GEOCHEMISTRY IN MINERAL EXPLORATION

Second Edition

GEOCHEMISTRY IN MINERAL EXPLORATION

Second Edition

ARTHUR W. ROSE

Professor of Geochemistry, Pennsylvania State University, University Park, Pennsylvania, U.S.A.

HERBERT E. HAWKES

Consultant, Tucson, Arizona, U.S.A.

JOHN S. WEBB

Professor of Applied Geochemistry, Imperial College of Science and Technology, London, U.K.

1979

ACADEMIC PRESS

LONDON * NEW YORK * TORONTO * SYDNEY * SAN FRANCISCO

A Subsidiary of Harcourt Brace Jovanovich, Publishers

ACADEMIC PRESS INC. (LONDON) LTD
24–28 Oval Road,
London NW1

U.S. Edition published by
ACADEMIC PRESS INC.
111 Fifth Avenue
New York, New York 10003

British Library Cataloguing in Publication Data

Rose, A W
Geochemistry in mineral exploration—2nd ed.
1. Geochemical prospecting
I. Title II. Hawkes, H E
III. Webb, John S
622′.1 TN270 79–50532

Cased edition ISBN 0-12-596250-9
Paperback edition ISBN 0-12-596252-5

Printed in Great Britain by
JOHN WRIGHT & SONS LTD, at THE STONEBRIDGE PRESS, BRISTOL

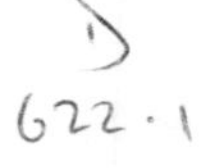

Preface to the Second Edition

The first edition of "Geochemistry in Mineral Exploration", published by Harper and Row in 1962, was offered as an introduction to geochemistry as applied in exploration for solid minerals. It was addressed to four principal groups of readers: (i) to the student as an introductory textbook; (ii) to the practising exploration geochemist as a source book and reference to the literature; (iii) to the specialized research worker in allied fields, such as environmental or agricultural geochemistry; and (iv) to non-specialized earth scientists who want a source of general information on exploration geochemistry.

The same may be said about this, the second edition. The present volume, however, differs from the first edition in a number of respects. Chemical principles, chemical equilibria, and chemical phenomena relevant to dispersion of elements are strongly emphasized, with the thought that an understanding of these principles will lead to valid new exploration methods and more accurate appraisal of exploration data. Two completely new chapters have been added, on vapor geochemistry and statistical methods. Most other chapters have been thoroughly revised and updated, based on the extensive literature published since the first edition. In particular, the previous single chapters on primary dispersion and on principles of secondary dispersion have each been enlarged into two chapters. Only brief discussions are offered of other fields of applied geochemistry, such as petroleum exploration, agriculture, public health, environmental problems, and forecasting of earthquakes and volcanic eruptions.

So far as practicable, the three approaches to geochemical exploration, namely basic principles, characteristics of geochemical anomalies, and practical surveying techniques, are treated in separate chapters or sections of the book. Thus, the student can learn basic principles in Chapters 6, 7, 8 and 9, and Sections IV of Chapter 2, I of Chapter 3, IV of Chapter 4, and I and II of Chapter 5, whereas Chapters 13 and 16 and Sections VI of Chapter 5, III and V of Chapter 17, and VIII of Chapter 18 are primarily aimed at the prospector or supervisor of a project.

The many friends and colleagues who helped with the preparation of the first edition, and who are listed in the preface of that volume, are again gratefully acknowledged. For this edition, special thanks for critically reviewing individual chapters are due to Fred Ward (Chapter 3), Edward Ciolkosz (Chapter 7), Margaret Hinkle and John Lovell (Chapter 18), and Richard Howarth (parts of Chapter 19). We are indebted to Martin Hale for a review of the entire manuscript and for many helpful suggestions. Numerous other colleagues and students too numerous to acknowledge individually have contributed ideas and stimulation. For care and patience in typing several drafts of the manuscript, we thank Margaret Biggers at Pennsylvania State University. Completion of the text would not have been possible without the sabbatical leave provided to Rose by Pennsylvania State University and the facilities made available to him at the Imperial College.

Although we sincerely hope that this edition is free of inaccurate statements and faulty judgment in interpreting data, we know that this is too much to hope. For these shortcomings, we offer our sincere apologies.

A. W. ROSE, H. E. HAWKES, and J. S. WEBB
August 1979

Contents

PREFACE . . . v

Chapter 1

INTRODUCTION

I. Geochemistry in Mineral Exploration . . . 2
II. Scales and Sampling Media in Geochemical Exploration . . . 2
A. Reconnaissance surveys . . . 2
B. Detailed surveys . . . 4
III. History . . . 5
IV. Present Status . . . 6
V. Literature . . . 9
VI. Related Fields of Applied Geochemistry . . . 11

Chapter 2

BASIC PRINCIPLES

I. The Geochemical Environment . . . 13
II. Geochemical Dispersion . . . 16
III. Geochemical Mobility . . . 18
IV. Geochemical Reactions . . . 18
A. Dispersion of elements under deep-seated conditions . . . 20
B. Mobility under surficial conditions . . . 23
V. Association of Elements . . . 25
VI. Patterns of Geochemical Distribution . . . 28
A. Normal background values . . . 30
B. Statistical distribution of background values . . . 32
C. The geochemical anomaly . . . 34
VII. Principles of Interpretation . . . 35

Chapter 3

PRINCIPLES OF TRACE ANALYSIS

I.	Mode of Occurrence of Trace Elements in Solids	44
	A. Stability of important minerals	46
II.	Preparation of Samples	46
III.	Decomposition of Samples	49
	A. Volatilization	51
	B. Fusion	51
	C. Vigorous acid attack	51
	D. Attack by weak aqueous extractants	52
	E. Oxidation – reduction agents	53
IV.	Separation	53
	A. Separation in liquid phase	53
	B. Separation in solid phase	54
V.	Estimation	54
	A. Colorimetry	54
	B. Emission spectrometry and spectrography	56
	C. Atomic absorption	57
	D. X-ray fluorescence spectrometry	58
	E. Fluorimetry	58
	F. Radiometric methods	58
	G. Electroanalytical methods	59
	H. Chromatographic methods	59
	I. Mass spectrometry	60
	J. Other methods	60
	K. Units for reporting chemical analyses	61
VI.	Reliability of Geochemical Analyses	61
	A. Precision	62
	B. Accuracy	65
	C. Routine checks of precision and accuracy	66
VII.	Analytical Procedures	66
VIII.	Selected References on Techniques of Chemical Analysis	70

Chapter 4

PATTERNS OF DEEP-SEATED ORIGIN—ORE TYPES, GEOCHEMICAL PROVINCES, AND PRODUCTIVE PLUTONS

I.	Classification of Mineral Deposits	72
	A. Ore-forming processes	72
	B. Geochemical recognition of ore types	72
II.	Productive Environments of Deep-seated Origin	77
III.	Geochemical and Metallogenic Provinces	78

IV. Processes Forming Productive Plutons 84
V. Ores Related to Productive Plutons 87
A. Pegmatites 87
B. Magmatic copper–nickel sulfides 88
C. Ores associated with granitic rocks 89
D. Hydrothermal tin deposits 90
E. Molybdenum, tungsten, and gold deposits 93
F. Base-metal sulfide deposits 93
G. Porphyry-copper deposits 94
H. Volcanogenic massive sulfides 95

Chapter 5

EPIGENETIC ANOMALIES IN BEDROCK

I. Formation of Diffusion Aureoles 98
II. Formation of Leakage Anomalies 101
III. Zoning in Epigenetic Orebodies and Aureoles 104
IV. Examples of Diffusion Aureoles 107
V. Examples of Leakage Anomalies and Zoning 109
A. Polymetallic vein and replacement deposits 109
B. Porphyry-copper and molybdenum deposits 114
C. "Epithermal" deposits of Au–Ag–As–Sb–Hg 118
D. Volcanogenic massive-sulfide deposits 119
E. Mississippi-valley type lead–zinc deposits 120
F. Leakage anomalies in specific types of material . . . 121
VI. Geochemical Rock Surveys 121
A. Orientation surveys 122
B. Collection and processing of samples 124
C. Preparation of maps and handling of data 125
D. Interpretation of data 126

Chapter 6

WEATHERING

I. Nature of Weathering 128
II. Weathering Processes 129
A. Physical weathering 130
B. Chemical weathering 131
C. Biologic agents in weathering 133
III. Factors Affecting Weathering Processes 134
A. Resistance of minerals in weathering 134
B. Permeability 136

C. Climate . . . 137
D. Relief and drainage . . . 138
IV. Products of Weathering . . . 138
A. Residual primary minerals from the parent rock . . . 138
B. Secondary minerals . . . 139
C. Soluble products . . . 146
D. Residual structures and textures . . . 148
V. Selected References on Weathering . . . 148

Chapter 7

SOIL FORMATION

I. Soil Profile Development . . . 150
II. Factors Affecting Soil Formation . . . 154
A. Parent material . . . 154
B. Climate . . . 154
C. Biologic activity . . . 157
D. Relief . . . 158
E. Time . . . 161
III. Classification of Soils . . . 162
A. The Soil Taxonomy system . . . 162
B. Old classification of soils . . . 163
IV. Soils of Humid Regions . . . 170
A. Podzolic and related soils of temperate forested regions . . . 170
B. Tropical soils . . . 171
C. Intrazonal hydromorphic soils . . . 173
V. Soils of Subhumid and Arid Regions . . . 174
A. Grassland soils . . . 175
B. Desert soils . . . 175
C. Soils of arctic areas . . . 176
D. Intrazonal saline and alkali soils . . . 178
VI. Mountain Soils . . . 178
VII. Selected References on Soil Formation . . . 179

Chapter 8

CHEMICAL EQUILIBRIA IN THE SURFICIAL ENVIRONMENT

I. Composition of Natural Waters . . . 181
II. Eh–pH Relationships . . . 181
A. Hydrogen-ion concentration . . . 182
B. Oxidation–reduction potential . . . 184
C. Eh–pH diagrams . . . 185

D. Forms of sulfur and carbon in solution 188
III. Formation of Complexes 191
IV. Solubility of Minerals 192
V. Adsorption and Ion Exchange on Colloidal Particles 195
A. Ion exchange in clay minerals 196
B. Adsorption on surfaces 197
VI. Organic Matter 206
A. Simple organic compounds 206
B. Humic substances 208
VII. Electrochemical Dispersion 212
VIII. Selected References on Chemical Equilibria in the Surficial Environment 214

Chapter 9

MECHANICAL AND BIOLOGICAL DISPERSION IN THE SURFICIAL ENVIRONMENT

I. Mechanical Factors 216
A. Simple gravity movement 216
B. Dispersion in ground water 220
C. Mechanics of dispersion in surface water 224
D. Dispersion by glaciers 231
E. Dispersion by wind action 235
F. Dispersion by animal activity 235
G. Dispersion in permafrost areas 236
II. Biological Factors 236
A. The effect of vegetation 236
B. The effect of microorganisms 238
III. The Influence of Environment on Dispersion 238
A. Climate 238
B. Relief 239
C. Rock types 239
D. Life processes 240
E. Time 240
IV. Selected References on Dispersion in the Surficial Environment . 241

Chapter 10

SURFICIAL DISPERSION PATTERNS

I. Classification of Surficial Dispersion Patterns 242
II. Syngenetic Patterns 244
A. Clastic patterns 244

B. Hydromorphic patterns 245
C. Biogenic patterns 245
III. Epigenetic Patterns 245
A. Hydromorphic patterns 245
B. Biogenic patterns 246
IV. Partition of Dispersed Constituents Between Liquid and Solid Media 246
A. Liquid media 246
B. Solid media 248
V. Extractability of Metal from Clastic Samples 249
VI. Contrast 250
VII. Form of Surficial Patterns 252
A. Clastic patterns 252
B. Hydromorphic patterns 254
C. Biogenic patterns 256
VIII. Anomalies Not Related to Mineral Deposits 256
A. High-background source rocks 257
B. Contamination 258
C. Sampling errors 259
D. Analytical errors 260
IX. Suppression of Significant Anomalies 260

Chapter 11

ANOMALIES IN RESIDUAL OVERBURDEN

I. Anomalies in Gossans and Leached Outcrops of Ore . . . 261
II. Syngenetic Anomalies in Residual Soil 266
A. Mode of occurrence 266
B. Form and magnitude of the anomaly 269
C. Homogeneity 278
D. Variations with depth and soil type 281
III. Hydromorphic Anomalies in Residual Soil 285
IV. Anomalies Not Related to Mineral Deposits 285

Chapter 12

ANOMALIES IN TRANSPORTED OVERBURDEN

I. Common Features of Anomalies in Transported Overburden . 288
A. Syngenetic anomalies 289
B. Epigenetic anomalies 289
C. Anomalies not related to mineral deposits 290
II. Glacial Overburden 290
A. Syngenetic (clastic) anomalies 291

B. Hydromorphic anomalies 300
C. Biogenic anomalies 304
D. Superjacent anomalies of complex origin 305
E. Schematic models of dispersion in glaciated areas . . 306
III. Organic Deposits 306
IV. Colluvium and Alluvium 311
A. Syngenetic anomalies 313
B. Epigenetic anomalies 314
V. Other Types of Transported Overburden 317

Chapter 13

GEOCHEMICAL SOIL SURVEYS

I. Orientation Surveys 320
A. Residual soil 321
B. Transported overburden 325
C. Contamination 325
D. Choice of procedures 328
II. Field Operations 329
A. Sampling pattern 329
B. Sampling procedure 333
C. Locating and identifying samples 336
D. Sample preparation and analysis 337
E. Preparation of geochemical maps 339
III. Interpretation of Data 341
A. Estimation of background and threshold values . . . 341
B. Recognition of non-significant anomalies 342
C. Distinction between lateral and superjacent anomalies . . 344
D. Appraisal of anomalies 345
IV. General Procedure for Follow-up 346

Chapter 14

ANOMALIES IN NATURAL WATERS

I. Mode of Occurrence of Elements 348
II. Factors Affecting Composition of Natural Waters 351
A. Composition of rain water 351
B. Reaction with soil and rock 352
C. Regional and climatic factors 354
D. Age of ground water 355
E. Oxidation–reduction 355
F. Adsorption 355

G. Mixing of waters 358
H. Background content of natural waters 358
III. Persistence of Anomalies 358
A. Contrast at source 358
B. Decay by dilution 361
C. Decay at precipitation barriers 363
D. Re-solution at precipitation barriers 368
IV. Time Variations 368
V. Ground-water Anomalies 373
VI. Stream-water Anomalies 377
VII. Lake-water Anomalies 380
VIII. Anomalies Not Related to Mineral Deposits 381

Chapter 15

ANOMALIES IN DRAINAGE SEDIMENTS

I. Spring and Seepage Areas 383
II. Active Stream Sediments 392
A. Mode of occurrence 392
B. Contrast 394
C. Decay patterns 399
D. Anomaly enhancement 406
E. Effect of precipitation barriers 411
F. Lateral homogeneity 413
G. Vertical homogeneity 415
H. Time variations 415
III. Flood-plain Sediments 416
IV. Lake Sediments 419
A. Mode of occurrence 420
B. Homogeneity and contrast 421
C. Decay patterns 424
V. Drainage Anomalies Not Related to Mineral Deposits 426
VI. Marine Sediments 427

Chapter 16

GEOCHEMICAL DRAINAGE SURVEYS

I. Orientation Surveys 431
A. Water 432
B. Drainage sediment 433
C. Contamination 434

II. Choice of Material to be Sampled 435
A. Water 435
B. Seepage soils 436
C. Stream sediments 437
D. Heavy minerals 438
E. Organic sediment 439
F. Lake sediment and water 439
III. Sample Layout 440
A. Ground-water patterns 441
B. Drainage channel patterns 442
IV. Collection and Processing of Samples 444
A. Water 445
B. Stream and seepage sediment 447
C. Lake sediment 449
D. Heavy minerals 450
V. Preparation of Maps 450
VI. Interpretation of Data 453
VII. Follow-up Techniques 454

Chapter 17

VEGETATION

I. Uptake of Mineral Matter by Plants 457
A. Plant nutrition 457
B. Availability of elements in soil 458
C. Reactions in the root tips of plants 458
D. Movement and storage within plants 463
II. Biogeochemical Anomalies 464
A. Variation between plant species 464
B. Variation between plant parts 464
C. Depth of root penetration 466
D. Variation with other factors 467
E. Contrast 468
F. Homogeneity 470
G. Form of anomalies 470
III. Biogeochemical Surveying Techniques 471
A. Orientation survey 473
B. Choice of sampling medium 473
C. Collecting and processing of samples 474
D. Choice of analytical method 474
E. Interpretation of data 475
F. Advantages and disadvantages of biogeochemical surveys . 475
IV. Geobotanical Indicators 476

A. Indicators of ground water 478
B. Indicators of saline deposits 479
C. Indicators of hydrocarbons 479
D. Indicators of rock types 481
E. Indicators of ore 481
V. Geobotanical Surveying Techniques 484
VI. Remote Sensing of Geobotanical Anomalies 485

Chapter 18

VOLATILES AND AIRBORNE PARTICULATES

I. Properties of Gases and Aerosols 490
II. Units of Measurement 491
III. Source of Naturally Occurring Gases 492
A. Atmospheric gases 492
B. Deep-seated gases 494
C. Radiogenic gases 495
D. Biogenic gases 495
E. Gases generated in sulfide deposits 495
F. Atmospheric particulates 497
IV. Reactivity of Gases 498
V. Movement of Gases 500
A. Diffusion 500
B. Water transport 502
C. Transfer between water and vapor phases 503
D. Vapor transport 504
E. Relative effects of infiltration versus diffusion 505
VI. Geochemical Behavior of Selected Gases 506
VII. Applications of Vapor Geochemistry 511
A. Exploration for sulfide deposits 511
B. Exploration for uranium 512
C. Exploration for petroleum 513
D. Location of buried faults 514
E. Forecasting of earthquakes 514
F. Forecasting of volcanic eruptions 514
G. Location of geothermal areas 514
VIII. Vapor Survey Techniques 514
A. Soil air 516
B. Water 516
C. Gas traps 517
D. Atmospheric air 517

Chapter 19

COMPUTERIZED DATA-HANDLING AND STATISTICAL INTERPRETATION

I. Data-handling by Computer 520
A. Recording of data 520
B. Production of tables, histograms, and summaries . . 521
C. Computer-generated maps 521
D. Contouring of concentrations 523
E. Moving averages and trend surfaces 523
II. Multivariate Statistical Methods 525
A. Regression analysis 525
B. Discriminant analysis 527
C. Factor analysis 531
D. Cluster analysis 533

Chapter 20

GEOCHEMICAL METHODS IN MINERAL EXPLORATION

I. Optimization of Exploration 535
II. Planning of Exploration 536
A. Selection of professional leadership 536
B. Selection of areas 538
C. The exploration sequence 538
III. Choice of Exploration Methods 540
A. Target size 543
B. Property control 544
C. Reliability of method 544
D. Cost 544
E. Value of expected ore 545
F. Comparison of programs. 545
G. The role of geochemistry in an exploration system . . 545
IV. Organization and Operations 546
A. Field operations 546
B. Laboratory 547
C. Supervision 548

APPENDIX. Geochemical Characteristics of the Elements 549

REFERENCES 583

SUBJECT INDEX 637

Chapter 1

Introduction

The period from 1940 to the present time has witnessed an unprecedented scale of activity in prospecting for mineral deposits. In the first half of this period, the mining industry became aware that it could not depend indefinitely on discoveries brought to its attention by relatively untrained prospectors. Most mining companies developed a more consciously organized approach to exploration and prospecting, and focused greater attention on development of geologic guides to ore. In addition to this geologic activity, an intensified effort was directed toward the development and application of new techniques of mapping geological structures and of detecting ore concealed beneath a mantle of organic debris, soil, or barren rock. Included here were photogeological, geophysical, and geochemical techniques. These techniques became increasingly attractive not only because they were able to detect concealed orebodies but also because they were cost-effective, particularly when applied to the exploration of large areas.

In recent years, geochemical and geophysical techniques have developed to the point of routine use in nearly all mineral exploration programs, and now account for a major proportion of exploration spending in the predrilling stage. The methods have been justified by a large number of discoveries in which geochemical and/or geophysical methods have been an essential link in the chain of discovery. Many techniques for detection of concealed orebodies are now available, and methods for more complex and deeper ores are under active development.

I. GEOCHEMISTRY IN MINERAL EXPLORATION

Geochemical prospecting for minerals, as defined by common usage, includes any method of mineral exploration based on systematic measurement of one or more chemical properties of a naturally occurring material. The chemical property measured is most commonly the trace content of some element or group of elements; the naturally occurring material may be rock, soil, gossan, glacial debris, vegetation, stream or lake sediment, water, or vapor. The purpose of the measurements is the discovery of abnormal chemical patterns, or geochemical anomalies, related to mineralization.

Geochemical methods of exploration should be viewed as integral components of the arsenal of weapons available to the modern prospector. They are distinguished from other methods of gathering geological data only because they call for certain specialized techniques. The goal of every exploration method is, of course, the same—to find clues that will help in locating hidden ore.

II. SCALES AND SAMPLING MEDIA IN GEOCHEMICAL EXPLORATION

Geochemical surveys are usually classified as reconnaissance surveys or detailed surveys, according to their purpose and scale (Fig. 1.1). They can also be classified according to the material sampled in the survey, that is, soil, vegetation, sediment, water, rock, or vapor. Some of these media are more effective or more efficient in reconnaissance surveys, others are more suitable in detailed surveys.

A. Reconnaissance Surveys

In a reconnaissance survey, the purpose is to search a relatively large area for indications of ore. The spacing and type of samples are chosen to detect but not necessarily outline favorable provinces, districts, or orebodies with as few samples or at as low a cost as possible. The aim in a reconnaissance survey is to determine the mineral possibilities of a relatively large area, eliminate the barren ground, and draw attention to local areas of interest. Areas of tens to thousands of square kilometres are evaluated during reconnaissance surveys, often with no more than one sample per 1 km^2 to 100 km^2.

Of the geochemical ore guides, recognition of geochemical provinces shows the most promise as a method of appraisal of very large regions. If the probability of ore occurrence can be correlated with the average abundance of the

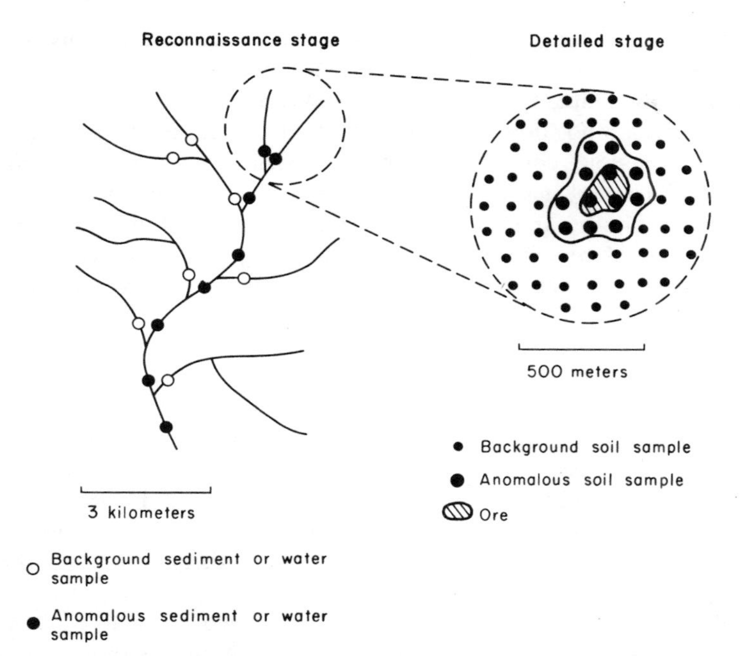

Fig. 1.1. Reconnaissance exploration using stream sediment or water, as contrasted to detailed exploration using soils.

element in average rocks of the region, or with the abundance in certain specific types of igneous or sedimentary rocks, then widely spaced samples of rock, sediment, or soil can be collected and analyzed to focus attention on regions with above-average mineral potential.

In more detailed reconnaissance surveys, samples of sediment from streams and lakes, and water from streams, lakes, springs, seepages, and wells have become the most effective and widely used media. This is particularly true of stream sediment, which is applicable to most metals. Exploration based on water samples is necessarily limited to the more readily soluble elements. Ease of operation and low cost per unit area, combined with a relatively high degree of reliability, place stream-sediment surveys in an outstanding position as a reconnaissance tool. In glaciated areas, dispersion of visible boulders or of chemically detectable traces of metals in till have been successfully used to detect orebodies.

Residual soil anomalies related to the weathering of suboutcropping deposits are normally of very local extent. Nevertheless, despite the consequent high cost per unit area, soil surveys have been widely used in reconnaissance because of their relatively high dependability. In the U.S.S.R., systematic soil sampling is used as a matter of routine in all areas that are being mapped on the regional scale, either geologically or geophysically. More usually, however, soil sampling is applied in detailed surveys in areas where the ground is already known to be promising.

The chemical composition of plants or the distribution of plant species that prefer anomalous soil have possible applications in reconnaissance. Plants or plant assemblages uniquely related to ore can sometimes be identified visually or photographically from the air, with resulting economy in coverage of large areas. Airborne devices that sample and analyze the airborne particles shed by vegetation are now in development for reconnaissance surveys.

B. Detailed Surveys

In detailed surveys, the purpose is to outline mineralized ground and to pinpoint the mineralized source with the greatest possible precision, preparatory to physical exploration by trenching, drilling, or underground work. In order to localize the bedrock source, a relatively close sample spacing, normally between 1 m and 100 m, is usually required. On grounds of cost alone, therefore, the application of detailed surveys is mostly restricted to limited areas of particular interest, selected on the basis of other geochemical, geological, or geophysical information. To be most useful, the anomalies should be well defined and should occur in close proximity to the ore.

Systematic sampling of residual soil in search of superjacent anomalies (anomalies directly overlying ore) has been outstandingly successful in detailed exploration. The method has been widely used under a variety of conditions for locating suboutcropping deposits of many kinds, appraising geophysical anomalies and other features of interest, tracing vein extensions, and in property examinations. The points in favor of residual-soil sampling are its dependability and simplicity.

Much the same can be said of soil surveys as a method of locating superjacent anomalies in transported overburden. Here, the dependability factor is much lower, inasmuch as deposits concealed by a blanket of transported cover are not so consistently accompanied by readily detectable superjacent anomalies. Nevertheless, the dependability can often be improved by deep sampling, though with a consequent steep rise in cost. It may, therefore, be more profitable to confine soil surveying in areas of transported overburden to small target areas already delimited by other cheaper methods, such as seepage reconnaissance or airborne geophysical surveys.

Plant sampling is, in general, applicable mainly in environments where for one reason or another soil sampling is unsatisfactory. A variety of problems in recognizing and collecting a single species throughout a survey area, in variability of metal content with age and time of year, and in processing vegetation make the routine use of vegetation surveys generally less desirable and more costly than soils. The main circumstances in which plant sampling may be preferable are in winter in snow-covered areas, in areas where soil is absent, and in areas where plant roots penetrate deeply into a layer of transported material.

In rocks, anomalies formed by leakage or diffusion of metal away from an orebody constitute targets for detailed exploration. Leakage anomalies may be sought by sampling the cap rocks or the residual soil derived therefrom. Anomalies developed in the wallrock of the ore are particularly applicable in detailed surveys underground, where rock sampling may be used to detect nearby ore during crosscutting and drilling.

Recent experimental surveys of soil gases and atmospheric constituents show promise as a technique for detecting ore through thick overburden, but many problems of sampling, analysis, and interpretation remain to be solved, and the conditions under which vapor surveys are preferable to other types of surveys have yet to be established.

III. HISTORY

The principles of geochemical prospecting are as old as man's first use of metals. The earliest prospector soon learned that the environment of many mineral deposits was marked by certain conspicuous and diagnostic features. Fragments of fresh or weathered ore might be scattered about the surface of the ground near its source in the bedrock. Smaller fragments of similar material could be seen in the sediments deposited by streams draining the mineralized area. The prospector found that by searching for such material and then following the trail of increasing concentrations, he was often led to the bedrock source of the metals. Here he was guided only by what he could see with the unaided eye. However, basically, he was following dispersion patterns that are strictly analogous to those used in modern geochemical methods.

Trace-element analysis as a supplement to visual observations began to find its place in exploration technology in the middle 1930s. Inspired by the classical researches of Goldschmidt, Fersman, and Vernadsky in the field of fundamental geochemistry, research workers in Scandinavia and the U.S.S.R. independently carried out successful experiments using spectrographic trace analysis of soils and plants as a prospecting method (Hawkes, 1976b; Hawkes and Webb, 1962). In the 1940s, with continually improving methods of wet chemical analysis as well as spectrography, geochemists in the U.S.A. and Canada continued with the development of progressively more economical and effective methods of geochemical prospecting. In the 1950s, the growth of active research spread to Great Britain and thence to other countries in western Europe, accompanied by a concomitant activity in their overseas territories, notably Africa and the Far East. In addition, experimental work was initiated in Australia, India, and many other parts of the world, until today there are relatively few mineral fields in which no geochemical studies of any sort have been made.

The evolution of the various techniques is by no means complete, and not all are in the same stage of development. Geochemical soil surveys are undoubtedly the most advanced and were first applied on a commercial scale in the 1930s. Vegetation surveys have been utilized for about the same period. Geochemical drainage surveys on a routine basis did not come into their own until the middle 1950s, but are now the most widely used reconnaissance method. Exploration based on systematic sampling of weathered or fresh rock was first undertaken on a selective basis in the 1950s. By the 1960s, some types of rock survey had been developed to the point of routine application, particularly in the U.S.S.R. Measurement of gases in soils and the atmosphere is still in the experimental stage.

IV. PRESENT STATUS

Geochemical methods of prospecting have been credited in the literature with a number of mineral discoveries. Almost invariably, geochemical methods were used in conjunction with other geological or geophysical exploration methods, so that an assignment of "credit" to a single method is not possible. However, as an indication of the contribution of geochemical methods, and as a reference to a few of the case histories of successful exploration, Table 1.1 lists some discoveries in which geochemical surveys are reported to have been important. Other discoveries, credited with a total of $2{\cdot}5 \times 10^9$ tons of porphyry copper ore, 10^8 tons of 2·6% Ni, and 60×10^6 tons of 11% Zn and 2·7% Pb, are reported by Webb (1973).

Perhaps a better index of the status of the method is the spectacular growth of activity in the field over the past 30 years, and the inclusion of geochemical methods in most major exploration programs. Growth and scale of operations may be interpreted as an indication of the overall opinion of the exploration profession, based on practical experience under many conditions and in many parts of the world.

Over the years, the U.S.S.R. has probably seen the largest volume of both geochemical surveying and research in geochemical methods of exploration of any country. The work is being carried out in many state organizations and

Table 1.1
Some discoveries in which geochemical exploration played an important part

Deposit and location	Type	Method	Reference
Agricola Lake, N.W.T., Can.	Zn–Cu–Pb–Ag	Lake sed.	Cameron, 1975 a
Mt Pleasant, N.B., Can.	Sn, W, Mo	Str. sed.	Hawkes *et al.*, 1960
Murray, N.B., Can.	Cu–Zn–Pb	Str. sed.	Fleming, 1961
Panama, Argentina, Malaysia, Ecuador	Cu, Mo	Str. sed.	Lepeltier, 1971, 1974
Howards Pass, Yukon, Can.	Cu	Str. sed.	Ainsworth *et al.*, 1977
Kazakhstan, U.S.S.R.	20 deposits	Str. sed.	Beus and Grigorian, 1977, p. 169
Gortdrum, Ireland	Cu–Ag	Str. sed., soil	Thompson, 1967
Kalengwa, Zambia	Cu	Str. sed., soil	Ellis and McGregor, 1967
Butiriku, Uganda	Carbonatite	Str. sed., soil	Reedman, 1974
Flat Gap, Tenn., U.S.A.	Zn	Soil	Hoagland, 1962
Fitula, Zambia	Cu	Soil	Cornwall, 1969
Navan, Ireland	Pb–Zn	Soil	O'Brien and Romer, 1971
Casino, Yukon, Can.	Cu, Mo	Soil	Archer and Main, 1971
Lady Loretta, Australia	Zn–Pb–Ag	Soil	Cox and Curtis, 1977
Yellow Cat, Utah, U.S.A.	U	Plants	Cannon, 1964
S.W. Wisconsin, U.S.A.	Zn	Spring water	DeGeoffroy *et al.*, 1967
Wyoming, U.S.A.	U	Ground water	Denson *et al.*, 1959
Cortez, Nev., U.S.A.	Au	Rock	Erickson *et al.*, 1966b

universities. In 1955, by order of the Ministry of Geology and Conservation, systematic geochemical surveys of the soil and weathered rock were made compulsory in all geological organizations for exploration work on any scale (Solovov, 1959, p. 10). Much of this work has been concentrated in the desert and mountainous terrain of Central Asia, where soil survey techniques seem to be especially applicable. More than 60 million geochemical samples were collected in Kazakhstan in the period 1947–1969 (Beus and Grigorian, 1977, p. 169). Sampling continues at the level of millions of samples per year (Boyle, 1976).

In the Western world, replies to a questionnaire sent out by the Association of Exploration Geochemists indicate that 90% of the 150 companies replying make frequent use of geochemical exploration, and spend on an average about

Table 1.2
Some principal Western organizations conducting research and development of geochemical exploration methods in 1977

Australia	Commonwealth Scientific and Industrial Research Organization (CSIRO)	N. Ryde, N.S.W.
	University of N.S.W.	Kensington, N.S.W.
Canada	Geological Survey of Canada	Ottawa, Ont.
	University of British Columbia	Vancouver, B.C.
	Queens University	Kingston, Ont.
	University of New Brunswick	Fredericton, N.B.
	Barringer Research, Inc.	Toronto, Ont.
Finland	Geological Survey of Finland	Espoo
France	Bureau de Recherches Géologique et Minières (BRGM)	Orléans
Germany	Technische Hochschule	Aachen
	Federal Institute for Geosciences and Natural Resources	Hannover
New Zealand	Massey University	Palmerston North
Norway	Geological Survey of Norway	Trondheim
Sweden	Geological Survey of Sweden	Stockholm
U.K.	Imperial College of Science and Technology	London
	Institute of Geological Sciences	London
U.S.A.	U.S. Geological Survey	Denver, Colo.
	U.S. Department of Energy	Grand Junction, Colo.
	Pennsylvania State University	University Park, Penn.
	Colorado School of Mines	Golden, Colo.
	University of Georgia	Athens, Georgia

12% of their exploration budgets for this purpose (Lakin, 1973). Approximately 8 000 000 samples per year were being collected in the early 1970s, plus about 10 000 000 samples in the U.S.S.R. (Webb, 1973). About 50% of the samples were soil, 20% stream sediment, 30% rock, 2% water, and ½% vegetation. The major elements of interest were Cu, Pb, Zn, Ag, Au, Ni, Mo, and U. A very large reconnaissance survey for U is currently under way in the U.S.A., involving water and sediment samples, and similar surveys are in progress in other countries.

In the Western countries, organized research programs have been undertaken in a large number of university and government departments, the more important of which are listed in Table 1.2. Most universities are now teaching geochemical exploration either as separate courses or as parts of courses in economic geology (Carpenter, 1975).

Another indication of the activity and status of geochemical exploration is provided by the formation of the Association of Exploration Geochemists in 1970 and its subsequent growth. The Association now has about 500 members, sponsors a journal (*Journal of Geochemical Exploration*), and publishes newsletters for the members and occasional special publications, including periodic bibliographies of the literature of geochemical exploration. The Association grew out of International Symposia on Geochemical Exploration held in Ottawa in 1966 and Golden, Colorado, in 1968, and has since sponsored symposia every two years, in Toronto in 1970, London in 1972, Vancouver in 1974, Sydney in 1976, and Denver in 1978.

V. LITERATURE

The literature on geochemical exploration is increasing at a rapid rate. Bibliographies have been compiled for the periods before 1952 (Harbaugh, 1953), 1952–1954 (Erickson, 1957), 1955–1957 (Markward, 1961), 1965–1971 (Hawkes, 1972a), 1972–1975 (Hawkes, 1976a), and 1976 (Hawkes, 1978). The publication rate for the Western world is approximately 400 papers per year for the period 1972–1975, based on the papers cited by Hawkes (1976). A hundred or so additional papers per year appear in Russian and a few more in oriental languages. Technical journals and publications that carry papers on geochemical exploration and allied topics are listed in Table 1.3.

Previous major textbooks on geochemical exploration in English are by Hawkes and Webb (1962) and Levinson (1974). These have been translated into several other languages. Other relatively comprehensive books are by Granier (1973, in French), Polikarpochkin (1976, in Russian), Ginzberg (1960, in Russian translated into English), and Siegel (1974). More specialized books include those by Beus and Grigorian (1977), mainly on the use of rock samples; by Antropova (1975, in Russian) on the form of elements in dispersion haloes; by Belyakova *et al.* (1963, translated from Russian) on

Table 1.3
Principal journals and other publications carrying articles on geochemical exploration

Journal of Geochemical Exploration
Transactions of the Institution of Mining and Metallurgy (Section B, Applied Earth Sciences)
Economic Geology
Canadian Mining Journal
Akademiya Nauk SSSR Doklady (translated to English)
Geokhimiya (partly translated to English in *Geochemistry International*)
Geochimica et Cosmochimica Acta
Geological Survey of Canada Papers, Bulletins
U.S. Geological Survey Bulletins, Circulars, Professional Papers, *Journal of Research*
International Geochemical Exploration Symposia
First (Ottawa; Cameron, 1967)
Second (Golden; Canney, 1969)
Third (Toronto; Boyle, 1971a)
Fourth (London; Jones, 1973a)
Fifth (Vancouver; Elliot and Fletcher, 1975)
Sixth (Sydney; Butt and Wilding, 1977)
Seventh (Denver; Watterson and Theobald, 1979)
International Geological Congresses
XX (Mexico; Lovering *et al.*, 1958, 1959, 1960)
XXI (Copenhagen; Marmo and Puranen, 1960)
XXIV (Montreal, 1972)
Conferences on Prospecting in Areas of Glaciated Terrain
Jones (1973b)
Jones (1975)
Institution of Mining and Metallurgy (1977)
Proceedings of United Nations Conferences on the Peaceful Uses of Atomic Energy
First (1955, Vol. 6)
Second (1958, Vol. 2)
U.N. Seminars on Geochemical Exploration Methods and Techniques
Mineral Resources Development Series, No. 21, 1963
Mineral Resources Development Series, No. 38, 1970

hydrogeochemical methods; by Brooks (1972), Malyuga (1964) and Kovalevskiy (1974, in Russian) on the use of vegetation; and by Kvalheim *et al.* (1967) on geochemical prospecting in Fennoscandia. Reviews and summaries have been published by Boyle (1967), Andrews-Jones (1968), Boyle and Smith

(1968), Boyle and Garrett (1970), and Bradshaw *et al.* (1972), and annual summaries appear in the journal *Mining Engineering*. Geochemical exploration methods for petroleum are summarized by Siegel (1974, Ch. 9).

VI. RELATED FIELDS OF APPLIED GEOCHEMISTRY

Mineral exploration is not the only field where the science of geochemistry in its various ramifications has served to solve some of the practical problems confronting the human race. In fact it does not even represent the first large-scale application of geochemistry, as agricultural scientists have depended on the principles of geochemistry as the basis of their studies of plant and animal nutrition, a field that historically served as both the stimulus and the source of analytical technology for the earliest experiments in exploration geochemistry. Other, newer fields of applied geochemistry have been developing their own specialized concepts and technologies which conceivably could contribute to the future development of geochemical exploration methods. For this reason, it is profitable for the exploration geochemist to maintain active channels of communication with workers in these allied fields either through personal contact or through the literature. Following are the fields of applied geochemistry that are particularly active at the present time, together with references that may serve as an introduction to the appropriate specialized literature.

Agricultural geochemistry is concerned with the geologic sources of the major and minor elements that are necessary in plant and animal nutrition. The *Journal of Soil Science* and the *Journal of the Soil Science Society of America* are leading publications in this field.

Environmental geochemistry is concerned with the pollution of the surficial environment by industrial waste, including metals and the radioactive by-products of nuclear reactors. *Environmental Science and Technology*, *Science of the Total Environment*, *Environmental Research*, and *Environmental Pollution* are leading journals, although articles are scattered throughout a wide literature.

Geochemistry in public health is concerned with minor elements in foods and drinking water and their relation to human health. Among the major publications in this area are the proceedings of conferences on Trace Substances in Environmental Health and the journal *Interface*.

Geological applications of geochemistry include the uses of vapor-phase emanations in locating faults and forecasting earthquakes and volcanic eruptions, and the use of geochemical methods to aid geological mapping. These topics are discussed in Chapter 19 of this volume.

Marine geochemistry is concerned with the chemistry of the ocean waters and sediments. Devices developed for the remote sensing of the composition of the sea floor are of particular interest to exploration geochemistry. Publications in this field appear in *Marine Technology* and *Marine Pollution Bulletin*.

Chapter 2

Basic Principles

Geochemistry, as originally defined by Goldschmidt and summarized by Mason (1958, p. 2), is concerned with (i) the determination of the relative and absolute abundance of the elements in the earth, and (ii) the study of the distribution and migration of the individual elements in the various parts of the earth with the object of discovering principles governing this distribution and migration. In more recent years, a third concern has been added to these two, namely the application of geochemical principles and information in solving human needs. At present, geochemical exploration and environmental geochemistry are the major fields in which geochemistry is applied.

Although the methods and results of geochemical exploration are largely descriptive, with products usually taking the form of geochemical maps, the optimum planning and interpretation of geochemical surveys must be based on a knowledge of the distribution and abundance of elements, and on the principles that govern the migration of elements. This chapter discusses some of the basic principles and concepts useful in many facets of geochemical exploration.

I. THE GEOCHEMICAL ENVIRONMENT

Geologically and geochemically, the earth is a dynamic system in which materials are moved from place to place and changed in form and composition by a variety of processes, including melting, crystallization, erosion, dissolution, precipitation, vaporization and radioactive decay. Although the detailed behavior of matter in this system is extremely complex, the geochemical environment, as defined by pressure, temperature, and availability of the most abundant chemical components, determines the stability of mineral and fluid phases at any given point. On the basis of gross differences in pressure, temperature, and chemistry, the geochemical environments of the earth can be classified into two major groups: deep-seated and surficial.

The deep-seated environment extends downward from the lowest levels reached by circulating surface water to the deepest level at which normal rocks can be formed. Magmatic and metamorphic processes predominate in this zone. It is an environment of high temperature and pressure, restricted circulation of fluids, and relatively low free-oxygen content. Volcanic phenomena, hot springs, and similar features can generally be included with the deep-seated environment, in view of the temperature and source of material. The terms "hypogene", "primary", and "endogenic" have been used by some workers to refer to phenomena in this environment. However, primary implies a sequence which is not necessarily followed, and hypogene has been more commonly applied to ores than to large-scale geochemical processes. Endogenic, used in the U.S.S.R., is too easily confused with its converse, exogenic. In this book, the term "deep-seated" will be used in referring to environments and locations. Primary will refer to the stage of ore formation, as discussed further below.

The surficial environment is the environment of weathering, erosion, and sedimentation at the surface of the earth. It is characterized by low temperatures, nearly constant low pressure, free movement of solutions, and abundant free oxygen, water, and CO_2. The terms "supergene", secondary", and "exogenic" have also been used to refer to processes in this environment, but surficial seems preferable when considering environments. Secondary is used for processes acting on already formed orebodies, as discussed later.

The movement of earth materials from one environment to another can be conveniently visualized in terms of a closed cycle, as illustrated in Fig. 2.1. Starting on the right-hand side of the diagram and moving clockwise, sedimentary rocks are progressively metamorphosed as they are subjected to increasing temperature, pressure, and increments of new materials from outside the system. They may eventually attain a state of fluidity such that on recrystallization they can differentiate into various kinds of igneous rocks and hydrothermal extracts. When erosion brings the resulting suite of rocks into the surficial environment again, the component elements are redistributed by weathering agencies. A new series of sedimentary rocks is then deposited, and the cycle is closed. Figure 2.1 is, of course, highly simplified, as in reality large parts of the cycle may be missing in any given case, as suggested by the arrows in the center of the diagram. It is quite normal, for example, for sedimentary sandstone and shale to be exposed to weathering and erosion with-

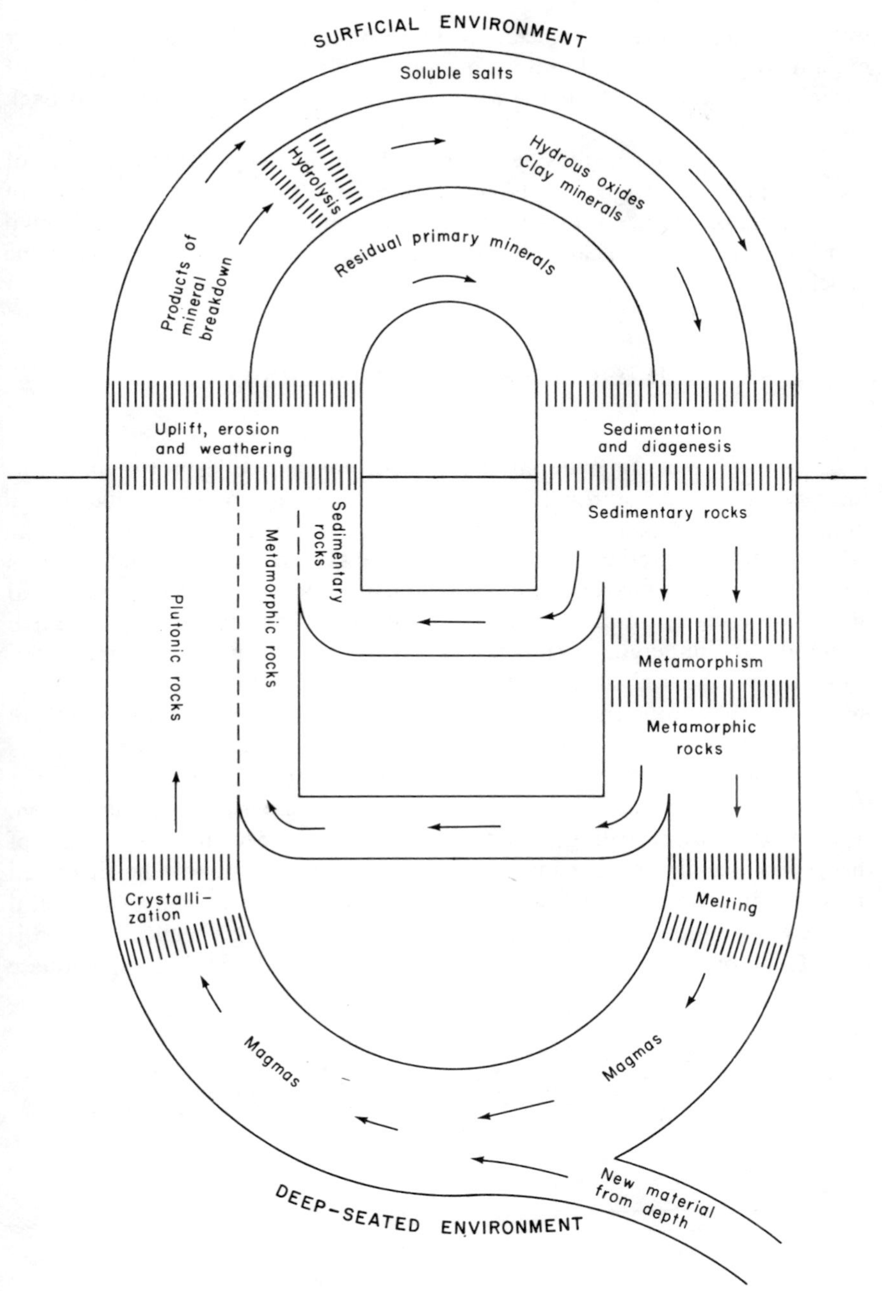

Fig. 2.1. The geochemical cycle.

out remelting or even significant metamorphism. Furthermore, this major cycle incorporates several important minor cycles, such as the circulation of carbon from the air into living plants, animals, organic deposits, and back into the air again.

The major geochemical cycle embraces both the deep-seated processes of metamorphism and igneous differentiation, and the surficial processes of weathering, erosion, transportation, and sedimentation. The horizontal division in Fig. 2.1 indicates the boundary between these two sectors of the geochemical cycle.

II. GEOCHEMICAL DISPERSION

A given small mass of material in the earth normally does not maintain its identity as it passes through the major transformations of a geochemical cycle, but rather tends to be redistributed, fractionated, and mixed with other masses of material (Fig. 2.2). This process, in which atoms and particles move to new locations and geochemical environments, is called geochemical dispersion. Nearly all dispersion occurs in dynamic systems in which earth materials are undergoing changes in chemical environment, temperature, pressure, mechanical strain, or other physical conditions. The rocks or minerals stable in one environment and the grains or atoms contained in them are released to be dispersed by either chemical or mechanical processes.

Dispersion may be the effect of exclusively mechanical agencies, such as the injection of magma or the movement of surficial material by glacial action. Apart from alluvial sorting of clay and sand, purely mechanical processes of dispersion usually involve mixing but not differentiation of the dispersed materials into specialized fractions. In contrast, chemical and biochemical processes commonly create fractions of widely differing chemical composition. The more mobile fractions tend to leave their original host if adequate

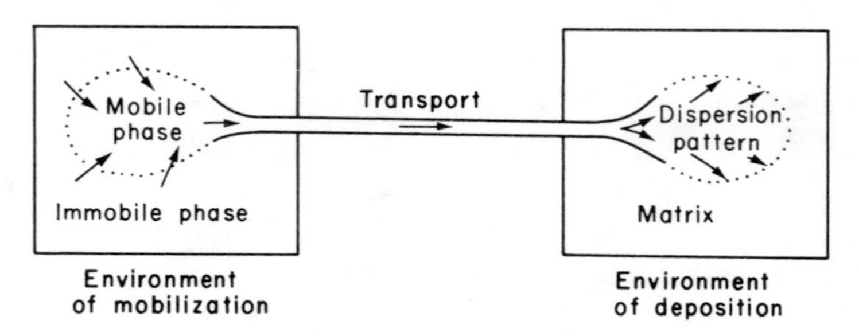

Fig. 2.2. The dispersion process.

channel-ways and chemical or physical gradients are available. When the mobile phase enters a new environment, a part of the introduced material may be deposited.

Dispersion may be either deep-seated or surficial, according to the geochemical environment in which it occurs, and primary or secondary, according to whether it occurs during the formation of the ore deposit or during a later stage, as discussed by James (1967). Around magmatic and most hydrothermal deposits, primary dispersion occurs in the deep-seated environment, and secondary dispersion in the surficial environment. However, because an increasing number of ore deposits are now considered to be formed in the surficial environment, a clear distinction is needed between the environment of ore formation and dispersion (deep-seated vs surficial), and the stage (primary vs secondary). Primary dispersion includes all processes leading to emplacement of elements during the formation of an ore deposit, no matter how the orebody was formed. Secondary dispersion applies to the redistribution of the primary patterns by any later process, usually in the surface environment. For ores formed by hydrothermal solutions at depth in the earth, the two sets of terms have parallel meanings; however, for syngenetic sedimentary deposits, primary patterns would be formed during sedimentation (Table 2.1). Secondary patterns would be formed at a later time if the

Table 2.1
Examples of dispersion in different environments and stages

Stage	Environment: Deep-seated	Environment: Surficial
Primary	Diffusion of metals into wallrock around hydrothermal deposit during ore deposition	Precipitation of traces of metals on sea floor near a volcanogenic deposit
Secondary	Diffusion of metals from ore deposit undergoing metamorphism	Weathering of sulfide ore deposit

deposit was exposed to weathering or if the deposit was metamorphosed after formation. Although the terminology may be confusing at times, it is important to distinguish between the environment and the time at which the processes occur, because these determine the characteristics of the resulting dispersion.

In deep-seated dispersion, the channel-ways and sites of redeposition are generally the fissures and intergranular openings of deep-seated rocks. Surficial dispersion, on the other hand, takes place at or near the surface of the earth, where patterns are formed in the fissures and joints of near-surface rocks, in the unconsolidated overburden, in streams, lakes, vegetation, and even in the open air.

The exploration geochemist is searching for traces of material that have been dispersed away from the orebody he is seeking. These dispersion patterns may form a considerably larger target area than the ore itself, so that a correspondingly lower sample density is required for discovery. In addition, the exploration geochemist is concerned with patterns of element distribution in normal rocks because he must be able to distinguish these normal patterns from anomalous patterns associated with an orebody.

III. GEOCHEMICAL MOBILITY

Fundamentally, the response of an element to dispersion processes is governed by its mobility, that is, the ease with which it may be dispersed relative to the matrix of other materials surrounding it. In some environments, mobility depends on the mechanical properties of the mobile phase, for instance on factors such as the viscosity of magmas and solutions, or the size, shape, and density of clastic grains in a flowing stream. The relationship between mobility and chemical differentiation, on the other hand, is more involved. Here, the prime factors are the chemical stability of the elements in immobile solid phases relative to the coexisting mobile fluid phases. To understand these processes, a digression into some chemical principles is necessary.

IV. GEOCHEMICAL REACTIONS

One important step in understanding any chemical process causing dispersion of an element is to write one or more balanced chemical reactions for the process. Because of the complexity of earth materials, an accurate and complete reaction sometimes cannot be written; nevertheless, just an attempt to do so usually increases one's understanding of the process.

In writing such reactions, two features of the process must be kept in mind. First, the mineral or fluid phase containing the element in question must be specified, as must the form of the element in the phase. The phase may be solid, liquid, or gaseous, and the element may occur as either a major essential element in the phase, or as a minor or trace component not essential to the existence of the phase. Second, many reactions in complex earth materials involve components other than the elements of specific interest. These components must be recognized and included if a correct reaction is to be written.

If the element of interest occurs as a major component of a mineral, then it may become mobile by dissolution, melting, volatilization, or some similar process. The most common type of reaction is dissolution of the mineral to form ions in aqueous solution. For instance, if we are concerned with the mobility of Zn as $ZnCO_3$ in near-surface waters, the simplest reaction would be:

$$ZnCO_3(s) = Zn^{2+}(aq) + CO_3^{2-}(aq). \qquad (2.1)$$

Although this reaction is chemically balanced, it does not express the real chemical reaction in most geochemical environments. In the majority of near-surface waters, the CO_3^{2-} ion is much less abundant than HCO_3^- or dissolved H_2CO_3 (Garrels and Christ, 1965). In view of this, a more realistic reaction may be written:

$$ZnCO_3(s) + H^+(aq) = Zn^{2+}(aq) + HCO_3^-(aq). \qquad (2.2)$$

From this equation, we see that the dissolution process depends on hydrogen ions (pH) and is not a simple dissociation into ions. The solubility of $ZnCO_3$ therefore is strongly dependent on the pH of the solution in contact with it.

Although for many reactions we can observe or infer the reactants and products, for others some reactants or products are difficult to ascertain, especially those involved in processes in the deep-seated environment. For example, in what form does Cu occur in a hydrothermal fluid? The discussions by Garrels and Christ (1965) are very helpful in writing common types of reactions in the surficial environment. Helgeson (1969) and Barnes and Czamanske (1967) give useful information for the deep-seated environment. However because of the variety of environments involved and the rapid advance in understanding of geochemical processes, a good grasp of geochemical principles is essential.

In addition to their occurrence in pure compounds such as $ZnCO_3$, trace elements also commonly occur as trace components of minerals and fluids. Crystalline solid solutions and exchange sites on clays or the surfaces of colloidal particles are examples of phases that may contain trace elements of interest. For instance, if Zn occurs in solid solution in calcite, the behavior of Zn may depend on the following reaction:

$$ZnCO_3(\text{calcite}) + Ca^{2+}(aq) = CaCO_3(s) + Zn^{2+}(aq). \qquad (2.3)$$

The $ZnCO_3$ in this reaction is not a pure phase but is specified as a component in calcite. The content of Zn in the calcite is seen to depend on the properties of the host $CaCO_3$ and on Ca^{2+} and Zn^{2+} in solution.

The chemical environment can be expressed in terms of the chemical activities of chemical species. The chemical activity (a) is the effective concentration of the species in the environment in question. In aqueous solutions, the activity of a constituent is approximately equal to its concentration in moles (gram-atomic weights) per liter. In gases, activity of a constituent is approximated by its "partial pressure". In solid solutions or mixtures of liquids, activity of a constituent is related to its mole fraction (moles of constituent per total moles of all substances in a given volume). For more details, see Garrels and Christ (1965).

For any chemical reaction, an equilibrium expression involving the chemical activities of the species taking part can be written and used to estimate the stability of the phases involved in the reaction. Equilibrium expressions for the three reactions discussed above are:

$$K_1 = \frac{a_{Zn^{2+}} \cdot a_{CO_3^{2-}}}{a_{ZnCO_3}} \tag{2.1a}$$

$$K_2 = \frac{a_{Zn^{2+}} \cdot a_{HCO_3^-}}{a_{ZnCO_3} \cdot a_{H^+}} \tag{2.2a}$$

$$K_3 = \frac{a_{CaCO_3} \cdot a_{Zn^{2+}}}{a_{ZnCO_3} \cdot a_{Ca^{2+}}}. \tag{2.3a}$$

In eqns 2.1a and 2.2a, a_{ZnCO_3} is approximately one because the $ZnCO_3$ is nearly pure, but in eqn 2.3a, a_{ZnCO_3} is much less than one because $ZnCO_3$ is a trace constituent in calcite. The value of *K*, the equilibrium constant, defined as the product of the activities when the reaction is at equilibrium, is dependent on temperature. For many reactions, *K* can be estimated from data in the literature, and concentrations can be converted to activities so that equilibrium expressions can be used to evaluate the stability of reactants vs products in a particular environment. Even though quantitative results cannot be obtained, the form of the equilibrium expression gives useful insights into the nature of the reaction and the effects of changes in one parameter, such as pH, on other parameters.

A. Dispersion of Elements Under Deep-seated Conditions

As a magma crystallizes, trace elements are partitioned between the crystals and the silicate melt. For instance, for a divalent element such as Zn^{2+}, the partition into pyroxene can be expressed by a reaction such as:

$$Zn^{2+}(\text{melt}) + CaMgSi_2O_6(\text{pyroxene}) = CaZnSi_2O_6(\text{pyroxene}) + Mg^{2+}(\text{melt}). \tag{2.4}$$

The equilibrium expression, which is analogous to eqn 2.3a, can be rearranged to give:

$$\frac{a_{CaZnSi_2O_6}}{a_{CaMgSi_2O_6}} = K\frac{a_{Zn^{2+}}}{a_{Mg^{2+}}}. \qquad (2.4a)$$

If concentrations are substituted for activities, then K is the distribution coefficient (MacIntire, 1963). Assuming that activities in the solids are approximately proportional to mole fractions, the above equation indicates that the Zn/Mg ratio in the pyroxene is related to the Zn/Mg ratio in the melt, multiplied by K. If K is large, Zn is concentrated in the pyroxene during crystallization; if K is one, the pyroxene and melt have the same Zn/Mg; if K is less than one, then Zn/Mg is low in the pyroxene compared to the melt. A low value of K, corresponding to low Zn in the pyroxene, results in Zn remaining in the magma (as far as reactions involving pyroxene are concerned). Zinc is then mobile, that is, it remains in the mobile fluid phase.

The value of K in such reactions is dependent on the relative energy involved in substituting Zn for Mg in the pyroxene and in the melt. Factors affecting the energy are the relative ionic sizes, ionic charges, coordination,

Table 2.2

Index of ionic replacement of common cations[a]

Tl^{+}	0·03	Cu^{2+}	0·14	Be^{2+}	0·24
K^{+}	0·03	Co^{2+}	0·14	Nb^{4+}	0·28
Ag^{+}	0·04	Ni^{2+}	0·14	W^{4+}	0·28
Na^{+}	0·06	Mg^{2+}	0·14	Mo^{4+}	0·28
Cu^{+}	0·06	Th^{4+}	0·16	Ti^{4+}	0·28
Ba^{2+}	0·07	U^{4+}	0·19	Al^{3+}	0·35
Pb^{2+}	0·08	Zr^{4+}	0·20	Ge^{4+}	0·46
Ca^{2+}	0·09	Sc^{3+}	0·20	Si^{4+}	0·48
Mn^{2+}	0·13	Fe^{3+}	0·22	As^{5+}	0·60
Zn^{2+}	0·14	Cr^{3+}	0·22	P^{5+}	0·62
Fe^{2+}	0·14				

[a] Source: Green (1959).

and bonding characteristics. In Table 2.2, the ionic radius, charge, coordination number, and electronic configuration of cations have been combined into an "index of ionic replacement" (Green, 1959, p. 1155). Elements with similar values of this index tend to have distribution coefficients close to unity and to substitute extensively for each other in crystals. Thus, Mn, Zn, Co, and Ni tend to occur in the Fe–Mg sites of ferromagnesian minerals, and U and Th substitute for Zr in zircon.

If an aqueous fluid separates from a magma, elements are partitioned between the melt and the aqueous fluid according to the relative energy with which they are held in the two phases. For such situations, reactions of the type:

$$CO_2(\text{melt}) = CO_2(\text{aq}) \tag{2.5}$$

can be written for which

$$K = \frac{a_{CO_2}^{\text{aq}}}{a_{CO_2}^{\text{melt}}} \tag{2.5a}$$

where K in this case is a distribution coefficient. If the value of K is large (as it is for CO_2 in most types of magmas), the constituent is concentrated in the aqueous hydrothermal solution and acquires greater mobility as a result. The components H_2O, CO_2, H_2S, SO_2, F, Cl, B, and in some environments, Cu, Pb, Zn, Au, and other metallic elements, become mobile in this type of reaction.

The stability of an element in a mobile phase may be greatly affected by the formation of complex ions or molecules made up of more than one element. Common complexes includes species such as H_2S and HCO_3^-, as well as MoO_4^{2-}, $CuCl_2^-$, $ZnCl_4^{2-}$, $Hg(HS)_3^-$, and other metal-bearing species. Because most of the metal in solution may occur in such a complex, reactions must be written in terms of the most abundant species, or the concentration of the simple ion must be known in terms of the concentration of the complex. For relatively insoluble minerals, variability in concentration of the complexing element (O, S, Cl, H, etc.) may have a strong influence on the mobility of the element.

In metamorphic processes, a pore fluid rich in water furnishes a mobile phase into which some constituents of the rock may be extracted by solution of minerals or by exchange in and out of solid solutions. Migration of this pore fluid furnishes mobility for these constituents, which migrate with the pore fluid until a change in physical or chemical conditions causes precipitation or incorporation into solid phases.

Empirically, information on the mobility of elements in the deep-seated environment may be gained by observing their impoverishment or enrichment in a specific mineral species formed at different stages of a deep-seated differentiation cycle. Thus the higher Sn, Sr, and Li content of micas in late pegmatite differentiates as compared with the same species of mica in the parent igneous mass suggests that these elements are relatively mobile under the prevailing conditions. The occurrence of an element as a characteristic constituent of deposits formed from fluids of various kinds, either vapors, supercritical solutions, or liquid aqueous solutions, may also be an indication of mobility. In this group are the elements of complex pegmatites, hydrothermal vein deposits, juvenile water, and gaseous emanations.

B. Mobility Under Surficial Conditions

Mobility in the surficial environment is dominated by transport in aqueous solutions. An approximate guide to mobility in such solutions is given by the ionic potential, which is equal to the ionic charge divided by the ionic radius (Fig. 2.3). Elements with low ionic potential (Ca, Na) are soluble as simple cations; those with very high ionic potential attract oxygen ions and form soluble oxy-anions (PO_4^{3-}, SO_4^{2-}, MoO_4^{2-}). Elements with intermediate ionic potential are generally immobile because of very low solubility and strong adsorption to surfaces (Al, Ti, Sn). Transition elements with incomplete inner electron shells (Fe, Cu, Cr, Ag, and others of the middle of the periodic table) tend to be less soluble and more strongly adsorbed than non-transition ions of similar charge and ionic radius. Differences in valence state (Fe^{2+} vs Fe^{3+}) also result in differing mobility.

Where the identity of the stable mineral species and of the soluble ionic phases is known, and where, at the same time, the thermodynamic constants for these species have been determined, it is possible to compute the relative solubility and hence the mobility of minor elements in natural surface waters. Such information is available for a wide variety of minerals and aqueous species (Robie *et al.*, 1978; Garrels and Christ, 1965; Sillen and Martell, 1964), and the solubility of minor elements can be predicted for many surficial geochemical environments. Effects of coprecipitation and adsorption by hydrous Fe- and Mn-oxides, clay, and organic matter are less well understood, but great advances are currently being made (Chapter 8).

A quantitative estimate of relative mobility under surficial conditions is difficult, but an empirical estimate can be obtained by comparing coexisting mobile and immobile phases. In particular, an approximation can be made by comparing the composition of natural waters and the rocks or soils with which they are in contact. Another method is to compare the soil with the rock from which it has developed. Perel'man (1967) has used the first method to estimate mobilities of elements, using the coefficient of aqueous migration (K), equal to the content of the element in the dissolved solids of a surface or ground water divided by its content in the associated rock:

$$K = \frac{100M}{a \cdot N} \tag{2.6}$$

where M is the concentration of the element in the drainage water (in mg/liter), a is the total mineral residue in the water (%), and N is the concentration of the element in the rocks (%). Mobility obviously depends on the

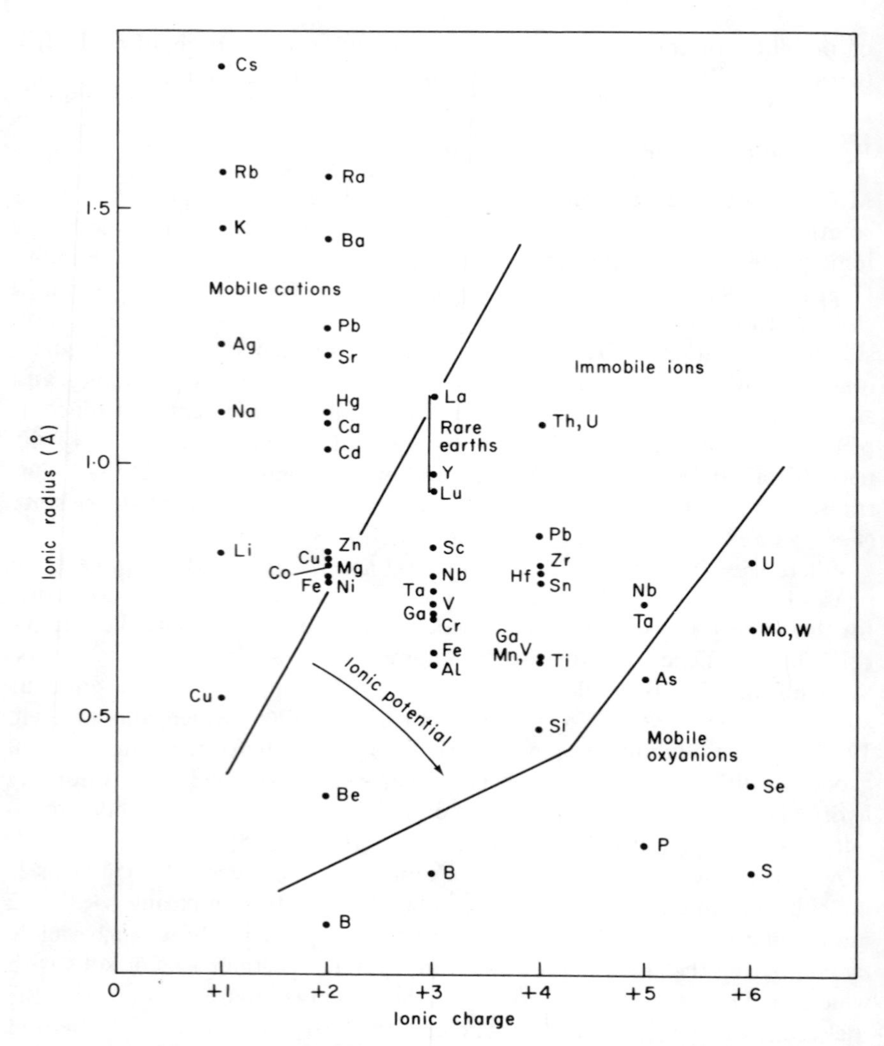

Fig. 2.3. Mobility in the surficial environment as a function of ionic charge and ionic radius. (Data from Whittaker and Muntus, 1970.)

chemical conditions, so at least the general conditions must be specified. As yet, quantitative data are not readily available for many environments, but Table 2.3 summarizes information based on coefficients of aqueous migration, plus practical experience in exploration geochemistry. In normal surface weathering, conditions would fall in the oxidizing, pH 5–8 column of Table 2.3; near an oxidizing sulfide orebody, conditions would be much more acid

Table 2.3
Mobility of elements in surficial environments[a]

Relative mobility	Oxidizing (pH 5–8)	Oxidizing (pH<4)	Reducing
Highly mobile ($K>10$)	Cl, Br, I, S, Rn, He, C, N, Mo, B (Se, Te, Re ?)	Cl, Br, I, S, Rn, He, C, N, B	Cl, Br, I, Rn, He
Moderately mobile ($K = 1$–10)	Ca, Na, Mg, Li, F, Zn, Ag, U, V, As (Sr, Hg, Sb ?)	Ca, Na, Mg, Sr, Li, F, Zn, Cd, Hg, Cu, Ag, Co, Ni, U, V, As, Mn, P	Ca, Na, Mg, Li, Sr, Ba, Ra, F, Mn
Slightly mobile ($K = 0{\cdot}1$–1)	K, Rb, Ba, Mn, Si, Ge, P, Pb, Cu, Ni, Co (Cd, Be, Ra, In, W ?)	K, Rb, Ba, Si, Ge, Ra[b]	K, Rb, Si, P, Fe[c]
Immobile ($K<0{\cdot}1$)	Fe, Al, Ga, Sc, Ti, Zr, Hf, Th, Pa, Sn, rare earths, Pt metals, Au (Cr, Nb, Ta, Bi, Cs ?)	Fe, Al, Ga, Sc, Ti, Zr, Hf, Th, Pa, Sn, rare earths, Pt metals, Au, As[b], Mo[b], Se[b]	Fe, Al, Ga, Ti, Zr, Hf, Th, Pa, Sn, rare earths, Pt metals, Au, Cu, Ag, Pb, Zn, Cd, Hg, Ni, Co, As, Sb, Bi, U, V, Se, Te, Mo, In, Cr (Nb, Ta, Cs ?)

[a] Based on data from Perel'man (1967), Chapter 14, the Appendix, and other sources.
[b] In presence of limonite.
[c] Intermediate oxidation potential.

but still oxidizing; and in swamps, waterlogged soils, and similar organic-rich environments, conditions would be reducing. An oxidizing orebody in a carbonate environment would approximate the near-neutral oxidizing situation. Note that certain elements, including As and Mo, are strongly scavenged from solution in Fe-rich environments, either by formation of Fe compounds or by adsorption on Fe-oxides.

V. ASSOCIATION OF ELEMENTS

In a geochemical survey an element measured in order to detect an orebody is termed an indicator element. In a majority of situations, the indicator element is an economically valuable component of the ore being sought, for example, Cu for Cu ores or U for U ores. However, if the valuable component is difficult to analyze, immobile, or yields data which are difficult to interpret, another element associated with the ore may be more useful. Such an element is called a pathfinder element. Before deciding to use a pathfinder element, the reason for association of the pathfinder element with ore and the range of geochemical environments in which the association occurs should be examined.

Elements tend to be associated because of similar relative mobility in a group of geological processes. Within the range of environments in which the elements are associated, the ratio of the two elements remains relatively constant, so that high contents of one element are accompanied by high values of the other, and vice versa. This relationship is illustrated in Table 2.4. Some

Table 2.4
Examples of As content of soil as a pathfinder for Au ore[a]

Position on traverse	As content of soil (ppm)	Au content of soil (ppm)
300 N.	40	0·1
50 N.	100	0·1
10 N.	480	1·0
0	1000	3·0
15 S.	560	1·0
50 S.	190	0·5
300 S.	80	1·0
400 S.	10	0·25

[a] Source: James (1957).

elements maintain characteristic associations throughout a wide range of geological conditions. Others may travel together during most processes of the deep-seated environment but part company in the surficial environment. Still others are characteristic of very specific plutonic rocks and associated oxide ores, of sulfide ores, or of certain kinds of sedimentary deposits. The presence of one member of the association suggests the probable presence of other members.

Useful pathfinder elements are those with more desirable geochemical or analytical properties than the principal ore metal being sought. Some examples are given in Table 2.5; a more extensive list is presented in Table 4.2. For instance, an element with high mobility in the surficial environment may allow more efficient detection of an ore composed of immobile elements if it is consistently associated with such ores. Molybdenum is more mobile than

Table 2.5
Some common geochemical associations of elements[a,b]

Group	Association
Generally associated elements	*K*–Rb
	Ca–Sr
	Al–Ga
	Si–Ge
	Zr–Hf
	Nb–Ta
	Rare earths, La, Y
	Pt–Ru–Rh–Pd–Os–Ir
Plutonic rocks	
General association (lithophile elements)	*Si–Al–Fe–Mg–Ca–Na–K*–Ti–Mn–Zr–Hf–Th–U–B–Be–Li–Sr–Ba–P–V–Cr–Sn–Ga–Nb–Ta–W–the halogens–rare earths
Specific associations	
Felsic igneous rocks	*Si–K–Na*
Alkaline igneous rocks	*Al–Na*–Zr–Ti–Nb–Ta–F–P–rare earths
Mafic igneous rocks	*Fe–Mg–Ti*–V
Ultramafic rocks	*Mg–Fe*–Cr–Ni–Co
Some pegmatitic differentiates	Li–Be–B–Rb–Cs–rare earths–Nb–Ta–U–Th
Some contact metasomatic deposits	Mo–W–Sn
Potash feldspars	*K*–Ba–Pb
Many other potash minerals	*K–Na*–Rb–Cs–Tl
Ferromagnesian minerals	*Fe–Mg*–Mn–Cu–Zn–Co–Ni
Sedimentary rocks	
Fe-oxides	*Fe*–As–Co–Ni–Se
Mn-oxides	*Mn*–As–Ba–Co–Mo–Ni–V–Zn
Phosphorite	*P*–Ag–Mo–Pb–F–U
Black shales	*Al*–Ag–As–Au–Bi–Cd–Mo–Ni–Pb–Sb–V–Zn

[a] Source: Goldschmidt (1954), Krauskopf (1955), and Boyle (1974).
[b] For additional association in orebodies see Table 4.2.

Cu in most surficial environments and can be used as a pathfinder for porphyry-copper deposits which typically contain some Mo. In other cases, the pathfinder may produce clearer and less ambiguous patterns of high values than more abundant elements of the ore, as in the use of Cu as a pathfinder for Ni–Cu ores, which typically occur in Ni-rich rocks that furnish high Ni values whether mineralized or not. Thirdly, the pathfinder element may be more easily detected than the element being sought, as in the case of As as a pathfinder for Au deposits. The essential characteristic is that the pathfinder element should have a consistent relationship to mineralization. In this regard, note that pathfinders are normally usable only for certain types of ore or geochemical environments. For instance, Mo is not a pathfinder for all Cu deposits, only for porphyry-copper deposits. The most widely used pathfinders are listed in Table 2.6.

Table 2.6
Pathfinders

Pathfinder element	Material sampled	Ore type
As	Wallrock, residual soil, stream sediment	Vein-type Au ore
Hg	Wallrock and soil	Complex Pb–Zn–Ag ores
Se	Gossan, residual soil	Epigenetic sulfides
Ag	Residual soil	Ag-bearing Au ore
Mo	Water, stream sediment, soil	Porphyry-copper deposits
SO_4	Water	Sulfide deposits

VI. PATTERNS OF GEOCHEMICAL DISTRIBUTION

In any given area, the geographic distribution of a given element in rocks, soils, and other materials is a response to the sum total of all the processes concerned in the movement of earth materials. In many instances, this distribution reflects simply the distribution of lithologic units (Fig. 2.4). Other processes of the deep-seated environment, such as hydrothermal alteration, may modify the basic relation to rock types. Weathering, erosion, and a variety of other processes in the surface environment will further blur and modify the patterns of the deep-seated environment.

The recognition of patterns related to ore is the aim and function of geochemical exploration. If the effects related to ore are intense and relatively local, the distinction from more normal processes is simple; on the other hand, if the effects are weak and related to ore in a complex manner, the interpretation of geochemical data is correspondingly more difficult and complex. For effective recognition of patterns related to ore, it is necessary first to determine the background of the indicator elements in unmineralized materials.

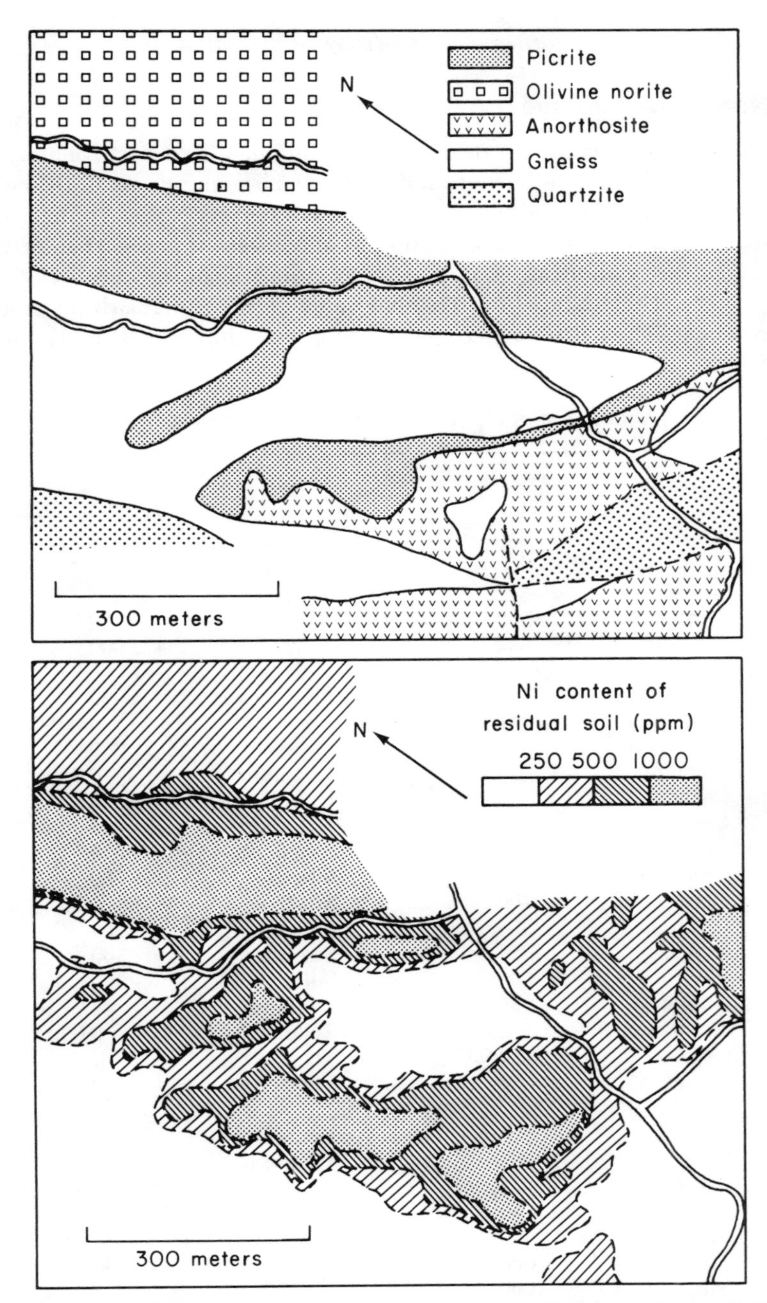

Fig. 2.4. Relationship between geology and the pattern of Ni in residual soil, Nguge region, Tanzania. (Colluvial and alluvial overburden occur flanking the main rivers.) (After Coope, 1958.)

A. Normal Background Values

The normal abundance of an element in unmineralized earth materials is commonly referred to as background. For any particular element, the normal abundance is likely to differ considerably from one type of earth material to another. Furthermore, the distribution of an element in any particular earth material is rarely uniform. Thus it is usually more realistic to view background as a range rather than as an absolute value, even in a relatively uniform environment. The nature of the environment itself, however, can also have a

Table 2.7
Abundance of elements in average crustal rocks (ppm)[a]

Element	Abundance	Element	Abundance
Aluminum	81 000	Mercury	0·02
Antimony	0·1	Molybdenum	1·5
Arsenic	2	Nickel	75
Barium	580	Niobium	20
Beryllium	2	Oxygen	473 000
Bismuth	0·1	Palladium	0·01
Boron	8	Phosphorus	900
Bromine	1·8	Platinum	0·005
Cadmium	0·1	Potassium	25 000
Calcium	33 000	Rhenium	0·0006
Carbon	230	Rubidium	150
Cerium	81	Scandium	13
Cesium	3	Selenium	0·1
Chlorine	130	Silicon	291 000
Chromium	100	Silver	0·05
Cobalt	25	Sodium	25 000
Copper	50	Strontium	300
Fluorine	600	Sulfur	300
Gallium	26	Tantalum	2
Germanium	2	Tellurium	0·002
Gold	0·003	Thallium	0·45
Hafnium	3	Thorium	10
Indium	0·1	Tin	2
Iodine	0·15	Titanium	4400
Iron	46 500	Tungsten	1
Lanthanum	25	Uranium	2·5
Lead	10	Vanadium	150
Lithium	30	Zinc	80
Magnesium	17 000	Zirconium	150
Manganese	1000		

[a] Sources: Green (1959), Taylor (1964), Wedepohl (1969–1978), and Appendix data (average of contents in granite and mafic rocks).

marked influence on distribution, because under certain conditions some elements may be enriched and others impoverished. Consequently, whatever type of sample is involved, the range of background values should be determined or at least reconsidered whenever a new area is studied.

As a guide to the general level of background values expected in rocks, data on the composition of average igneous rocks are helpful (Table 2.7, Fig. 2.5). The composition of many types of rock, however, differs significantly

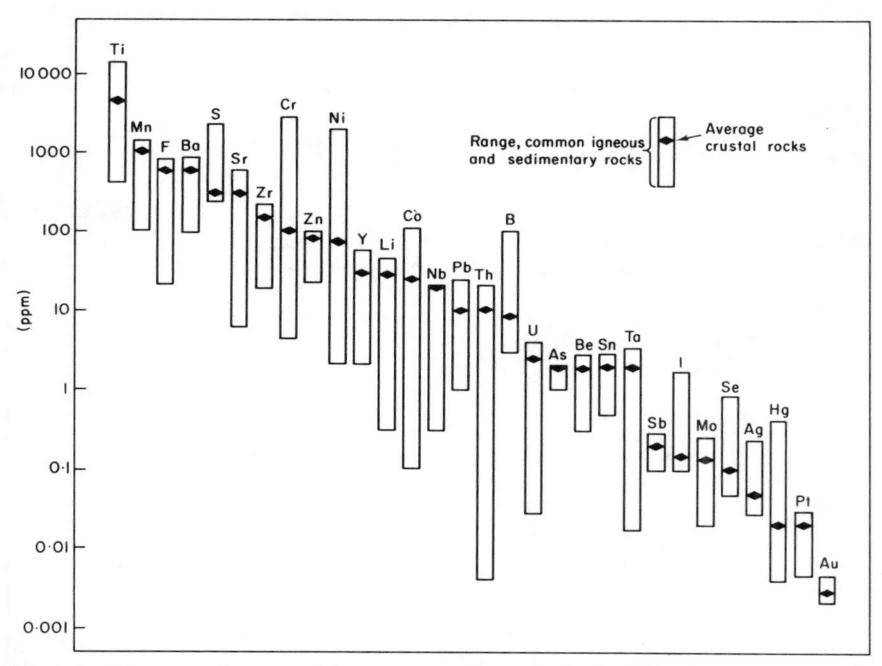

Fig. 2.5. Average and range of the content of the principal minor elements in normal rocks. (Data from Appendix.)

from the average, for both minor elements and trace elements. For instance, ultramafic rocks are characterized by an unusually high content of Cr, Ni, and Co, and some granitic rocks contain higher than average amounts of Li and Rb. In addition to the data for average igneous rocks, Fig. 2.5 also shows the range for selected minor elements covered by the average compositions of six principal rock types (ultramafic, mafic, and granitic igneous rocks; sandstone, limestone, and shale).

Background in soils is subject to appreciable variation, according to soil type and soil horizon, particularly in well-differentiated profiles characterized by marked enrichment of some constituent such as Fe-oxide or organic matter. The range of values and extremes actually observed in normal soils in one comprehensive survey of soil composition is given in Fig. 2.6. In glacial deposits, materials from different sources tend to be homogenized by mixing, but for most elements, background approximates to that in the parent rocks.

Present data concerning the normal minor-element content in different mineral species, water, and vegetation are either inadequate or show too wide a variation to be usefully summarized in table form. Some data are given in the Appendix and in Chapter 14, but for more detailed information the reader is referred to references cited by Rankama and Sahama (1950), Fleischer (1954), Goldschmidt (1954), Green (1959), Wedepohl (1969), and Turekian (1977).

B. Statistical Distribution of Background Values

Studies of the frequency distribution of elements in rocks and other natural materials suggest that in many cases the log-transformed concentrations are distributed approximately normally (Ahrens, 1954, 1957), as indicated in

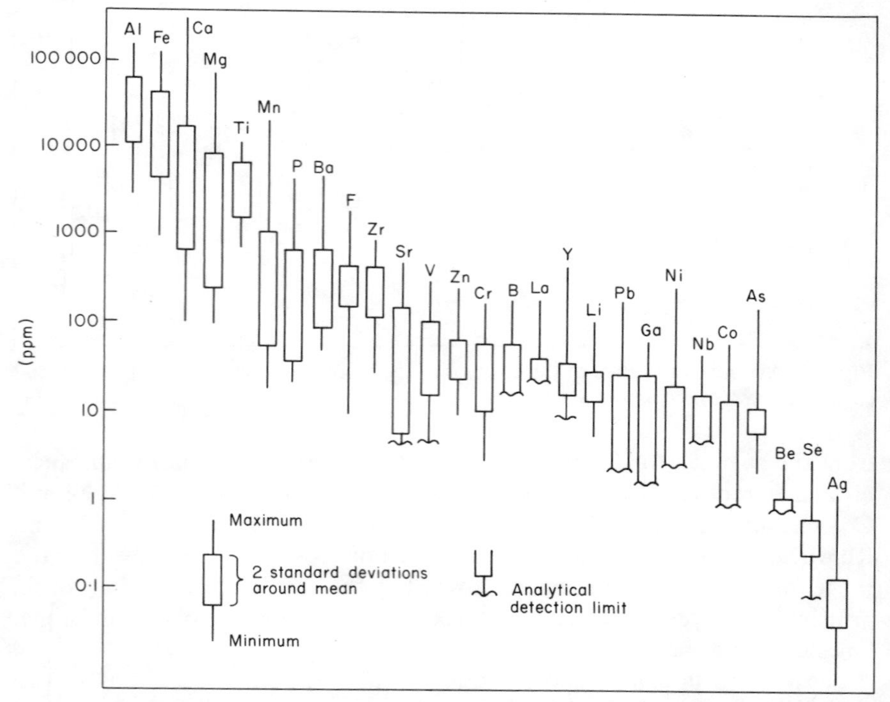

Fig. 2.6. Range of the content of major and trace elements in soil. (Based on data from Connor and Shacklette, 1975.)

Fig. 2.7. A great deal of controversy was aroused by this generalization (Shaw, 1961), and, as a result, it is clear that although many trace and minor elements are distributed approximately log-normally, neither this distribution nor any other is universally applicable. Because of the variety of processes affecting geological materials, and the variety of materials included in any group of items considered as a population, data cannot always be fitted by a specific frequency distribution. For instance, Oertel (1969) shows that some sets of data are fitted better by gamma or beta distributions than by log-normal functions, and Tolstoy *et al.* (1965) find that Pearson type I and IV curves fit better than log-normal curves. Govett *et al.* (1975) suggest that elements in closely defined populations are normally rather than log-normally distributed.

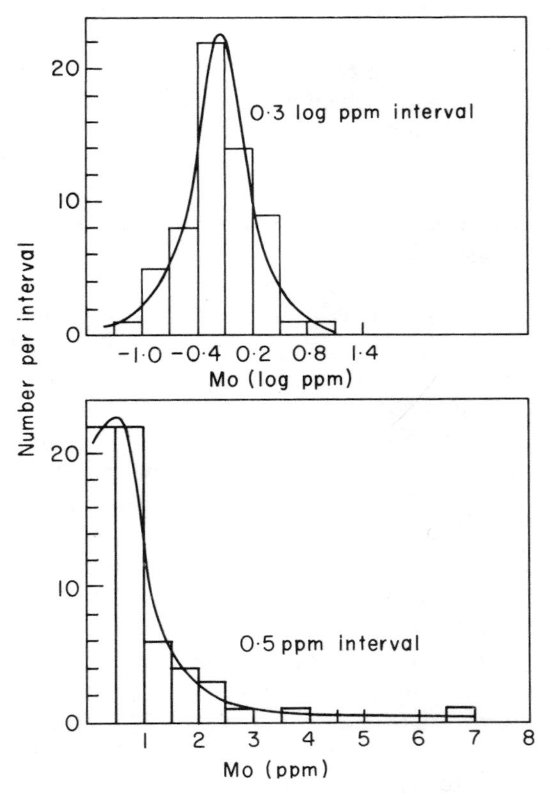

Fig. 2.7. Plot illustrating approximately log-normal frequency distribution of Mo in granite (top), contrasted with marked skewness of raw data (bottom). (After Ahrens (1954). Reprinted with permission from *Geochim. Cosmochim. Acta*, © Pergamon Press Ltd, 1954.)

Whether or not the hypothesis of log-normal distribution lacks rigor and generality, the fact remains that most sets of data encountered in exploration appear more nearly log-normal than normal, and that useful results can be obtained by assuming log-normal behavior (Tennant and White, 1959; Miesch, 1977; Sinclair, 1974, 1976). For most sets of geochemical data, little appears to be gained by use of more complex distributions.

C. The Geochemical Anomaly

By definition, an anomaly is a deviation from the norm. A geochemical anomaly, more specifically, is a departure from the geochemical patterns that are normal for a given area of geochemical environment. Strictly speaking, an ore deposit, being a relatively rare or abnormal phenomenon, is itself a geochemical anomaly. In addition, the recognizable patterns of geochemical dispersion related either to the genesis or to the erosion of the ore deposit are anomalies.

Anomalies that are related to ore and can be used as guides to ore are termed significant anomalies. Such anomalies usually consist of values that are higher than background; "negative" anomalies are only rarely useful in searching for ore. Unfortunately, high contents of indicator elements can arise from non-economic mineralization or by geological or geochemical processes unrelated to ore; such superficially similar anomalies unrelated to ore are termed non-significant anomalies.

The threshold is the concentration of an indicator element above which a sample is considered anomalous. In the simplest case, the threshold is the upper limit of normal background fluctuations; any higher values are anomalies, and lower values are considered background. In more complex cases, two or more types of threshold values may be recognized. Sometimes the anomalies related to ore may be set in a background of higher than normal values, representing a geochemical relief consisting of (i) a low-lying plain of regional background, separated by a regional threshold from (ii) a plateau of higher values related to extensive feeble mineralization or dispersion, from which rise (iii) the anomalies most closely related to ore, defined by a local threshold (Fig. 2.8). Recognition of regional and local thresholds can be extremely important in prospecting as it may then be possible to limit the detailed search for peak anomalies to the plateaus of high values defined by preliminary wide-spaced regional sampling.

In other instances, data from orientation surveys at several known deposits may lead to setting a threshold well above the level of normal background fluctuations in order to separate significant anomalies from those related to weak mineralization or unusual geochemical features. In any case, the threshold is set at a level that is most convenient and efficient in detecting and delimiting ore.

The contrast of an anomaly expresses its strength as a ratio to the average background or the threshold. Both background and threshold are used by different workers in the denominator of this ratio, and it should be clearly stated which reference level is used. For most purposes, a ratio of anomaly to background is simpler and more easily comparable.

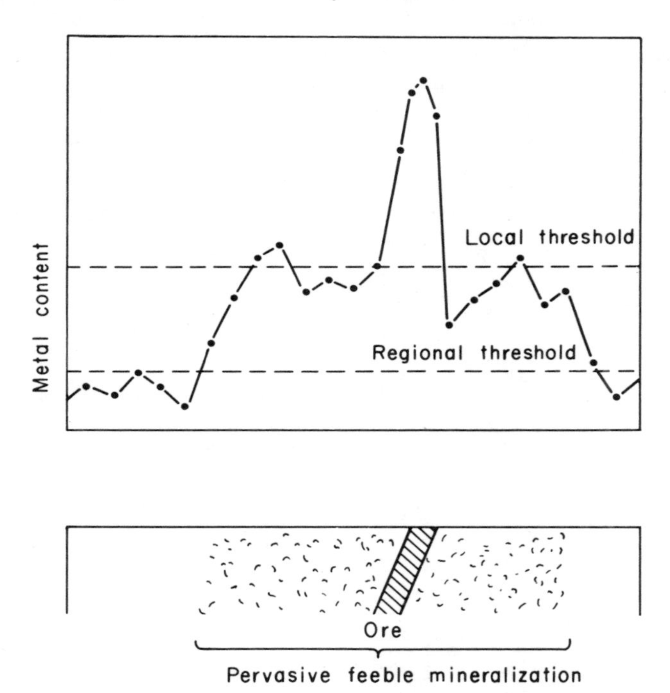

Fig. 2.8. Diagram illustrating local and regional threshold values.

VII. PRINCIPLES OF INTERPRETATION

Effective interpretation of geochemical data involves consideration of multiple populations of data. In data from a survey, a log-normally distributed population of background samples may constitute one population. Samples near ore and affected by dispersion involving the ore can be considered a second population. Samples related to certain rock types or unusual aspects of the environment may define additional populations. The samples related to ore usually have an approximately log-normal frequency distribution but with a higher mean and different standard deviation than the background samples. The background and ore populations commonly overlap, so that a completely satisfactory discrimination of background and ore samples is impossible (Fig. 2.9).

If the survey is to be effective, at least some of the samples near ore must contain anomalous values that can be distinguished from the background population with moderate to high probability. If the threshold is set too high (threshold A, Fig. 2.9), some orebodies will be missed. If the threshold is set too low (threshold C), time and money will be wasted in following up non-significant anomalies. The more the populations overlap, and the more numerous the individual populations among the samples, the greater is the difficulty in selecting an effective threshold.

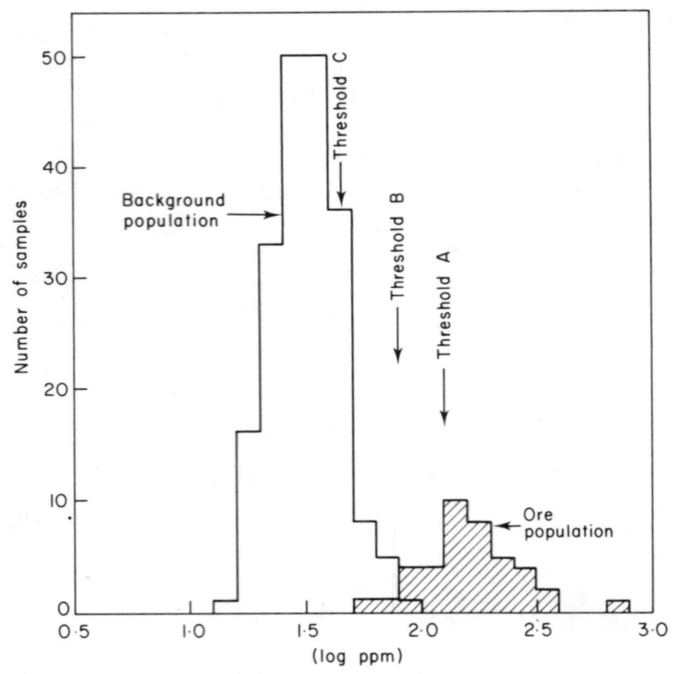

Fig. 2.9. Overlap of values in background and ore populations (derived as random samples from two log-normally distributed populations).

In order to separate anomalous samples from background, a number of methods can be used, depending on the amount of data, purpose of the survey, knowledge of the area surveyed, and economic consequences of the selection. Methods for selection of a threshold include (i) comparison with data from the literature, (ii) graphical discrimination of the numerous background values from the smaller proportion of anomalous values in an upper tail or second peak on a histogram of the data, (iii) calculation of the threshold from the mean plus 2 (or 3) standard deviations, (iv) plotting the cumulative frequency on log probability paper and partitioning into anomalous and background populations, (v) recognition of clusters of anomalous samples when the data are plotted on a map, and (vi) comparison with the results of an orientation survey. In addition, combinations of these methods and more complex versions involving statistical treatment or use of multiple variables have been utilized and will be described later.

The simplest method, useful if only a few analyses are available from a new area, is a simple comparison of the content of the indicator element with data in the literature for similar material (see Appendix). Samples that are anomalous on the basis of this criterion may be worth further study, sampling, and follow-up. However, care must be taken to allow for unusual features of the geochemical or geological environment. The conclusions from this simple procedure should be regarded as preliminary and very tentative.

If limited numbers of analyses are available from an area, the first step is to plot a histogram (Table 2.8, Fig. 2.10) showing the frequency of values in successive classes (small ranges of concentration). The class interval should be selected to give 10–20 classes (Lepeltier, 1969), preferably with at least five

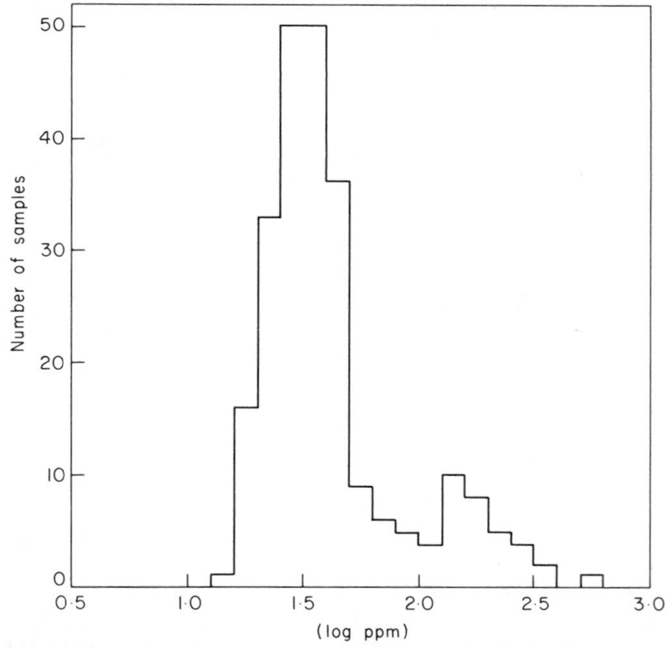

Fig. 2.10. Histogram plotted from data of Table 2.8, and equal to the sum of the two populations in Fig. 2.9.

Table 2.8a
Partial list of data for Fig. 2.10 and Table 2.8b

Concentration	Log concentration	Concentration	Log concentration
28	1·45	45	1·66
25	1·39	32	1·51
42	1·62	29	1·46
25	1·40	29	1·46
33	1·51	37	1·57
120	2·07	18	1·26
40	1·60	49	1·69
31	1·49	132	2·12
15	1·17	.	.
28	1·45	.	.
		.	.

Table 2.8b
Classification of data for combined populations of Fig. 2.9

Log ppm	Number of samples	Cumulative number	Cumulative percent
1·10–1·20	1	240	100·0
1·20–1·30	16	239	99·6
1·30–1·40	33	223	93·0
1·40–1·50	50	190	79·2
1·50–1·60	50	140	58·3
1·60–1·70	36	90	37·5
1·70–1·80	9	54	22·5
1·80–1·90	6	45	18·8
1·90–2·00	5	39	16·3
2·00–2.10	4	34	14·2
2·10–2·20	10	30	12·5
2·20–2·30	8	20	8·3
2·30–2·40	5	12	5·0
2·40–2·50	4	7	2·9
2·50–2·60	2	3	1·2
2·60–2·70	0	1	0·4
2·70–2·80	1	1	0·4

samples in many of the classes in order to minimize statistical fluctuations. Examination of the histogram usually shows a large peak representing background, plus a tail or second peak at higher values (Fig. 2.10). More symmetrical peaks may be produced after conversion of the data to logarithms.

A value near the upper end of the large background peak may be useful as a first approximation to the threshold, but some of the additional methods described below should usually be tried before settling on this value.

A modification of this procedure is to assume a threshold at a specified number of standard deviations above the mean. This procedure is most useful in the early stages of major surveys for data sets that are approximately normally distributed and appear to contain very few significant anomalies, as for many reconnaissance geochemical surveys. For a normally distributed population of this type, the 2·5% of the samples with concentrations greater than the mean plus 2 standard deviations include most of the significant anomalies and only a small proportion of the background population. However, as pointed out by Cachau-Herreillat (1975), because only the characteristics of the background population are considered, the selection of 2 (or 3) standard deviations is arbitrary and not of general applicability. For detailed surveys in which an appreciable proportion of ore-related samples is likely, this method is clearly not appropriate..

A fourth method of determining a threshold uses a plot of the cumulative frequency distribution on probability paper (or log probability paper), as suggested by Tennant and White (1959). The resulting graph is partitioned into two or more populations representing background, anomalies, and other geochemical effects, as described by Sinclair (1974, 1976). Preferably, the percentages in each successive class are cumulated from the highest class down in order to give maximum emphasis to the higher values (Lepeltier, 1969). The cumulative percent scale on probability paper is graduated so that a normal distribution plots as a straight line (Fig. 2.11). A mixture of two normal populations plots as nearly linear segments separated by curved segments containing an inflection point. Log-normal distributions plot similarly if a log scale is used for concentration, or if the data are log-transformed before being

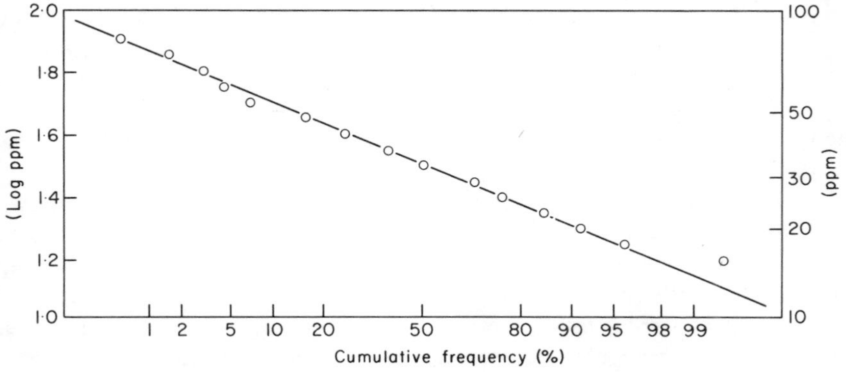

Fig. 2.11. Log probability plot of background population of Fig. 2.9.

plotted on an arithmetic scale. The result for a combination of two log-normal populations is an S-shaped curve (Fig. 2.12). The inflection point gives an estimate of the proportions of the two populations in the mixture; for instance, proportions of 16% and 84% are indicated from the inflection point shown on the figure. The curve may then be partitioned into two curves by the procedure described in Table 2.9 (Sinclair, 1974, 1976).

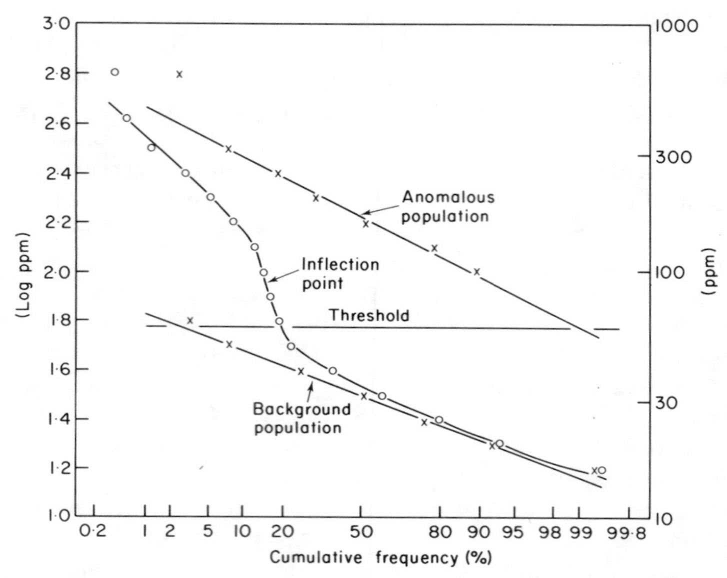

Fig. 2.12. Log probability plot of combined populations of Fig. 2.9, plotted from data of Table 2.8. The two populations have been resolved by the method of Sinclair (1974).

Assuming that the partition of the frequency distribution into background and anomalous populations is correct, the consequences of choosing various threshold values can now be evaluated. For instance, selecting a threshold value of 60 ppm (1·78 log ppm), 99% of the ore-related samples are correctly identified as anomalous and 1% are missed. Of the background samples, 97·5% are correctly identified as not indicative of ore, and 2·5% are mistakenly selected for further follow-up, as illustrated in Table 2.10. The two kinds of mistakes are analogous to the type I and type II errors recognized by statisticians (Koch and Link, 1970, p. 107). Because the costs of follow-up work are generally small relative to the value of an ore deposit, the threshold in geochemical work is usually set so that nearly all the ore-related anomalies will be correctly recognized. However, the economic and other consequences of this decision should be carefully considered in any specific case.

Table 2.9
Procedure for partitioning a single population into two normally distributed populations[a]

(i) Convert the data to logarithms if necessary.
(ii) Classify by concentration into 10–20 groups, with at least 5 samples in most of the groups.
(iii) Compute cumulative percentages, starting at the high concentrations, ending with 100% at low concentrations (Table 2.8b).
(iv) Plot the cumulative percentages against the lower class limits on probability paper as indicated in Fig. 2.12, and draw a smooth curve through the points.
(v) If the data plot is an S-shaped curve as in Fig. 2.12, select the inflection point of the curve (16% in Fig. 2.12). This point defines the approximate proportions of the two populations. If shape is more complicated, see Sinclair (1976).
(vi) For class 1, compute $P_A = P_1/f_A$, where P_1 is the observed cumulative percentage for class 1, and f_A is the proportion of samples falling in population A (0·16 in example). For example, $P_A = 0{\cdot}4\%/0{\cdot}16 = 2{\cdot}5\%$.
(vii) Plot a new point at P_A and the lower limit of class 1. Continue step 6 until the inflection point is reached or until newly calculated points markedly diverge from a straight line.
(viii) Draw a line to fit the points plotted in step (vii). This line defines the anomalous population.
(ix) Compute analogous points starting at low concentrations, using f_B (proportion of samples in population B, as determined by inflection point), and cumulative percentages computed as $100-P$. Draw line for background population.
(x) Check the calculated populations by combining them in the proper proportions, using $P_m = f_A P_A + f_B P_B$, where P_m is the cumulative percentage in the mixture. Plot the values of P_m and compare with original points. Repeat using a slightly different inflection point if necessary.

[a] After Sinclair (1974, 1976).

Table 2.10
Results of classifying samples as anomalous vs background

Assigned class	Real condition: Ore present	Real condition: Ore not present
Anomalous	Correct decision	Useless follow-up, type II error
Background	Ore missed, type I error	Correct decision

A fifth method of selecting a threshold value is to plot the data on a map, and contour or otherwise separate the higher values from the low values. A cluster of high values has a low probability of occurring by chance and may reflect a mineralized area. By raising and lowering the threshold value, a narrow range can frequently be found in which the anomalous area changes from one or a few clusters of samples into a wide dispersion of isolated anomalous values. The threshold should be set to furnish the largest possible clusters without producing a large number of isolated highs. Careful consideration of the distribution of known mineral occurrences, rock types, and geochemical environments should accompany this method. This method is most applicable to surveys of moderate detail, but not to many reconnaissance surveys.

A sixth method of determining the threshold is applicable if an orientation survey on known mineral occurrences in a similar geochemical and geological environment has been carried out, as it should be for any new region or type of survey. The threshold should be set lower than the content of the indicator element in most or all of the samples associated with ore.

In a typical large geochemical survey, all of the methods described here should be utilized or at least considered. A threshold derived from a histogram or log probability plot should be tested by mapping the anomalies (method 5) and adjusted up or down according to the clustering of values and the results of orientation studies. Although statistical techniques may help in presenting and analyzing geochemical data, they cannot provide the interpretation. A reliable interpretation requires a combination of geological and geochemical experience, plus an appreciation of the economic aspects of further exploration. Pure mathematical analysis, therefore, is not likely to replace the subjective interpretive talents of the exploration geologist for some time to come.

Chapter 3

Principles of Trace Analysis

The practical effectiveness of any geochemical method of exploration depends in large part on the availability of an analytical procedure that is properly suited to the problem at hand. The procedure must be sensitive enough to detect elements present in very small concentrations; it must be reliable enough that the chances of missing an important anomaly are negligible; and, for most types of survey, it must be economical enough that very large numbers of samples can be processed. Added desirable features are simplicity of techniques, so that the analysis can be entrusted to relatively untrained personnel, and portability of equipment, so that the analytical laboratory can, if desirable, be set up near field operations.

Four steps are involved in almost every trace analysis procedure. First, the sample must be treated to prepare it for transport, storage, and subsequent steps in the analytical procedure. Then the sample must be partially or completely decomposed, so that the element to be determined is released in a form that can be easily manipulated. The third step is the separation of the element from other constituents that might interfere with later measurements. And, finally, the quantity of the element present must be estimated. In some trace-analysis systems, as in X-ray spectrography, one or two of the intermediate steps may be omitted. Also, in some systems, such as cold-extraction colorimetry and emission spectrography, several of the steps may be carried out essentially simultaneously as a single process.

In this chapter, the mode of occurrence of trace elements in natural materials is discussed as a basis for understanding the behavior of elements both during chemical treatments in the laboratory and geochemical processes in nature. The techniques of sample preparation, decomposition, separation, and estimation are briefly reviewed, with emphasis on the methods used in geochemical exploration. Finally, methods of determining and expressing precision and accuracy of analyses are discussed.

For more comprehensive summaries of chemical methods, see Wainerdi and Uken (1971), Stanton (1966, 1976) and other references at the end of the chapter.

I. MODE OF OCCURRENCE OF TRACE ELEMENTS IN SOLIDS

The form in which trace elements occur in samples of soil, stream sediment, rock, or other materials determines their chemical behavior in response either to the natural environment or to laboratory procedures. Four major types of occurrence are (Fig. 3.1):

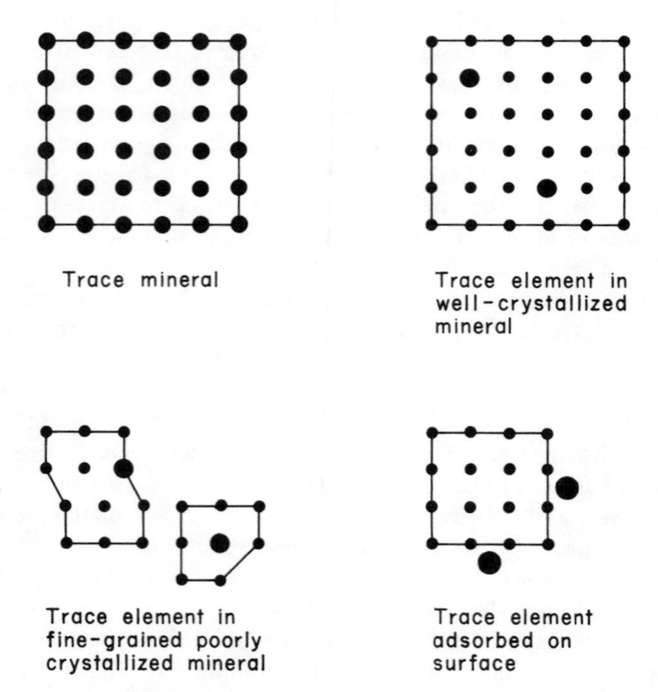

Fig. 3.1. Schematic diagrams of four modes of occurrence of trace elements. (●) Atom of trace element; (•) atom of major element.

(i) As a major element in a trace material, such as Pb in anglesite ($PbSO_4$), Cu in malachite ($Cu_2CO_3(OH)_2$), or Au as the native metal. The mobility of trace elements in these forms is dependent mainly on simple solubility and solution chemistry, or on physical processes transporting the grains.

(ii) As a trace constituent in the crystal structure of a well-crystallized mineral, such as Zn in magnetite, Pb in K-feldspar, or Cu in biotite. Well-crystallized minerals are most commonly formed by igneous, metamorphic, and hydrothermal processes. The behavior of a trace element occurring in this form depends primarily on the properties of the host mineral. If and when this host is destroyed or decomposed, then the trace element will be governed by simple solution chemistry and solubility.

(iii) As a trace element in a poorly crystallized material, or occluded as a trace mineral in such a phase, or adsorbed on such a material and trapped by further precipitation. Such materials are commonly formed in the surface environment. Examples are Co or Cu in Fe–Mn-oxides, Zn in the strongly bonded octahedral sites of a montmorillonite clay, and Hg in organic compounds. Basically, the controls on the behavior of such trace elements are the same as in (ii), but because of the poorly crystalline nature of the host, the trace elements tend to be more accessible to the surrounding solutions than those in coarser-grained, more nearly perfect lattices of minerals formed in igneous, metamorphic, or diagenetic environments. In spite of this, partial dissolution of the host or a strong acid attack is usually required to liberate trace elements in this form.

(iv) As a trace element adsorbed on the surface of a colloidal particle of Fe–Mn-oxide, clay or organic material, or in the exchange layer of a clay mineral. Elements in these sites are controlled mainly by ion exchange equilibria. Even minor changes in composition of surrounding solutions may liberate the trace element to solution.

From the above classification, it is clear that a trace element enclosed by a host mineral that is inert in either the natural or laboratory environment will remain in the solid phase and can only be affected by physical processes, such as sorting by size or density of the particle. In contrast, a trace element occurring as a trace mineral or as an adsorbed ion may be directly taken into solution by appropriate changes in the surrounding solution. Trace elements in the lattice of major minerals can go into solution only after decomposition of the host.

A. Stability of Important Minerals

The minerals formed during weathering are generally fine grained, poorly crystalline, and have a large surface area, and, as a result, may be dissolved or modified by relatively mild chemical treatments. The silicates of igneous and metamorphic rocks (and their detrital grains in sedimentary rocks, soils, and stream sediment) generally require more drastic chemical treatment for decomposition. Trace minerals have a wide range of properties, ranging from phases soluble in water to zircon and other heavy resistant minerals.

The stability of Fe–Mn-oxides is a function of acidity (pH) and oxidation potential (Eh), as illustrated in Fig. 3.2. Oxides of both Fe and Mn are dissolved by either acid or reducing solutions. The Mn-oxides are more easily dissolved than the Fe-oxides. By an appropriate choice of acids or reducing agents, Fe- or Mn-oxides or both may be selectively dissolved from samples of soil or stream sediment, leaving more-resistant minerals behind. Conversely, if the chemical conditions are kept within or near the field of stability of the Fe- and Mn-oxides, trace metals enclosed in them should be unaffected. Similarly, organic materials and sulfides are stable only under reducing conditions and may therefore be dissolved by strongly oxidizing treatments (Fig. 3.2). Clays, micas, feldspars, and mafic silicates are only slowly decomposed by weak acids or bases because silica and alumina are relatively insoluble in such solutions (Fig. 3.3). Many of these silicates are only slowly decomposed even in hot concentrated acids and therefore are relatively unaffected by conditions that dissolve Fe- or Mn-oxides.

II. PREPARATION OF SAMPLES

Samples of naturally occurring material almost always require some kind of preparatory treatment before they are ready for chemical analysis. The purpose of such treatment may be (i) to put the sample into a form that can be readily transported and stored, (ii) to homogenize the sample so that subsample variations are minimized, or (iii) to effect a preliminary separation of the elemental constituents according to their occurrence in the different kinds of particle. All such treatments must be designed to avoid contamination of the samples, and if large numbers of samples are to be processed, they must be handled with minimum of time and effort. Also, the fewer steps and manipulations involved in preparation and determination, the less chance there is for contamination and error, other factors being equal. General aspects of sample preparation are discussed by Lavergne (1965).

Water is normally removed from samples before shipment to the laboratory. Water is not only an unnecessary component of clastic or organic

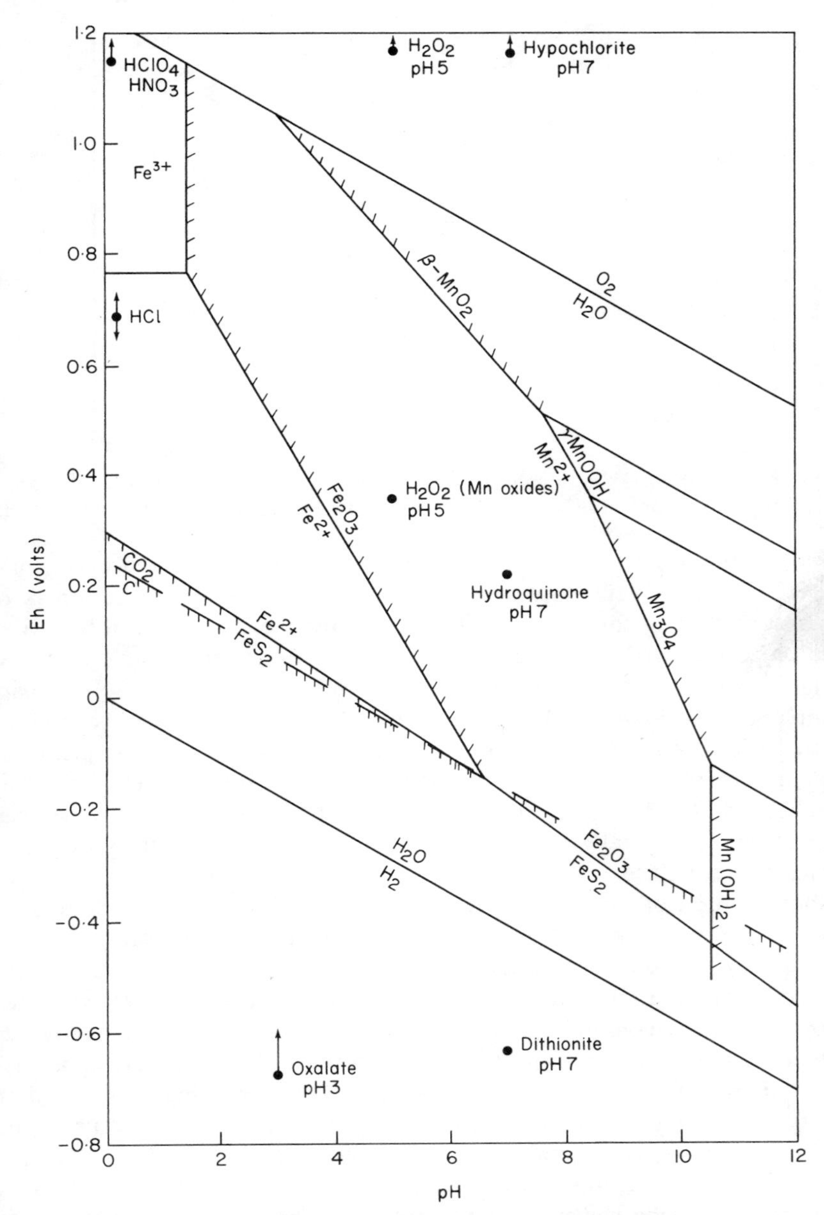

Fig. 3.2. Stability of Fe-oxide, Mn-oxide, pyrite and organic matter (C) as a function of Eh and pH, as compared to conditions imposed by several types of selective leach solutions. Solid phases are stable on the hatched sides of boundaries.

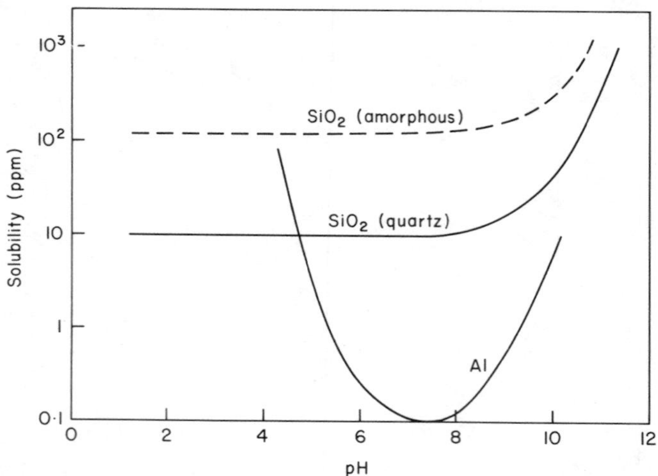

Fig. 3.3. Solubility of aluminosilicates vs pH, as indicated by solubility of Al and Si. (After Krauskopf (1967), p. 168, © McGraw-Hill Book Co., Inc., and Parks (1972).)

samples, but it can also actively interfere with subsequent processing in a variety of ways. Wet samples tend to cause decay of paper containers or corrosion of metal containers. Wet clastic material cannot be readily pulverized or sieved. Furthermore, analytical results are normally desired on a dry-weight basis, so that the sample must be dried before weighing. Soil or plant samples are most commonly dried either in the sun or in drying ovens set up at the field camp. Samples to be analyzed for readily extractable metals or similar components should not be completely dried, because their form may be changed. The inorganic constituents of water samples may require concentration before analysis, either by evaporation in the laboratory or by ion exchange in the field.

Pulverization of samples of rock or vegetation by crushing, grinding, or chopping serves in part to increase the surface area of the sample that is exposed for subsequent chemical attack and in part to homogenize the sample.

Rock samples are usually crushed to 6–10 mm in a steel-plate jaw crusher with only minor contamination by Fe, Mn, Cr, and similar ferroalloy elements. If contamination from these elements must be avoided, crushing with large blocks of the sample itself has been used. Grinding from 1 cm to smaller sizes may be carried out with steel-plate pulverizers or disc mills, which usually add considerable steel and associated Mn, Mo, Cr, V, Ni, and related elements. Ceramic-plate pulverizers (Barnett *et al.*, 1955), or ceramic or tungsten-carbide ball mills or disc mills may be used to avoid ferroalloy contamination. Rock samples are commonly ground to pass 60–200 mesh (250–74 μm).

Soils and stream sediments are usually sieved before analysis. This sieving has two purposes: (i) rejection of coarse fragments of quartz, organic litter, and other materials relatively poor in trace metals, and (ii) production of a fine-grained homogeneous product from which representative subsamples may be conveniently taken. Sizes of sieves should be specified by the dimension of the sieve opening, because sieves of the same nominal mesh may differ in opening by as much as a factor of two.

Mineral separations of various other kinds may be made on clastic samples prior to chemical analysis. These separations take advantage of the principal physical properties of the minerals, particularly the density, magnetic susceptibility, or electrical properties.

When small portions of pulverized rock, soil, or sediment are removed from a larger volume for further processing or analysis, care must be taken that these samples are representative. Relationships between grain size, content of element, and volume of material required to maintain specified precision are discussed by Gy (1966), Ottley (1966), Nicholls (1971), and Engels and Ingamells (1970). Sample splitters or quartering methods should be used in dividing larger samples, and segregation of minerals should be avoided by thorough mixing.

III. DECOMPOSITION OF SAMPLES

Many methods of extracting trace metals from rock, soil, sediment, and other materials are used in geochemical exploration. The user must choose a method giving optimum contrast between anomalies and background, within the constraints of cost, time, equipment, and subsequent analytical steps. In many geochemical surveys, contrast can be improved by selective extraction of only certain forms of the element, rather than the total content of the element. Methods of selective extraction are based on differing modes of occurrence of the element (Section I of this chapter).

The terms "total" ,"readily extractable", "hot extractable", "cold extractable", and similar expressions are widely used in geochemical exploration but have no well-defined meanings. "Total" frequently refers to decomposition by fusion or treatment by hot concentrated acid, which usually extracts 80–100% of heavy metals from most samples, but may extract less than 50% from some minerals and types of samples. Such methods are better termed "near-total". Alternatively, "total" may refer to analyses by emission spectrography, X-ray fluorescence or neutron activation, or decomposition by HF combined with strong acids, which give essentially total metal content for most natural materials. "Hot extractable" generally refers to treatment with hot acid in concentrations of 0·1—1 M. "Cold extractable" (contracted to cx, as in cxCu) generally refers to treatment with buffer solutions of pH 4—9, possibly combined with complexing agents such as dithizone (diphenyldithiocarbazone) or EDTA (ethylene diamine tetraacetic acid), at ambient temperatures.

In view of the many possible treatments, no concise nomenclature seems possible, and the specific procedure used should be specified whenever possible, and as closely as possible.

Both time and temperature, as well as chemical stability, affect the rate of decomposition of minerals in a sample being subjected to selective extraction or total dissolution. In general, a reaction such as ion exchange or dissolution proceeds at a relatively rapid initial rate and then slows down as equilibrium is approached or the phase is nearly exhausted (Fig. 3.4). Rates of

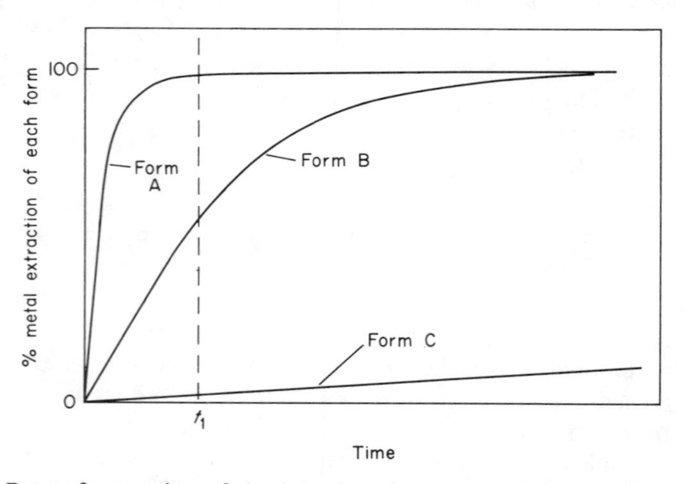

Fig. 3.4. Rate of extraction of three modes of occurrence of a metal in a sample. At the time t_1, metal in form A is essentially all extracted, and is effectively separated from form C, but not from form B.

reaction also tend to increase with temperature. For example, Ellis *et al.* (1967) show that rates of extraction can increase by factors as great as five as temperature is increased from 30 °C to 50 °C. Different minerals or phases react or dissolve at differing rates, so that in favorable cases, one reaction, such as dissolution of Fe-oxides, may be nearly complete when another, such as breakdown of clays, is barely started. Fine grain size also tends to increase the rate of reaction. In designing and using selective extractions, it is rarely possible to separate modes of occurrence completely, but adjustment of the time, temperature, extractant, and other conditions of extraction may allow nearly complete separation of components. For new types of samples or new extractants, experiments should be conducted to establish optimum conditions for selective extraction of the desired component.

The characteristics of some common methods of extraction and sample decomposition are summarized below.

A. Volatilization

The sample may be decomposed by volatilization in the extreme heat of an electric discharge (emission spectrography), releasing essentially the total content of the element, or in the more adjustable heat of an electric furnace (distillation of Hg) or induction furnace. The volatilized material may be determined directly by optical or other methods, or trapped by condensation or absorption in water for later determination. Methods using furnaces are capable of selectively extracting and measuring only certain forms of the element (Köksoy *et al.*, 1967; Watling *et al.*, 1973; Meyer and Leen, 1973) and will probably be more widely used in the future.

B. Fusion

An effective method of attack is fusion of the sample with an inorganic salt that melts at a reasonably low temperature but, at the same time, is capable of a vigorous attack on the sample. Fusion in potassium bisulfate ($KHSO_4$) has been commonly used because, in contrast to most other fluxes, the fusion can be carried out in Pyrex test tubes. Some fluxes, in addition to having a high temperature, attack the sample analogously to an acid, as in the case of $KHSO_4$; others are alkaline (Na_2CO_3) or oxidizing (Na_2O_2). Experiments with $KHSO_4$ suggest that it effectively decomposes clays, Fe–Mn-oxides, and to a lesser extent micas and felspars but not chain-structure silicates (Harden and Tooms, 1964; Foster, 1973). Lithium- or Na-borate or -metaborate decomposes nearly all common minerals and can be used in graphite crucibles (Medlin *et al.*, 1969; Ingamells, 1970; Cremer and Schlocker, 1976). In favorable cases the molten borate fusion can be poured directly into an aqueous solution, but in other cases a separate dissolution of the solidified melt is required. Non-oxidizing fusions may not release metal from organic-rich materials (Stanton *et al.*, 1973).

C. Vigorous Acid Attack

Hot concentrated acids are widely used as agents for decomposing samples; HNO_3, $HClO_4$, HCl, H_2SO_4, and HF have all been used. An acid attack by relatively concentrated HNO_3, HCl, $HClO_4$, or mixtures of these with each other and other acids decomposes Fe- and Mn-oxides, clay minerals, carbonates, and some other silicates, especially olivine, but trace elements in pyroxenes and amphiboles are only partly released (Foster, 1971, 1973). Mixtures containing $HClO_4$ appear to release the highest proportion of trace elements, probably because of the high temperature attainable and the oxidizing effect of this acid. Use of $HClO_4$ requires destruction of readily oxidizable organic material either by prior roasting, or oxidation by HNO_3 or other oxidizing agent to avoid explosions (Everett, 1967). Complete breakdown of silicates and most other common minerals is achieved with a combination of HF and another acid, using Teflon (fluorocarbon) beakers or, in favorable cases, test tubes (Foster, 1973). In general, oxidizing acids are best for decomposing sulfides and organic matter, but non-oxidizing acids are better for dissolving Fe- and Mn-oxides (Peachey, 1976), as might be inferred from the stability of these phases in terms of pH and oxidation potential (Fig. 3.2).

Care must be taken not to lose elements by volatilization from concentrated acids; for instance, As, Sb, Cr, Se, Mn, Re, Ge, Mo, and other elements can be lost during perchloric acid digestions (Hoffman and Lundell, 1939; Chapman *et al.*, 1949). Loss of volatile metal–organic compounds during oxidation of organic matter is also possible (Gorsuch, 1959).

D. Attack by Weak Aqueous Extractants

The trace metal content in exchange sites of clays and on the surfaces of colloidal particles can be released by displacing an exchange reaction of the type

$$Tr\text{–}X + Me^{2+} = Me\text{–}X + Tr^{2+} \tag{3.1}$$

to the right, where Tr is the trace cation, Me a major cation (both assumed to be divalent for purposes of illustration), and X the substrate (clay or colloidal particle). The H^+ of dilute or weak acids, such as HCl, HNO_3, or acetic acid, acts as the major cation Me and displaces the trace metal. In near-neutral buffer solutions, such as Na-acetate, the high content of Na^+ acts as the major cation. Organic complexing agents, such as citrate, tartrate, EDTA, and dithizone, decrease the content of free trace cation (Tr^{2+}) by complexing or chelating action, and thereby promote release of the trace element. Associated alkali, H^+, or other cations also aid the exchange. Some of these reagents also dissolve carbonates and other readily soluble minerals, but silicate lattices are little affected.

E. Oxidation–Reduction Agents

Hydroxylamine, dithionite, oxalate, and hydroquinone act as selective reducing agents for Fe- and Mn-oxides (Rose, 1975; Chao, 1972; Chao and Sanzolone, 1973). In near-neutral solutions, these reagents have little effect on clay lattices and other silicates. Similarly, hydrogen peroxide, hypochlorite, chlorate, and nitric acid act as oxidizing agents for sulfides and organic material (Lynch, 1971; Olade and Fletcher, 1974; Chao and Sanzolone, 1977). Ashing or treatment with strong acids may be necessary to extract metal chelated by organic matter (Chowdhury *et al.*, 1972). Several authors report on specific treatments necessary to deal with organic-rich samples (Horsnail, 1975; Maynard and Fletcher, 1973; Bradshaw *et al.*, 1974). The use of H_2O_2 in Na-carbonate solution to dissolve uraninite and HCl–NaCl to dissolve galena has been reported by Beus and Grigorian (1977).

The rate of dissolution of various phases can be important in designing analytical procedures and understanding the occurrence of elements. Ellis *et al.* (1967) illustrate some applications of this approach to geochemical exploration.

IV. SEPARATION

Once the trace element under study has been released from the sample, it may be necessary to separate it from interfering elements liberated from the sample at the same time. Separation is especially needed prior to most colorimetric methods, but it not usually required before determination by atomic absorption. The process of separation may also aid by concentrating the element in the separate phase, thus increasing the effective sensitivity of the overall method. Where the element is present in extremely low concentrations, as is often the case in natural waters, enrichment of the element is often necessary for its determination.

A. Separation in Liquid Phase

Many separations in trace analysis involve liquid–liquid solvent extraction permitting the transfer of dissolved material between two immiscible liquid phases, usually water and an organic liquid. Solvent extraction requires vigorous mechanical shaking of the system to emulsify the two component phases and thus increase the effective surface area across which the transfer of dissolved components takes place. Separation may be facilitated by the use of appropriate complexing agents that modify the solubilities of the trace element in the two phases. Solvent extraction is the basis of many of the colorimetric tests that have been extensively employed in geochemical prospecting.

Paper chromatography (adsorption on paper in a flowing capillary film) and other types of chromatography can also be used to separate elements.

B. Separation in Solid Phase

A solid phase containing the desired elements may be separated from a liquid phase by ion exchange or by precipitation. These methods may be applied to samples of natural water or to aqueous solutions obtained by some of the sample extraction procedures mentioned earlier.

V. ESTIMATION

Table 3.1 lists the principal methods of estimation and the elements for which they are especially well suited. Some of the advantages and disadvantages of these methods in geochemical prospecting work are mentioned in the following summary.

A. Colorimetry

The formation of colored compounds in solution by reaction of an element with a specific chemical reagent is the basis of colorimetry. Quantitative estimation of the element is possible if the intensity of the color, as measured by the absorption of light over a small range of wavelengths, is proportional to the concentration of the colored compound. A few colorimetric reagents form colored compounds only with one element, but most react with several elements, so preliminary exclusion of interfering elements by complexing or separation is usually required. Very commonly, the colored complex is extracted from aqueous solution into an organic solvent, thereby concentrating it and separating it from many interferences. Dithizone is a common colorimetric reagent which can be used to separate and estimate many elements by varying the pH of extraction and adding complexing agents for unwanted elements (Hawkes, 1963; Sandell, 1959). The advantages of colorimetry in geochemical exploration are the simplicity, low cost, and portability of the equipment, and the ease of training unskilled personnel in its operation. The disadvantages are the inability to determine more than one element at a time, and the sensitivity of many reagents to interferences and aberrant chemical conditions. Colorimetric methods are available for nearly all elements of interest in geochemical exploration.

Table 3.1
Methods of estimation and minor elements for which they are commonly used

Element	Colorimetry	Emission spectrometry d.c. arc	Emission spectrometry Plasma	Radiometry	Atomic absorption	X-ray fluorescence
Antimony	×		×			
Arsenic	×		×			
Barium	×	×	×		×	
Beryllium	×	×	×	×		
Bismuth	×		×		×	
Boron			×	×		
Cadmium	×	×	×		×	
Chromium	×	×	×		×	×
Cobalt	×	×	×		×	
Copper	×	×	×		×	
Fluorine[a]	×					
Gold					×	
Iron	×		×		×	×
Lead	×	×			×	
Lithium		×	×		×	
Manganese	×	×	×		×	×
Mercury	×				×	
Molybdenum	×	×	×		×	
Nickel	×	×	×		×	×
Niobium	×					×
Platinum	×	×				
Rare earths		×	×			×
Rubidium					×	×
Selenium	×		×			
Silver	×	×	×		×	
Strontium		×	×		×	×
Sulfur	×					×
Tantalum	×					×
Thorium				×		×
Tin	×	×	×			×
Titanium	×	×	×		×	×
Tungsten	×					×
Uranium[b]				×		
Vanadium	×	×	×		×	×
Zinc	×		×		×	×

[a] Most commonly used method is specific ion electrode.
[b] Commonly used method is visible fluorescence.

B. Emission Spectrometry and Spectrography

Almost all elements, when vaporized and ionized in the intense heat of an electric discharge or other source of energy, emit radiation of characteristic wavelengths in the visible and ultraviolet range as a result of electrons refilling the outer electron orbitals (Fig. 3.5). The element can be identified by the wavelength emitted, and the quantity of the element can be determined by the intensity of the light.

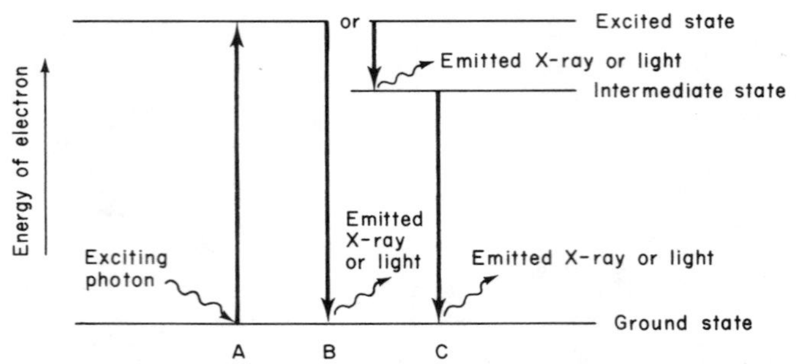

Fig. 3.5. Electronic energy changes in methods involving absorption and emission of radiation. (A) atomic absorption, absorption of energy by outer electron; (B) and (C) emission of energy from excited electron in emission spectrography (outer electron, visible or ultraviolet light) or X-ray fluorescence (inner electron, X-ray wavelength).

Intensity may be measured on photographic plates or directly by electronic photometers, the latter allowing a high degree of automation. In the past, the main source of energy was a d.c. or other arc in which a small amount of sample was burned as a powder or occasionally as a liquid. Feeding the sample to the arc as a powder on a tape has been widely used for geochemical samples (Danielson *et al.*, 1959). Recently, plasmas (gases highly excited by radio-frequency induction or other means) have been used as a source, with the sample introduced as a solution. The high temperature of the plasma and the homogeneity of solutions minimize the previous problems caused by instabilities in the d.c. arc and poorly reproducible sampling of a powder, and also improve detection limits. The advantages of emission spectrometry are the large number of elements that can be determined simultaneously, the low detection limits for most elements, and the low unit cost for large-scale operations. The disadvantages are the high cost of equipment and the need for a highly trained operator. Poor reproducibility has also been a problem with conventional electric arc sources. The improvements in reproducibility and detection limits with plasma sources, along with a demand for multi-element determinations, seem likely to lead to increased use of this technique in the future.

C. Atomic Absorption

Uncharged atoms in a vapor state are capable of absorbing photons of light having energy appropriate for exciting the outer electrons (Fig. 3.5). In atomic absorption, a sharp spectral line of the element to be analyzed is generated in a source lamp and passed through the vaporized sample. The wavelength absorbed allows specific identification of the element, and the proportion of light absorbed is a measure of the concentration of the element in the light path (Price, 1972). Dissolved samples can be aspirated into a flame where the atomic vapor is formed (Fig. 3.6), or solids or liquids can be heated

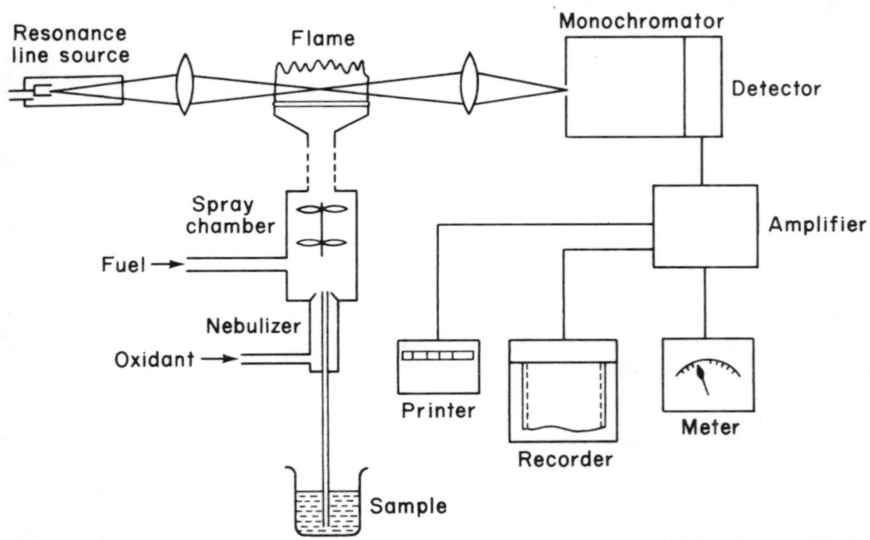

Fig. 3.6. Practical system for atomic absorption spectrometer. (After Price, 1972.)

in a furnace and the vapors passed into the light path. Atomic absorption, mainly using dissolved samples, has become the most widely used analytical method in recent years because of the low detection limits for most elements of interest, the specificity for individual elements, the opportunity to determine several elements on one solution, and the relatively inexpensive and simple equipment required. Methods of analysis are summarized by Perkin-Elmer (1976), Wainerdi and Uken (1971), Price (1972), and Dean and Rains (1975). A minor disadvantage for most conventional equipment is the limitation to one element at a time. Recent experiments with selective volatilization from solid samples in furnaces indicate very low detection limits for some elements, allow inferences on the form of elements in samples, and have potential advantages analogous to other methods of partial extraction (Meyer and Leen, 1973).

D. X-Ray Fluorescence Spectrometry

The inner electrons may be activated by an X-ray beam in such a way that fluorescent X-rays of a wavelength characteristic of the activated element are emitted. Advantages of X-ray fluorescence include multi-element capability, high precision, and low cost per determination. Disadvantages are high cost of equipment and detection limits of about 50 ppm, inadequate for many elements of interest in mineral exploration. However, nearly all elements with an atomic weight greater than Mg can be detected, many at levels lower than by emission spectrometry (Levinson and dePablo, 1975; Leake *et al.*, 1969; Leake and Aucott, 1973).

Normal laboratory X-ray fluorescence units use an X-ray tube as the source of exciting radiation, but because gamma rays and X-rays are identical, a gamma-emitting radioactive source can also be used to excite fluorescent radiation. Portable units using a radioactive source have been designed for evaluation of outcrops and drill holes (Bowie *et al.*, 1965; Kunzendorf, 1973; Wollenberg *et al.*, 1971; Gallagher, 1970).

E. Fluorimetry

Samples containing U, when fused with a suitable flux and cooled, emit a visible luminescence under ultraviolet activation. Under properly controlled conditions, the luminescence is quantitatively proportional to the amount of U present down to extremely low concentrations. This effect may be measured either visually by comparison with standards, or instrumentally with photoelectric devices. It is one of the most widely used and simplest methods of determining traces of U (Smith and Lynch, 1969).

F. Radiometric Methods

The elements U, Th, and K and some of their decay products may be detected and measured by their natural radioactivity. Non-radioactive elements may be converted into radioactive elements by bombardment with neutrons or other atomic particles (neutron-activation analysis). The identity of the radioactive element may be determined by the energy of the emitted gamma rays or other radiation, and the amount in a sample can be estimated from the intensity of the radiation. Extremely low detection limits are attained for many elements, and for some elements, the determination can be made directly on rock powders without chemical processing (Gordon *et al.*, 1968; Plant *et al.*, 1976). Disadvantages of neutron activation are the high cost of the equipment and general unavailability of a neutron source, the hazards of dealing with radiation, and the need for highly trained operators.

Portable equipment for evaluation of outcrops and drill holes has recently been developed, using ^{252}Cf and other neutron-emitting elements as sources (Moxham *et al.*, 1972).

G. Electroanalytical Methods

Measurements of electrical potentials and electrical currents in solutions have several applications in analyzing for trace elements. Measurement of pH by the potential of a glass electrode and Eh by potential of a platinum electrode are the best known procedures of this group. Analogous methods using electrodes sensitive to F^-, Cl^-, Cu^{2+}, Pb^{2+}, and other ions are presently available (Durst, 1969; Van Loon *et al.*, 1973; Plüger and Friedrich, 1973; Friedrich *et al.*, 1973). Specific-ion electrodes are easy to use and have sensitivities generally less than 1 ppm, but can be subject to interferences from other ions. The electrodes measure the chemical activity of the ions rather than their total concentration in solution, so that careful calibration and selection of the chemical medium are important. The specificity and simplicity of specific-ion electrodes are likely to lead to more widespread use of this technique.

In polarography, a gradually increasing voltage is applied across a pair of electrodes in a solution. When the voltage at the cathode is sufficient to reduce an ion to metal or to a lower valence state, a current is produced in the circuit by electrons transferred between the electrodes and the dissolved ion. The magnitude of this current is proportional to the concentration of the ion. Polarography has high sensitivity and can be carried out with very small amounts of solution. The equipment is relatively inexpensive. However, interferences between elements are common so that chemical separations and carefully controlled conditions are usually required. For this reason, polarography has not been widely used in geochemical exploration.

Other electroanalytical methods include the measurement of dissolved oxygen in water by electrical measurement of the current derived from decomposition of oxygen diffusing through a membrane selectively permeable to oxygen, and measurement of conductivity of solutions as a means of estimating the content of ionized solutes in water.

H. Chromatographic Methods

The adsorption or partition of solutes between a flowing fluid and a stationary sorbent is the basis of chromatographic methods. As the solvent flows, the less strongly sorbed atoms and molecules tend to be carried along by the solvent at a faster rate than the more strongly sorbed species. In gas–liquid chromatography, the solvent is a gas and the sorbent is liquid coating the surface of solid particles in a tube or column. In gas–solid chromatography, the same configuration is used, except that the sorbent is a surface-active solid. In both types, the sample containing the volatile compounds is injected into the column, carried along by an inert gas, and separated into atomic and molecular components after traversing a suitable length of column. The separated components are detected by measurement of thermal conductivity or ionizability of the gas stream, or they are passed into a mass spectrometer. The amplitude of the peak can be used to determine the amount of each constituent. Gas chromatography is most widely used in separating organic compounds. Similar methods are available for determining gases such as CH_4 and SO_2. The method is extremely sensitive, but some preliminary treatment of samples is usually necessary.

In paper chromatography, the liquid flows by capillary action through an adsorbing paper. Any contained solutes are separated according to their strength of adsorption on the paper. Similarly, in thin-layer chromatography, the liquid flows through a thin layer of solid sorbent coated on a plate of glass or other inert material. Extremely small amounts of inorganic or organic constituents can be separated, identified, and estimated by application of appropriate tests to the chromatogram. For the most part, this method is qualitative to semi-quantitative.

I. Mass Spectrometry

The mass spectrometer consists of an evacuated chamber into which a sample of gas is injected and in which the molecules or atoms are ionized. The ions are accelerated in a strong electric field and then passed through a magnetic field, which separates them according to their ratio of mass to charge. Molecules differing by one atomic mass unit can be easily separated. The ions are then recorded electronically or photographically. The intensity of the electrical or photographic signal at a specific mass is proportional to the abundance of molecules or atoms of that mass. The mass spectrometer is extremely sensitive, but is expensive and requires highly skilled maintenance.

Mass spectrometers are used in isotopic studies (because they can resolve individual isotopes of the same element), in organic geochemistry, and in gas analysis.

J. Other Methods

A wide variety of other methods has been used for geochemical analyses, including gravimetry (used in fire assay for Au, Ag, and Pt metals), turbidimetry (absorption or scattering of light by a suspended precipitate, used for $BaSO_4$), spot tests (formation of a colored spot), and flame spectrometry (analogous to emission spectrometry but using a flame for excitation instead of an arc or plasma). For more detailed discussions of these methods, see the references at the end of the chapter.

K. Units for Reporting Chemical Analyses

The common units for reporting trace element analyses are parts per million (ppm) by weight. For example, 20 ppm Cu indicates that the sample contains 20 grams of Cu in 10^6 grams of sample, or 20 micrograms (20 μg) per gram. Other common units for solid samples are per cent (%) and parts by billion (ppb), where a billion is 10^9, following the American usage. To convert from percent to ppm, multiply by 10^4; that is, 1% = 10 000 ppm. To convert from ppm to ppb, multiply by 10^3; that is, 1 ppm = 1000 ppb.

In water samples, the sample is usually measured by volume rather than by weight, and results are commonly reported in milligrams per liter (mg/l) or micrograms per liter (μg/l). For dilute solutions with a density of 1 g/ml, mg/l = ppm, and μg/l = ppb, by weight.

For a discussion of units for gas samples, see Chapter 18.

VI. RELIABILITY OF GEOCHEMICAL ANALYSES

Reliable analyses are required if significant anomalies and patterns are to be detected with any degree of confidence. The reliability of analyses may be conveniently evaluated in terms of precision and accuracy. Precision is a measure of the reproducibility of replicate determinations without regard to how close their average is to the true value. Accuracy deals with the closeness of the measured value to the true value. The difference is illustrated in Fig. 3.7.

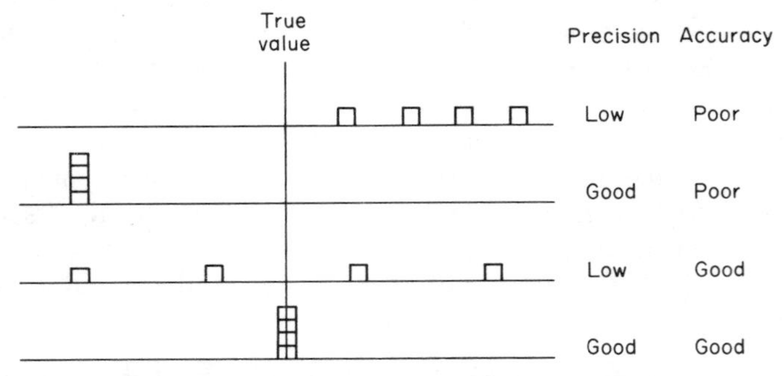

Fig. 3.7. Possible combinations of precision and accuracy based on four analyses of a sample. In case 3, only the average may be termed accurate.

In most geochemical surveys, precision is more important than accuracy, because if precision is poor, weak anomalies cannot be recognized and spurious anomalies may be created. However, if results are to be compared with other surveys or with data from the literature, then accurate results are also needed. Inaccurate results may take the form of a constant bias or a bias that varies from sample to sample (Miesch, 1967).

A. Precision

The precision of an analytical method can be determined by multiple analyses of a single sample or, preferably, by multiple analyses of many samples from the area being surveyed. The successive determinations of an element in a sample normally differ and produce a spread of values which can be shown as histogram (Fig. 3.8).

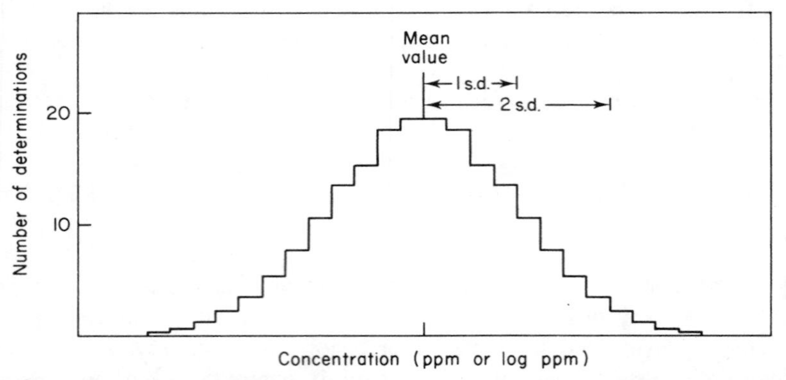

Fig. 3.8. Normal or Gaussian frequency distribution of 100 determinations of a sample.

The determinations commonly approximate to a normal (Gaussian) distribution (for major elements) or a log-normal distribution (for trace elements). If the errors are normally (or log-normally) distributed, then statistical techniques can be used to express the confidence that the content of the element is greater than or less than a certain value, such as the threshold, or the probability that the content falls within a certain range. The improvement in precision from multiple determinations on a sample and the results of other manipulations of the data can also be estimated.

Using multiple determinations of a single sample, the standard deviation (s_a) of an analytical method is calculated as

$$s_a^2 = \frac{\sum(x^1-\bar{x})^2}{n-1} = \frac{1}{n-1}\left[\sum x(_i^2) - \frac{(\sum x^1)^2}{n}\right] \quad (3.2)$$

where x_i is the ith determination of an element on a sample, and $\bar{x}$ is the mean (or average) of n determinations on a sample. If the values are normally distributed, then 68% of the determinations are expected to fall within 1 standard deviation of the mean, 96% within 2 standard deviations, and other proportions as given by the normal distribution (Fig. 3.8).

The mean of several determinations is obviously more precise than any individual determination. For normally distributed determinations (or for the logs of log-normally distributed determinations), the standard deviation of the mean ($s_{\bar{x}}$, also called the standard error) is

$$s_{\bar{x}} = \frac{s_a}{\sqrt{n}}. \tag{3.3}$$

For example, the standard deviation of the mean of duplicate analyses of a sample is $0 \cdot 7 s_a$. Duplicate analyses are relatively effective in improving the precision of reported analyses and, in addition, serve to detect the occasional large errors caused by mistakes in procedure, misnumbering, and other blunders. Additional replications can further improve the precision, but the improvement for each additional determination becomes progressively smaller.

Analytical error commonly varies with concentration, so that replicate analyses of one sample do not allow an estimate of precisions for samples of different concentration. For instance, at 100 ppm, the standard deviation might be 10 ppm, but at 20 ppm, it might be 4 ppm. Replicate determinations for several samples of differing concentrations are therefore desirable. Computation and expression of precision depend on how precision varies with concentration. A rigorous treatment is not usually possible, and the results must be regarded as approximations.

The precision is often approximately proportional to the concentration; that is, at 10 ppm, the standard deviation is 2 ppm; at 100 ppm, it is 20 ppm; at 1000 ppm, it is 200 ppm. In such a case, the coefficient of variation (C.V.), defined as $100 s_a/\bar{x}$, is a convenient measure of precision, 20% in the case of the example above. Alternatively and more precisely, the data may be converted to logarithms, giving a constant standard deviation of about 0·088 log ppm for the above examples. Note that the conversion from coefficient of variation to standard deviation of logs is not exact, because one standard deviation on either side of a mean of 2·0 (100 ppm) gives 82–123 ppm, which is different from the 80–120 ppm obtained using the coefficient of variation. For this reason, all precise calculations on data with this type of error should be conducted using logarithms.

If the standard deviation (or log standard deviation) is constant, it may be estimated from replicate analyses of a series of k different samples:

$$s_a^2 = \frac{(n_1-1)s_1^2+(n_2-1)s_2^2+\ldots(n_k-1)s_k^2}{n_1+n_2+\ldots n_k-k+2} \tag{3.4}$$

where n_i is the number of determinations of sample i, s_i is the standard deviation of determinations of sample i, and s_a is the average standard deviation for the entire group of samples. For the special case of a series of samples each analyzed twice:

$$s_a^2 = \frac{1}{k}\sum_{i=1}^{k}\frac{(x_{1i}-x_{2i})^2}{2} \tag{3.5}$$

where k is the number of samples, and x_{1i} and x_{2i} are the first and second determinations on sample i (Garrett, 1969).

In the more general case, if the standard deviation varies with concentration for both the raw data and the logged data, the standard deviation and number of replicates for individual samples may be listed in a table or shown graphically in a plot of s vs $\bar{x}$ (Fig. 3.9), so that analytical precision may be estimated for any sample. Simple methods for determining precision using duplicate analyses are presented and evaluated by Thompson and Howarth (1976).

The total analytical standard deviation may be subdivided into variability introduced by various stages of analysis, such as sieving, subsampling of powders, decomposition, and final determination steps. These standard deviations are related as follows:

$$s_t^2 = s_1^2+s_2^2+s_3^2+\ldots s_n^2 \tag{3.6}$$

where s_t is the total standard deviation of the final analyses, and s_1, s_2, s_3, s_n are the standard deviations introduced by single stages. Procedures for obtaining these estimates are discussed by Bennett and Franklin (1954, pp. 402—427).

The precision of geochemical analyses is commonly stated as the standard deviation, giving the range around the mean within which 68% of the determinations are expected to fall, or as 1·96 or 2·00 standard deviations within which 95% or 96%, respectively, of the determinations are expected to fall. Which of these alternative expressions is chosen should be clearly stated. For some geochemical work, coefficients of variation of 10–50% are common, especially for field methods, but recent atomic-absorption methods in fixed laboratories are capable of giving a coefficient of variation in the range 3–10%.

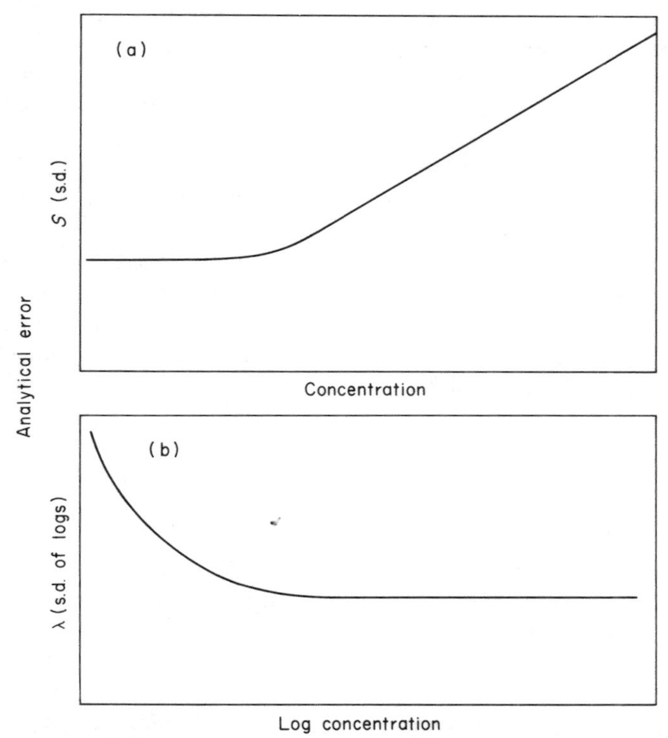

Fig. 3.9. (a) Typical relation of error (standard deviation of replicate analyses) and concentration. (b) Typical relation of standard deviation of logarithms of replicate analyses (λ) and log concentration.

B. Accuracy

Ideally, in order to evaluate the accuracy of a geochemical analysis, the true content of an element in some sample must be known and compared with determinations by the analytical method in question. The simplest type of inaccuracy is a constant deviation or constant percentage deviation of the determinations from the true value. In more complicated cases, the error depends on the form of the indicator element or on the content of some other element. If the error is constant or a constant percentage, the result may be very usable within the group of samples analyzed, but comparison with values from the literature or with other surveys may be misleading. With increasing use of computerized data banks and comparison of data from different sources, verification of the accuracy of geochemical data is becoming increasingly important. In large surveys, minor changes in analytical method during the survey may make it impossible to interpret the data properly.

At present, the simplest method for estimating accuracy is analysis of one or more reference samples carefully analyzed by other laboratories. A listing of such samples is given by Flanagan (1974, 1976). Compositions of these samples have been compiled by Flanagan (1973, 1974, 1976), and Allcott and Lakin (1975, 1978), but additions and improvements to the analytical data appear periodically. In exploration geochemistry, the samples USGS-GCX-1 to -6 are usually the best reference standards because of their relatively high contents of most elements of interest.

Other methods of evaluating accuracy include the analysis of synthetic samples made up to contain known amounts of elements, and the analysis of unknown samples spiked with known amounts of the element in question. Flanagan (1974) lists some synthetic samples available on request. All of these methods are capable of detecting a simple bias in the analytical method, but not all are capable of detecting bias related to differing behavior of elements in different chemical forms.

C. Routine Checks of Precision and Accuracy

In any geochemical project, the precision and accuracy of the analyses must be evaluated and monitored during the progress of the project. For small projects, one or more samples may be split and included for analysis under new numbers. In addition, one or more reference samples of known content should be analyzed.

In large projects, check samples should be included with each batch of samples or at a regular interval. Replication of every 10th or 20th sample, submitted under a different number, allows a measure of routine precision, as described by Garrett (1969). In addition, a reference sample, or possibly a composite of samples from the project, should be analyzed regularly. Plant *et al.* (1975) describe procedures used in monitoring precision and accuracy in a large geochemical survey and some errors uncovered by the procedure.

VII. ANALYTICAL PROCEDURES

Selecting an analytical system that combines the most suitable procedures for the preparation and decomposition of the sample, and for the separation and estimation of the desired constituents must be based on an appraisal of a number of factors. The principal of these are (i) the mode of occurrence of the element in the material sampled, (ii) the number and nature of the elements to be determined, (iii) the sensitivity and precision required, and (iv) the economics and logistics of the operation. The optimum combination of the first three factors should be determined by an orientation survey in known mineralized areas (Section I in both Chapters 13 and 16), or by information from previous surveys in similar terrain.

The economics of an operation make themselves felt where the capital investment of a large sum of money for an expensive instrument is required. For very large projects, many of the instrumental methods of estimation are far cheaper per unit than the methods that do not require any substantial investment. Thus, if hundreds of samples per day are to be processed continuously over a long period of time an X-ray spectrometer or a radiometric instrument could easily be amortized by the savings in cost per sample.

The logistics of the operation are involved where the problem of prompt reporting of analytical data and hence the desirability of a field analytical laboratory arises. Many of the expensive instruments that offer a real economy in cost per unit determination are too heavy or delicate to transport to the field. The need for a high-capacity power supply for many of the instrumental methods may also limit their applicability.

The analytical procedures that have found the widest acceptance in geochemical prospecting are listed in Tables 3.2, 3.3, and 3.4. In general terms, however, each geochemical survey presents its own analytical requirements. It cannot be emphasized too strongly that the optimum procedure in each case can only be devised in close collaboration between the geologist and the analyst, each with a full appreciation of the other's problems.

Table 3.2
References to principal analytical procedures used in routine geochemical exploration[a]

Element	Reference and type of method[b]
Ag	Ward *et al.*, 1969 (AA); Chao *et al.*, 1971 (AA)
As	Ward *et al.*, 1963 (COL); Aslin, 1976 (AA); Harms and Ward, 1975 (AA); Terashima, 1976 (AA)
Au	Ward *et al.*, 1969 (AA)
B	Stanton, 1976 (COL)
Ba	Stanton, 1976 (AA)
Be	Patten and Ward, 1962 (FL); Stanton, 1966 (COL)
Bi	Ficklin and Ward, 1976 (AA)
C	Dean, 1974 (G)
Cd	Ward *et al.*, 1969 (AA); Gong and Suhr, 1976 (AA)
Cl	Kesler and Van Loon, 1973 (E)
Co	Ward *et al.*, 1969 (AA); Stanton, 1966 (COL)
Cr	Ward *et al.*, 1963 (COL); Stanton, 1966 (COL)

Table 3.2—continued

Element	Reference and type of method[b]
Cold-ext. metals	Hawkes, 1963 (COL); Stanton, 1976 (COL)
Cu	Ward *et al.*, 1969 (AA); Ward *et al.*, 1963 (COL)
F	Kesler *et al.*, 1973b (E); Ficklin, 1970 (E)
Hg	Ward *et al.*, 1969 (AA); Jonasson *et al.*, 1973 (AA)
I	Ficklin, 1975 (E)
In	Hubert and Lakin, 1973 (AA)
Major elements	Medlin *et al.*, 1969 (AA); Shapiro, 1975 (AA)
Mn	Medlin *et al.*, 1969 (AA); Stanton, 1976 (AA)
Mo	Ward *et al.*, 1963 (COL); Kim *et al.*, 1974 (AA); Nakagawa *et al.*, 1975 (AA)
Nb	Greenland and Campbell, 1974 (COL)
Ni	Ward *et al.*, 1969 (AA)
P	Peachey *et al.*, 1973 (COL)
Pb	Ward *et al.*, 1969 (AA); Ward *et al.*, 1963 (COL)
Pt–Pd	Dorrzapf and Brown, 1970 (SP); Stanton, 1976 (COL).
Rb	Medlin *et al.*, 1969 (AA); Stanton, 1976 (AA)
Re	Mahaffey, 1974 (COL)
S	Searle, 1968 (furnace and titration); Shapiro, 1973 (turbidimetric)
Sb	Ward *et al.*, 1963 (COL); Welsch and Chao, 1975 (AA); Aslin, 1976 (AA)
Se	Crenshaw and Lakin, 1974 (FL)
Sn	Ward *et al.*, 1969 (COL); Welsch and Chao, 1976 (AA); Goodman, 1973 (XRF)
Sr	Stanton, 1976 (AA)
Te	Ward *et al.*, 1969 (AA); Watterson and Neuerberg, 1975 (AA); Chao *et al.*, 1978 (AA)
Th	Aruscavage and Millard, 1972 (NA); Stanton, 1976 (COL)
Tl	Hubert and Lakin, 1973 (AA)
U	Smith and Lynch, 1969 (FL); Ostle *et al.*, 1972 (NA)
V	Ward *et al.*, 1963 (COL)
W	Ward *et al.*, 1963 (COL); Quin and Brooks, 1972 (COL)
Zn	Ward *et al.*, 1969 (AA); Ward *et al.*, 1963 (COL)

[a] For sample preparation, see Wainerdi and Uken (1971) and LaVergne (1965); for decomposition and separation, see Stanton (1966); for general information, see Stanton (1976).
[b] AA, atomic absorption; COL, colorimetric; E, electrode; FL, fluorimetry; G, gravimetric; NA, neutron activation; SP, emission spectographic; XRF, X-ray fluorescence spectrometry.

Table 3.3
Analytical methods specifically adapted to water

Constituent	Method	Reference
pH	Electrode	Wood, 1976
Eh	Electrode	Wood, 1976
Specific conductivity	Electrode	Wood, 1976
Alkalinity (HCO_3^-, CO_3^{2-})	Titration	Wood, 1976
Dissolved oxygen	Electrode	Wood, 1976
Major cations	Atomic absorption	Brown *et al.*, 1970
Cd, Co, Cu, Pb, Mn, Ni, Ag, Zn	Atomic absorption–extraction	Brown *et al.*, 1970, Appl. Geochem. Res. Gr., 1975
F	Electrode	Brown *et al.*, 1970
As	Colorimetric	Brown *et al.*, 1970
Cl	Titration	Brown *et al.*, 1970
SO_4	Colorimetric	Brown *et al.*, 1970
	Turbidimetric	Ward *et al.*, 1963
	Atomic absorption	Meyer and Peters, 1973
Mo	Colorimetric	Nakagawa and Ward, 1975
U	Neutron activation	Reimer, 1975
	Fluorimetric	Smith and Lynch, 1969
Se	Colorimetric	Brown *et al.*, 1970
Heavy metals (field)	Colorimetric	Huff, 1948

Table 3.4
Methods for analysis of gases and volatile species[a]

Constituent	Method	Reference
Sulfur gases	Absorption colorimetry	West and Gaeke, 1956; Meyer and Peters, 1973
	Gas chromatography	Banwart and Bremner, 1974
	Infrared spectrometry	Adams, 1976
	Correlation spectrometry	Newcombe and Millan, 1970
CO_2	Infrared spectrometry	Liptak, 1974, p. 338
	Absorption in solution	—
	Thermal conductivity	—
	Gas chromatography	Smith, 1977
He	Mass spectrometry	Friedman and Denton, 1976
O_2	Fuel cell detector	Commercial instrumentation
	Paramagnetic detector	Commercial instrumentation
	Polarographic detector	Commercial instrumentation
	Gas chromatography	Smith, 1977
CO	Infrared spectrometry	Altshuller, 1976
	Gas chromatography	Smith, 1977
Hg	Atomic absorption	Ward, 1970; Barringer, 1966
	Gold microbalance	Bristow, 1972
	Gold resistivity	McNerney *et al.*, 1972
Rn	Radioactivity	Dyck, 1968a
		Allen, 1976

[a] General reference: Katz (1977).

VIII. SELECTED REFERENCES ON TECHNIQUES OF CHEMICAL ANALYSIS

General	Yoe and Koch (1957)
Colorimetric techniques (with specific procedures for geological and biological problems)	Sandell (1959)
Instrumental techniques	Wainerdi and Uken (1971)
Analysis of water	Brown *et al.* (1970)
Analysis of soils and plant material	Piper (1950)
Methods used in exploration geochemistry	Stanton (1966, 1976), Ward *et al.* (1963, 1969), Ward (1975)
Precision of analyses	Thompson and Howarth (1976)

Chapter 4

Patterns of Deep-seated Origin—Ore Types, Geochemical Provinces, and Productive Plutons

The processes forming mineral deposits in the deep-seated environment create many types of geochemical patterns and anomalies that are useful in mineral exploration. Some deposits are accompanied by extensive anomalies that constitute the enriched source material from which ore was concentrated. Other anomalies result from dispersion of metals away from ore during the ore-forming process. In addition, some types of ore are accompanied by certain assemblages of elements, by which the type of ore can be identified. Because of the dependence of anomalies and patterns on the nature of the ore-forming process, a brief initial section of this chapter deals with types of ores and the processes forming them.

Anomalies of deep-seated origin may be divided into two groups by their age relation to host rocks. Anomalous volumes of rock ranging in size from single plutons to entire geological provinces generally acquired their content of anomalous elements as part of the process forming the enclosing rock. Such anomalies may be termed syngenetic, indicating that the anomalous metal was emplaced during the formation of the enclosing rock. The major part of this chapter discusses the geochemical provinces and productive plutons making up this first group of anomalies.

In contrast, in a second group of anomalies the anomalous elements were introduced into pre-existing rock, or the original rock was modified in mineralogy or other properties. These anomalies are termed epigenetic, indicating an origin later than the enclosing rock. Epigenetic patterns tend to be related to fractures or other superimposed structures, because such structures commonly control the introduction of material. The best-developed epigenetic patterns are relatively local in extent, forming haloes of a few meters to a few hundred meters in extent around orebodies. Epigenetic anomalies are discussed in Chapter 5.

I. CLASSIFICATION OF MINERAL DEPOSITS

A. Ore-forming Processes

A knowledge of ore-forming processes and of the geological, geochemical, and mineralogical characteristics of orebodies is increasingly important in mineral exploration. Orebodies formed by different processes have differing grade, mineralogy, areal extent, and geological relations. Geochemical differences in pathfinder elements and in character of primary dispersion aureoles are also recognized. Models of the geological, geochemical, and other characteristics of the desired type of orebody are used to plan exploration, including geochemical surveys, and to interpret the results of surveys. For example, porphyry-copper deposits typically have an areal extent of two square kilometers or more, so that samples of soil, rock, or stream sediment may be widely spaced. Molybdenum is a valuable pathfinder element for porphyry-copper deposits. In contrast, volcanogenic massive-sulfide deposits of Cu, Zn, and Pb are much smaller and hence generally require closer sample spacing. Molybdenum is not a useful pathfinder element for these deposits.

Table 4.1 summarizes the various ore-forming processes and the types of deposit formed. They are divided into chemical processes of the deep-seated environment (class I), chemical processes of the surface or near-surface environment (class II), and mechanical processes (class III). Each of these classes is subdivided according to the phases involved (magma, hydrothermal fluid, surface waters, etc.) and the location and geological association of the process.

B. Geochemical Recognition of Ore Types

Mineral exploration in recent years has been increasingly focused on particular types of ore with high economic value and favorable mineralogy, geological relations, size, and grade. Upon discovery of mineralization in a drill hole or isolated outcrop at an early stage of exploration, it may be difficult to determine whether the ore is the desired type. A number of geochemical techniques aid in classifying the mineralization so that a decision can be made regarding further exploration of the site.

Table 4.1
Classification of mineral deposits

I. *Deposits produced by chemical processes of concentration at elevated temperatures within the earth or at the sea floor*
   A. In magmas (magmatic deposits)
      1. By concentration of crystals from magma (chromite and magnetite of Bushveld complex)
      2. By separation of immiscible sulfide or oxide liquids from magma (Cu–Ni at Sudbury, Ont.; Ti at Allard Lake, Que.)
      3. By crystallization of unusual magmas
         (a) Carbonatites (Nb at Oka, Que.; Cu and phosphate at Palabora, South Africa)
         (b) Pegmatites (Nb–Ta in Nigeria; mica at Petaca, N. Mex.; Li at Kings Mtn, N. Car.)
   B. From hot aqueous fluid formed within the earth (hydrothermal deposits)
      1. Deposited within the earth and associated with intrusive igneous bodies or volcanic centers
         (a) Disseminated sulfides in and adjacent to igneous bodies (porphyry–Cu–Mo deposits of Bingham, Utah)
         (b) Contact metasomatic replacement of carbonate rocks (skarn deposits of Fe at Iron Springs, Utah; Cu–Pb–Zn at Central District, N. Mex.)
         (c) Vein and replacement deposits
            (i) In and adjacent to granitic intrusions (Sn–Cu at Cornwall, England)
            (ii) Peripheral to granitic intrusions (Cu at Magma, Ariz.; Pb–Zn–Ag of Central District, N. Mex.; Pb–Ag of Coeur d'Alene, Idaho)
            (iii) Associated with volcanic centers and hot spring systems on land (Ag at Pachuca, Mexico; Au at Carlin, Nev.)
            (iv) Cu associated with basaltic volcanism (northern Michigan Cu, Mich.)
      2. Deposited within the earth but with no obvious relation to igneous activity
         (a) Pb–Zn sulfide deposits in carbonate rocks (Mississippi Valley deposits, U.S.A.)
         (b) U deposits in sandstones (Colorado Plateau, U.S.A.)
         (c) Cu deposits associated with red sediments (Nacimiento, N. Mex. and White Pine, Mich.)
      3. Deposited on the sea floor by fluids from hot springs
         (a) Massive Fe-sulfides with base and precious metals, in association with volcanism (volcanogenic massive sulfides, Kuroko deposits, Japan)
         (b) Base-metal sulfides unrelated to volcanism (Cu at Ducktown, Tenn.)
         (c) Extensive Fe- and Mn-rich deposits with associated Au and other metals (as in premetamorphic carbonate beds at Homestake, S. Dak.)
   C. By regional or dynamic metamorphism
      1. By redistribution of chemical constituents (talc and tremolite deposits, concentration of Au at Homestake, S. Dak.)
      2. By recrystallization (garnet, kyanite)

Table 4.1—continued

II. *Deposits formed by chemical processes of concentration at or near the surface of the earth at low temperatures*
- A. By weathering and related processes on land
  1. By leaching of soluble constituents to leave residual concentrations (bauxite, Fe-, Mn-, and Ni-rich laterites)
  2. By supergene enrichment of sulfides (Cu at Miami, Ariz.)
  3. By evaporation of pore waters from soil (U in caliche at Yeelerie, Australia)
- B. By precipitation in lakes and oceans
  1. By evaporation of water (evaporites, gypsum, halite, borates)
  2. By chemical changes in solution
     - (a) Precipitation of limestones and dolomites
     - (b) Unusual precipitates (Fe formation, Mn nodules, phosphates, base-metal sulfides)
     - (c) By biological processes and diagenesis
       - (i) Accumulation of plant debris (coal)
       - (ii) Formation of liquid and gaseous products from plant and animal debris (oil and gas deposits)
       - (iii) Conversion of sulfates to native sulfur (sulfur deposits)

III. *Deposits produced by mechanical processes of concentration*
- A. Concentration by size in flowing water (gravels, sands, clays)
- B. Concentration of dense minerals by flowing water (placer deposits of Au, Pt, Sn, diamond)

Isotopic studies of Pb in ores and rocks are one means of identifying types of ore (Cannon *et al.*, 1971; Doe and Stacey, 1974). Of the four isotopes of lead, ^{206}Pb and ^{207}Pb are formed by radioactive decay of U, ^{208}Pb is formed by radioactive decay of Th, and ^{204}Pb is non-radiogenic. Because of these relations, the ratios of the Pb isotopes in galena or other ore and gangue minerals can provide information on the chemical environment or source of Pb in the material. Major base-metal ores, especially the volcanogenic massive-sulfide ores, have Pb isotopes consistent with derivation from average rock in the lower crust or upper mantle. Small deposits tend to have Pb deviating from this average deep-seated Pb. In the Mississippi Valley Pb–Zn deposits, high proportions of U- and Th-derived Pb are found, apparently because the Pb in these deposits was leached from sedimentary or igneous rocks of the upper crust. In some districts or regions, several types of Pb of different origin can be distinguished, or the Pb isotopes are zoned in isotopic ratios about a central area of large orebodies (Stacey *et al.*, 1968;

Cannon *et al.*, 1971). Another application is suggested in the MacArthur River area, Australia, where pyrites from pre-ore barren black shales have distinctly different Pb isotope ratios from the similar black shales in which the ore occurs (Gulson, 1977).

Sulfur from igneous sources usually has an ^{34}S content near that of S from meteorites, which is the standard reference material from which deviations in $^{34}S/^{32}S$ are measured ($\delta^{34}S$ values in ‰, or parts per thousand). Sea-water sulfate has $\delta^{34}S$ of about +20 to +25‰ or contains 2–2·5% more ^{34}S than meteoritic S. Sulfides in sedimentary environments have highly variable but usually negative $\delta^{34}S$ values. The values in hydrothermal sulfides are usually near 0% but depend on the isotopic composition, pH, and temperature of the ore fluid, and on the minerals deposited. Ohmoto (1972) and Rye and Ohmoto (1974) describe methods for estimating the sulfur isotopic composition of hydrothermal ore fluids and thereby distinguishing sulfides of different origin.

Associations of elements that may be useful in ore-typing are listed in Table 4.2. As indicated in the table, many elements are associated with a wide variety of types of ore. In addition, some elements listed are not always associated with a given type of ore. However, used in conjunction with geological and mineralogical data, the trace-element associations can be of considerable aid in recognizing ore types. Undoubtedly many additions and improvements will be made to this listing.

The content or ratio of elements in minerals can also be used to identify types of ores. For instance, the Re content of molybdenites from porphyry-copper deposits is much higher than in other types of Mo-bearing deposits (Giles and Schilling, 1972). Sedimentary pyrite tends to have $Co < Ni$, in contrast to most pyrite of hydrothermal origin, which generally has $Co > Ni$ and commonly has $Co > 100$ ppm (Fleischer, 1955; Loftus-Hills and Solomon, 1967). Sedimentary pyrite also tends to have low Se content ($S/Se > 30\,000$, Edwards and Carlos (1954)). Antweiler and Campbell (1977) suggest that Au in different types of ores has distinctive contents of Ag, Cu, and other elements, and that the ratios in placer gold can be used to identify porphyry-copper and other types of sources.

Relations of trace element content to size of orebody have also been suggested. For example, the concentration of U, Y, Na, Fe, Zr, Mn, Ca, and Ni in sandstone-type U deposits in the Morrison Formation on the Colorado Plateau shows a well-defined statistical correlation with the size of the deposit (Miesch *et al.*, 1960).

Roedder (1977) summarizes some differences in character of fluid inclusions in different types of ores; inclusions at porphyry-copper deposits contain halite crystals or high gas content; those in Mississippi Valley deposits have high salinity; those in epithermal Au deposits have low salinity.

Table 4.2
Associated elements (pathfinders) useful in ore-typing[a]

Type of deposit	Major components	Associated elements
Magmatic deposits		
Chromite ores (Bushveld)	Cr	Ni, Fe, Mg
Layered magnetite (Bushveld)	Fe	V, Ti, P
Immiscible Cu–Ni-sulfide (Sudbury)	Cu, Ni, S	Pt, Co, As, Au
Pt–Ni–Cu in layered intrusion (Bushveld)	Pt, Ni, Cu	Cr, Co, S
Immiscible Fe–Ti-oxide (Allard Lake)	Fe, Ti	P
Nb–Ta carbonatite (Oka)	Nb, Ta	Na, Zr, P
Rare-metal pegmatite	Be, Li, Cs, Rb	B, U, Th, rare earths
Hydrothermal deposits		
Porphyry copper (Bingham)	Cu, S	Mo, Au, Ag, Re, As, Pb, Zn, K
Porphyry molybdenum (Climax)	Mo, S	W, Sn, F, Cu
Skarn-magnetite (Iron Springs)	Fe	Cu, Co, S
Skarn–Cu (Yerington)	Cu, Fe, S	Au, Ag
Skarn–Pb–Zn (Hanover)	Pb, Zn, S	Cu, Co
Skarn–W–Mo–Sn (Bishop)	W, Mo, Sn	F, S, Cu, Be, Bi
Base-metal veins	Pb, Zn, Cu, S	Ag, Au, As, Sb, Mn
Sn–W greisens	Sn, W	Cu, Mo, Bi, Li, Rb, Si, Cs, Re, F, B
Sn-sulfide vein	Sn, S	Cu, Pb, Zn, Ag, Sb
Co–Ni–Ag vein (Cobalt)	Co, Ni, Ag, S	As, Sb, Bi, U
"Epithermal" precious metal	Au, Ag	Sb, As, Hg, Te, Se, S, U
Mercury	Hg, S	Sb, As
Uranium vein	U	Mo, Pb, F
Copper in basalt (L. Superior type)	Cu	Ag, As, S
Volcanogenic massive-sulfide Cu	Cu, S	Zn, Au
Volcanogenic massive–sulfide Zn–Cu–Pb	Zn, Pb, Cu, S	Ag, Ba, Au, As
Au–As-rich Fe formation	Au, As, S	Sb
Mississippi Valley Pb–Zn	Zn, Pb, S	Ba, F, Cd, Cu, Ni, Co, Hg
Mississippi Valley fluorite	F	Ba, Pb, Zn
Sandstone-type U	U	Se, Mo, V, Cu, Pb
Red-bed Cu	Cu, S	Ag, Pb
Calcrete U	U	V
Sedimentary types		
Copper shale (Kupferschiefer)	Cu, S	Ag, Zn, Pb, Co, Ni, Cd, Hg
Copper sandstone	Cu, S	Ag, Co, Ni

[a] Some data from Beus and Grigorian (1977, p. 232) and Boyle (1974).

II. PRODUCTIVE ENVIRONMENTS OF DEEP-SEATED ORIGIN

The association of certain kinds of ore deposit with specific types of plutonic rock has been known and used by exploration geologists for many decades. Familiar examples include the association of cassiterite with potassic granites, ilmenite with anorthosites, chromite with ultramafic rocks, and nickeliferous sulfides with both ultramafic and mafic rocks. Similarly, provinces with an unusual abundance of certain types of ore are known. Examples are the Sn ore of south-east Asia and the porphyry-copper province of the south-western U.S.A. The regions or rock types with which ores are preferentially associated may be termed productive, in the sense that exploration in these environments is likely to be more productive of new ore than exploration in a region picked at random. Although the productive environments given as examples above are recognizable by simple examination of mineralogy or distribution of producing deposits, other productive environments may be detected only by systematic sampling and chemical analyses of rocks or other natural materials. Geochemical detection of such productive environments can be an effective method of reconnaissance exploration when a new region or district is being sought, or when the types of ore most likely to occur in a given region are being evaluated.

The syngenetic productive environments described here originate either prior to or during ore formation. In many of the productive environments, anomalous metal content probably formed a source from which the ore has been concentrated. The metal-rich source rock can still be recognized because it was only partially or locally depleted to form ore. In other localities, the indicator elements are pathfinders for a valuable element and reflect the operation of an ore-forming process. For example, unusual amounts of sulfur or chloride may have promoted the solubility of metals in hydrothermal fluids, or large amounts of As, Se, or other elements may have accompanied the ore element. Again, relics of the original large volume of anomalous material can serve as guides to ore.

Productive environments range in size from the continents and ocean basins to an individual ore zone, but, for convenience in this chapter, they are grouped into geochemical provinces of regional extent and local productive environments the size of a mining district. Other types of syngenetic productive environments exist but have received very little study. An example in a metamorphic environment is Homestake, South Dakota, where Au originally disseminated in a Precambrian sedimentary Mn–Fe formation has been concentrated into quartz veins during meramorphism to form the ore (Rye and Rye, 1974). The remnants of Fe formation in this and other similar areas still contain anomalous Au and may be useful as a guide to Au ore (Ridler and Shilts, 1974).

III. GEOCHEMICAL AND METALLOGENIC PROVINCES

A geochemical province may be defined as a relatively large segment of the earth's crust in which the chemical composition is significantly different from the average. On the broadest scale, continents and ocean basins clearly differ in composition. Smaller geochemical provinces exist within the continents because of primordial differences in the continental crust or later formation by large-scale igneous or sedimentary processes. Geochemical provinces are most clearly manifested by suites of igneous rocks, all members of which are relatively rich or relatively impoverished in certain chemical elements. Weathering and erosion of these unusual igneous rocks may incorporate unusual abundances of elements in the sedimentary rocks of the region. Conversely, assimilation of unusual crustal materials by magmas may result in igneous rocks of distinctive composition. The enriched or depleted rocks of an area may have a variety of ages. In fact, one of the criteria of a bona fide geochemical province is that the characteristic chemical peculiarities should be recognizable in rocks covering a substantial period of time.

Large areas of the earth's crust may also be characterized by an unusual abundance of ores of a particular metal or of a particular type of ore. Areas of this kind, known as metallogenic provinces, have long been the object of speculation by geologists (Turneaure, 1955). Examples are the Cu-producing areas of Chile and Peru, the Sn fields of north-western Europe, and the U fields of the Canadian Shield. Figure 4.1 shows the location of the great Cu mines that define the Cu province of the south-western U.S.A. These classic examples of metallogenic provinces are defined on the basis of known mines and their mineral production statistics.

Metallogenic provinces can originate in several ways. To form ores, one or more processes of concentration must have extracted metals from a relatively large volume of source material and concentrated it into a small volume, the orebody. The formation of ore may thus be promoted by an unusually rich source operated on by an ordinary process, or by the operation of unusually efficient ore-forming processes on a normal source, or by some combination of these two. Following this model, a metallogenic province with many orebodies can be either (i) a geochemical province, where the absolute concentration of metal is higher than normal, or (ii) a region where the physical and chemical environment promoted concentration of ore elements into deposits. A metallogenic province formed by the first process is simply one manifestation of a geochemical province. Such provinces can be detected and delineated by wide-spaced rock or drainage sampling. Particularly in inaccessible or poorly explored terrain, the delineation of geochemical provinces can provide useful guidance in the reconnaissance stages of exploration.

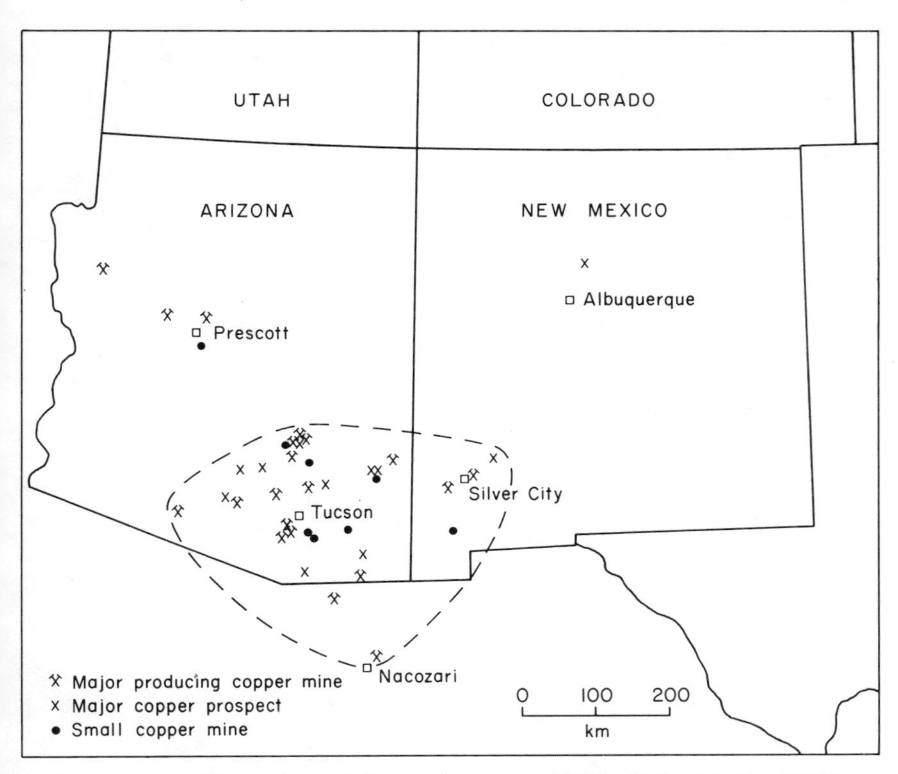

Fig. 4.1. The Cu province of the south-western U.S.A. (After Noble, 1974.)

One example of a geochemical province is the Bear Province of the northwestern Canadian Shield (Cameron and Allan, 1973). Uranium in lake sediments in this area averages four times as high as in the adjacent Slave Province (Fig. 4.2). Composite rock samples show similar differences in the rocks of the two regions. Uranium ore has been extracted at the Port Radium and Rayrock mines, making the Bear Province a metallogenic as well as geochemical province.

Uranium provinces also appear to exist in western U.S.A. and in central Europe. In western U.S.A., silicic volcanic rocks in the Basin and Range and Rocky Mountain region have higher contents of U than in regions nearer the coast (Coats, 1959), and U-rich volcanic ash may have been the source for some U accumulations in sandstones and coals. These features may reflect the enrichment of U and other "incompatible elements" in late differentiates of the alkali-rich igneous rocks characteristic of the interior regions of continents. In the Erzgebirge of Saxony, Silesia, and Czechoslovakia, uraniferous granite, uraniferous pegmatite, and pitchblende veins are closely associated.

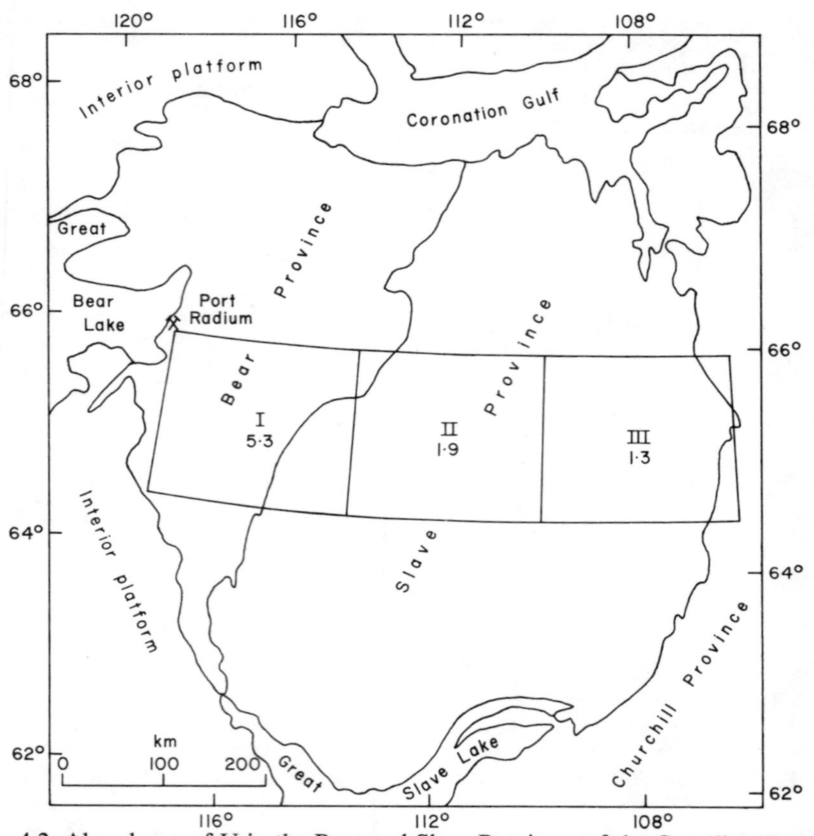

Fig. 4.2. Abundance of U in the Bear and Slave Provinces of the Canadian Shield. (After Cameron and Allan, 1973.) The numbers are mean ppm U in 1200–1300 lake sediments for each of three quadrangles, and illustrate the much higher content of U in the Bear Province.

In general, well-defined U geochemical provinces appear to be a very useful guide to U ores.

Tin provinces have received widespread attention, undoubtedly because of the very striking concentration of Sn deposits in the extensive yet well-defined Sn fields of the world (Schuiling, 1967). Ahrens and Liebenberg (1950) report that mica from pegmatite dikes near Sn veins in South West Africa contains from 10 to 100 times as much Sn as mica from similar dikes in areas where no Sn mineralization is known. Goldschmidt (1954, p. 393) states:

> The absence of workable tin deposits in large regions of the earth, which are also characterized by the scarcity of even small amounts of tin minerals, seems to be followed by a scarcity of tin as a trace element as indicated by spectroscopic observations on magmatic rocks.

In South Africa, Viljoen *et al.* (1970) suggest that relatively high contents of Au in the ancient mafic and ultramafic rocks of the Onverwacht series and similar sequences were remobilized into minable veins during intrusion of granitic rocks. These concentrations in turn were eroded and segregated into the placer ores of the Witwatersrand district.

Regional stream-sediment surveys (Armour-Brown and Nichol, 1970) in Zambia show that mining districts are surrounded by large regions of slightly anomalous metal content, indicating the existence of geochemical provinces associated with the Cu, Co, and Sn deposits of this region (Fig. 4.3). Chromite-bearing regions in Sierra Leone were detectcd by stream sediments collected at one sample per 200 km^2 (Garrett and Nichol, 1967).

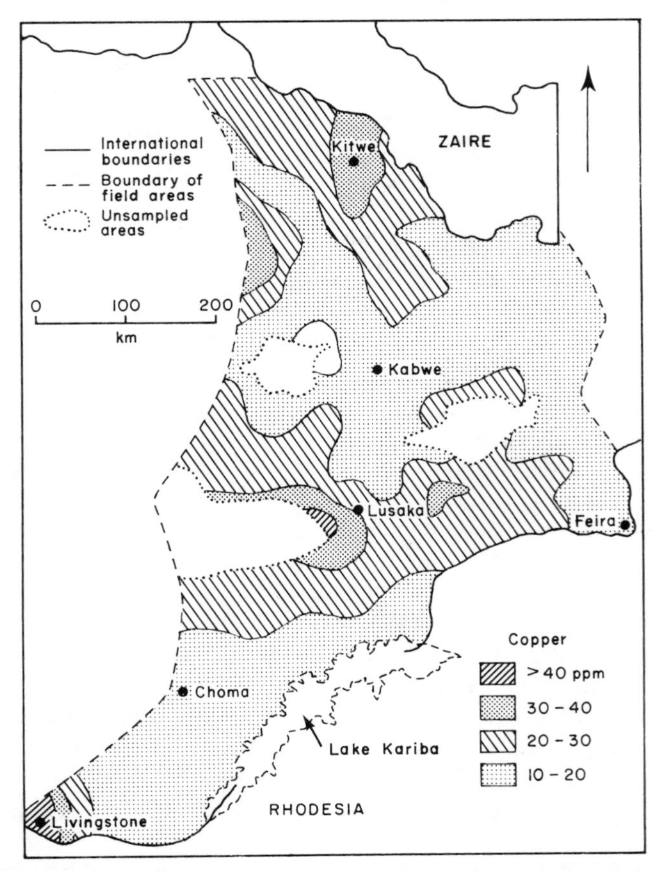

Fig. 4.3. Copper content of stream sediments collected at one sample per 200 km^2 in Zambia showing anomaly related to Copperbelt (Kitwe area) and mines in the Lusaka area. (After Armour-Brown and Nichol, 1970.)

The anomalies described above are so extensive that they cannot be attributed to dispersion from single ore deposits or ore districts. Two kinds of sources are suggested: (i) extensive clusters of small non-economic deposits accompanied by their primary and secondary dispersion haloes (Fig. 4.4), and (ii) regionally elevated contents of the indicator elements in one or more rock types that are common in the region. In the Bear Province of Canada the U anomalies were associated with Proterozoic granite, but in addition,

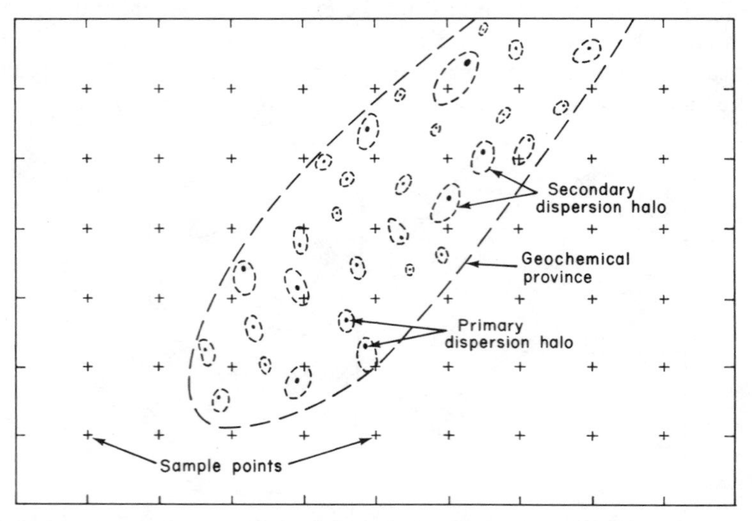

Fig. 4.4. Geochemical province composed of a cluster of primary and secondary dispersion haloes, and sampling grid, illustrating anomalous but highly variable results of wide-spaced sampling. (After Cameron and Hornbrook, 1976.)

the strongest anomalies were near lineaments and faults, suggesting that epigenetic U concentrations may have contributed to the patterns in lake sediment. In some areas, analysis of bedrock or soil showed that the anomaly reflects high values in rocks of the area (Armour-Brown and Nichol, 1970; Garrett, 1975, p. 385). but in other areas no such anomalies in rock or soil could be demonstrated (Garrett and Nichol, 1967).

The contrast of the regional sediment anomalies is relatively weak compared to contrast near orebodies, and the anomalies are relatively inhomogeneous. In Zambia, the Copperbelt was outlined in smoothed data by Cu values of 30–40 ppm compared to background values of 10–30 ppm. In the Bear Province, lake sediments average 5 ppm U compared to 1·3 ppm in the adjacent unmineralized Slave Province. In part, the low contrast of the Zambian data results from deletion of obvious anomalies followed by smoothing using a moving average.

Higher than normal variability in the content of the indicator elements is a characteristic of many metallogenic provinces. This may be expressed qualitatively as geochemical "relief", as illustrated by the cxHM content of stream sediments in the New Brunswick–Gaspé area (Fig. 4.5). In fact, most regional anomalies tend to be more variable than can be accounted for by normal sampling and analytical errors. Because of this variability, some system of regional averaging may help to outline the limits of a regional anomaly. Higher than normal standard deviations for groups of samples have also been suggested as a means of recognizing geochemical provinces.

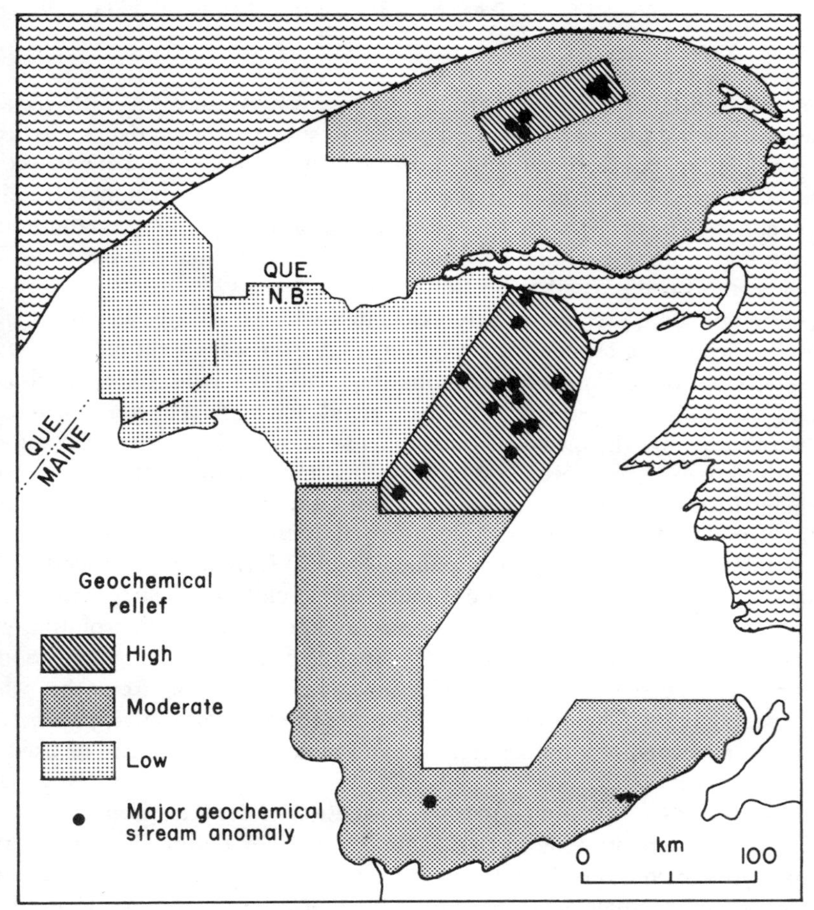

Fig. 4.5. Map of New Brunswick–Gaspé area showing relative geochemical relief. Low relief, cxHM values rarely exceed 5 ppm; moderate relief, values up to 20 ppm common; high relief, values of 40–200 ppm in scattered areas. (After Hawkes, 1976b.)

Regional zoning of minor elements in epigenetic sulfide minerals has been described within the metallogenic province of western U.S.A. (Burnham, 1959). The Sn, Ag, and combined Co–In–Ni–Ag–Sn content of chalcopyrite and sphalerite apparently define three broad zones that parallel both regional tectonic features and previously defined metallogenic provinces. The zonal pattern is attributed to regional differences in the chemical composition of deep-seated source materials.

IV. PROCESSES FORMING PRODUCTIVE PLUTONS

The association of ore with certain plutons, for instance Sn with certain granites, and not with all plutons of superficially similar characteristics, has been a long-standing puzzle to students of ore deposits. A higher than average content of the ore element in these productive intrusions has been an attractive explanation. The observation that granites with associated Sn ores commonly have a high Sn content supports this concept, as do similar observations for other metals, as discussed in Section V of this chapter.

Although some ores are clearly associated with igneous bodies unusually enriched in the ore element, the relation is far from consistent and is poorly understood. Possible processes creating the relation include: (i) separation of metal-rich fluids or crystals directly from metal-rich magma during crystallization and cooling, (ii) origin of both magma and ore fluid from a common metal-rich parent or source at depth, (iii) leaching or segregation of the ore element from an already consolidated body of metal-rich igneous rock, and (iv) introduction of metal from an external source into a large volume of consolidated igneous rock, accompanied by local formation of orebodies. In processes (i) and (ii) the anomalous metal in the pluton is clearly syngenetic; in process (iii), a close tie exists between ore and the metal-rich pluton, but the leaching might have occurred long after the cooling of the igneous body. In process (iv), the anomaly is epigenetic, but distinction from the other processes may be difficult in practice. Despite these uncertainties, productive plutons are recognized in many regions and in favorable cases can be used as a guide to ore.

In processes (i), (ii), and (iii) above, formation of ore was promoted by an initial enrichment of metal in the magma, source, or pluton. Because of the initial enrichment, the final magmatic or hydrothermal ore-concentrating process could produce a higher grade of ore or a larger volume of ore than would the same process operating on less enriched source material. Most concentrating processes involve partition of the ore element between two phases, such as magma and ore fluid, magma and crystal, or crystal and ore fluid. The chemical relationships are thus expressed by partition equations like eqn 2.4a.

As an example to clarify the factors leading to development of ore from metal-rich plutons, consider the case of Zn in a magma and its extraction to form hydrothermal Zn deposits.

Zinc probably exists in the magma largely as Zn^{2+} ions weakly bonded to O^{2-}, SiO_4 polymers, and other complexes. Zinc has an ionic radius and charge similar to Mg^{2+} and Fe^{2+}, and it may substitute for these elements in olivine, pyroxene, amphibole, and biotite. The partition of Zn between magma and early-crystallizing clinopyroxene (px) can be expressed as:

$$Zn^{2+}(\text{magma}) + CaMgSi_2O_6(\text{px}) = Mg^{2+}(\text{magma}) + CaZnSi_2O_6(\text{px}). \quad (4.1)$$

If the Zn and Mg occur as ideal solutions in the magma and the pyroxene (or if the deviations from the ideal for both Zn and Mg are similar, as seems possible in this case), then the equilibrium expression for this reaction (eqn 2.4a) may be rearranged and simplified to

$$\frac{Zn}{Mg}(\text{px}) \simeq K \frac{Zn}{Mg}(\text{magma}) \quad (4.2)$$

where the Zn/Mg values are atomic ratios, or

$$N_{Zn}(\text{px}) = K \frac{Zn}{Mg}(\text{magma}) N_{Mg}(\text{px}) \quad (4.3)$$

where N indicates the atomic fraction or mole fraction of the element in the Mg–Fe sites of the mineral. From these expressions, it is apparent that if Zn/Mg is unusually high in the magma, then Zn/Mg will also be high in the pyroxene (if K is constant, and the assumptions about ideality are correct). Analogous reactions may be written for other mafic minerals crystallizing from the magma. A high Zn content is thus expected in pyroxenes or other mafic minerals crystallizing from a Zn-enriched magma.

As complicating factors in the above relationship, note that an increase in K, which might be caused by a change in temperature, could result in a high content of Zn in the pyroxene formed from a magma of normal Zn content, as could changes in the Mg content of the magma or the pyroxene. Finally, changes in the character of the magma or other substitutions in the pyroxene might modify the relationships of concentration to activity in the magma and the pyroxene. In view of these complications, the Zn content of an early crystallizing mineral can be regarded as only an approximate indicator of the initial Zn content of the magma.

If a hydrothermal solution later separates from this magma, analogous equations can be used for the partition of Zn and other elements between magma and hydrothermal solution. The process has been discussed in some detail by Holland (1972), who found experimentally that Zn probably occurred as the non-ionized complex $ZnCl_2^0$ in the chloride solutions of his experiments. The partition equation into an NaCl-rich aqueous phase can be expressed as:

$$Zn^{2+}(\text{magma}) + 2NaCl^0(\text{aq}) = ZnCl_2^0(\text{aq}) + 2Na^+(\text{magma}). \tag{4.4}$$

The equilibrium expression for this reaction is:

$$K = \frac{a_{ZnCl_2^0(aq)} \cdot a^2_{Na^+(mag)}}{a^2_{NaCl^0(aq)} \cdot a_{Zn^{2+}(mag)}} \tag{4.5}$$

where (aq) refers to the hydrothermal solution and (mag) to the magma. The Zn concentration in the hydrothermal solution is then expressed as:

$$a_{ZnCl_2^0(aq)} = K \frac{a_{Zn^{2+}(mag)} a^2_{NaCl^0(aq)}}{a^2_{Na^+(mag)}}. \tag{4.6}$$

Although the activities in this expression are not easily converted to concentrations, the expression clearly indicates that Zn in the hydrothermal solution increases proportionally to the first power of Zn in the magma, but to the second power of NaCl concentration in the hydrothermal solution. As shown by Holland, the NaCl content of the solution is dependent mainly on the chloride content of the magma. Thus, although Zn in hydrothermal solutions depends on the Zn in the magma, it depends even more strongly on the Cl content of the magma. This feature has led to studies of the Cl content of biotites and apatites, in hopes of recognizing Cl-rich intrusives (Stollery *et al.*, 1971; Kesler *et al.*, 1975a; Kesler *et al.*, 1975b; Parry and Jacobs, 1975). A further complication is variation in the water content of the magma and the timing of its release from the magma. Because a low content of magmatic water results in high chloride content of the hydrothermal fluids (reflecting a high Cl/H_2O ratio in the magma), the solutions separated from a magma of low water content can have unusually high Zn content. Given all these complexities, the many exceptions to simple correlations between Zn in magmatic minerals and Zn in ores are not surprising.

Further complications exist for elements occurring in variable oxidation states or as complexed ions in the magma. For instance, Cu probably occurs as Cu^+ (or as Cu–S complexes) in most magmas, but may occur as Cu^{2+} in relatively oxidizing magmas and in some magmatic minerals. The equilibrium expressions in these cases must include additional components to accomplish the oxidation–reduction process and balance the reactions. The simple correlation of Cu in the igneous minerals with Cu in associated ores is thus disrupted.

At the present time, the identification of productive intrusions by their metal content relies on empirical observations that ores of certain types and regions are associated with metal-rich igneous bodies. More general use of productive plutons as a guide to ore will require a better understanding of the many complexities indicated above.

V. ORES RELATED TO PRODUCTIVE PLUTONS

In general, the clearest and best-developed examples of ores associated with productive plutons are furnished by the relatively simple magmatic deposits (pegmatites and segregations of molten sulfide). Among hydrothermal ores, productive plutons appear to be more easily recognized for Sn, W, Be, U, Li, and other lithophile elements than for Cu, Pb, Zn, Au, Ag, and other chalcophile elements.

As suggested by the equilibrium expressions of the preceding section, the use of ratios of trace elements to major elements may compensate for some of the complexities in the chemical relations. So far, very little use has been made of such ratios, which can correct at least in part for the effects of differentiation. For example, the ratio Zn/Mg might compensate for differing degrees of differentiation in samples from several plutons.

A. Pegmatites

The clearest examples of ores associated with metal-rich igneous bodies are the pegmatites. For instance, in pegmatites containing economic deposits of pollucite (a Cs mineral), the lepidolite has a much higher content of Cs than in other rare-metal pegmatites, as shown on Fig. 4.6 (Gordiyenko, 1973). Analysis of easily recognized lepidolite can thus serve as a guide to pegmatites of interest for pollucite, which is difficult to recognize and may occur in an unexposed zone of the pegmatite.

Similar results have been obtained using feldspars and muscovite enriched in Be, Nb, Rb, Cs, and Tl as a guide to pegmatites containing beryl and Nb–Ta minerals (Heinrich, 1962; Shmakin, 1973). Ratios such as Rb/Ba in feldspar or mica have been suggested as an index of extreme differentiation and rare element enrichment. High Li values in feldspar and mica can be used as a guide to pegmatites with Li minerals. The content of Ba, Rb, Pb, Li, Cs, and other elements in muscovitized wallrock near pegmatites has been used to detect unexposed muscovite pegmatite (Shmakin *et al.*, 1971).

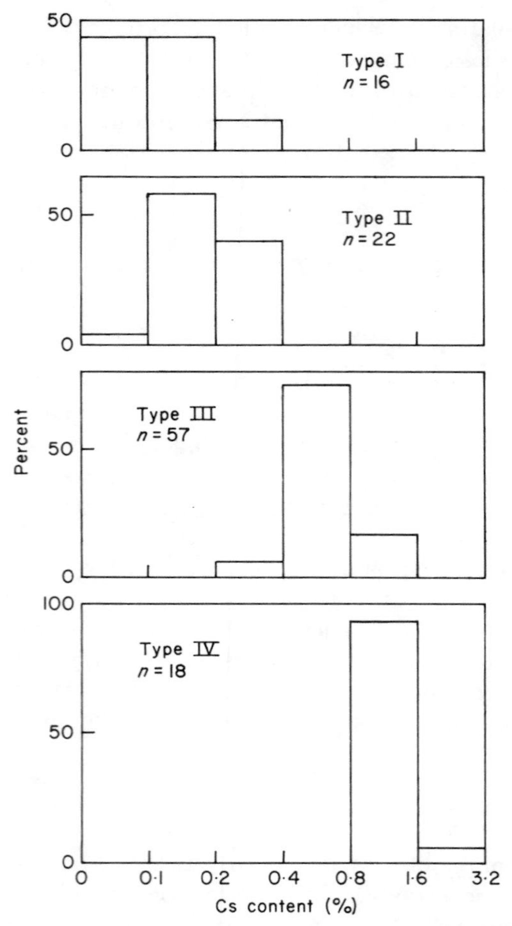

Fig. 4.6. Cesium content of lepidolites in four types of pegmatites. (I) Pegmatite districts lacking pollucite and other Cs minerals, but with Li–Be–Ta minerals. (II) Pegmatites lacking pollucite but other pegmatites of district contain pollucite. (III) Pegmatites with small amounts of pollucite. (IV) Pegmatites with economic amounts of pollucite. (After Gordiyenko, 1973.)

B. Magmatic Copper–Nickel Sulfides

Several studies of magmatic Cu–Ni-sulfide deposits show that associated ultramafic rocks have anomalous metal and S content. For instance, Cameron *et al.* (1971) treated 1079 samples of ultramafic rock from 61 localities in the Canadian Shield with a sulfide-selective leach (ascorbic acid and hydrogen peroxide). Leachable Ni and Cu and total S were considerably enriched in the ultramafic bodies having associated Ni–Cu ore, as compared to barren intrusions. The anomalies can be explained by incomplete separation and settling of sulfide droplets exsolved from a sulfur-rich magma, or by introduction of S to the igneous rock, causing sulfidation of Cu and Ni. A statistically determined function of Ni, Cu, and S contents (discriminant function) was developed by Cameron *et al.* (1971) to distinguish productive and barren intrusions with greater certainty than any single element. Relatively high Ni and S values and an S/Ni ratio near one are characteristics of serpentinites near Ni-sulfide deposits (Hausen *et al.*, 1973).

In a group of mafic and ultramafic bodies in the U.S.S.R., the content of Ni varies in a regular way with SiO_2, but the intrusions with Ni–Cu deposits have anomalously high Ni content for their SiO_2 content (Nyuppenen, 1966). The Cr/V ratio of the productive intrusions is also high (1·3–180) relative to normal gabbros ($<1{\cdot}1$) (Sesterenko and Smirnova, 1964).

The Pd/Ir ratio of mineralized ultramafic rocks in Western Australia is relatively low and may furnish an ore guide (Keays and Davison, 1976).

C. Ores Associated with Granitic Rocks

Tauson and Kozlov (1973) have classified granitic rocks into five types and listed criteria for distinguishing the types based on their trace element abundance:

(i) Plagiogranites, formed by differentiation of gabbroic magmas.
(ii) Ultrametamorphic granites, formed by partial melting during metamorphism in the crust.
(iii) Palingenic granites, formed by complete remelting of granitic rocks.
(iv) Plumasitic granitic rocks with high Li, Rb, Sn, and F, formed as late differentiates of the above types (possibly combined with crystal assimilation).
(v) Agpaitic granitic rocks with alkali-rich mafic minerals (aegerine, riebeckite).

Tin, W, Mo, U, and other rare elements are concentrated into deposits associated with the plumasitic series, which can be recognized by abundance and ratios of trace elements. Niobium, Ti, P, and other elements occur in carbonatites associated with rocks of the agpaitic series. The use of trace-element "signatures" to distinguish series or types of igneous rocks with similar major-element content may be useful in reconnaissance selection of favorable provinces, and in separating favorable and unfavorable igneous bodies within a mineralized region.

D. Hydrothermal Tin Deposits

Productive intrusions for Sn deposits have been the focus of a great deal of attention, especially in the U.S.S.R. and Australia. Most intrusions associated with Sn deposits have an Sn content appreciably higher than the normal value of about 3 ppm for granites (Fig. 4.7). Most are also plumasitic and are relatively enriched in Rb, F, Be, Li, B, and other elements concentrated in

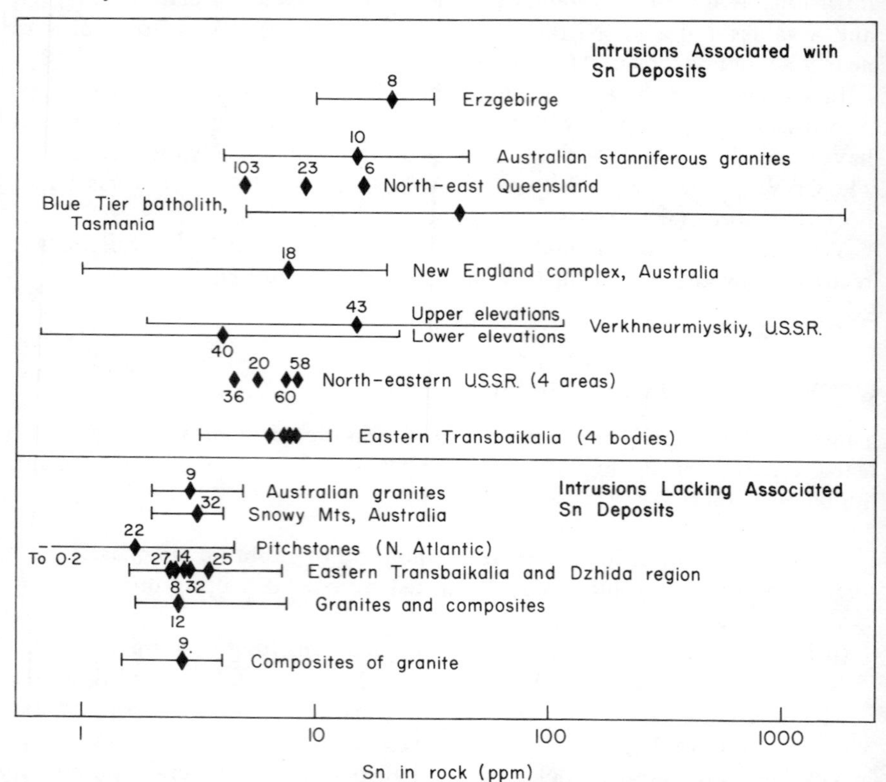

Fig. 4.7. Tin content of igneous rocks, showing higher Sn in intrusives associated with Sn deposits. Diamond is mean value, bar shows range, and figures are number of samples.

residual granitic magmas (Sheremet *et al.*, 1973; Kozlov *et al.*, 1975; Groves, 1972; Sheraton and Black, 1973). The productive intrusions are typically highly differentiated alkali-rich granites with SiO_2 contents of 70–77%, very low MgO and CaO, and relatively Fe-rich mafic minerals, usually biotite. In at least some cases, high initial $^{87}Sr/^{86}Sr$ ratios indicate that at least part of the magma was derived from the crust, either as a primary magma formed from crust or by assimilation (Sheraton and Black, 1974, pp. 336, 338; Jones *et al.*, 1977). The K/Rb ratio is typically very low (35–150 compared to normal values of 150 or more), Mg/Li is very high, and Zr/Sn is very low (Beus, 1969). The Sn-bearing granites usually have a lower F/Li ratio than normal granites (Bailey, 1977). Comparison using these ratios is preferable to absolute contents of Rb and Li, because the ratios allow comparison of rock series at different stages of differentiation. If sphene is present, the intrusion is probably non-productive, apparently because Sn^{4+} (radius 0·77 Å) readily substitutes for Ti^{4+} (0·69 Å) and is firmly bonded instead of being enriched in the later differentiates. The high resistance of sphene to later hydrothermal leaching of Sn may also be a factor.

Another possible guide to a productive intrusion for Sn deposits is a high content of F and H_2O in the magma, as indicated by the proportion of F in the OH sites of hydrous minerals and the abundance of OH-bearing minerals. The F content is important because of the likelihood that Sn is transported in hydrothermal solutions as an Sn–(F, OH) complex ion (Barsukov and Kurilchikova, 1966). The common occurrence of F minerals with cassiterite deposits tends to support the existence of such an F-bearing complex in hydrothermal solutions (Barsukov and Volosov, 1967). A high water content would promote the separation of a hydrothermal fluid from the magma, and the resulting partition of Sn and F into this solution.

Within a productive granite, Sn is typically concentrated in biotite, where Sn replaces Fe^{2+} (0·69 Å) and Mg^{2+} (0·80 Å). As a generalization, the Sn content of biotite (and other Fe–Mg and Ti minerals) is relatively high in most granites associated with Sn deposits, as shown in Fig. 4.8, but some exceptions and a large range of values have been found. Possible causes for these irregularities can be identified, but to date very little has been done to evaluate them and to develop methods of correcting for the effects. The following discussion suggests some causes for the lack of consistency. Similar considerations affect the use of other trace elements in minerals as guides to productive intrusions.

Given a magma with a certain content of Sn, the concentration of Sn in the biotite will depend partly on the amount of biotite in the rock. The given amount of Sn may occur at a relatively low concentration in a large percentage of biotite, or at a relatively high concentration in a small amount of biotite, in both cases mixed with relatively Sn-poor quartz and feldspars (Barsukov and Durasova, 1966). In another granite, the same total amount of Sn might be divided between biotite and sphene, or biotite and hornblende, so that the concentration of Sn in the biotite would be less. The content of Sn in biotite is thus a function of the biotite-forming elements (K, Fe, Mg, Al, Si, OH, F) in the magma as well as the Sn content.

The oxidation state of Sn in magmas is uncertain, but at least some probably occurs as Sn^{4+}. The substitution of Sn^{4+} for Mg^{2+} and Fe^{2+} requires

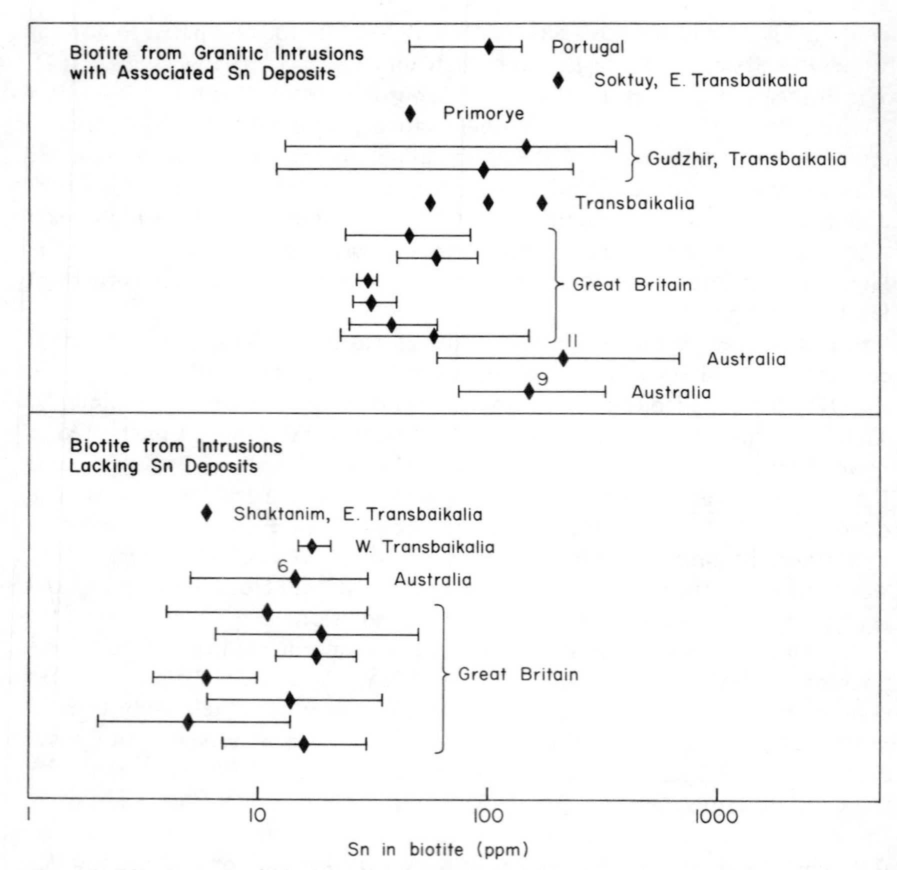

Fig. 4.8. Tin content of biotite, showing relation to tin ores. Symbols as in Fig. 4.7.

a compensating substitution of Al^{3+} for Si^{4+}, O^{2-} for OH^-, Li^+ for Mg^{2+}, or similar substitution. A possible exchange reaction (which would go to the right only to a minor extent) is

$$KMg_3AlSi_3O_{10}(OH)_2 + Sn^{4+} + O^{2-} + Al^{3+} \\ = KSnMg_2Al_2Si_2O_{11}(OH) + Mg^{2+} + OH^- + Si^{4+} \quad (4.7)$$

where the ions occur in the magma, and the "Sn-phlogopite" is in solid solution in the biotite. Because balancing substitutions are required, the content of Sn in biotite will depend in a complicated way on the availability of Al, Si, Mg, Fe, O, OH, F, and other elements as well as on the Sn content of the biotite.

A further possible problem is post-magmatic leaching of Sn from biotite or addition of Sn to biotite. In some districts, the original high Sn content of a biotite has been reduced by hydrothermal leaching, and the resulting Sn apparently contributed to an orebody (Nedashkovsky and Narnov, 1968; Tischendorf, 1969). Finally, the analysis for Sn must be on a pure biotite, because some Sn-rich rocks contain cassiterite, which obviously increases the apparent content of a biotite if it is included in the analysis.

The data in the literature clearly indicate that most Sn districts can be detected by high Sn content of at least some of the igneous rocks of the area and by relatively high Sn contents in the biotite. Although much remains to be learned about the details of the association, the technique appears to be useful in reconnaissance exploration for Sn districts. Further research might clarify some of the complexities described above. Bolotnikov and Kravchenko (1970) suggest that highly variable Sn content may be more significant than the absolute content of Sn in the intrusives, and also demonstrate a considerable increase in Sn content in the upper portions of igneous body.

E. Molybdenum, Tungsten, and Gold Deposits

In western Canada, plutons with associated Mo, W, and Au mineralization have anomalous contents of Mo, W, and Hg (Garrett, 1971a, 1974a) and also a high variability of these elements. However, Russian and Australian workers do not find higher than normal contents of Mo and W in intrusions associated with Mo and W deposits (Ivanova, 1963; Sotnikov and Izyumova, 1965; Sheremet *et al.*, 1973; Flinter *et al.*, 1972), although the intrusions do tend to be enriched in Rb, Li, F, Sn, and related elements.

F. Base-metal Sulfide Deposits

High contents of several chalcophile and halogen elements have been reported in plutons associated with hydrothermal Pb–Zn–Cu deposits. In the northern Cordillera of Canada, Garrett (1973) found that the content of Cu, Pb, and Zn, after statistical correction for the effects of differentiation, was anomalously high in granitic rocks associated with base- and precious-metal deposits and prospects. Slawson and Nackowski (1959) report that the Pb content of the potash feldspars from quartz monzonites in areas of Pb mining is higher than in other areas (Table 4.3), and Hamil and Nackowski (1971) found high values of Zn in magnetite of Zn districts. However, Pb and Zn were low in biotites from the Pb and Zn districts. In Yugoslavia, Pb in feldspars of igneous rocks in the Dinaride zone, which contains Pb deposits, is higher than in the adjacent Balkanide zone, a Cu province (Cuturic *et al.*, 1968). In contrast, ore metals were depleted in magnetite and silicates near base-metal veins in a granitic intrusion. Although some correlations may exist, they are not yet clear enough for confident use in exploration.

Table 4.3
Lead content of potash feldspars from plutonic rocks associated with Pb deposits in the western U.S.A.[a]

District	Lead production	Number of samples	Pb (ppm) Range	Pb (ppm) Mean
Bingham, Utah	Major	22	11–126	61
Park City, Little Cottonwood, Utah	Major	21	10–85	47
Tintic, Utah	Major	10	10–44	29
Robinson, Nev.	Minor	25	9–37	14
Iron Springs, Utah	None	3	12–18	15
Background Pb content of K-feldspar[b]	—	—	—	25

[a] Source: Slawson and Nackowski (1959). [b] Wedepohl (1956).

Biotites and apatites of some intrusions associated with base-metal mineralization have weakly anomalous Cl and F, but the anomalies are not generally strong enough to be easily useful in exploration (Parry and Jacobs, 1975; Kesler *et al.*, 1975a). Postmagmatic leaching may have destroyed the original patterns, because micas are found to exchange F and Cl with surrounding solutions relatively readily. Apatite is more resistant to this process and has not been thoroughly investigated.

G. Porphyry-copper Deposits

The close association of porphyry-copper deposits with certain granodioritic intrusions has suggested to many workers that these igneous bodies are chemically distinct from normal granodiorites, either in terms of unusually high or low content of Cu, or high content of Cl, which might determine the partition of Cu into hydrothermal fluids (Stollery *et al.*, 1971). However, research has not revealed any consistent, simple relationship. High contents of Cu in biotite associated with porphyry-copper deposits are clearly demonstrated for intrusions in Bingham, Ely, Park City, and other districts in Utah and Nevada (Parry and Nackowski, 1963), for districts in southern Arizona (Lovering *et al.*, 1970; Graybeal, 1973), and for the Boulder batholith near Butte, Montana (Al-Hashami and Brownlow, 1970), but in several of these districts the Cu is observed to occur as chalcopyrite inclusions. At the Patagonia and Ajo districts in Arizona, Cu, Zn, and Mn show consistent partitioning between biotite, hornblende, and chlorite, suggesting equilibrium between the minerals and magma during the magmatic stage (Graybeal, 1973). However, in the Patagonia stock, the content of Cu in biotite varies from 15 ppm to 3600 ppm, high values occurring in early magmatic biotite and low values in late magmatic biotite that may have been formed from a magma depleted in Cu by loss of a hydrothermal fluid or complexing of Cu. Copper in biotite gives higher contrast between mineralized and non-mineralized intrusions than total Cu in the rocks, suggesting that Cu in biotite may produce more reliable anomalies.

In contrast to these examples of positive relations of Cu-in-biotite to ore, Kesler *et al.* (1975a) found that biotites from intrusions in about 25 porphyry-copper districts in North America had a slightly lower average content of Cu than biotites from about ten non-mineralized districts, although very high individual values occur only in the mineralized districts. Probably these samples include both early magmatic and late magmatic biotite of widely varying Cu content, as described by Graybeal (1973). In the Boulder batholith, Al-Hashami and Brownlow (1970) found much higher Cu contents than Kesler *et al.* (1975a), but noted that fine inclusions of chalcopyrite were present in the high-Cu samples. Microprobe studies showed that Cu in biotite from intrusions associated with the Ray and Sierrita deposits in Arizona was concentrated in chloritized biotite, and specifically at the boundary of chlorite and biotite within the grains (Banks, 1974). He suggested that the anomalous Cu was introduced after the magmatic stage. Rehrig and McKinney (1976) substantiated Banks' results, and showed that a vermiculite-like material formed from the biotite during weathering contained the very high Cu contents. The high Cu content in the biotites studied by them was evidently introduced by supergene solutions. None of the biotites analyzed by microprobe in the districts studied by Banks and by Rehrig and McKinney contained more than 100 ppm Cu, in contrast to the values up to 3600 ppm inferred by Graybeal (1973) to occur as an equilibrium product of magmatic crystallization. Thus, although anomalous amounts of Cu clearly occur in some "biotite" of porphyry-copper intrusions, the relation to ore is not completely consistent, and the anomalies in many cases arise by hydrothermal and surficial processes rather than igneous processes. Application of this method will require careful examination of the origin and character of the sample that is analyzed.

Rutile in porphyry-copper intrusions has a high (Cr + V)/(Nb + Ta) ratio and contains 100–400 ppm Cu (Williams and Cesbron, 1977). Apatite in these intrusions has an orange fluorescence, and mafic minerals have rims with lower Fe/Mg than centers of grains, possibly indicating that the magmas were unusually oxidized (Mason, 1979).

H. Volcanogenic Massive Sulfides

Some types of ore are clearly related to certain series of igneous rocks. For instance, Cameron (1975b) and Wolfe (1975) show that volcanogenic massive-sulfide deposits of Pb–Zn–Cu occur in calc-alkaline volcanic sequences (andesite–dacite–rhyolite) in the Canadian Shield and elsewhere, but not in tholeiitic volcanic series (basaltic with a trend toward Fe-enrichment). In contrast, the large mafic layered intrusive bodies, such as the Bushveld and Stillwater, are formed from tholeiitic magma.

The volcanic sequences hosting volcanogenic massive-sulfide ores of Cu–Zn appear to have slightly higher contents of Zn and distinctly higher contents of sulfur than unmineralized sequences (Davenport and Nichol, 1973; Cameron, 1975b; Wolfe, 1975; Garrett, 1975). In addition, the variability of Zn, Cu, and S within such rock bodies in mineralized districts is much higher than normal. Whether these patterns are syngenetic or epigenetic is not established, but the anomalous values seem to extend for distances of miles around the known deposits.

Chapter 5

Epigenetic Anomalies in Bedrock

During deep-seated mineralizing processes, escape and leakage of elements from the orebody and the channels conducting ore-forming fluids creates a wide variety of chemical, mineralogical, and isotopic aureoles in the surrounding rocks. These anomalies are termed epigenetic because they are superimposed on the pre-existing country rock.

Such anomalies are most extensively developed near hydrothermal deposits and channel-ways, because the low viscosity of the hydrothermal fluids permits their penetration along fractures and pores in the country rock. Effects include addition of anomalous amounts of elements near hydrothermal channels, hydrothermal alteration of minerals in the wallrock, and leaching of elements along portions of the path of the ore-forming fluid. Gradients in temperature, oxidation state, and many other effects can be involved in forming aureoles (Fyfe and Kerrich, 1976). These chemical and mineralogical aureoles surrounding the ore and the hydrothermal channels provide an enlarged target for exploration.

Two mechanisms for epigenetic emplacement of indicator elements can be distinguished:

(i) Diffusion of dissolved metals through stationary pore fluid into wallrock adjacent to a vein or other zone of high metal content, with subsequent precipitation and adsorption in the wallrock. The resulting aureoles are diffusion aureoles.

(ii) Flow of fluid through veins, fractures, and pore spaces within the rock, with subsequent precipitation or adsorption of metals. This type of transport is termed infiltration, and the resulting anomalies are leakage anomalies.

In practice, anomalies combining both these effects are observed, and the two mechanisms should be considered as end members the effects of which cannot always be clearly distinguished.

I. FORMATION OF DIFFUSION AUREOLES

No matter what other processes are occurring in a solution, the ions and other dissolved constituents undergo random atomic motions and thereby tend to diffuse through the solution toward regions of lower concentration. The rate of diffusion is extremely slow, so that if a fluid is flowing at even 0·001 mm/s (32 m/yr), the effects of diffusion are negligible compared to transport by flow (Fig. 18.5). However, unless a high pressure gradient or an outlet to a nearby relatively permeable zone exists, flow through fine pores and fractures will be very slow and diffusion may dominate. If thousands or millions of years are available, the resulting dispersion by diffusion can be significant.

Diffusion can be described mathematically by Fick's first and second laws:

$$F = D\frac{\partial C}{\partial x} \tag{5.1}$$

$$\frac{\partial C}{\partial t} = D\frac{\partial^2 C}{\partial x^2} \tag{5.2}$$

where C is the concentration of the diffusing species, t is time, x is distance along the concentration gradient, D is the diffusion coefficient, and F is the flux of the diffusing component (mass per unit time across unit area). Given a specific set of initial conditions, these equations can be solved to give the pattern of concentrations as a function of time and location. For instance, assume a vein through which an ore fluid containing 100 ppm Pb (C_0) is flowing (Fig. 5.1). Before the fluid starts flowing, the pore fluid of the surrounding rock is assumed to contain no Pb, giving the initial pattern of concentration shown for 0 years (t_0) in Fig. 5.2.

At times after t_0, eqn 5.2 may be solved to give

$$C = C_0\left(1 - \text{erf}\,\frac{x}{\sqrt{(4Dt)}}\right) \tag{5.3}$$

where C is the concentration at a distance x from the vein, C_0 is the constant concentration in the fissure, and erf denotes the "error function" of the quantity $x/\sqrt{(4Dt)}$. Tables of the error function can be found in handbooks and texts on heat flow and diffusion (e.g. Ingersoll *et al.*, 1954). Figure 5.2a

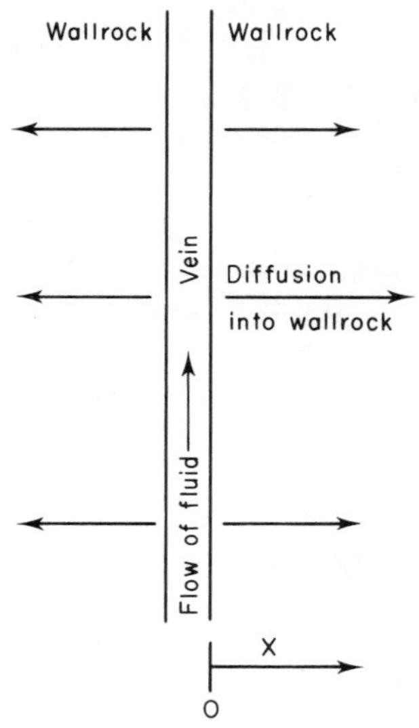

Fig. 5.1. Schematic illustration of the formation of a diffusion aureole fed by the flow of ore solution in a vein. "X" indicates coordinate system for Fig. 5.2.

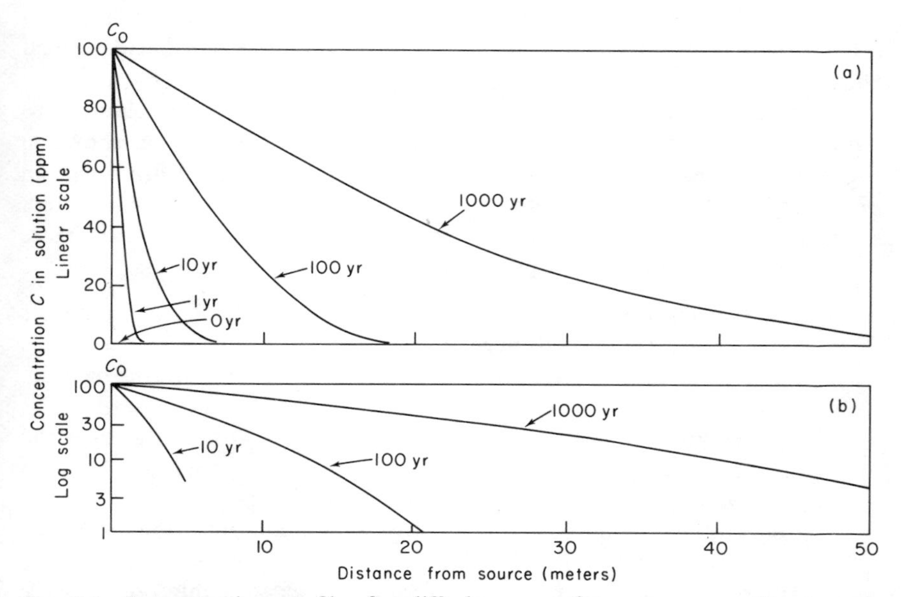

Fig. 5.2. Concentration profiles for diffusion away from a source with constant concentration, plotted on a linear concentration scale (a), and on a log scale (b). $C_0 = 100$ ppm; $D = 10^{-4}$ cm^2/s; no interaction with rock.

illustrates the pattern of concentration in the pore fluid as a function of time for this situation. The value of D (10^{-4} cm^2/s) is approximately correct for diffusion through a liquid at temperatures of 300–500 °C. After a year, Pb is detectable (1 ppm) about 2 m from the vein, after 10 years at about 7 m, after 100 years at about 20 m, etc.

In nature, several factors would decrease the distance of diffusion:

(i) In the pores of a rock, the diffusing element must follow a tortuous path, so that the effective rate would be of slower than shown, amounting to a lower value of D.

(ii) If we are to observe a halo in the rock, some Pb must be adsorbed or precipitated, and a given level of concentration in solution would advance a shorter distance from the vein. If the Pb is rapidly and completely precipitated from the solution, a sharp metasomatic front between replacement ore and wallrock would be formed; if the Pb is adsorbed in amounts proportional to concentration in solution, or reacts with the rock at a rate that is slow relative to the diffusion rate, then a pattern similar to but steeper than that of Fig. 5.2 will result.

The preceding discussion assumes that the fluid flowing through the vein at $x = 0$ maintains constant composition and constant physical properties. In most hydrothermal deposits the minerals are deposited in a sequence, indicating that the fluid changed composition with time. Any such changes will modify the pattern of metal distribution in the aureole from that described above.

The nature of eqn 5.3 is such that if the axes of Fig. 5.2a are changed to $\log C$ vs x, the plot is approximately (though not exactly) linear, as shown in Fig. 5.2b. This nearly linear relation does not necessarily hold if adsorption or precipitation is occurring, although in many cases it will be approximated.

From the above discussion, we can conclude that the nature of an aureole formed by diffusion and adsorption–precipitation will depend on:

(i) The concentration of the diffusing element at the source; a higher value of C_0 (concentration in the fissure) will result in higher values throughout the profile; variations of C_0 with time will also have an effect.

(ii) The length of time that diffusion operates; the longer the time the greater the extent of the aureole.

(iii) The nature of reaction with the wallrock; a reactive wallrock will result in narrow aureoles but relatively high concentrations, whereas a weakly reactive wallrock will allow the formation of wide aureoles.

(iv) The porosity and permeability of the wallrock; a highly porous rock with connected pores will tend to have wider aureoles than one with low porosity.

(v) The value of the diffusion constant for the chemical species and conditions; in general, small ions and high temperatures will tend to produce wider aureoles.

The theory of formation of aureoles of this type is discussed in more detail by Lavery and Barnes (1971), Wehrenberg and Silverman (1965), Golubev and Garibyants (1971), and Golubev and Beus (1970).

II. FORMATION OF LEAKAGE ANOMALIES

Ore-stage material, either in the form of visible minerals or trace elements that must be determined chemically, may be dispersed into massive rocks and in fracture zones above, below, and adjacent to hydrothermal ore deposits. In the literature of exploration geochemistry, dispersion of this type has been called leakage, following the concept that some of the fluid and metal has leaked upward or outward from the ore deposit or main channel of ore fluid (Fig. 5.3). Some leakage anomalies may represent precipitates from nearly spent mineralizing fluids moving upward after the major part of their metallic load has been left behind. If the deposit is blind (unexposed), this type of anomaly can be an excellent guide to the existence of ore. In others, the

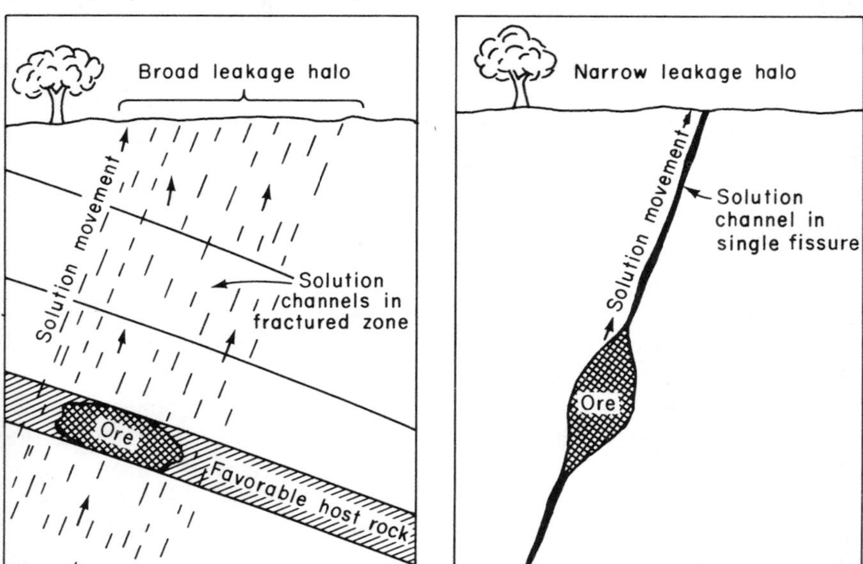

Fig. 5.3. Diagram illustrating relation of ore-solution channels to width of leakage halo.

mineralizing solutions may have been on their way to deposit an orebody that has long since been eroded. Leakage anomalies entirely similar to those related to ore may also mark the passage of non-productive hydrothermal solutions that lacked either the opportunity or the potential to deposit large concentrations of metal.

The location, dimensions, and intensity of leakage anomalies depend on:

(i) The path of flow of the ore fluid.
(ii) The amounts of indicator elements in the ore fluid.
(iii) The controls of precipitation, adsorption, and other processes transferring indicator elements from the hydrothermal fluid to the rock.

The flow path of a hydrothermal fluid is usually determined by zones of fracturing or porous rock, because flow is far more rapid along these permeable zones than along grain boundaries or pores in massive rocks. For example, calculations show that a single crack 1 mm wide through a slab of massive rock increases the rate of fluid flow by a factor of 10^5–10^6, assuming the typical permeability for massive granite. For this reason, leakage anomalies are usually better developed in fractured rocks than in more massive rocks.

Most hydrothermal fluids apparently flow in an upward direction, as a result of the lower density of the hot fluids compared to that of normal ground waters and the elevated pressures generated at depth. However, horizontal or even downward flow is not precluded if local permeable channels are so inclined, or if the fluid has unusual properties.

Fluids depositing the common base-metal ores are generally thought to contain between 1 ppm and 1000 ppm metal (Barnes and Czamanske, 1967). In comparison, normal surface and ground waters contain only about 0·01 ppm of Cu, Pb, and Zn. Even if the ore-depositing process is 99·9% efficient, it may still leave spent ore fluid with more metal than normal waters. In addition, the ore fluid may be hot and contain high levels of Cl, F, As, Hg, B, Rb, and other constituents. Thus, it would be surprising if the spent ore fluids left no trace whatever of their passage through rocks.

Elements may be transferred from the fluid to the solid phases by adsorption, ion exchange, or precipitation. The extent of adsorption depends on the surface area of the solid, the concentration of the adsorbed element in solution, and the strength of interactions between the dissolved element and the solid. Fine-grained or porous materials of large surface area may adsorb large amounts of other substances from solution. At low concentrations in solution, the amount adsorbed is usually proportional to the concentration. Ion exchange depends on the exchange capacity of the solid, on the concentration ratio of the elements being exchanged, and on the affinities of the dissolved elements for the exchange sites on the solids.

Precipitation from solution may be caused by changes in temperature, pressure, pH, oxidation state, and other chemical disturbances resulting from interaction with wallrock, boiling and mixing with ground water or other fluids (Barnes, 1973). The pattern and intensity of leakage anomalies will reflect the distribution of these processes and may be controlled by rock type, permeability, and other geological factors. In many respects the leakage anomalies and the resulting metal dispersion may be modeled by the relationships developed by chemists for chromatography, in which chemical species are transported through an adsorbent by a flowing fluid.

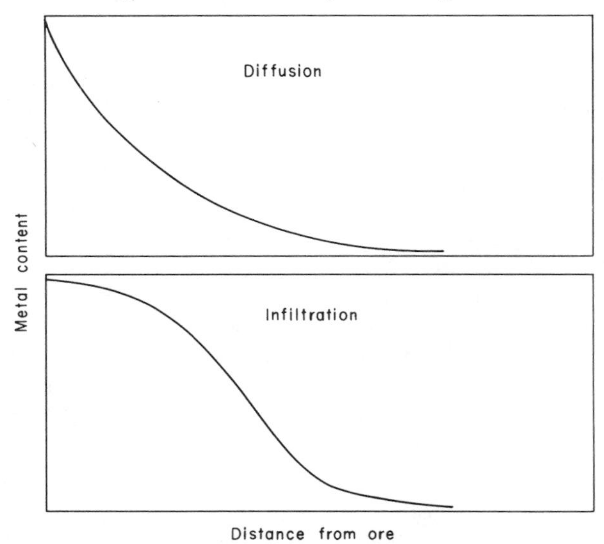

Fig. 5.4. Theoretical shapes of dispersion patterns formed by diffusion and infiltration.

Leakage anomalies differ from diffusion aureoles in their extent, their control by fracturing and permeability, and their pattern of decay away from the source of fluid. Leakage anomalies can extend for hundreds of meters from ore, whereas diffusion anomalies are limited to about 30 m. For diffusion aureoles, the plot of concentration vs distance from source is concave up (Fig. 5.2), but for leakage anomalies, this plot is usually S-shaped, concave down near the ore and concave up at greater distances from it (Fig. 5.4). However, the pattern can be strongly influenced by permeability and rock type.

The vertical extent of leakage haloes above ore varies widely depending on the permeability and structural controls for the deposit and the haloes. Beus and Grigorian (1977, p. 154) indicate that primary haloes extend 200–800 m above most types of hydrothermal deposit controlled by steeply dipping structures, and that even deeper blind orebodies may be detected in favorable cases. Where the ore and dispersion patterns are controlled by horizontal structures or bedding, the vertical extent may be much smaller but the lateral extent of the anomalies is correspondingly increased.

III. ZONING IN EPIGENETIC OREBODIES AND AUREOLES

In many epigenetic ores and aureoles, ratios of pairs of elements change gradually and progressively with location as a result of changes in the ore-forming fluid and the conditions of deposition. This zoning of elements is observed both in the ores and in the aureoles near ore. Ratios of metals can provide a means of deducing the direction towards ore, regardless of erratic changes in absolute content of elements, and a means of distinguishing the roots of orebodies from anomalies overlying ore.

Zoning of elements emplaced by deep-seated processes occurs at several scales. Within large mining districts, orebodies in different parts of a district may differ in metal ratios in a progressive way, commonly symmetrically around a center. Similarly, ore in large individual orebodies and ore shoots is commonly zoned laterally and vertically. Finally, the metals introduced into wallrock aureoles are usually zoned vertically and laterally relative to the orebody and the hydrothermal channels (Fig. 5.5). The following discussion covers mainly the last type of zoning. The first two, which share many of the characteristics of zoning in wallrock aureoles, are discussed by Park and MacDiarmid (1975) and by many papers on individual ore deposits.

Zoning in hydrothermal aureoles can also be classified according to direction (Beus and Grigorian, 1977). Zoning along the direction of flow of the ore-forming fluids is termed axial. Zoning outward from ore into wallrock, in a direction normal to the hydrothermal flow direction, is termed transverse. At most orebodies the pattern of zoning in these two directions is similar, but the extent may differ greatly, and at some deposits the sequence of elements differs.

Observations of axial or vertical zoning in leakage haloes around hydrothermal orebodies, based on studies of a large number of deposits of different types (Grigorian, 1974; Beus and Grigorian, 1977), shows the following average sequence of metals:

Shallowest *Deepest*
Ba–(Sb, As, Hg)–Cd–Ag–Pb–Zn–Au–Cu–Bi–Ni–Co–Mo–U–Sn–Be–W.

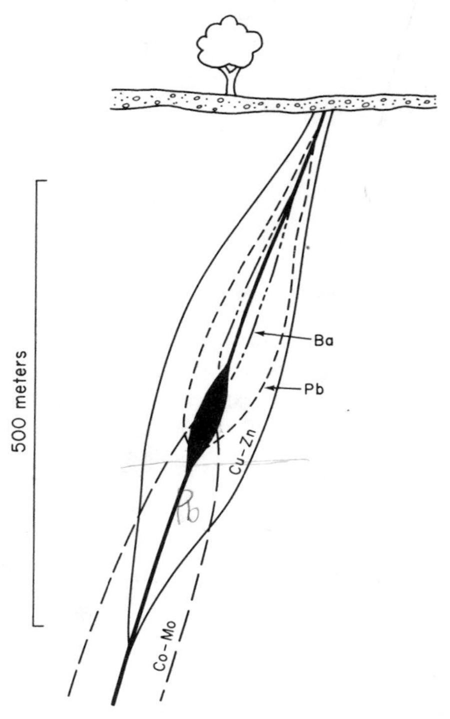

Fig. 5.5. Leakage haloes around a pyritic Cu deposit. (After Ovchinnikov and Baranov, 1972.)

Selenium and Te probably belong in the shallow zone, based on results of Gott and Botbol (1973) and Chaffee (1976a). At specific deposits, deviations from this sequence are found but generally do not involve discrepancies of more than one or two positions in the series (Table 5.1). Sequences of transverse zoning differ between types of deposit, but the major element of the ore usually forms one of the widest haloes (Beus and Grigorian, 1977, p. 127).

If the anomaly is formed by diffusion, the parameters controlling the relative behavior of elements are the diffusion coefficients and the tendency for adsorption or precipitation to take place. If the anomaly is formed by infiltration, then chemical reactions control the relative position of elements. These factors vary with temperature, pressure, chemical composition of the solution (which controls the form of elements in solution, the mineralogy of precipitates, and the nature of interaction with solids), and the identity and characteristics of solid phases. Because the zoning sequence is similar for many deposits, the chemical properties of the elements are inferred to exert a larger influence on zoning than the host rock, temperature, pressure, and fluid composition.

Table 5.1
The lateral and vertical zoning of haloes in different types of deposit[a]

Deposits	Transverse (lateral) zoning[b]	Axial (vertical) zoning[c]
Lead–zinc deposits in skarns	Ba, *Zn*, *Pb*, As, Ag, Cu, Sb	Sb, Cu, As, Ba, Ag, *Pb*, *Zn*, Cu
Lead–zinc deposits in acid effusive rocks	*Pb*, Ba, *Zn*, Ag, Cu, As, Co	As, Ba, Ag, *Pb*, *Zn*, Cu, Co
Skarns–scheelite deposits	*W*, Mo, Cu, Ba, Zn, Pb	Ba, Pb, Zn, Cu, *W*, Mo
Gold–quartz deposits	*Au*, As, Bi, Ag, Pb, Sb, Cu, Be, Mo, Co, Zn	Sb, As, Ag, Pb, Zn, Cu, Bi, Mo, *Au*, Co, Be
Copper–gold deposits	*Au*, *Cu*, Mo, Ag, As, Sb	Sb, As, Ag, *Cu*, Mo, *Au*
Copper–bismuth deposits	*Cu*, *Bi*, Pb, Ag, As, Ba, Zn, Co	Ba, Ag, Pb, Zn, *Cu*, *Bi*, Co
Uranium–molybdenum deposits	*U*, *Mo*, Pb, Cu, Zn, Ag	Ag, Pb, Zn, Cu, *Mo*, *U*
Mercury deposits	*Hg*, As, Ba, Cu, Pb, Zn, Ni, Ag, Co	Ba, *Hg*(?), Sb, As, Ag, Pb, Zn, Cu, Ni, Co
Sulfide–cassiterite deposits	Ag, Zn, Pb, *Sn*, Cu, Mo	Ag, Pb, Zn, Cu, Mo, *Sn*
Stratiform lead–zinc deposits	Ag, *Pb*, Cu, As, Ba, Co, *Zn*, Ni	As, Ba, Ag, Cu, *Pb*, *Zn*, Co

[a] After Ovchinnikov and Grigorian (1971).
[b] Elements are given in decreasing order of the width of their haloes in cross-section; for example, in Pb–Zn deposits in skarns, Sb has the narrowest halo, whereas Ba has the most extensive.
[c] Reading from the left to the right, the indicators of the supra-ore parts grade or pass downward to the indicators of the sub-ore parts of an ore zone. Italic indicates major economic elements of ore.

The models discussed in this and the preceding sections assume that the composition of the hydrothermal fluid remains constant with time. If the fluid changes with time, either gradually or abruptly, additional complications can be expected in the zoning sequence, especially if later fluids are controlled by different fractures than earlier fluids.

Discrepancies from the zoning sequence described above may be caused by differences in the mineralogical form of indicator elements. For example, if Cu occurs both as chalcopyrite and tennantite in the ore and dispersion halo, then Cu may have two peaks in the zoning sequence. Occurrence of Sn as both cassiterite and stannite has similar effects. Another source of discrepancy is the presence of two or more stages of mineralization introduced at different times.

IV. EXAMPLES OF DIFFUSION AUREOLES

Natural anomalies exhibiting the features of diffusion aureoles have been found in many localities. For instance, in the Tintic district, Utah, aureoles of Pb, Zn, and Cu extend up to 30 m from veins in quartz monzonite, but aureoles in limestone and dolomite are limited to 2–3 m in width (Fig. 5.6). The aureoles show an approximately logarithmic decay. In the limestone country rock at Gilman, Colorado, careful sampling failed to reveal any aureole (Engel and Engel, 1956), but in most other studies, aureoles have been

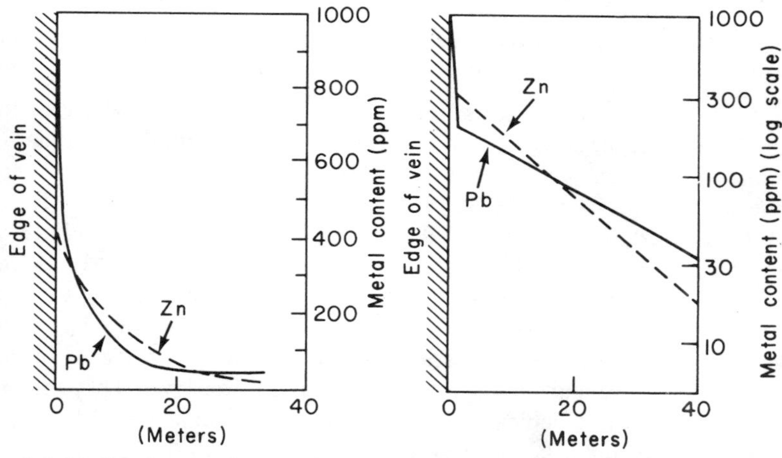

Fig. 5.6. Wallrock aureole as defined by Pb and Zn content of quartz monzonite, Tintic district, Utah. (After Morris, 1952.)

found. The aureoles are typically wider in igneous and metamorphosed silicate rocks than in carbonates.

In the Wisconsin–Illinois district, Zn aureoles around Zn orebodies extend up to 60 m from the deposits (Lavery and Barnes, 1971). Narrow aureoles of low concentration and steep gradient are found around the numerous minor fractures through the limestone, but broader anomalies with a large range of values and gentler gradient are found associated with large orebodies (Fig. 5.7). Lavery and Barnes suggest that the broad haloes result from a longer period of flow of ore solution in forming the orebodies.

Around some veins, the width of aureoles differs for different metals, probably because of differences in diffusion coefficient and in interaction with the wallrock. For instance, Au and Ag form regular aureoles extending up to 50 m from veins at Searchlight, Nevada, but base metals are anomalous only to 20 m from the same veins (Bolter and Al-Shaieb, 1971). One vein in

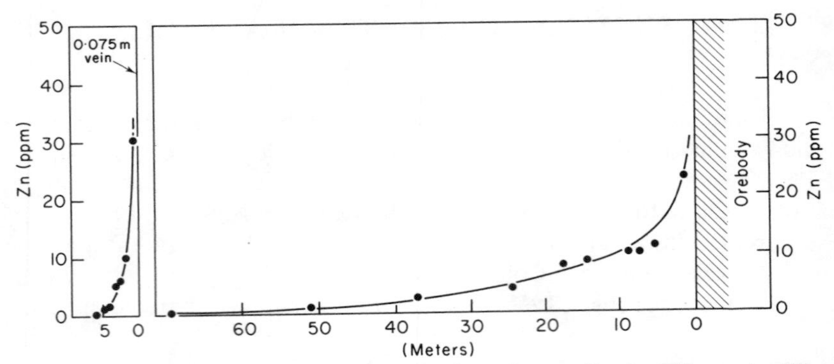

Fig. 5.7. Diffusion aureoles around an orebody and a small vein, Wisconsin–Illinois Zn district. (After Barnes and Lavery, 1977.)

this district has a much larger aureole on the hangingwall of the vein than in the footwall, perhaps as a result of some flow of solution through the rock during the mineralizing process. Anomalies for F extend to 10 m from veins in limestone in Derbyshire, U.K., but only to 3 m for Pb and Zn (Ineson, 1970). In the Park City district, Utah, Au–Ag aureoles are widest adjacent to the most Au–Ag-rich ores in the eastern part of the mine, and Mn aureoles are widest in the western part where Mn is most abundant in the veins (Al-Shaieb and Bolter, 1973). Reinking *et al.* (1973) describe wide haloes above zones with little mineralization, but narrow haloes adjacent to veins along which replacement bodies were formed at a lower level. They attribute the narrow haloes to silicification and sealing by quartz acquired by the ore solution while forming the replacement orebody.

In Rhodesia, James (1957) studied the distribution of As in the wallrock of shear zones containing Au and As veins. The As content of unweathered rock collected from underground workings showed very marked pseudo-logarithmic decay curves extending into the wallrock from the shear zones, as shown in Figs 5.8 and 5.9. At the Bell Mine, where the country rock is sandstone, the aureole is only 8 m wide, whereas at Motapa, in a greenstone

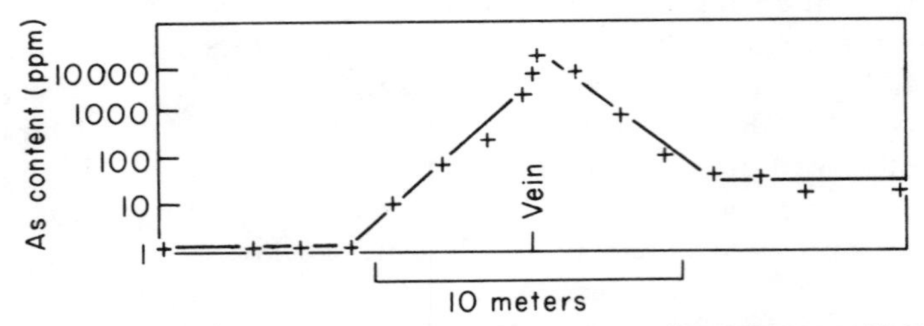

Fig. 5.8. Wallrock aureole as defined by As content of sandstone adjoining Au vein, Bell Mine, Rhodesia. (After James, 1957.)

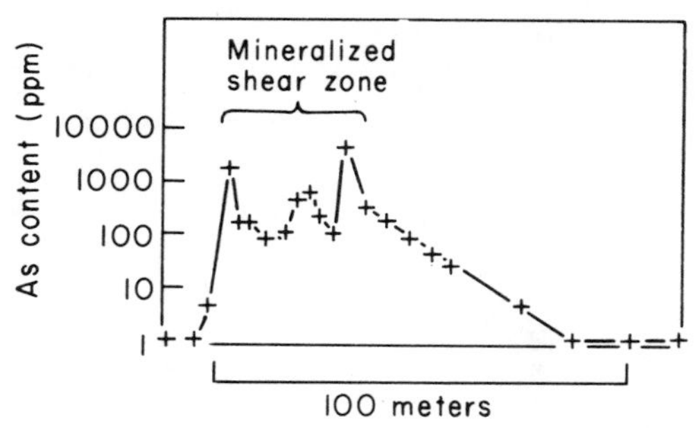

Fig. 5.9. Wallrock aureole as defined by As content of greenstone adjoining shear zone, Motapa Gold Mine, Rhodesia. (After James, 1957.)

country rock, the aureole is almost 70 m wide. Probably this difference is related to either the chemical reactivity or the permeability of the two rock types, together with the duration of the period of hydrothermal activity.

Stoll (1945) found increased concentrations of Be in the wallrocks of New England pegmatites. He postulated that the Be was added to the wallrock by solutions emanating laterally from the main channels through which the pegmatitic liquids were flowing.

V. EXAMPLES OF LEAKAGE ANOMALIES AND ZONING

A. Polymetallic Vein and Replacement Deposits

In the U.S.S.R., extensive studies have been made of the distribution of chalcophile metals in rocks around veins, replacements, and skarns containing a variety of metals (Grigorian, 1974; Ovchinnikov and Baranov, 1972; Khetagurov *et al.*, 1970; Beus and Grigorian, 1977). Haloes of anomalous metal content were found around all the ores tested. At many deposits of small to moderate size, the total amount of metals disseminated in the wallrocks is considerably larger than the amount in the orebodies. Haloes of anomalous metal content characteristically extend for distances of meters to hundreds of meters from the ore. Anomalies are detected adjacent to feeding structures beneath the deposit (subore anomalies), in rocks adjacent to the deposit, and in rocks near the controlling structures above the deposit (supra-ore anomalies) (Fig. 5.5). The metals are vertically zoned, so that at a Cu deposit, Co and Mo may have their highest concentrations and widest haloes along the feeding structures beneath the deposit, Cu its widest halo at the level of the deposit, Zn and Pb just above the deposit, and As, Sb, and Ba completely above the deposit. Figure 5.10 illustrates similar anomalies at a Cu–Pb–Zn deposit in Sweden. Such anomalies are useful for detection of blind orebodies at depth, and serve to enlarge the exploration target in areas of partial cover by soil and other overburden.

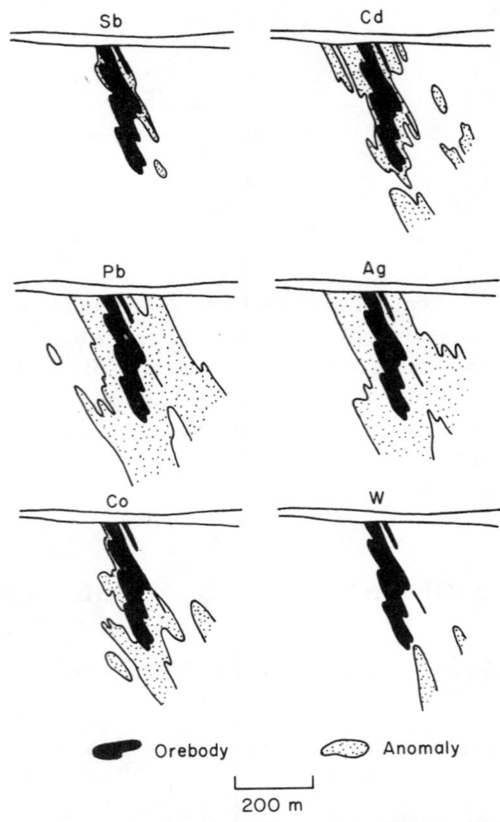

Fig. 5.10. Anomalies for trace metals at Harpenberg Pb–Zn–Cu–Ag skarn deposit in Sweden. (After Beus and Grigorian, 1977.)

Ratios of metals widely separated in the zoning sequence are useful in identifying vertical zoning and the level of a particular sample in the zoning sequence. For instance, the Ag/Cu ratio or the Pb/Bi ratio can be used to compare samples with differing amounts of total sulfides. As an extension of this concept, Grigorian recommends the use of "multiplicative ratios" such as $(Ag \times Pb \times Zn)/(Cu \times Bi \times Co)$, because this type of ratio smooths out erratic analytical and sampling errors, and allows a better comparison of different districts by minimizing local differences in the zoning sequence for one or two elements.

The best samples for defining leakage haloes consist of several meters of continuous drill core or chips of outcrops over a distance of several meters. Results from several such samples may be averaged to smooth out the local variations. Samples of soil may also be used, but the effects of leaching and enrichment in the soil profile must be allowed for, especially if several elements are being compared.

In the Coeur d'Alene district, a comprehensive study of leakage anomalies has shown many of the same features reported in the Russian deposits (Gott and Botbol, 1973). This study has helped to interpret the structure of the district, and has suggested extensions of the district. Silver in rock and soil outlines areas of shallow Pb–Zn deposits discovered in outcrop, but not the deposits of the Lucky Friday area which are found at 700 m depth. In contrast, anomalous Te values are not found near the known deposits but are found above the deep deposits in the Lucky Friday area. The Te anomalies appear to mark a zone many hundreds of meters above the productive Pb–Zn deposits. Similarly, high contents of Cd and high values of Cd/Zn are also indicative of ore at depth. A vertical zoning sequence of (Te, Mn)–Cd–Ag–(Zn, Pb) may be inferred from the data given by Gott and Botbol (1973).

At the Gregory mine, Derbyshire, a Pb and Zn anomaly in shallow residual soil occurs over a vertical fissure vein in limestones situated beneath 200 m of shales and sandstones, as illustrated in Fig. 5.11. The anomaly is absent

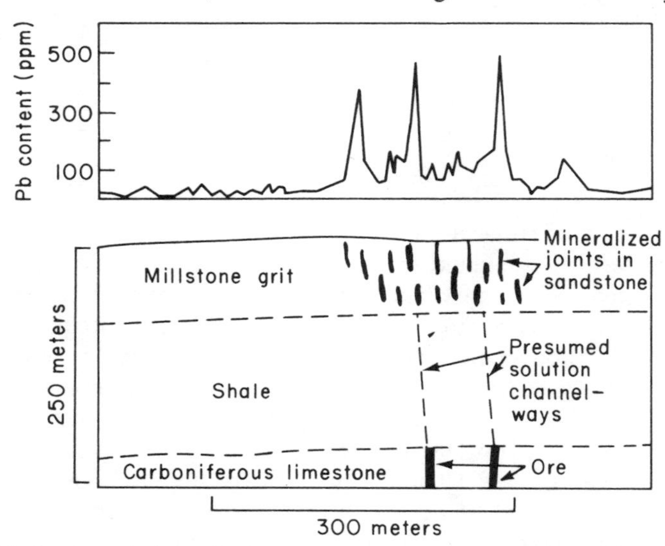

Fig. 5.11. Leakage halo as defined by the Pb content of residual soil overlying a base-metal deposit at a depth of 200 m, Gregory Mine, Derbyshire. Data on −80-mesh fraction. (After Webb, 1958b.)

where the fracturing of the overlying cap rock was not sufficiently extensive to allow the upward migration of the ore solutions.

A comparison of aureoles in fractured and massive rocks shows that the patterns in fractured rock are more irregular but broader than those in massive rock, presumably because flow of solution was guided and promoted by the fractures (Table 5.2).

Table 5.2
Primary Pb and Zn aureoles in massive and fractured wallrocks alongside galena–sphalerite veins in Derbyshire, U.K.[a]

Distance from vein (m)	Metal content of limestone wallrock (ppm)			
	Massive wallrock		Fractured wallrock	
	Pb	Zn	Pb	Zn
0	1600	1900	600	34 000
1·5	600	850	1700	500
3·0	230	180	1600	800
4·5	220	220	1400	900
6·1	120	260	1300	900
7·6	60	80	3500	1700
9·1	—	—	1200	400
12·2	—	—	1250	850
15·2	—	—	130	60
18·3	—	—	440	170
21·3	—	—	1400	600
24·4	—	—	750	1250
27·5	—	—	250	70
45·7	30	140	1200	1000

[a] Source: Webb (1958b).

The distribution of Hg near at least some base- or precious-metal deposits exhibits a pattern of maximum enrichment in the zone above the ore. For instance, Crosby (1969) and Chan (1969) describe haloes of Hg extending several hundreds of meters from ore in the Coeur d'Alene district, and high concentrations in hood-shaped zones above blind ore shoots. In the Pachuca–Real del Monte Ag district in Mexico, Hg forms strong anomalies above blind orebodies at depth but is absent from wallrock immediately adjacent to veins (Friedrich and Hawkes, 1966).

Near the Sn deposits at Altenberg in the Erzgebirge of central Europe, leakage haloes of Ga and Mo extend 100–200 m laterally from the greisen and Sn ore, anomalies in Bi, Sn, and Li extend to 500 m from the ore, and all except Li show strong vertical zoning and disappear a short distance below the level of the greisen (Fig. 5.12). In Czechoslovakia, Sn, Rb, W, and Mo

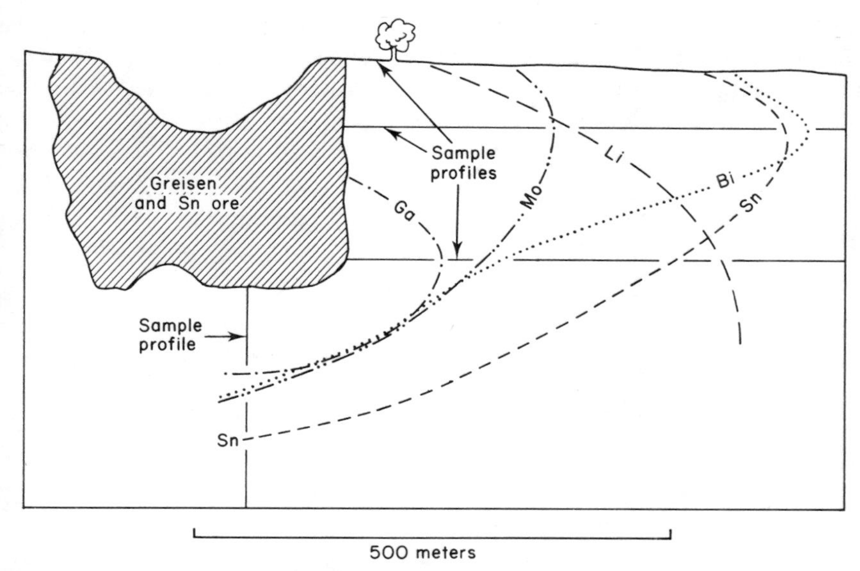

Fig. 5.12. Horizontal and vertical extent of Sn, Li, Bi, Ga, and Mo anomalies around the Sn deposit of Altenburg, East Germany. Broken lines indicate threshold contours, equal to mean plus 3 standard deviations. (After Tischendorf, 1973.)

give useful anomalies above greisen-type Sn deposits; and Sn, Cu, As, Zn, and B give useful anomalies above Sn-bearing skarns (Pokorny, 1975).

At Johnson Camp near Willcox, Arizona, Cooper and Huff (1951) have described a broad leakage halo in the fractured rock 130 m above the concealed Moore orebody. The ore is localized in one bed of a moderately dipping sequence of partly silicated calcareous rocks. Chip samples of rock collected at close intervals across outcrop surfaces were composited in sections of 17 m or more. The heavy-metal content of samples collected in this manner along a series of traverses showed an anomalous area similar in shape to the horizontal projection of the ore deposit, but displaced about 70 m to one side (Fig. 5.13).

In the Tintic district of Utah, complex Pb–Zn–Ag ore occurs in limestone of Paleozoic age overlain by barren rhyolite that pre-dates the mineralization (Lovering *et al.*, 1948; Bush and Cook, 1960). Surface studies of the rhyolite over known ore disclosed extensive areas of argillic and pyritic alteration. The alteration haloes are so extensive, however, that they are of limited use in guiding diamond-drill exploration. Sampling and chemical analysis of the altered rhyolite disclosed a well-defined heavy-metal anomaly that coincides closely with the up-rake projections of one of the principal ore shoots. This anomaly corresponds to the surface trace of the channels through which the nearly spent mineralizing solutions escaped. Even though the rhyolite was not a favorable host for ore deposition, enough metal was deposited along the solution channels to be detectable by trace analysis.

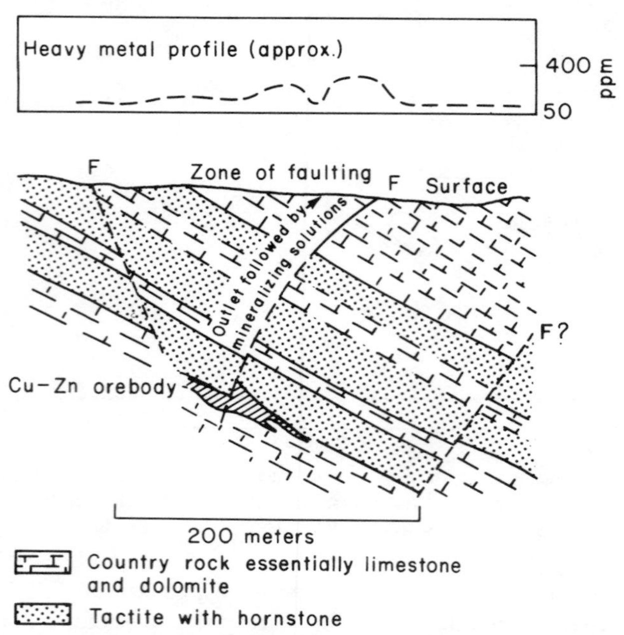

Fig. 5.13. Leakage halo as defined by the heavy-metal content of rocks exposed at the surface, Johnson Camp, Arizona. F = Pre-mineral fault zones. (Based on Figs 2, 3, and 5 of Cooper and Huff, 1951.)

B. Porphyry-copper and Molybdenum Deposits

At porphyry-copper deposits, not only have Cu- and Fe-sulfides been added to large volumes of rock in the form of veinlets and disseminated grains, but the ore and a large volume of surrounding rock have typically undergone extensive hydrothermal alteration to new minerals, with accompanying addition and depletion of both major and trace elements (Lowell and Guilbert, 1970, Rose, 1970). A central zone of pervasive potassic alteration, in which orthoclase and biotite are formed at the expense of plagioclase and mafic minerals, is typically coextensive with the central, most Cu-rich parts of the deposit, and is usually centered on an intrusive complex (Fig. 5.14). Peripheral to and cutting across the potassic zone are zones of sericitic alteration in which the rock is converted to sericite, pyrite, and, at some deposits, clay. Outside these zones of strong alteration, the rock is typically propylitized (altered to chlorite, calcite, clay, albite, pyrite, etc.). The potassic and sericitic zones are normally highly shattered and veined by quartz, sulfides, and other minerals. In the propylitic zone, widely spaced large veins are usually present but shattering is less developed.

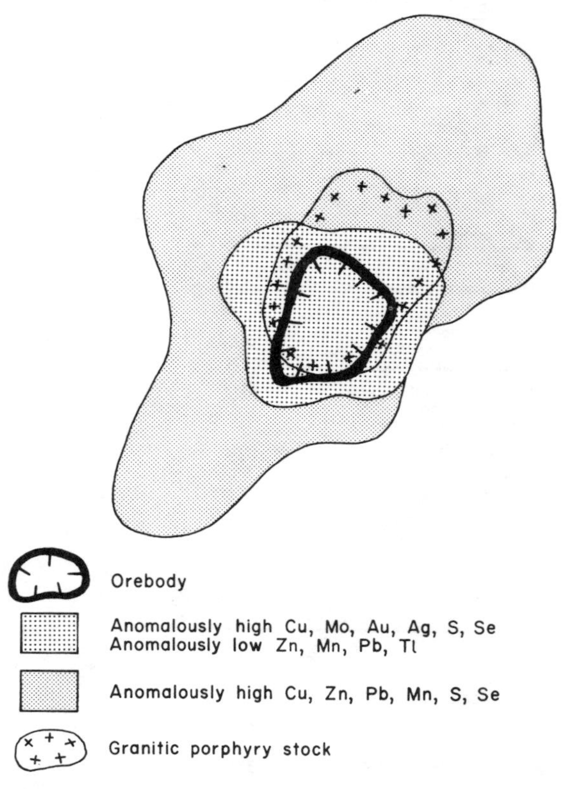

Fig. 5.14. Schematic plan of trace metal zoning at porphyry-copper deposits.

Recent research suggests that the central early-formed parts of the ore and alteration may be produced by hydrothermal solutions derived from magmatic sources, but that sericitic and peripheral alteration and mineralization may be caused by convecting ground water set in motion by the heat of the igneous activity, and by mixing of magmatic and ground waters. The chemical effects thus extend over the very large volumes of rock involved in the convection cells, and include elements supplied both from igneous sources and the surrounding country rock.

Patterns of enrichment and depletion in lithophile elements are related to mineralogical changes. In the potassic zone, K, Cu, S, and other metals are usually added, and Ca is depleted, usually along with Na and Mg. Similar

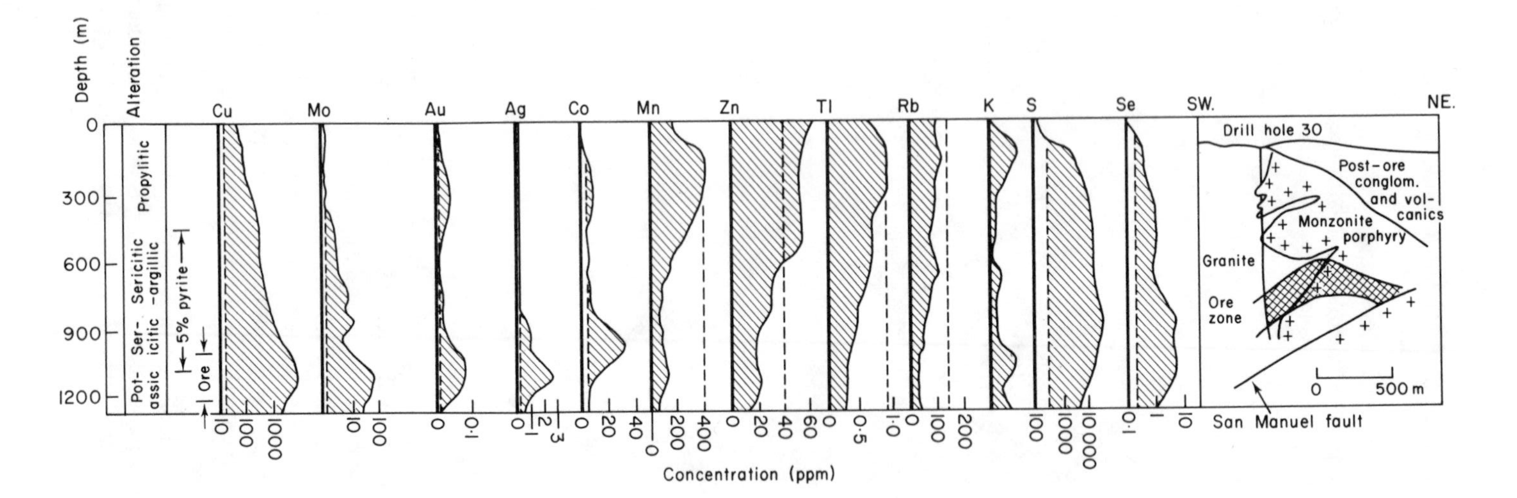

Fig. 5.15. Lateral zoning of elements in drill hole 30, Kalamazoo porphyry-copper deposit, Arizona. Dashed lines are inferred background values. Deposit has been faulted and tilted about 70° to the NE. (After Chaffee, 1976a.)

but more extreme changes in chemical composition are found for sericitic alteration, but in the propylitic zone, additions and depletions of major elements are generally small, except for H_2O, CO_2, and S. Patterns of lithophile trace elements, such as Rb, Sr, Li, Ga, and V, tend to be similar to the major elements they resemble (K, Ca, Mg, Al, Fe, respectively). Strontium is depleted and Rb enriched in most ore zones of porphyry-copper deposits (Armbrust *et al.*, 1977; Gunton and Nichol, 1975; Chaffee, 1976a). Both major and trace elements can be useful guides to zoning and alteration in porphyry-copper districts, provided that changes in background related to rock type are taken into account. In most cases, the patterns of lithophile elements and alteration can be efficiently mapped by observations of the mineral assemblages (Rose, 1970), supplemented by petrographic work and instrumental analysis of rock samples (Lyon and Tuddenham, 1959; Hausen and Kerr, 1971).

Extensive anomalies in chalcophile trace elements are recorded for porphyry Cu–Mo districts. At the San Manuel–Kalamazoo deposit in Arizona, Chaffee (1976a) found the following patterns of lateral zoning in rock samples from two long drill holes through the tilted deposit (Fig. 5.15):

Anomalously high in ore zone:	Cu, Mo, Au, Ag, Co, K, S, Se, Ba, B
Anomalously high in halo zone:	Cu, Zn, Fe, S, Se
Anomalously low in ore zone:	Mn, Zn, Tl, Rb, Li, Na
Anomalously low in halo zone:	Tl, Li, Na.

Work at other deposits suggests that anomalously high contents of Cu, Mo, Au, and S are typical of the ore zone at porphyry-copper deposits, as are high values of Cu, Zn, Pb, and S in the halo zone (Fisher, 1972; Jambor, 1974; Wargo, 1964; Theodore and Blake, 1975; Drewes, 1973; Chaffee, 1976b, 1977; Olade, 1977). At porphyry-molybdenum deposits, Mo anomalies extend well outside the ore zone, and Zn anomalies are found in the halo zone (Theobald and Thompson, 1961; Tauson and Petrovskaya, 1971; Woodcock and Carter, 1976). The contrast of these anomalies, especially in the halo zone, is greatly increased by analysis of selected material from veins, iron-stained rock, or highly altered rock, as shown by several of the above workers. Anomalies in Te, Hg, Au, and Ag are found in selected samples of veinlets and Fe-stained rock from the peripheral zones around the Ely porphyry-copper deposit (Gott and McCarthy, 1966). Similar patterns have been detected in altered zones not containing known major deposits, so that not every such anomaly is related to an ore deposit (Drewes, 1967, Moore *et al.*, 1966; Marsh and Erickson, 1975).

Vertically above the blind Red Mountain Cu deposit in Arizona, Corn (1975) reports very low Zn values near the surface, but very high Zn values in deep drill holes above the zone of ore-grade Cu, suggesting that the lateral

Zoning patterns discussed above also occur in the vertical dimension but may not extend all the way to the surface. Of the radioactive elements (K, U, and Th), only K shows patterns related to ore (Davis and Guilbert, 1973).

Porphyry-copper, epithermal precious-metal, and other hydrothermal deposits associated with hypabyssal igneous bodies are surrounded by extensive areas in which the isotopic ratios $^{18}O/^{16}O$ and D/H are much lower than in the original rocks (Taylor, 1974). This variation in the isotopic composition presumably results from interaction of the rocks with circulating heated waters of surface origin (meteoric waters). The surface waters are relatively low in ^{18}O and D compared to rocks and waters of the deep-seated environment, and they deplete these isotopes in the rocks through which they circulate. These haloes of low $^{18}O/^{16}O$ and D/H, which extend for several kilometers from the ores, can serve as an indication of an extensive hydrothermal circulation system. The extent of change in $^{18}O/^{16}O$, in conjunction with knowledge of the ^{18}O content in local ground water, can allow a semiquantitative estimate of the amount of water that has circulated through the system (Taylor, 1974). The passages through which the fluids flowed can be considered as possible ore-controlling structures. The center of such a hydrothermal system is normally an igneous intrusive body, which contributed heat and possibly part of the fluid in the convection system.

C. "Epithermal" Deposits of Au–Ag–As–Sb–Hg

Vein and replacement deposits of these metals are common in volcanic areas of the Cordilleras in North America (Carlin, Goldfield, Getchell in Nevada, some ores in San Juan Mts, Colorado), in parts of the U.S.S.R., New Zealand, and elsewhere. The association with recent volcanic activity, the fine-grained character of the minerals, the limited vertical range of mineralization, and the character of fluid inclusions in minerals suggest near-surface deposition in an environment similar to some hot springs. Extensive haloes of As, Sb, Pb, Zn, Cu, Hg, Tl, W, and Bi form around and above these deposits (Polikarpochkin *et al.*, 1965a; Akright *et al.*, 1969; Erickson *et al.*, 1964b; Burbank *et al.*, 1972; Ewers and Keays, 1977).

At the Baley Au deposits in eastern Transbaikalia, the Au-bearing sections of the veins have anomalies in As, Sb, Ag, Cu, and Hg (Polikarpochkin *et al.*, 1965b). The anomalies widen and coalesce upward so that a zone more than 300 m wide is anomalous near and above the ore, and can be detected at least 250 m above the zone of exploitable ore. Copper, Pb, and As are not anomalous immediately adjacent to the veins, but become anomalous farther away, both upward and laterally. Gold and Ag form continuous anomalies from the vein outward. The zones of maximum content of As, Sb, Cu, and Hg are well above the zones of maximum Au and Ag, so that ratios of elements can be used to determine whether ore lies at depth or has already been eroded away.

D. Volcanogenic Massive-sulfide Deposits

A large group of base-metal deposits, including the Kuroko deposits of Japan and deposits in eastern Canada, Cyprus, and Spain, are inferred to have formed by deposition from hydrothermal solutions that emerged on the sea floor (Lambert and Sato, 1974; Jenks, 1975). Three types of material can be deposited by the hydrothermal activity (Fig. 5.16): a stockwork and

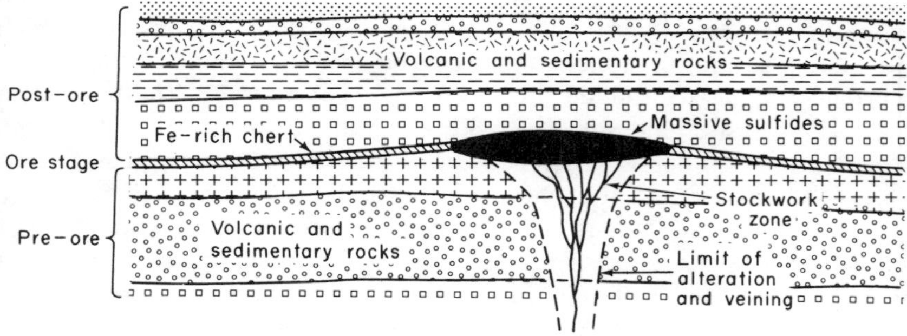

Fig. 5.16. Generalized relationship of massive-sulfide ore to Fe formation, stockwork zone, and post-ore volcanic and sedimentary rocks.

replacement zone of veins and alteration along the fissures carrying the solutions to the sea floor, a lens of massive sulfides precipitated on the sea floor, and sometimes an extensive layer of cherty iron formation apparently resulting from wide dispersion of some constituents.

In the stockwork zone, sulfide veins and replacements are accompanied by alteration to quartz, sericite, and chlorite. Calcium and Na are generally depleted, as at the porphyry-copper deposits, and Mg and Fe are generally added (Goodfellow, 1975; Bennett and Rose, 1973). These alteration effects may extend for many hundreds of meters from the center of the stockwork, and can provide a guide to ore, even where later metamorphism has obscured the original mineralogy. Anomalies in Cu, Pb, Zn, and other heavy metals accompany the altered rocks. At many Russian deposits, these anomalies extend into unaltered pre-ore rocks (Ovchinnikov and Baranov, 1972). At Mattabi, Cu and Zn anomalies extend at least 100 m beneath the deposit (Franklin *et al.*, 1975). However, Govett (1972), Lambert and Sato (1974), and Thurlow *et al.* (1975) indicate that in Canadian and Japanese deposits the haloes extend for only a few meters to about 30 m from the ore in the footwall rocks, although local high and low values are found over a larger area. Baranov *et al.* (1972) found that much of the anomalous metal is in pyrite, either as inclusions or in solid solution, and that the contrast and extent of anomalies are enhanced by analysis of pyrite concentrates. Of course, anomalies in the pre-ore footwall rocks are only of value as a guide to ore in areas of steeply folded rocks. If the deposits are correctly interpreted as sea-floor precipitates, the more extensive weak anomalies reported in the hangingwall of the ores (Whitehead and Govett, 1974) must be secondary and result from later dispersion during diagenesis, metamorphism, or weathering.

Extensive anomalies have been reported in the cherts, tuffs, and Fe formation deposited at the same time as the sulfide lens. Ridler and Shilts (1974) describe a prospecting system based on the concept that these Fe–Mn-rich "exhalites" associated with volcanogenic massive-sulfide deposits have high contents of Au and base metals that can be used as a guide to the nearby occurrence of massive-sulfide ore. Near the McArthur Pb–Zn deposits in Australia, extensive haloes of K, Zn, Pb, As, and Hg are described in Fe-rich dolomites and tuffs deposited at the same time as the ore (Lambert and Scott, 1973). Modern versions of such haloes are described in the Red Sea brine area by Holmes and Tooms (1973). Anomalies extending several hundred meters from ore are reported near the Kuroko deposits in Japan (Lambert and Sato, 1974). Such haloes in the contemporaneous rocks constitute a basis for detection of deposits formed by hot springs on the sea floor.

A slightly different kind of halo is described at the Tynagh deposit in Ireland (Russell, 1975). In this case the ore deposit replaced permeable limestones adjacent to a fault at a shallow depth beneath the sea floor, but apprently some ore fluids continued up the fault zone and entered the sea, producing a metal-rich iron formation and limestone with anomalous Mn content, the latter extending as far as 7 km from the deposit. Similarly, Lur'ye (1957) reported high contents of Pb and Zn in limestones stratigraphically above the Mirgalimsai Pb–Zn deposits in the U.S.S.R. and suggests that a similar process furnished the metals to the sea.

E. Mississippi-valley Type Lead–Zinc Deposits

An example of a leakage anomaly is seen in the increase in the Mg–Ca ratio in calcareous rocks over blind ore, as demonstrated by Agnew (1955) in the Gutenberg limestone in the south-western Wisconsin Pb–Zn district. A halo of pyrite occurs in the flat-lying sedimentary rocks overlying Zn ore in the Shullsburg district of south-western Wisconsin (Kennedy, 1956). Lead and Zn also occur in the pyritized cap rock, but in such small quantities that their distribution is more easily determined by chemical analysis.

F. Leakage Anomalies in Specific Types of Material

Because rocks and minerals vary greatly in both background values and ability to precipitate or incorporate minor elements, the collection and analysis of specific types of material is sometimes advantageous. For instance, Baranov *et al.* (1972) found that Cu, Pb, Zn, and other chalcophile metals were concentrated in pyrite from rocks near pyritic ores. Analysis of pyrite concentrates separated from the rock produced anomalies with greater contrast and with three times greater lateral extent than analysis of bulk rock samples.

Jasperoid, or highly silicified rock formed by replacement of limestone and other rocks, has been used as a special sampling medium in western U.S.A. The jasperoids typically weather to prominent iron-stained outcrops and are thus easily recognized and sampled. Jasperoids from the vicinity of hydrothermal ore typically contain much higher Fe, Ag, Bi, Cu, Hg, Mo, Pb, Te, and Zn than barren silicified zones, which may also occur in the same district (Lovering and Hamilton, 1963; Lovering *et al.*, 1966).

In the northern Michigan Cu district, leakage anomalies around the native Cu deposits are not detectable in the basaltic host rocks, but calcite veins cutting the basalts have higher Mn content near ore, and can be used as a guide to favorable areas (Ruotsala *et al.*, 1969).

In several districts in Nevada, Erickson and coworkers have sampled limonite, small veins, and altered zones, and have demonstrated extensive dispersion of many chalcophile elements in the vicinity of ore and mineralization (Erickson *et al.*, 1961, 1964a, 1964b; Marsh and Erickson, 1975).

VI. GEOCHEMICAL ROCK SURVEYS

Surveys of the distribution of elements in samples of bedrock are most useful in detailed exploration for individual orebodies and ore shoots, and in reconnaissance exploration for productive igneous bodies. Surveys to detect geochemical provinces have also been carried out, but are better accomplished by drainage surveys rather than rock surveys. A major advantage of rock surveys is the direct measurement of deep-seated dispersion patterns unaffected by complications of the surficial environment. Rock surveys are finding increasing application in many phases of mineral exploration.

The dimensions of the various types of anomalies developed in the deep-seated environment range from hundreds of kilometers to a few centimeters, and their applicability varies accordingly, Anomalous plutons range from many kilometers to tens of meters in size, and are mainly useful in the detection and recognition of mineralized districts. Large leakage haloes and hydrothermal alteration patterns with dimensions of several kilometers to a few meters are useful in surveys of mineralized districts in order to locate small areas that may offer exceptional promise for intensive exploration. Detection of blind orebodies using this method is especially promising. Leakage haloes and diffusion aureoles with dimensions of a few meters to a few tens of meters can assist in finding individual orebodies, especially using samples from underground workings and drill cores. Zoning patterns can be used to evaluate the direction of ore and whether ore is likely to be found at a deeper level or has already been eroded away.

The type of ore being sought is an important consideration in planning rock geochemical surveys. Geochemical provinces and productive plutons are much better developed for lithophile elements such as U, Sn, and Li than for chalcophile elements (Cu, Pb, Zn, Ag). Plutons giving rise to magmatic ores (immiscible Cu–Ni-sulfides, pegmatites) are much more easily detected geochemically than those giving rise to hydrothermal deposits. On the other hand, leakage anomalies are much better developed for hydrothermal deposits. Because of these differences, orientation surveys of known mineralized districts and background areas are usually necessary before starting a rock survey. A careful study of the literature for information on similar types of deposits is also desirable.

A. Orientation Surveys

The purpose of an orientation survey is to provide the technical information on which to base the complete survey. One or more known deposits of the type being sought, preferably located in the area to be surveyed, are sampled in detail to obtain data for design of an optimum procedure for the complete survey and interpretation of its results. Background samples of typical rocks from unmineralized portions of the area must be collected for comparison with the anomalous samples. The details of the orientation survey will vary with the type of ore being sought, the scale and stage of exploration, the type of material that can be easily sampled, and other factors. Table 5.3 summarizes the type of information collected in a typical orientation survey for rock sampling.

The type and amount of sample and the method of sample collection should be established during the orientation stage. Indicator elements are usually relatively homogeneously distributed in dispersion patterns of syngenetic or diffusion origin, and relatively small specimens of rock may be adequate. In contrast, leakage anomalies are commonly localized along fractures, and the metal values are very erratically distributed at the scale of hand specimens. In such cases, analysis of an aggregate of small chips from randomly chosen

Table 5.3
Check list of factors to be optimized by an orientation survey preparatory to rock sampling

(i) Type of sample (rock, vein material; comparison with soil or drainage samples)
(ii) Size and character of sample (single large chunk, many small chips, channel sample, length of drill core)
(iii) Best indicator elements (ore element, pathfinder elements, related major elements for use in ratios)
(iv) Applicability of separated minerals or types of material (sulfides, limonite, biotite, calcite, etc.)
(v) Effects of weathering, rock type, hydrothermal alteration and other geological variables on background and contrast of anomalies
(vi) Shape, extent, and homogeneity of anomalies, and reproducibility of anomalies from a single site
(vii) Method of sample decomposition and analysis (total analysis, sulfide- or oxide-selective leach, acid digestion)
(viii) Sources of contamination (metal from collecting and crushing equipment, dust, drill steel, circulating waters, smelter fumes)

sites over several square meters of outcrop is generally preferable to analyses of single pieces of rock. Composite samples from many meters of drill core should be used if available. If leakage haloes are being sought, vein materials, limonite, or other special types of samples should be collected in both mineralized and unmineralized areas. The contrast of anomalies in different types of samples can be compared at this stage, as described in Section I of Chapter 13.

A relatively comprehensive selection of elements should be tested during the orientation stage inasmuch as regions and districts differ in the applicability of possible pathfinder elements. The literature on similar surveys and geochemical theory can suggest possible elements to include, but, because the processes and geological environments are diverse, tests on a wide selection of elements at this stage are desirable in order to recognize unexpected associations.

The optimum method of decomposing and analyzing samples should be investigated in order to develop a method that is reliable and gives the optimum contrast and extent of anomaly. Sulfide-selective leaches, mineral separations, and other methods of sample treatment may be compared between background and mineralized areas to determine optimum techniques. These methods can greatly increase the extent and contrast of anomalies (Beus and Grigorian, 1977, pp. 76–90). In addition, the analysis of mineral separates may aid in understanding the geochemistry of indicator elements and thereby improve interpretation of anomalies.

The effects of varying rock type, weathering, hydrothermal alteration, man-made contamination, and other geological variables should be tested in the orientation stage, so that the routine sampling procedure may be designed to avoid problems from these sources or to note them when they occur. Suites of samples, each strictly comparable as regards rock type and degree of weathering and alteration, should be collected from selected mineralized and background areas. It is advisable to collect, if possible, samples from several areas of known mineralization in order to test the consistency of any apparent relationships that may be disclosed.

It is essential that the results of orientation for leakage dispersion should be interpreted in terms of the geology in three dimensions and care taken to discriminate wherever possible between primary dispersion and secondary redistribution of metal by percolating meteoric waters. In the latter connection it is very helpful to determine the primary ratio of metals possessing markedly different mobilities in the zone of weathering. Thus, in supergene redistribution patterns, the ratio of Zn (mobile) to Pb (immobile) would be expected to show a marked increased compared to the Zn/Pb ratio in primary ore.

In orientation studies involving dispersion in wallrock aureoles, a series of samples that is closely spaced near the vein or orebody, and more widely spaced away from it, is usually most effective in determining the extent of the aureoles. The series of samples should extend well into unaltered rock, so that background levels are clearly established. Inclusion of more than one rock type in a single sample should be avoided. Needless to say, it is essential that the geology of the sample traverse be logged in complete lithological and structural detail. Samples may be conveniently collected along crosscuts near ore or taken from drill cores. The former generally gives a more complete section but removal of surface contamination may be troublesome.

In many areas, poor exposures militate against systematic rock sampling at the surface. However, the primary pattern, whatever its nature, may be reflected in the soil, particularly if the overburden is residual. It is common practice, therefore, to compare the anomalies obtained by rock and soil sampling during the course of orientation. The procedure for examining metal distribution in the soil and the factors that need to be taken into account are similar to those outlined later in connection with detailed soil surveys.

B. Collection and Processing of Samples

Many aspects of sample collection are similar to those used in soil and drainage surveys, and many of the procedures described in Section II of Chapter 13 and Section IV of Chapter 16 can be adopted. In particular, the procedures for selecting sampling patterns and for locating and identifying samples are applicable to detailed surveys using rocks. The sample spacing for detailed surveys seeking veins or small orebodies should be close enough that three to six samples fall within an anomaly and define a clear trend toward the ore.

The method for selection of sample sites will vary with the scale of the survey, the type of samples to be used, and the availability of outcrops, drill core or other material. In reconnaissance for productive plutons or similar features, several samples are usually collected from each pluton or body of interest. In detailed surveys, a regular grid or spacing along traverses is normally preferred. Since large areas of complete outcrop are rare, rock samples cannot usually be collected on a rigid grid or a rigid spacing along traverses. If exposures are relatively abundant, the closest outcrop to the grid point may be sampled. If no site exists within some specified distance (say half or a quarter of the sample spacing), then no sample is collected. Alternatively, the area or pluton to be sampled may be divided into blocks and one or more samples selected from each block. It is important that samples be selected in as unbiased a fashion as possible so that they are representative of the area rather than units especially resistant to weathering or preferentially exposed along streams. However, if outcrop is poor, the use of samples of residual soil in place of rock should be considered.

Rock samples are usually broken from the outcrop with a hammer as large as can be conveniently carried and used in the field. The sample should be collected from a typical part of the outcrop (which is not necessarily the portion most easily broken off). Weathered or contaminated surface material should be removed in the field if the orientation survey has indicated that such features complicate interpretation. Cloth or heavy plastic bags are usually the most convenient containers, except for small chip-samples for which paper envelopes may be used.

Considerable care must be taken in the preparation of rock samples in order to avoid contamination. Drillcore may require sand-blasting to remove surficial smears of metal from the drill. Samples from underground workings and outcrops may be coated with debris from blasting, smelter fumes, or stains and films from migrating surface or ground waters. Before grinding the samples, all such surficial contamination should be removed by breaking or diamond-sawing off the affected portions. If this is impossible or impractical, such materials should be tested separately to evaluate the effect of such contamination on the results of the survey. Because minor contamination of metal from the hammer used to collect rock samples is difficult to avoid, the effects of such contamination should also be evaluated.

C. Preparation of Maps and Handling of Data

The procedures for preparation of maps are generally similar to those used for soils and drainage samples, as described in Section II.E of Chapter 13

V of Chapter 16. Data maps should show individual sample element contents. Information on rock types, structures, and ı should be plotted on the map or should be available at the comparison. From the data maps, interpretation maps depicting increasing concentration levels with symbols of increasing size and darkness may be used to bring out anomalies, or the data may be contoured. In large surveys, plotting of maps and handling of data may be carried out with a computer, as described in Section I of Chapter 19.

D. Interpretation of Data

Because of the variety of scales and purposes for bedrock surveys, no uniform procedure for interpretation can be suggested. The following comments apply mainly to detailed surveys for epigenetic (diffusion and infiltration) anomalies. Surveys for productive plutons are still in the research stage. Aspects common to most types of detailed surveys are (i) selection of anomalous areas, (ii) recognition of non-significant anomalies, and (iii) relation of anomalies to the location of possible ore.

The general principles for recognition of anomalies, using background and threshold values, are discussed in Chapter 2, and these methods may be applied to most rock surveys. However, because many detailed surveys are conducted in areas of extensive mineralization where a high proportion of samples may be anomalous, calculation of the threshold from the mean plus 2 standard deviations is not generally recommended. If sufficient samples are available, different thresholds may be calculated for different rock types. A further aid if rock types differ in threshold is the conversion of analyses to "background units" by dividing each analysis by the background value for the rock type represented by that sample so that the strength of anomalies in different rock types may be more easily compared.

The erratic nature of element contents in bedrock aureoles has led to the common use of smoothing techniques such as moving averages (Chapter 19; Beus and Grigorian, 1977, p. 75). If the samples are spaced closely enough, the trends and anomalies recognizable in the smoothed data are much more reliable than in the raw data. Products or ratios of elements can also be used in smoothing data, because some effects of local differences in amount of epigenetic minerals are removed by use of the ratios.

Non-significant anomalies include those due to analytical and sampling errors, human contamination, and (in surveys seeking epigenetic anomalies) rock types with consistently anomalous metal content. If sampling or analytical error is suspected, the samples in question should be re-collected, re-prepared, or re-analyzed. Possible contamination of anomalous rock types can be checked by reference to field notes, examination of hand speci[illegible] coarsely crushed splits of the samples, or investigation in the field.

Zones of extensive weak mineralization lacking economic concentrations of metal give rise to another type of non-significant anomaly. Grigorian (1974) suggests that such non-significant anomalies may be distinguished from significant anomalies related to ore by the following features:

(i) Widely scattered small centers exist, rather than one or a few intense anomalies.
(ii) A multiplicative ratio of the principal ore components (numerator) divided by associated trace metals (denominator) does not show clear trends across possible zones of mineralization.
(iii) Vertical zoning in the metal anomalies is lacking.

Leakage anomalies produced by hydrothermal processes are strongly controlled by structure and permeability, so that ore is not necessarily located directly beneath a surface anomaly. Careful geological mapping followed by tracing of the anomalies in drill cores may be necessary to determine the location of blind ore at depth.

Distinction between supra-ore anomalies indicative of blind orebodies, anomalies around exposed ore, and sub-ore anomalies in the roots of eroded orebodies is an important problem. For many types of hydrothermal ore deposits, the zonal relations of elements may be used to make this distinction. For example, if Cu ore is being sought, the presence of anomalies in Co, Mo, and Cu at the surface would suggest that the samples were in or near the roots of eroded ore, and that if good ore is not present at the surface, then none is likely in depth. In contrast, an anomaly in Pb, Zn, Cu, and Sb would suggest that a blind orebody may occur in depth. Orientation studies on known orebodies within the district are very desirable to confirm that the general zoning patterns of Table 5.1 are applicable in the district in question.

Chapter 6

Weathering

The preceding chapters were concerned with geochemical dispersion patterns formed at depth. By uplift and erosion, however, the rocks and minerals typical of the deep zone may be brought into the vastly different environment prevailing near the surface of the lithosphere. Deep-seated dispersion patterns are modified, and new patterns are formed by surficial processes.

The surficial environment is characterized by low temperature and pressure, and high concentrations of water, free oxygen, and carbon dioxide. Most of the minerals that are formed under deep-seated conditions are not stable at low pressures and temperatures. When rocks of igneous or metamorphic origin are exposed at the surface, therefore, the original materials tend to be reconstituted to new forms that are stable under the new conditions. The term weathering includes all the processes of reconstitution that take place in the near-surface zone.

I. NATURE OF WEATHERING

Weathering may be defined as the breakdown and alteration of materials near the earth's surface, to form products that are more nearly in equilibrium with the atmosphere, the hydrosphere, and the biosphere (Reiche, 1950; Ollier, 1969). This definition, taken literally, would include not only the primarily inorganic processes commonly associated with weathering, but the processes of soil formation where organic activity dominates the scene, together with the mechanical processes by which the products of weathering and soil formation are eroded, transported, and redeposited elsewhere, processes in which water is the dominant agent.

In a more restricted sense, weathering can be better defined as the "change of rocks from the massive to the clastic state" (Polynov, 1937). It is the phase dominated by the initial physical and chemical changes arising in response to the demand for equilibrium in the surface environment. The process begins with a progressive disintegration and decomposition of rock material *in situ*. The product of this decay is a mixture of resistant primary minerals with a suite of new mineral constituents that are stable in the new environment. Together these form the mantle of unconsolidated material overlying the solid rock, known as the regolith.

The process of weathering is followed by the processes of soil formation, erosion, transportation, and sedimentation, all of which together comprise the surficial geochemical cycle illustrated in Fig. 2.1. Although these processes are often intimately associated both in their operation and their effect, clarity of presentation requires that they be considered separately. Following their sequence in the geochemical cycle, weathering and soil formation are reviewed first, followed by erosion, transport, and deposition. No attempt is made here to give more than an outline of the basic principles, a general understanding of which is essential for the efficient planning and interpretation of geochemical surveys. Soil formation is treated in rather more detail in Chapter 7, because geologists are generally less familiar with this aspect of the surficial environment which is of so much importance in geochemical exploration. For further information on weathering and related topics, the reader should consult the selected references given at the end of this and the following two chapters.

II. WEATHERING PROCESSES

Three main types of weathering may be distinguished: physical, chemical, and biological. Physical processes include all those that cause rock disintegration without chemical or mineralogical changes. This disintegration increases the reactive surface area and thus facilitates the decomposition of rocks by chemical reaction with the abundant water, oxygen, and carbon dioxide of the surface environment. Biologic activity contributes either directly or indirectly to both physical and chemical weathering. All these processes usually take place side by side, though their relative importance varies according to environment. Thus, in the extremes of arid deserts and arctic conditions, and in many areas of mountainous relief, physical disintegration is usually the dominant mechanism of rock decay. Under most other climates, chemical attack is by far the dominant factor in controlling the nature of the weathering products at all depths within the zone of weathering. By contrast, the principal domain of biologic activity is restricted to the near-surface zone of soil formation.

A. Physical Weathering

The first step in the physical disintegration of massive rock following uplift and erosion is the development of a lacework of cracks and joints. These cracks increase in abundance as the surface is approached.

Simple unloading of deep-seated rocks allows expansion toward the free surface. The resulting differential stresses lead to rupture of the rock by sheeting parallel to the surface, or spalling of chips and blocks. Rupture along the boundaries of individual crystals and grains also takes place because of the differing rates of expansion of different minerals and different crystal orientations.

Incipient cracks may also be formed by differential thermal expansion and contraction along different crystal directions resulting from extreme heating by forest fires (Blackwelder, 1927) or the periodic burning of vegetation for agricultural purposes in many parts of the world. Normal diurnal changes in temperature are generally thought to be inadequate to cause rupture in a massive rock (Griggs, 1936).

As soon as the very first cracks are formed in a rock, a number of forces may work together to increase the width of the cracks and thus cause further rupture. Even cracks of microscopic dimensions provide access to aqueous solutions. In temperate and arctic climates, the expansion of water on freezing can furnish very high stresses that lead to further extension of the crack, as well as additional fracturing around pores filled with water. In arid regions, crystallization of salt can widen and extend cracks. Roots grow into cracks and widen them.

Chemical reactions of water and its dissolved solutes also can lead to physical breakdown of the rock. One of the principal disruptive effects of the introduction of water into the rock comes from the volume increases caused by hydrolysis and other forms of incipient alteration of the minerals adjoining the crack. The net effect is overall expansion and hence a progressive widening of the cracks. Repeated wetting and drying accentuates the effects of these processes.

In the erosional cycle, still other processes lead to physical disintegration of rocks and access to other agents of weathering. Gravity causes collapse of steep undercut faces. Abrasion by glacial ice, by wind-blown particles, and by stream-borne particles acts to physically disintegrate the solid materials at the surface.

B. Chemical Weathering

By comparison with physical processes, the chemical agents of weathering are capable of much more powerful attack on rocks and their constituent minerals. In extreme cases the resulting changes in composition, properties, and texture may be such as to obliterate almost completely the original nature of the parent material. Thus, under appropriate conditions, coarsely crystalline silicate rocks can be reduced to an ultra-fine complex of clay minerals; limestones and dolomites may be completely leached except for a fractional residuum of insoluble material; and considerable thicknesses of hydrated Fe- and Al-oxides (laterite and bauxite) or $CaCO_3$ (calcrete or caliche) may be developed at the expense of a variety of different rocks.

At all levels and in all environments, chemical weathering depends on the presence of water and the solids and gases dissolved therein. All minerals are more or less soluble even in pure water. The presence of dissolved oxygen, carbon dioxide, and humic complexes greatly increases the corroding power of natural solutions. Rain water contains small but significant quantities of dissolved oxygen and carbon dioxide, together with chlorides and sulfates derived from the ocean and man's domestic and industrial activities. Percolation through the soil adds humic compounds, further carbon dioxide, and many other products of organic origin. Ground waters can acquire diverse new constituents liberated from the rocks undergoing decomposition or by mixing with thermal waters rising from the depths (Table 6.1).

The principal types of chemical reactions associated with weathering are hydration, hydrolysis, oxidation, and simple solution.

1. Hydration and hydrolysis

Hydration (absorption of water) and hydrolysis (chemical reaction to produce or consume H^+ or OH^- ions) are commonly regarded as the most important chemical reactions involved in rock decomposition. Hydration implies absorption of water molecules into the crystal structure of a mineral. A simple example is the transformation of anhydrite ($CaSO_4$) to gypsum ($CaSO_4 \cdot 2H_2O$). More commonly, hydration occurs in conjunction with other processes of chemical weathering, such as the conversion of aluminosilicates to clays, or Fe minerals to hydrated ferric oxides, with the result that the solid products of weathering contain more water than the initial minerals.

In hydrolysis, an Al- or Fe-bearing silicate is typically converted to a clay or Fe-oxide, accompanied by release of cations and incorporation of H^+. An example is the reaction of albite with weak acid to form clay, silica, and Na^+ ions:

$$\underset{\text{Albite}}{2NaAlSi_3O_8} + 2H^+ + H_2O = \underset{\text{Kaolinite}}{Al_2Si_2O_5(OH)_4} + 4SiO_2 + 2Na^+. \quad (6.1)$$

Table 6.1
Composition of some natural waters

Constituent	Content in rain water (ppm) (1)	Content in ground water (ppm) (2)	(3)	(4)	(5)	Content in thermal water (ppm) (6)
Ca^{2+}	0·1–10	30	40	12	19	4
Mg^{2+}	⩾0·1	31	22	6·6	5·1	2
Na^+	⩾0·4	279 (Na + K)	0·4	7·2	4·4	48
K^+	⩾0·03		1·2	3·1	3·2	30
SiO_2	—	13	8·4	38	13	303
HCO_3^-	⩽1·0	445	213	85	39	—
SO_4^{2-}	2·0	303	4·9	4·4	30	1100
Cl^-	0·5	80	2·0	1·2	5·8	5
Others	0·7	18	6·5	0·7	16	348
Salinity	—	973	190	115	116	1850

(1) After Hutchinson (1957, p. 551), except HCO_3^- after Gorham (1955).
(2) Well in shale–sandstone formation, New Mexico (Hem, 1959, p. 88).
(3) Spring from dolomite, Tennessee (Hem, 1959, p. 83).
(4) Well in basalt, Washington (Hem, 1959, p. 55).
(5) Well in gneiss, Connecticut (Hem, 1959, p. 1154).
(6) Spring, Yellowstone National Park, Wyoming (White and Brannock, 1950).

Note that this reaction also involves hydration, as do many other hydrolysis reactions, as well as the dissolution of Na^+. The Al in this and many similar hydrolysis reactions may be released as a well-crystallized clay mineral or as a colloidal, poorly crystalline aluminosilicate, and the SiO_2 as colloidal silica, dissolved silicic acid (H_4SiO_4) or solid quartz, opal or other silica mineral. The released cation may be sorbed to the surface of the colloidal particles or released to solution. These adsorbed ions are then available for reaction and exchange with the constituents of passing solutions. Although hydrolysis can take place in pure water, the reaction is intensified in the presence of natural acids, of which the most common are carbonic acid and humic acids.

The hydrolysis reaction produces heat and may produce an increase in volume. The stress that results is one of the principal factors in the disintegration of rocks, and its effect may extend to appreciable depths, far below the range of most of the purely physical agents of disintegration.

2. *Oxidation*

Oxidation reactions are characteristic of the aerated environment of the weathering zone. The elements most commonly affected are Fe, Mn, and S, which occur as Fe^{2+}, Mn^{2+}, and S^{2-} in most deep-seated and some sedimentary rocks, but can be oxidized to Fe^{3+}, Mn^{4+}, and S^{6+} in the surficial environment. Other oxidizable elements include C, N, V, Cr, Cu, As, Se, Mo, Pd, Sn, Sb, W, Pt, Hg, and U. Oxidation reactions tend to be slow, and the presence of water catalyzes reactions involving gaseous oxygen. Products of oxidation are new minerals (Fe- and Mn-oxides) and dissolved constituents (SO_4^{2-}). Optimum conditions for oxidation naturally occur in the moist ground above the zone of permanent saturation, but in places, oxidizing waters can descend to depths far below the water table before their oxidizing power is consumed.

3. Simple solution

The simple solution of many minerals in the abundant water of the surface environment can be an extremely important factor under some conditions. The most spectacular example of solution is the formation of limestone caves by the solution of calcite in CO_2-bearing waters to form soluble calcium bicarbonate. Less spectacular is the slower release of silica and the common cations of K, Mg, Na, and Ca, especially during the hydrolysis of primary silicates.

C. Biologic Agents in Weathering

Organic processes not only constitute the principal genetic factor in soil formation but also play a significant part in rock decomposition and weathering.

Plants contribute to the physical weathering of rocks by widening the cracks into which they insinuate their roots. A vast amount of near-surface material may be mixed and sorted by worms, termites, and rodents. The resulting disaggregation and increased permeability facilitate the entry of air and water, thereby promoting more intense chemical weathering.

Biologic agents also contribute significantly to the chemical disintegration of rocks. The local conditions of extreme acidity generated at the root tips of plants can act as a powerful corrosive force in the chemical breakdown of rocks. Computations based on the silica content of tropical vegetation show that the rate at which plants remove silica from silicate minerals is adequate to account for a large part of the observed high mobility of silica in tropical weathering (Lovering, 1959). Plant respiration is a major factor in the biochemical cycle of oxygen and carbon dioxide, which are two of the most important reagents in chemical weathering, and transpiration of water by plants enriches the remaining pore solutions in solutes. Oxidation of Fe and S is catalyzed by bacteria, as is the fixation of nitrogen. Organic acids and complexing agents generated by decomposition of plant material in the upper soil horizons can contribute materially to reactions in the deeper zones of weathering and to solubility in natural waters.

III. FACTORS AFFECTING WEATHERING PROCESSES

The processes just outlined operate over the entire land surface and throughout a wide variety of environments. Although the processes are basically the same everywhere, local environmental conditions can have a considerable influence on the rate and type of weathering, and on the nature of the end products. The principal factors that condition the processes of weathering are: (i) the resistance to weathering of the primary rock-forming minerals, (ii) the grain size and texture of the primary rock, (iii) climate, with particular reference to temperature and rainfall, and (iv) topographic relief and drainage.

A. Resistance of Minerals in Weathering

Mineral species differ widely in their relative resistance to weathering processes. Strictly speaking, one must specify the chemical and physical conditions in order to rank minerals in resistance to weathering, but an order applicable for rock-forming silicates in temperate humid climates has been developed by Goldich (1938) and is summarized in Fig. 6.1.

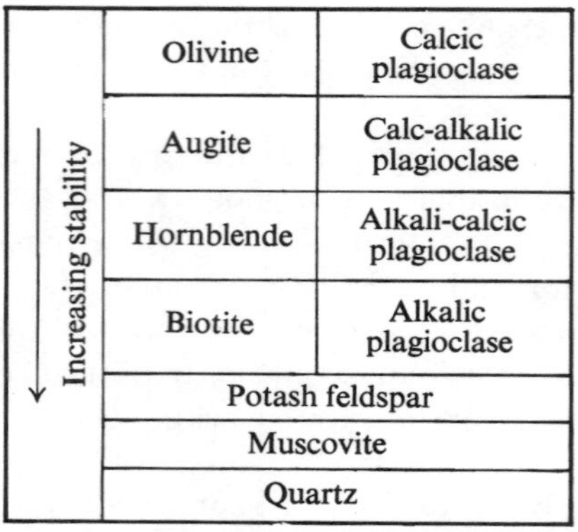

Increasing stability ↓		
	Olivine	Calcic plagioclase
	Augite	Calc-alkalic plagioclase
	Hornblende	Alkali-calcic plagioclase
	Biotite	Alkalic plagioclase
	Potash feldspar	
	Muscovite	
	Quartz	

Fig. 6.1. Relative stability of common rock-forming silicates in chemical weathering. (After Goldich, 1938.)

This arrangement is essentially the same as Bowen's reaction series, which sets out the order of progressive reaction during the course of magmatic crystallization (Bowen, 1922). Goldich's series indicates that the minerals that crystallized at the highest temperatures, under the most anhydrous conditions, are more readily weathered than those that crystallized last from the lower-temperature, more aqueous magmas. As a general rule, the closer the conditions of crystallization approximate to those now prevailing at the earth's surface, the more resistant is the mineral in the weathering environment. The Goldich stability series also applies to the same minerals when they are of metamorphic origin.

A similar sequence based on weathering of fine-grained minerals in soils is presented in Table 6.2, after Jackson *et al.* (1948). The order is similar to that in Fig. 6.1, but incorporates a wider range of minerals including some formed in the surface environment and common in sedimentary rocks. Not surprisingly, the sedimentary minerals originally formed in a surface environment are relatively stable when the sedimentary rock is exposed to weathering.

Table 6.2
Weathering sequence for clay-size minerals in soils and sedimentary deposits[a]

	Weathering stage	Clay-size minerals characteristic of different stages in the weathering sequence
Increasing stability ↓	1	Gypsum (halite, etc.)
	2	Calcite (dolomite, aragonite, etc.)
	3	Olivine–hornblende (diopside, etc.)
	4	Biotite (glauconite, chlorite, etc.)
	5	Albite (anorthite, microcline, etc.)
	6	Quartz
	7	Illite (muscovite, sericite, etc.)
	8	Intermediate hydrous micas
	9	Montmorillonite
	10	Kaolinite (halloysite)
	11	Gibbsite (boehmite, etc.)
	12	Hematite (goethite, limonite, etc.)
	13	Anatase (rutile, ilmenite, etc.)

[a] Source: Jackson *et al.* (1948). Reproduced with permission of the American Chemical Society.

The relative resistance to weathering of the constituent minerals of ore deposits is naturally a matter of prime importance in the development of geochemical anomalies. Generally speaking, the order of increasing susceptibility to decomposition by weathering appears to be oxides < silicates < carbonates and sulfides. Sulfide minerals are particularly vulnerable to oxidation and solution. The oxidation of pyrite and marcasite results in the formation of sulfuric acid, ferrous ions, and limonite by reactions such as the following:

$$2FeS_2 + 7O_2 + 2H_2O = 2Fe^{2+} + 4SO_4^{2-} + 4H^+ \tag{6.2}$$

$$2Fe^{2+} + 3H_2O + \tfrac{1}{2}O_2 = Fe_2O_3 \cdot H_2O + 4H^+. \tag{6.3}$$

In the resulting strongly acid environment, the gangue minerals and silicates of the wallrocks are attacked at a rate far in excess of that in a nonsulfide environment. The presence of abundant carbonate or other alkaline constituents in the gangue or wallrock naturally results in more rapid neutralization of the acid than is the case in a dominantly silicic environment.

An important factor in the oxidation of sulfides is the electrochemical reaction which develops in aggregates consisting of more than one electrically conducting sulfide (Gottschalk and Buehler, 1912; Sato, 1960). In these circumstances, the oxidation of one mineral is favored over that of others. Thus, in a deposit consisting of chalcopyrite and pyrite, the chalcopyrite is oxidized preferentially to the pyrite.

Some sulfides, notably pyrite, can show considerable variations in stability, apparently related to structural and compositional disorders in their crystal lattices. Other primary sulfides, galena for example, may be transformed on weathering to insoluble secondary minerals which coat the primary sulfides and thereby impede further alteration. A compact stable gangue can also provide a measure of protection for unstable minerals imbedded within it.

B. Permeability

The rate of weathering of a rock depends on how easily the reactive solutions can get to the sites of reaction within the body of the rock. This is strictly a matter of permeability. A fine-grained rock in which solutions can find their way along grain boundaries will be relatively permeable and will undergo rapid weathering because of the large surface area of the grains. Most rocks that meet this specification are clastic sedimentary rocks in which the constituent minerals are already in equilibrium with the surficial environment. On the other hand, many medium- to coarse-grained igneous and metamorphic rocks have micro-channel-ways following the grain boundaries, but fine-grained igneous rocks such as diabase are very rarely permeable. Thus it happens that coarsely granular rocks are often more susceptible to decomposition than fine-grained rocks, in spite of the apparently greater internal surface area of the latter. Permeability can also be furnished by fractures or by dissolution of minerals. Weathering often extends to greater depths within an epigenetic mineral deposit than in the enclosing rock, in part because vein structures commonly provide preferential channel-ways for circulating waters, and in part because of the effect of the acid generated by the oxidation of sulfide minerals. Fracturing on both the macro- and micro-scale usually results in a marked increase in the intensity and depth of chemical decomposition. As weathering proceeds, the selective removal of soluble carbonates and sulfides provides further access for weathering solutions. Primary wallrock alteration alongside mineral deposits is frequently attended by an increase in porosity, though silicification usually renders a rock more resistant to weathering. In general, the effects of oxidation and leaching of disseminated sulfides are more pervasive in sandstones than in shales, because of the greater permeability.

C. Climate

The principal climatic elements that bear on weathering are rainfall and temperature. Rainfall controls the amount of water available for chemical weathering, while temperature influences the rate of chemical reactions and, in particular, the rate of decomposition of organic matter. Temperature affects the availability of water by increasing evaporation at high temperatures or by freezing at low temperatures. Climate also controls the amount and type of vegetation which in turn provides the raw material for organic reagents. Chemical weathering is generally most intense in tropical areas of high rainfall and high temperature. The intensity of weathering is somewhat less in temperate climates of moderate rainfall and seasonal variations in temperature. An example of the contrast between the products of chemical weathering of dolerite in tropical and in temperate climates is given in Table 6.3. Under arid and arctic conditions, chemical weathering is at a minimum and physical processes are dominant. As a result of the extremes of aridity in some desert

Table 6.3
Changes in chemical weathering in temperate and tropical zones[a]

	Dolerite Staffordshire, England		Dolerite Bombay, India	
Constituent	Content in fresh rock (%)	Content in overlying clay (%)	Content in fresh rock (%)	Content in overlying laterite (%)
SiO_2	49·3	47·0	50·4	0·7
Al_2O_3	17·4	18·5	22·2	50·5
Fe_2O_3	2·7	14·6	9·9	23·4
FeO	8·3	—	3·6	—
MgO	4·7	5·2	1·5	—
CaO	8·7	1·5	8·4	—
Na_2O	4·0	0·3	0·9	—
K_2O	1·8	2·5	1·8	—
H_2O	2·9	7·2	0·9	25·0
TiO_2	0·4	1·8	0·9	0·4
P_2O_5	0·2	0·7	—	—
Total	100·4	99·3	100·5	100·0

[a] Source: Warth (1905).

climates, corrosion by wind-borne sand may be an important factor in physical weathering.

D. Relief and Drainage

In very mountainous terrain, physical erosion may cause the rock debris to be removed faster than it can be decomposed chemically. Chemical weathering here is confined largely to decomposition of eroded fragments as they are carried away, first by soil creep and subsequently by stream waters. Moderate to strong relief is commonly associated with extreme variability in the depth to the water table. In these areas, chemical decomposition is most active beneath the crests of ridges, where the water table tends to lie at maximum depth below the surface. A shallow water table and near-surface zones of permanent saturation are usually limited to the immediate vicinity of springs and drainage channels. The circulation of ground-water solutions is most vigorous in the areas between the ridge crests and the drainage channels.

Flat-lying terrain, by contrast, is characterized by less active erosion and by relatively sluggish ground-water movement. Swamp conditions are common where the precipitation is sufficient to maintain a high water table. Under such conditions, the low rate of erosion inevitably leads to the slowing down of ground-water movement and rock decomposition until eventually equilibrium is approached and the weathering process comes to a virtual standstill.

IV. PRODUCTS OF WEATHERING

The surficial geochemical anomalies on which most geochemical prospecting is based are composed of the weathering products of ores superimposed upon the weathering products of the normal unmineralized rocks in which the ore occurs. The products of weathering of both rocks and ores take three forms: residual primary minerals from the parent rock, secondary minerals resulting from the processes of weathering, and soluble material that can be removed by circulating water.

A. Residual Primary Minerals from the Parent Rock

As discussed previously, most igneous and metamorphic minerals are unstable in the zone of weathering, where pressure and temperature are much lower than when they were formed, and where water, carbon dioxide, and oxygen are more abundant. If water is percolating through the weathering rock, no minerals are truly stable because all minerals are soluble in at least trace amounts. However, quartz and many accessory minerals (zircon, Ti-oxides, tourmaline) are dissolved very slowly and persist while other minerals decompose more readily. Therefore, the slowly dissolved minerals become more abundant in highly weathered materials, and are mixed with the minerals newly formed by weathering.

Of the ore minerals, Au, Pt, cassiterite, columbite–tantalite, chromite, and beryl are among the most common representatives of the residual category. Minerals that are reasonably resistant chemically may yet be too friable or soft to withstand physical abrasion. Thus wolframite, scheelite, and barite tend to persist in the regolith as residual products of weathering but are quickly pulverized by abrasion during erosion and transport.

B. Secondary Minerals

On weathering, primary rock-forming silicate minerals tend to undergo both leaching and hydrolysis to form a characteristic suite of secondary minerals. The particle size of most of the secondary products is extremely fine, of the order of 2 micrometers (μm) or less (Fig. 6.2). In this range the individual crystals can be seen only with the aid of the electron microscope, and the identity of the minerals themselves can be determined only by special instrumental techniques, such as X-ray diffraction and differential thermal analysis.

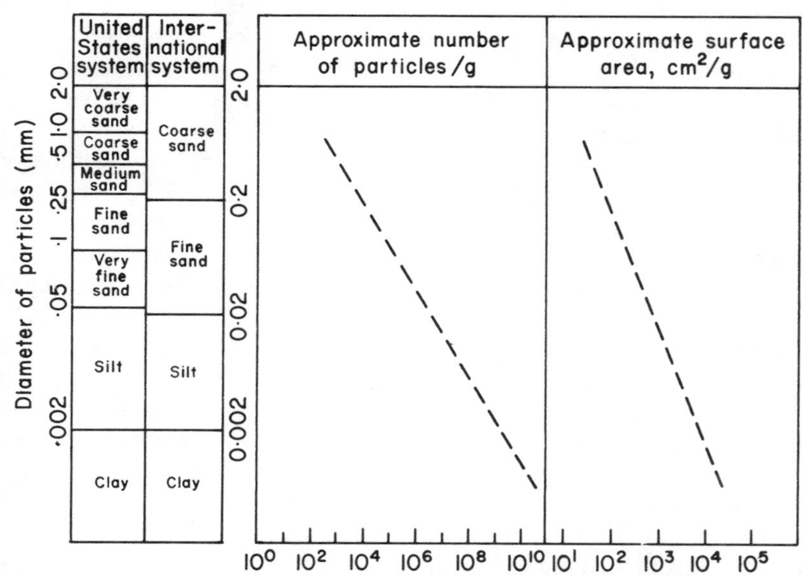

Fig. 6.2. Particle-size classes and some of their properties. (After Birkeland, 1974.)

Among these secondary weathering products are included clay minerals, Fe–Al-oxides, secondary ore minerals, and gossans, as discussed in the following sections.

1. Clay minerals

The term "clay" has been a source of considerable confusion. As defined here, a "clay" is simply a material of a certain specified grain size, with no implication as to its mineralogical constitution. The term "clay mineral", however, refers to a very specific group of "clay-sized" aluminosilicate minerals, as distinguished from "clay-sized oxides" which are limonitic and bauxitic minerals. The terminology is further confused by use of the term "colloid" to refer to any complex of fine-grained particles regardless of the state of dispersion of the fine-grained phase. In this book, the following definition will be followed: (i) "clay" refers to material containing particles sizes less than 2 μm; (ii) "clay minerals" refer to fine-grained hydrated aluminosilicate minerals; and (iii) "colloids" are fine-grained particles dispersed throughout water or some other fluid.

Almost all of the clay-sized weathering products are either silicates or oxides, hence the terms "silicate clays" and "oxide clays" used in the terminology of soil science for the two principal groups. The silicate clays are composed mainly of clay minerals (kaolinite, montmorillonite, etc.) and the oxide clays of the oxides of Fe, Al, and to a lesser extent Mn.

The clay minerals are abundant in most weathered material, and vary widely in chemical composition and ion-exchange capacity, so are worth consideration in some detail. The crystal structures of all the common clay minerals are characterized by a layered structure. Tetrahedral layers consist of linked $(Si,Al)O_4$ groups in which each Si^{4+} or Al^{3+} ion is surrounded by four O^{2-} ions in a tetrahedral configuration (Fig. 6.3a, b). Octahedral layers consist of Al^{3+}, Mg^{2+}, or Fe^{2+} ions surrounded by six O^{2-} or OH^- ions in an octahedral configuration (Fig. 6.3c). The identity of the mineral is determined by the sequence of layers, the elements present in the octahedral layers, the charge on the layers, and the nature of interlayer cations balancing the layer charge.

In kaolinite, $Al_2Si_2O_5(OH)_4$, one octahedral and one tetrahedral layer are bound together by sharing a common layer of O^{2-} and OH^- ions (Fig. 6.3d). The combined tetrahedral–octahedral layer is bound to other similar layers by hydrogen bonds and weak ionic forces. Serpentine has a similar structure, but the octahedral layer contains three Mg^{2+} rather than two Al^{3+}. Halloysite is like kaolinite except for a layer of water between the sheets. Kaolinite and serpentine are denoted as two-layer structures.

Most of the other common clay minerals have a three-layer structure, with an octahedral layer sandwiched between two tetrahedral layers (Fig. 6.3e).

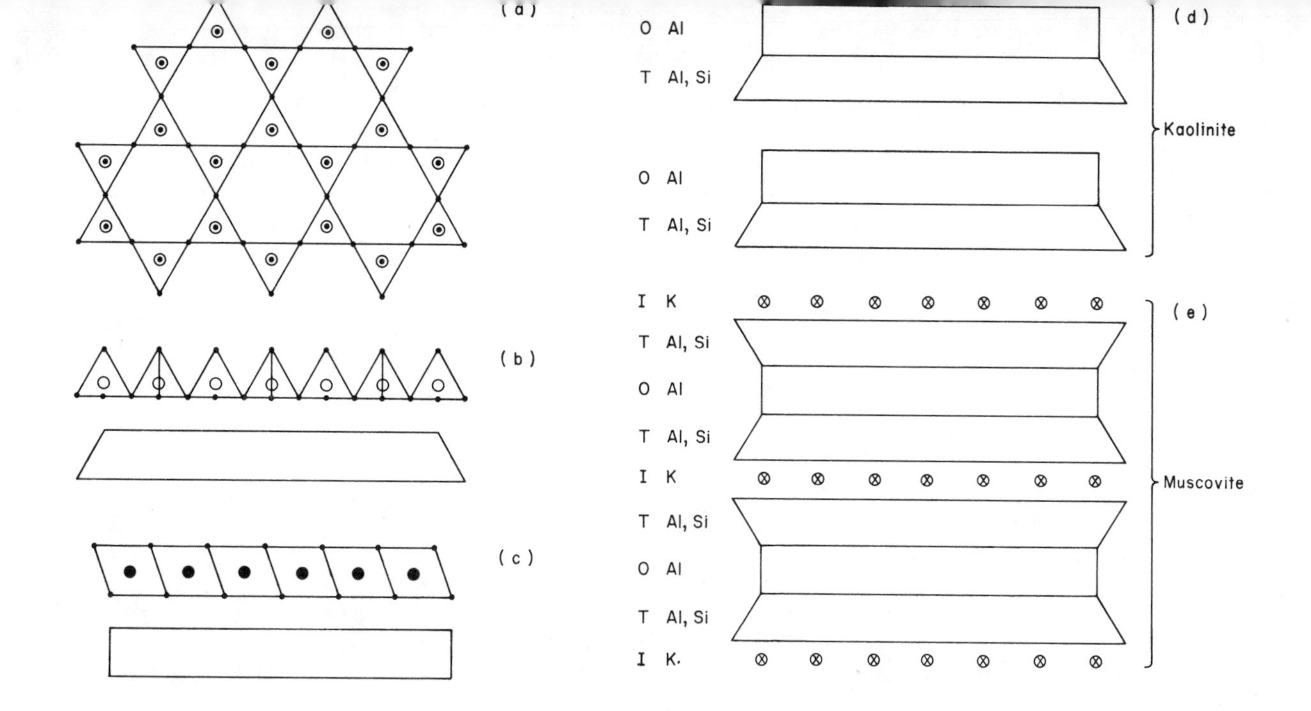

Fig. 6.3. Structure of clay minerals. (a) Top view of $(Si,Al)O_4$ groups in tetrahedral layer; (b) side view of tetrahedral layer and schematic representation; (c) side view of octahedral layer and schematic representation; (d) kaolinite structure, schematic representation; (e) muscovite structure, schematic representation. T, tetrahedral layer; O, octahedral layer; I, interlayer cations.

In muscovite, $K_2Al_4(Al_2Si_6)O_{20}(OH)_4$, one quarter of the tetrahedral sites contain Al^{3+} rather than Si^{4+}, leaving a net layer charge on the tetrahedral and octahedral layers of -2 per formula unit. This charge is balanced by K^+ between the layers (Fig. 6.3e). Fine-grained clay minerals deviating in only minor ways from the muscovite are called illite. In smectites (montmorillonite group), substitution of Mg^{2+}, Fe^{2+}, or other divalent cations for Al^{3+} in the octahedral sites, and Al^{3+} for Si^{4+} in the tetrahedral sites, leaves a net layer charge of up to $-0{\cdot}67$ per formula unit. This layer charge is balanced by Ca^{2+}, Na^+, and other ions between the layers. The layer charge and the properties of these ions are such that water easily enters the interlayer space and surrounds the interlayer cations. These structures are termed expandable because the number of interlayer water molecules is variable, and organic molecules can also enter the interlayer spaces, giving rise to an expanded thickness of the structure. Dissolved ions can easily enter the interlayer sites and displace the original ions, which are therefore called exchangeable. Vermiculites have a higher negative layer charge than smectites, because of greater substitution in the tetrahedral and octahedral layers. Larger amounts of exchangeable ions accompany the higher layer charge. Chlorites have a layer of $Mg(OH)_2$ separating the basic three-layer units.

In addition to the simple repeating units described above, the different types of structures may be interlayered. For instance, mica layers may alternate with smectite layers, or smectite with kaolinite layers. The interlayering may follow a repeating pattern or be random.

Figure 6.4 presents a scheme to illustrate how primary minerals change to successive secondary minerals with increasing intensity and duration of weathering. Throughout this process, the genesis of the clay minerals is accompanied by the removal of the relatively soluble elements K, Na, Ca, and Mg.

Kaolinite formation is favored by an acid environment, with free drainage, leading to a thorough leaching of Na, Ca, Mg, and K, as well as many trace constituents. It develops most readily, therefore, in relatively humid climates where rainfall exceeds evaporation and where downward percolation and lateral movement of underground water is active. Figure 6.5 shows the relation of the abundance of the kaolinite group and the other common clay mineral groups to rainfall in different environments and over different rock types. As this illustration shows, kaolinite is normally the commonest clay mineral in the zone of weathering.

Montmorillonite formation is favored by neutral to alkaline conditions and by the incomplete leaching of Na, Ca, Mg, K, and trace metals. The bases may be retained within the system as the result of either impeded drainage or excessive evaporation. Montmorillonites are, therefore, typical end products of weathering in waterlogged ground or in semi-arid climates. Characteristically, members of this group have the property of expanding considerably on wetting and thereby assist in maintaining a poorly drained environment. A high concentration of available Fe^{2+} and Mg^{2+} is naturally a strong predisposing factor in the formation of montmorillonite. Even in freely drained ground, montmorillonite may be a transient intermediate product in the course of the decay of the ferromagnesian constituents of mafic rocks to kaolinite. When drainage conditions change and leaching becomes effective, montmorillonite may alter to kaolinite.

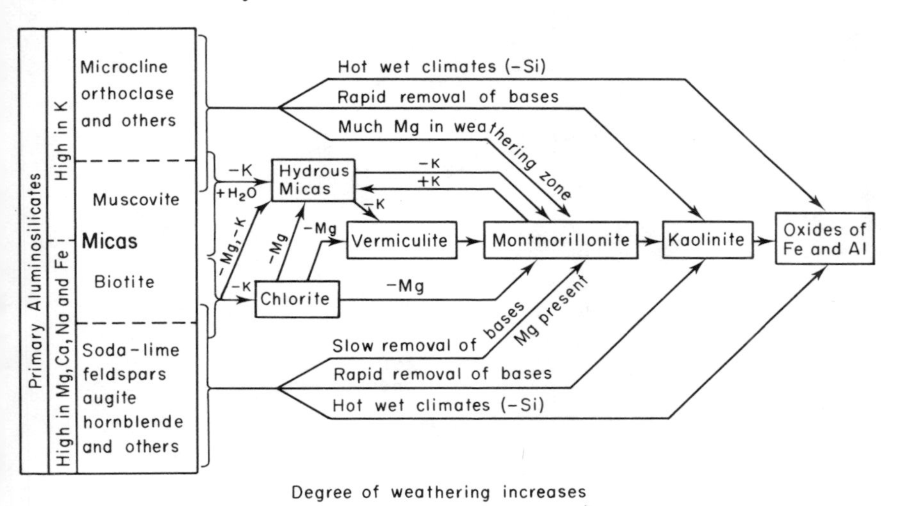

Fig. 6.4. Weathering products of primary minerals, and sequence of forming clays. (After Brady (1974). Reprinted with permission from "The Nature and Properties of Soils", 8th edn, © Macmillan Publishing Co., Inc., 1974.)

Illite, also known as hydrous mica, is similar structurally to montmorillonite in that unsatisfied charges within the layers are neutralized by interlayer cations. It is different, however, in that the interlayer bonding by K^+ is much stronger than the weaker ionic bonds in montmorillonite, with the result that interlayer expansion does not occur. Although illite is typically developed in deep-sea sediments, it also forms on land under appropriate conditions, as indicated in Fig. 6.4.

Magnesium-rich vermiculite is found most commonly in the weathering products of mafic rocks. Aluminum-rich vermiculite is a common weathering product of micas and other aluminosilicates in temperate humid climates. Some vermiculites show interlayer expansion similar to montmorillonite.

The clay-sized fraction of soils normally contains mixtures of the above principal clay minerals, together with representatives of some of the rarer species. Hybrid minerals intermediate between the well-defined clay-mineral

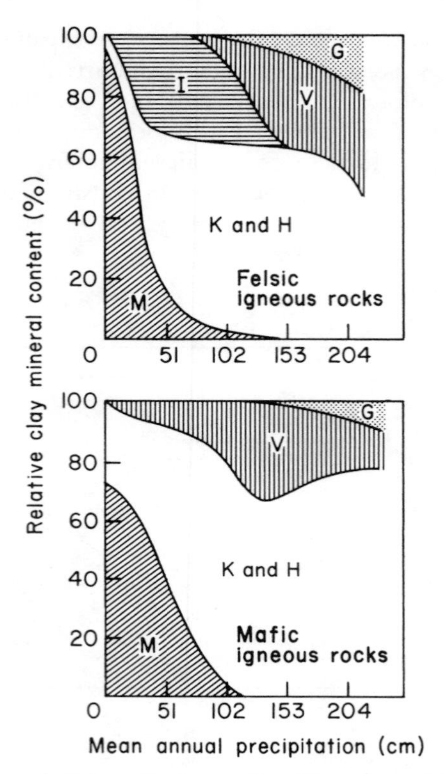

Fig. 6.5. Relative clay-mineral content as a function of precipitation for surface-soil samples, California. Mean annual temperature 10–15 °C. M, montmorillonite; K, kaolinite; H, halloysite; I, illite; V, vermiculite; G, gibbsite. (After Barshad, 1966.)

species, can also occur, and add materially to the complexity of soil mineralogy.

2. Iron and aluminum oxides

The oxides of Fe and Al, often lumped as "sesquioxides", are widely distributed among the products of weathering. The most important representatives of this group are limonite ($Fe_2O_3.nH_2O$), hematite (Fe_2O_3), goethite ($Fe_2O_3.H_2O$), diaspore ($Al_2O_3.H_2O$), and gibbsite ($Al_2O_3.3H_2O$). Most limonite is probably a mixture of geothite and hematite, with jarosite ($KFe_3(SO_4)_2(OH)_6$) occurring in the presence of oxidizing pyrite. Iron not required for clay formation is precipitated as a hydrated ferric oxide, particularly in oxidizing environments. Under reducing conditions, however, ferrous iron is liable to be removed in solution. Aluminum hydroxide may develop directly from mafic and intermediate rocks. Both hydrous Fe- and Al-oxides attain their maximum development under humid tropical climates, as suggested by the data of Fig. 6.6.

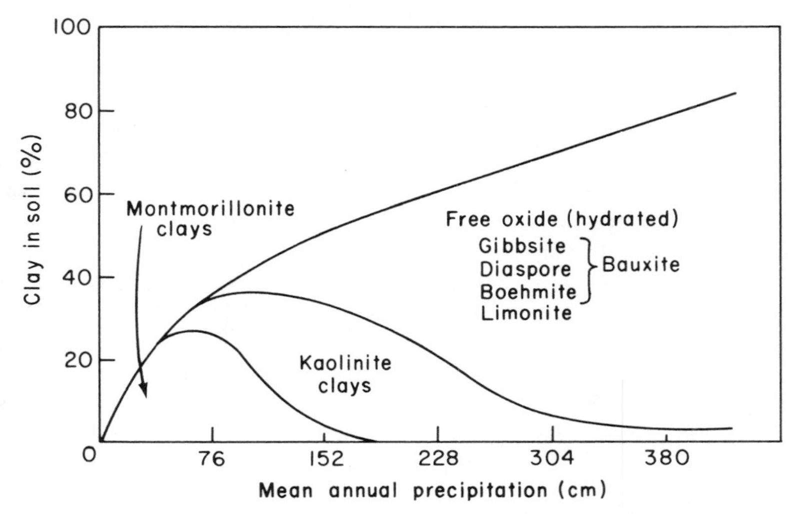

Fig. 6.6. Clay mineralogy as a function of precipitation under a continuously wet climate, Hawaii. (After Sherman, 1952.)

3. *Secondary ore minerals*

On weathering, sulfide ores yield a characteristic suite of secondary minerals, including not only hydrous Fe-oxides but a host of secondary metalliferous minerals. Unlike clay minerals, many of these are visibly crystalline and are not so readily dispersed. Different secondary minerals of the same metal may be formed under a variety of chemical conditions, thereby leading to a great diversity of mineral species. Thus, under certain conditions of Eh, pH, CO_2 pressure, Cu concentration, etc., malachite ($Cu_2CO_3(OH)_2$) is the principal product of weathering, whereas under other conditions, other oxidized Cu minerals may be dominant.

4. *Gossan*

Residual hydrous Fe-oxide derived from the oxidation of Fe-bearing sulfides and carbonates has long been of significance in mineral exploration. In the massive form, accumulations of hydrous Fe-oxide are known as gossan. Gossans are formed most commonly from the weathering of pyrite, marcasite, pyrrhotite, Cu–Fe-sulfides, arsenopyrite (FeAsS), siderite ($FeCO_3$), and ankerite ($Ca(Mg,Fe)(CO_3)_2$). The mineralogy of gossans derived from a dominantly sulfide ore is limited to species that are stable in contact with acid sulfate solutions. Later, as active oxidation recedes from the surface, percolating waters of more normal composition, in which bicarbonate is usually the predominant anion, may bring about further changes in composition. In all gossans, regardless of whether they were derived from sulfide or carbonate ores, the predominant minerals are limonite, quartz, and secondary silica. Depending on the parent material and the maturity of the gossan, other minerals may occur as accessory constituents, including a wide variety of sulfates, arsenates, carbonates, silicates, and many other secondary metalliferous salts; clay minerals derived from vein silicates or incorporated country rock may also be present.

Most of the metallic constituents of the ore that are dissolved in the course of weathering necessarily escape in ground and surface waters. Appreciable amounts may be adsorbed or coprecipitated with the ubiquitous hydrous Fe-oxides. Examples have been reported where the limonitic material of gossans contains up to several percent of Zn derived in this manner from sphalerite associated with the primary ore. The metal content of gossans and their significance in prospecting are considered more fully in Chapter 11.

C. Soluble Products

The soluble products of weathering consist of those constituents that are released by the decomposition of primary minerals and that are not required in the formation of insoluble secondary minerals. Analyses of ground and surface waters show that much of this material is removed in solution from the site of weathering (Table 6.1). In the general case, the soluble constituents reflect the composition of the parent rocks. Ca^{2+}, Mg^{2+}, and CO_3^{2-} are naturally the principal soluble products derived from the weathering of calcareous rocks. In contrast, siliceous rocks yield alkalis, alkaline earths, and colloidal silica resulting from hydrolysis of the primary silicates. The proportions of these constituents vary according to the felsic or mafic nature of the parent material. For example, Ca and Mg predominate in the case of mafic rocks, while granites, felsic schists, and argillaceous sediments yield higher proportions of K and Na. In general, Ca is more liable to be removed than is Mg, which may be strongly adsorbed by clays or incorporated in the structure of montmorillonite or chlorites. Potassium may be retained to a certain extent in illites, whereas Na tends to remain almost entirely in solution. Iron and Mn are appreciably soluble only under reducing conditions, although ferric hydroxide may in part be removed in stabilized colloidal solution.

Insofar as the solid products of weathering are controlled by climate and relief, so these factors will also modify the nature of the soluble products.

Thus, in arid climates, soluble salts may be precipitated by evaporation; under humid tropical conditions silica is rendered relatively mobile, and ultimately all but the hydrous oxides tend to be removed in solution. Relief, through its relationship with drainage, influences the solubility and removal of many metals. For example, under conditions of impeded drainage such as occur where ground water comes to the surface in topographically low situations, montmorillonite may form, thereby retarding the solution of Fe and Mg and favoring the retention of other cations by adsorption of the clay mineral.

Table 6.4
Composition of ground water draining sulfide ore deposits

Element	Content in water (ppm) (1)	(2)
Ca	68	260
Mg	41	49
Na	23	13
K	20	3·2
SiO_2	56	23
Al	433	12
Fe	2178	143
Cu	312	—
Zn	200	345
SO_4	6600	1650
Cl	0·1	3·7
Salinity	9990	2500

(1) Burra-Burra Mine, Ducktown, Tennessee.
(2) Victor Mine, Joplin district, Missouri. (Data from Emmons, 1917.)

Ores, on weathering, similarly yield soluble products that can escape from the immediate environment of the source if they are not trapped during the precipitation of secondary minerals. The relations here are more complex than in the case of the weathering products of silicate rocks because of the more diverse mineralogy and the more complex stability fields of the secondary minerals that make up the soluble end products. Many of the latter minerals are stable only in the presence of relatively concentrated solutions of the component elements. Consequently, if ground-water circulation is at all active, a substantial proportion of the metals will leave the site of oxidation in soluble form (Table 6.4). The many physical and chemical factors affecting the solubility of trace elements in the zone of weathering are considered in following chapters.

D. Residual Structures and Textures

Most igneous and siliceous sedimentary and metamorphic rocks do not usually appear to change greatly in volume as a result of weathering. Traces of bedding, foliation, and vein structures in the bedrock can often be traced well up into the residuum even after intense weathering. The texture of the original rock itself may also be preserved, despite the fact that alteration of the unstable primary minerals is essentially complete and the entire rock rendered loose and friable. Such weathered rock retaining traces of its original texture and structure is known as saprolite. Ruxton (1958), however, has demonstrated that under certain conditions appreciable compaction may take place as a result of subsurface erosion. The removal of carbonates from calcareous rocks results in a very substantial reduction in volume, and the resultant slumping causes partial or total obliteration of the original rock structures. In extreme cases, pure limestone and saline deposits may be completely removed by solution. In the upper layers of residual overburden, relic structures have been almost invariably destroyed by physical disturbance accompanying soil creep, frost heaving, and biologic activity. Compaction following eluviation of clays in soil formation is another common contributory factor in the eradication of relic structures.

The weathering of ore deposits composed predominantly of carbonate or sulfide minerals may be accompanied by cavitation and slumping. In contrast, a resistant quartz gangue favors the preservation of vein structures. Relic textures inherited from both of these types of primary ore material, however, are commonly preserved in limonite inherited from weathering sulfide and carbonate minerals. These cellular box-work pseudomorphs, as they are called, have received special study as a means of identifying the specific parent minerals originally present in the primary ore (Blanchard, 1968).

V. SELECTED REFERENCES ON WEATHERING

General review	Birkeland (1974) Ollier (1969)
Clay mineralogy	Grim (1953) Brown (1961)

Chapter 7

Soil Formation

The word "soil" means different things to different people. To the engineer it is the material that must support the structures and highways he wants to build. To the farmer it is the material on which plants grow. And to the geologist or prospector it is the material that covers the bedrock and must either be removed, penetrated, or interpreted in terms of the underlying rock or mineral deposit. Most investigations into the genesis and chemical properties of soils have been the work of agricultural scientists. This chapter summarizes those aspects of soil science as developed by agriculturalists that are of particular importance in understanding the movement and distribution of elements in soils, which are commonly sampled in detailed geochemical surveys, and which form much of the raw material for stream sediments.

For further information the reader should consult the selected references listed at the end of the chapter.

Soil has been defined by Joffe (1949, p. 41) as

> a natural body of mineral and organic constitutents, differentiated into horizons, of variable depth, which differs from the material below in morphology, physical make-up, chemical properties and composition, and biological characteristics.

As emphasized by Jenny (1941), the unique characteristic of soil lies in the organization of its constituents and properties into layers that are related to the present-day surface and that change vertically with depth. This is in direct contrast to the character of the parent material from which the soil is formed.

Biologic activity figures prominently in the soil-forming processes, and most soils are more or less fertile in that they can support plant growth to an extent far greater than the parent material. During weathering, rock decomposition and soil formation merge indistinguishably into one another. As a rule, they proceed simultaneously, the former paving the way for the latter. From this point of view, soil formation can be considered as an advanced stage of weathering.

I. SOIL PROFILE DEVELOPMENT

As previously stated, soils are characteristically organized into layers differing from each other and from the underlying parent material in their properties and composition (Fig. 7.1). Apart from differences in color and texture which aid recognition in the field, the properties of greatest significance that affect geochemical dispersion of the elements are pH, organic-matter content, clay-mineral type and assemblage, and the amount of Fe–Al–Mn-oxides.

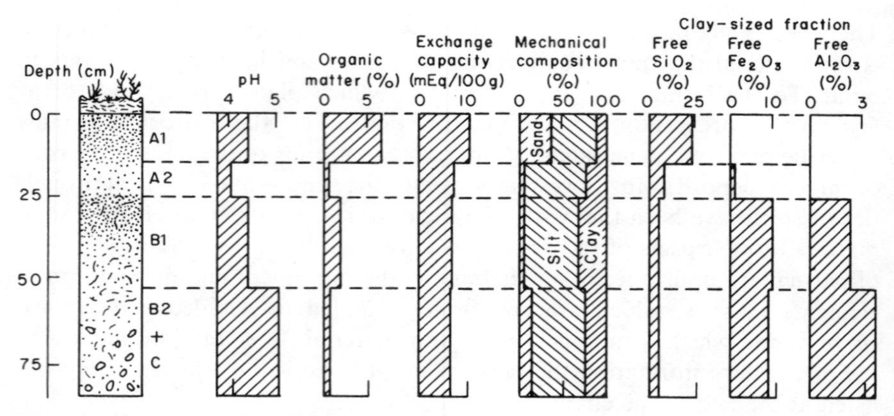

Fig. 7.1. Some variations in the physical and chemical properties of the different horizons comprising a soil profile. (Based on data for a tropical podzol by Hardon, 1936.)

The individual layers are referred to as soil horizons and may range from a few centimeters to a meter or more in thickness. Taken together, these horizons comprise the soil profile. In general terms, profile development is synonymous with soil formation. It is primarily the result of vertical (upward and downward) movement of material in solution and suspension, accompanied by a complex series of chemical reactions, many of which are organic in origin. Water is the essential medium in which this transfer and reconstitution takes place.

Soil profiles vary in make-up within wide limits according to their genetic and geographic environment. Most well-developed profiles, however, can be divided into four principal horizons. From the surface downward these are identified by the letters A, B, C, and R. The A and B horizons together constitute the solum, or "true soil", while the C horizon is the partly weathered parent material from which the solum has been derived by soil-forming processes, and the R horizon is the underlying rock material. A horizon of nearly pure organic matter (O or A0) may lie above the A horizon. A hypothetical soil profile is shown diagrammatically in Fig. 7.2.

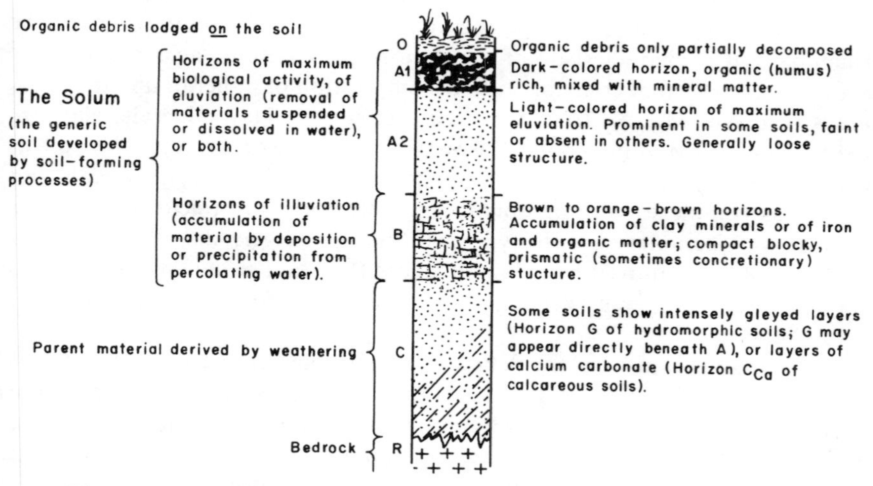

Fig. 7.2. Hypothetical soil profile showing the principal horizons.

The entire sequence need not always be represented. For instance, immature soils frequently lack a B horizon, or erosion may lead to truncated profiles, sometimes to the extent of exposing the C horizon. When studied in detail, each of the principal horizons may be further subdivided. These subdivisions are identified by combined letters and numbers thus: A1, A2, B1, B2, B3, and so on. Recognition of thess subdivisions, apart from the A1 and A2 horizons, is usually unnecessary in geochemical prospecting. The distribution of metals may vary markedly with major changes down the profile, however, and it is therefore important to distinguish the master horizons and to recognize immature and truncated profiles when these are encountered.

The A horizon is characterized by a process of partial leaching known as eluviation, a word derived from the Latin meaning "to wash out". Eluviation is accomplished by the downward percolation of water through the soil. Some constituents are removed as ions or molecules in true solution in the downward-moving water; others are removed as dispersed colloidal particles.

A major agent catalyzing the leaching of the A horizon and determining the character of the entire soil profile is humus, which is generally concentrated in the A horizon. Humus has been defined by Brady (1974, p. 14) as

> a complex and rather resistant mixture of brown or dark brown amorphous and colloidal substances modified from the original plant tissues or synthesized by various soil organisms.

It is the product of decay of surficial organic debris together with the decomposition products of roots within the uppermost soil horizons.

The organic acids and complexing agents generated in the humus by bacterial action, and the CO_2 generated by decay of humus promote the leaching that is characteristic of the A horizon. Carbonic acid and organic acids furnish H^+ that contributes to decomposition of minerals and displaces bases (Ca, Na, Mg, K) from the exchange sites of clay minerals, clay-size oxides, and organic materials. Bases move downward through the soil as dissolved ions; Fe and Al move as colloidal particles of clay minerals and oxides, as complexes with organic groups, or, in the most acid soils, as free ions or ions complexed with hydroxyl. Silica is largely dissolved as silicic acid (H_4SiO_4) or moved as colloidal silica. Resistant primary minerals and rock material undergoing decomposition tend to remain behind in the upper soil.

The accumulation of plant debris is responsible for the most commonly noted subdivisions of the A horizon into an organic surface horizon (A0 or O), a dark upper layer containing humus with mineral grains (the A1 horizon), and an underlying, strongly leached, light-colored horizon with little organic matter (the A2). The relative thicknesses of these horizons vary considerably according to the supply of organic debris, the rate at which it is decomposed, the effectiveness of leaching and eluviation, and the age of the profile. Both A1 and A2 can generally be discerned in mature profiles developed in forested areas under moist climates, but the A2 may be absent in grassland and dry regions, or in young soils.

Under moist conditions and free drainage, the more soluble constituents leached from the A horizon will descend to the water table and eventually pass with the ground water into the surface drainage. Some suspended matter may follow the same course. More usually, however, colloidal silicates, oxides, and organic material, as well as some dissolved constituents derived from the A horizon, are locally redeposited in the zone of accumulation or illuviation (from the Latin "to wash in"), constituting the B horizon. As a result, the B horizon tends to be enriched in clay and Al-oxides, and to assume a red or yellow-brown color in those profiles where illuviation also involves Fe-oxides. Under some conditions organic material may accumulate in the B horizon. In many soils, however, organic matter is likely to be completely broken down to CO_2 and H_2O in the A horizon. In some profiles, the B horizon may also gain material by precipitation of soluble material derived from underlying horizons by ground-water circulation, capillary rise of water, or evaporation of soil moisture. The level at which illuviation takes place is dependent on the decrease in the acidity of the soil water as it reacts with soil during downward percolation, and on increases in contents of dissolved solids. These changes flocculate the colloidal material and precipitate the oxides.

The C horizon consists of more or less weathered parent material for the overlying A and B horizons. It is important to appreciate that the parent material may be rock *in situ*, transported alluvial, glacial, or wind-blown overburden, or even soil of a past pedological cycle. As a rule, inorganic decomposition extends deeper than soil formation, and the C horizon can often be subdivided into zones of weathering that decrease in intensity with depth. Organic matter is at a minimum in the C horizon, which usually contains less clay and is lighter in color than the B horizon. Relic rock structures and textures are also more commonly preserved than in the overlying horizons.

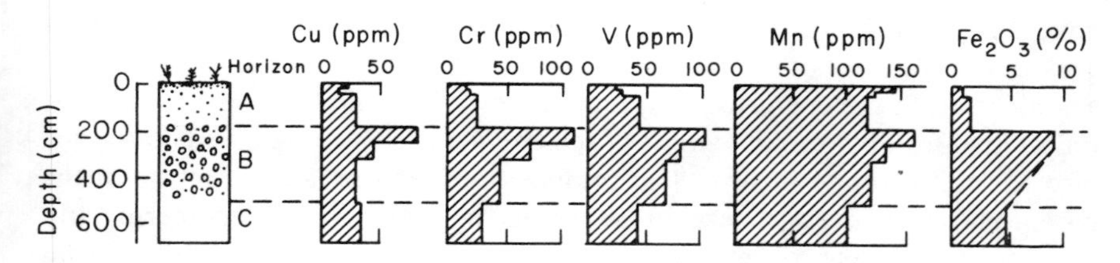

Fig. 7.3. Variation in metal content with soil horizon, latosol profile, Zambia. Alkalis increase with depth; Co and Ni show little change. Data on −80-mesh fraction. (Sampling by J. S. Tooms; analyses by J. D. Kerbyson, Geochemical Prospecting Research Centre, Imperial College, London.)

The importance of the soil profile in geochemical prospecting will be readily appreciated from the foregoing paragraphs. Metals indigenous to the parent material vary in their response during the development of the soil horizons. Soluble metals and those incorporated or adsorbed on clays and colloids are liable to be removed from the A horizon, whereas those contained in resistant primary minerals are liable to be enriched in that horizon. Metals taken up by deep-rooted plants will be returned to the surface in the organic debris, and their subsequent fate will depend on the stability of their organic compounds in the A1 horizon. Some of the metals which are removed from the A horizon may tend to accumulate along with hydrous Fe- and Mn-oxides or clays in the B horizon. A striking example of metal distribution developed over granite bedrock in the tropics is shown in Fig. 7.3.

II. FACTORS AFFECTING SOIL FORMATION

In nature, soil profiles may differ widely from the necessarily idealized picture conveyed in the preceding paragraphs. Their diversity arises from the fact that the direction and rate of horizon differentiation are governed by no less than five major factors; namely, parent material, climate, biologic activity, relief, and time.

A. Parent Material

The parent material contributes the raw material of the soil and is therefore a factor controlling the nature of the resultant soil. The effects of parent material are most evident in the deeper soil horizons, which grade into the parent material. The effects of the parent material are also evident in youthful soils that have not had time to develop fully, as in mountainous terrain, and in desert and arctic regions where chemical weathering is slow.

Parent material is also more obvious in areas where the surface horizons have been eroded away. In well-developed soils, effects of the parent materials are subordinate to other factors, but are usually recognizable by careful examination of the chemistry and mineralogy. Geologic features may also be reflected in residual soil in regions of relatively uniform climate and constancy of other soil-forming factors. The composition, texture, and structure of the parent material have a general influence on the rate at which weathering and soil formation take place. Thus, soils are developed more rapidly from permeable or readily decomposed rocks than from those which are relatively impermeable or resistant to decay. On the grand scale, however, climate appears to be more potent than geology in determining soil type.

Fortunately, even under conditions of the deepest weathering and diversity of soil type, it is often possible to detect variations in the underlying rocks, providing the overburden is residual. Thus, characteristic assemblages of resistant minerals and the mechanical composition of the soil may be diagnostic of certain parental rock types (Fig. 7.4). The original concentration of bases may influence the clay-mineral assemblage. Furthermore, the relative concentrations of trace elements in the soil may also provide a clue as to the bedrock geology (Fig. 7.5). This latter approach is, of course, the basis of geochemical soil surveys, because ores are a special type of parent material.

B. Climate

The significance of temperature and rainfall in rock decomposition has already been mentioned. These two climatic elements are also of the utmost importance in soil formation. It has long been recognized that vastly different soils may develop from similar parent materials according to the climatic environment. In general terms, the soils of humid regions tend to be the more thoroughly leached and to possess Fe-rich B horizons. Arid climates, on the other hand, militate against the formation of well-eluviated A horizons, while calcareous soils are characteristic of semi-arid, warm regions.

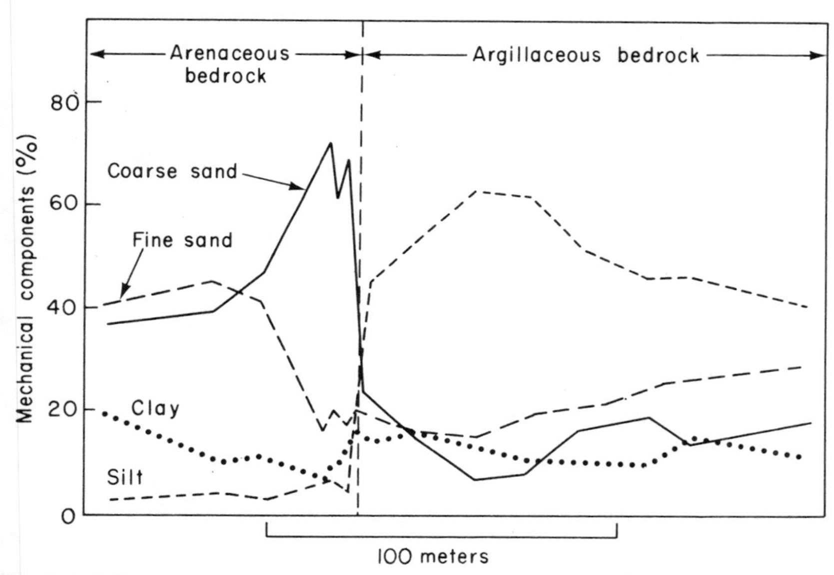

Fig. 7.4. Influence of bedrock on mechanical composition of the A-horizon soil, Mutupa, Zambia. (After Govett, 1958.)

The interrelations between rainfall and temperature govern the precipitation/evaporation ratio. This affects not only the amount of water and the depth to which it may percolate through the soil, but also the direction of movement. A high ratio of precipitation to evaporation favors downward percolation, whereas a low ratio favors evaporation of moisture from the soil and an upward rise from the underlying water table by capillary forces. Where the rainfall is markedly seasonal, the direction of soil-water movement may reverse periodically during the course of the year. Upward movement of water above the capillary fringe has been shown to be possible (Marshall, 1959). Calcium, Mg, and other soluble salts precipitated as carbonates or sulfates by evaporation thus may be a feature of soil formation under the influence of relatively dry climates. Colloidal materials may likewise rise toward the surface and coagulate to form a concretionary horizon near the upper level reached by the water table. Iron oxides and colloidal silica are the outstanding materials to accumulate in this way. As a result of

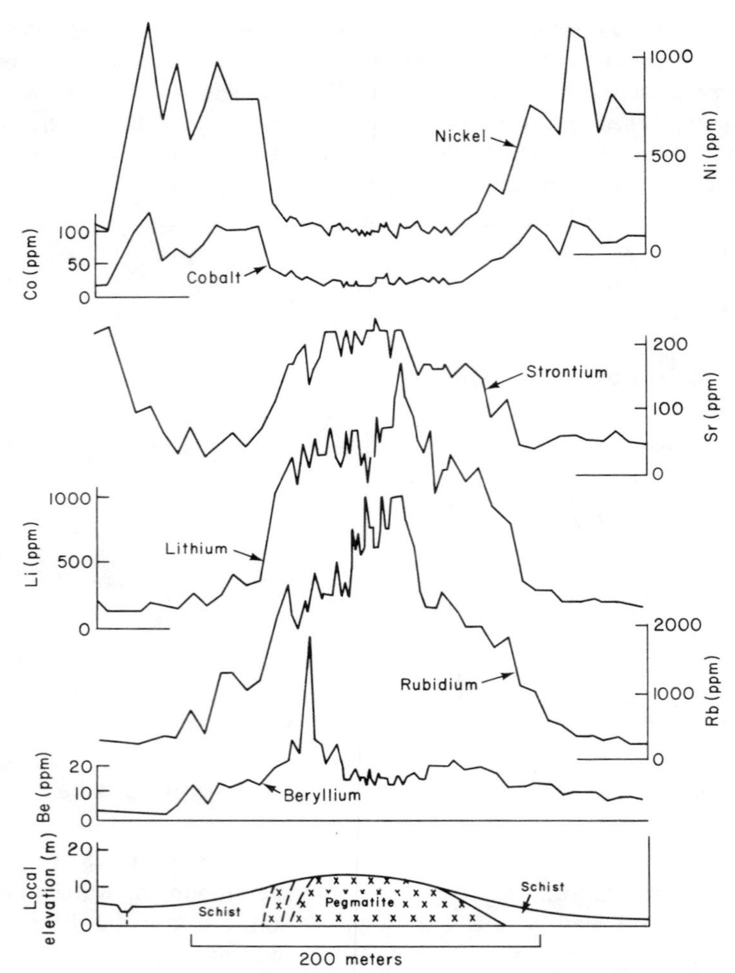

Fig. 7.5. Relationship between minor elements in residual soil and bedrock geology, Bepe pegmatite vein, Rhodesia. Data on −80-mesh fraction. (Spectrographic analysis by J. D. Kerbyson, Geochemical Prospecting Research Centre, Imperial College, London.)

incomplete leaching, the soils of arid and semi-arid areas tend to be alkaline, and thus favor the formation of montmorillonite-type clays with high exchange capacity. Neutral or acid soils with clays of low exchange capacity are more typical of humid regions.

Some soils inherit important features from past climates. For example, soils affected by frost-stirring during the Pleistocene may have much higher metal content in surface horizons than similar soils unaffected by frost effects.

In tropical areas, some thoroughly leached soils may have developed under conditions of much higher rainfall in the past.

In addition to their other properties, the soil types characteristic of different climates may differ widely in pH, clay-mineral assemblage, and content of Ca, Fe, and organic matter, all of which affect metal dispersion (Figs 7.6 and 7.7). Indirectly, therefore, climate has a bearing on the choice of the optimum geochemical techniques in different parts of the world.

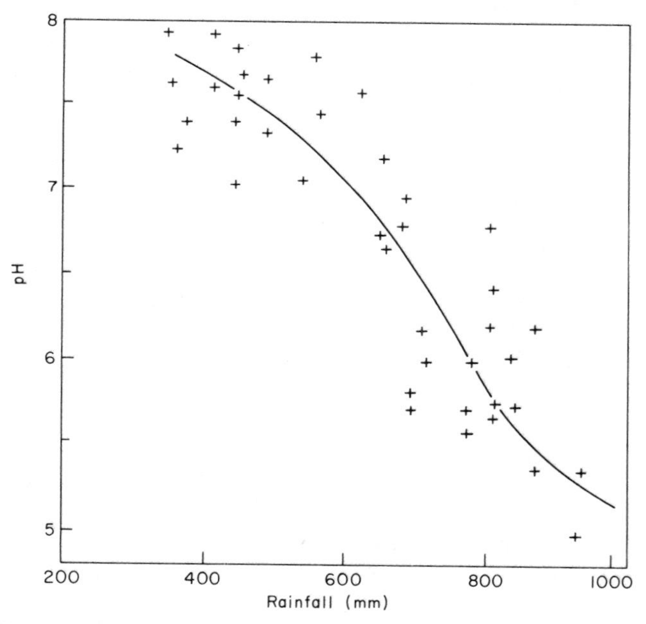

Fig. 7.6. Effect of rainfall on soil pH. (After Jenny, 1941.)

C. Biologic Activity

The biologic factor in soil formation is largely a function of vegetation plus the activity of microorganisms in decomposing the plant debris. These in turn are related to climate. In general terms, dry climate means less vegetation and less raw material for the formation of humus, and hence less soil-forming activity. Warm climate means increased biologic activity that more rapidly destroys the plant litter and hence restricts the accumulation of humus, but accelerates the action of the organic acids that are generated during the decay of the litter. Thus we should expect maximum accumulation of humus only in cool humid climates.

Another biologic factor in soil formation is the effect of different types of vegetation that yield different organic decay products which, in turn, affect profile development and soil properties (Fig. 7.8). Deep-rooted plants offset leaching to some extent by taking up elements which are returned to the surface soil when the plant dies or sheds its leaves. Plants also affect the all-important moisture regime in the soil by transpiring water and by shading the soil. Moreover, vegetation tends to protect soils from erosion and to limit disturbance by soil creep, thereby favoring the development of soil horizons. On the other hand, horizon differentiation may be retarded when the soil is mixed by root and animal activity.

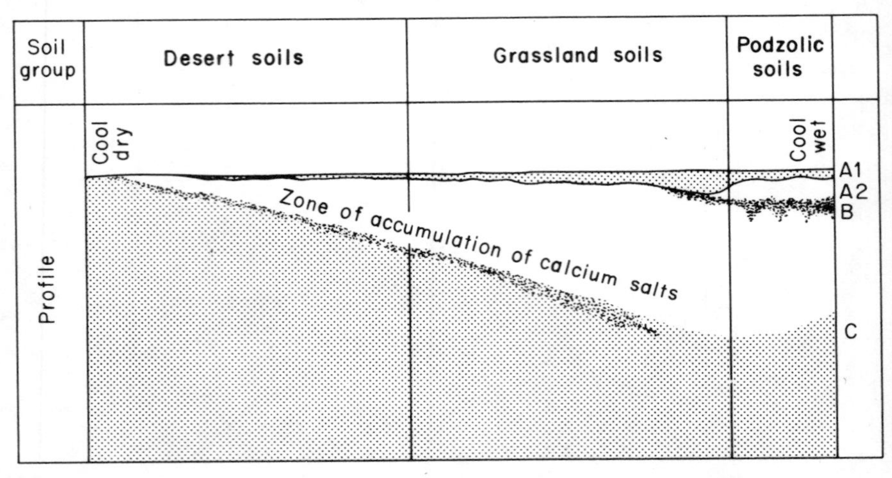

Fig. 7.7. Influence of climate on concentration of organic material and carbonate in soil profiles. (After Millar *et al.* (1958). Reprinted with permission from "Fundamentals of Soil Science", © John Wiley and Sons, Inc., 1958.)

D. Relief

Topographic relief influences soil formation through its relationships with ground-water levels, drainage, and erosion. The A and B horizons are most strongly differentiated in regions of moderate to high rainfall where there is free drainage and effective leaching. Such profiles develop most readily, therefore, on the interfluves in undulating country. In low-lying areas, the terrain may be saturated with water, almost to the surface. In such circumstances, a very different profile may develop, often consisting of an organic-rich surface layer, overlying a pallid or mottled subsoil, in which reducing conditions prevail and leaching is at a minimum. If water stands at the surface, peat may form. In semi-arid regions a temporary water table may exist alongside rivers in the wet season and induce the formation of saline profiles behind the levee where the water table lies near the surface (Fig. 7.9).

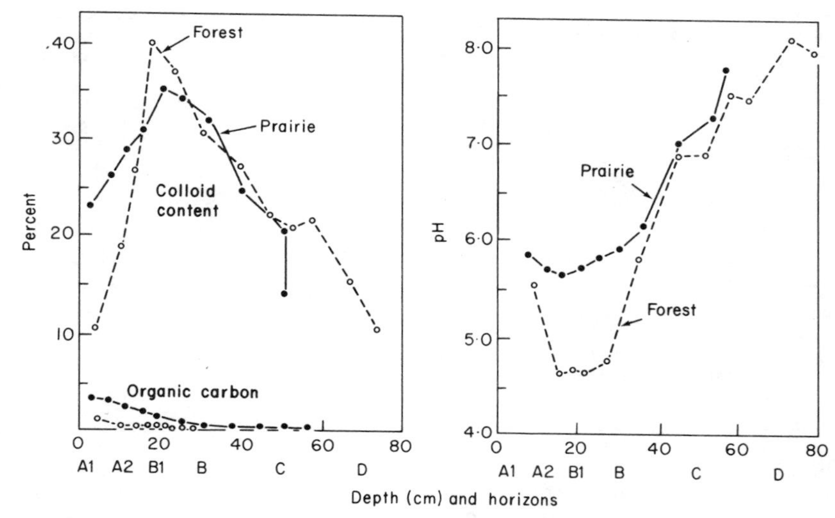

Fig. 7.8. Effect of vegetation on soil properties. (After Jenny, 1941.)

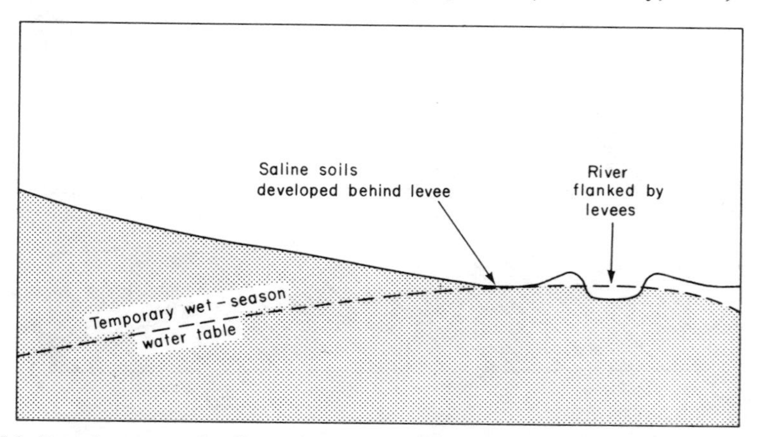

Fig. 7.9. Development of saline soils under the influence of ground water in an arid climate. (After Jenny (1950). Reprinted with permission from "Applied Sedimentation" (P. D. Trask, ed.), © National Academy of Sciences, 1950.)

Changes from one drainage condition to another are usually transitional and give rise to a related sequence of profile types which constitute a soil catena (Fig. 7.10).

The angle of slope affects drainage and erosion. In general, there is more rapid erosion, a greater volume of surface runoff, and less percolation on steep slopes than on gentle ones. Consequently, soils on steep slopes tend to be shallower, and show less distinct horizon development and a higher content

Granite hill with tors | Upper footslope | Lower footslope | Valley margin | Valley floor | Seasonal swamp

Dark gray loam | Brownish-red loam directly on granite | with iron concretions in subsoil | Gray sand with irregular iron concretions in subsoil | Hardpan soil not calcareous | Black sandy clay calcareous | Heaviest black clay

Fig. 7.10. A soil catena or repetitive sequence of soil types dependent on slope and drainage. (After Clarke (1938). Reprinted with permission from "The Study of Soil in the Field", © Clarendon Press, 1938.)

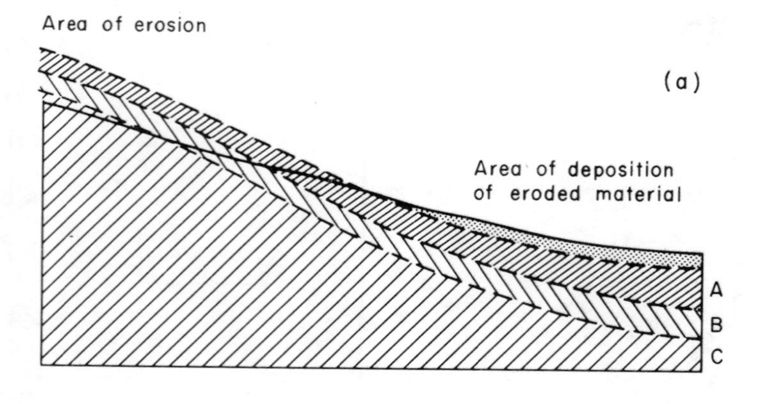

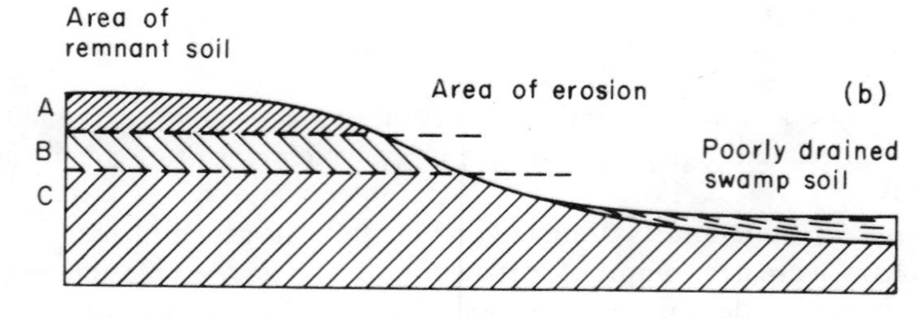

Fig. 7.11. Effect of erosion on soil type.

of stony material than those on gentler slopes. In some areas, an increased rate of erosion may result in truncated profiles on the upper slopes, accompanied by burial of the original profile at lower topographic levels (Fig. 7.11a). In other areas, truncated profiles may occur on the lower slopes (Fig. 7.11b). In areas of very high relief, the elevation determines the local climate, with the result that soil types characteristic of the different climates occur in topographically controlled patterns (Fig. 7.12).

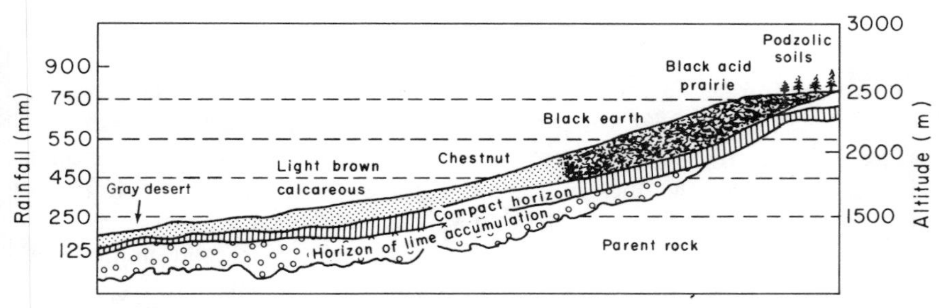

Fig. 7.12. Gradation of soil types from desert to humid mountain top, west slope of Big Horns, Wyoming. Scale of soil profiles is greatly exaggerated. (After Thorp, 1931.)

E. Time

Generally speaking, the accumulation of parent material by weathering takes longer than its differentiation into soil horizons. Given a moderate humidity and free drainage, faint A horizons may become apparent in weathering parent material after some decades. The development of a distinct B horizon normally needs a much greater length of time, often measured in centuries or even tens of thousands of years (Fig. 7.13). Soils are sometimes classified as

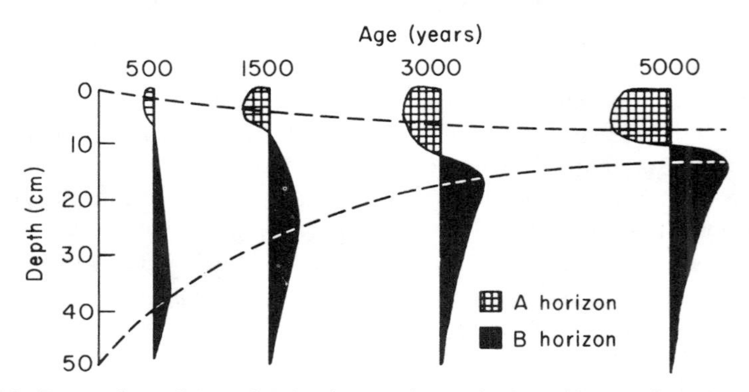

Fig. 7.13. Formation of A and B horizons of a podzol profile as a function of age. (After Jenny, 1941.)

juvenile or mature according to their state of development with respect to the present-day surface. Where erosion is active, soils will necessarily remain in a juvenile condition irrespective of time. If the time factor is adequate, and erosion proceeds no faster than soil formation, a mature profile will eventually result.

III. CLASSIFICATION OF SOILS

The exploration geochemist is concerned with classification of soils because of his need to recognize different types of dispersion during soil-forming processes, and to correlate these dispersion processes from one region to another. The older classifications, such as the older U.S. system (Thorp and Baldwin, 1938), were largely genetic, that is, the classifications were based on inferred genesis as controlled by climate, drainage, relief, parent material, and similar factors. More recent classifications have been descriptive, emphasizing the presence of certain chemical and physical features that can be identified unambiguously and that are chosen to reflect genesis and growth of crops. The most widely used example of such a classification is Soil Taxonomy of the U.S. Department of Agriculture (Soil Survey Staff, 1975), which incorporates a new set of terms and soil types based on quantitative measurement of the chemical and physical properties of soils. In some respects, the Soil Taxonomy system is not as easily used in geochemical exploration as the older approach, since, because the names are unfamiliar, the classification requires some effort to understand; also laboratory tests are necessary to classify soils with certainty. However, many of the terms and definitions are tied to clearly defined features and processes, and thus are of potential value in understanding the nature and extent of chemical dispersion in soils. In addition, although many national systems of soil classification still are in use, the trend has been toward incorporation of many of the terms and concepts of the U.S. Soil Taxonomy system as, for example, in the classification used by the Food and Agricultural Organization (F.A.O.) of the United Nations in its soil map of the world (F.A.O., 1974). As a basis for understanding the soils literature of the present and future, and to indicate the potential relevance of the new classification in geochemical exploration, the Soil Taxonomy system is briefly presented in the following pages. Soils of the world are then discussed, using both the old classification and the Soil Taxonomy system.

A. The Soil Taxonomy System

At the broadest level of classification in Soil Taxonomy, soils are divided into ten Orders determined largely by the presence of one or more of the diagnostic horizons described in Table 7.1. The ten Orders are listed in Table 7.2, along with their diagnostic features. Equivalents in the old system are indicated in Table 7.3. The diagnostic horizons are the key to separation and classification of Orders, and are also of critical importance in recognizing the nature of trace-metal dispersion. For example, oxisols are highly leached soils, and alfisols are less leached than ultisols.

Orders are subdivided into Suborders on the basis of climate, drainage, or other distinctive features or properties. Names for Suborders are formed from a prefix denoting the "formative element" (Table 7.4) and a suffix denoting the Order. For example, an aquod is a water-saturated (aqu-) spodosol (-od), known in the old system as a ground-water podzol. About 50 Suborders are recognized. Successively more detailed subdivisions are Great Groups, Subgroups, Families, Series, and Phases. Great Groups are formed by adding a prefix to denote diagnostic horizons or variants of horizons. The names of Subgroups are formed by an adjective preceding the name of the Great Group to indicate its gradational position in relation to the neighboring Great Groups. Family names add more adjectives to indicate properties important to plant growth. For example, a Family name for a given soil might be fine-loamy mixed aquic hapludalf (a wet soil with a poorly leached clay-rich B horizon, formed in a humid region with minimal development of diagnostic horizons). The Series and Phase are respectively the geographical and textural names for soils that have been in current use for years.

B. Old Classification of Soils

In general, three categories of soils may be distinguished, azonal, zonal and, intrazonal. In azonal soils, the nature of the soil is determined mainly by the parent material, and only incipient horizons are present. Immature soils on recent alluvium, volcanic ash, wind-blown sands, and shallow bedrock are the major representatives of this group. These soils are the entisols and inceptisols of Soil Taxonomy and include the lithosols, regosols, and alluvial soils of the older classification. In zonal soils, a mature soil profile has developed, and the general characteristics of the soil are determined mainly by climate and vegetation. Podzols, chernozems, latosols, and desert soils in the old classification are zonal soils, in part equivalent to spodosols, mollisols, oxisols, and aridosols, respectively, in the Soil Taxonomy system. In the intrazonal soils, a local condition, such as drainage, is the dominant influence on the character of the soil, so that soils in different climatic regions resemble each other more closely than the zonal soils of the region in which they occur. For example, in humid regions, soils in localities with poor drainage develop a waterlogged reducing zone beneath the surface. Such reducing subsurface

Table 7.1
Some diagnostic soil horizons and properties[a]

A. Horizons formed at the surface (A horizons)

1. Histic A horizon: Peaty or mucky, organic matter >20%, usually water saturated. Diagnostic of histosols.
2. Mollic A horizon: A granular dark horizon >18 cm thick with organic matter >1% and base saturation >50% (see below); Ca is main exchangeable cation. Typical of grasslands. Diagnostic of mollisols.
3. Umbric A horizon: Similar to a mollic horizon but with base saturation <50%. Typical of forested areas.
4. Ochric A horizon: A relatively pale or thin surface horizon not qualifying as mollic, umbric, or histic; organic content <1%. Typical of arid regions and some forested regions.

B. Horizons formed beneath the surface (B horizons)

1. Argillic B horizon: Accumulation of illuvial clay, as indicated by clay coatings or markedly higher clay content than overlying horizons. Present in alfisols and ultisols.
2. Cambic B horizon: Texturally distinguishable from A horizon but lacking illuvial clay, Fe–Al-oxides, organic matter, or alkali enrichment. May be present in inceptisols and mollisols.
3. Natric B horizon: Like an argillic B horizon, but alkali-enriched; exchangeable Na >15% of total exchange capacity, columnar structure. Present in some aridosols and mollisols.
4. Oxic B horizon: Highly weathered, composed of kaolinite, Fe–Al-oxides, very low cation-exchange capacity (<15 mEq/100 g clay), no clay coatings. Diagnostic of oxisols.
5. Spodic B horizon: Accumulation of organic matter and Al-oxides with or without Fe-oxides and without clay coatings, usually beneath albic horizon. Diagnostic of spodosols.

C. Other horizons and properties

1. Albic horizon: Subsurface horizon from which clay and Fe-oxides have been leached, leaving a strongly bleached horizon, usually beneath an organic-rich A horizon and above a spodic B horizon (A2 horizon of podzols).
2. Calcic horizon: Accumulation of appreciable carbonate.
3. Gypsic horizon: Accumulation of gypsum.
4. Salic horizon: Accumulation of salts more soluble than gypsum.
5. Gley horizon: A blue-gray subsurface waterlogged horizon, reducing, and usually mottled with spots of brown to yellow.
6. Ferric properties: With red mottles and concretions, and low cation-exchange capacity.
7. Base saturation: The fraction of exchange sites on clays or other colloidal particles that is occupied by Ca, Mg, Na, and K; the remaining sites are occupied mainly by H and Al. Base saturation is a measure of the degree of weathering and leaching. It tends to decrease with increased weathering and leaching.
8. Aridic moisture regime: Profile not moist throughout for more than two consecutive months, nor moist in some part for more than three.

[a] Source: Soil Survey Staff (1975), Dudal (1968), Young (1976), and Dregne (1976).

zones are known as gley horizons and are characterized by a pallid blue-gray mottled or streaked appearance, with Fe predominantly in the ferrous state. These soils are denoted by gley, gleyic, ground-water, or similar terms in the old classification and by the prefix aqu- in Soil Taxonomy. Similarly, in arid regions, low or poorly drained areas develop salt-rich or saline soils, denoted as solonchak or solonetz in the old system and by prefixes natr- and sal- in Soil Taxonomy.

Table 7.2
Soil Orders, according to Soil Taxonomy[a]

Soil order and suffix to form name of suborder	Latin or other root	Description
Entisol (-ent)	Rec*ent*	Soils lack significant profile development, consist of little modified parent material
Inceptisol (-ept)	L. *inceptum*, beginning	Soils with weakly developed horizons of alteration of parent material, but not accumulation
Spodosol (-od)	L. *spodos*, wood ash	Soils with a spodic B horizon
Alfisol (-alf)	Al–Fe	Soils with an argillic or natric B horizon and a base saturation >35%; moist part of the time; lack calcic, gypsic, or mollic horizons
Ultisol (-ult)	L. *ultimus*, last	Soils with an argillic B horizon and base saturation <35%; warm most of the time
Oxisol (-ox)	*Ox*ide	Soils with an oxic B horizon
Aridosol (-id)	L. *aridus*, dry	Soils with an aridic moisture regime, an ochric A horizon and high base saturation; may have calcic, gypsic, argillic, or natric B horizons
Mollisol (-oll)	L. *mollis*, soft	Soils with a mollic A horizon and high base saturation in B horizon
Vertisol (-ert)	L. *verto*, change	Soils with a high content of swelling clay, forming wide cracks when dry into which surface soil falls or is washed
Histosol (-ist)	L. *histos*, tissue	Soils with a histic (peaty) A horizon (organic soils)

[a] Source: Soil Survey Staff (1975).

The relation between zonal soils and climate is illustrated in Figs 7.14 and 7.15, and is listed in Table 7.3. Major groups (desert, grassland, podzolic, tropical, tundra) are determined by climate and vegetation. Within each of the broad zones, other soils are normally present in small areas. Figure 7.16 shows idealized profiles of some of the major soil groups.

Table 7.3
Approximate correlations[a] of soil Orders, and their relation to climate and local conditions

	U.S.D.A. (1975)	Old system or subdivision	F.A.O. (1974)
Zonal soils			
1. Soils of the cold zone	Inceptisols, mollisols (cryo- or aq- prefix)	Tundra (gleyic) soils	Gleysols
		Arctic brown soils	
2. Podzolic and related soils of temperate and tropical forested regions	Spodosols	Podzols	Podzols
		Brown podzolic soils	
	Alfisols	Gray-brown podzolic soils	Luvisols
		Gray podzolic soils	
		Non-calcic brown soils	
		Degraded chernozems	
	Ultisols	Red-yellow podzolic soils	Acrisols
	Oxisols	Ferrallitic soils[b]	Ferralsols
		Ferruginous soils[b]	
		Ferrisols[b]	(Nitosols)
3. Soils of semi-arid, subhumid, and humid grasslands	Mollisols	Degraded chernozems	Phaeozem
		Prairie soils	
		Reddish prairie soils	
		Chernozem	Chernozem
		Chestnut soils	Kastanozem
		Reddish chestnut soils	
		Brown soils (in part)	
	Vertisols	Grumusols	Vertisols

4. Light-colored soils of arid regions	Aridosols	Brown soils (in part)	Xerosols
		Reddish-brown soils	
		Sierozem soils	
		Desert soils	Yermosols
		Red desert soils	
Intrazonal soils			
1. Hydromorphic soils of marshes, swamps, flats, and seepages	Aqu- prefix	Humic-gley soils	Gleysols
		Alpine meadows oils	
		Low humic-gley soils	
		Planosols	
		Ground-water podzols	
		Ground-water latosols	
	Histosols	Bog soils	Histosols
		Half-bog soils	
2. Halomorphic (saline and alkali) soils of imperfectly drained arid regions	Sal-, natr- prefixes	Solonchak	Solonchak
		Solonetz	Solonetz
3. Calcimorphic soils	Rend- prefix	Brown forest soils	
		Rendzina	Rendzina
Azonal soils			
1. Soils on little-weathered parent material	Entisols	Lithosols	Lithosols
		Regosols	Regosols
		Alluvium	Fluvisols
2. Soils with only incipient horizons	Inceptisols		Andosols
			Cambisols

[a] After Aubert and Tavernier (1972), Young (1976), Brady (1974), and Dudal (1968).
[b] After Young (1976).

Table 7.4
Some prefixes for formative elements used in naming Suborders in the Soil Taxonomy[a]

And-	Soils from vitreous volcanic parent materials
Aqu-	Soils that are wet for long period
Arg-	Soils with a horizon of clay accumulation
Bor-	Soils of cool climates
Fibr-	Soils with minimal decomposition
Hum-	Soils with organic matter
Ochr-	Soils with little organic matter
Orth-	Soils typical of their order
Psamm-	Sandy soils
Sapr-	Soils with extreme decomposition
Ud-	Soils of humid climates
Umbr-	Soils with dark surface layer reflecting much organic matter
Ust-	Soils of dry climates with summer rains
Xer-	Soils of dry climates with winter rains

[a] Source: Soil Survey Staff (1975).

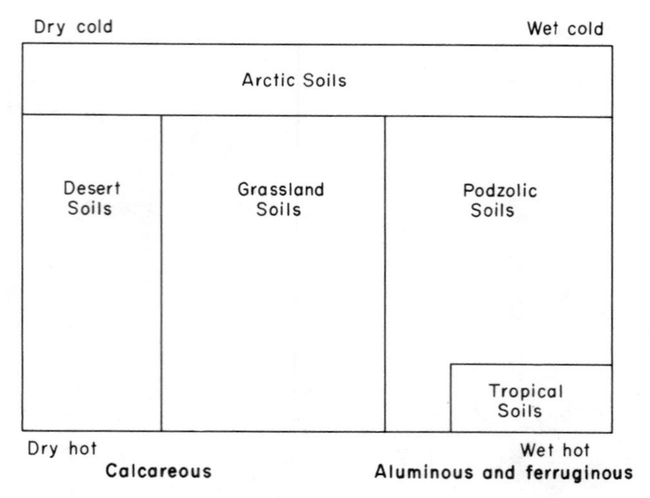

Fig. 7.14. Classification of soils in relation to climate.

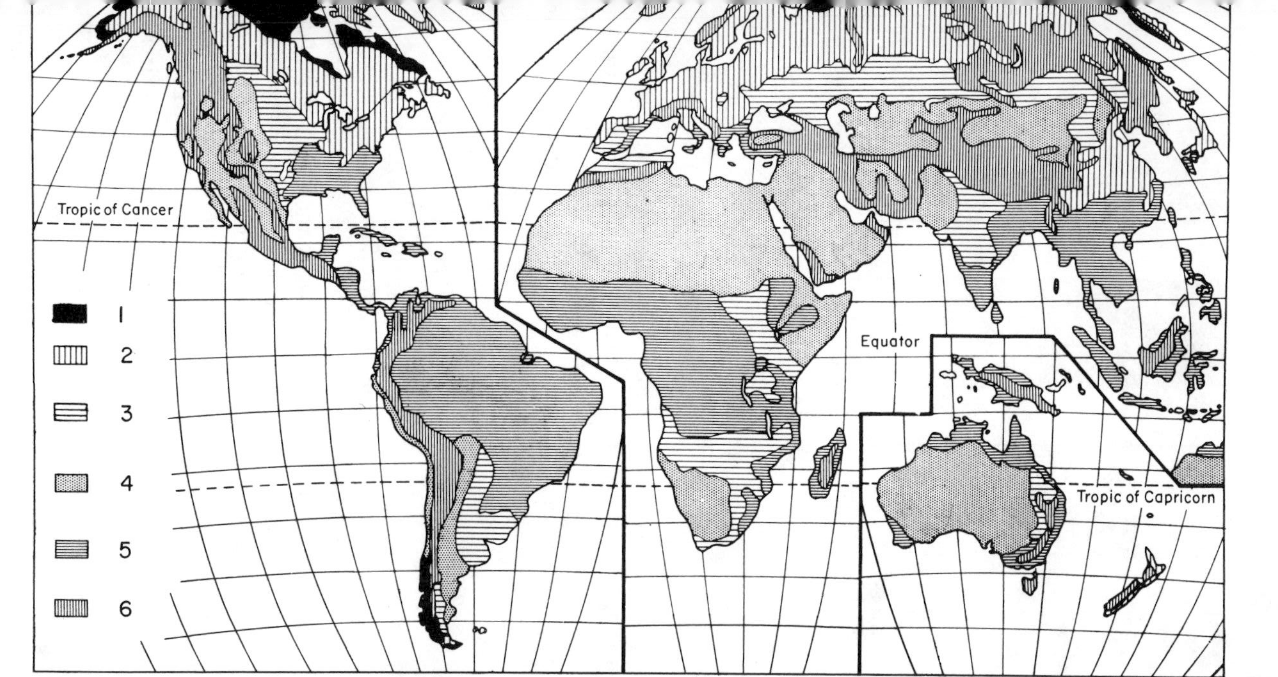

Fig. 7.15. Map of the world, showing six broad soil zones. Each zone generally has similar processes of horizon differentiation prevailing over it. Important areas of organic soils and other intrazonals are omitted, as well as very important bodies of alluvial soils along great rivers such as the Mississippi, Amazon, Nile, Niger, Ganges, Yangtze, and Yellow. (1) Arctic soils: dwarf shrub- and moss-covered soils of frigid climates. (2) Podzolic soils: forested soils of humid, temperate climates; includes many areas of organic soils. (3) Grassland soils: grass-covered soils of subhumid temperate climates; includes some soils of wet–dry tropical savannas such as black and dark gray clays. (4) Desert soils: sparsely shrub- or grass-covered soils of arid, temperate, and tropical climates; includes large areas of lithosols and regosols. (5) Tropical soils: forested and savanna-covered soils of humid and wet–dry tropical and subtropical climates. (6) Mountain soils: stony soils (lithosols) with inclusions of one or more of the above soils, depending on climate and vegetation, which vary with elevation and altitude. (After Simonson, 1957.)

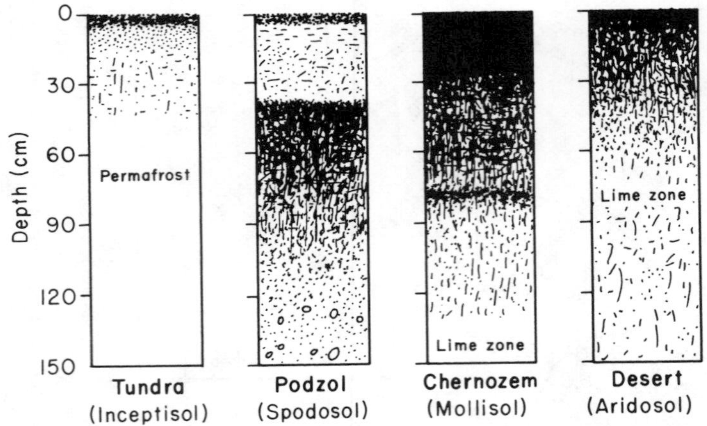

Fig. 7.16. Idealized profiles of four zonal soils. (After Winters and Simonson, 1951.)

IV. SOILS OF HUMID REGIONS

The humid zone is characterized by an annual rainfall greater than about 60 cm. Vegetation ranges from forest and grasses in temperate and tropical regions to low-order plants and shrubs in the colder belts. Mature soils are generally characterized by an accumulation of Fe- or Al-oxides or clays in some part of the profile and the absence of calcareous horizons.

A. Podzolic and Related Soils of Temperate Forested Regions

In well-drained areas of coniferous forest in the northern portions of North America and Eurasia, a surface horizon of partly decomposed organic matter is immediately underlain by a pale strongly leached (albic) A2 horizon. Organic matter with variable amounts of Al–Fe-oxides and clay accumulates in the B horizon of these podzols (spodosols). Farther south, under deciduous or mixed deciduous–coniferous vegetation, a distinct organic-bearing mineral-rich A1 horizon is underlain by a similar A2 and a B horizon with either clay, Al–Fe-oxide, or organic illuviation in the podzolic soils. Leaching by organic acids and the chelating ability of the organic matter are apparently responsible for the strongly leached A2 horizons in these soils.

By virtue of the intense leaching in the A horizon, metal dispersion patterns are commonly sought in the underlying B horizon. As a rule, this horizon is encountered within 30 cm to 60 cm of the surface and is, therefore, conveniently situated for most prospecting purposes. It is possible that the distribution of some metals during soil formation may vary according to whether the profile is that of a podzol or of a podzolic soil. Lead is a case in point where marked surface enrichment has been noted under deciduous cover but not where the vegetation is coniferous (Webb, 1958b).

B. Tropical Soils

The zonal soils of the tropics and subtropics usually have a red to yellow color in subsurface horizons and an enrichment of Fe and Al in the B horizon. These soils, formerly termed latosols, develop at relatively high temperatures under moderate to high rainfall (>60 cm/yr) where the profile is freely drained, at least in its upper part. Tropical soils are discussed by Young (1976) and Mohr *et al.* (1972).

The older classifications do not clearly distinguish types of soils in the tropics, but in recent years, two major soil types have been clearly differentiated (Young, 1976). The ferruginous soils (alfisols or ferric luvisols) occur in areas having savanna vegetation and a markedly seasonal distribution of rainfall in the range 60–120 cm/yr. The parent material is generally a felsic to intermediate igneous rock or its metamorphic equivalent. The surface A1 horizon is normally reddish brown and sandy, with a small amount of organic matter. A paler but still red to yellow A2 horizon may be present beneath the surface horizon. The blocky red to yellowish B horizon exhibits distinct illuviation of clay, as evidenced by clay coatings or "skins" on surfaces in the B horizon. The clay fraction is composed mainly of kaolinite with a low cation-exchange capacity, but a high proportion of this capacity is occupied by bases (Ca, Na, Mg, or K), indicating only moderate leaching of the profile and incomplete weathering of the parent materials.

Ferrallitic soils (oxisols or ferralsols) are found in two environments: in tropical rainforests with a rainfall greater than about 175 cm/yr, and on ancient plateau surfaces receiving lower rainfall. They are characterized by extreme and complete weathering of the parent material to kaolinite and Fe- and Al-oxides. The surface sandy to loamy A horizon contains small to moderate amounts of humus and may be underlain by a lighter colored A2 horizon. The B horizon is rich in Fe-oxides and clays. The soil has very low cation-exchange capacity, much of which is occupied by H and Al, rather than bases. Iron-rich layers or concretions that harden on drying may be present to form a layer of laterite, which is discussed further below. Beneath flat surfaces with moderate rainfall, the zone of weathered rock beneath the B horizon may be as much as 100 m thick. The deep weathering may reflect a former climate with a higher rainfall in some such localities. On steep slopes in tropical rainforests, bedrock may be present only a few meters beneath the surface.

A third common type of tropical soil develops on mafic rocks, which tend to produce a clay-rich profile. The ferrisols (alfisols or ultisols, depending on base saturation) or reddish brown lateritic soils of the older classification have a thick clay-rich C horizon with clay coatings on surfaces, overlain in the A horizon by a darker, sandy clay with 2–5% organic matter (Young, 1976). Concretions of Fe-oxide may be present. The clay fraction is mainly kaolinite and Fe-oxide with a base saturation less than 50%, and a low to moderate cation-exchange capacity. These soils develop in rainforests and rainforest–savanna transition zones. With lower rainfall (70–150 cm/yr) in savanna zones, the zonal soils on mafic rocks may have a brown color with distinct enrichment of clays in the B horizon and appreciable amounts of weatherable minerals in the profile. Cation-exchange capacity is moderate to high because montmorillonite is typically present, and base saturation is 50–100%.

In many respects, there is a considerable measure of similarity between tropical and temperate forest soils. They are both strongly leached and show accumulation of Fe–Al-oxides somewhere in the profile. The basic differences appear to lie mainly in the rate and intensity of weathering rather than in any differences in the processes involved. Under tropical temperatures, decomposition of aluminosilicates and organic matter proceeds faster, and there is a higher turnover of organic matter than is the case in more temperate climates. The soil temperature is consistently high in the tropics, and the oxidation state may be higher because of the smaller amount of organic matter. In tropical areas with a wet and dry season, the soil may dry out completely, in contrast to the consistently moist condition of soils in temperate humid regions.

Considerable attention has been given to the accumulations of Fe- and Al-oxides in tropical soils. If these accumulations harden irreversibly on drying, they are termed laterites. Some laterites have been used as low-grade Fe ore, and they can be difficult to distinguish from gossans over sulfide ore. At the other extreme, similar accumulations rich in alumina constitute the important bauxitic ores of aluminum. Other types of hardened subsurface regolith materials (duricrusts) include silcrete, composed of silica, and calcrete or caliche, formed of carbonate.

A typical laterite contains the normal pale-reddish A horizon underlain by a hard, red or brown material rich in Fe-oxides, possibly accompanied by Al-oxides. The hard Fe-rich material may occur as closely spaced concretions, as a massive zone, or as a combination of the two. This laterite zone commonly overlies a mottled zone and a pallid zone, which extend to depths as great as 100 m. In the mottled zone, the weathered parent material contains red, orange, or purple mottles separated by paler, soft, clay-rich material. The mottled zone usually grades downward into the pallid zone, consisting of white kaolinitic clay crushable in the hand.

The origin of laterites is controversial, despite a great deal of attention in the literature (McFarlane, 1976). One widely accepted theory attributes the profile to a fluctuating water table producing an alternation of wet reducing conditions, during which Fe is mobilized upward from the pallid and mottled zones, and dry oxidizing conditions, during which Fe is fixed as hematite or goethite. However, the amount of Fe in many laterite profiles is too great to be explained by the relatively minor depletion of Fe from the mottled and pallid zones (McFarlane, 1976). Other workers suggest leaching of Fe from large volumes of previously overlying weathered and eroded soil, lateral movement of Fe in ground water flowing along the surface of the water table, and residual accumulation by leaching of all other soil constituents. Downward migration of the laterite horizon by solution and reprecipitation seems necessary in the formation of extensive continuous laterites, so that Fe can accumulate from a large thickness of parent material. Iron is clearly mobile during the formation of laterites, probably in the periodic reducing environment, and is able to replace other minerals and to grow to relatively large crystals and coatings. Undoubtedly, several of the above mechanisms operate to varying extents in different areas. In some regions, erosion of the overlying soil horizons has left fossilized laterites exposed at the surface over large areas.

Tropical soils share with temperate forest soils the feature of extensive leaching from the surface horizons and accumulation of elements in the B horizons. If the soil profile is thin, sampling of the B horizon is generally preferred, but in thicker profiles, sampling of shallow soil in the A horizon is generally more practical. The possibility of extensive migration, both lateral and vertical, in forming the laterites must be recognized. Methods of distinguishing laterites from gossans over sulfides are discussed in Chapter 11.

C. Intrazonal Hydromorphic Soils

In the present context, the most important intrazonal modifications are those resulting from impeded drainage. Lack of aeration is mainly responsible for the characteristic profile developed in waterlogged soils. Although they naturally occur more frequently in humid regions, they may also develop under comparatively dry climates. Their common distinguishing features are: a grayish to black surface horizon grading sharply into a pale, bluish-gray subsoil, often with rusty streaks, mottling, or concretions. The necessary waterlogged condition may arise in the presence of a shallow water table or impeded drainage due to an impervious subsurface layer. For this reason, ground-water or hydromorphic soils, as they are called, are often developed in depressions and areas of ground-water seepage. Here the entire profile is commonly subjected to alternate oxidation and reduction as the shallow water table rises and falls according to the season of the year. Plant growth is restricted to shallow-rooted species, commonly grasses, appropriate to the local climate. Breakdown of humus is retarded when the ground is waterlogged, and consequently, organic matter tends to accumulate in the surface horizon, even under tropical climates.

Predominantly reducing conditions are reflected in the pallid, bluish-gray colors of the subsoil where Fe tends to remain in the ferrous state; such horizons are known as gley horizons. Rusty mottling and Fe concretions testify to some oxidation in the zone of ground-water fluctuation near the surface. In drier climates, calcareous nodules may be deposited in the subsoil. As a rule, leaching is restricted by a sluggish flow of water which is dominantly in a lateral direction. Under these conditions, clay minerals of the montmorillonite type tend to develop. Where organic acids are developed in quantity, however, kaolinitic clays will predominate. Mohr has drawn attention to the rapidity with which parent material may be kaolinized beneath some tropical swamps, where the pH may be as low as 3·0.

If water stands at the surface for much of the year, thick deposits of partially decayed organic matter may accumulate, Peat soils (histosols) of this type are more common in temperate and cool climates, and cover extensive areas in northern Europe and America. Peats can also occur in the tropics as, for example, in mangrove swamps.

As already mentioned, impeded drainage in the surface soil may result from the presence of an impervious horizon. Thus, hydromorphic soils can develop at times over basic rock, such as norite, the weathering of which may produce an abundance of clay. Or, the parent material itself may be clay-rich and impermeable, and for this reason, waterlogged soils are common on glacial overburden in northern latitudes. An impervious layer may also be developed by the extreme accumulation of clay in the B horizon of an originally well-drained profile.

Under some northern, temperate conditions, a hard impermeable horizon of Fe-oxide (placon) develops at depths of a few centimeters to several tens of centimeters (Tilsley, 1977) and separate zones of drastically different chemical composition. The zone above the placon may undergo intense leaching or be the site of a perched water table. Anomalies may be weak or lacking in the surface horizons compared to the deeper horizons beneath the placon.

V. SOILS OF SUBHUMID AND ARID REGIONS

dic soils form in areas with an annual rainfall that is generally ut 75 cm. A thin forest cover may be present in the wetter parts, but more commonly the vegetation, if any, includes certain shrubs, grasses, and desert plants appropriate to local conditions. In the absence of a plentiful supply of water, chemical weathering and leaching are restricted. Most profiles are characterized by the development of calcareous horizons. In others there may be little or no horizon differentiation or, indeed, no breakdown of the primary rock-forming minerals.

A. Grassland Soils

Grassland and analogous soils (mollisols) are developed in temperate and tropical regions under a rainfall ranging from approximately 23 cm to 45 cm. Typically, the profile consists of a dark A horizon overlying lighter-colored parent material. The B horizon is usually indistinct. Concretions and earthy accumulations of calcium carbonate are commonly present below the A horizon. In temperate zones the A1 horizon is high in organic matter but is less so in the tropics. By virtue of the low rainfall, leaching and eluviation of clay are not so marked as in the soils of humid regions. Nevertheless, some material is dissolved in the A horizon and may be reprecipitated, mostly as carbonate, in the lower horizons. The typical accumulations of calcium carbonate are mostly formed in this way. The depth to the upper limit of the calcareous horizon increases with annual rainfall (Fig. 7.17). The lower limit of this horizon usually indicates the maximum depth to which rain water percolates before being dissipated by evaporation or by transpiration from plants.

In contrast to the foregoing, carbonates may also accumulate in the soil by evaporation of ground-water solutions in the capillary fringe immediately above the water table. Given a slowly falling ground-water level, the calcareous horizon may then attain a very considerable thickness. Accumulations of indurated calcium carbonate are often known as caliche or calcrete.

Other things being equal, calcareous horizons can form only on terraces and interfluves providing the underlying rock is lime-bearing. Under the influence of a shallow water table, however, the soil of semi-arid regions may contain calcium carbonate irrespective of the nature of the immediate underlying parent material.

Grassland soils are usually neutral or slightly alkaline in reaction, whereas the tropical analogues are more strongly alkaline and tend to be richer in montmorillonitic clays. The high pH and high base-exchange capacity of the soil are not conducive as a rule to the ready dispersion of soluble ore metals.

B. Desert Soils

Every gradation exists between the grassland soils of semi-arid regions and the soils of truly arid deserts. Extreme shortage of moisture greatly restricts

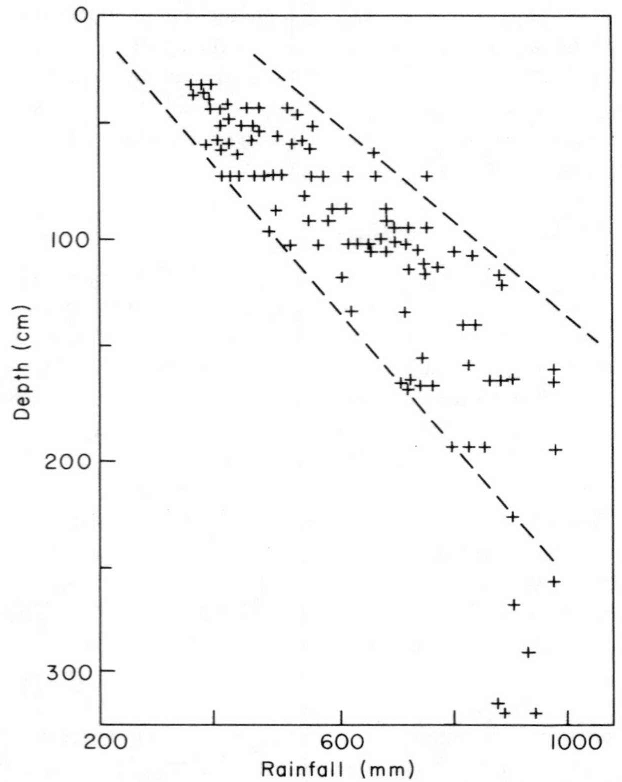

Fig. 7.17. Relationship between depth of lime accumulation and rainfall. (After Jenny, 1941.)

weathering, leaching, and plant growth. As a result, desert soils usually possess indistinct shallow profiles. Organic matter is invariably low, and the predominant soil color is gray or reddish. There may be some loss of carbonate from the A horizon and a slight accumulation of clay in the B horizon. Generally speaking, however, there is a high proportion of relatively fresh undifferentiated parent material throughout the profile. Under such circumstances, the most effective agent of secondary metal dispersion is likely to be through plant species whose deeper roots are capable of penetrating to ground water level.

C. Soils of Arctic Areas

The arctic zone is located north of the tree line, where vegetation consists mainly of grasses, moss, heath, herbs, and lichens with small shrubby trees. Permafrost is present beneath the surface in many areas and restricts the vertical movement of water through the soils. The northern parts of this zone are essentially a desert, with evaporation exceeding precipitation. Important factors involved in soil formation in this environment are production of organic matter, leaching from the surface horizons, formation of a gley zone beneath the surface, frost action, and evaporation and transpiration of soil moisture (Tedrow, 1977).

A uniform soil classification is even farther from acceptance in this environment than in others. The Soil Taxonomy system has not been widely applied in this region, so correlation of older names is not attempted in this section.

Plant growth is slow in arctic regions, but the rate of decay is also slow, so that a layer of organic matter tends to build up in those parts of the tundra zone receiving adequate precipitation. The subsurface layers of this organic mat tend to become humified. If the soil is well drained, colloidal organic material tends to be transported down into the underlying mineral soil to produce a dark to pale brown A1 horizon. Soils of this type are called arctic brown soils. Some leaching of Ca, Na, Mg, and K from the surface horizons occurs but, in most localities, is confined to the top few centimeters to tens of centimeters.

Development of a subsurface gley horizon is typical where drainage is poor as a result of shallow permafrost or underlying clay-rich materials. This blue-gray, waterlogged, mottled, reducing horizon may occur immediately beneath the surface organic horizon, or a thin organic soil may separate the two. Soils with a distinct development of a subsurface gley horizon over permafrost are generally termed tundra soils. It should perhaps be mentioned that although such tundra soils are essentially hydromorphic in that they owe their characteristic features, in part at least, to impeded drainage arising from a frozen subsoil, they are usually regarded as zonal soils since their distribution is dependent on a regional climatic control.

Frost-heaving, frost-stirring, and related cryogenic processes are widespread in northern zones and constitute a major factor in disrupting and destroying soil profiles. The net result is a vertical stirring of the soil, leaving bare patches of nearly raw parent material, separated by accumulations of rock and vegetation on which some soil formation has taken place. The resulting frost boils, stone circles, and stone stripes have dimensions of from a few tens of centimeters to tens of meters. For geochemical purposes, these processes are important because they bring relatively unweathered parent material to the surface for easy sampling.

In the far north, evaporation exceeds precipitation and moisture is lost from the soil to leave saline conditions and precipitation of salts and efflorescences. Such conditions are usually accompanied by loss of fines from the surface by wind action, leaving a pavement of small stones (desert pavement).

D. Intrazonal Saline and Alkali Soils

These soils (a type of aridosol) occur most commonly under arid conditions, but they may also be found in semi-arid and subhumid regions. They represent modified desert and grassland soils and are characterized by the accumulation of soluble salts of Mg or Na. In some instances, K salts may be present in appreciable amounts. There are two principal profile types, known as saline, white alkali, or solonchak soils, and alkali, black alkali, or solonetz soils. Both varieties are usually found in depressions where they originate by evaporation under shallow ground-water conditions. They are, therefore, the counterpart of hydromorphic soils in more humid climates. The salt content of these soils may represent the soluble products of hydrolysis of silicates or the vestiges of former seas or salt lakes.

The typical salts of solonchak are sulfates and chlorides of Na, Ca, and Mg. These often appear as a white efflorescence at the surface, which is liable to be dissolved and temporarily washed downward when it rains. Solonetz, on the other hand, is characterized by Na-carbonate as the predominant salt and a dark-colored B horizon which is strongly alkaline in reaction. A rather bleached A2 horizon is commonly present. By comparison, solonchak is only moderately alkaline.

All gradations exist between these two soil types which are believed to be closely related. Thus, solonetz is assumed to be a development from solonchak as a result of the soil clay becoming dominated by exchangeable Na, so that when drainage is improved, both clay and organic matter move downward to give the dark compact B horizon.

VI. MOUNTAIN SOILS

Some of the principal features of mountain soils (inceptisols and entisols) have been mentioned earlier when considering relief as a factor in soil formation. In general terms, relief and altitude combine to impart specific modifications on the zonal soil type most nearly characteristic of the local climatic conditions. Where erosion is active, the soil tends to be maintained in a juvenile state of development with thin, indistinct horizons containing a high proportion of partially weathered rock debris. Such skeletal soils are known as lithosols and rightly belong to the azonal group. Where erosion is retarded, horizon differentiation takes place and bears the stamp of the local climate and vegetation. The resulting profile possesses in greater or lesser degree the characteristics of the appropriate zonal soil type.

In geochemical prospecting, the most important features of many mountain soils include (i) active movement of material by soil creep; (ii) relatively little leaching in some environments or accelerated oxidation in others, according to the relative rates of chemical weathering and erosion; (iii) the accumulation of transported eroded material (colluvium) at the base of slopes.

VII. SELECTED REFERENCES ON SOIL FORMATION

Short summaries	Jenny (1950)
	Simonson (1957)
General texts	Brady (1974)
	Birkeland (1974)
Tropical soils	Young (1976)
	Mohr *et al.* (1972)
Trace elements in soils	Vinogradov (1959)
Desert soils	Dregne (1976)
Arctic soils	Tedrow (1977)
Soil classification	Soil Survey Staff (1975), F.A.O. (1974)

Chapter 8

Chemical Equilibria in the Surficial Environment

As rocks weather, their substance is made available for erosion and dispersion away from the place of origin. Selective solution, deposition, or sedimentation during the course of erosion and dispersion can result in a far-reaching redistribution of the products of weathering. Redistribution is governed by the chemical and physical properties of the various dispersed constituents and of the media within which they move. An understanding of the chemical equilibria that determine the behavior of the chemical elements in the surficial environment is clearly essential for effective geochemical prospecting.

The products of weathering at the earth's surface are partitioned between the relatively immobile solid phase that constitutes the regolith, and the mobile fluid phase made up of free-flowing underground and surface waters. The solid phase is composed of the insoluble products of weathering, which are dispersed by the relatively slow mechanical movement of clastic fragments. Weathering products that either are water soluble or that occur in forms that can be readily suspended and swept along by flowing water are dispersed more rapidly.

The mobility of the chemical elements in the surficial dispersion cycle is therefore largely controlled by their solubility in water. Some elements, such as the Si of detrital quartz grains, characteristically occur as components of insoluble minerals. Others, such as Na or Cl, are almost always found in the water-soluble phase. Most elements, however, fall somewhere between these extremes. Either they are partitioned between the two forms, or are soluble under some conditions and insoluble under others. Empirical observations on the mobility of elements are reviewed in Chapter 2 and in the Appendix. The following sections summarize some of the factors of the chemical environment that have an effect on the partition of elements between the more mobile aqueous phase and the less mobile solid phase.

I. COMPOSITION OF NATURAL WATERS

The normal content and range of major and minor elements in natural fresh water are illustrated in Fig. 14.5 and Table 14.1. The total salinity (dissolved solids) depends on both climate and rock type. It ranges from a minimum of perhaps 10 ppm in areas of cold climate and high rainfall to several per cent in warm, desert environments. The presence of calcareous or other readily soluble rocks naturally tends to raise the total salinity in all environments. The bulk of the solids dissolved in water normally consists of dissociated ions, principally Na^+, K^+, Mg^{2+}, Ca^{2+}, Cl^-, SO_4^{2-}, and HCO_3^-. To these should be added soluble but undissociated H_4SiO_4, the stable colloidal suspensions of hydrous Fe-oxide, and the dissolved and colloidal organic matter. Small to moderate amounts of dissolved gases (CO_2, H_2S, CH_4, N_2, O_2) are also present (see Chapter 18). The relative proportions of the major constituents depend to a large extent on the rocks through which the water percolates. High relative and absolute amounts of Ca^{2+}, Mg^{2+}, and CO_3^{2-} with associated high pH values occur in waters derived from the chemical weathering of calcareous rocks. High SO_4^{2-} concentrations ranging up to thousands of ppm may occur in areas of oxidizing sulfide minerals. Extremely high contents of Ca^{2+} and SO_4^{2-}, and sometimes Na^+, K^+, and Cl^-, may result from the leaching of saline deposits.

Variations in the major constituents affect the dispersion of the minor elements in fresh water primarily by providing reactants, either for the precipitation of insoluble minerals or for the formation of soluble complex ions with the minor elements. These effects are considered separately in the following sections.

II. Eh–pH RELATIONSHIPS

Perhaps the most important of all the factors governing the solubility of a given element in water is the relationship between the concentration of hydrogen ions in the solution, expressed as pH, and the oxidation potential, or Eh.

A. Hydrogen-ion Concentration

This parameter is customarily expressed as the negative logarithm of the hydrogen-ion activity, for which the symbol "pH" is used. Most natural waters have a pH between about 4 (acid) and 9 (alkaline). Near oxidizing sulfides, concentrations of sulfuric acid may be high enough to produce a pH of 2 or even less, and conditions more alkaline than pH 9 occur in some desert soils and lakes.

Most metallic elements are highly soluble in acid solution but can be precipitated as oxides or hydroxides by increasing the pH. For alkali and most alkaline-earth elements, the pH of precipitation is above the range of natural waters, but many higher-valent or transition elements are precipitated in the pH range of natural waters. For example, Cu^{2+} may precipitate as CuO by the following hydrolysis reaction:

$$Cu^{2+}(aq) + H_2O(l) = CuO(s) + 2H^{2}(aq) \qquad (8.1)$$

for which the equilibrium constant is

$$K_{CuO} = \frac{a^2_{H^+}}{a_{Cu^{2+}}} = 10^{-7 \cdot 35} \qquad (8.2)$$

where a, the chemical activity in solution, is approximately equal to the concentration in moles per liter. By taking logs of this equation and re-arranging:

$$\log a_{Cu^{2+}} = 2 \log a_{H^+} + 7 \cdot 35 \qquad (8.3)$$

or, since $pH = -\log a_{H+}$:

$$\log a_{Cu^{2+}} = 7 \cdot 35 - 2 \text{ pH}. \qquad (8.4)$$

The solubility of Cu may then be plotted against pH, as in Fig. 8.1. At pH 7, the solubility of Cu^{2+} is $10^{-6.65}$ mole/liter or $10^{+0.35} \times 10^{-7.0}$, equal to $2 \cdot 2 \times 10^{-7}$ mole/l $\times$ 63·5 g/mole $= 14 \times 10^{-6}$ g/l, or 14 μg/l (14 ppb), which is approximately the Cu content of many natural waters (see the Appendix).

In the simplest case, the solubility of CuO determines the maximum amount of Cu in solution; any larger amounts will tend to precipitate as CuO. This simple relation may be modified by other ions that may either form complex ions of higher solubility, such as $CuCl^-$, or that react with Cu^{2+} to form compounds with a lower solubility than CuO, such as CO_2 in the precipitation of malachite. In addition, varying degrees of supersaturation may exist.

Figure 8.1 shows that Cu is very soluble under acid conditions, but that under neutral to alkaline conditions it is insoluble. In agreement with this generalization, high Cu contents are found in acid water near oxidizing Cu orebodies, but Cu contents in normal ground-waters with pH of 7·0–8·5 are very low, usually 10 ppb or less.

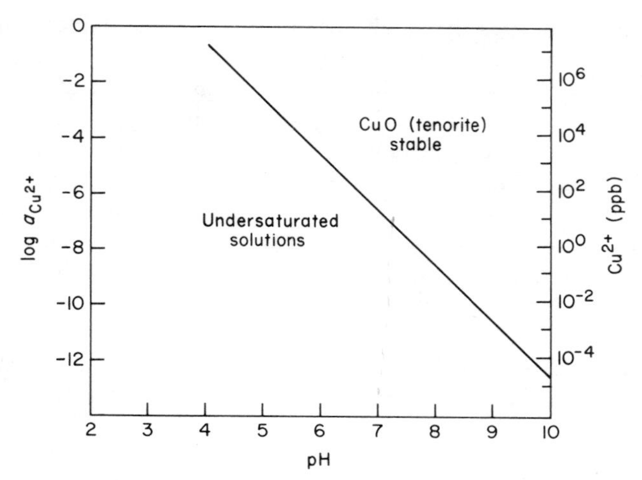

Fig. 8.1. Solubility of Cu^{2+} as a function of pH, showing limitation by CuO.

Table 8.1
Solubility of metal oxides and hydroxides[a]

Compound	Ion	Log K[b]	Saturation pH (10^{-6} M ion)
BeO	Be^{2+}	−6·4	6·2
$Mg(OH)_2$	Mg^{2+}	−16·75	11·4
$Mn(OH)_2$	Mn^{2+}	−15·30	10·65
$Fe(OH)_2$	Fe^{2+}	−11·67	8·83
$Co(OH)_2$	Co^{2+}	−12·37	9·18
NiO	Ni^{2+}	−12·46	9·23
CuO	Cu^{2+}	−7·35	6·67
ZnO	Zn^{2+}	−11·54	8·77
CdO	Cd^{2+}	−15·12	10·56
HgO	Hg^{2+}	−2·49	4·25
PbO	Pb^{2+}	−12·72	9·36
Ag_2O	Ag^+	−12·57	12·28
Al_2O_3	$Al(OH)_2^+$	—	4·74
FeOOH	Fe^{2+}	−44·2	1·3

[a] Source: Wagman *et al.* (1968), Truesdell and Jones (1974), and Langmuir and Whittemore (1971).

[b] Reactions are of the type:

$$2Me^+ + H_2O = Me_2O + 2H^+, \quad K = a_{H^+}^2/a_{Me^+}^2$$
$$Me^{2+} + H_2O = MeO + 2H^+, \quad K = a_{H^+}^2/a_{Me^{2+}}^2$$
$$Me^{2+} + 2H_2O = Me(OH)_2 + 2H^+, \quad K = a_{H^+}^2/a_{Me^{2+}}^2$$
$$2Me^{3+} + 3H_2O = Me_2O_3 + 6H^+, \quad K = a_{H^+}^6/a_{Me^{3+}}^2.$$

Table 8.1 summarizes the solubilities of some simple oxides which can be precipitated by changes in pH. As can be seen from the pH values at which 10^{-6} mol/l may be dissolved, the oxides and hydroxides of Mg^{2+}, Ca^{2+}, Mn^{2+}, Cd^{2+}, and Ag^{+} are relatively soluble in the pH range of natural waters, and, for practical purposes, precipitation of oxides and hydroxides does not limit the solubility of the metal. For Fe^{2+}, Co^{2+}, Ni^{2+}, Zn^{2+}, and Pb^{2+}, the oxides and oxyhydroxides may limit the solubility in a few very alkaline environments. In the case of Be^{2+}, Cu^{2+}, Hg^{+}, and Al^{3+}, the oxides and hydroxides exert a very important control in natural waters. For other ions and elements, such as Fe^{3+}, Si^{4+}, Ti^{4+}, Zr^{4+}, V^{3+}, Cr^{3+}, and Sn^{4+}, the solubility of the simple ions is so low as to be of no significance in natural waters. These relationships are closely related to the overall mobility of the elements in surface processes (Section II of Chapter 2).

B. Oxidation–Reduction Potential

A second very important control on mobility of elements is the concentration of electrons in the environment, called the oxidation–reduction or redox potential. This factor is important because many elements occur in more than one valence or oxidation state, and the properties of elements both in solutions and solids change considerably from one valence to another. For instance, Fe occurs in minerals and natural waters as Fe^{2+} and Fe^{3+}, and more rarely as metallic Fe. Other elements with more than one valence state in nature include C, N, S, V, Cr, Mn, Cu, As, Se, Mo, Sn, Sb, W, Hg, and U.

For many purposes, the concentration of electrons is most conveniently measured and expressed as the Eh, a voltage measured between a platinum electrode and a standard hydrogen electrode immersed in the solution (Langmuir, 1971; Garrels and Christ, 1965). In the convention generally used in geochemistry, a high Eh indicates an oxidizing system, and a low Eh a reducing system. Many chemists use the same scale but with the signs reversed.

Measurement of Eh and chemical calculations involving Eh can be illustrated by a simple cell involving a Cu electrode placed in a solution of 1 M Cu^{2+} and 1 M H^{+} (or, more precisely, a solution with unit activity of Cu^{2+} and H^{+}). A hydrogen electrode (platinum in contact with hydrogen at 1 atm is also placed in the solution. The measured voltage between the electrodes is 0·337 V, produced by the following reactions at the electrodes:

$$\text{copper electrode:} \quad Cu^{2+} + 2e^{-} = Cu \tag{8.5}$$

$$\text{hydrogen electrode:} \quad H_2 = 2H^{+} + 2e^{-}. \tag{8.6}$$

The electrons furnished by the reaction of H_2 to H^{+} flow through the connector and are utilized to reduce Cu^{2+} to metallic Cu.

If the measured concentration of Cu^{2+} in solution differs from 1 M, the measured Eh follows the relation:

$$\mathrm{Eh} = E^0 + \frac{RT}{nF} \ln \frac{1}{a_{Cu^{2+}}} \tag{8.7}$$

where E^0 is the standard reaction potential of the cell (0·337 V in this case) with all reactants and products in their standard states, R is the gas constant (1·98 cal/deg-mole), T is the absolute temperature (°K), n is the number of electrons transferred in the oxidation–reduction reaction (two in this case), and F is the Faraday (23·06 cal/V). For the general oxidation reaction involving reactants B and C forming products D and E:

$$b\mathrm{B} + c\mathrm{C} = d\mathrm{D} + e\mathrm{E} + ne^- \tag{8.8}$$

$$\mathrm{Eh} = E^0 + \frac{RT}{nF} \ln \frac{a_\mathrm{D}^d . a_\mathrm{E}^e}{a_\mathrm{B}^b . a_\mathrm{C}} = E^0 + \frac{RT}{nF} \ln Q \tag{8.9}$$

where a is the chemical activity of the species and Q is the reaction quotient. In the example above, Cu metal is in its standard state, so $a_{Cu} = 1$ and:

$$Q = \frac{a_{Cu}}{a_{Cu^{2+}}} = \frac{1}{a_{Cu^{2+}}}.$$

At 25 °C and using logs to the base 10:

$$\mathrm{Eh} = E^0 + \frac{0{\cdot}0592}{n} \log Q. \tag{8.10}$$

Figure 8.2 shows the application of these equations to the solubility of Cu. Under reducing conditions, the solubility of Cu is very low, and Cu is in solution as Cu^+, but under more oxidizing conditions, solubility increases and Cu^{2+} predominates.

At the surface of the earth, atmospheric oxygen is the most abundant oxidizing agent. Environments with access to the atmopshere or to downward percolating waters saturated with oxygen are normally oxidizing. As access to free or dissolved oxygen decreases because of depth below the surface or impeded flow of oxygenated surface waters, the environment becomes progressively less oxidizing and more reducing. Carbonaceous material produced by plant or animal life furnishes the principal reductant in the surficial environment. Where carbonaceous matter has accumulated and access to the atmosphere is restricted, strongly reducing conditions are found.

C. Eh–pH Diagrams

Because Eh and pH are such important variables, many aspects of the mobility of elements are clearly expressed in Eh–pH diagrams, using these

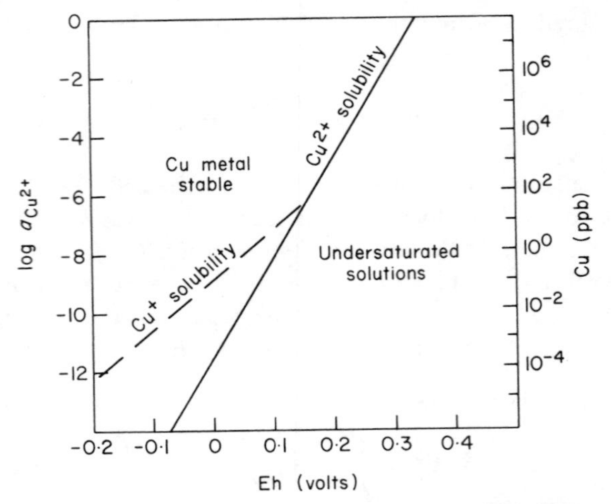

Fig. 8.2. Solubility of Cu^{2+} as a function of Eh, showing limitation by Cu metal.

two variables as axes. The simplest diagrams are those involving H, O, and the element in question. Boundaries between solid phases can be calculated from equations like the following for the Cu system:

$$2Cu + H_2O = Cu_2O + 2H^+ + 2e^- \tag{8.11}$$

$$E^0 = 0{\cdot}470 \text{ V} \tag{8.12}$$

$$\text{Eh} = 0{\cdot}470 + \frac{0{\cdot}059}{2} \log a^2_{H^+} = 0{\cdot}470 - 0{\cdot}059 \text{ pH}. \tag{8.13}$$

Values for E^0 can be calculated from the relationship $E^0 = \Delta G^0_r/nF$, using free-energy values (ΔG^0) tabulated by Wagman *et al.* (1968, 1969, 1971), Truesdell and Jones (1974), Naumov *et al.* (1974), and other reference books. Garrels and Christ (1965) describe methods for calculating Eh–pH diagrams and show examples of their use.

A diagram for the system Cu–O–H is shown in Fig. 8.3. A unique line separates pairs of solids, but an arbitrary concentration of the dissolved species must be selected to locate boundaries between solids and dissolved species. A value of 10^{-6} M is conventionally chosen for most major and minor elements, because solubilities greater than this value generally are significant geochemically. However, for the less-abundant elements, values of 10^{-8} M and even less may be appropriate. Equations 8.4 and 8.7, as graphed in Figs 8.1 and 8.2, are used to calculate the limit of significant Cu^{2+} in the figure.

The Eh and pH of some natural environments are summarized in Figs 8.4 and 8.5. The upper slanting line marks conditions oxidizing enough to decom-

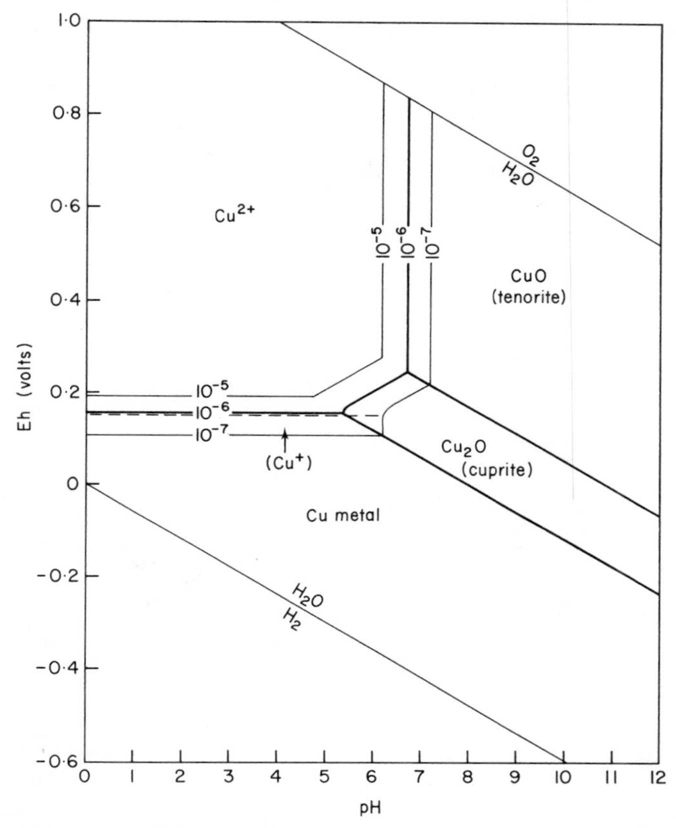

Fig. 8.3. Eh–pH diagram for system Cu–O–H at 25 °C and 1 atm. Lines labeled 10^{-5}, 10^{-6}, 10^{-7} indicate molarity of dissolved Cu. The dashed line separates fields of dominantly Cu^{2+} from dominantly Cu^+.

pose water to O_2 (1 atm) and H^+. The lower slanting line is a similar limit for conversion of H^+ to H_2 (1 atm). As noted previously, natural environments containing organic matter and isolated from the atmosphere fall near the lower limit of water stability. These include waterlogged soils, euxinic marine, and similar environments. Where atmospheric oxygen is readily available, as in rain, streams, and surface ocean-water, measured Eh and pH values fall in the upper part of the diagram. Actually, if the water is saturated with oxygen at 0·2 atm, the oxidation potential is very close to the upper limit of water stability, but both dissolved and gaseous oxygen react so slowly that neither Eh electrodes nor most chemical reactants can sense this higher oxidation potential. Instead, they sense less-oxidizing conditions created by the intermediate product H_2O_2 (Sato and Mooney, 1960). In

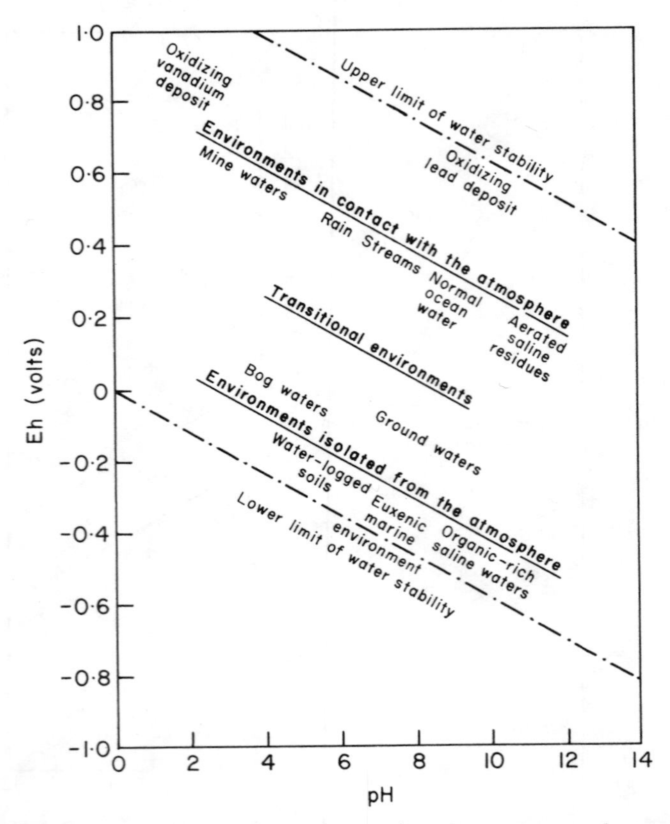

Fig. 8.4. Approximate position of some natural environments as characterized by Eh and pH. (From Garrels, 1960.)

general, the measurement of Eh is fraught with great difficulties in most natural waters, and an accurate value is very difficult to obtain (Morris and Stumm, 1967).

D. Forms of Sulfur and Carbon in Solution

Among the relatively abundant components of natural waters are sulfur and carbon. As illustrated in Figs 8.6 and 8.7, the form taken by these elements in solution varies widely, depending on the pH and Eh. The mobility of many other elements is strongly dependent on the abundance and form of C and S, because of the insolubility of many carbonates, sulfides, and sulfates.

The major dissolved species for sulfur are SO_4^{2-}, HSO_4^-, H_2S, and HS^-. In addition, S is relatively insoluble in acid solutions and may precipitate as the element under intermediate oxidation conditions. The size of the wedge-shaped field of native S decreases with decreasing dissolved S; at 10^{-4} M dissolved S ($\Sigma S = 10^{-4}$), the field extends only to about pH 3. In the H_2S(aq) field, the pressure of H_2S(g) in atmospheres is approximately equal to the activity of H_2S(aq) in solution, and the gas tends to escape to the atmosphere, making such solutions unstable unless confined. In typical fresh waters, dissolved S is in the range 10^{-4}–10^{-3} M, but near oxidizing sulfide deposits or in brines, much higher values are normal.

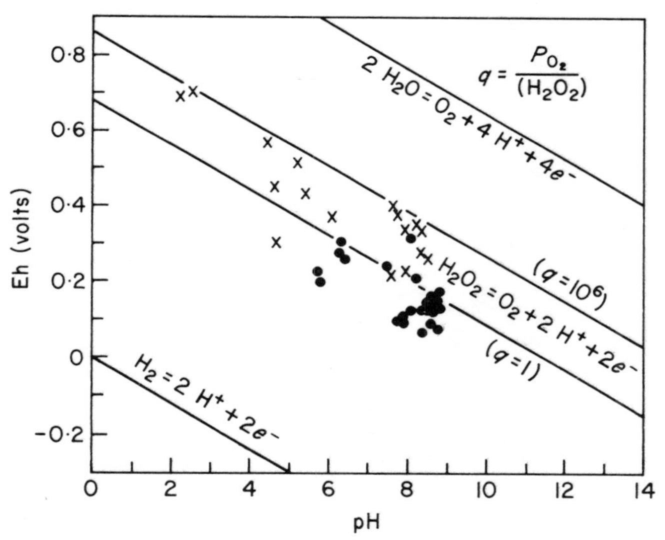

Fig. 8.5. Eh and pH values of mine waters. (×) Oxidized ore zone; (●) primary ore zone. (From Sato and Mooney, 1960.)

In the oxidizing region, the stable carbon species are $H_2CO_3^0$, HCO_3^-, and CO_3^{2-}. The relative amounts of each of these species can be evaluated from the equilibria:

$$CO_2(g) + H_2O = H_2CO_3; \qquad K_{CO_2} = 10^{-1 \cdot 47} \quad (25\ ^\circ C) \qquad (8.14)$$

$$H_2CO_3^0 = H^+ + HCO_3^-; \qquad K_1 = 10^{-6 \cdot 4} \quad (25\ ^\circ C) \qquad (8.15)$$

$$HCO_3^- = H^+ + CO_3^{2-}; \qquad K_2 = 10^{-10 \cdot 3} \quad (25\ ^\circ C) \qquad (8.16)$$

The pressure of CO_2 in the atmosphere is normally about $10^{-3.5}$ atm, from which the amounts of each of the above species in aerated waters can be calculated for any given pH. In soils and most ground waters, P_{CO_2} is higher, commonly from $10^{-1.0}$ atm to $10^{-2.5}$ atm, as a result of oxidation of organic matter. Under reducing conditions, graphite and organic C compounds

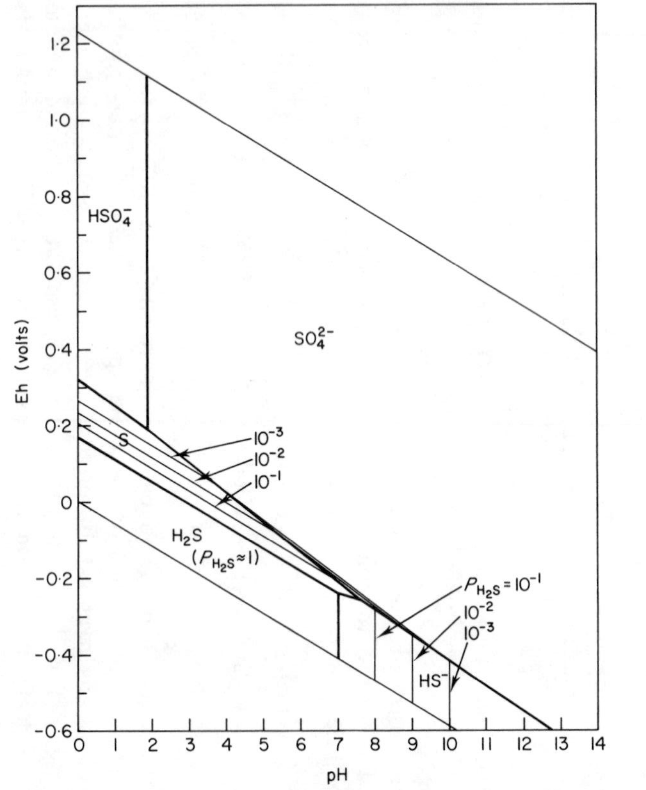

Fig. 8.6. Eh–pH diagram for sulfur species. Light labeled contours indicate pressure of H_2S in atm. Total dissolved sulfur = 10^{-1} M.

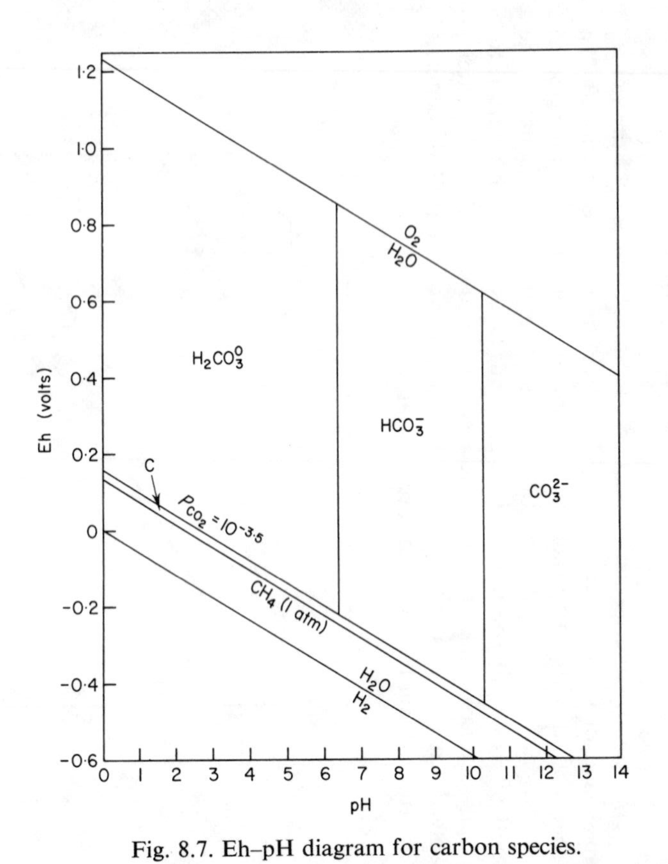

Fig. 8.7. Eh–pH diagram for carbon species.

become stable (in contrast to their metastability in contact with normal air), and species such as methane have appreciable equilibrium pressures, as evidenced by the release of bubbles of methane ("swamp gas") from organic muck in lakes and swamps.

The oxidation–reduction reactions of S and C species tend to be very sluggish, but bacteria and other organisms are able to catalyze the reactions and use the energy in their metabolism. The abundance of the reduced species tends to depend on the activity of bacteria as much as any other factor.

III. FORMATION OF COMPLEXES

Although most metallic elements occur in acid solution as simple ions (Cu^{2+}, Pb^{2+}, Ag^{+}), a wide variety of complex hydroxy- and oxy-ions are formed in more alkaline solutions. Complexes may also be formed in solutions containing other anions, especially those with which the cation forms a compound of low solubility. For example, Pb forms the complexes $PbOH^{-}$, $Pb(OH)_2^0$, $Pb(OH)_3^-$, $PbCO_3^0$, $PbCl^+$, $PbCl_2^0$, $PbCl_3^-$, $PbHS^+$, and others. Figure 8.8 illustrates the proportions of different forms of Pb in solution, considering only the complexes with OH^- and CO_3^{2-}, for two values of P_{CO_2}. The partition between two forms is found from equilibrium expressions such as:

$$Pb^{2+} + OH^{-} = PbOH^{+} \tag{8.17}$$

$$K_{PbOH^+} = \frac{a_{PbOH^+}}{a_{Pb^{2+}} \cdot a_{OH^-}} = 10^{6.29}. \tag{8.18}$$

In the complexes described above, the anion forms only one bond with the cation and is termed unidentate. However, some more complex anions can occupy several of the coordination sites around a metal cation. For example, the oxalate ion is bidentate:

```
O=C—O
  |    \
  |     Me.
  |    /
O=C—O
```

Formation of such multi-dentate complexes is termed chelation, and the resulting affinity between the ions tends to be stronger than in unidentate complexes.

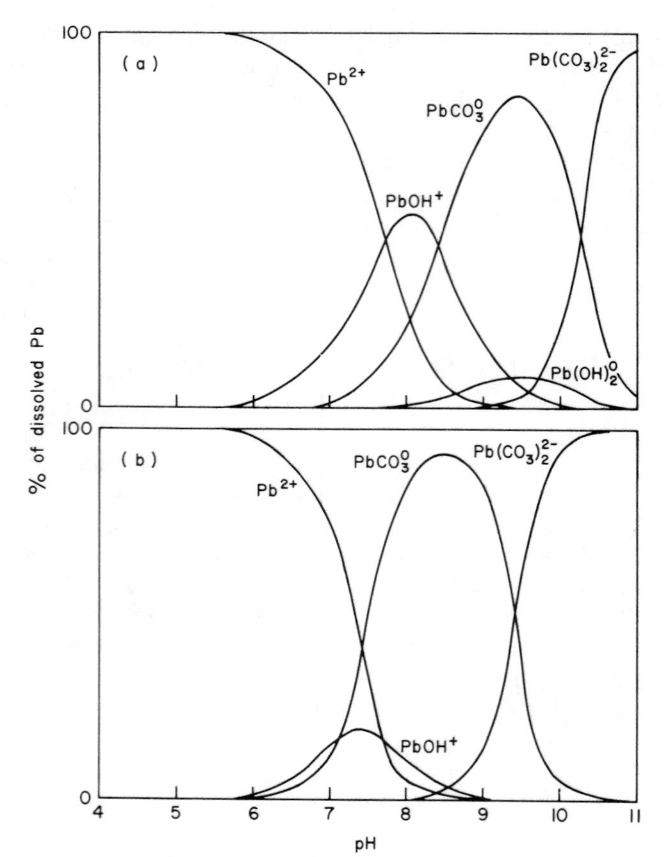

Fig. 8.8. Distribution of Pb species in solution as a function of pH at 25 °C. (a) with $P_{CO_2} = 10^{-3.5}$ atm; (b) with $P_{CO_2} = 10^{-2.2}$ atm.

IV. SOLUBILITY OF MINERALS

The solubility of minerals places an important restriction on the mobility of elements, as indicated in the previous section. If perfect chemical equilibrium always existed, the solubility would be an absolute upper limit for the concentration of an element in a natural water. However, supersaturation and related non-equilibrium phenomena can lead to concentrations higher than the equilibrium values. Supersaturation is most evident for minerals of very low solubility. For example, the low solubility of hematite and goethite can limit Fe concentrations to very low levels in oxidizing waters, but in some waters, observed Fe concentrations are higher than the equilibrium solubility by a factor of more than 10^7 (Langmuir and Whittemore, 1971). Instead of the stable goethite or hematite, these waters precipitate a metastable amorphous Fe-oxyhydroxide with much higher solubility, which only slowly recrystallizes to the more stable crystalline forms, with a concomitant decrease in Fe solubility. The very fine grain size of most Fe-oxyhydroxides also tends to increase their solubility. In contrast to the large factors of supersaturation observed for insoluble minerals, the more soluble minerals tend to nucleate more readily and recrystallize to coarse crystals so that supersaturation is not so marked. On the other hand, an Fe mineral with an equilibrium solubility of 10^{-10} M that is supersaturated by a factor of 10^3 is only supersaturated by 5·6 ppb, but Ca that is 10% supersaturated at 10^{-3} M is supersaturated by 4000 ppb. Thus, for many geochemical purposes, the solubilities are an effective limit on the mobility of elements in natural waters.

A wide variety of minerals other than metals, oxides, and hydroxides can limit the mobility of elements. For example, malachite ($Cu(CO_3)_2(OH)_2$) is more common than tenorite (CuO) in the oxidized zone of Cu deposits, and anglesite ($PbSO_4$) and cerussite ($PbCO_3$) are common oxidized Pb minerals. In reducing environments, sulfides are highly insoluble. For trace elements, the least soluble mineral may be rare or unrecognized, so that a very complete knowledge of possible solids is necessary to predict limits on mobility. The true limit of equilibrium solubility can be far lower than the inferred one, if the correct mineral phase is not known, but the least soluble known mineral does give an upper limit.

The concentration level at which a precipitation of a mineral can control an element depends on the concentration of all components of the mineral, so that the solubility product must be used to determine the equilibrium concentration of the element. For example, for the reaction

$$PbSO_4(s) = Pb^{2+} + SO_4^{2-} \tag{8.19}$$

$$K_{PbSO_4} = a_{Pb^{2+}} . a_{SO_4^{2-}} = 10^{-7.75}. \tag{8.20}$$

A typical content of SO_4^{2-} in fresh waters is 30 mg/l, or 3×10^{-4} mole/l. At this level of sulfate, the maximum solubility of Pb is given by:

$$a_{Pb^{2+}} = \frac{10^{-7.75}}{3 \times 10^{-4}} = 6 \times 10^{-5} \simeq 12 \text{ ppm}. \tag{8.21}$$

Considering that the normal content of Pb in natural waters is 0·5–6 ppb (0·0005–0·006 ppm), as indicated in the Appendix, natural waters are undersaturated in anglesite by a factor of about 1000.

However, anglesite is not the least soluble Pb mineral at all values of pH. As illustrated in Fig. 8.9, cerussite is less soluble at pH values of about 6·5 and above. The precise relations are determined by the content of carbonate species and the amount of sulfate, as well as the pH. In normal stream waters at pH 8 and higher, Pb values are limited to about the observed range of background values. In most other environments, much higher Pb values are possible, but note that dissolved Pb contents in oxidizing ores are limited to about $10^{-6.5}$ M, and the dilution with normal waters will rapidly deplete this to background levels away from the orebody.

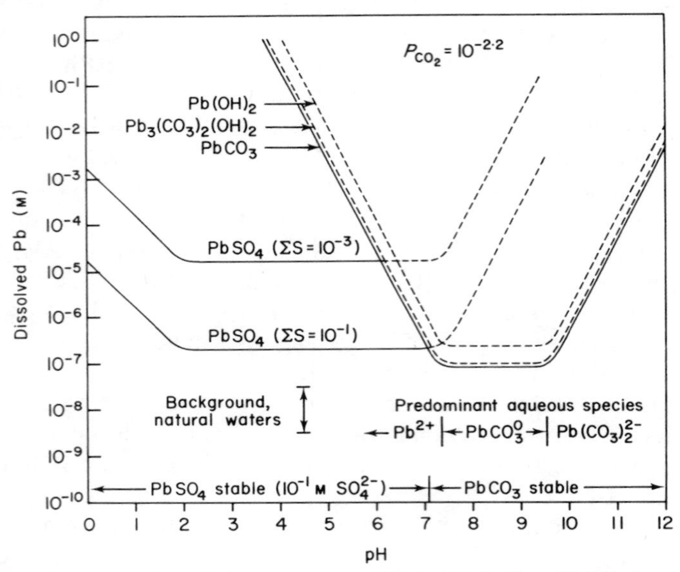

Fig. 8.9. Solubility of minerals in the system Pb–O–H–S–C at 25 °C, 1 atm under oxidizing conditions, $P_{CO_2} = 10^{-2.2}$ atm (typical of soil and ground water) at two levels of total sulfur. $PbCO_3$ (cerussite) and $PbSO_4$ (anglesite) are stable phases; $Pb_3(CO_3)_2(OH)_2$ (hydrocerussite) is nearly stable. Dashed lines show metastable solubilities.

The effects of S, C, and other constituents can be depicted on Eh–pH diagrams by selecting a value for the total content of the additional element, as was done for Fig. 8.9. For example, choosing a value of 10^{-3} M for total dissolved S, and a CO_2 pressure of $10^{-2.2}$, Fig. 8.10 shows the fields of Pb mobility (Pb $> 10^{-6}$ M) and Pb solubility as a function of Eh and pH. Note the large area in which PbS is very insoluble in the reducing portion of the diagram, as well as the area of $PbCO_3$. If the dissolved Pb content for mobility is decreased to 10^{-8} M, for example, the fields of the solids will contract and the fields of the ions will increase.

The mobility of some trace elements may be controlled in indirect Eh–pH effects. Hem (1977) showed that the Ag and Cu contents of many natural waters were consistent with precipitation of Ag and CuO at the Eh and pH conditions set by equilibrium between Fe-oxides and dissolved Fe^{2+}. This redox couple is common in many waters of intermediate oxidation state.

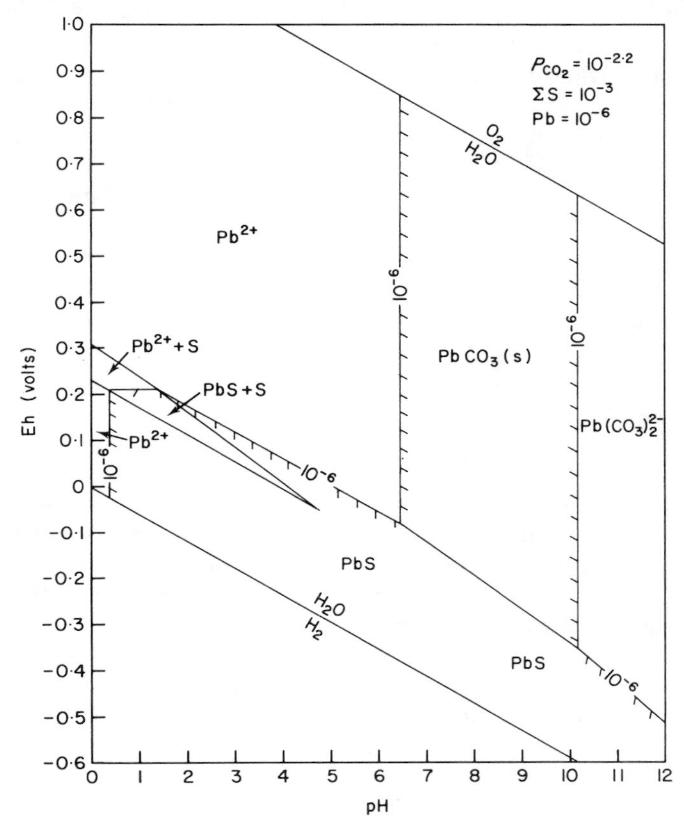

Fig. 8.10. Eh–pH diagram for the system Pb–O–H–S–C, $P_{CO_2} = 10^{-2.2}$ atm (oxidizing region only), $\Sigma S = 10^{-3}$ M, 25 °C, 1 atm. In hatched area, Pb is relatively immobile.

V. ADSORPTION AND ION EXCHANGE ON COLLOIDAL PARTICLES

Another major process governing the mobility of trace elements in the surficial environment is adsorption and ion exchange onto colloidal particles, which have diameters in the range 10–10 000 Å (10^{-7}–10^{-4} cm), and consequently have a large surface area for adsorption (Fig. 6.2). Iron and manganese oxides and hydroxides, organic matter, clays, and silica are the most common natural materials occurring as colloidal particles. Initially, the ions adsorbed on surfaces are in active equilibrium with the surrounding solution, but with time, the adsorbed ions may move into lattice sites or become overgrown by the solid. Under favorable conditions small amounts of colloidal materials can scavenge important amounts of dissolved elements from solution. Upon flocculation of the colloidal particles, the originally dissolved element is immobilized in soil or stream sediment or as a coating on a mineral grain (Jenne, 1977).

Most adsorption and exchange is caused by electrical charges on the surfaces or in the lattices of colloidal particles. Two types of charge distribution may be distinguished: materials having fixed charge, primarily the clay minerals; and those of fixed surface-potential, such as the Fe- and Mn-oxides (von Olphen, 1963).

A. Ion Exchange in Clay Minerals

Montmorillonite and vermiculite clay minerals have an inherent excess of negative charge in their silicate layers because of substitution within the silicate lattices. Illites and mixed-layer clay minerals share this type of charged lattice, but the amount of charge (exchange capacity) is considerably less than for montmorillonites and vermiculites. Cations enter the interlayer exchange sites in clay minerals only by exchanging for a cation already in the interlayer sites in order to maintain an electrically neutral state of the entire clay particle. Exchange of ions follows the following pattern (Wahlberg *et al.*, 1965):

$$\text{Ca–clay} + \text{Zn}^{2+} = \text{Zn–clay} + \text{Ca}^{2+} \tag{8.22}$$

$$\text{Ca–clay} + 2\text{Na}^{+} = \text{Na}_2\text{–clay} + \text{Ca}^{2+} \tag{8.23}$$

where Ca–clay, Zn–clay, and Na_2–clay indicate the clay mineral with Ca, Zn or Na in the exchange sites. In natural waters, clay minerals contain a mixture of ions in the interlayer sites.

Although the ratios of ions in the exchange sites can be estimated from the equilibrium constants for reactions like those listed above (Maes *et al.*, 1975), the amounts of various ions held in exchange sites are also dependent on the exchange capacity. Some values of exchange capacity for clay minerals and other natural materials are listed in Table 8.2. Montmorillonite, vermiculite, and organic matter have much larger exchange capacities than other materials, and if present, dominate the ion-exchange properties of the sediment or soil. A commonly observed sequence of increasing replacing power of the common simple ions on clays is

$$Li^+ < Na^+ < K^+ < Rb^+ < Cs^+ < Mg^{2+} < Ca^{2+} < Ba^{2+} < H^+.$$

In general, ions of higher valence are more strongly held in exchange sites than those of low valence, with the prominent exception of hydrogen ion.

Table 8.2
Exchange capacity of some clay minerals and common soils[a]

Minerals and soils	Cation-exchange capacity (mEq/100 g)[b]
Kaolinite	3–15
Halloysite	5–50
Montmorillonite	80–150
Illite	10–40
Chlorite	10–40
Vermiculite	100–150
Organic fraction of soils	150–500
Podzolic soils (U.S.A.)	5–25
Chernozem (U.S.S.R.)	30–60
Black cotton soil (India)	50–80
Latosol (Zambia)	2–10
Gley soil (Zambia)	15–25

[a] Source: Joffe (1949), Grim (1953), Mohr and Van Baren (1954), Tooms (1955), and Govett (1958).
[b] Milliequivalents per 100 g.

B. Adsorption on Surfaces

The surface charge of oxides, hydroxides, organic materials, and other types of colloids with constant surface potential results from ionization of surface atoms, or from chemical adsorption of dissolved ions onto the surface. For example, the OH on the surface of a Mn-hydroxide tends to dissociate in a manner analogous to other weak acids:

$$Mn\text{—}OH^0 = Mn\text{—}O^- + H^+(aq) \qquad (8.24)$$

where the Mn is part of the mineral structure and OH is on the surface. The surface thus acquires a negative charge (Parks, 1965). By dissociation of adsorbed water molecules, oxides form charged surfaces in a similar manner.

The extent of dissociation of surface OH or H_2O depends on pH, as shown in Fig. 8.11, and as predicted from eqn 8.24. For Mn-oxides, dissociation at pH 7 is extensive, and the surface of most natural colloidal Mn-oxides and -hydroxides has a strong negative charge. Under very acid conditions, an excess of H^+ is adsorbed to the surface, and it has a positive charge. At about pH 3, the surface is uncharged. This pH is termed the zero point of charge (ZPC). The lack of charge near this pH allows particles to closely approach each other and flocculate into larger particles which settle out. Because the charge on the surface is controlled by H^+ and OH^- in solution, these ions are termed potential-determining ions.

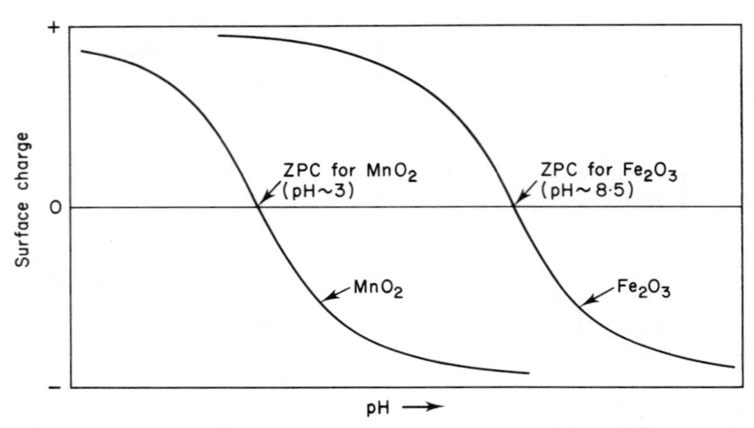

Fig. 8.11. Schematic diagram of surface charge on Mn- and Fe-oxides, as a function of pH.

For Fe-oxides, the ZPC is 6·5–8·5, so that natural Fe-oxides are usually positively charged (Parks and de Bruyn, 1962). Organic colloidal particles are usually negatively charged. Adsorption onto colloidal oxides, hydroxides, silica, and organic materials is also controlled by the number of charged sites, and charge balance may not be maintained. For example, adsorption of Co^{2+} and other divalent cations onto Mn-oxides takes place by a one-to-one exchange of ions (Murray, 1975a):

$$\text{Mn–OH}^0 + \text{Co}^{2+}(\text{aq}) = \text{Mn–O–Co}^+ + \text{H}^+(\text{aq}). \qquad (8.25)$$

The surface charge thus depends on the Co^{2+} concentration in solution, as well as on H^+ (Murray, 1975b), and Co is another potential-determining ion.

Adsorbed ions can occupy several types of sites on or near a charged colloidal particle. In solution, most ions are accompanied by a shell of water molecules held by electrostatic attraction for the H_2O dipole (Fig. 8.12a). Charged colloidal particles are surrounded by a cloud of oppositely charged hydrated ions (counterions) attracted by the charged surface (Fig. 8.12b). Exchange of these electrostatically attracted ions with the surrounding solution takes place rapidly and easily, with only minor discrimination between ions. The attraction of such ions by predominantly electrostatic forces is

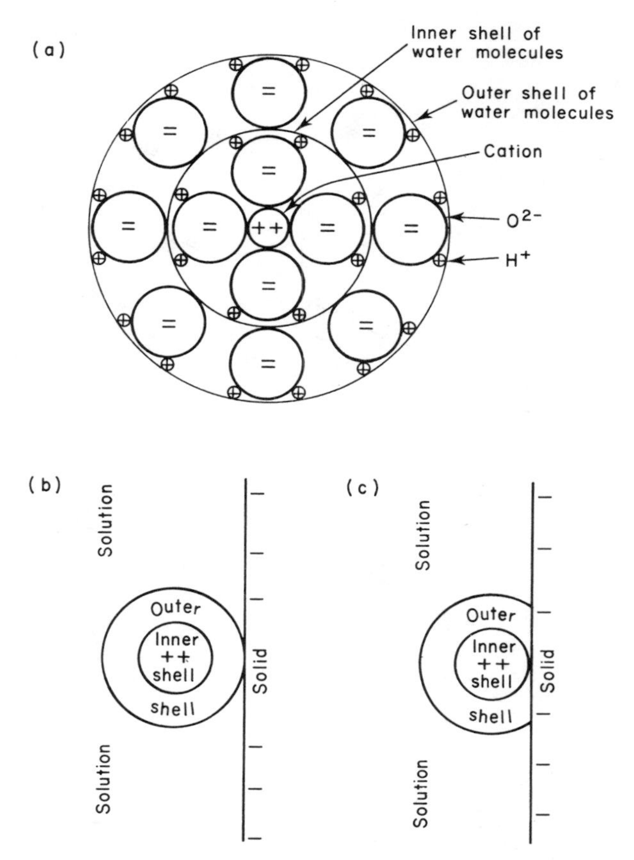

Fig. 8.12. (a) Schematic diagram of dissolved cation with strongly bonded inner shell of H_2O molecules and weakly bonded outer shell; (b) weak, non-specific adsorption in which attraction is predominantly electrostatic; (c) stronger, specific adsorption in which outer H_2O molecules and possibly some inner H_2O molecules are displaced, and chemical bonds may be operative.

termed non-specific adsorption. At the zero point of charge, these ions are not adsorbed, as in the case of Mg^{2+} on Mn-oxide (Fig. 8.13a).

Some ions, expecially those of higher valence or more complex electronic structure, such as the transition metal ions, are more strongly attracted to the surface of colloids and lose part or all of their envelope of water molecules (Fig. 8.12c). Chemical bonds specific to the ions and the surface may be operative. This process is termed specific adsorption. Co^{2+}, Mn^{2+}, Cu^{2+}, and, to a lesser degree, other divalent ions show this behavior on Mn-oxide, as demonstrated by significant adsorption at the zero point of charge (Fig.

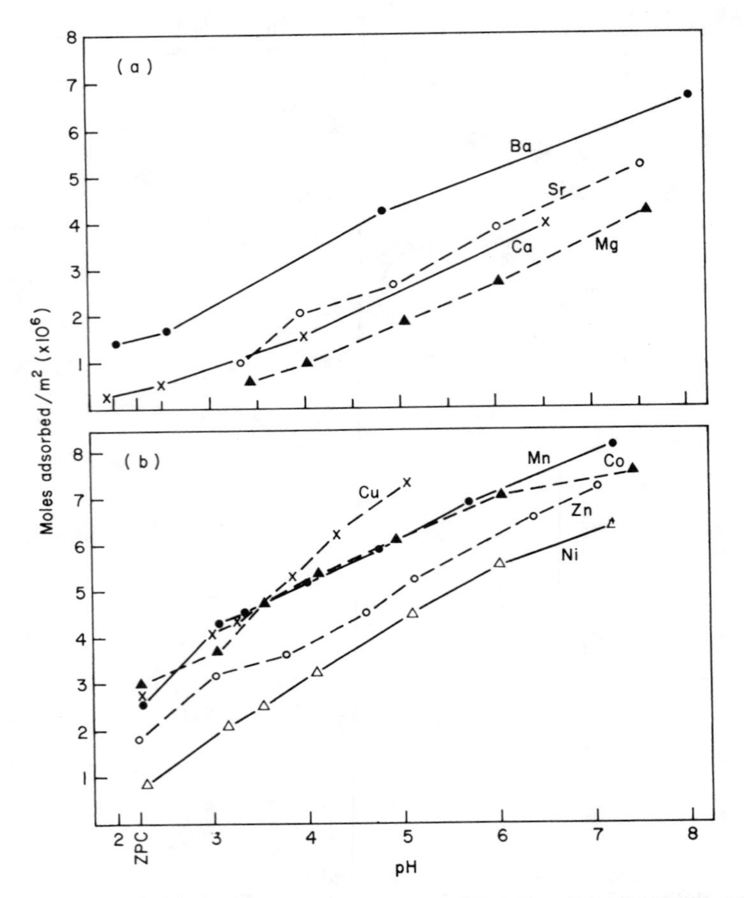

Fig. 8.13. Adsorption of divalent cations on MnO as a function of pH. (a) Alkaline earths; (b) transition metals. Cation concentration = 10^{-3} M. (After Murray (1975a). Reprinted with permission from *Geochim. Cosmochim. Acta*, © Pergamon Press Inc., 1975.)

8.13). The amount adsorbed can be expressed by the equilibrium constant for the adsorption reaction; for example, for eqn 8.25:

$$K^{Co}_{MnO_2} = \frac{a_{R-Co^+} \cdot a_{H^+}}{a_{R-H^0} \cdot a_{Co^{2+}}} \tag{8.26}$$

where a_{R-Co^+} is the activity of adsorbed Co on the surface of the MnO_2, and a_{R-H^0} is the activity of surfaces with undissociated OH on their surface. If the activity of MnO_2 with undissociated OH on its surface is considered to be relatively constant, then eqn 8.26 may be rearranged to give:

$$\log \frac{a_{Co^{2+}}}{a_{R-Co^+}} \simeq \log \frac{m_{Co^{2+}}}{m_{Co(ads)}} \simeq \log a_{H^+} + \log K' \qquad (8.27)$$

where $m_{Co}{}^{2+}$ and $m_{Co(ads)}$ are the molar concentrations of dissolved and adsorbed Co. Experimental data commonly follow this type of relation (Fig. 8.14), which is called a Kurbatov plot.

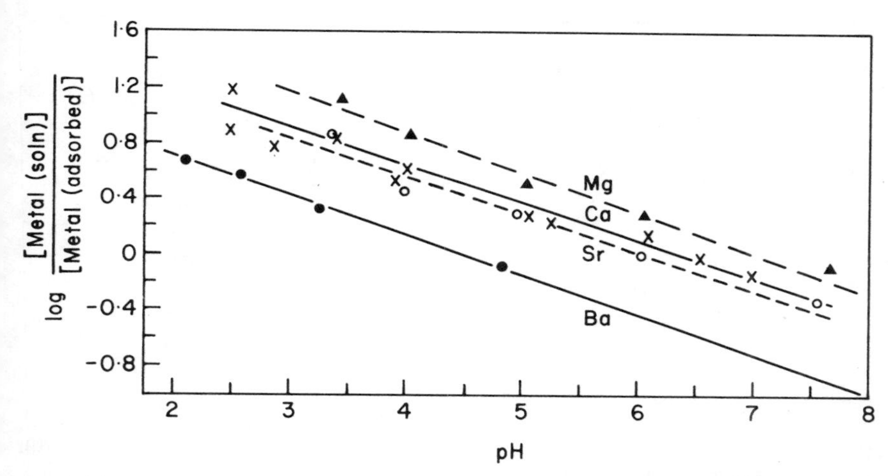

Fig. 8.14. Kurbatov plots for the alkaline earth metal ions on Mn-oxide. Metal concentrations = 10^{-3} M. (After Murray (1975a). Reprinted with permission from *Geochim. Cosmochim. Acta*, © Pergamon Press Inc., 1975.)

As indicated by the above relationships, the amount of adsorbed Co, which at low concentrations is proportional to a_{R-Co^+}, is dependent on the following factors:

(i) The activity of Co^{2+} ions in the surrounding solution (taking account of any complexing).
(ii) The pH of the surrounding solution.
(iii) The energy of bonding, as expressed by the equilibrium constant.
(iv) The number or density of adsorption sites, as expressed in the example by a_{R-H^0}.
(v) In more complex cases, the activity of other ions competing with Co^{2+} and H^+ for the sites.

The energy of adsorption varies from one ion to another, as suggested by greater adsorption for some elements than others, given similar concentrations in solution (Fig. 8.13). The relative strength of adsorption for some elements on Mn-, Fe-, and Al-oxides and hydroxides is summarized in Table 8.3. In general, divalent ions are more strongly sorbed than univalent ions, and transition metals more than non-transition metals. Well-crystalline materials may show different adsorption preferences than amorphous or poorly crystalline solids (Kinniburgh *et al.*, 1976).

Table 8.3
Relative affinity for adsorption of some cations[a]

	Mn-oxides	Amorphous Fe-oxides	Goethite	Amorphous Al-oxides	Humic substances (1)	Humic substances (2)
Greatest	Cu^{2+}	Pb^{2+}	Cu^{2+}	Cu^{2+}	Ni^{2+}	Cu^{2+}
	Co^{2+}	Cu^{2+}	Pb^{2+}	Pb^{2+}	Co^{2+}	Ni^{2+}
	Mn^{2+}	Zn^{2+}	Zn^{2+}	Zn^{2+}	Pb^{2+}	Co^{2+}
	Zn^{2+}	Ni^{2+}	Co^{2+}	Ni^{2+}	Cu^{2+}	Pb^{2+}
	Ni^{2+}	Cd^{2+}	Cd^{2+}	Co^{2+}	Zn^{2+}	Ca^{2+}
	Ba^{2+}	Co^{2+}		Cd^{2+}	Mn^{2+}	Zn^{2+}
	Sr^{2+}	Sr^{2+}		Mg^{2+}	Ca^{2+}	Mn^{2+}
	Ca^{2+}	Mg^{2+}		Sr^{2+}	Mg^{2+}	Mg^{2+}
Least	Mg^{2+}					

[a] Source: Mn-oxides, Murray (1975a); amorphous Fe- and Al-oxides, Kinniburgh *et al.* (1976); goethite, Forbes *et al.* (1976); humic substances, (1) Schnitzer and Hanson (1970), and (2) Gamble and Schnitzer (1973).

All ions in solution compete for adsorption sites on the surfaces of colloidal particles, and the equilibrium population of adsorbed ions is a composite of many adsorption reactions like eqn 8.25. At near-neutral pH, where H^+ is at low concentrations, Ca^{2+} or Na^+ may be the most important ions competing with trace elements for adsorption to the surface of particles, and exchange reactions may be more conveniently written with Ca^{2+} rather than H^+ as the competing ion. Complex ions may also be adsorbed. There is some evidence that hydroxy complexes of ions (such as $PbOH^+$) are more strongly adsorbed than simple ions (Stumm and Morgan, 1970).

Experiments show that specific sorption is generally not completely reversible. The amount of an ion adsorbed follows the simple relations described above as a cation is added to the solution, but when the amount in solution is decreased, only part of the expected amount is desorbed (Fig. 8.15). Ions initially in exchange sites apparently become permanently fixed in oxides and other phases. Possible causes for this behavior are diffusion of the adsorbed ion into a lattice site, recrystallization of the oxide to leave the adsorbed ion within the lattice rather than on its surface, and loss of coordinating water molecules so that the ion is directly bonded to the surface. Drying of colloidal phases is one way of accomplishing the last process. The adsorption properties of dried samples may be very different from the original natural sample.

Among the factors determining the abundance and nature of adsorption sites are the mineralogical identity of the particle and its crystallinity. Amorphous oxides generally have the largest surface area per unit mass. The several polymorphs of Mn-oxide and -hydroxide (pyrolusite, birnessite, todorokite, etc.) differ in adsorption properties. Further differences are expected from the non-stoichiometric nature of many oxides. For example, Murray (1975a) experimented with a hydrous phase of stoichimetry $MnO_{1\cdot93}$. Recrystallisation and dissolution–precipitation of colloidal Fe–Mn-oxides can have major effects on adsorbed elements (Jenne, 1968).

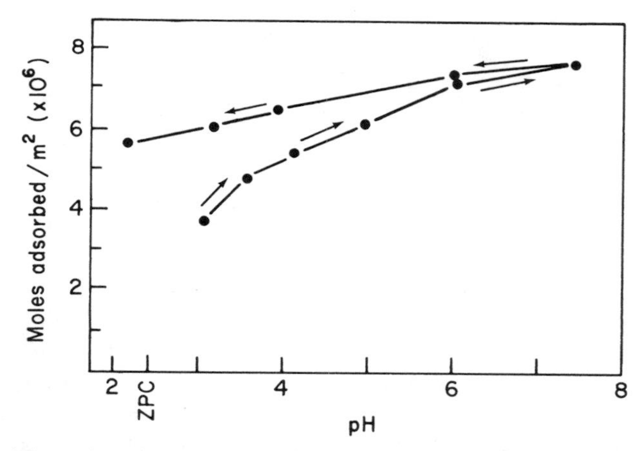

Fig. 8.15. Adsorption–desorption results with changing pH for Mn-oxide exposed to solutions of 10^{-3} M Co. (After Murray (1975a). Reprinted with permission from *Geochim. Cosmochim. Acta*, © Pergamon Press Inc., 1975.)

In clay minerals, dissolved ions may be adsorbed to sites on the edges of the sheets. The amount adsorbed to these sites is subordinate to interlayer adsorption in montmorillonites and vermiculites, but predominates over interlayer adsorption in kaolinites and some illites. Figure 8.16 shows the extent of adsorption of Cu on kaolinite, presumably on the broken bonds of the edges of sheets. A dependence on both pH and Cu content of the solution is observable. Adsorption of Cu and Co on illites by this process is extensive enough that most of these trace elements can be transported in adsorbed form in a solution with 0·5% suspended illite (O'Connor and Kester, 1975).

Table 8.4 shows the relative enrichment of certain minor elements in Fe- and Mn-oxides. In a parallel manner, the limonite of gossans and hydrous oxide precipitates in surface drainage channels tends to become preferentially

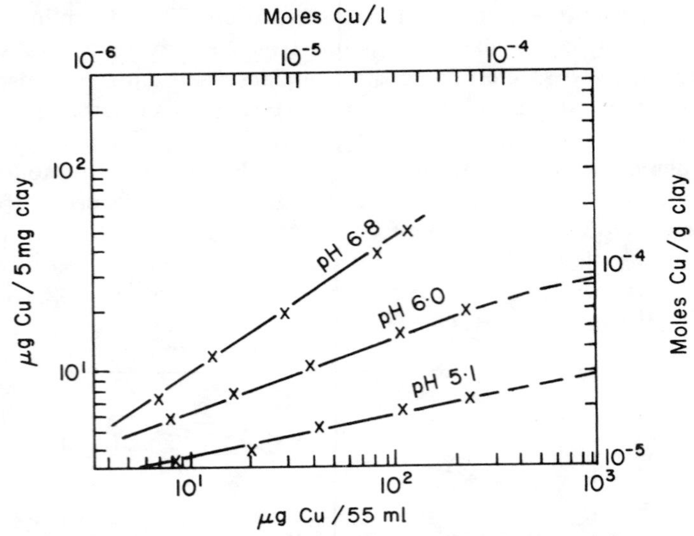

Fig. 8.16. Adsorption of Cu on kaolinite as a function of pH and Cu concentration of the solution. (After Heydemann (1959). Reprinted with permission from *Geochim. Cosmochim. Acta*, © Pergamon Press Inc., 1959.)

enriched in many minor elements. Figure 8.17 shows the relation of Pb and Zn to Fe in the active sediments of the Colden River, Isle of Man. No known sulfide mineralization occurs in this drainage basin, so that the Pb and Zn now occurring with the Fe almost certainly represent enrichment from normal background concentrations. The same association of many minor elements with Fe and Mn is observed in soils. Figure 8.18 shows the correlation of Co with Mn in soils over unmineralized rocks in Zambia. Jenne

Table 8.4
Concentrations of some minor elements in iron oxide and manganese oxide sediments[a]

Element	Content in average crustal rocks (ppm)	Content in Fe-oxide sediments (ppm)	Content in Mn-oxide sediments (ppm)
As	2	10–700	70
Ba	580	90–370	1000–7000
Cu	50	180	2000–20 000
Mo	1·5	—	300–3000
Ni	75	20–2000	1600–2000
Se	0·1	0·5–5·0	—

[a] Source: Krauskopf (1955) and Table 2.7.

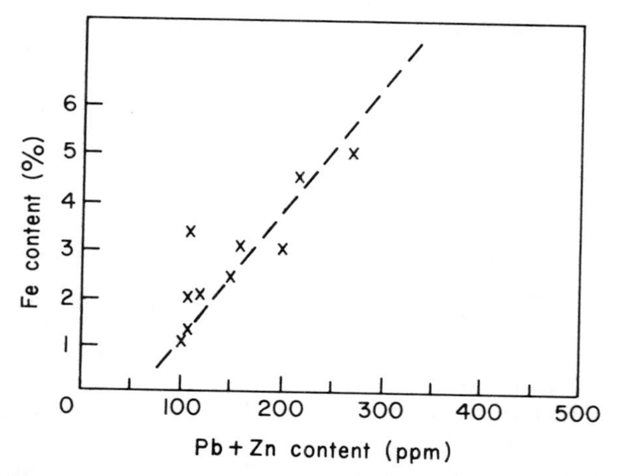

Fig. 8.17. Relation of Pb and Zn to Fe in modern stream sediments of the Colden River, Isle of Man. Data on −80-mesh fraction. (From de Grys, 1959.)

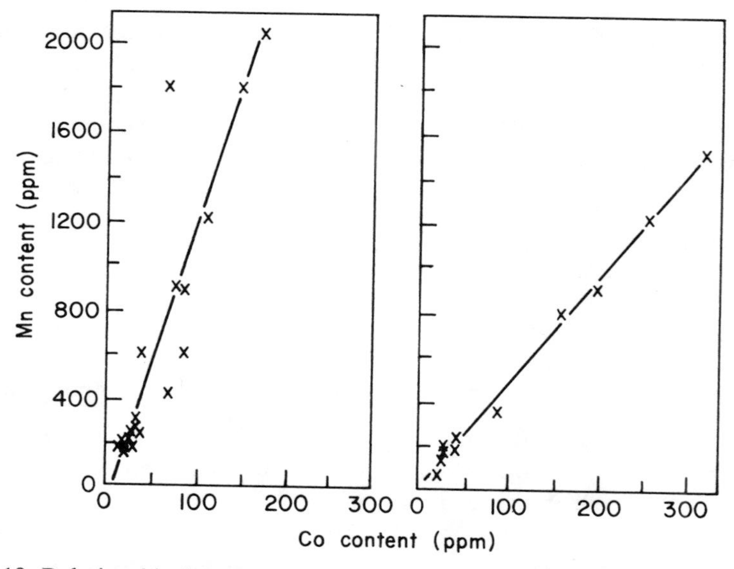

Fig. 8.18. Relationship between Mn and Co in soils from Zambia. Data on −80-mesh fraction. (After Jay, 1959.)

(1968) has reviewed evidence showing that Fe- and Mn-oxides are major controls on the content of Co, Ni, Cu, and Zn in soils and waters.

VI. ORGANIC MATTER

Organic matter is a major constituent of some soils, stream sediments, and natural waters, and is present in at least traces in most such materials. Because of its unusual properties compared to other natural materials, even trace amounts of organic materials can have important effects on the chemistry of trace elements. These effects include complexing of trace ions by dissolved organic matter, resulting in increased mobility of the element; adsorption or formation of organic compounds, resulting in immobilization; and reduction to lower valence states, with resulting changes in chemical properties.

Organic matter normally constitutes less than 10% by weight of soils and stream sediments, but reaches much higher abundances in some environments, especially in the A horizon of some soils and in relatively stagnant or waterlogged zones, including the bottom sediment of many lakes. Natural waters contain 0·1–200 mg/l of dissolved organic C, averaging about 10 mg/l, compared to an average of 120 mg/l of inorganic solutes (Stumm and Morgan, 1970, p. 283; Beck *et al.*, 1974; Livingstone, 1963; Reuter and Perdue, 1977).

Many types of organic compounds have been recognized in natural materials. The distribution and nature of these is evidently very complex (Swain, 1970; Cranwell, 1975). Major types of natural organic compounds include the following:

(i) carbohydrates
(ii) hydrocarbons
(iii) alcohols, organic acids, fats, waxes
(iv) amino acids, proteins
(v) humic substances
(vi) porphyrins, vitamins, pigments, and other complex compounds.

Most of these compounds are found in or associated with living organisms. The exceptions are the humic substances which are degraded and modified from organic matter in living organisms.

A. Simple Organic Compounds

Some basic building blocks for organic materials are illustrated in Fig. 8.19. Carbohydrates with the formula $C_xH_{2y}O_y$ include sugars, such as glucose, and larger molecules, such as cellulose, a polysaccharide. Hydrocarbons composed only of C and H include aliphatic (chain-type) compounds like methane, ethane, etc., and aromatic (ring-type) compounds, like benzene. Alcohols have an —OH substituted for an —H, as in methyl alcohol and phenol. Organic acids contain the carboxyl group (—COOH). Examples are acetic, oxalic, and citric acids. Amino acids contain an amine group ($—NH_2$) and a carboxyl group attached to an organic radical of simple or complex form.

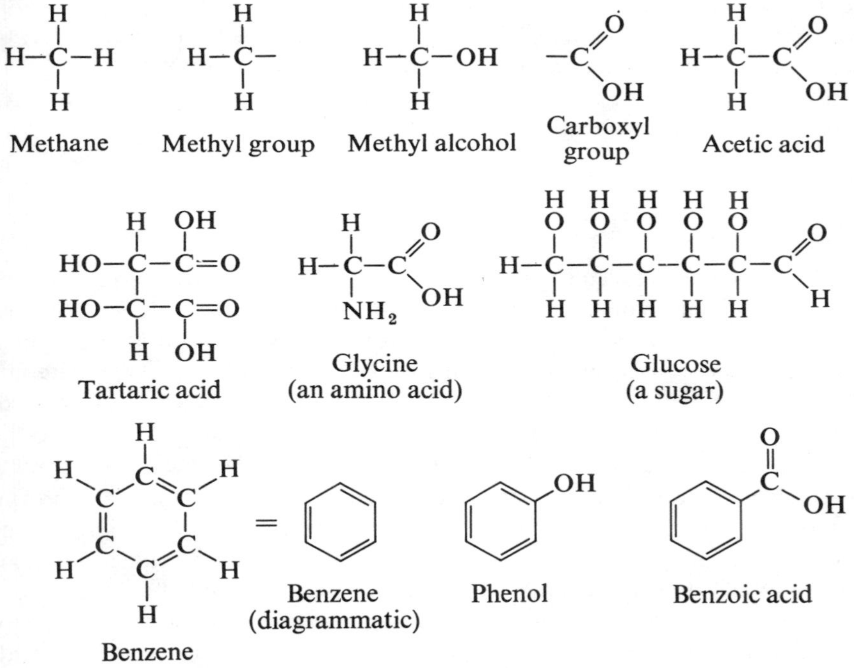

Methane — Methyl group — Methyl alcohol — Carboxyl group — Acetic acid

Tartaric acid — Glycine (an amino acid) — Glucose (a sugar)

Benzene — Benzene (diagrammatic) — Phenol — Benzoic acid

Fig. 8.19. Some organic compounds and groups.

Metals may be bound to organic matter in several ways. Organic acids containing —COOH, —OH, or similar groups may form organic salts in which the metal occupies the place of the ionizable H^+. This type of binding is generally of moderate strength. The metal may be bonded directly to carbon atoms, forming organometallic compounds, or to N, O, P, or S or other electron-donating atoms in an organic compound. Bonding is generally strong in such compounds. In many organic compounds, the metal is chelated, that is, it is bonded to two or more sites on the organic molecule, as for Fe in hemoglobin, and various metals in EDTA complexes (Section III of Chapter 3). Chelated metals are generally very strongly bound (Martell, 1971). Metals may also be electrostatically attracted to the vicinity of charged colloidal particles. This attraction is usually relatively non-selective and weak.

Some organic acids have considerable chelating power for metals. Silicate minerals are dissolved and mobilized much more rapidly in solutions of citric, salicylic, tannic, and other acids than in inorganic acids of the same pH (Huang and Keller, 1971). Similar increases in solubility of Cu, Pb, Zn, Fe, and Al are reported (Ong *et al.*, 1970). Lichens acquire nutrients from their substrate by attack with a complex organic acid, and fungi produce citric and oxalic acids. The concentrations of organic acids in pore solutions of soils range from about 10^{-3} M for simple organic acids such as acetic, to 10^{-5} M for aromatic and amino acids (Stevenson and Ardakani, 1972). The movement of Fe and Al out of the surface horizons of soils probably involves chelation by organic acids.

B. Humic Substances

Although much detailed chemical work has been done on traces of the above types of organic compounds in natural materials, the bulk of organic material in most soils, sediments, and natural waters is probably composed of humic substances. However, because of their molecular complexity, these materials have proven less easy to understand. Humic substances are dark-colored, acidic, complex organic materials of high molecular weight that lack the specific chemical and physical characteristics of simple organic compounds (Schnitzer, 1976). They form by the degradation of plant materials and from synthetic activities of microbes (Gamble and Schnitzer, 1973). They range in molecular weight from several hundred to perhaps several thousand (compared to less than a few hundred for compounds discussed above), and contain predominantly aromatic (ring-type) components. Humic substances are divided into three groups: humic acids, which are soluble in dilute alkali (NaOH) but are precipitated on acidification of the extract; fulvic acids, which remain in solution on acidification of the alkaline extract; and humin, which is the humic material insoluble in both dilute base and acid. The three types are similar chemically, with the fulvic acid tending to be lower in molecular weight and having a higher content of —COOH, —OH, and —C=O groups than the other types. Figure 8.20 illustrates a partial chemical structure of a fulvic acid, based on the types of compounds found on partial breakdown of samples. The abundant —COOH and —OH groups give the humic and fulvic acids their acid properties, and are also active in adsorbing, complexing, and chelating metals. When dissolved in water, the fulvic and humic components are typically negatively charged (Gamble and Schnitzer, 1973). Humic substances are a complex mixture of components, so that

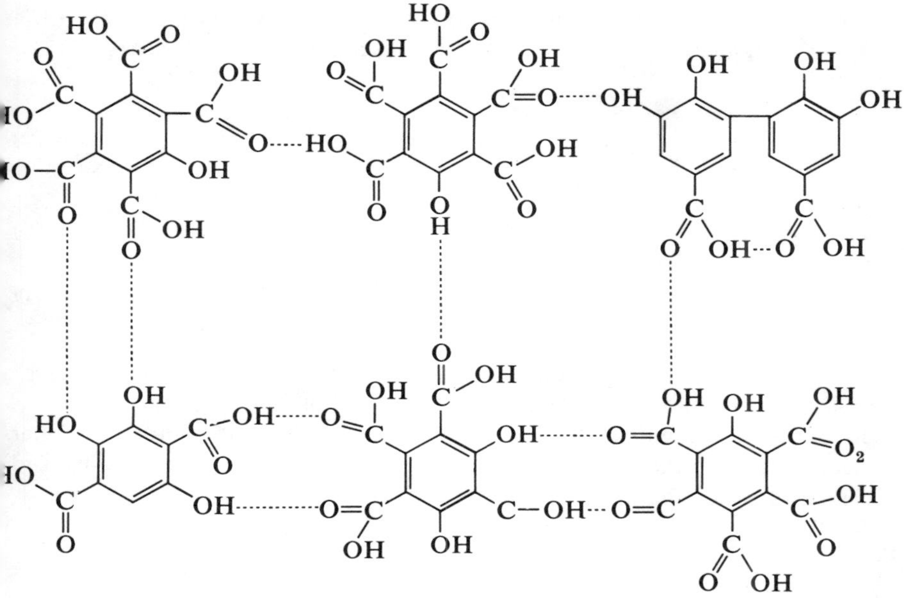

Fig. 8.20. A partial chemical structure for a fulvic acid. (After Gamble and Schnitzer (1973). Reprinted with permission from "Trace Metals and Metal–Organic Interactions in Natural Waters" (P. Singer, ed.), © Ann Arbor Science Publishers, 1973.)

chemical and physical properties are those of mixtures rather than simple compounds.

Humic substances are estimated to constitute 60–80% of the dissolved organic matter in river water, and to occur in concentrations of 5–150 mg/l, equivalent to molar concentration of about 10^{-5}–10^{-4} (Reuter and Perdue, 1977). The humic substances in water exist in particles ranging from about 10^{-3} μm to greater than 1 μm in diameter. The smaller particles can be considered as dissolved, but an arbitrary under limit must be set to distinguish these from colloidal materials. Particles less than 0·45 μm are usually considered dissolved, because easily usable filters are available at this size, but many colloidal materials are included as dissolved using this definition. A variety of processes, including decreases in pH and increases in total ionic concentrations, or in concentrations of cations capable of neutralizing the charge on the complex, can decrease solubility of the colloidal material and cause precipitation. Humic materials can also occur as coatings on particles of clay, soil, or sediment, and because of their large surface area, have effects out of proportion to their mass.

Reactions of humic and fulvic acids (HA) with metals are probably of the following types:

$$HA^0 + Me^{2+} = MeA^+ + H^+ \tag{8.28}$$

$$HA^0 + MeA^+ = MeA_2^0 + H \tag{8.29}$$

or the overall reaction

$$2HA^0 + Me^{2+} = MeA_2^0 + 2H^+ \tag{8.30}$$

$$K = \frac{a_{MeA_2} \cdot a_{H^+}^2}{a_{HA}^2 \cdot a_{Me^{2+}}} \tag{8.31}$$

where Me^{2+} is a divalent ion in solution, and the HA and MeA_2 are large molecules of humic acid which dissociate in dilute solution and behave as dissolved constituents at low values of metal content and acidity. At higher metal contents, the humic units tend to flocculate into large chains and networks, becoming immobilized.

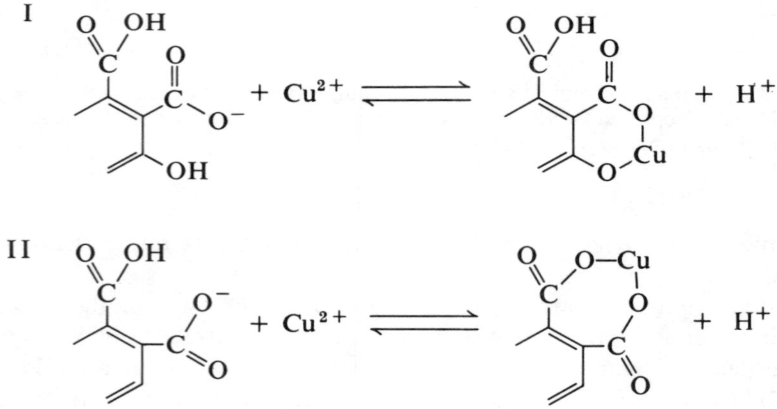

I [structure: O, OH, C, O, C, O⁻, OH] $+ Cu^{2+} \rightleftharpoons$ [structure: O, OH, C, O, C, O, Cu, O] $+ H^+$

II [structure: O, OH, C, O⁻, C, O] $+ Cu^{2+} \rightleftharpoons$ [structure: O, O—Cu, C, O, C, O] $+ H^+$

Fig. 8.21. Bonding of metal in humic substances by chelation. (After Gamble and Schnitzer (1973). Reprinted with permission from "Trace Metals and Metal–Organic Interactions in Natural Waters" (P. Singer, ed.), © Ann Arbor Science Publishers, 1973.)

Experimental work indicates that metals are bound in humic substances both by chelation (multiple bonds) and by complexing to a single site on the molecules. A probable type of chelation reaction is indicated in Fig. 8.21. The relative strength of bonding varies in the following order for typical humic substances (Gamble and Schnitzer, 1973):

$$Cu > Ni > Co > Pb = Ca > Zn > Mn > Mg.$$

The above sequence represents the average relative bond strength on some fulvic acids, but because the metals are bound in many different types of sites on the organic compounds, no single equilibrium constant can describe the process; the values are merely averages for a specific level of metal content. Individual sites on a humic substance may have very different affinities for metals, depending on the character of the compound, the pH, the presence of competing ions, the content of the ion in question, and other factors.

Chlorophyll and hemoglobin are examples of organic compounds in which metals are strongly chelated by four organic groups. Porphyrins containing V, Ni, and other elements have similar structures in which the metals are strongly bonded and effectively immobilized. Organic compounds containing U are thought to form strong bonds with this element in some environments (Schmitt-Colerus, 1967, 1969).

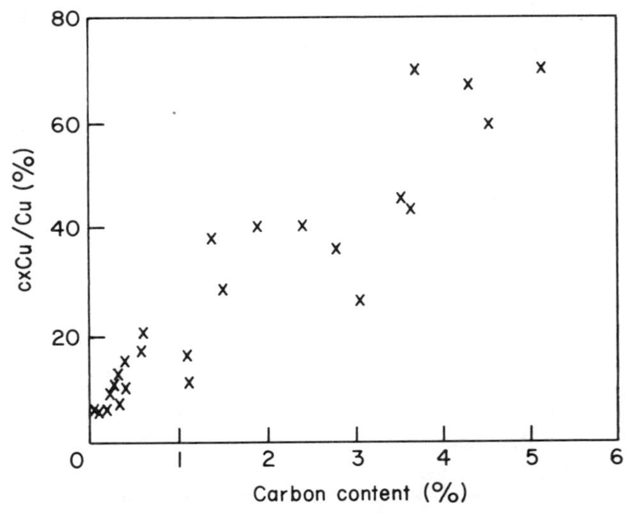

Fig. 8.22. Relation of carbon content to extractability of Cu in soils from Zambia. Data on −80-mesh fraction. (After Govett, 1960.)

As an example of the effects of complexing and immobilization by organic matter, Fig. 8.22 shows the clear relationship of extractable Cu to organic content of soils in Zambia. In stream sediments, U shows a similar correlation with organic content (Rose and Keith, 1976). The amount of adsorbed U on the organic matter of stream sediments, peats, and other natural materials follows the general relation

$$\text{U (organic matter)} = 10\,000\text{U (water)}$$

as determined by Szalay (1964) and illustrated in Fig. 8.23.

The opposite effect, namely mobilization of elements by formation of humic complexes, has also been documented (Rashid and Leonard, 1973; Bondarenko, 1968). In most instances this mobilization occurs by complexing of the metal, thereby increasing its solubility, but Ong and Swanson (1969) report that colloidal Au acquires a coating of organic material which transforms it into a stable colloid which is then capable of mobility. In typical natural waters, about half of the Cu^{2+} is expected to occur in complexes with fulvic acid, even given the competition by major cations (Reuter and Perdue, 1977).

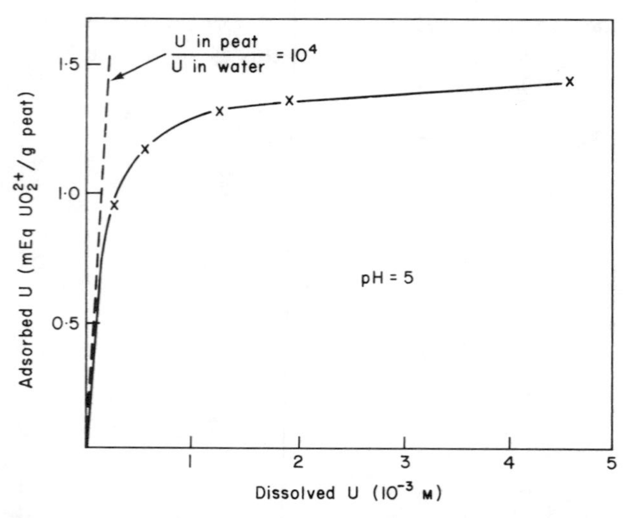

Fig. 8.23. Adsorption of U to peat, showing strong adsorption at low concentrations of U, with an enrichment of 10 000 times in the peat. At higher concentrations, all the adsorption sites are filled. (After Szalay (1964). Reprinted with permission from *Geochim. Cosmochim. Acta*, © Pergamon Press Inc., 1964.)

VII. ELECTROCHEMICAL DISPERSION

In general, ground waters and pore solutions at depth in the earth are relatively reducing compared to oxygenated waters near the surface. As a result, a vertical gradient in electrochemical potential (Eh) exists near the surface of the earth. The steepest gradient is commonly near the water table, above which gaseous oxygen may be present in pore spaces.

Sulfide ore minerals, graphite, and a few other materials of metallic luster are good conductors of electrons compared to silicates, carbonates, and most oxides. These natural conductors tend to "short-circuit" the usual gradients in electrochemical potential. Near orebodies, the high conductivity of the ore depletes electrons from the deep environment and supplies them to

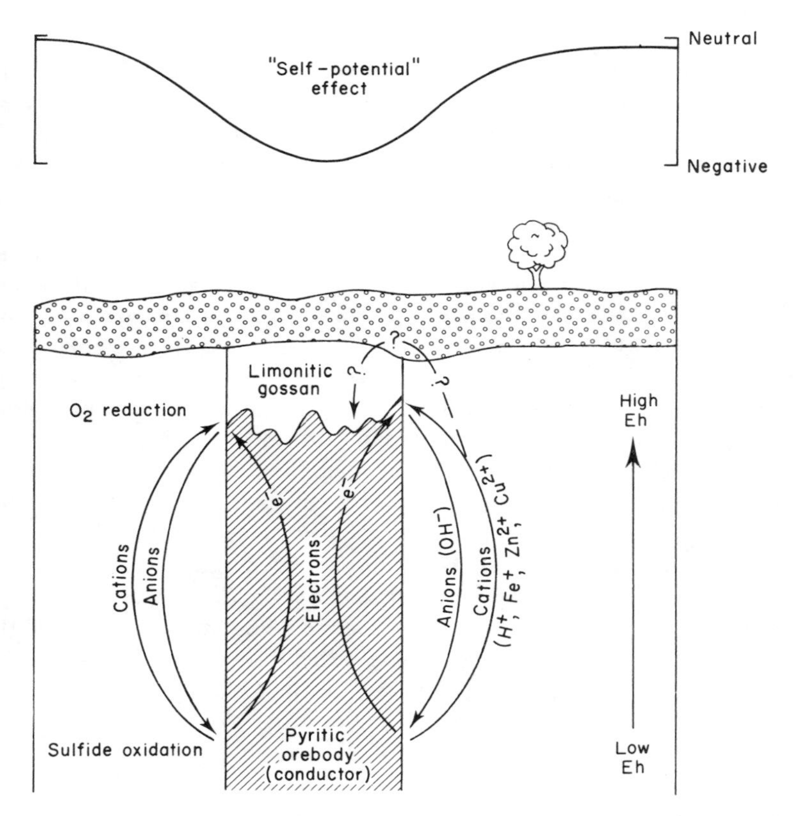

Fig. 8.24. Schematic diagram of oxidizing orebody, showing flow of electrons and ions.

near-surface regions (Sato and Mooney, 1960). The top of an orebody is thus negatively charged relative to its surroundings, and the bottom is positively charged. Figure 8.24 illustrates the flow of electrons and ions in and near a conducting orebody, where the orebody acts as a conductor connecting two electrochemical cells of differing oxidation potential. Oxidized substances approach the upper end of the orebody, receive electrons, and are reduced. Sato and Mooney (1960) suggest that the most common reactions here are:

$$O_2 + 2H^+ + 2e^- \longrightarrow H_2O_2 \longrightarrow H_2O + \tfrac{1}{2}O_2 \qquad (8.32)$$

$$Fe^{3+} + e^- \longrightarrow Fe^{2+}. \qquad (8.33)$$

At the deep end of the orebody, reduced substances are oxidized by loss of electrons to the orebody:

$$Fe^{2+} \longrightarrow Fe^{3+} + e^-. \qquad (8.34)$$

Metallic cations such as Pb^{2+}, Zn^{2+}, or Cu^{2+} may be released at depth in the ore by this process. In order to maintain charge balance at the two ends of the conductor, a current equal to the electronic current must be carried by migration of ions. Cations must migrate upward and anions downward. Govett (1973a) and Bølviken and Logn (1975) have suggested that around massive base-metal sulfide deposits, part of this current is carried by base-metal cations (Pb^{2+}, Zn^{2+}, Cu^{2+}) which are electrochemically driven upward from the ore where they may be fixed by adsorption or precipitation and give rise to anomalies, which are observed to occur even in post-ore till and alluvial deposits above massive sulfides. However, Webber (1975) suggests that the rates of ionic migration by this electrochemical process are much slower than solution flow, and that the mechanism may not be very effective. Although electrochemically driven dispersion clearly occurs, further research is needed to clarify the magnitude of such dispersion.

Electrochemical effects are also important at a more detailed scale during oxidation of sulfide ores. The minerals stable at the lowest Eh values, such as sphalerite and galena, tend to be oxidized preferentially relative to those stable at higher Eh, such as pyrite and chalcopyrite (Shvartsev, 1976).

The use of electrical surveys to identify conducting minerals and map their distribution has attracted considerable recent attention. Electrochemical effects constitute at least part of the basis for these surveys, which are therefore closely related to geochemical exploration (Angoran and Madden, 1977).

VIII. SELECTED REFERENCES ON CHEMICAL EQUILIBRIA IN THE SURFICIAL ENVIRONMENT

General	Garrels and Christ (1965) Stumm and Morgan (1970) Mortvedt *et al.* (1970) Hem (1970) Marshall (1964)

Adsorption	James and MacNaughton (1977) Parks (1975)
Metal complexes	Baes and Mesmer (1976) Sillen and Martell (1964, 1970) Smith and Martell (1976) Stumm and Morgan (1970, Ch. 6)

Chapter 9

Mechanical and Biological Dispersion in the Surficial Environment

★★★★★★★

The chemical equilibria that were discussed in the last chapter determine whether a given element or group of elements is dispersed in the surface environment principally in solution in natural waters, or as components of the clastic fragments of soils and stream sediments. This chapter reviews the specific mechanisms whereby this complex of fluid and solid materials actually moves under the influence of various mechanical and biological factors.

I. MECHANICAL FACTORS

Dispersion by movement of clastic fragments is largely restricted to the surface of the regolith where erosion is in active progress. The principal force responsible for mechanical dispersion is gravity, acting either directly on soil and loose debris or through the media of flowing water and ice. Wind action and animals are contributory agents and locally may even assume a dominant role. Mechanical dispersion by volcanism has no general significance in prospecting and is not considered here.

A. Simple Gravity Movement

Under the influence of gravity, solid components of the overburden tend to move downhill either by slow lateral creep or by more rapid landsliding. A classification of these phenomena, based on the kind and rate of movement, is presented in Fig. 9.1.

Even on moderate to gentle slopes, there is a continual, imperceptible flow of rock debris and soil in the downslope direction. In general terms, the rate of lateral movement within the regolith progressively increases from bedrock to surface. As a result, vestigial rock structures in the moving overburden tend to bend over in the direction of movement, as illustrated diagrammatically in Fig. 9.2. The distribution of residual weathering products related to an underlying metalliferous deposit may be similarly affected, as illustrated by the data of Fig. 9.3.

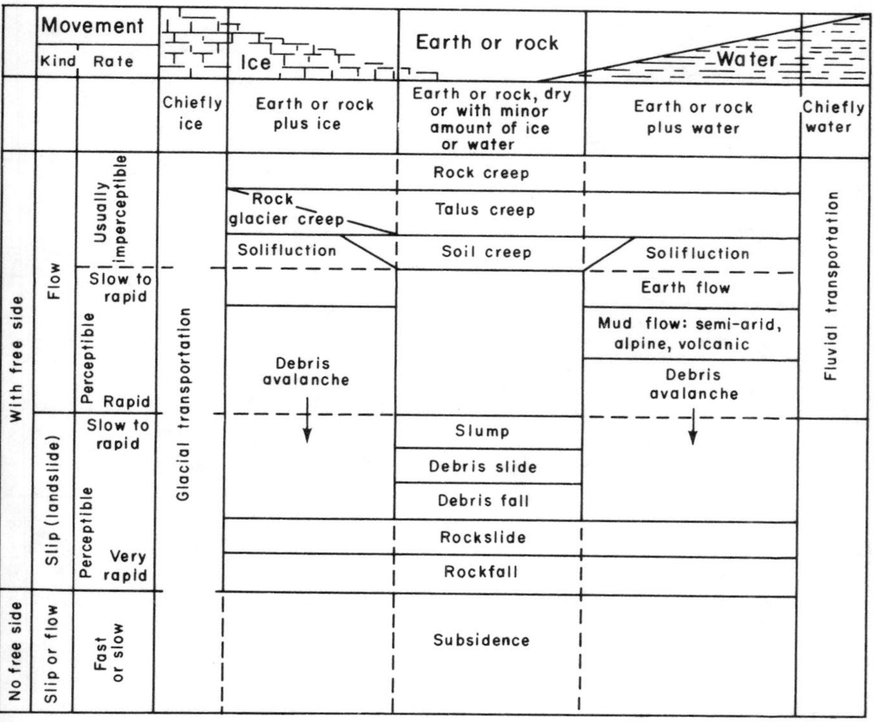

	Movement: Kind	Movement: Rate	Ice		Earth or rock	Water	
			Chiefly ice	Earth or rock plus ice	Earth or rock, dry or with minor amount of ice or water	Earth or rock plus water	Chiefly water
With free side	Flow	Usually imperceptible	Glacial transportation		Rock creep		Fluvial transportation
				Rock glacier creep	Talus creep		
				Solifluction	Soil creep	Solifluction	
		Perceptible: Slow to rapid				Earth flow	
				Debris avalanche		Mud flow: semi-arid, alpine, volcanic	
		Rapid				Debris avalanche	
	Slip (landslide)	Perceptible: Slow to rapid			Slump		
					Debris slide		
					Debris fall		
					Rockslide		
		Very rapid			Rockfall		
No free side	Slip or flow	Fast or slow			Subsidence		

Fig. 9.1. Classification of mechanical factors in the dispersion of weathering products. (After Sharpe, 1938.)

The presence of vegetation tends to stabilize the overburden and so reduce the rate of creep. In those areas where the vegetation is dense and shallow rooted, maximum movement may in fact take place immediately below the surface layer of matted roots.

In temperate and humid climates creep is facilitated by the lubricating effect of soil moisture. Alternate freezing and thawing of interstitial water, or alternate wetting and drying of the soil, tend to facilitate the downslope movement of clastic material.

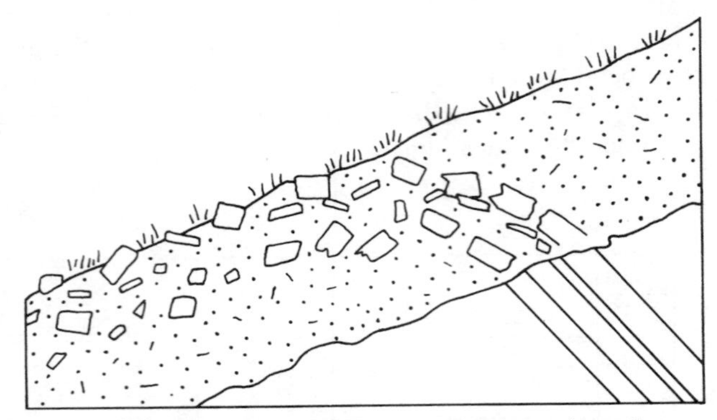

Fig. 9.2. Dispersion of resistant rock fragments by soil creep.

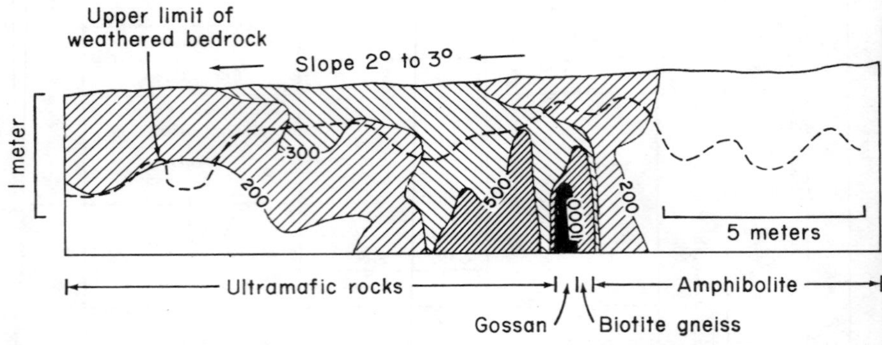

Fig. 9.3. Section showing predominantly mechanical dispersion of Cu in soil near Cu-rich gossan, Magogophate, Botswana. Data on −80-mesh fraction. (After Coope, 1958.)

Accelerated flow due to abundant interstitial water is known as solifluction. With increasing mobilization, a slurry of unconsolidated material mixed with water may develop, which can take the form of a rapidly moving mud flow. In northern climates, the high water content necessary for solifluction and mud flowage may be provided by the melting of frost crystals formed from upward-moving capillary soil moisture. In spring thaws, vast amounts of overburden may thus be mobilized on slopes as gentle as 1–3 degrees.

Whenever the load exceeds the internal strength of the soil or rock resting on a sloping surface, differential flow gives way to catastrophic displacement or landsliding of very large segments of unconsolidated cover. When movement of this kinds take place, geochemical soil anomalies may be completely disrupted, displaced, or buried, as illustrated in Fig. 9.4, Thus, landsliding can have a most important bearing on geochemical interpretation, particularly in mountainous terrain.

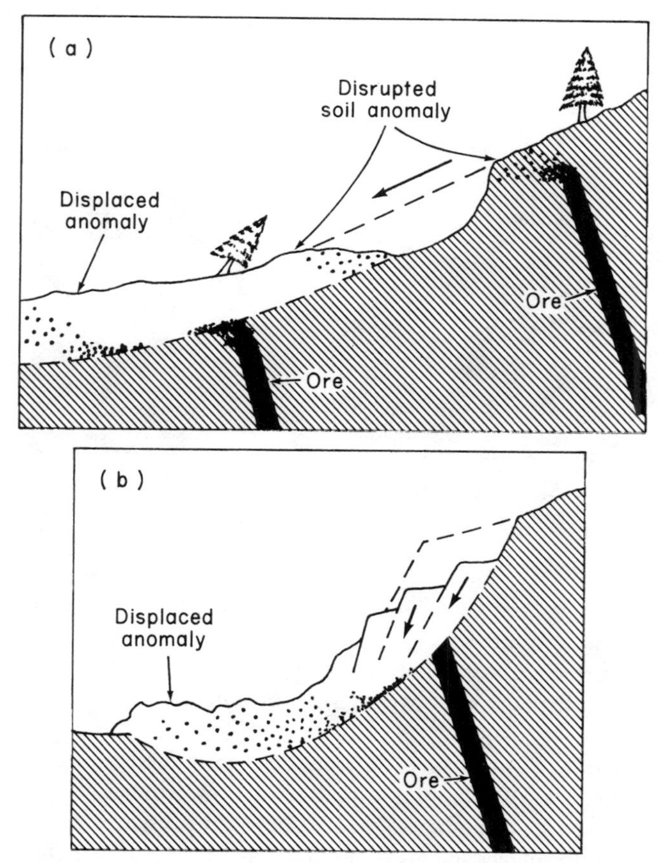

Fig. 9.4. Disruption and displacement of geochemical soil anomalies by (a) landsliding and (b) slumping.

Fragmental material that has moved down a mountainside by creep and landsliding will tend to accumulate at the foot of the slope. Very considerable thicknesses of transported material, or colluvium, may be built up in this way. The accumulation of colluvium is important in geochemical prospecting in that (i) the colluvium represents to some extent a composite sample of the overburden covering the slopes above, and (ii) residual soil anomalies in the base of slope zone may be effectively concealed (Fig. 9.5). In deeply dissected areas undergoing vigorous erosion, material moving down the valley slopes may feed directly into stream courses, to be channeled away through the surface drainage system.

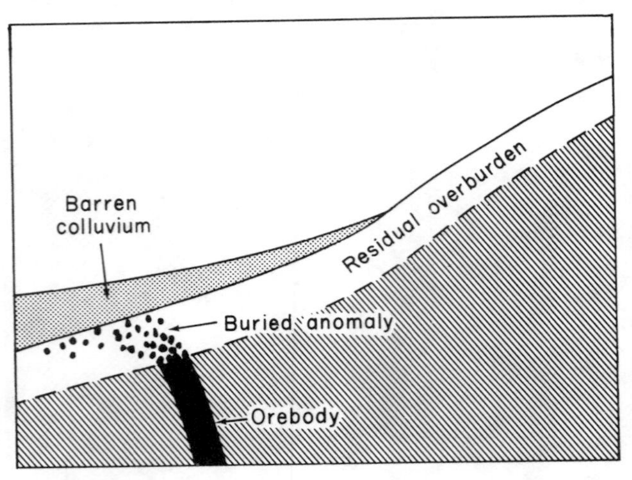

Fig. 9.5. Burial of anomaly in residual overburden beneath footslope colluvium.

B. Dispersion in Ground Water

The bulk of the water that falls to the ground as rain moves relatively rapidly either as ground water in the fractures or pore spaces of rocks and soils, or as surface runoff. This free-flowing water provides the principal vehicle for the dispersion of weathering products at and near the earth's surface.

The movement of subsurface water under the influence of gravity or of a hydrostatic head is a fairly simple matter of flow from a higher to a lower level, or from an area of higher to an area of lower pressure. Rain water soaks into the soil and replaces the air in the pore spaces of the surficial material or rocks. The upper unsaturated section is known as the zone of aeration, and the water contained in it as vadose water. If the pore spaces extend to sufficient depth and precipitation exceeds evaporation,the rain water eventually reaches the water table, which is defined as the surface below which the air in the pore spaces has been completely replaced by water. The ground water, or water below the water table, then tends to move downward and laterally in the direction of the easiest means of escape. It may emerge at the surface as springs and as seepages along the banks or in the beds of streams. The pattern of ground-water flow beneath the water table is such that a certain proportion of the water coming to the surface as seeps and springs may have come from substantial depth, as indicated by the arrows in Fig. 9.6. Springs and seepages near major streams may discharge water that has circulated to appreciable depths from recharge areas outside the immediate surface drainage, as illustrated by Toth (1963) and Williams (1970). Permeable units bounded by less permeable units also channel the flow of ground water. Under ordinary circumstances, a swamp, spring, or stream is in effect an "outcrop" of the water table. In climates with sufficient rainfall, the water table slopes in essentially the same direction as the surface of the ground, and its contours are more or less parallel with topographic contours. This simple pattern of lateral ground-water flow may be complicated by the upward movement of water under artesian pressure.

The gravity movement of ground water is impeded by the material through which it flows. Consequently, during a period of heavy rainfall, fresh rain water entering the underground reservoir of the ground-water system cannot escape as fast as it is added and tends to pile up. The effect of this piling up is to raise the water table so that higher escape channels are available, and new springs and seepages become active, as illustrated in Fig. 9.6. Conversely,

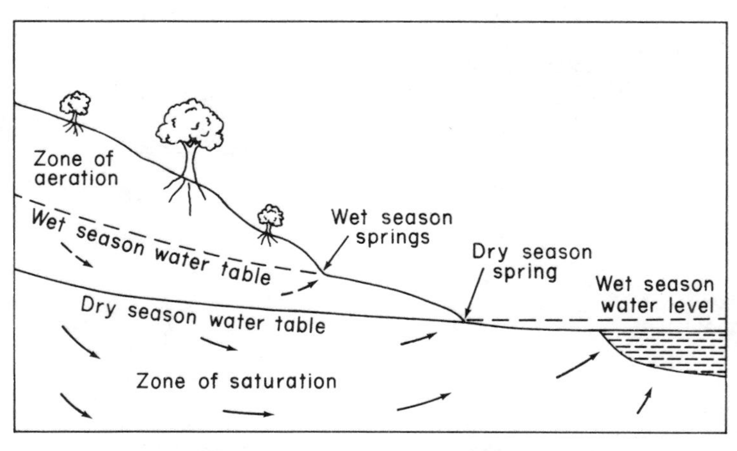

Fig. 9.6. Effects of seasonal rise and fall of water table.

during a period of light rainfall, the water table will be lowered, and springs will progressively dry up.

In areas of crystalline or metamorphic rocks which have negligible pore space, a true water table may not exist unless the cover of surficial material is deep enough. Here, rain water that has soaked into the soil tends to flow along channels on the surface of bedrock. When a channel of this kind crosses a permeable zone of fracturing or shearing, the water may enter the fractures to emerge again at the surface of bedrock at a lower elevation.

Water and the salts dissolved in it can move upward from the water table against the force of gravity as a result of capillary forces. The experimental upper limits of capillary rise range from several centimeters in coarse sand to ten meters or so in clays. The rate of movement varies markedly in an inverse relationship to particle size, and more than a year may be required to attain the full height in fine-grained material. The rate of capillary rise is further

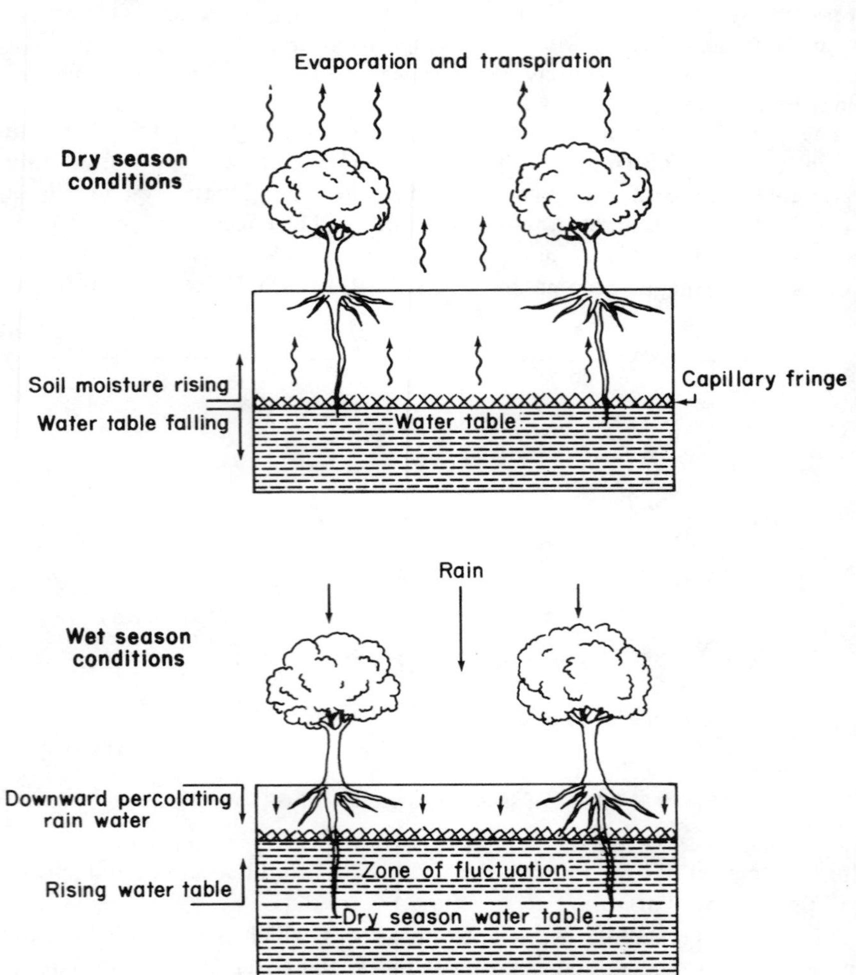

Fig. 9.7. Seasonal movements of soil moisture where rainfall penetrates to water table.

diminished as the content of dissolved salts is increased, although the magnitude of the effect is dependent on the composition of the solutes. Experiments have shown, however, that water and dissolved salts can readily move upward for substantial distances through unsaturated soils above the capillary fringe (Marshall, 1959). The mechanism, at least in part, appears to involve migration of water and dissolved ions on the surface of minerals. These effects are most conspicuous in the zone of aeration between the water table and the land surface. The net result of movement of this kind against the force of gravity is commonly a dispersion of the dissolved constituents both upward and laterally in the direction of decreasing concentration.

Under an appropriate seasonal climate, upward dispersion of this kind

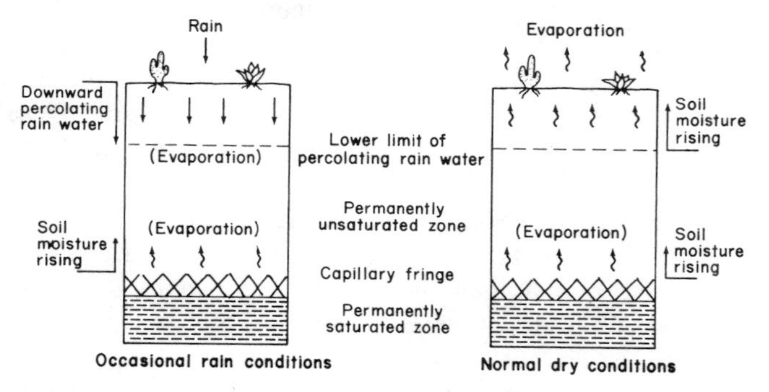

Fig. 9.8. Movements of soil moisture where rainfall is insufficient to penetrate to water table.

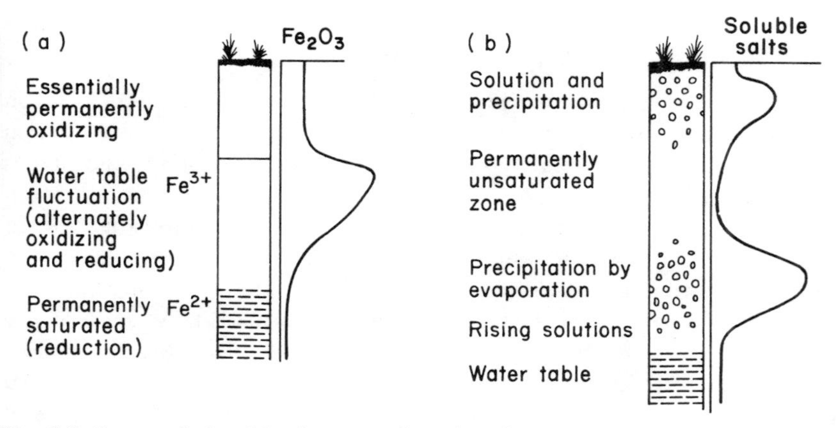

Fig. 9.9. Some relationships between the subsurface water regime and the redistribution of soluble constituents. (a) Accumulation of lateritic B horizon under a warm, humid, seasonal climate and shallow water table (see Fig. 9.7 for moisture regime). (b) Accumulation of soluble salts under a semi-arid climate and deep water table (see Fig. 9.8 for moisture regime).

may be assisted by the movement of material associated with a periodically rising and falling water table. Above the water table, moisture moves downward as rain water soaks in from above, while at the same time the groundwater level is rising; in dry periods, on the other hand, these directions of movement are reversed as moisture rises from the falling water table in order to replenish the water removed nearer the surface by evaporation and transpiration (Fig. 9.7). The pattern of movement is naturally modified in those areas where the rainfall is insufficient to saturate the ground right down to the water table. Here, although there is a seasonal reversal in the direction of flow of soil moisture in the surface horizons, the movement of water at the water table is always upward (Fig. 9.8). The overall effect of this rather complex "pumping" action is to assist in spreading soluble constituents upward from their source in the bedrock for distances depending on the local environment. In addition to dispersions formed by adsorption of the solute ions in the matrix through which the solutions move, accumulations of salts may develop by precipitation in zones where evaporation or oxidation is active (Fig. 9.9).

Special conditions influence the flow of subsurface water in the immediate vicinity of drainage channels which are underlain by permeable material. Here a substantial proportion of the water contained in a stream or river system will move as underflow through the permeable stream bed. The partition of water between channel flow and underflow varies greatly depending on the gradient of the stream and the width, depth and permeability of the bottom material at any given point.

C. Mechanics of Dispersion in Surface Water

The dispersion of solid weathering products by surface water takes place in three main environments: on the land surface by runoff or sheetwash, in stream water, and in the relatively quiet environment of swamps and bodies of standing water.

The same general principles of erosion, transport, and deposition apply in all three environments. The amount of material that can be picked up is governed primarily by particle size, the rate of flow, and turbulence. Erosion is only effective so long as the water remains underloaded. As soon as the load capacity is reached, no further net erosion is possible. If the rate of flow or turbulence decreases in a fully loaded water, then material will be deposited until the load capacity at the new energy level is attained. It is important to appreciate that each of the foregoing relationships refers only to a specific particle size. Thus, a moving body of water may be fully loaded with respect to one particle size while at the same time being capable of picking up and carrying more particles of another size.

Material may be transported in suspension, or by jumping and rolling along the bottom (saltation). Large fragments may also glide along over smaller rounded particles, which act as ball bearings. Transport, whether by suspension or saltation, is greatly enhanced by turbulence, which is itself a function of volume and velocity of the water and the roughness or irregularity of the surface over which it is flowing. The distance that material may travel before being deposited depends on the turbulence, the load, the ratio of settling velocity to the velocity of flow, and the depth of water. The settling velocity is a function of the size of the particles and their relative densities. It is also influenced by shape of the particles, inasmuch as flat particles tend to settle more slowly than rounded grains of equivalent size. Provided velocity and turbulence remain reasonably uniform, the particles deposited will tend to be well graded in size. The energy relationships between velocity and turbulence are such that sediments deposited from slowly moving water tend to be more poorly sorted than those deposited from fast-moving water.

1. Dispersion by surface runoff

The eroding and transporting power of surface runoff depends on a number of factors, notably the slope of the land, the nature of the overburden, the amount and intensity of the rainfall, and the proportion that soaks into the soil. The last is greatly influenced by the presence of vegetation, rubble, or other obstacles to effective sheetflow, as well as by the permeability of the overburden. Under favorable conditions, very considerable masses of material may be eroded and transported by simple surface runoff, accompanied by the development of extensive alluvial fans downslope from the outcrop areas. Channeling of runoff may result in catastrophic erosion by gulleying. Under most conditions, however, the effect of sheetwash is limited to the selective erosion and transport of the finest products of weathering, namely clays and finely divided Fe- and Al-oxides.

2. Dispersion in stream water

Streams acquire their load by erosion of the banks, scouring of the stream bed and by direct contributions from sheetwash and soil creep. Catastrophic floods occurring at relatively infrequent intervals probably account for a very large part of the total mass of material moved by streams. During such flood periods, very large boulders may be moved several miles in a matter of hours. Flood deposits may subsequently be sorted out and the components redistributed in an orderly manner during quieter periods of stable runoff conditions. Then the clay minerals and other fine products of weathering will move predominantly in suspension at the same speed as the water, while the coarser material moves by saltation and gliding in the bed of the stream at a somewhat slower speed.

Resistant primary minerals, notably quartz and the heavy accessory minerals, together with partially weathered rock fragments, are the dominant constituents of the sediment in fast-flowing streams. Micaceous minerals also occur, but because of their flat habit are more readily carried in suspension. Some of the stable secondary minerals are also coarse grained and may travel with the stream-bed material. Chief among these is secondary Fe-oxide, which is particularly common in areas of concretionary laterite. Manganese oxides and a variety of secondary ore minerals also may be dispersed in the sand and silt fraction.

During the course of transport along the stream bed, the particle size is liable to be progressively reduced by chemical and physical disintegration. Minerals and rocks differ in their physical resistance to abrasion. According to experiments carried out by Friese (1931), mafic rocks are more readily worn away by running water than felsic rocks, and metamorphic rocks resist abrasion better than do igneous rocks of similar mineralogical composition. Friese also determined the relative rate of disintegration of primary minerals under the abrasive action of flowing water. He found that minerals could be arranged in the following order of increasing transportation resistance: galena, quartz, zircon, ilmenite, sphalerite, cassiterite, magnetite, wolframite, rutile, chromite, pyrite, and tourmaline. This sequence does not necessarily represent the relative distance the minerals will travel from their source, as no account was taken of the relative rate of chemical decomposition of the minerals or the effect of shape of the mineral grains and the resulting differences in the bouyant effect of running water. Experiments by Kuenen (1959), however, indicate that purely mechanical abrasion of sand-size grains of quartz, limestone, and feldspar is entirely negligible in river transport. A combination of chemical attack and mechanical disintegration is probably needed to effect any appreciable reduction in the particle size of the finer alluvial fractions. If this is generally true, then effective comminution during fluviatile transport depends primarily on the stability of the mineral species involved.

Erosion and transportation naturally predominate over deposition in the upper reaches of a stream. As turbulence decreases in the quieter and flatter reaches, suspended material begins to settle out, and movement by saltation diminishes. Downstream, deposition continues on an increasing scale as the load capacity progressively falls. Considerable thicknesses of sediment may build up in the lower reaches, where the stream often flows in a channel carved through its own alluvium. As the water at this stage is usually fully loaded, erosion of the banks is balanced by deposition in the channel. At times of flood, erosion is enhanced and material previously deposited is liable to be picked up, eventually to settle down again further downstream. If the stream should overspill its banks, the sudden loss of velocity causes suspended material to be deposited on the flood plain flanking the main stream channel. Natural levees of the coarser sediment tend to build up nearest the channel, while the material laid down at greater distances is mostly finer grained.

Except where the flow is strong and uniform, river and flood-plain sediments will be poorly sorted. Changes in velocity and direction may result in gross variations in grain size, both laterally and in depth. The characteristic distribution of coarse and fine material around bends in the channel is a well-known feature related to velocity of flow. As a result of seasonal changes in the flow, the sediment is liable to be re-sorted and redistributed, sometimes resulting in stratification and cross-bedding. During the course of repeated erosion and redeposition, gravity sorting may result in the concentration of heavy minerals at critical points along the stream bed.

Introduction of material from a second source can be an important factor in locally modifying the overall tendency toward a progressive decrease in average grain size downstream. Part, at least, of the load carried by tributary streams is usually dumped at the confluence as velocity is lost on entering the main stream.

Ore elements undergoing dispersion in the solid products of weathering may travel preferentially either in the coarse or in the fine fraction, depending on the individual characteristics of the elements in question. The factors controlling grain-size distribution of minor elements during successive stages of erosion, transport, and deposition can have a marked bearing on technique and interpretation of geochemical drainage surveys. Furthermore, stream sediment may be reworked and redistributed consequent on rejuvenation of the drainage system by uplift. Drainage sediments dating from an earlier erosional cycle may be preserved as high-level terraces, or in abandoned channels remaining after river capture or even past glaciation. It is important, therefore, to recognize any aspects of the local geomorphological history which may have influenced the course of sedimentation in the past.

3. Deposition in swamps

Swamps are characterized by extremely sluggish flow and a dense mat of vegetation. Surface waters entering such an environment tend to lose practically all their load by a combination of deposition and filtration. Most of the dispersion processes in a swamp environment are chemical, involving either solubilization or precipitation of elements that react with organic matter. At the outlet of a swamp, mechanical erosion may once again become effective as flow increases in the vicinity of the outlet stream.

4. Dispersion in lakes

Lakes originate by a variety of processes, including glaciation, faulting, volcanic activity, landslides, solution of limestone, and human activities.

Most of these processes produce only a few lakes in a given region and are not of significance in geochemical exploration. However, lakes formed by the erosive and depositional effects of continental glaciation are a major surface feature of the Canadian Shield and similar areas. The remainder of this section applies to lakes in glaciated regions, where lake water and sediment have become important sampling media.

Lakes in the shield areas generally occupy rock basins eroded out by the ice and may be fed by one or more streams, plus direct precipitation and inflow of ground water. Water may flow out by a surface stream or discharge beneath the surface through a boulder field or other surficial deposit. In some lakes, evaporation is an important mechanism of water loss. Lakes form an efficient trap for all but the finest particulate material.

The nature of the water and sediment in a lake is affected by climate, depth of water, amount and source of water flowing through the lake, vegetation in the drainage basin, and geology of the area. Combinations of these variables result in lakes with a wide variety of characteristics (Hutchinson, 1957, 1967; Levinson, 1974; Dean and Gorham, 1976).

One of the most important factors determining the nature of lake water and sediment is thermal stratification of the water (Fig. 9.10). Lakes deeper than about 10 m and cooling below 4 °C during the winter tend in summer to have a layer of warm low-density water at the surface. This layer is termed the epilimnion. It is is mixed and stirred by wind and currents, and may be diluted by runoff, or concentrated by evaporation. Below the epilimnion, usually at a depth of 8–15 m, the temperature decreases rapidly to about 4 °C (39 °F), at which temperature water has its maximum density. Below this level is the hypolimnion, which is relatively stagnant most of the year, although it may be fed by ground water and undergo slow circulation. In winter, because surface water colder than 4 °C is also less dense than the deep water, stratification is also present (if the climate is cold enough) and mixing is inhibited. However, in the spring, when the surface layer is warmed to 4 °C, the layering is no longer stable. Wind and currents can mix the layers during this period, which is termed the spring overturn. A similar overturn may occur during cooling of the surface layer during the fall. Lakes overturning twice per year are termed dimictic, those overturning once are monomictic, and those overturning only occasionally are oligomictic. Because water in lakes is layered during most or all of the year, the chemical composition of the surface and bottom water may differ markedly. During periods of overturn, large lateral inhomogeneities may exist because of incomplete mixing.

A second major control on the nature of lake waters and sediments is the oxidation potential of the lake, especially the hypolimnion and sediment. The epilimnion is exposed to the air and is mixed, so that it is normally well

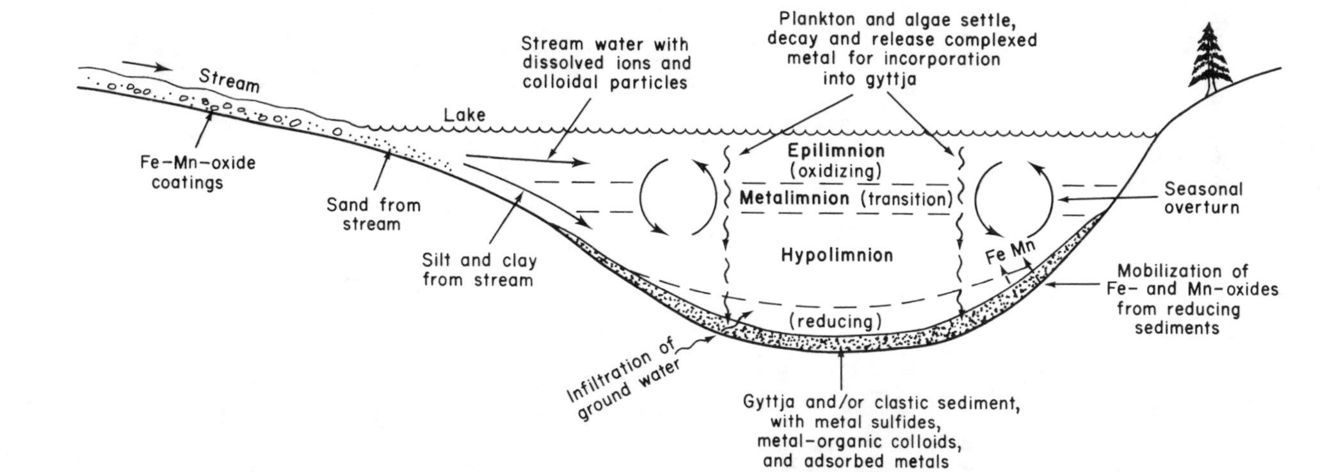

Fig. 9.10. Dispersion of metals in stratified glacial lake containing layers of organic-rich ooze (gyttja). (After Timperley and Allan, 1974.)

oxygenated. The hypolimnion can acquire oxygen only during periods of overturn. If vegetation is abundant in the drainage basin, or if nutrients are available in the lake waters for growth of algae and plankton, large amounts of organic matter enter the lake or are produced in it and sink to the bottom. Decay of this vegetation may consume most or all of the oxygen in the deep waters, so that organic material is preserved on the lake bottom, and reducing conditions prevail. Such a situation is common in lakes of areas of mixed deciduous and coniferous forests, especially shallower lakes with only a limited amount of oxygenated deep water. Trace elements in the decayed organic matter are released to the deep water and after an overturn are again available to plant and animal life. Lake waters in sedimentary terrain tend to be higher in nutrients and produce more organic material than lakes in igneous and metamorphic terrain,which tend to have water with low dissolved solids content. Lakes of volcanic belts tend to have higher dissolved solids and higher organic productivity than lakes in granitic areas. A covering of ice also tends to limit access of oxygen and leads to reducing conditions, even in shallow lakes.

In non-stratified lakes, or those with waters of relatively low dissolved solids content and limited vegetation, the entire lake and the surface of the lake sediment may be oxidizing, although reducing conditions may prevail at depth in the sediment. If the oxidizing–reducing boundary occurs at or just below the sediment surface, Fe–Mn-oxide nodules may grow at the surface by mobilization of Fe and Mn in the reduced zone and precipitation at the oxidizing surface.

Precipitation of carbonate minerals can be an important process in lakes draining carbonate-bearing areas, especially if evaporation significantly exceeds rainfall or waters are warm in the summer, so that CO_2 is lost. These conditions are found in the more temperate lakes.

Three types of sediment are distinguished in glaciated lakes (Jonasson, 1976). Inorganic sediments are common in the more northern, sparsely forested regions where lakes are relatively oxidizing and organic activity is minimal compared to the influx of sediment or the precipitation of carbonate. These sediments consist of sand, silt, marl, and clay, and, in Canada, occur near shores and inflow areas of southern lakes as well as throughout the northern lakes. Organic gels (also called gyttja) are predominant in the deep waters of lakes in organic-rich environments in forested areas. These gels become fluid when disturbed, have a density of 1·5–2 g/cm^3, typically contain 10–50% organic matter and detectable H_2S, and can form layers up to 10 m thick on the bottom of lakes. They are formed by coagulation and settling of colloidal organic material and fine particulates. The gel dries to a hard dark cake with conchoidal fracture. Typical rates of accumulation of deep sediment are 4–6 cm/100 yr (Allan and Timperley, 1975). Intermediate between the above two types of sediments are organic sediments, which are mixtures of the two types with appreciable amounts of coarser organic debris. These sediments are found near shore in shallow water, especially near inflow areas.

D. Dispersion by Glaciers

Scandinavia and Canada, together with adjoining areas, have been periodically covered by ice during the last million years of geologic time. As a result, the processes of weathering and dispersion in these glaciated areas are materially different from those in areas of residual weathering.

Erosion and deposition by glaciers involves a wide variety of processes. A continental glacier erodes its base by plucking blocks bounded by fractures, by abrasion between fragments and bedrock, and by freezing of water around material at the base (Goldthwaite, 1971). Some water is present at the base of most glaciers, even where the surface temperature is below freezing, as a result of melting induced by geothermal heat, frictional heat, and pressure-induced melting. Freezing and thawing at the base of the ice mass probably facilitate a great deal of the erosion by glaciers (Price, 1973). In polar glaciers, the ice may be frozen to the bedrock and much less erosion occurs. Near the front of an advancing glacier, the debris picked up by the glacier may be dominantly outwash, soil and older till, but nearer the source area, bedrock is usually actively eroded, at least on subglacial highs. In mountain glaciers, avalanches and rock falls from adjacent slopes add debris to the margin of the glacier.

Within a continental glacier or a large mountain glacier, most of the debris is contained within the basal few meters to tens of meters of ice. Overlying ice contains only small amounts of particulate material. The particles entrained in a glacier tend to decrease in size down-glacier as a result of abrasion and crushing, and approach a terminal size determined by the mineralogy and texture of the rock. Dreimanis and Vagners (1971) show that particle sizes in till are bimodal, with high abundances of coarse clasts acquired from bedrock, and of sand, silt, and clay formed by attrition of the clasts. Igneous and metamorphic rocks tend to produce terminal sizes in the sand range, limestones produce silt sizes, and shales produce clay sizes. The abundance of recognizable fragments derived from a given location decreases by attrition down-glacier, and the proportion of fines increases accordingly.

Major types of sediment deposited by glaciers include till, ice-contact stratified drift (eskers, kames), and proglacial stratified drift (glaciofluvial sediments, lake sediments) (Nichol and Björklund, 1973; Price, 1973). Wind-blown dust (loess) is also common near glaciers because of the large amounts of rock flour and the high winds characteristic of glacial areas. Because of the disrupted drainage, lakes and bogs are widespread. A wide variety of deposits thus obscure the bedrock in glaciated areas.

Till, which is deposited directly from ice (Fig. 9.11), is recognized by its lack of sorting and lack of regular bedding. The till may range in thickness from one pebble to many tens of meters. The terms "ground moraine" and "unstratified glacial drift" are also applied to till, although with less precision. The following generalizations about till are suggested by Goldthwaite (1971):

(i) The major volume of till is deposited in the outer third of the area covered by ice from a particular glacial episode.
(ii) Till becomes generally thinner as one proceeds from the edge to the center or source of ice movement, or topographically from valley floors to high bedrock knobs formerly covered by ice.
(iii) Most tills are deposited late in the glacial cycle after erosion of the underlying bedrock is largely completed.
(iv) Multiple till layers, deposited by multiple periods of glaciation or by multiple advances during a single period, are common.
(v) Till is a mixture of many materials over which the ice has passed, including soil, organic matter, and bedrock, but the lowest part of the till at any spot is most like the underlying bedrock or overburden.

Till has commonly been divided into a lower compact basal or lodgement till, which is deposited beneath the ice, and an upper looser ablation till, deposited by surficial melting of the ice (Fig. 9.11). Basal till is believed to be deposited by accumulation behind obstacles on the glacier floor, by shearing over basal debris-rich bodies of ice, and by melting of basal layers of ice by geothermal or frictional heat. Basal till usually contains a moderate to high content of clay. The compaction of this layer and the orientation of pebbles have been attributed to compression and shearing by the overlying ice, although some orientation may be inherited from the ice. Ablation till generally has a lower content of fines, because the clays are washed out during the melting process. Poorly developed layering, sometimes contorted, may be formed in ablation till by accumulation of sediment in hollows on top of the glacier, followed by melting of the underlying ice and resulting slumping. Mixing by frost-heaving can also affect the surface layers of till.

A variety of water-deposited sediments form within and along the margins of glaciers. Eskers are narrow sinuous ridges of sand, silt, and gravel thought to be deposited in streams flowing within or at the base of glaciers. Lengths range up to hundreds of kilometers, although only short segments are thought to form at any one time (Lee, 1965), and lengths greater than a few kilometers are rare. The deposits are stratified and cross-bedded, and usually show evidence of slumping by melting of the enclosing ice. Detritus is derived from upstream, from within the ice, or from the basal layer. Kames are

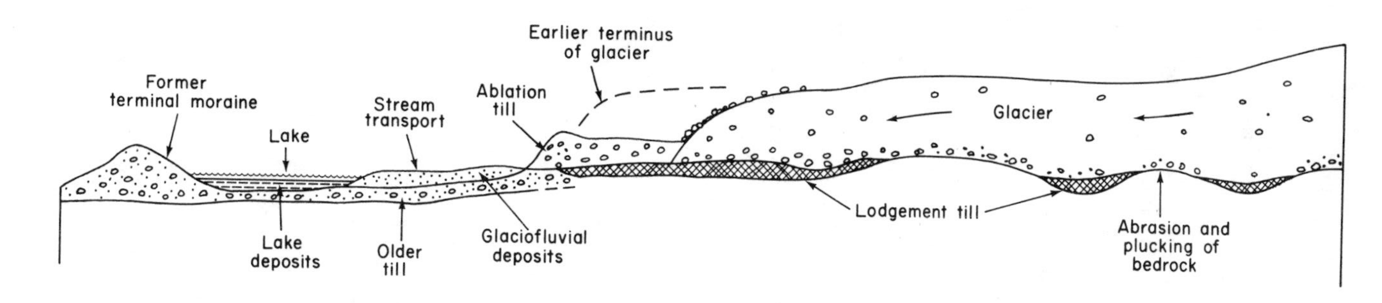

Fig. 9.11. Diagrammatic representation of glacial deposits during waning stages of glaciation.

mounds or terraces of stratified debris deposited in depressions in a glacier or along its margin.

The large streams flowing out of glaciers can cover extensive areas with glaciofluvial deposits (outwash). In addition, the glacier may cut off normal drainage channels to form lakes, sometimes many tens and even hundreds of kilometers in dimensions. The fine stratified sediments deposited in these lakes may be tens of meters thick, as in the Clay Belt of Ontario. Analogous deposits occur where glacially derived sediment entered a marine environment.

The distances over which glacially eroded material has been moved by the ice can be very great. Fragments of native Cu derived from the Keweenaw Peninsula of northern Michigan have been found 950 km distant in southern Illinois, and a boulder of a distinctive norite gneiss from near the city of Quebec has been picked up in southern Ohio some 1150 km away (Flint, 1957). The bulk of the material in glacial tills, however, is of relatively local origin. Chamberlin (1883) estimated that 90% of the material in the moraines of southern Wisconsin traveled less than 1·6 km, and studies of the glacial boulders in northern New Brunswick show that the most abundant rock type is a fairly reliable indication of the lithology of the bedrock directly beneath.

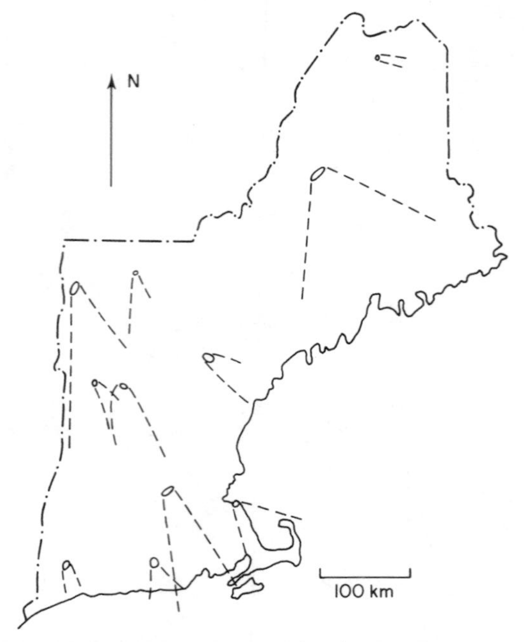

Fig. 9.12. Fan-shaped trains of glacial boulders in New England. (After Flint, 1971.)

Bayrock and Pawluk (1967) in Alberta, and Larsson and Nichol (1971) in Ireland, showed that most till was locally derived. Thus, although a few glacial erratics have traveled a very great distance from the source, the overwhelming majority of pebbles and boulders contained in till are of local origin.

The direction of movement of continental ice during a single glacial epoch is generally outward from a central area in the heart of the ice mass. At any one locality, however, the direction may fluctuate somewhat around a mean value, with the result that glacial erratics are commonly distributed in the form of a fan spreading from the source in the direction of ice movement, rather than as a linear pattern. The prerequisite for the formation of a simple fan-shaped pattern of glacial dispersion is the existence of a relatively small outcrop of a unique, mechanically resistant, and readily identified rock type. In New England alone, well-defined glacial fans have been mapped in association with 14 occurrences of distinctive rock types, some of which are shown in Fig. 9.12.

E. Dispersion by Wind Action

Erosion and transport by wind action is naturally most effective in arid or semi-arid regions, where there is little or no vegetation to protect the surface. Under favorable conditions, vast tracts of land have been covered by wind-blown desert sand and by loess, which is generally considered to be mainly wind-sorted material of glacial origin.

In addition to obvious wind-blown accumulations, there is a certain proportion of aeolian material in all superficial deposits. As a rule, the amount is relatively small and is not an important factor in geochemical prospecting. However, detectable aeolian dispersions of material derived from weathering ore deposits have been reported on rare occasions. Locally too, wind-blown dust and fumes from mine dumps and smelter stacks can interfere in geochemical surveys.

F. Dispersion by Animal Activity

Burrowing species, particularly worms and termites, are responsible for bringing vast amounts of weathered material to the surface. For example, in the Zambian Copperbelt, individual termitaries of up to 300 tons are distributed at a density of 0·4 termitaries per hectare, representing a depth of 7–15 cm of soil over the land surface as a whole. As a result of this animal activity, the metal content of surface horizons may be enriched, permeability of the overburden increased, and a loose aggregate prepared, facilitating lateral dispersion by gravity, water, and wind.

G. Dispersion in Permafrost Areas

The former concept that permaforst precluded oxidation and weathering in the frozen layers is now known to be incorrect. Surface and ground waters in permafrost regions contain anomalous amounts of heavy metals and other elements, indicating that oxidation and leaching are occurring at significant rates (Shvartsev, 1972; Jonasson and Allan, 1973; Levinson, 1974). Cameron (1977a) found intense oxidation and large volumes of subsurface water flow near a mineralized zone in Northwest Territories, Canada. Some of the dispersion of solutes undoubtedly occurs by movement of water through "windows" where permafrost is lacking, but some dispersion takes place through continuous permafrost. Although most of the water in frozen ground is in the form of ice, a small amount remains liquid, apparently because its high content of dissolved solids lowers its freezing point. This water may migrate upwards by capillary action or expansion during freezing, or it may flow downwards by gravity. Diffusion in this aqueous phase or along grain boundaries can also disperse the dissolved constituents. Jonasson and Allan (1973) show that the metals migrate upwards into snow, where they form geochemical anomalies. Because of the dry climate of many arctic areas, efflorescences of soluble salts form in some localities during the winter.

When the snow and surface soil thaw in the late spring and summer, the metals accumulated in these materials are rapidly flushed out into the drainage system. Unusually high and variable metal values can be measured in water during this season.

II. BIOLOGICAL FACTORS

Although life processes are often considered as something separate from strictly physical and mechanical processes, they still must abide by the same laws of physical chemistry that govern the solubility and hence the mobility of the chemical elements. At the same time, living organisms can cause effects that are quite different from those that would be expected from purely inorganic mechanisms. Of particular importance in this respect are plants and microorganisms.

A. The Effect of Vegetation

Living vegetation has a profound effect on the dispersion of weathering products. The uptake of a given element by the root system of a plant is a function of the relative solubility of the element in the soil solution, as modified by the extremely corrosive environment created by the plant in the vicinity of its root tips. The biogenic processes whereby elements may be solubilized from relatively stable mineral phases and ingested into the plant's circulatory system vary with different species of plant. These extremely complex processes are discussed at greater length in Chapter 17 in connection with plant nutrition.

The net effect of these combined inorganic and organic factors is an uptake of substantial quantities of inorganic matter which is then distributed in greater or less amounts through the body of the plant. As the leaves and other plant organs fall to the ground and decay, rain water leaches out the more soluble constituents. The bulk of the soluble products of plant decay are normally removed in ground and surface water. A part, however, may again be taken up by living plants or reprecipitated with Fe, Mn, and Al in the B horizon of many soils. The less soluble constituents released by plant decay tend to remain in the humus layer, wherein ions may also be retained by adsorption on organic matter. This effect, as originally pointed out by Goldschmidt (1937), is cumulative, and over the years very appreciable enrichment may take place. The entire sequence of processes is referred to as the biogeochemical cycle (Fig. 9.13).

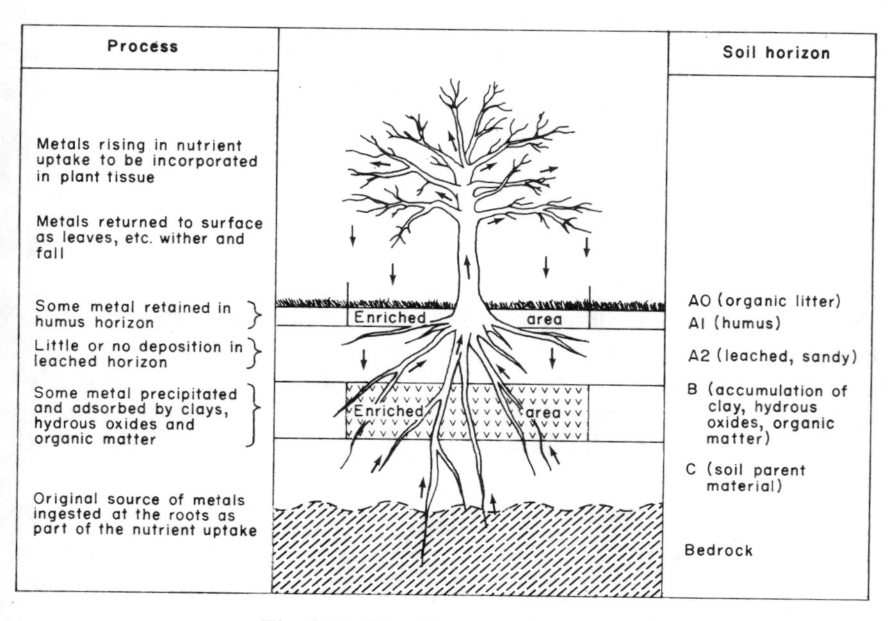

Fig. 9.13. The biogeochemical cycle.

B. The Effect of Microorganisms

Bacteria, algae, fungi, and other microorganisms exert a powerful influence on metal dispersion in the soil and in surface drainage systems. Apart from their vital role in the production of humus, microorganisms are intimately associated with other important redox reactions. Bacterially induced oxidation may lead to the precipitation of Mn- and Fe-oxides and also the production of soluble sulfates from sulfur and certain sulfides (Rudolfs and Heilbronner, 1922). Experiments have shown that under certain conditions microorganisms can precipitate native Cu and Cu-sulfide by reduction of Cu-bearing solutions (Lovering, 1927; Arkhangel'sky and Soloviev, 1938). It is important to realize that the effect of bacteria on these reactions is primarily catalytic. Reactions that are possible and the compounds that can be formed are restricted to those consistent with the pH and Eh of the environment.

There is also evidence that microorganisms are capable of assimilating and concentrating certain metals. In many cases it is not clear, however, whether the metal is in fact assimilated within the living organism or whether it is absorbed on the external mucilage or fixed in metabolic products. Riley (1937) has shown that much of the total Cu in lake waters in Connecticut is tied up by organisms in the plankton. *Spirogyra*, a fresh-water alga, growing in mine drainage water carrying 16 ppm total heavy metals, has been shown to contain 2900 ppm Zn, 6600 ppm, Pb and 930 ppm Cu in the dried algal material (Cannon, 1955). Under some conditions, soil microorganisms may be responsible for reducing the availability of metals for plant nutrition. An example of Zn deficiency caused either by assimilation or fixation in metabolic products is described by Hoagland *et al.* (1937).

III. THE INFLUENCE OF ENVIRONMENT ON DISPERSION

From the foregoing it will readily be appreciated that environment necessarily exercises a very strong influence on metal dispersion. As with weathering and soil formation, the principal environmental factors are those relating to climate, relief, rock types, life processes, and time. The close interaction between these fundamental factors has been stressed earlier and need not be repeated here.

A. Climate

Climate affects dispersion mainly through its control of the moisture regime, vegetation, and soil type. In arid regions, paucity of water and vegetation are reflected in the subordinate role of chemical as compared to mechanical dispersion. Ground water usually lies far below the surface, but some metals, particularly those which form soluble anion complexes such as Mo and W, may be brought up by certain deep-rooted plant species. Although chemical dispersion assumes greater importance as the rainfall increases, calcareous soils characteristic of semi-arid and subhumid climates provide a poor environment for the dispersion of soluble elements on account of high pH and the presence of abundant lime. Humid tropical or temperate climates, on the other hand, provide optimum conditions for chemical dispersion. The relative importance of mechanical dispersion is largely determined by the vegetation and relief. In the colder regions chemical reactions are slowed down by the low temperature and low organic activity.

B. Relief

Through its bearing on erosion and ground-water movement, relief exerts a most powerful effect on the dispersion of weathering products. In flat-lying terrain, the rate of mechanical dispersion is restricted, and although chemical weathering may extend to great depths, dispersion of the soluble products can be accomplished only slowly by the sluggish flow of surface and ground waters. With increasing topographic relief, more vigorous flow results in ready dispersion of soluble material. At the same time, however, the effectiveness of mechanical erosion is also increasing, and in mountainous terrain surface material may be removed faster than potentially soluble material can be released by weathering. Thus, while moderate relief undoubtedly promotes extensive chemical dispersion, the balance may swing in favor of mechanical dispersion in areas of strong topographic relief. By virtue of its relationship with ground-water level, the relief also influences the dispersion of soluble metals which may be precipitated in seepage areas at points where the water table crops out at the surface.

C. Rock Types

The prime importance of the geologic environment lies in its influence on the composition and the freedom of movement of ground-water solutions both through the rock and through the overburden. Dispersion of many semi-mobile elements such as Cu and Zn is more restricted in an alkaline, calcareous environment than it is in the more acid conditions commonly associated with relatively unreactive siliceous rocks. The high content of Ca^{2+} and HCO_3^- characteristic of ground waters in calcareous terrain can also effectively restrict the dispersion of Mo by precipitation of $CaMoO_4$. Elements such as As and Mo that tend to be precipitated in the presence of

Fe will be preferentially retained in the Fe-rich soils derived from mafic rock. Permeability and the degree of fracturing of the bedrock may impede or facilitate flow of ground-water solutions. The permeability of the overburden, largely related to clay content, is primarily dependent on the nature of the parent material. This relationship is witnessed by the clay-rich residual soils derived from limestones, argillaceous and mafic rocks, as contrasted with the more permeable residua associated with arenaceous and felsic rocks. Equally variable is the permeability of transported overburden ranging from fine-grained glacial boulder clay and lake or flood-plain clay sediments to more permeable coarse-grained alluvial deposits.

D. Life Processes

The distribution and activities of plants, animals, and microorganisms are largely controlled by climate, topography, and drainage. Although the modifying influence of organisms is evident in metal distribution in most environments, the influence of biological processes attains predominant geochemical significance only under certain rather specialized conditions. Thus, plants may play the leading role in bringing metal to the surface in areas of transported overburden and in "living" desert regions. The extremely widespread dispersions of metal in soil in parts of the Central African peneplain are considered to be due in part to metal taken up by the vegetation from moving ground-water solutions (Tooms and Webb, 1961). The most significant accumulations of metals by microorganisms take place as a rule in swamps and lakes wherein the Eh and pH are in part controlled by the microbial population, which in turn reflects the anaerobic conditions and temperature in these relatively stagnant environments.

E. Time

Where the environment predisposes a slow rate of weathering and erosion, or a slow rate of transport, appreciable time is clearly essential for widespread dispersion to take place. In residual overburden the maturity of metal dispersion patterns probably parallels that of the soil profile. Other things being equal, immature dispersion patterns tend to be narrower and more intense than do the mature patterns. The time required for soluble metals to move up into transported overburden must clearly depend on a number of factors, such as the rate of oxidation, the rise and fall of ground water, and the rate at which metal can be transferred to the surface soil by the local vegetation. From a few empirical observations it appears that appreciable dispersion in transported overburden can take place, under some conditions at least, in a few hundred years.

By comparison with the soil environment, the rate of change of conditions is much greater where weathering products are being transported in the subsurface and surface drainage system. Very little is known about the rate at which reactions involved in chemical dispersion tend to come into equilibrium. Answers to this problem could have a material bearing on metal dispersion in sediments and waters, particularly where there are seasonal variations in the nature and amount of eroded material and in the metal content of the transfluent solutions.

IV. SELECTED REFERENCES ON DISPERSION IN THE SURFICIAL ENVIRONMENT

Hydrology	Davis and DeWeist (1966)
Glaciology	Nichol and Björklund (1973)
	Goldthwaite (1973)
Biogeochemistry	Brooks (1972)

Chapter 10

Surficial Dispersion Patterns

★★★★★★★★

The characteristics of surficial dispersion patterns are a natural consequence of the dynamic processes of dispersion discussed in the previous chapter. These processes and the wide range of environments in which they operate are complex in the extreme, and the resultant patterns of redistributed material show a corresponding diversity in origin, mode of occurrence of their constituents, and physical form.

The present chapter reviews the more important general features of surficial dispersion patterns. Later chapters present more detailed discussions of the various kinds of anomalous patterns, with special emphasis on their significance in mineral exploration.

I. CLASSIFICATION OF SURFICIAL DISPERSION PATTERNS

Genetically, surficial dispersion patterns may be classified according to (i) time of formation, relative to the host matrix, and (ii) mode of formation. This system of classification is employed because correct recognition of the time and mode of formation of a geochemical pattern provides the only sure foundation for its interpretation in terms of bedrock geology.

On the foregoing basis, therefore, patterns introduced or deposited at the same time as the host matrix are classified as syngenetic, while those which were introduced into the matrix after its formation are distinguished as

Genetic classification		Dispersion process	Principal transporting agent	Matrix	Mode of occurrence of dispersed elements	Form of dispersion pattern
Syngenetic patterns	Clastic	Weathering *in situ*		Weathered rock Residual overburden Gossan	Resistant primary and secondary minerals; minor constituents of clay minerals and secondary hydrous oxides	Superjacent patterns
		Movement of solid particles by:	Gravity	Residual overburden Gossan Colluvium		Fans and asymmetrical superjacent patterns
			Ice	Moraine		Fans
				Glaciofluvial deposits		Trains and irregular patterns
			Water	Sheetwash deposits		Fans
				Stream sediment		Trains
				Lake sediment		Delta fans
			Wind	Aeolian deposits		Fans
	Hydro-morphic	Movement of solutions	Ground water	Ground-water solution	Soluble salt complexes and sols	Fans
			Surface water	Surface-water solution		Trains.
				Precipitates and evaporite deposits	Precipitated salts	Lateral patterns
	Biogenic	Plant metabolism	Uptake by living plants	Living-plant tissue	Metallo-organic compounds	Superjacent and lateral patterns
				Organic debris		
Epigenetic patterns	Hydro-morphic	Movement of solutions followed by precipitation	Ground water	Any clastic overburden	Ions sorbed on clay minerals, hydrous oxides and organic matter; ions coprecipitated and occluded in hydrous oxides; metallo-organic compounds; precipitated salts	Superjacent patterns; fans
				Soils of seepage areas		Lateral patterns
			Surface water	Stream sediments		Trains
	Biogenic	Plant metabolism followed by redistribution of organic decomposition products	Nutrient solutions; soil moisture	Any clastic overburden		Superjacent and lateral patterns

Fig. 10.1. Classification and general characteristics of the principal types of surficial dispersion patterns.

epigenetic. Patterns may be further classified as (i) clastic, where the dispersion is mainly by movement of solid particles; (ii) hydromorphic, where the dynamic agents are aqueous solutions; and (iii) biogenic, where the patterns are the result of biological activity. The principal characteristics of the various kinds of surficial dispersion patterns are summarized in Fig. 10.1.

Because of the fact that surficial patterns are often the net result of a combination of processes, a genetic system of classification such as the one proposed cannot be applied too rigidly. Careful studies of the chemical and physical characteristics of a dispersion pattern may, however, lead to a fairly reliable prognosis as to its source, even where its genetic history has been relatively complicated.

II. SYNGENETIC PATTERNS

Syngenetic patterns may be either clastic, hydromorphic, or biogenic. For the most part, these patterns are relatively simple to interpret, inasmuch as the history of the matrix provides a direct link between the pattern of dispersed metal and its primary source in the bedrock.

A. Clastic Patterns

Residual soil, colluvium, alluvial sediments, and glacial till are the common media for virtually all clastic patterns. In residual soil, they reflect the distribution of the elements in the underlying bedrock surface; the bedrock patterns of distribution of the immobile elements are usually more faithfully preserved than those defined by the more mobile constituents, which are subject to leaching and redistribution in the weathered residuum. Syngenetic patterns in colluvium and glacial moraine also point to a metal source at the bedrock surface, although here the relationship between the dispersion pattern and the bedrock source may be complicated by substantial lateral movement and by the vagaries of glacial transport and redeposition.

Clastic patterns in stream sediment result from the erosion and alluvial transport of metal-rich overburden. Here the relationship between the anomaly and the bedrock source may be complicated by the previous dispersion history of the metal in the overburden. If the sediment anomaly is derived directly from the erosion of a residual soil anomaly, the bedrock source of metal will occur in the immediate vicinity of the soil anomaly. If the sediment anomaly comes from the erosion of a seepage soil anomaly, however, the bedrock source must be sought up the slope of the water table from the seepage anomaly.

B. Hydromorphic Patterns

The soluble load of ground and surface water is another kind of syngenetic dispersion pattern. Here the matrix is water rather than soil or some other solid-phase material. As with syngenetic soil anomalies, determining the source of the metal-rich matrix, in this case water, goes a long way toward determining the bedrock source of the anomalous constituents. For this reason, it seems more realistic to think of dispersion patterns in water as syngenetic, rather than epigenetic as proposed by Ginzburg (1960, p. 155).

Massive chemical precipitates and evaporite deposits would be properly classified as syngenetic patterns of hydromorphic origin. Where the precipitates are not massive but are disseminated in the interstices of a clastic matrix (e.g. soil or stream sediment), the pattern would be regarded more appropriately as epigenetic. An important example of syngenetic patterns of this kind would be those formed in and contemporaneously with extensive deposits of bog Fe and Mn ores.

C. Biogenic Patterns

An anomalous concentration of readily available metals in the soil will normally be reflected by an anomalous metal pattern in the plants growing on the soil. Plant anomalies are syngenetic in that they were formed contemporaneously with the growth of the plant itself. Similarly, dispersion patterns in organic debris derived solely from the accumulation of partially decomposed plant material are syngenetic. Biogenic patterns of syngenetic origin reflect in part the composition of local bedrock or overburden and in part the composition of circulating ground-water solutions.

III. EPIGENETIC PATTERNS

Epigenetic patterns of dispersed materials introduced subsequent to the matrix are necessarily the result of either hydromorphic or biogenic processes. For this reason, they are most commonly defined by the semi-mobile elements, such as Zn and Cu, that can be both readily dissolved and readily precipitated with local changes in the environment.

A. Hydromorphic Patterns

Natural aqueous solutions normally leave a pattern of precipitates of one kind or another in the matrix through which they flow. The resulting epigenetic dispersion patterns are superimposed on indigenous (syngenetic) patterns originally present in the matrix, whether rock or soil of either residual or transported origin. Hydromorphic anomalies of this type are always particularly well developed wherever the local environment is especially favorable to precipitation. Such conditions are common in spring or seepage areas and in organic swamps. The source of hydromorphic anomalies naturally lies upstream, updrainage, or in depth, according to the route followed by the metal-bearing solutions.

B. Biogenic Patterns

With the decay of the living or dead organic matter serving as the host for a plant anomaly, a major part of the mineral content may be leached away. A certain fraction of the mineral matter released by the decay may, however, be retained in the soil where it forms an epigenetic anomaly of biological origin. Biogenic patterns of this kind can develop in either residual or transported overburden. Their relationship to the bedrock source is naturally the same as that of the particular syngenetic vegetation patterns from which they are immediately derived.

IV. PARTITION OF DISPERSED CONSTITUENTS BETWEEN LIQUID AND SOLID MEDIA

The behavior of the individual chemical elements in the cycle of surficial dispersion is conditioned throughout by their relative mobility. Immobile elements tend to lag behind with the clastic products of weathering, while the mobile elements tend to travel away from the site of weathering as part of the soluble load of ground and surface water. It is important to realize, however, that during the course of dispersion there is usually interchange to a greater or less extent between solutions and the solid phase with which they are in contact. This interchange is illustrated schematically in Fig. 10.2.

A. Liquid Media

A dispersed constituent will be relatively mobile when it is carried in natural waters as a component of stable solutes or suspensoids. Where the constituent is carried in the water in a form that is less stable and hence more prone to reaction with the matrix in contact with it, the chances of forming a precipitate are higher, and the effective mobility is lower. The partition of an element between any two competing phases such as natural waters and the surrounding clastic matrix depends not only on its relative stability in the two

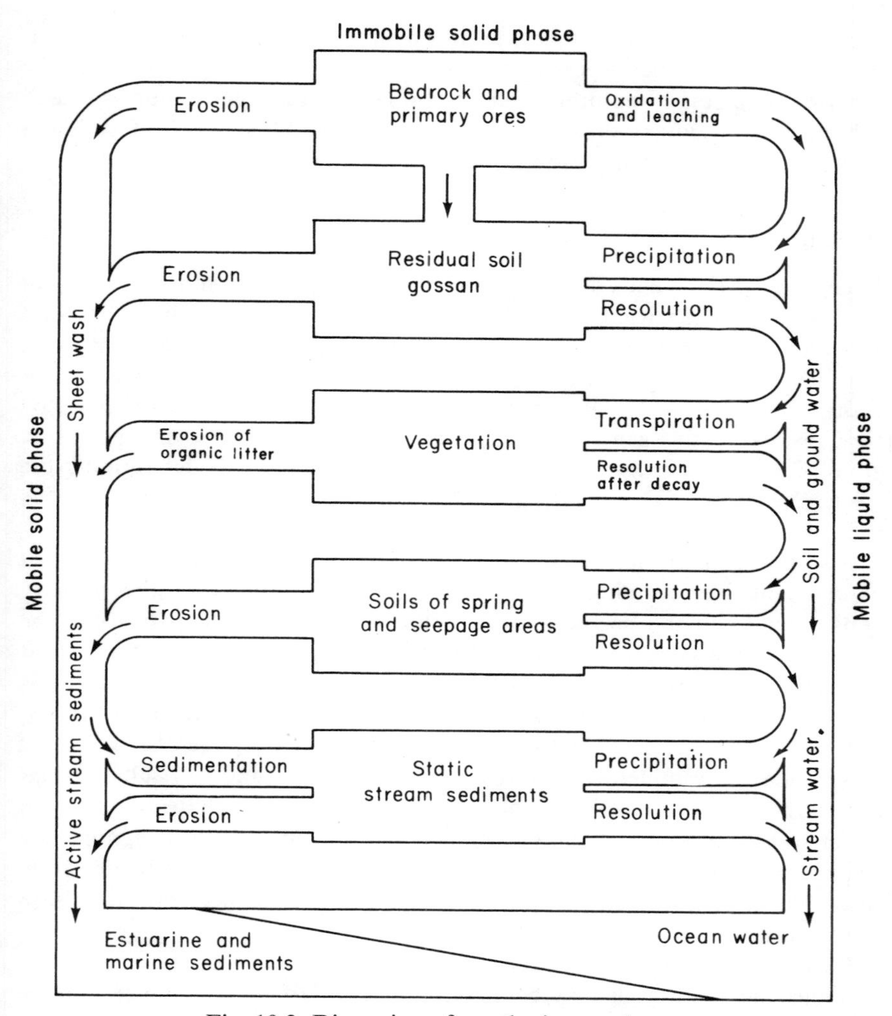

Fig. 10.2. Dispersion of weathering products.

phases but also on the speed of the reaction that leads from the less stable to the more stable form. The reactivity of the minor constituents of natural waters thus depends on their mode of occurrence—whether they are carried with highly reactive components of the water or with components that are either stable or that do not react rapidly.

The ionic constituents of water are generally by far the most reactive. Ions are free to react very rapidly either to form precipitates or to enter into ion-exchange reactions on the surface of charged solid particles. It is the ionic constituents of water, therefore, that most commonly take part in the formation of hydromorphic dispersion patterns in soils and stream sediments. The other components of natural water, including dissolved gases, undissociated soluble organic and inorganic matter, and dispersed material of colloidal dimensions, are by comparison far less reactive.

B. Solid Media

Except where the metals occur in distinctive coarse-grained minerals, it is not always easy to tell what mineral or non-mineral phases of the matrix are serving as hosts for the trace elements. Nevertheless, indirect evidence that may be useful as a guide to the probable partition of metals in residual or transported overburden and in stream sediments can often be obtained by (i) considering the general geochemical behavior of the elements in the zone of weathering, as reviewed in the Appendix, and (ii) determining the relative proportions of metal released by different chemical extractants, as indicated below.

Resistant primary minerals include some ore and gangue minerals as well as resistant rock-forming minerals. The relative stability of the primary constituents and the degree of weathering are the controlling factors. The chemical and physical properties characteristic of the mineral species involved determine the methods by which they may be detected. If chemical analysis is required, it is usually necessary to resort to a vigorous attack by strong, hot acids or fluxes, in order to break down the primary minerals.

Secondary ore minerals include a variety of oxides and oxy salts. Oxides and carbonates are mostly soluble in cold, weak acids; sulfates range from water-soluble to those which, in company with secondary silicates, normally require strong or hot acids for solution. Phosphates are likely to be relatively insoluble, so that very strong, hot acids or fluxes may be necessary to release the metals.

Clay minerals make up the bulk of the solid breakdown products of the rock-forming silicates. Metals may be incorporated within the clay-mineral lattice or adsorbed in exchange positions on the particle surface. Complete extraction of lattice-held metal, which is commonly of residual origin, requires destruction of the mineral, which can be accomplished by treatment with strong, hot acids or by fusion. Metal adsorbed on clay minerals is indicative of the ions dissolved in the solutions with which they are in contact. Adsorbed ions are loosely held and can normally be released by leaching with cold aqueous extractants. With time, adsorbed metal may become incorporated within the lattice and will then be less readily extracted.

Secondary hydrous oxides of Fe and Mn may be derived from the weathering of rock-forming minerals as well as from primary ore minerals. Important amounts of many soluble ore metals may be coprecipitated, occluded, or adsorbed with the hydrous oxides. Metal held in this way may be either residual or acquired from soil moisture or ground-water solutions. The readiness with which the metal may be extracted varies greatly according to the nature of the bonding and the condition of the oxide host. Freshly precipitated or adsorbed metals may be readily extracted by cold reagents. The proportion of readily extractable metal tends to decrease with time and with progressive dehydration and crystallization of the host, and can then be released only by hot acids or fluxes.

Organic matter may contain appreciable amounts of many metals. The greater part of this metal has usually been introduced by ground water, stream water, or decaying vegetation. The bonding is extremely variable and complex, ranging from simple adsorbed ions to metallo-organic compounds and metal incorporated in the structure of living organisms. The adsorbed fraction, as might be expected, is readily extracted by cold aqueous solutions, though often less readily than from clays. Complete release of the more firmly bonded metal generally requires complete destruction of the organic matter. This is usually accomplished by ashing or wet oxidation.

V. EXTRACTABILITY OF METAL FROM CLASTIC SAMPLES

It will readily be appreciated from the foregoing that the syngenetic or epigenetic character of a dispersion pattern may be recognized by the partition of the metal between the various solid phases, and that the mode of occurrence of the metal may in fact be suggested by the relative extractability of the metal in different reagents. Broadly speaking, the epigenetic components of a sample tend to be more readily extractable in weak aqueous reagents than syngenetic components.

By convention, the content of a metal that can be extracted from weathered rock, overburden, or stream sediment, by weak chemical reagents (e.g. cold HCl or citrate solutions) is referred to in the literature of geochemical prospecting as readily extractable or cold-extractable metal. These terms are conveniently abbreviated "cxMe" or, in the case of a specific metal such as Cu or Zn, "cxCu" or "cxZn". In this terminology, "cxMe/Me" refers to the fraction of the total metal content of a sample that is soluble in weak chemical extractants, and is usually expressed as a percentage.

Conclusive evidence of syngenetic clastic origin is, of course, the presence of residual or detrital grains of primary ore minerals. However, with unstable ore minerals, most of the metal of clastic patterns, including anomalous metal, is contained in secondary minerals, principally in the clay minerals and hydrous oxides. Insofar as these minerals have resisted the chemical attack of natural solutions during weathering and possibly also a cycle of erosion and transport, they usually tend to be relatively resistant to attack and solution by laboratory extractants. As a result, the cxMe/Me ratio in syngenetic clastic anomalies tends to be low. Exceptions to this rule occur, particularly where weathering is incomplete, where soluble secondary minerals are present, and in poorly drained material where leaching may not have been effective.

In epigenetic patterns where the introduced metal is of hydromorphic origin and formed by the relatively recent precipitation of soluble material from ground or surface waters, the content of readily extractable metal is maintained by active exchange with the metal in solution. In consequence the cxMe/Me ratio tends to be high. The original metal content of the matrix occurs in the same forms as in purely syngenetic patterns and is usually strongly bonded.

A more complex situation exists in biogenic patterns. Here, part of the introduced metal is held in the same manner as in hydromorphic patterns while part is more firmly bonded in residual organic compounds that differ widely in their solubility in standard extractants. The cxMe/Me ratio varies accordingly but is typically higher than for syngenetic patterns and is usually less than for hydromorphic patterns in the same general environment.

VI. CONTRAST

The contrast in metal content between surficial geochemical anomalies and normal background is dependent on a number of factors. These include (i) the primary contrast between ore and country rock, (ii) the relative mobility of elements in the dispersion environment, and (iii) dilution with barren material. Primary contrast varies widely for the different metals and classes of mineral deposits (Table 10.1). In clastic anomalies, primary contrast is preserved to a greater extent by immobile elements such as Sn and Be than by more mobile elements such as Zn and Cu, which are more susceptible to leaching. Even with the most mobile elements the degree of leaching is, of course, determined by the intensity of weathering, the rate of flow of water, the pH, and the many other factors that play a part in the formation of dispersion patterns. These same factors influence the contrast shown by hydromorphic anomalies. In waters, contrast is also a function of mobility in that, other things being equal, the highest contrast is shown by elements possessing the greatest mobility. Mobile elements that are susceptible to

Table 10.1
Average contrast between the metal content of marginal ore and unmineralized rock

Principal metals	Content in igneous rocks (A) (ppm)[a]	Content in workable ore (B) (ppm)[b]	Contrast (ratio B/A)
Chromium	2000[c]	250 000	125
Cobalt	25	5000	200
Copper	50	10 000	200
Gold	0·003	10	3300
Iron	46 000	300 000	6
Lead	10	50 000	5000
Manganese	1000	250 000	250
Molybdenum	1·5	5000	3300
Nickel	160[d]	15 000	95
Silver	0·05	500	10 000
Tin	2	10 000	5000
Tungsten	1	5000	5000
Vanadium	150	25 000	170
Zinc	80	80 000	1000

[a] Table 2.7 except as noted.
[b] Figures for Cr, Cu, Au, Fe, Pb, Mn, Ni, Ag, Sn, and Zn from Fleischer (1954).
[c] Average ultramafic rock (Vinogradov, 1956).
[d] Average gabbro (Vinogradov, 1956).

precipitation with moderate yet critical changes in the chemical and biological environment tend to give the best contrast in hydromorphic soil and sediment anomalies.

Contrast in plant anomalies is dependent to a degree on the contrast in the metal available in the soil of the root zone. The availability of an element to plants is often an approximate reflection of its mobility in the more general sense. In biogenic soil anomalies, on the other hand, contrast is governed by the metal content of the present vegetation, subject only to modification to a greater or less extent by accumulation or leaching of the biogenic metal during the course of soil formation.

Dilution consequent on mixing with barren material is essentially a local problem, although the effect is liable to be more pronounced in some environments than in others. Generally speaking, rapid dilution of an anomaly with distance from the source is more serious with a vigorous dispersion mechanism and with an abundant supply of barren material at the point of origin of the dispersion pattern. Thus, dilution and a resulting reduction in contrast is characteristic of glacial and fluvial dispersion and is less severe with residual soils in flat terrain or on ridge crests.

Ideally, the aim of a geochemical exploration survey is to detect only those patterns of metal derived from mineral deposits. If it is possible to exclude that part of the total metal content that is not related to the mineralization, then anomaly contrast will be greatly enhanced. Selective extraction is most likely to be successful in mapping epigenetic patterns, both hydromorphic and biogenic, where the introduced metal is often more readily extractable than the original component. Even in syngenetic anomalies, however, the metal derived from ore minerals may occur in a different form than that from the country rock, though for many metals the partition is usually less marked or is not so readily detectable by fractional analysis.

VII. FORM OF SURFICIAL PATTERNS

Classification of surficial patterns according to their geometrical shape and location with respect to the bedrock source is helpful not only in laying out sampling programs but also in interpreting the data in terms of the probable cause of anomalies. The various characteristic forms that result from the different modes of formation of dispersion patterns are illustrated in Figs 10.3–10.8.

The terminology used is purely descriptive. Patterns developed more or less directly over the bedrock source are said to be superjacent, as distinct from lateral patterns that are displaced to one side and entirely underlain by barren bedrock. Superjacent patterns more or less symmetrically disposed about the source are termed haloes. Directional movement during dispersion results in asymmetry, the pattern then taking the form of a fan spreading outward from the source, or a train if dispersion takes place along a restricted channel-way.

For complete description it is necessary also to take into account the distribution of anomalous metal values within a pattern. Anomalies are said to be intense if values rise sharply to well-defined peaks, or diffuse if the pattern is more subdued and does not show any pronounced focal point. The homogeneiety of an anomaly is also determined by the regularity with which values are distributed within the pattern.

A. Clastic Patterns

The form of clastic anomalies of syngenetic origin depends very much on the dispersion medium. Superjacent patterns are typical of residual overburden.

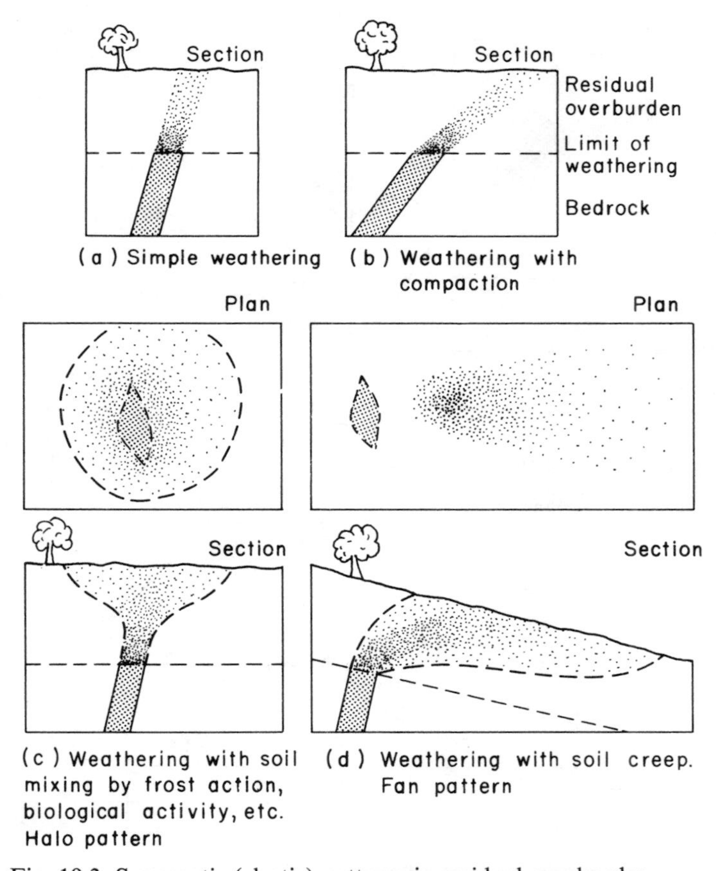

Fig. 10.3. Syngenetic (clastic) patterns in residual overburden.

Lateral patterns may result from compaction during weathering, though as a rule the amount of displacement is small. Gravity creep causes distortion, leading in extreme cases to well-developed fans extending downslope from the deposit. Fans are also characteristic of syngenetic patterns in glacial till and aeolian deposits, the apex of the fan lying near the deposit and spreading out in the direction of ice or wind movement. In deposits of alluvium laid down by sheetwash flowing across unrestricted pediment areas, the anomalies again take the form of fans, spreading out from the bedrock deposit or from the point where the surface water leaves a restricted channel. Where the drainage channel is well defined throughout its course, the alluvial anomaly is a linear train. At the point of entry into a lake, fans may again develop in deltaic sediments.

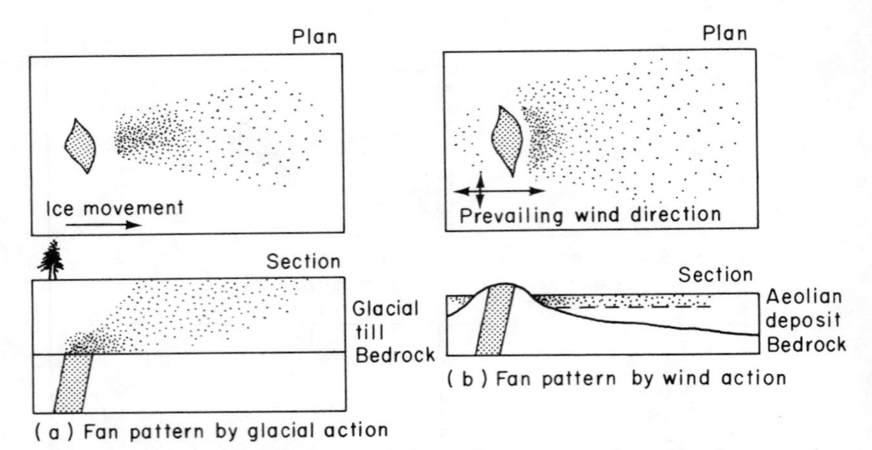

Fig. 10.4. Syngenetic (clastic) patterns in transported overburden.

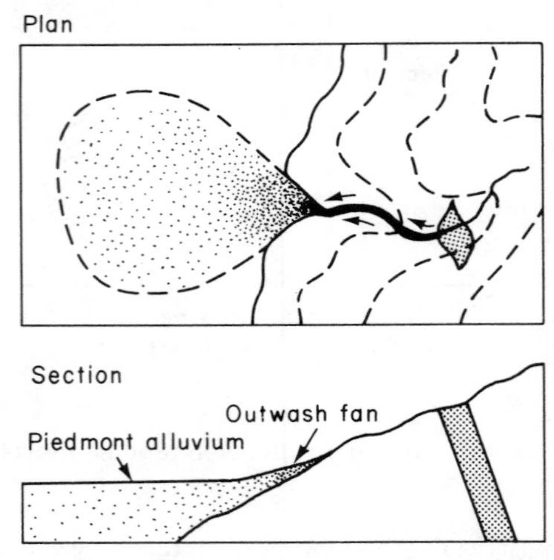

Fig. 10.5. Syngenetic (clastic) pattern in outwash fan and piedmont sheetwash alluvium.

B. Hydromorphic Patterns

The form of hydromorphic patterns depends first on the local flow pattern of the solutions. Linear dispersion trains result where the flow is strongly channeled, as in surface drainage patterns. Ground-water patterns, on the other hand, tend to be more nearly fan-shaped with local modifications resulting from preferential flow along bedrock channel-ways or permeable horizons in the overburden. Superjacent hydromorphic patterns, sometimes

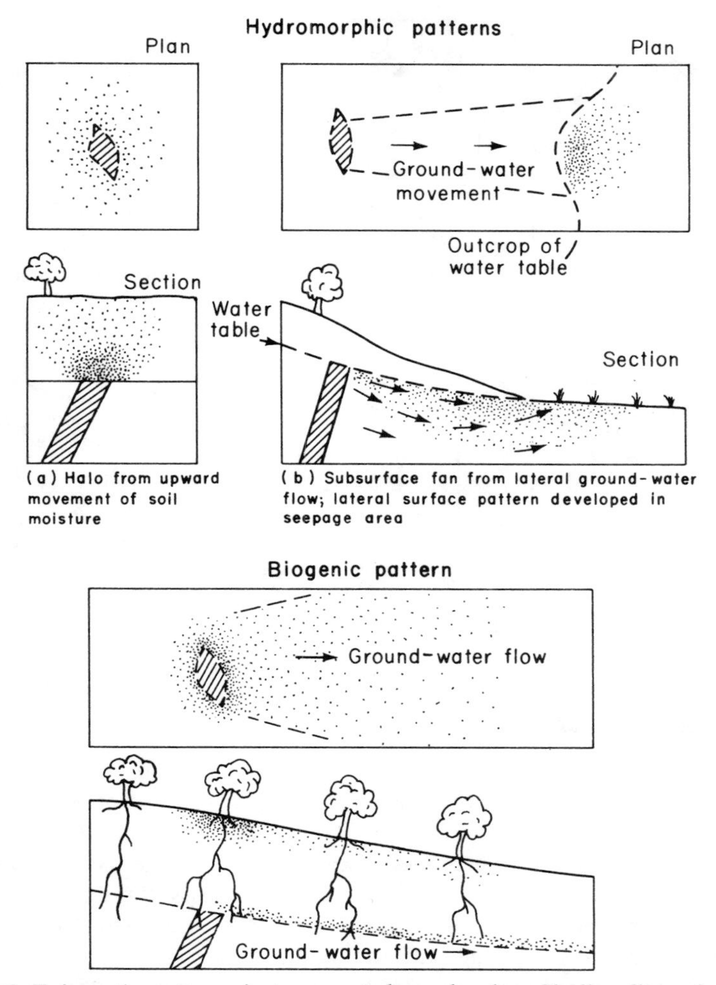

Fig. 10.6. Epigenetic patterns in transported overburden. Similar dispersion may also contribute to patterns in residual overburden.

precipitated from ground-water solutions in the overburden above a concealed deposit, range in form from haloes to fans, according to the amount of lateral flow. The form of hydromorphic patterns may be further complicated by the uneven distribution of local environments that favor precipitation. This effect is particularly noticeable in the lateral patterns developed in seepage zones, where the solutions may be canalized by the bedrock topography and precipitation is governed by the local distribution of appropriate Eh–pH conditions, organic matter, and clay minerals.

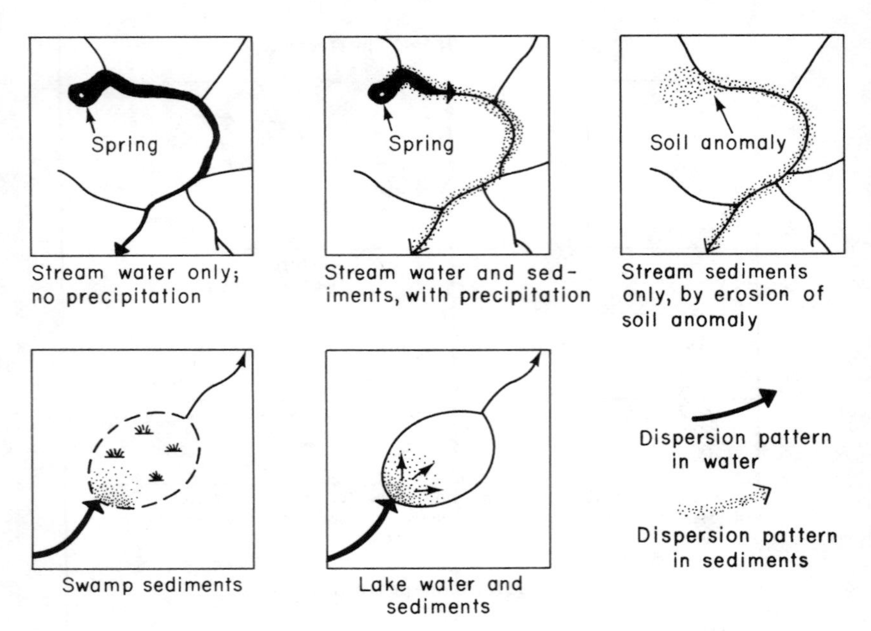

Fig. 10.7. Principal types of dispersion patterns in surface drainage.

C. Biogenic Patterns

The form of biogenic anomalies in plants and in the soil beneath the plants is determined by the pattern of available metal in the root zone. Biogenic anomalies may range, therefore, from superjacent to lateral, from haloes to fans, and even to trains, where metal-bearing ground water is canalized along a subsurface drainage system. In view of the complexity of the biogeochemical cycle, however, biogenic anomalies are often less well defined than the parent syngenetic or hydromorphic patterns in the root zone. Strong haloes may form, however, when the plants are rooted directly in an underlying mineral deposit.

VIII. ANOMALIES NOT RELATED TO MINERAL DEPOSITS

Most geochemical surveys disclose a bewildering array of anomalies, or departures from the geochemical patterns that are considered normal for the survey area. One of the most critical and often one of the most difficult

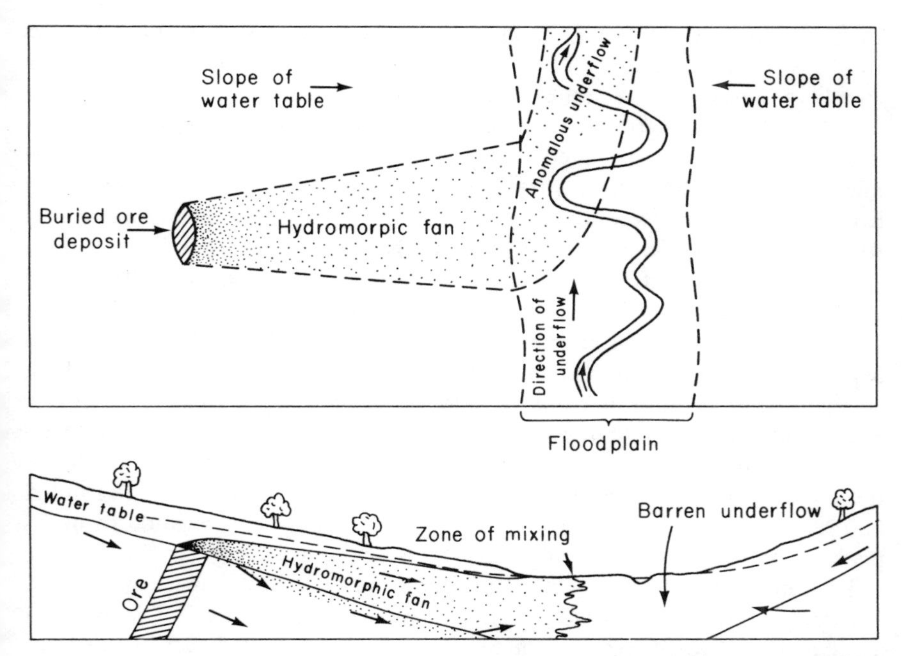

Fig. 10.8. Relation of ground-water anomalies on valley slopes to anomalous underflow.

tasks is that of discriminating between anomalies that should be followed up and anomalies that are of no economic significance. Non-significant anomalous patterns usually fall into one of three main types.

(i) Patterns related to certain rocks that are characterized by a relatively high background metal content.
(ii) Anomalies due to contamination as a result of man's activities.
(iii) Apparent anomalies resulting from sampling or analytical errors.

These misleading patterns will be discussed in more detail in later chapters; it suffices here to review briefly some of their general characteristics, for comparison with those of patterns related to mineral deposits.

A. High-background Source Rocks

Many kinds of rocks are characterized by relatively high concentrations of many of the same elements that occur in ore deposits, but that have no genetic relation to the ore. Surficial dispersion patterns developed from the weathering of these high-background rocks may show many of the features of patterns that are derived from ores. Discriminating between non-significant anomalies resulting from high-background rocks and significant anomalies resulting from ore deposits can be an extremely difficult problem. Fortunately, many metals in high-background rocks occur in a different form and are accompanied by different associated elements or primary minerals than the same metals in ores. Where this contrast in primary mineralogy and associations is carried over into the secondary patterns, it may be possible to develop criteria for screening out anomalies due only to high-background rock.

The family of ultramafic rocks, including peridotite, serpentine, and kimberlite, is probably the most spectacular example of a high-background source rock. These rocks are typically very much enriched in Cr, Ni, Co, and Mg. The weathering product of ultramafic rocks commonly contains a high concentration of montmorillonite, and thus has a high exchange capacity and possibly also a high content of readily extractable cations. The association of the four elements, Cr, Ni, Co, and Mg, normally carries through into the surficial dispersion patterns and serves as a key to the source. The stunted nature of vegetation growing on "serpentine soils" also is a guide to ultramafic rocks.

The family of mafic rocks, including gabbro, basalt, and diabase or dolerite, is characterized by a relatively high content of Fe, Ti, and Cu. The high pH associated with weathering calcareous rocks may restrict dispersion of their metal content to the extent that apparently anomalous patterns occur in the residuum. Acid from the oxidation of pyrite-rich rocks such as pyritic shale may have the reverse effect and cause the accelerated leaching of metals from rocks of normal composition and the resulting development of anomalous hydromorphic patterns unrelated to ore.

Less common high-background rocks that should be kept in mind in sorting out anomalies are some black shale (As, Cu, Pb, Mo, Ag, U, V, Zn), phosphorite (P, V, U, Mo, Zn), saline deposits (SO_4), and carbonatite (Zr, Nb, rare earths).

B. Contamination

Probable sources of metal contamination arising from human activity are many and varied. The most common are mine dumps, old mine workings, smelting operations, metal-rich agricultural chemicals, road metalling, industrial and domestic fumes, effluent, and waste of many kinds.

Dispersion is normally by gravity movement of solid particles, wind-blown material or in aqueous solutions, while plants may take up contaminating metal at any stage of its dispersion. Contamination patterns may thus form in any type of clastic, hydromorphic, or biogenic environment.

In clastic patterns, the modes of occurrence of exotic contaminating metal are usually very different from those of the natural metal, although the distinction may not be so readily made when the contaminating source derives from the soil of ancient mining activity. In most hydromorphic and biogenic dispersion patterns, however, it is extremely difficult to tell whether the metal came from a natural or an artificial source.

Initially the form of contamination patterns is conditioned by the geometrical shape of the source area. On dispersion away from the source of contamination, fans and trains may be developed which, in the case of windblown material, stream sediment, and aqueous solutions, may be very extensive. The outstanding feature of contamination, however, is the fact that it almost invariably originates at the surface of the ground. As a result, soil patterns are most strongly developed in, and in many cases confined to, the surface horizon, in contrast to natural superjacent soil patterns of clastic and hydromorphic origin. Apart from this, however, the surface origin of contamination need not lead to any dissimilarities with hydromorphic and biogenic patterns, either in the overburden or in the drainage system.

Nevertheless, despite the limitations and difficulties sometimes imposed by contamination, it has rarely, to the authors' knowledge, presented an insuperable problem in geochemical surveys, even in well-populated or intensively prospected areas.

C. Sampling Errors

Spurious anomalies related to sampling errors may be more difficult both to detect and to guard against, particularly in a routine survey where sampling is usually carried out by relatively untrained labor. For the most part, they arise from the collection of samples which, though superficially similar to the main body of samples, are enriched in metal by some natural process unrelated to ore. Natural processes of weathering, erosion, and surficial dispersion not uncommonly result in patterns of enrichment of ore metals in areas of background composition that can readily be confused with significant anomalies related to ore deposits. Natural enrichments of background metal commonly occur in the organic matter of humus horizons, in the limonitic B horizon of podzols, and in clastic material that for one reason or another has a high exchange capacity. Enrichments are characteristic of seepage areas and any other points along the drainage pattern where conditions favor the preferential accumulation of metal. Enrichment of metals in plants also may take place for a variety of reasons, all unrelated to the amount of metal in the supporting soil. As a rule, apparent anomalies arising from the inadvertent collection of naturally enriched material may tend to be related to some recognizable geomorphological feature of the environment, such as the topography, in which case they are readily recognized for what they are. The need to collect material that is strictly comparable in all respects and to note all changes in the sample environment that may possibly affect the dispersion of both background and anomalous metal cannot be overemphasized.

D. Analytical Errors

Anomalous patterns of no significance whatever may appear in geochemical data as the result of errors in analytical technique. Such patterns, if suspected, can be eliminated simply by a repeat analysis of the samples in question. Isolated erratic values are immediately suspect and should be rechecked. Apparent patterns arising from analytical bias may be recognized by their association with groups or batches of samples or with individual analysts. Such patterns may also be eliminated by analysis of the entire batch of samples in random order (Chapter 13, Section II.E). A commonly used method of protection against analytical bias is a system of routine repeat analysis of samples selected at random from previous batches.

IX. SUPPRESSION OF SIGNIFICANT ANOMALIES

Some of the same factors responsible for anomalies unrelated to mineral deposits may also operate in the reverse direction, with the result that significant anomalies in the vicinity of a metal-rich bedrock source may be missed. Thus during soil formation, for example, the A horizon may be leached so effectively that no anomalous values may be detectable in this horizon, whereas the underlying horizons may show strong anomalies. Samples intended to represent material from hydromorphic soils may have mistakenly been taken from freely drained ground, and hence miss a seepage anomaly. Not uncommonly, anomalies are overlooked because of gross failures in recognizing the nature of the matrix. Transported loess, alluvium, or moraine may be mistaken for residual soil. The absence of anomalous patterns in material that is assumed to be residual can be misinterpreted as an indication of the absence of an underlying metal-rich source.

Chapter 11

Anomalies in Residual Overburden

The nature of the parent material from which a residual soil has been formed is not always apparent from a cursory inspection of the color or texture of the soil. As a result, many ore deposits have undoubtedly escaped detection because there is no readily visible indication in the residual overburden. Even where the bedrock is exposed, weathering and leaching may still obscure evidence of the primary metal content.

Vestigial traces of the chemical and mineralogical characteristics of the original parent material, however, will normally be retained more or less *in situ* in the residuum even where the more obvious criteria are lacking. Vestigial features resulting from simple weathering in place of a metal-rich parent material provide an extremely direct and straightforward geochemical guide to buried ore. For this reason, geochemical anomalies in weathered rock and residual soil have become widely used and have proven successful under almost all conditions.

The simple residual pattern is often complicated, however, by hydromorphic patterns of metal superimposed on the residual material by precipitation from metal-rich ground waters draining the mineralized ground. These patterns may be formed at a considerable lateral distance from the deposit. The picture may be further complicated by anomalous concentrations of metal that bear no relation at all to the occurrence of ore.

I. ANOMALIES IN GOSSANS AND LEACHED OUTCROPS OF ORE

Residual limonite and other products of the weathering of Fe-bearing sulfides and carbonates have long been one of the most reliable guides to ore in areas of deep weathering and residual cover. However, because weathering tends to obscure the original chemistry, mineralogy, and texture, confusion is frequently possible between the weathering products of ore and those of ordinary rocks. Trace-element studies can help to distinguish weathered ore from similar material such as lateritic limonites, bog-iron ores, and weathering products of barren hydrothermal pyrite, syngenetic pyrite, and Fe-carbonates. Such studies can be made on samples from outcrops, shallow pits, drill holes, and float from soil. In addition, a knowledge of weathered ore is helpful in interpreting data from soils over orebodies where the concentrated weathering products are modified and diluted by soil-forming processes.

The texture and mineralogy of the weathering products of ores depend on many factors, including the texture and mineralogy of the primary ore, the degree of oxidation and leaching, the nature of the country rock, and the permeability of the country rock and ore. Gossans, or relatively massive concentrations of limonite, may be developed from sulfide-rich veins or massive-sulfide ores. Where non-sulfide gangue minerals predominate in the primary ore, the weathering product may be oxidized and leached rock with scattered limonite and leached cavities. Such material is commonly referred to simply as leached ore, or if it overlies enriched sulfides at a porphyry-copper deposit, as leached capping. The abundance and texture of the limonite grains are a guide to the abundance and identity of the original sulfides. The mineralogy of the limonite, as expressed by its color, can help to indicate the primary mineralogy and conditions of oxidation. However, a good deal of experience is needed for confident visual interpretation of limonites, even within a particular district. Visual interpretation of the mineralogy and textures of limonites in leached outcrops is thoroughly discussed by Blanchard (1968) and McKinstry (1948, pp. 268–276).

Pyrite oxidizes to form an Fe-oxide or -hydroxide plus sulfuric acid (Section III of Chapter 6). The initial solid product is usually an amorphous ferric hydroxide, perhaps $Fe(OH)_3$. This original amorphous material gradually crystallizes to goethite or hematite with a coarsening of the crystal size with time (Langmuir and Whittemore, 1971). Hematite and goethite are nearly equally stable at ambient temperatures, so that the species formed and the rate of crystallization depend on the composition of the surrounding fluid.

Trace metals tend to be strongly adsorbed by the initial amorphous to very fine-grained precipitates. As these precipitates recrystallize and coarsen in grain size, their affinity for adsorbing ions decreases, but some previously adsorbed ions may be incorporated into the crystals. The amount adsorbed is decreased by very acid solutions, so that under very acid conditions in pyritic cappings, only traces of the valuable metals may remain in the capping. At less acid to neutral pH values, most heavy metals are retained, at least in part, either as trace minerals (anglesite, malachite) or as adsorbed or coprecipitated elements in limonites, as discussed in Section IV of Chapter 8. In general, Pb, Cu, and Ag are strongly adsorbed by limonite, and Zn less so. Heavy elements occurring as anions (MoO_4^{2-}, AsO_4^{2-}, SeO_4^{2-}) are also strongly adsorbed to limonite, especially under acid conditions, and tend to be retained in cappings.

The pH of pore fluid in an oxidizing ore depends on the concentration of sulfuric acid, which in turn depends on the amount of pyrite in the ore, and the reactivity of associated gangue minerals and country rock. Quartz, chalcedony, kaolinite, alunite, and sericite are relatively unreactive in acid solutions, so that pH values below 2 are not unknown. Carbonates rapidly consume any acid produced by oxidation of nearby pyrite, and feldspars, mafic minerals, and most other silicates react with acid solutions to maintain only weakly acid pH values.

Geochemical studies of gossans have been successfully used in exploration for Ni–Cu ores in the lateritic terrain of Western Australia (Cox, 1975). The extreme lateritic weathering in that region obscures even the parent rock identity and has produced many types of Fe-rich materials easily confused with the gossans over pyrrhotite–pentlandite ores.

Initial geochemical work in Australia showed that gossans over many orebodies could be distinguished from "pseudo-gossans" by their very high Ni content (Fig. 11.1 and Table 11.1). However, in some localities, Ni was enriched or leached during weathering, so that additional criteria were necessary (Butt and Sheppy, 1975). High Cu contents were favorable indications in most localities, but exotic Cu mobilized from nearby sedimentary units and

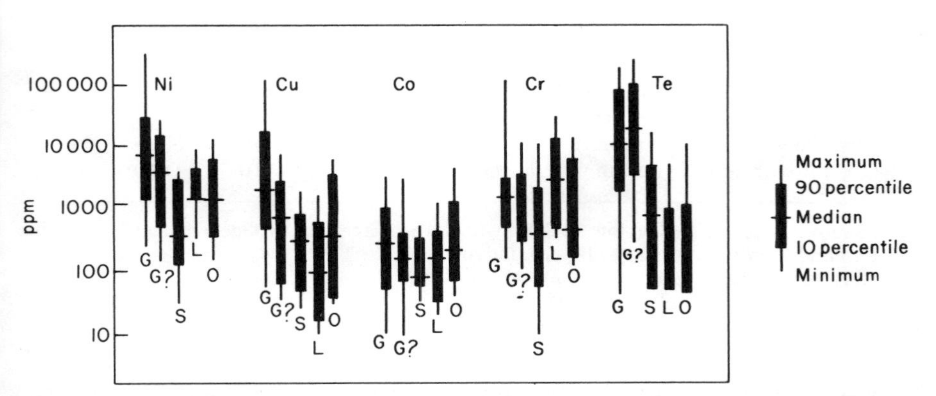

Fig. 11.1. Composition of gossans and pseudo-gossans from the Archaean of Western Australia. G, gossans over Ni ore; G ?, gossans probably over Ni ore; S, pseudo-gossan over sulfide-rich sediment or tuff; L, lateritic weathering product of ultramafic rock; O, gossans over non-nickeliferous sulfide accumulations. (After Bull and Mazzucchelli, 1975.)

Table 11.1
Analyses of limonites showing contrasting ore-metal contents in gossans and pseudo-gossans

Metal content of gossan and pseudo-gossan, Borneo[a]

Sample	Cu (ppm)	Co (ppm)	Ni (ppm)
Sulfide gossan	11 000	500	50
Pseudo-gossan derived from ultramafic rock	110	200	2000
Gossan of unknown derivation but probably related to sulfides	3200	25	130

Metal content of gossan and pseudo-gossan, Western Australia[b]

Sample	Ni (ppm)	Cu (ppm)	Zn (ppm)	Pb (ppm)	Mn (ppm)	Cr (ppm)
Gossan over Ni–Cu ore	7600	3000	90	80	340	360
Laterite over ultramafic rock	1200	20	20	20	1200	3400
Laterite over shale	1975	40	590	<20	1850	350

Metal content of gossan derived from mineralized and unmineralized sulfide veins, Sierra Leone[c]

Sample	Pb (ppm)	Mo (ppm)	Au (ppm)	Ag (ppm)
Quartzose gossan (mineralized)	900	600	20	6·6
Gossan stringers (mineralized)	400	400	0·3	13
Massive gossan (mineralized)	300	10	Nil	Nil
Gossan with muscovite (unmineralized)	50	10	Trace	Trace

[a] Samples supplied by Geological Survey of Borneo; analyses by Geochemical Prospecting Research Centre, Imperial College, London.
[b] Source: Clema and Stevens-Hoare (1973).
[c] Source: Geological Survey of Sierra Leone.

adsorbed by limonites locally led to confusion. High Cu and Ni combined with low Mn and Cr were also favorable indications of ore in most cases (Clema and Stevens-Hoare, 1973). Comprehensive studies show that Ir, Pd, and Te, which are highly enriched in the ore relative to normal rocks and are relatively immobile during weathering, can be used to evaluate limonite-rich materials of uncertain origin in this province (Travis *et al.*, 1976; Bull and Mazzucchelli, 1975; Wilmshurst, 1975).

Leached cappings overlying porphyry-copper deposits can also be evaluated by combined mineralogical and trace element studies. Hydrothermal alteration products are important in determining the behavior of Cu in these cappings. "Reactive" cappings containing feldspars, chlorite, biotite, or montmorillonite, when accompanied by relatively low amounts of pyrite, neutralize most of the acid produced during weathering, and Cu is precipitated as malachite, chrysocolla, or other oxidized minerals within the capping. Oxidized ore at Ajo, Arizona, is a good example of this combination of alteration and sulfide abundance (Gilluly, 1946). In more typical cappings derived from ore with moderate contents of pyrite and small amounts of reactive minerals, most of the Cu is leached downwards, there precipitating to form an enrichment zone, but anomalous amounts remain in limonite of the capping. With abundant pyrite and non-reactive gangue, nearly all Cu is leached, to leave less than 100 ppm Cu in capping overlying high-grade enriched ore, as at the Ruth orebody, Ely, Nevada. However, in this and other cappings, distinctly anomalous amounts of Mo are retained in the capping. In Puerto Rico, Learned and Boissen (1973) show that Au and Mo are retained in soils from which nearly all Cu is leached. In capping containing only traces of limonite, the metal content of the limonite can be approximated by leaching with dithionite or other Fe-selective leaches (see Section III of Chapter 3) and determining ratios of metal to Fe in the leachate.

The metal content of gossan-like materials from two other areas (Table 11.1) illustrates the kind of differences expected between true and false gossans. According to Sindeyeva (1955), gossans derived from epigenetic sulfides are characterized by a high content of Se. Within a district of relatively uniform weathering and soil type, the ratio of ore metals in gossans may provide a rough guide to the ratios of ore metals in the underlying ore. For example, in the Nyeba district of Nigeria, gossan collected over a known Pb–Zn deposit contained 4000 ppm Pb and 500 ppm Zn, but at a prospect shown by a soil survey to contain Zn but no Pb, the gossan contained 50 ppm Pb and 8000 ppm Zn (Hawkes, 1954).

So-called transported gossans have an origin somewhat different from the residual gossans discussed above. Some transported gossans are simply colluvial accumulations of fragments of normal gossan that have moved down the slope from the site of weathering. Another variety of transported gossan of an entirely different origin is effectively a fossil spring or seepage deposit, where at one time Fe-rich ground water resulting from the leaching of Fe minerals has precipitated massive limonite at or near the surface. The composition of lateral hydromorphic gossans of this type will also tend to reflect the metal content of the parent mineral deposit, although differences in the relative mobilities of the ore metals will show up more strongly than with residual gossans, where transport in ground-water solutions is not an important factor.

II. SYNGENETIC ANOMALIES IN RESIDUAL SOIL

The goal of a geochemical soil survey is recognition of the primary distribution pattern of selected elements in underlying parent rock. In residual soils, this primary pattern is generally expressed in a recognizable fashion, but superimposed on it are the effects of many surface processes. Some of these processes tend to make the soil more homogeneous, and thereby tend to subdue the primary pattern (frost, plants, animals, gravity, local solution, and redeposition), and others tend to form vertically differentiated horizons (soil formation). This dynamic system is further complicated by removal of elements during weathering and soil formation (leaching by rain water, uptake by plants), and by addition of elements (deposition from ground water, addition from decaying vegetation, airborne dust, or solutes in rain).

The effects of all these factors on the formation of anomalies in residual soil are best understood by consideration of some basic physical and chemical characteristics of anomalies in soil. The most important of these characteristics are (i) the mode of occurrence of the indicator element, (ii) the form and magnitude of the anomaly, (iii) the homogeneity of the pattern, and (iv) the variation with depth.

A. Mode of Occurrence

Elements derived from the weathering of both ore and unmineralized rock occur in soils in a manner dependent on the chemical properties of the element, its mode of occurrence in the parent material, and the chemistry and mineralogy of the soil. A few metals, such as W, Sn, and Au, are normally retained in the soil as components of resistant primary minerals. Most metals in well-developed soils occur predominantly as secondary ore minerals or as firmly bonded components of clay minerals and hydrous Fe- and Mn-oxides. Many elements in poorly developed soils occur as partially weathered grains from the parent material. A relatively small part of the metal from the parent may be held in organic material or as ions adsorbed on clays, colloidal Fe-Mn-oxides, and organic material. The ore minerals that are relatively resistant to chemical breakdown in the environment of weathering include cassiterite, wolframite, scheelite, columbite–tantalite, pyrochlore, diamond, Au, Pt, beryl, chromite, ilmenite, corundum, and cinnabar. Common resistant gangue minerals are quartz, tourmaline, garnet, magnetite, barite, and fluorite. Less resistant minerals may also persist where mechanical weathering predominates over chemical weathering, as is often the case in arid or very cold climates and on steep slopes undergoing active erosion. The grain size will depend on the original particle size and the degree to which this may have been reduced by solution or abrasion during weathering. Commonly an appreciable proportion of the metal held in resistant minerals occurs in the intermediate to coarse fraction, as suggested by the data for immobile metals in Table 11.2.

Secondary ore minerals occur in some soils over ore as a result of the weathering of chemically unstable primary ore minerals. The formation of limonite from Fe-bearing sulfides and Mn-oxides from Mn-carbonates are simple examples. Malachite ($Cu_2CO_3(OH)_2$), anglesite ($PbSO_4$), ferrimolybdite ($Fe_2(MoO_4)_3.8H_2O$), and carnotite ($K_2(UO_2)_2(VO_4)_2 . nH_2O$) are other examples in which specific conditions or chemical constituents are required for their formation. Such conditions or constituents may occur only in very special geological environments. For example, Cu-carbonates are not stable in normal acid soils of humid regions, but have been noted in soil at the Bushman Mine in Botswana, where alkaline calcareous soils have developed in an arid climate. Similar occurrences have been noted in freely drained latosolic soils derived from calcareous bedrock in central Africa. Primary and secondary Cu-sulfides occur along with secondary oxidized Cu minerals in the Philippines where soil has been derived directly from the zone of secondary enrichment which has been exposed by recent erosion. In Sierra Leone, high concentrations of As and Mo in soils probably occur largely as Fe-arsenate and Fe-molybdate, respectively (Mather, 1959).

Most mobile or semi-mobile metals in freely drained residual soils overlying oxidizing ores are firmly bound in clays and Fe–Mn-oxides. Under these conditions, semi-mobile trace metals tend to be concentrated in the silt- and clay-size fractions of the soil, as suggested by the data for Cu, Co, and Mo in Table 11.2. Selective chemical extractions of soils over and near Zn ores in limestones subjected to weathering in a humid subtropical climate in Tennessee show that about 50% of the Zn occurs in hydrous Fe-oxides (White, 1957). Less than 5% was found to be readily extractable, and the remaining 40–50% is assumed to occur in the lattice of clay minerals. Similar proportions were found in both anomalous and background soils.

Distinct enrichment of anomalous Cu in ferruginous concretions of the B horizon was noted by Tooms and Webb (1961) in Zambian latosols. The concretions contained up to 1290 ppm Cu, compared to 800 ppm in the surrounding matrix. Copper and Co were also enriched in Mn-oxides in these soils. Jay (1959) has demonstrated a correlation of Co with Mn in these same soil profiles (Fig. 8.18). Lead shows a tendency to accumulate with Fe or Mn

Table 11.2
Distribution of mobile and immobile metals in different size fractions of anomalous residual soil[a]

Size fraction	Mesh (B.S.S.)	μm	Immobile metals remaining in resistant primary minerals (ppm)				Residual portion of mobile metals remaining in secondary soil minerals (ppm)				
			Cr (Rhod.)	Sn (Malaysia)	Be (Rhod.)	Nb (Zambia)	Zn (U.K.)	Cu (Zambia)	Co (Zambia)	Ni (Tanzania)	Mo (Sierra Leone)
	10	1980									
			—	150	80	—	—	—	—	—	—
Coarse sand	20	894									
			3000	2100	70	500	200	20	6	600	20
	35	471									
			4000	6700	25	800	240	30	4	600	30
	80	186									
Fine sand			10 000	4000	20	1500	180	40	4	600	70
	135	104									
			11 000	2400	15	2300	300	25	6	600	90
	200	76									
Very fine sand			—	1000	15	1800	300	35	10	—	—
	—	20									
Silt			8000	175	10	1700	500	170	24	1200	100
	—	2									
Clay			—	45	3	250	1500	500	40	—	—

[a] Source: Geochemical Prospecting Research Centre, Imperial College, London.

in subtropical soils (Webb, 1958a; Ledward, 1960). Lead and Zn are concentrated to levels of 1·5% and 0·5% in Fe-oxides of soils over granites in France (Leduc and Boucetta, 1971).

The proportion of readily extractable metal (cxMe) in the A and B horizons of freely drained residual soils rarely exceeds 5–10% of the total metal and in some localities is 1% or less. The low percentage of readily extractable metal reflects the relative stability of minerals exposed to constant leaching by downward-percolating rain water. Higher proportions of cxMe can usually be attributed to hydromorphic dispersion or incomplete weathering. Figure 11.2 illustrates profiles of total Cu and cxCu for residual soil over ore in Zambia, and for soil bearing hydromorphically introduced Cu. Total Cu at the surface differs relatively little, but the proportion of cxCu/total Cu differs markedly. This difference in form of anomalous metal furnishes an important guide to the source of anomalies in soils.

Adsorption of metal on organic matter in the A0 and A1 horizons accounts for an important fraction of the total metal in some soils. In such cases, the ratio of cxMe to total metal, though erratic, is commonly higher than in the underlying soil horizons.

In perennially moist soil horizons, where metal has been introduced or redistributed by dispersion in ground water or soil moisture, very significant proportions will be cold extractable (Fig. 11.2c). inasmuch as the bulk of such metal is held by adsorption on clay-size particles and organic matter or as minor constituents coprecipitated with hydrous Fe- and Mn-oxides.

B. Form and Magnitude of the Anomaly

1. Intensity and contrast of anomalies

In a crude way, the contrast of anomalies in soil tends to reflect the primary contrast between ore and unmineralized host rock. This tendency is subject to so many extraneous factors, however, that a reliable quantitative correlation between the tenor or ore and either contrast or intensity of anomalies in overlying soil can rarely be established. In the first place, the background metal content is likely to vary as the wallrocks change their character. Secondly, anomaly contrast in the soil can at best reflect the primary contrast at depth only insofar as the latter is preserved in the weathered parent material of the soil. Consequently, the contrast at the surface can vary greatly, even along the strike of the same orebody, solely as an effect of variations in wallrock lithology and the degree of oxidation and leaching at the present-day level of the suboutcrop. To these factors must be added variations related to the overburden, particularly with regard to topography, drainage, and the type, depth, and maturity of the soil profile, all of which may vary markedly over

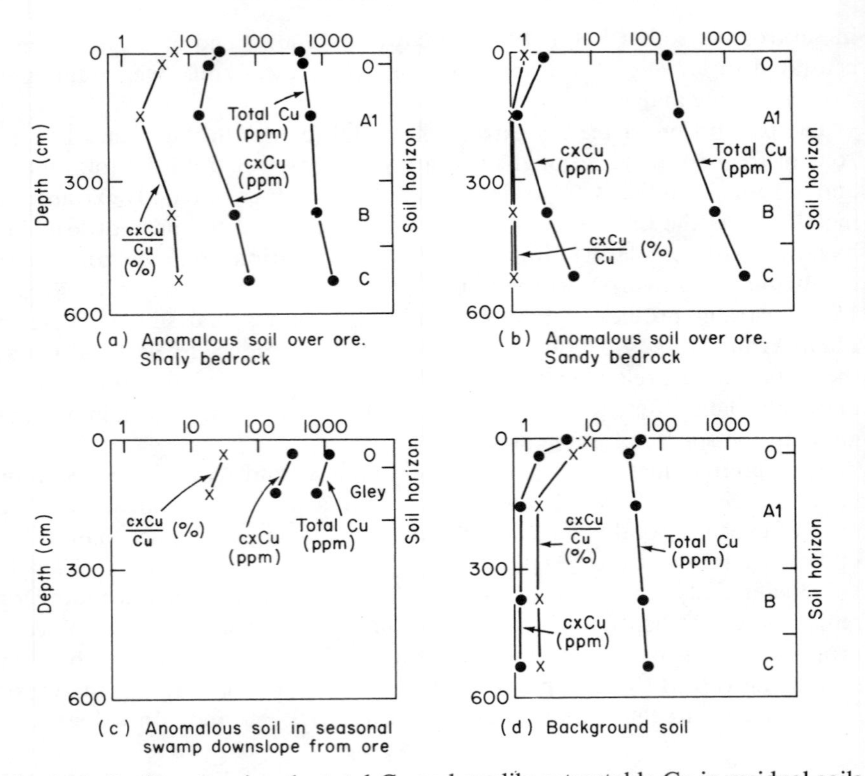

Fig. 11.2. Profiles showing the total Cu and readily extractable Cu in residual soils over ore, compared to background soil and soil with hydromorphically introduced Cu. Note differences in proportion of cxCu/total Cu, and more extensive leaching in sandy soils. (After Webb, 1958a.)

short distances. Differences between the tops, upper slopes, and lower slopes of hills are to be expected, as might be predicted from the discussion of factors affecting soil formation in Chapter 7.

Where all the modifying factors are essentially constant, however, the anomaly contrast may reflect roughly the size and tenor of underlying mineralization. Figure 11.3 shows an example of such a correlation in areas of similar bedrock geology and soil conditions.

Relatively immobile elements such as Sn, Nb, and Cr suffer little leaching, and to this extent the primary contrast tends to be preserved in the soil. In places, contrast for such elements may even be enhanced by residual accumulation of metal in the soil due to solution or washing out of non-metalliferous matrix material. Enrichment of this type can be seen over a carbonatite deposit in Zambia, where solution of the carbonate has resulted in accumulation of stable pyrochlore in the soil (Table 11.3).

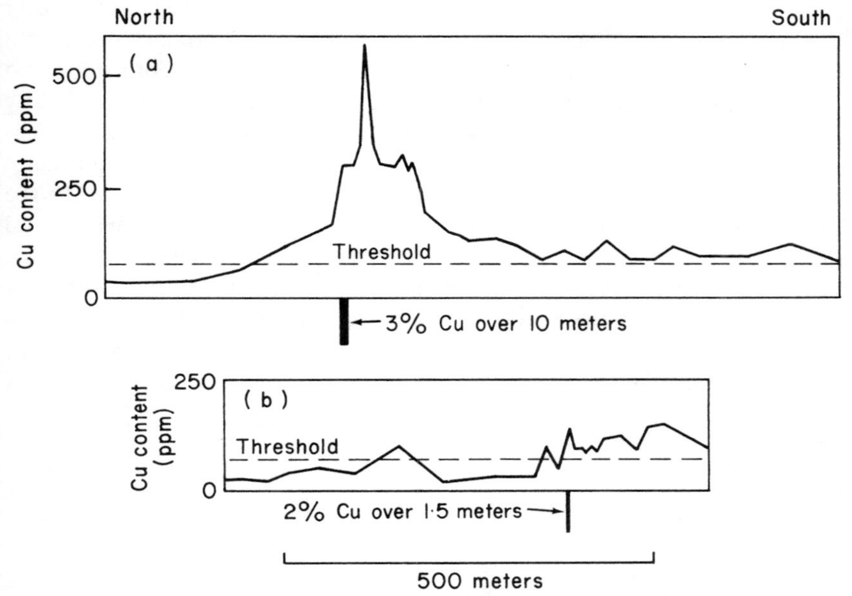

Fig. 11.3. Relationship between soil anomalies and tenor of underlying mineralization at Baluba, Zambia. Note greater intensity of anomaly over wider higher-grade ore zone. Data on −80-mesh fraction. Mineralization: disseminated chalcopyrite and pyrite in argillaceous ore horizon; overburden, freely drained, residual latosol, 7–14 m thick; relief: flat peneplain, gentle slope to the south, water table 2–7 m from the surface; sample depth: 45 cm (A2 horizon); analysis: total Cu, bisulfate fusion, dithizone/benzene reaction. (After Tooms and Webb, 1961.)

Table 11.3
Accumulation of Nb as residual pyrochlore in residuum from carbonatite, Kaluwe, Zambia[a]

Depth (cm)	Description	Nb (ppm)
0–7		1600
7–15	Brown residual soil	1000
15–30	↓	800
30–45		600
45–60		500
60–90	Gray–brown calcareous residuum, highly watered	400
90–120	↓	200
120–150		200
150–180	Weathered carbonatite	400

[a] Source: Watts *et al.* (1963); data for −80-mesh fraction.

With mobile elements such as Zn, U, Mo, and Cu, the contrast and intensity of anomalies are usually much less than the values in the primary ore. Metal contents in soil over ore are typically reduced by 50–95% or more compared to the grade of primary ore, whereas background values are only slightly affected (Huff, 1952). In general, the less mobile elements tend to give anomalies with higher contrast and intensity than more mobile elements. Figure 11.4 illustrates the preservation of the primary contrast for Pb, compared to the suppression of contrast for Zn and Cu in residual soils over a base-metal deposit in North Carolina. Figure 11.5 shows the characteristic difference between an intense, sharp Pb anomaly and a weaker, flatter Zn anomaly.

The extent to which the contrast is suppressed or enhanced depends on the total effect at any given point of all the factors influencing mobility and the length of time they have been operative. Briefly, mobility and consequent low

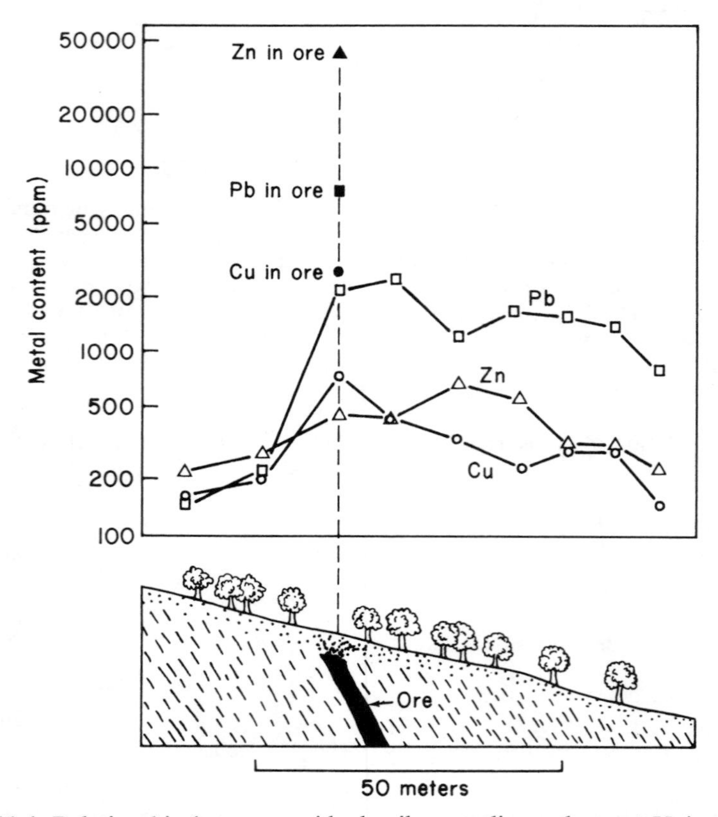

Fig. 11.4. Relationship between residual soil anomalies and ore at Union Copper Mine, North Carolina. Data on −12-mm fraction. (Data from Huff (1952) and Hawkes (1957).)

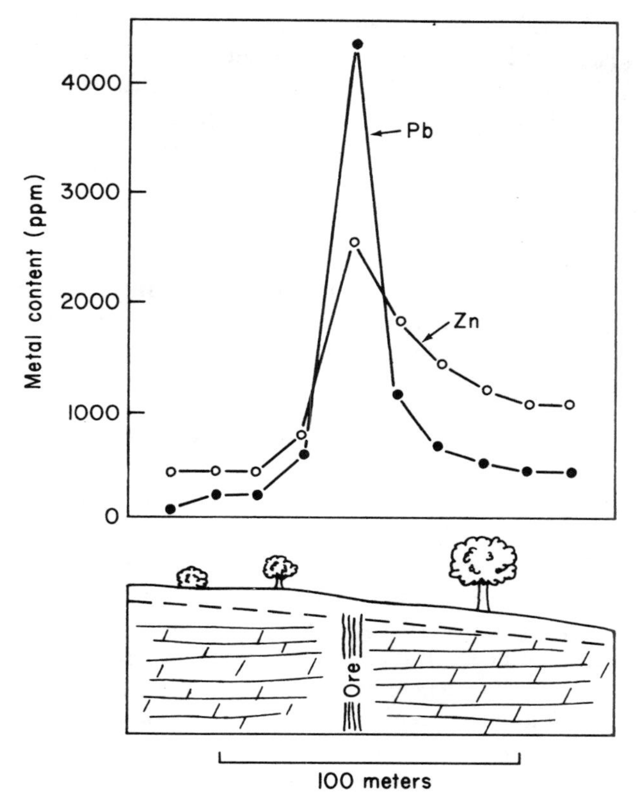

Fig. 11.5. Relationship between residual soil anomaly and Pb–Zn vein at Porter's Grove, Wisconsin. Data on −2-mm fraction. Mineralization: fracture-fillings of coarse galena and sphalerite in dolomite; overburden: shallow, freely drained residual podzol; relief: gentle. (After Huff, 1952.)

contrast are promoted by a low content of adsorbents (organic matter, Fe- and Mn-oxides, clay), a siliceous soil, and relatively acid soil (for cations). Leaching is also greater in areas of high rainfall, free drainage, deep chemical weathering, mature soils, and old stable land forms. Conversely, leaching tends to be minimal in immature, alkaline, or calcareous soils, and those with a high content of adsorbents such as Fe- and Mn-oxides, clays, and organic matter. Shallow chemical leaching, low rainfall, poor drainage, and active erosion are associated with relatively high contrast in soil anomalies.

Many exceptions to these general trends may be noted. For instance, anomaly contrast for Mo is weak in alkaline soils because of the high mobility of Mo at high pH values. Similar effects are to be expected for Se, As, and other elements occurring as anions. The primary contrast may even be enhanced in the soil if the leached zone has been removed by erosion, and the present-day suboutcrop is in the zone of secondary enrichment. Another special case is the marked enrichment of Cu noted in partially leached soils derived from the weathering of calcareous rocks under tropical conditions; in extreme cases, 1–2% Cu occurs in soil derived from bedrock containing only about 0·2% Cu.

2. Width of anomalies

The lateral extent, or width, of a residual soil anomaly reflects the width of the primary halo around ore, plus the dispersion generated in weathered rock during oxidation and weathering, plus the dispersion in the soil. The extent of dispersion during weathering prior to incorporation of metal in soil can be appreciable, as demonstrated by Scott (1975) for a pyritic Cu–Zn orebody in metamorphosed aluminosilicate rocks in South West Africa (Fig. 11.6). At this locality, oxidation of pyrite creates pH values as low as 3·5, allowing Cu to migrate away from the orebody until the acid is neutralized by reaction with the wallrocks. Only 7% of the original Cu remains in the oxidized orebody, the rest having migrated into the country rock. The resulting halo of rock with anomalies of four times background extended 50–100 m from the vein. Further dispersion occurred in the surface soil.

The importance of bedrock type in controlling leaching and dispersion is well illustrated in Zambia, where broad, diffuse anomalies that may only rise to 1·5 times threshold are characteristic of an arenaceous ore horizon, whereas an anomaly related to equivalent mineralization in argillaceous host rock is usually narrower and shows a better-defined peak rising to five to ten times threshold over the ore suboutcrop. The anomaly becomes even more intense where the ore horizon is overlain by poorly drained organic-rich soil, wherein lateral dispersion is retarded by metal adsorption on organic matter. Anomalies formed under these three conditions are illustrated in Fig. 11.7.

3. Distortion of anomalies

Symmetrical haloes of dispersed metal are characteristic of level terrain. On sloping ground, however, anomalies generally tend to fan out asymmetrically in the downslope direction (Figs 11.4, 11.5, and 11.8). Downslope distortion is usually the effect of physical movement of metal-bearing particles by soil creep, aided by solution and precipitation from laterally moving subsurface waters in the case of soluble metals. Relative mobility is therefore important in the development of asymmetrical anomalies. Figure 11.5 shows the difference between the dispersion of Pb and Zn, where the lower mobility of Pb results in its being dispersed mainly mechanically by soil creep, whereas the dispersion of Zn has been aided by chemical solution and precipitation.

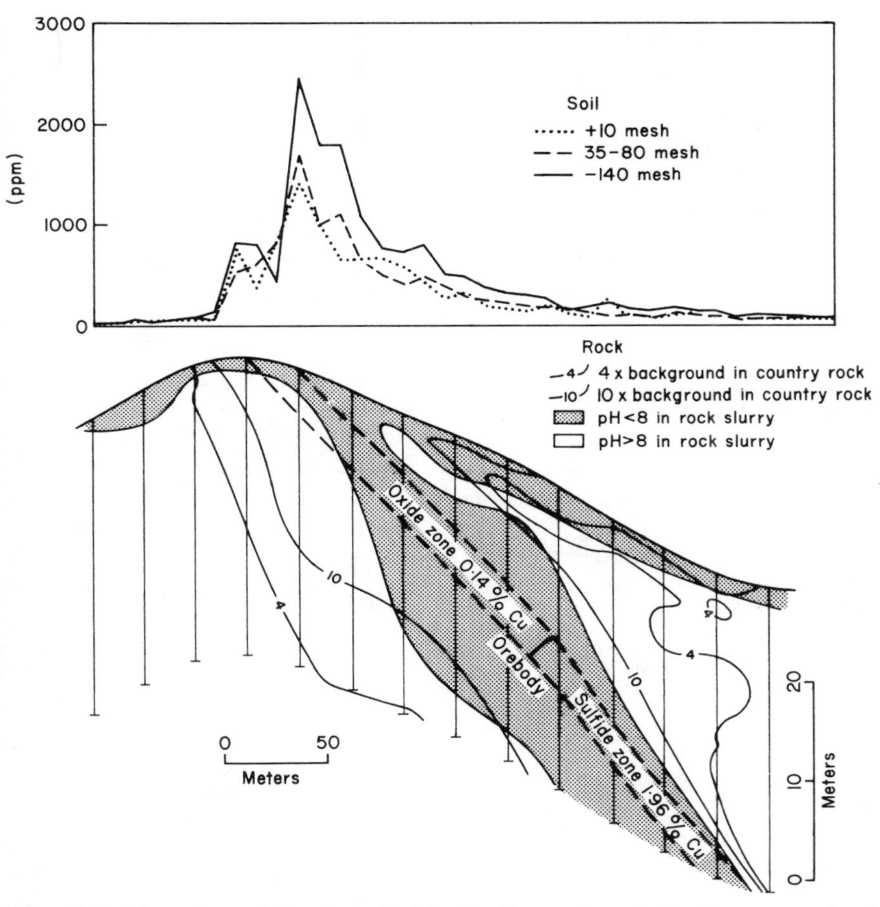

Fig. 11.6. Dispersion of Cu from pyritic Cu–Zn ore into bedrock as a result of weathering in the Windhoek district, South West Africa. (After Scott (1975), as determined by a series of drill holes and Cu in the resulting soil.)

As a rule, the steeper the slope, the greater is the distortion. In extreme cases the anomaly may persist for 100 m or more to the foot of the slope and even beyond into the colluvium (Fig. 11.9). However, the relation between relief and the degree of distortion is not always a simple one. For example, in the Ruwenzori Mountains in Uganda, very intense anomalies may occur over the suboutcrop, but extend for only relatively short distances downslope despite gradients of more than 35° (Fig. 11.10). The reason for this appears to be that erosion is keeping pace with oxidation, but conditions are such that metal is being leached from the soil as it moves down the slope (Jacobson,

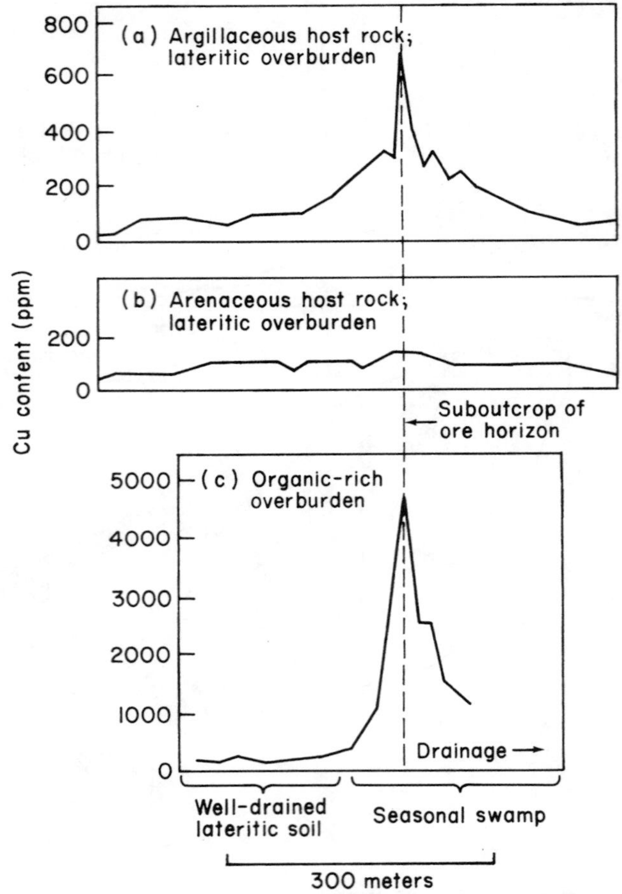

Fig. 11.7. Variation in soil anomalies according to host rock and drainage conditions, Zambia. Samples: A2 horizon in (a) and (b), G horizon in seasonal swamp in (c); sample depth 45 cm; analysis: total Cu, bisulfate fusion, dithizone/benzene determination; cxCu, cold ammonium citrate (pH 2), dithizone/white spirit determination. (After Webb (1958a) and Tooms and Webb (1961).)

1956). In contrast, if given enough time, a very appreciable degree of distortion can take place on very gentle slopes. For example, on the mature peneplain of the Zambian Copperbelt, anomalies in residual soil may extend for 600 m and more on slopes of only 1–2° (Fig. 15.4). The extensive downslope distortion of these anomalies is believed to be due in part to the development of biogenic anomalies over areas of metal-rich ground waters draining the mineralized rock (Tooms and Webb, 1961). Considerable distortion purely by

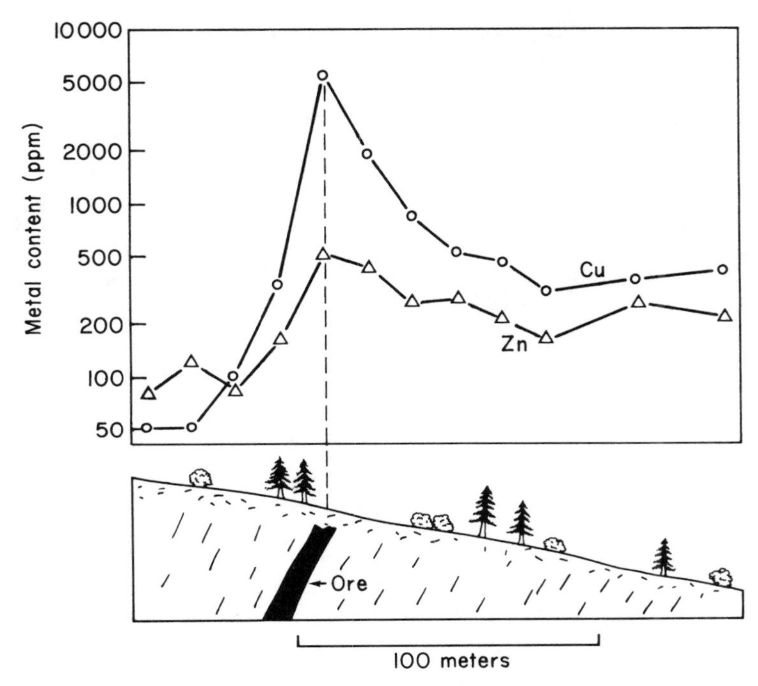

Fig. 11.8. Relationship between residual soil anomalies and ore at Malachite Copper Mine, Colorado. Data on −12-mm fraction. (Data from Huff, 1952.)

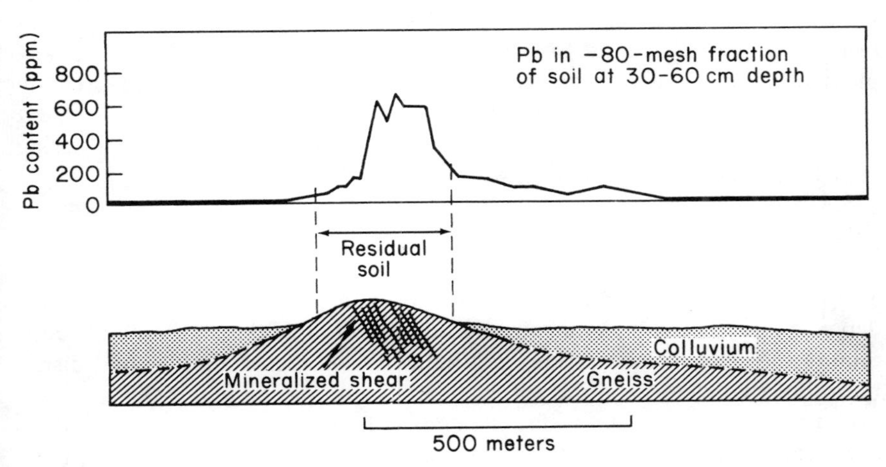

Fig. 11.9. Lateral dispersion of Pb from area of mineralized outcrop into soil and colluvium, Mpanda, Tanzania. (After Ledward, 1960.)

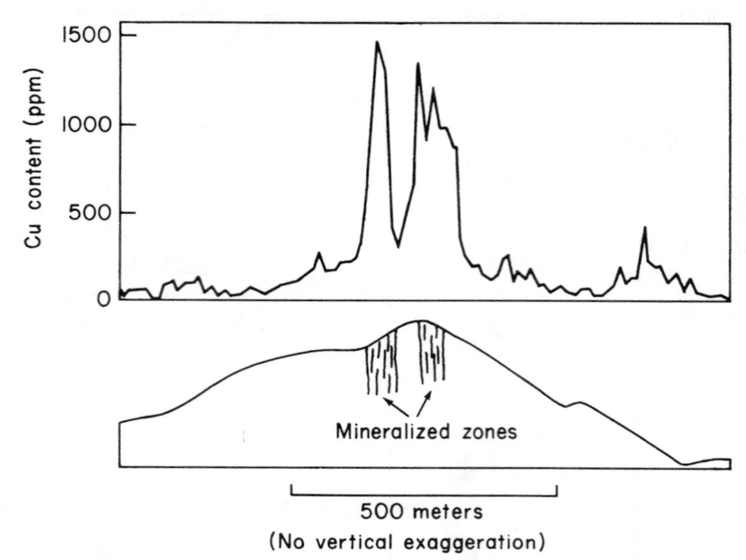

Fig. 11.10. Copper anomalies in near-surface soil, Ruwenzori Mountains, Uganda. Data on −80-mesh fraction. (After Jacobson, 1956.)

soil creep can also take place on very old peneplain surfaces if conditions are unfavorable for ground-water dispersion (Fig. 11.11).

Where anomalies are distorted by soil creep, it is not uncommon to find that the anomaly peak in the near-surface soil is also displaced downslope from the suboutcrop of the ore (Fig. 11.12). The amount of displacement naturally tends to increase with angle of slope and depth of overburden, but in practice is rarely found to exceed 20 m and is often much less. Displacement of the peak may also occur where the trace of an inclined orebody projects up into the overburden. A case of this kind where the displacement is further accentuated by compaction of the overburden during weathering of the limestone host rock is shown in Fig. 11.13.

C. Homogeneity

The homogeneity or regularity of residual anomalies in well-developed soils depends directly on the grain-size and distribution of the soil minerals containing the element in question, and indirectly on the occurrence and distribution of that element in the parent bedrock and/or country rock.

Ore metals which occur as discrete grains of resistant primary or secondary minerals scattered through the soil often show irregular anomalies. On the other hand, much more homogeneous anomalies are characteristic of elements that are pervasively dispersed throughout the body of the soil as

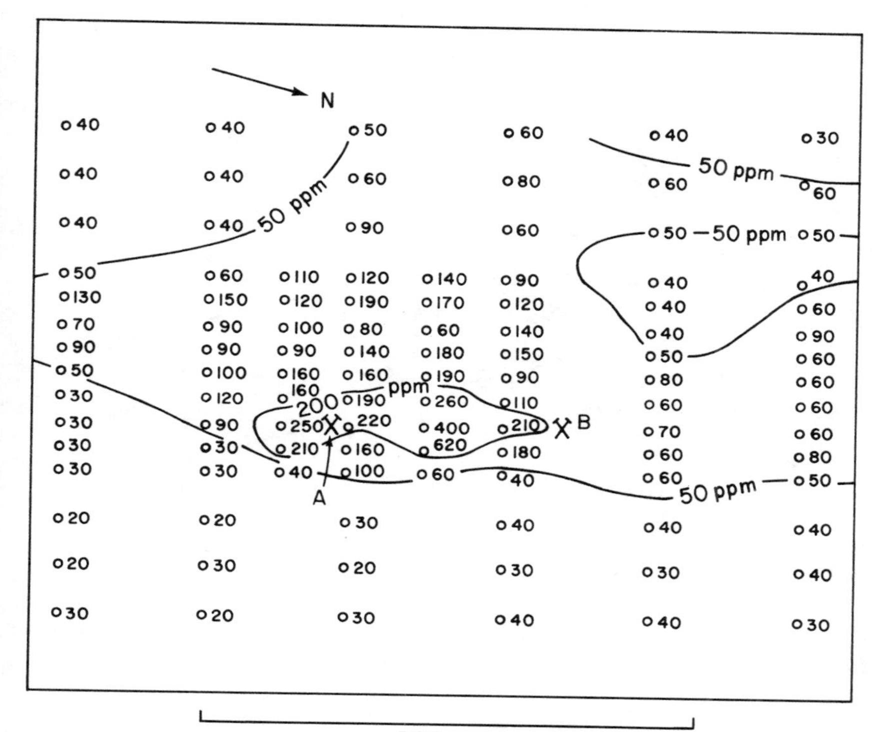

Fig. 11.11. Distortion of Cu anomaly in residual soil on ancient land surface despite alkaline environment and flat-lying terrain, Bushman Mine, Botswana. Mineralization: Cu-bearing quartz breccia exposed at A and B; climate: subhumid; overburden: residual soil, pH 8·0; terrain: essentially flat, general drainage direction is westward; sample depth: 23 cm; analysis: total Cu in −80-mesh fraction. (After Webb, 1958a.)

minor constituents of the fine-grained clay minerals and hydrous oxides. Thus many of the relatively immobile elements, such as Sn, Be, and Pb, tend to give less homogeneous anomalies than Cu, Zn, Co, and other relatively mobile elements. Under certain conditions even the mobile elements may give irregular anomalies if they are preferentially enriched in relatively coarse secondary minerals, concretions, or segregations spotted through the soil.

Despite the very thorough mechanical and chemical homogenization that often occurs in the upper soil horizons, excessively irregular patterns in the bedrock may still persist in the overburden, even in the case of well-dispersed mobile elements. An example of a broad but inhomogeneous Pb anomaly representing the summation of a closely related series of smaller anomalies, each of which by itself may be homogeneous in terms of metal distribution in

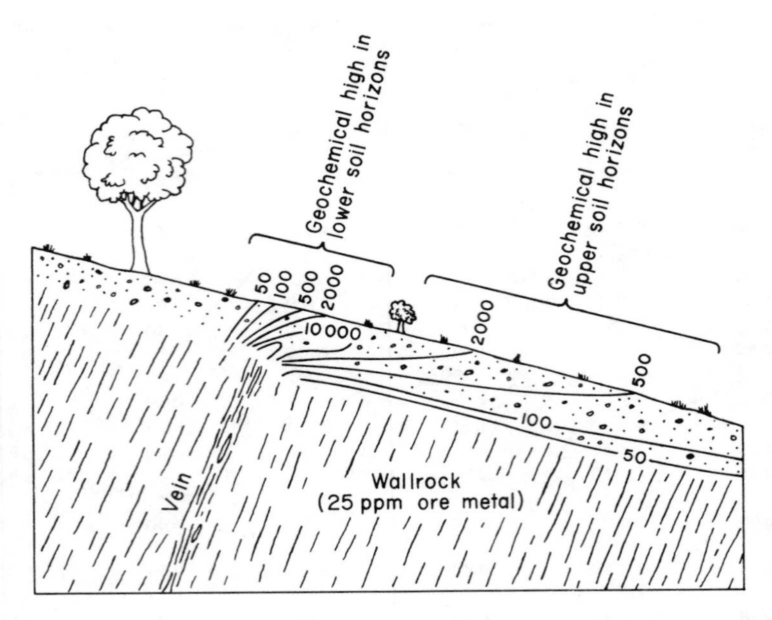

Fig. 11.12. Diagrammatic cross-section showing residual soil anomaly resulting from purely mechanical dispersion. (After Huff, 1952.)

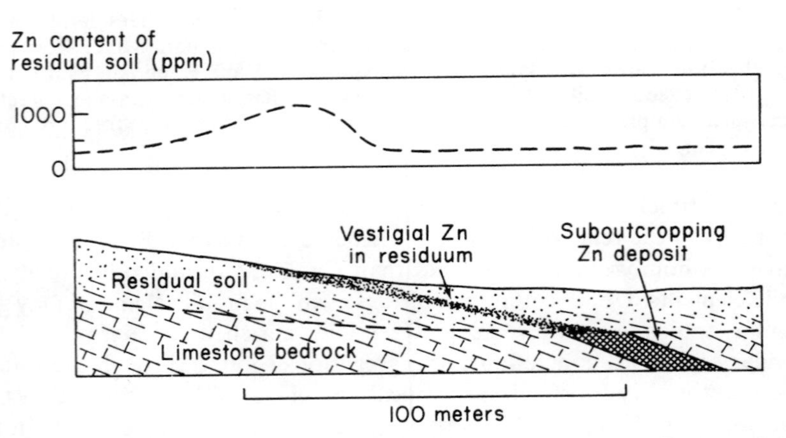

Fig. 11.13. Cross-section showing displacement of a residual soil anomaly as a result of compaction during weathering of a gently dipping deposit, Tennessee. Mineralization: sphalerite replacing limestone; climate: temperate, seasonal; relief: moderate; overburden: deep residual kaolinitic soil; sample depth: 1·3 m; analysis: total Zn in −80-mesh, bisulfate fusion. (After Hawkes and Lakin, 1949.)

the individual samples, is illustrated in Fig. 5.11. Here the inhomogeneity of the soil anomaly reflects only an inhomogeneous distribution of metal in the bedrock. This is a very different type of inhomogeneity from that described in the previous paragraph, where smoother anomalies could be obtained merely by modifying the sampling technique.

D. Variations with Depth and Soil Type

Variations in the metal content of the overburden with depth below the surface can come about in three ways: (i) by preservation of the trace of an inclined ore deposit projecting up into the overburden, (ii) by physical distortion resulting from soil creep, and (iii) by soil formation and profile development.

The Zn anomaly shown in Fig. 11.13 is an example of the first type. Similar occurrences are also commonly observed in the C horizon of tropical latosols in Zambia. Preservation of the inclined attitude can only be expected below the zone of soil mixing and will only assume practical importance in areas of deep and intense weathering, where determination of such variations in metal content with depth may aid in locating the suboutcrop of the primary ore.

In the zone of active soil creep, the metal content varies with depth in a characteristic manner according to the position of the sampling point with respect to the suboutcrop (Fig. 11.12). Well downslope from mineralization, the metal content of the soil progressively decreases with depth; on approaching the ore the vertical metal profile shows a maximum at an intermediate depth which becomes progressively deeper until, immediately over the deposit, the metal content increases continuously from surface to bedrock. From Fig. 11.12 it will also be seen that the anomaly becomes narrower and more intense with depth. In practice, the ideal sequence is liable to be more or less obscured by slumping, soil homogenization, and the development of soil horizons. Nevertheless, where soil creep is active in moderate to deep overburden, the general trend is usually discernible (Fig. 9.3), and the pattern of metal variation down the soil profile can often aid in locating the suboutcrop where conditions are appropriate.

Variation in the distribution of metal with depth is also a natural consequence of weathering and soil formation. The resulting patterns of redistribution in the soil profile depend primarily on the dispersion mechanism, the behavior of the individual metals, and the soil type.

In immature soils, profile variations are not usually important, although under appropriate conditions metals such as Pb, Cu, and Zn, all of which tend to be enriched in humus, may start to accumulate in the organic topsoil. With progressive differentiation of the soil horizons, the pattern of metal redistribution becomes more pronounced.

In mature, freely drained soils, many mobile metals tend to become impoverished in the A horizon and enriched in the B horizon; enrichment in the organic A0 and A1 horizons continues erratically. Immediately over the suboutcrop of a deposit the metal content is usually greatest in the C horizon. By contrast, background profiles and anomalous profiles at a distance from the suboutcrop of the ore may show a relative enrichment of metal in the B horizon. This type of redistribution is commonly developed in podzols and podzolic soils where the dominant agent is downward-percolating rain water. Table 11.4 shows a consistent increase in metal content with depth in an

Table 11.4
Metal content of anomalous and background profiles of Alamance silt loam, North Carolina[a]

Soil horizon	Average depth (cm)	Description	Anomalous profile[b] Pb (ppm)	Cu (ppm)	Zn (ppm)	Background profile[b] Pb (ppm)	Cu (ppm)	Zn (ppm)
A1	0–1	Humus	440	150	260	100	20	160
A2	1–5	Gray silt loam	840	300	300	190	24	140
B1	5–40	Red to yellow silty clay	1000	380	280	230	34	140
B2	40–75	Same, mottled	1300	750	410	370	57	160
C	>75	Weathered rock	1700	1100	440	180	59	110

[a] Source: Hawkes (1957); data for −80-mesh fraction.
[b] Each figure is the average of five samples collected within a radius of 1.3 m; values rounded to two significant figures.

anomalous soil, and a slight enrichment in the B horizon in a corresponding background soil. The accumulation of metal in the B horizon has presumably been derived from residual metal leached or washed out of the A horizon together with any metal that may have been brought to the surface by plants and subsequently leached from decaying vegetable matter.

Similar trends are often observed in latosols, where the enrichment of metals that tend to be precipitated with Fe and Mn may be very pronounced in the ferruginous B horizon. In this case, in addition to material derived from the overlying A horizon, ground-water solutions also contribute important amounts of metals dissolved from the underlying rock. Typical accumulation of Cu and Pb with Fe in the ferruginous concretionary B horizon and the enrichment of Pb in organic matter are shown in Table 11.5. Variation in the form of Cu anomalies in the different horizons of latosolic soils in Zambia is shown in Fig. 11.14.

Table 11.5
Examples of marked differentiation of anomalous metal in the different soil horizons of latosol profiles, central Africa[a]

Soil horizon	Description	Profile near Cu deposit Cu (ppm)	Profile over Cu deposit Cu (ppm)	Profile near Pb deposit Pb (ppm)
A1	Humic topsoil	130	300	350
A2	Sandy subsoil	160	300	140
B	Compact with ferruginous concretions	400	1000	880
C	Weathered parent material	200	2000	170

[a] Data for −80-mesh fraction.

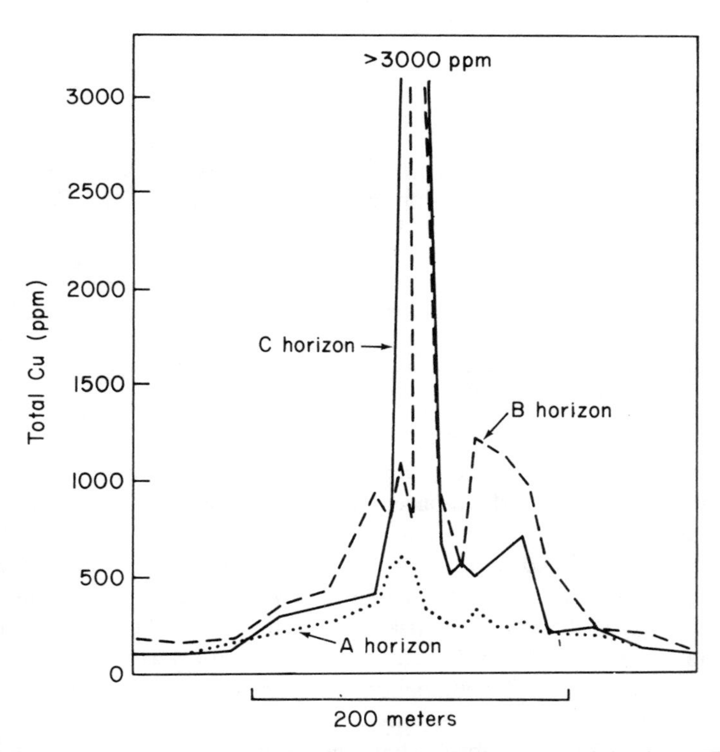

Fig. 11.14. Variation in the form of soil anomalies in different soil horizons, Zambia. (After Tooms (1955) and Tooms and Webb (1961).)

In relatively arid environments soluble metals may be expected to be enriched in horizons formed by evaporation of (i) downward-percolating rain water or (ii) ground-water solutions rising into the capillary fringe zone above the water table. For most semi-mobile metals, however, redistribution in desert environments is likely to be relatively restricted in view of the characteristically high pH. No reliable data are available concerning metal distribution in thick deposits of massive caliche (in Europe, "calcrete") developed in truly residual overburden. Here, too, mobility is likely to be restricted, with resulting retention or even enrichment of metals in the zone of precipitation. Metal distribution observed in massive caliche deposited in overburden of possibly colluvial origin is considered in the next chapter.

The foregoing has been concerned with freely drained profiles. Where drainage is impeded, the metal distribution in normal hydromorphic soils shows a tendency toward enrichment in the organic-rich A1 horizon. Directly over mineralization, however, the metal content of hydromorphic soils generally increases progressively with depth (Table 11.6).

Table 11.6
Profile distribution of residual anomalous metal in a poorly drained soil, Zambia[a]

Soil horizon	Description	Background profile Cu (ppm)	Profile near Cu deposit Cu (ppm)	Profile over Cu deposit Cu (ppm)
A1	Organic topsoil	90	360	3250
G	Gley subsoil	80	125	4200

[a] Data for −80-mesh fraction.

In permafrost areas, Pitulko (1968) reports differences between soil profiles in frost boils and in the adjacent vegetation-covered areas. Frost boils show no development of soil horizons, and metal content gradually increases with depth in profiles over ore. In the adjacent vegetation-covered areas, mobile metals are depleted from near-surface organic-rich horizons, accumulated in underlying brown clays at 0·5–0·8 m depth, and depleted in underlying yellow clays just above the permanently frozen layer. Profiles for other permafrost regions are not available to determine whether this type of profile is general in permafrost areas. In any case, sampling of frost boils produces well-defined anomalies over ore (Allan and Hornbrook, 1971), especially for the less mobile metals (Cameron, 1977a).

III. HYDROMORPHIC ANOMALIES IN RESIDUAL SOIL

Lateral hydromorphic anomalies may be developed in residual soil by precipitation of metal from anomalous ground-water solutions draining a mineralized area. They are most commonly found in seepage areas down the slope of the water table from the bedrock source, at any point where the metal-bearing ground waters approach close to the surface. General aspects of seepage anomalies are considered in detail in Chapter 10 and, more specifically, in Chapter 15; transported gossans, which are effectively no more than fossil seepage anomalies, are mentioned in Section I of this chapter. Suffice here to compare the features of superjacent and lateral anomalies in residual overburden in terms of their mutual relationship.

Seepage anomalies are, of course, restricted to the group of semi-mobile metals that are soluble in the local ground waters but that are commonly precipitated in the soil of the seepage area. In marked contrast to the metals contained in syngenetic anomalies in residual soil, a substantial proportion of the anomalous metal of seepage anomalies is loosely bonded and therefore readily extractable. In Zambia, for instance, the proportion of total Cu that is cold extractable ranges from 20% to 80% in seepage anomalies as compared to less than 10% in residual anomalies in freely drained soil (Webb and Tooms, 1959). The cxMe/Me ratio may therefore be used as a criterion for differentiating syngenetic and hydromorphic soil anomalies under these conditions. However, if the ground water recedes and the soil becomes freely drained, both total metal and the cxMe/Me ratio will tend to decline with time as the extractable metal is both removed by leaching and fixed in the course of the drying out of the ground. The background content of mobile metals in seepage soils tends to be higher and to show a greater range of fluctuation than in residual freely drained soils. Thus, both the threshold value and the contrast of threshold to background for seepage anomalies tend to be higher than for syngenetic soil anomalies. Nevertheless, providing the ore is actively undergoing oxidation and leaching, the seepage anomaly may often show a greater anomaly contrast, particularly for cold-extractable metal, than does the residual anomaly associated with the same deposit (Table 11.7). This comes about because hydromorphic anomalies are continually being reinforced by precipitation of metal, whereas in residual anomalies the metal is being lost by leaching.

IV. ANOMALIES NOT RELATED TO MINERAL DEPOSITS

Occasionally variations in the bedrock lithology can result in residual patterns simulating those related to mineral deposits. In the Kilembe area of Uganda, for example, the soil over diabase dikes may contain up to 250 ppm Cu and 140 ppm Ni, as compared with the normal background of 50 ppm Cu and 20 ppm Ni over granulites and gneisses (Jacobson, 1956). The Cu anomalies related to Cu–Co mineralization may be distinguished, however, by their low Ni content and high Co/Ni ratio. In Zambia, anomalous soils associated

Table 11.7
Contrast for residual and seepage anomalies in areas underlain by Katanga sediments, Zambia[a]

Location of anomaly	Metal	Average threshold (ppm)	Peak anomaly (ppm)	Contrast
Anomaly in residual soil overlying ore deposit	Cu	75	480	6·4
	cxCu	5	20	4·0
Lateral seepage anomaly 2000 feet downslope from ore	Cu	100	2100	21·0
	cxCu	20	780	39·0

[a] Source: J. S. Tooms (personal communication); data for −80-mesh fraction.

Table 11.8
Metal content of residual overburden derived from kimberlite and other rocks, Sierra Leone[a]

	Depth (ft)	Kimberlite	Mafic schist	Felsic schist	Granite
Nickel (ppm)	2	150	160	30	20
	4	260	180	30	20
	8	500	200	30	—
Cobalt (ppm)	2	20	10	<10	10
	4	60	80	<10	<10
	8	50	10	<10	—
Chromium (ppm)	2	340	900	60	50
	4	400	1050	110	55
	8	1000	1650	35	—
Cold-extractable zinc (ppm)	2	7	1	2	1
	4	5	1	1	1
	8	12	1	<1	—
Base exchange capacity (mEq/100 g clay)	2	8·4	3·6	2·7	6·8
	4	12·0	2·0	1·5	5·7
	8	28·8	2·9	0·5	—

[a] Source: Webb (1958a); data for −80-mesh fraction.

with Cu deposits carry more Co than Ni, whereas the reverse holds over gabbroic rocks, although the Cu anomalies in both cases may be identical at 150 ppm compared with the normal background of 20–70 ppm Cu.

An interesting example has been recorded from Sierra Leone, where kimberlite and mafic schist can readily be distinguished from quartzose bedrock by differences in the Co, Ni, and Cr content. In this area, the kimberlite soil is further characterized by a relatively high cxZn content reflecting the high base-exchange capacity due to the presence of montmorillonitic clay minerals (Table 11.8).

Chapter 12

Anomalies in Transported Overburden

★★★★★★★★

Over large areas of the earth, the bedrock is blanketed by relatively recent deposits of glacial debris, alluvium, colluvium, peat, wind-borne material, or volcanic debris. This cover of transported overburden effectively prevents any direct observation of mineral deposits that occur at the surface of bedrock. Just as with residual overburden, studies of the distribution of traces of ore metals in the surficial cover may provide clues to the presence of concealed ore.

Geochemical anomalies in transported overburden may be either syngenetic or epigenetic. A syngenetic anomaly is, by definition, an integral part of the matrix, formed at the same time as the deposit of transported material in which it occurs. An epigenetic anomaly, on the other hand, is a dispersion pattern introduced subsequent to the deposition of the matrix. Syngenetic and epigenetic anomalies may, of course, occur together and be mutually superimposed.

I. COMMON FEATURES OF ANOMALIES IN TRANSPORTED OVERBURDEN

The geochemical anomalies developed in transported surficial cover have certain features in common that are independent of the type of transported overburden.

A. Syngenetic Anomalies

Syngenetic clastic patterns developed in transported overburden are the effect of purely mechanical movement of solid particles. Irrespective of the nature of the impelling force, be it ice, water, or wind, the general pattern of movement of particulate matter from its source in the bedrock to the site of deposition is the same. As a consequence of this common pattern of movement, all syngenetic anomalies in transported cover tend to be more or less elongated in the direction of movement. This simple pattern due to purely mechanical dispersion may be complicated by a certain amount of local redistribution and reconstitution by ground water or soil moisture that may have taken place during or subsequent to the deposition of the matrix.

As a rule, the concentration of ore materials in the overburden is greatest in the immediate vicinity of the source and decays rapidly with distance as a result of dilution by barren material. Although rapid decay with distance is the rule, special environmental factors in the formation of some kinds of mechanical dispersion patterns may cause feeble traces to persist for very substantial distances.

The ore metals in syngenetic anomalies occur partly as resistant fragments of rocks and of primary or secondary ore minerals, and partly as reconstituted material that has been dissolved and locally reprecipitated. Determination of the metal content of resistant minerals usually requires rigorous chemical extraction. In contrast, the reconstituted material is likely to consist largely of limonitic precipitates and adsorbed ions, both of which are more readily extractable than the resistant material.

Near the source, the metal content of overburden usually increases with depth. If the overburden has been thoroughly mixed, the pattern may decay relatively smoothly with distance from the source. On the other hand, if the overburden is deposited as a series of layers, the syngenetic pattern may persist only in certain horizons, often the bottommost. Syngenetic anomalies in transported overburden can be derived from a given orebody only while that orebody is exposed to physical erosion. Once the orebody is covered, succeeding events producing transported overburden will not produce syngenetic anomalies related to that orebody. For this reason, syngenetic anomalies are often lacking in the surface layers of transported overburden, although they may be present in deeper layers.

B. Epigenetic Anomalies

Hydromorphic and biogenic patterns in transported overburden show no special features beyond those already discussed in previous chapters. However, epigenetic anomalies assume special importance in transported cover, in view of the fact that syngenetic patterns are not as consistently associated with concealed mineralization as they are in residual overburden. Thus hydromorphic and biogenic anomalies in transported cover may be the only usable geochemical guides to buried ore. In desert areas of low water table where the formation of near-surface hydromorphic dispersion patterns is inhibited, the only surface indication of buried ore may be biogenic anomalies resulting from the uptake of metal by deep-rooted plants.

C. Anomalies not Related to Mineral Deposits

Extensive surficial deposits of transported material commonly represent a more or less homogenized sample derived from all the rocks in the source area. As a result, anomalies traceable to a specific rock type are less developed than in residual overburden. Except where unusual rocks occur as major units, or where the overburden has not moved very far, local patterns not related to ore are more likely to be the result of sorting during the course of transport and deposition than of variations in the composition of the source rocks. Epigenetic patterns, on the other hand, may reflect the presence of quite small-scale lithological units, particularly where the introduced metal is derived from the immediately underlying bedrock.

II. GLACIAL OVERBURDEN

The development, detection, and interpretation of geochemical anomalies in glacial overburden are complicated by the diverse nature and origin of glacial deposits, as reviewed in Chapter 9. Sands, clays, gravels, and morainal deposits of varying composition and permeability may be intermixed or stratified. Over short distances the cover may vary in depth from a few centimeters to many tens of meters. Eroded material, scoured and plowed up from the surface over which the ice moved, may have been transported for distances ranging from a few meters to many kilometers from its place of origin. The glacial debris may have been further subjected to water transport and sorting, and the resultant glaciofluvial deposits distributed over wide areas as outwash fans and lake-bed sediment. The direction of ice movement, which is the principal factor conditioning the form of a glacial anomaly, may change during the course of a single period of glaciation. Complexities in the local deposits and in the local glacial history naturally lead to a corresponding complexity in the dispersion processes and the resulting geochemical patterns. Any or all of these factors add materially to the difficulties of conducting geochemical surveys in glacial terrain. Therefore, a first step in geochemical studies of glaciated areas is to determine the type of glacial deposits present, either by field studies or reference to the literature.

Because of the large area obscured by glacial overburden, numerous tests, research projects, and orientation studies have focused on geochemical dispersion in glaciated areas. Books and symposia dealing specifically with geochemical exploration in glaciated areas include Kvalheim (1967), Jones (1973b, 1975), Institution of Mining and Metallurgy (1977), Bradshaw (1975), and Kauranne (1976). These publications give many details of research and exploration in glaciated areas, and the successes and problems encountered.

Both syngenetic and epigenetic anomalies are useful in glaciated areas. Syngenetic anomalies formed by clastic dispersion of grains of ore minerals are most useful in till (ground moraine). Epigenetic anomalies, formed by hydromorphic and biogenic dispersion, may be present in any type of glacial, glaciofluvial, or glaciolacustrine deposit.

A. Syngenetic (Clastic) Anomalies

The particle size of the eroded ore material may range from large boulders down to the finest clays. Patterns formed by both boulders and small particles are useful in exploration.

Glacially transported boulders are characteristically distributed in fan-shaped patterns extending outward from the bedrock source in the direction of ice movement. Boulder fans have been observed to extend for many kilometers from the source (Figs 9.12 and 12.1). More commonly, a well-defined boulder fan cannot be traced for more than one or two kilometers (Fig. 12.2). Boulder tracing by prospectors has been a very successful method of prospecting for many years, especially in Fennoscandia (Hyvarinen *et al.*, 1973). One recent development is the training of dogs to smell weathering ore boulders (Nilsson, 1973). The dogs can detect boulders buried a meter or more below the surface, and are used to extend and refine the boulder trains detectable by a human prospector.

Fine-grained metal-bearing particles or "micro-boulders", too small or too decomposed to be identified by eye, may likewise be dispersed mechanically to form fan-shaped patterns of abnormally high metal content that are more or less coextensive with boulder fans (Figs 12.2–12.4). The homogeneity of such anomalies is dependent on the grain size of the particles and the degree of mixing or sorting that took place during glaciation. As a rule, fine-grained material gives a more homogeneous pattern than coarser fragments. The homogeneity of a micro-boulder pattern may be modified by chemical solution of metals from the larger ore boulders and local reprecipitation in the fine-grained matrix.

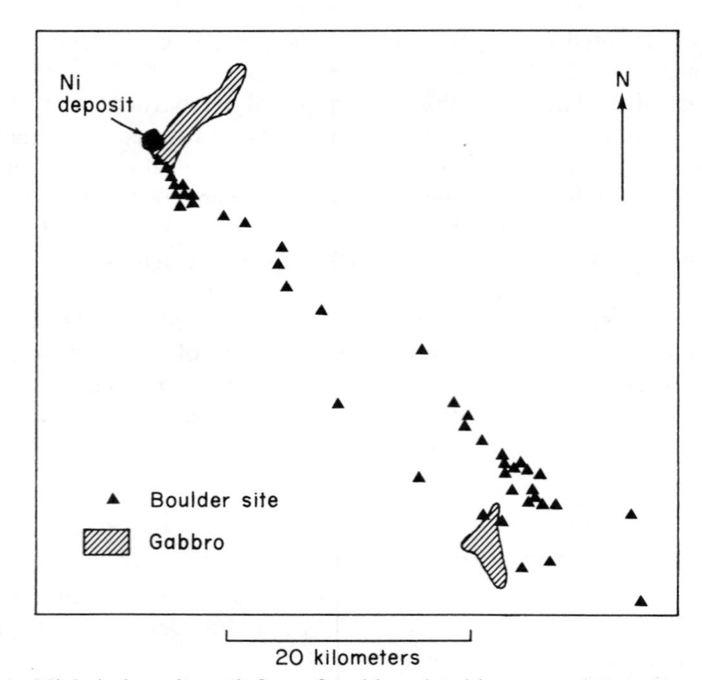

Fig. 12.1. Nickel deposit and fan of gabbro boulders containing disseminated sulfide, Storbodsund, Sweden. (After Grip, 1953.)

Detectable micro-boulder anomalies cannot usually be followed as far as boulder fans. In surface soils, they have rarely been observed extending more than a kilometer or so from the source. However, the effects of large bodies of unusual rock, such as ultramafic intrusions, can be detected for tens of kilometers (Shilts, 1973) and some anomalies in basal till, especially in the heavy mineral fraction, also extend many kilometers, as discussed below.

In till less than about 1 m thick, an anomaly is generally found in the A and B horizons of the soil directly over the ore, with little or no lateral displacement (Govett, 1973b; Morissey and Romer, 1973). In single thicker layers of till up to about 5 m depth, an anomaly is generally detectable at the surface, but may be offset down-ice by tens or hundreds of meters (Fig. 12.5). Profiles through single layers of till show that the highest metal contents typically are in basal till lying directly on the orebody (Figs 12.6 and 12.7), and that the anomaly approaches the surface and becomes more diffuse in a down-ice direction (Figs 12.7–12.9). In a general way, the intensity of anomalies diminishes with thickness of till and distance of lateral transport, but the extent of diminution varies widely depending on the friability of the ore mineral, the subglacial topography, and other variables controlling

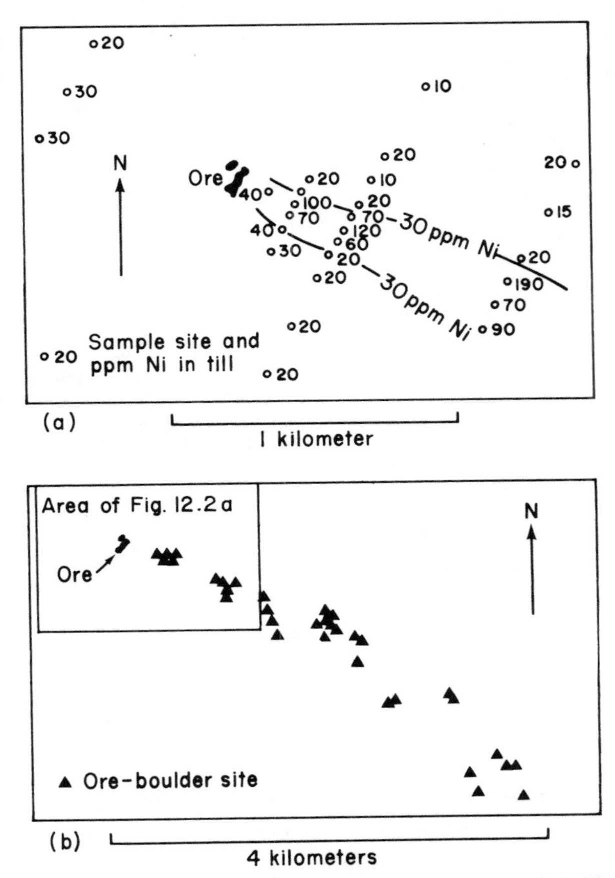

Fig. 12.2. (a) Nickel content of −0·012-mm fraction of glacial till near Makola Ni-sulfide deposit, Finland. (b) Discovery sites of ore boulders from Makola Ni-sulfide deposit, Finland. (After Kauranne, 1959.)

glacier flow. Bradshaw (1975) summarizes data on the size and intensity of anomalies in the Canadian Shield. As till becomes thicker, anomalies tend to become patchier as well as weaker. Spotty weak anomalies are also characteristic of diffuse non-economic mineralization in bedrock (Morissey and Romer, 1973).

If multiple layers of till overlie bedrock, anomalies in surface soil are generally weak or lacking. If present, anomalies show complex patterns of offsets from the ore because of differing directions of ice movement during the different glacial episodes. In overburden containing interlayers of stratified drift, anomalies are generally lacking, or if present, are of hydromorphic

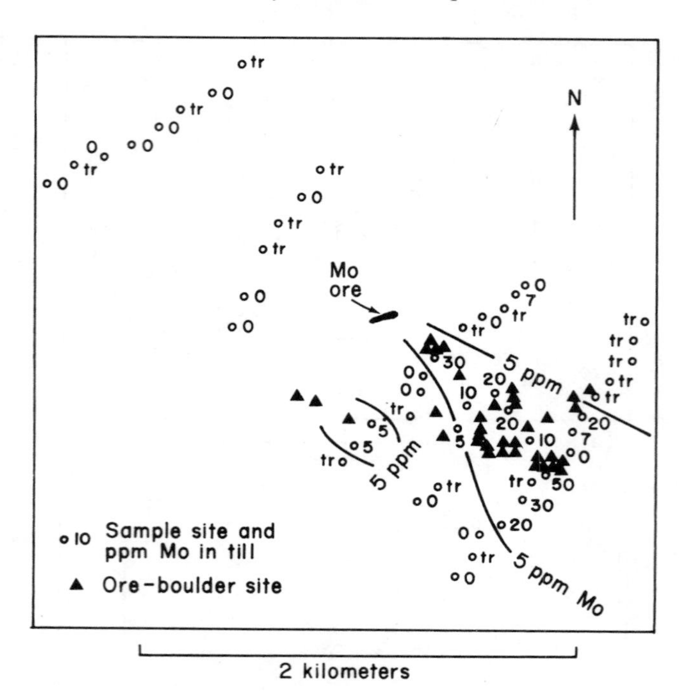

Fig. 12.3. Comparison of Mo content of the −0·012-mm fraction of glacial till with the distribution of ore boulders, Susineva, Finland. (After Kauranne, 1958.)

origin. An important example is the Clay Belt of Ontario and Quebec, where many meters to tens of meters of glaciolacustrine clay overlie till and bedrock in an area of important mineral deposits (Gleeson and Cormier, 1971). In this and similar areas, sampling of deep till is usually necessary to detect anomalies.

Several methods of sampling deep overburden have been developed and are in widespread use in exploration of glaciated areas. Sampling equipment and methods are discussed further in Chapter 13. Lodgment till from just above bedrock is usually the most readily interpreted material, but deep samples at a constant distance below the surface have also proven useful. In addition to chemical analyses of the deep till samples, examination of rock fragments from the bedrock or from basal till gives valuable geological and mineralogical information. Observation of sulfides or other primary ore minerals demonstrates a clastic origin for the anomaly and indicates that the source lies up-ice. Mineralogical observations also allow a decision on whether an electromagnetic geophysical anomaly results from sulfides or graphite.

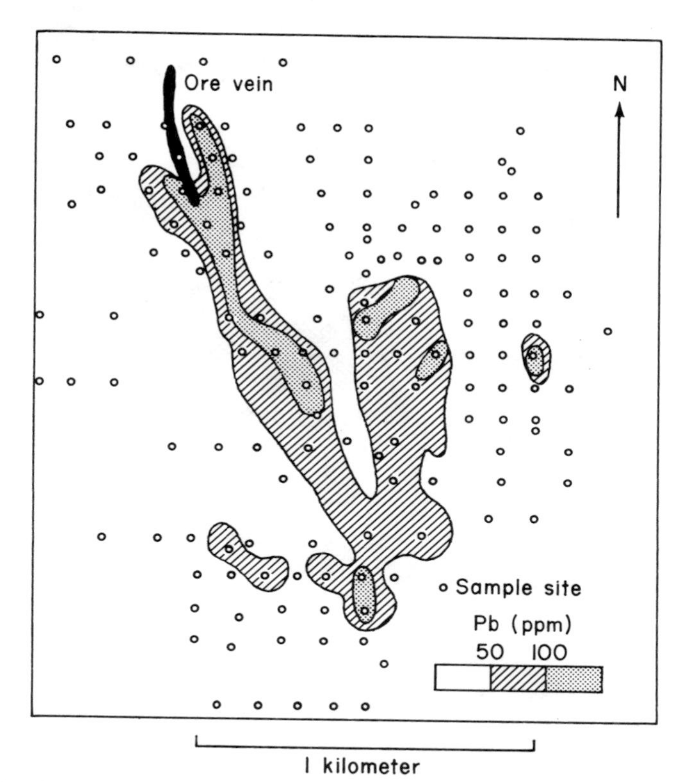

Fig. 12.4. Distribution of Pb in fine-grained fraction of glacial moraine at Korsnäs, Finland. (After Hyvarinen, 1958.)

Both total till and fractions of till have been used in studies of deep till. The sand and silt fraction of basal till is distinctly anomalous near most orebodies. If the basal till is unoxidized, as it is in many areas, then sand- and silt-size grains of sulfides and other heavy ore minerals are usually responsible for the anomaly. An increase in contrast and a more extensive anomaly can be obtained by discarding the clay fraction and concentrating the heavy minerals from the sand and silt fraction with heavy liquids (Gleeson and Cormier, 1971; Garrett, 1971b; Shilts, 1971). With the aid of this technique, anomalies are observed to extend at least 10 km down-ice from the Kidd Creek massive-sulfide body in Ontario, whereas in surface samples no anomaly is detectable because of a thick clay layer (Skinner, 1972). High values of sulfide-extractable base metals may give similar results in basal till.

In oxidized till, the metals originally present as sulfides have generally been dispersed hydromorphically for at least short distances, and adsorbed or coprecipitated with clays or Fe–Mn-oxides. In this type of environment,

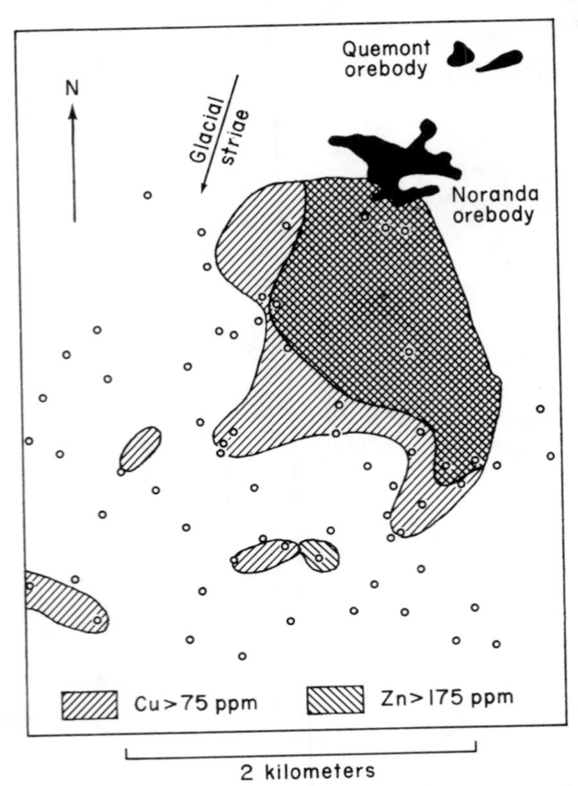

Fig. 12.5. Copper and Zn content of moraine near Noranda Cu deposit, Quebec. Data on −80-mesh fraction. (After Dreimanis, 1960.)

analysis of the clay fraction of the basal till may give better contrast. However, both background and anomaly values may vary considerably with the content of clay in the sample. Ridler and Shilts (1974) suggest that the clay fraction of oxidized till in frost boils gives optimum contrast in permafrost areas.

Some of the possible complexities in geochemical surveys in tills are illustrated in Fig. 12.8, from a Ni–Cu prospect in Finland (Nurmi, 1976). Ore boulders, geophysical measurements, and shallow till surveys led to trenching and drilling that disclosed the orebody. Later studies showed that the anomaly near the orebody was restricted to the lower till (2 m or deeper) but reached the surface about 150 m down-ice (Fig. 12.8). A deep depression in the bedrock a short distance down-ice from the orebody was partly filled with an older till, as indicated by pollen analyses. This older till was overlain by an apparently separate tongue of anomalous metal content. Sulfides were easily

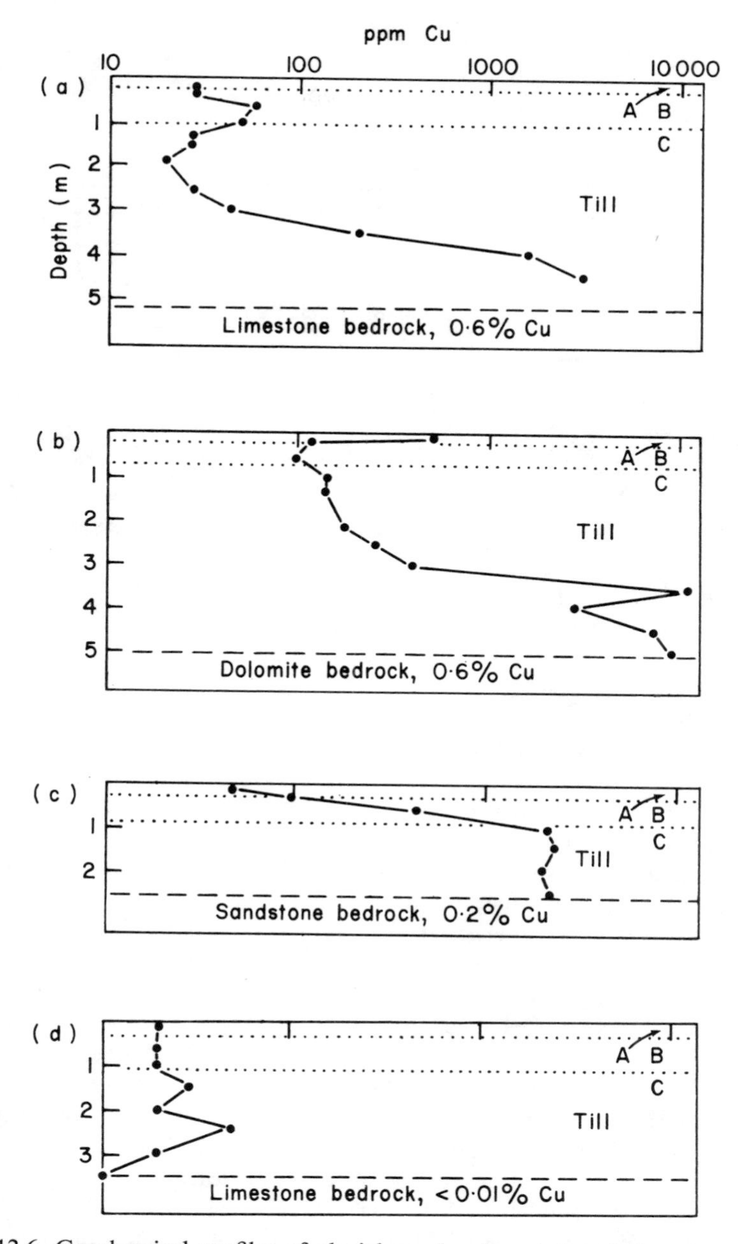

ig. 12.6. Geochemical profiles of glacial overburden at the Mallow Cu deposit, reland. Profiles (a), (b), and (c) are over mineralized bedrock, (d) over unmineral-zed bedrock. (After Wilbur and Royall, 1975.)

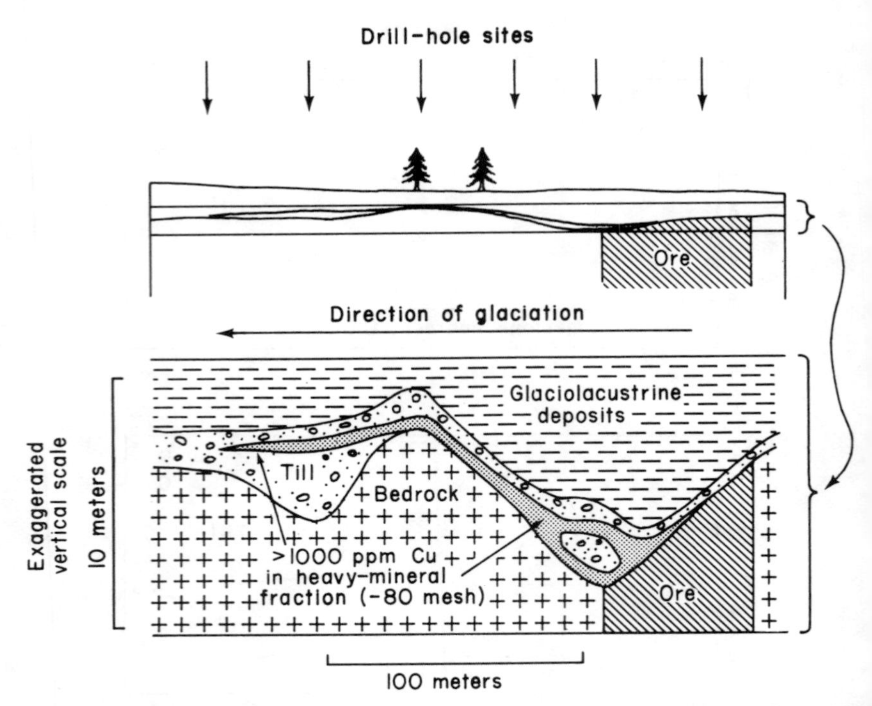

Fig. 12.7. Vertical profile of dispersion train over the Louvem massive-sulfide deposit, Quebec. (After Garrett, 1971b.)

detected in the till near the orebody, but not in the anomaly to the east. The irregularities of the more detailed profiles suggest the vagaries of glacial transport, and the types of problems that can be encountered in detailed work, especially if only a few samples are collected.

Syngenetic anomalies in eskers have also been investigated. In the Munro esker of Ontario distinct peaks in the abundance of dunite fragments, trachite fragments, and Au particles were found at distances of 3–13 km down-esker from their bedrock sources (Lee, 1965). The pattern of pyrope grains in the esker led to the discovery of a new kimberlite in the area (Lee, 1968). In a survey of two eskers in northern Ontario, Shilts (1973) obtained distinct anomalies in the −250-mesh fraction near a Cu–Ni prospect. Any original sulfides dispersed clastically had apparently been weathered and the products incorporated in the fine fraction, because anomalies in the heavy-mineral fraction were weak. Studies by Virkkala (1958) on eskers in Finland indicate that most esker material is derived from within a few kilometers of the sample site, in agreement with the above results. However, this conclusion may not be universal, because some esker-forming streams flow within or on top of the

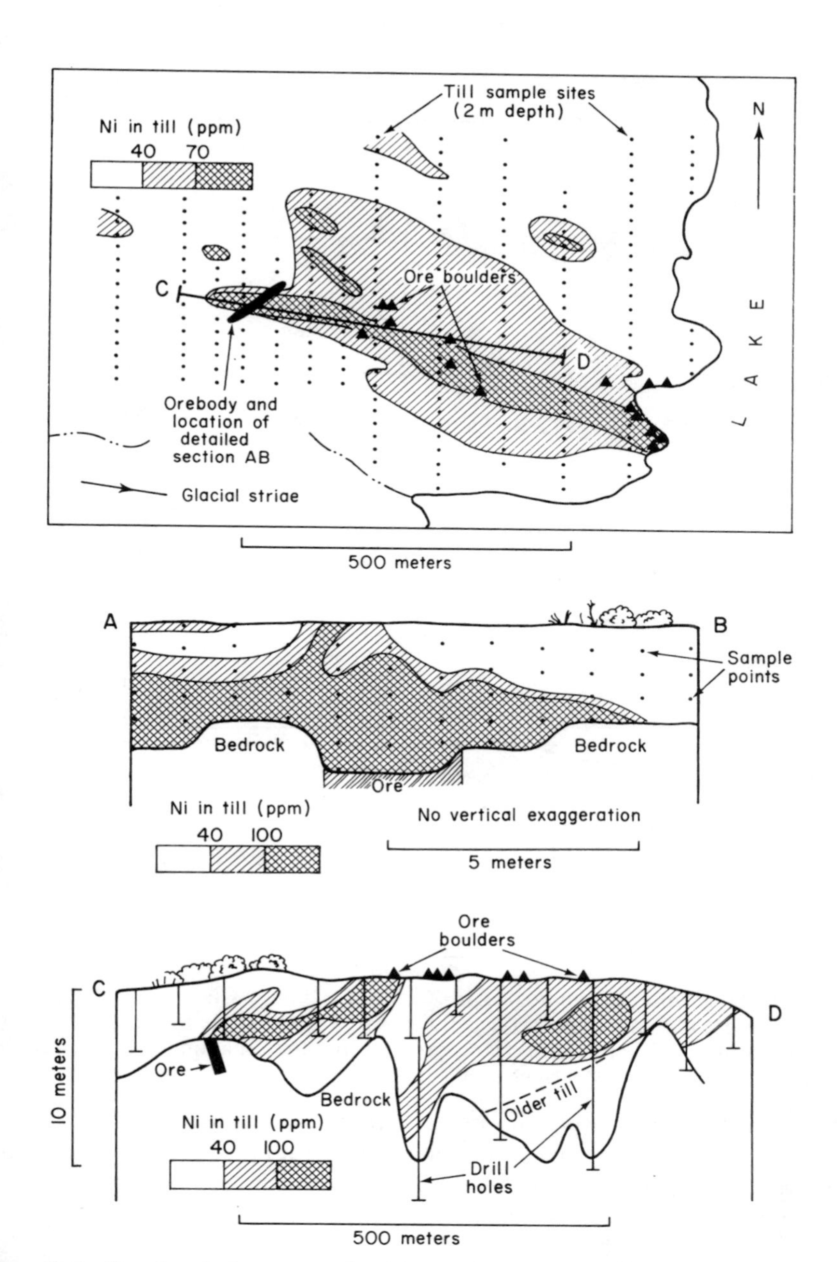

Fig. 12.8. Geochemical patterns in till near an Ni–Cu prospect, Finland. Plan shows Ni anomalies in <0·5 mm fraction at 2 m depth. Section AB shows detailed cross-section of Ni in till over ore. Section CD shows Ni content of till and sub-glacial topography parallel to ice movement. (After Nurmi, 1976.)

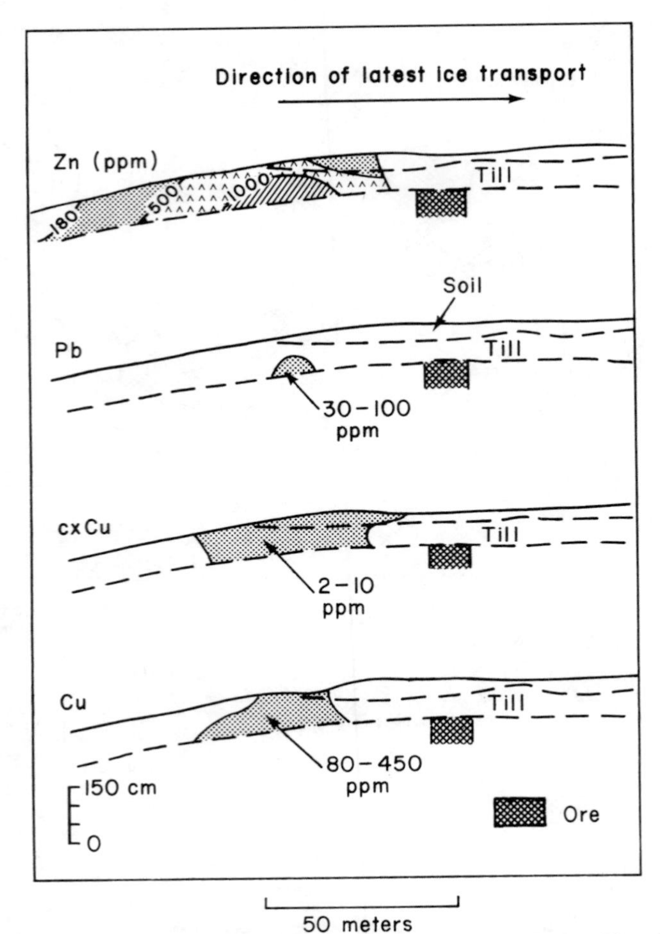

Fig. 12.9. Hydromorphic anomalies in Zn, Pb, Cu, and cxCu in till and soil near Cu–Pb–Zn ore at Søndre Gjeitteryggen, Norway. Note that the metals have moved in a direction opposite to glacial transport, and that the Zn anomaly is much more intense than Pb. (After Mehrtens *et al.*, 1973.)

glacier, where they can pick up material from a more distant source (Price, 1973). In general, eskers may be useful in regional studies, but are of questionable value for more detailed surveys because of their sparse occurrence and the uncertain source of detritus.

B. Hydromorphic Anomalies

A second mode of dispersion from bedrock into glacial overburden is by ground water. This mode of dispersion is effective only for the mobile elements. It furnishes a means of detecting orebodies by analysis of surface and near-surface samples of glacial overburden, and also leads to the development of anomalies in stream sediment, lake sediment, and waters in the vicinity of ore. Because of the poorly developed surface drainage of most glaciated terrain, the water table in many areas is shallow, seepages are common in lower slopes, and hydromorphic dispersion commonly carries metal to the surface.

Accurate interpretation of geochemical data on glacial overburden is not possible without some means of distinguishing clastic (syngenetic) from hydromorphic patterns. Three criteria are applicable: the form of the pattern, the mineralogy of the anomalous material, and the mobility of the elements.

The form of hydromorphic patterns is controlled by the path of groundwater flow away from the orebody, and by the distribution of adsorbing and precipitating materials. Because of the much higher permeability of glacial deposits than most bedrock, and because of the normally shallow water table in glaciated areas, ground water tends to flow within the overburden and emerges at the surface in seepages, springs, bogs, and swamps. The direction of hydromorphic dispersion is thus related to the direction of ground-water flow, and if this differs from the direction of ice movement, a distinction can be made between clastic and hydromorphic dispersion (Fig. 12.9). Immediately downslope from the suboutcrop of ore, hydromorphic anomalies are usually restricted to the bottom 1–2 m of overburden (Fig. 12.10). For example, Koehler *et al.* (1954) report high Co contents in relatively permeable deposits of sandy clay near the bottom of a 5 m blanket of glacial overburden at Cobalt, Ontario. Patterns of this kind may extend 100 m or more downslope from the source before emerging as anomalous springs and seepages.

Most of the hydromorphically dispersed metal is adsorbed and coprecipitated in Fe–Mn-oxides, clays, organic matter, and other fine-grained materials, as at Tverrefjellet, Norway, where Cu is concentrated with Fe-oxides cementing the basal till downslope from the ore (Mehrtens *et al.*, 1973). Changes in the abundance of host materials along the path of flow can significantly affect the intensity of anomalies. Similarly, a marked change in chemical conditions along the path of flow, forming a precipitation barrier, may lead to an intense anomaly. The change from subsurface to surface conditions commonly produces such an effect. As a rule, the strongest patterns are developed in silty overburden, where the permeability is adequate to permit access of solutions, but at the same time sufficient fine material is present to adsorb metal. In contrast, clastic anomalies related to sulfides are most intense in the dense sand and silt fraction, and in extractions by sulfide-selective leaches. Caution must be exercised in using these criteria, however, because weathering and oxidation of clastically dispersed minerals may produce anomalies in readily extractable metal or in the fine fraction.

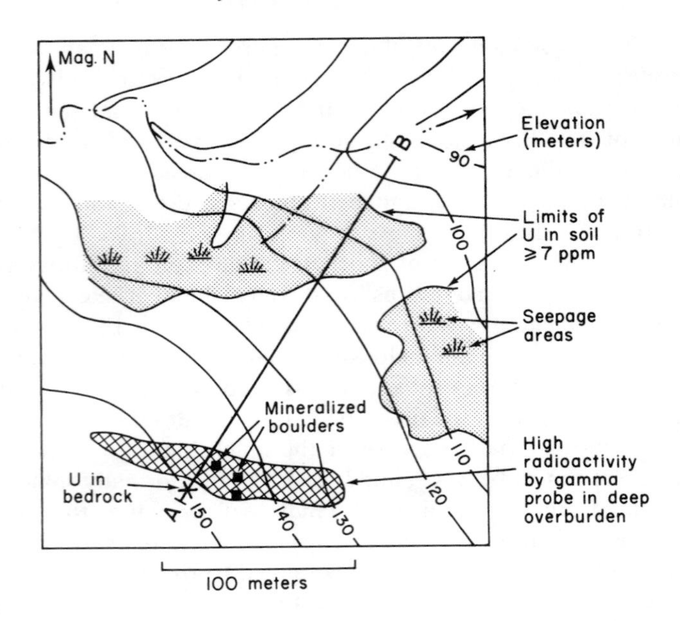

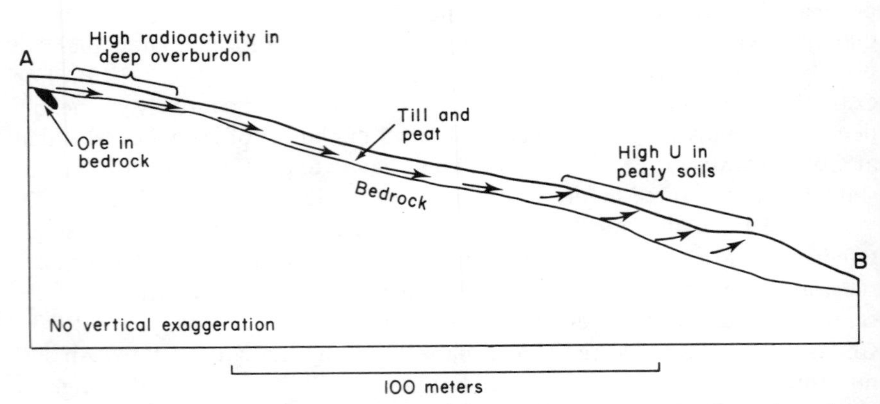

Fig. 12.10. Hydromorphic anomaly of U in soils of a seepage area downslope from a mineralized zone in bedrock, Scotland. (After Michie *et al.*, 1973.)

The mobile metals are obviously dispersed farther and form more intense hydromorphic anomalies than the metals of intermediate mobility (Figs 12.9 and 12.11). The immobile metals do not exhibit hydromorphic anomalies. The pattern of different metals is thus a guide to the origin of anomalies in glacial terrain. For example, Zn tends to form broader and more intense anomalies in glacial overburden than does Pb, and migrates farther from the orebody, as at Tynagh (Fig. 12.12). Conversely, the less mobile metals are

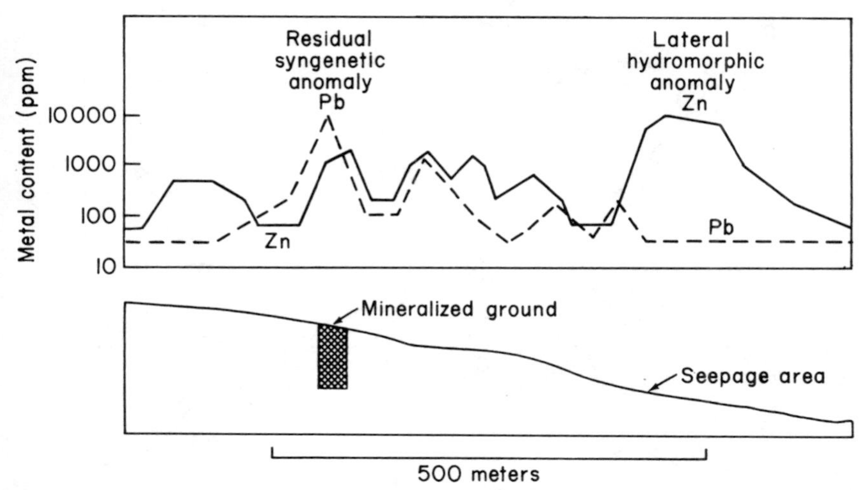

Fig. 12.11. Relationship between Pb and Zn in associated syngenetic and hydromorphic anomalies, Charlotte Prospect, New Brunswick. Data on −80-mesh fraction.

enriched near the mineralized zone in the residual material (Fig. 12.11). An additional means of confirming the hydromorphic nature of a currently forming anomaly is analysis of ground water in the vicinity of the anomaly.

The intensity of a hydromorphic anomaly depends in the first instance on the rate of oxidation of the primary ore minerals. It is a common observation that in many glaciated areas, bedrock deposits have been eroded by the ice down to the relatively compact and impermeable fresh primary ore. The slow rate of oxidation of such material militates against the formation of strong epigenetic anomalies in the glacial cover. However, fractured and broken ore may oxidize rapidly, and electrochemical effects (Govett, 1973a, Bølviken and Logn, 1975) may accelerate solution if the upper part of the ore is exposed to oxidizing conditions.

Seepage anomalies, or more strictly, lateral hydromorphic anomalies, are commonly the most intense of all the patterns occurring in glacial overburden. At Vangorda Mines, Yukon Territory, for example, a strong seepage anomaly in Zn was observed in a topographic depression 300 m from the nearest known ore, as indicated by drilling and gravity surveying (Fig. 12.13).

Recent investigations (Govett, 1973a) suggest that soils developed on transported overburden near sulfide orebodies have anomalously high electrical conductivity when slurried with a small amount of water. The anomalous conductivity apparently arises from an elevated concentration of exchangeable hydrogen ions in the soil (Govett and Chork, 1977). The high conductivity may occur directly over the ore, or offset on one or both sides of the

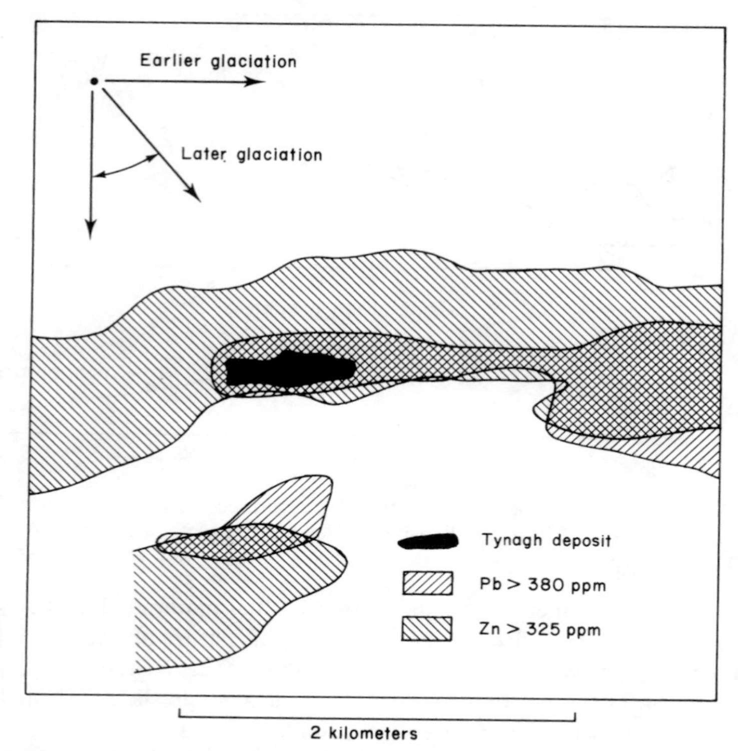

Fig. 12.12. Distribution of Pb and Zn in −80-mesh till-derived soil at 20–30 cm depth at the Tynagh deposit, Ireland. Both clastic and hydromorphic dispersion are responsible for the anomaly. (After Donovan and James, 1967.)

orebody. Unusually rapid diffusion of hydrogen ions released during oxidation of the ore may explain the anomalies. As yet, the reliability of conductivity anomalies in routine surveys remains to be established.

C. Biogenic Anomalies

Concentration of metal at the surface rather than at depth is reported over the Van Stone orebody in eastern Washington, where the glacial cover immediately overlying Zn ore in the bedrock ranged from 5–15 m in thickness (Cox and Hollister, 1955). Inasmuch as the anomalous pattern was most strongly developed in the top 30 cm and disappeared with depth, it is almost certainly of biogenic origin. Similarly, Boyle and Dass (1967) and Hornbrook (1971) found much higher contents of As and Ag in the A horizon than in the B horizon of till near Ag veins at Cobalt, Ontario. Anomalies formed by

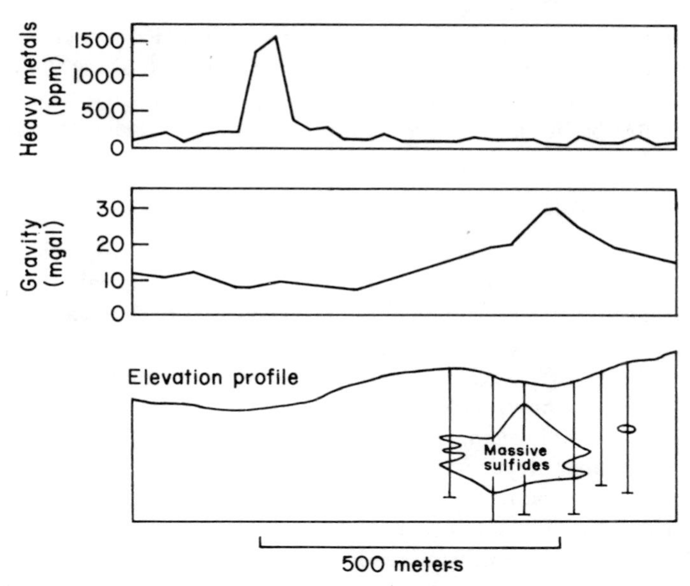

Fig. 12.13. Cross-section showing seepage anomaly in glacial overburden downslope from ore, Vangorda Mines, Yukon Territory. (After Chisholm, 1957.)

biogenic processes are rarely intense, however, compared to anomalies in residual soils.

The recognition of stronger anomalies in the A than in the B horizon at some localities has led to the suggestion that the A horizon should be preferred for geochemical surveys, especially in transported overburden (Boyle and Dass, 1967; Kokkola, 1977; Nuutilainen and Peuraniemi, 1977). However, care should be exercised before undertaking such surveys, because in some areas many false anomalies are disclosed in A-horizon soils (Bradshaw, *et al.*, 1972, p. 36; Brotzen, 1967), apparently because of local variability in the intensity of metal transport by biogenic processes. In addition, analytical complications may be encountered in high-organic materials (see Section III of Chapter 3).

D. Superjacent Anomalies of Complex Origin

Anomalous patterns of probable complex origin have been reported in morainal material immediately over mineralized bedrock in a number of areas. For example, at Tynagh, Ireland, both Pb and Zn anomalies extend considerably to the east of the orebody, apparently because of glacial smearing, but the Zn anomaly is much more extensive than that for Pb, suggesting hydromorphic dispersion (Fig. 12.12). Morissey and Romer (1973, pp. 49 and 50) describe soils weakly anomalous in Zn displaced along the direction of glacial transport, and a much stronger Zn anomaly in peaty soils displaced downslope but at nearly 90° to the direction of glacial transport.

E. Schematic Models of Dispersion in Glaciated Areas

Figure 12.14 summarizes in generalized form the types of dispersion and the resulting profiles of metal content for several types of glacial environment, after Bradshaw (1975). In thin local till, a syngenetic (mechanical) anomaly is detectable at the surface and extends down-ice. In overburden of remote origin (stratified drift), such an anomaly is only detectable in the basal till. Seepage anomalies may occur in both types of overburden if drainage is poor. Cold-extractable metal is much higher in hydromorphic anomalies than in clastic anomalies. Hydromorphic dispersion also furnishes anomalies in stream sediment, stream water, and lake sediment.

As shown on Fig. 12.14, anomalies of various types differ in metal distribution with depth and in ratio of readily extractable to total metal content. Clastic anomalies generally show low values of readily extractable metal (except where weathered) and increase in intensity downward at sites overlying or close to ore. Seepage (hydromorphic) anomalies have high ratios of readily extractable to total metal, and commonly decrease in intensity downward. Analysis of both readily extractable and total metal on samples collected in vertical profiles through the overburden thus allows distinction of the origin of most anomalies (Bradshaw *et al.*, 1974).

III. ORGANIC DEPOSITS

Organic deposits accumulate in any situation where vegetable matter is formed faster than it decomposes. In this class of overburden are included the peat, muck, and muskeg bogs of moist, cool climates, and the mangrove and other organic swamps of the tropics. Although such deposits are usually confined to swampy stagnant depressions, in certain environments they can occur also in upland regions. In particular, bogs and peatlands are abundant in glaciated areas. Essentially all organic deposits are water-saturated during formation, and receive only small amounts of detrital sediment. For the most part, organic deposits accumulate by the growth and decay of vegetation in place. Rarely, organic deposits are formed in basins of sedimentation by the accumulation of clastic plant debris and mineral matter washed in from the surrounding high ground.

The movement of water into and within a bog or peatland is probably the most important factor determining the chemistry and physical character of the peat. At one extreme are the ombrotrophic bogs, which receive most of their water supply from precipitation (Usik, 1969). As a result, the water in the bog is very low in dissolved solids and is relatively acid (pH 4–5). Trace-element contents of the peat are correspondingly low. Water movement through such bogs is generally downward. This type of bog is extensively developed in upland areas of Ireland (Horsnail, 1975).

At the other extreme is the minerotrophic bog, which receives most of its water supply from adjacent or underlying rock or overburden. Both dissolved solids and pH are higher in these waters, and the metal contents of the peat are correspondingly higher. Such bogs tend to occur in distinct basins and valleys, and have been termed basin bogs (Horsnail, 1975). Movement of water in these bogs is generally lateral.

Water can also move into minerotrophic bogs by upward migration, induced by capillary action, by evaporation, and by transpiration of the plants growing in the bog. Salmi (1967) suggests that upward movement of water is common in Finnish bogs during the relatively dry summer season.

The composition of waters entering minerotrophic bogs from the surrounding rocks and sediments depends on the Eh and pH of the water, and on the availability of elements in the surrounding rock or overburden. Relatively high pH values are generally observed over limestones, but in silicic rocks or near oxidizing pyrite, a relatively acid pH is typical. The oxidation potential of the water is similarly varied, depending on the history of the water. Acid oxidizing waters tend to bring in Cu, Zn, Pb, Ni, Co, and other elements soluble in such waters, to the extent that they are available in the surrounding material. Alkaline oxidizing waters may bring in U, Mo, V, and Se, but only traces of Cu and Pb. Under reducing conditions, Fe and Mn are mobilized, and Cu, U, and V are immobile.

As a result of the above processes of introduction and fixation of metal, bogs usually show both lateral and vertical variations of metal content. High values of many metals are found near the point where the water entered the peat (at the margin or base, Fig. 12.15). Other elements are concentrated at the top or base because of favorable pH, Eh, or biological factors. Studies by Salmi (1955, 1967), Gleeson and Coope (1967), Mitchell (1954), Hvatum (1965), and Tanskanen (1976) indicate that V, Ti, Al, Ni, and Cr tend to concentrate at the base of peat bogs, Mo, Fe, and Mn at the top, and Cu, Zn, Pb, and Co at differing locations in different bogs. Gleeson and Coope (1967) found the highest values in clay just beneath the peat.

In spite of the complexities of detail, bogs near ore are generally enriched in ore elements (Salmi, 1967; Gleeson and Coope, 1967). However, the ash of peats is generally enriched in metals relative to normal soils, so that they

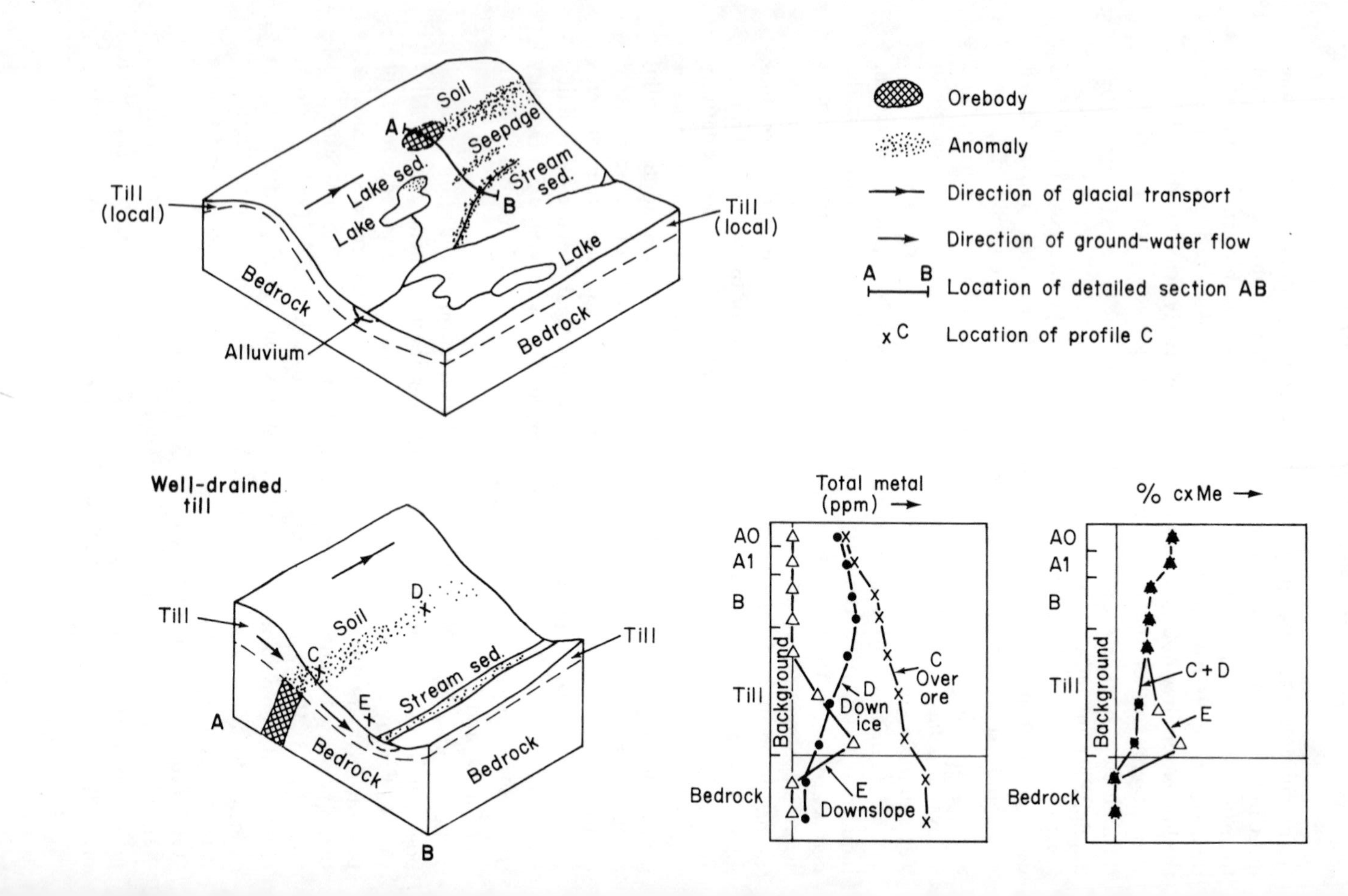
Soil
Seepage
Stream sed.
Lake sed.
Lake
Lake
A
B
Till (local)
Till (local)
Bedrock
Bedrock
Alluvium
Orebody
Anomaly
Direction of glacial transport
Direction of ground-water flow
Location of detailed section AB
Location of profile C
Well-drained till
Till
Till
Soil
Stream sed.
C
D
E
A
B
Bedrock
Bedrock
Total metal (ppm)
AO
A1
B
Till
Bedrock
Background
C Over ore
D Down ice
E Downslope
% cxMe
AO
A1
B
Till
Bedrock
Background
C + D
E

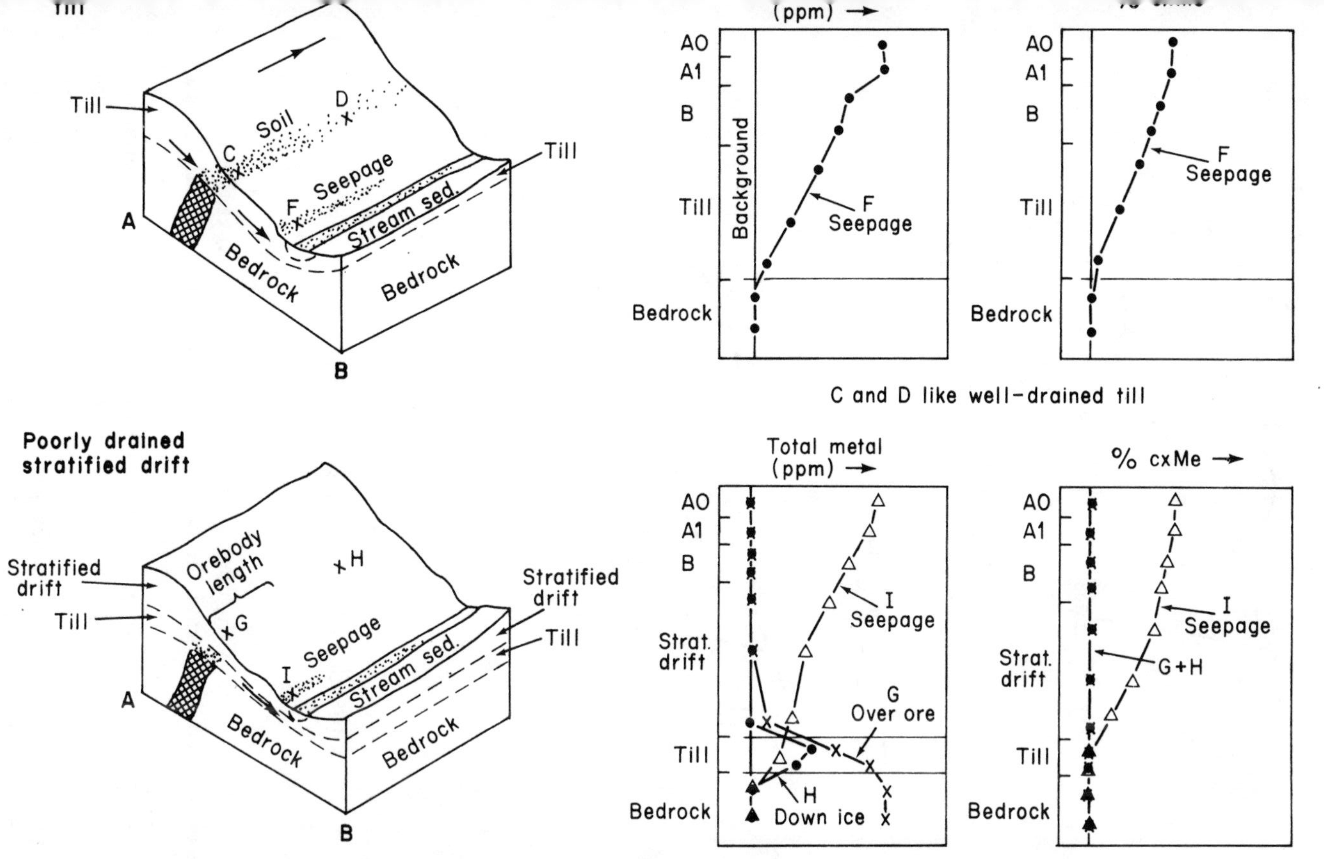

Fig. 12.14. Schematic diagrams showing dispersion of mobile metal in glacial overburden, and profiles of total and readily extractable metal in soil and glacial deposits. (After Bradshaw, 1975.)

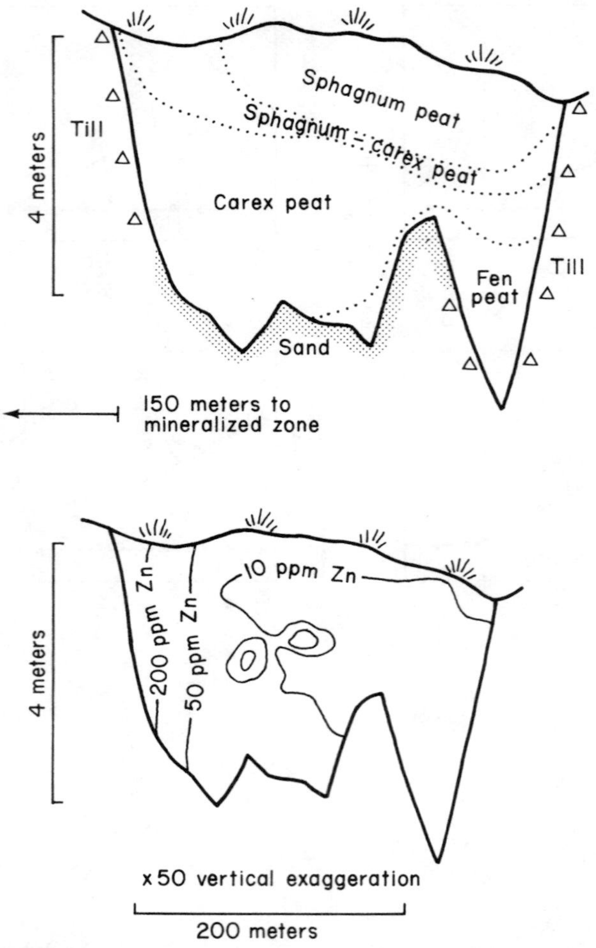

Fig. 12.15. Distribution of Zn and types of peat in a Swedish bog. (After Eriksson, 1973.)

cannot be lumped with other surficial materials in surveys. Peat sampling has most often been used in detailed follow-up surveys in glaciated areas. For example, Larsson (1976) describes the successful use of peat sampled during the winter season to follow-up stream sediment and geophysical anomalies (Fig. 12.16). Salmi (1955) found a close relationship between the distribution of Fe, Ti, and V in a bog and the presence of underlying Ti- and V-bearing magnetite ore (Fig. 12.17). In New Brunswick, Fe is enriched in peat and moss in a bog overlying pyritic sulfide ore in the Bathurst district (Hawkes and

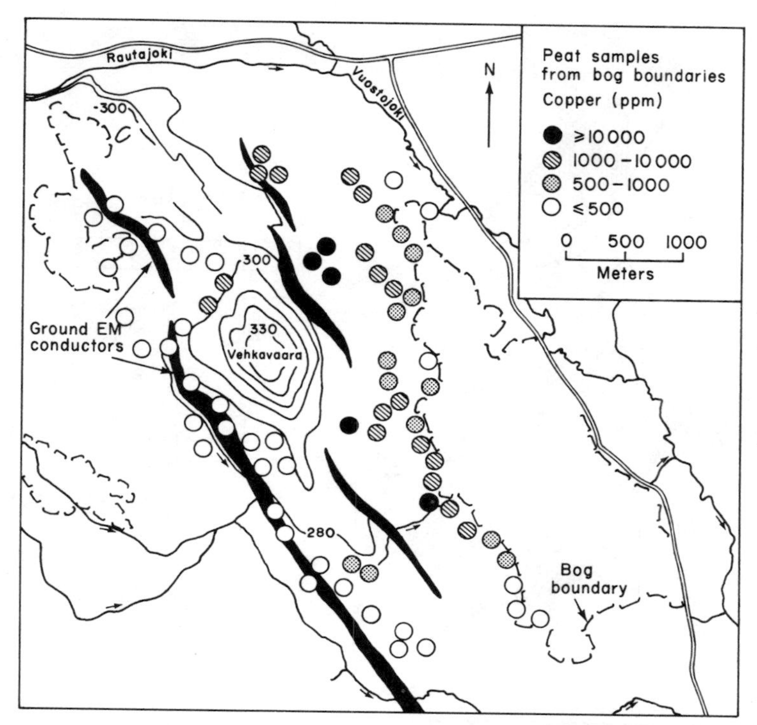

Fig. 12.16. Copper in peats near geophysical anomalies, Pajala district, northern Sweden. The westerly conductor is believed to be barren, the easterly mineralized. (After Larsson, 1976.)

Salmon, 1960). Eriksson (1976) describes the use of peat in reconnaissance sampling in Sweden, using two samples per bog, collected at the margin of the bog at the base of the peat.

IV. COLLUVIUM AND ALLUVIUM

Colluvium is generally taken to include all those deposits of local origin built up at the base of slopes as the result of gravity movement assisted by frost action, soil creep, and sheetwash. It ranges from poorly sorted angular rock fragments to fine clays, according to the nature of the bedrock material supplied from above. Although as a rule colluvial deposits are limited in extent, on ancient land surfaces they may cover very large areas, locally filling the old valleys to depths of 100 m or more.

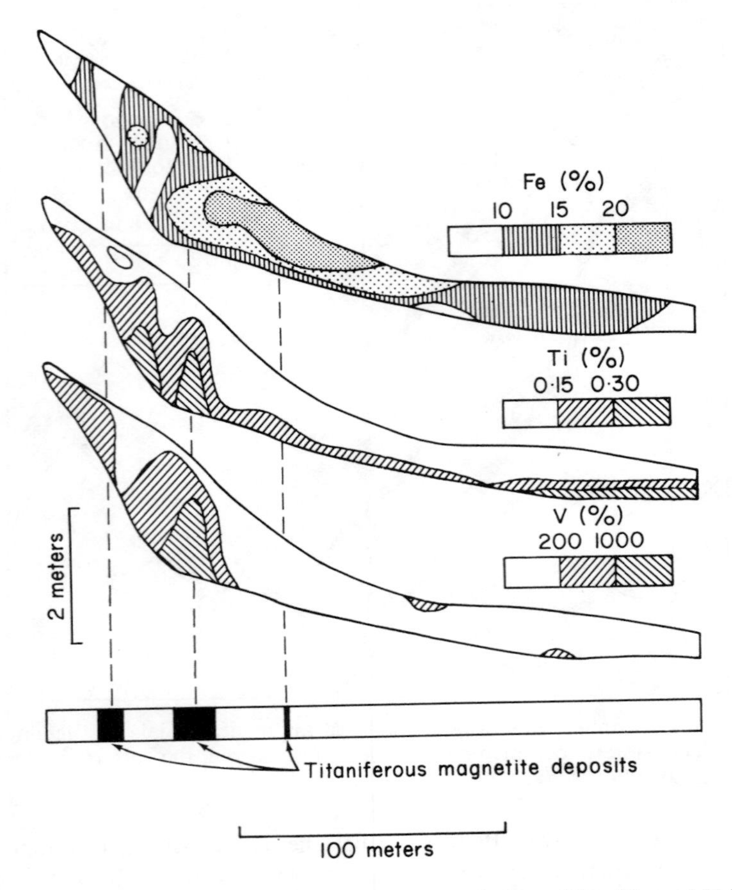

Fig. 12.17. Section through organic bog showing relation of Fe, Ti, and V in peat to underlying ore, Otanmäki, Finland. (After Salmi, 1955.)

Alluvial deposits include an equally wide range of material. They are, however, commonly better sorted and are more likely to be stratified than is colluvium. In confined valleys, alluvial deposits are necessarily restricted in their lateral extent, but in open valleys and on flat, mature surfaces, vast areas may be covered by thick accumulations of water-borne material.

In the Basin and Range Province of western U.S.A., extensive areas are mantled with alluvium derived from the neaby mountains (Fig. 12.18).

The anomalies formed in both kinds of overburden possess many features in common and are conveniently considered together.

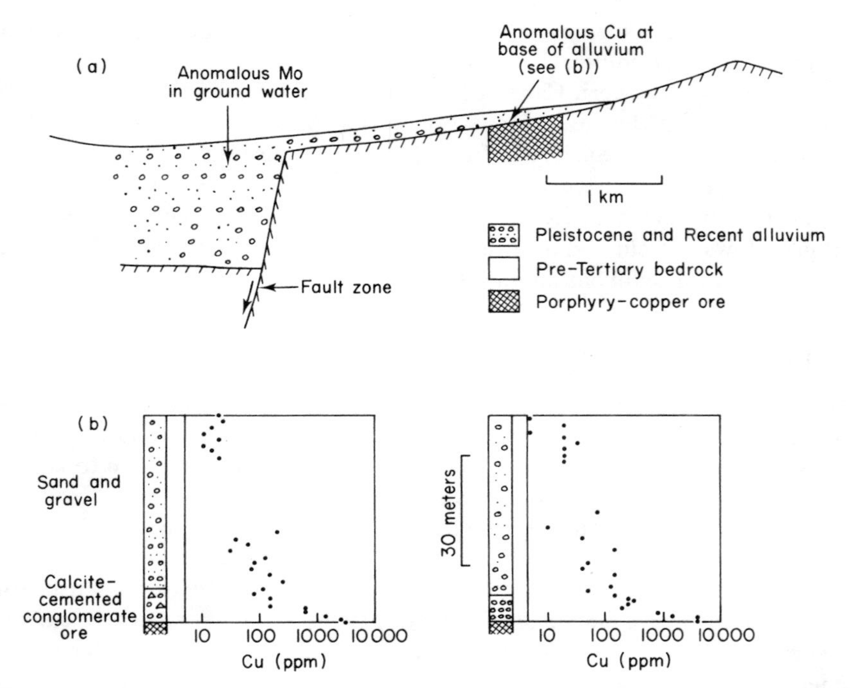

Fig. 12.18. (a) Schematic cross-section of geology near the Mission-Pima deposit, Arizona. (b) Copper content of alluvium in two drill holes overlying ore. (Data from Huff, 1970.)

A. Syngenetic Anomalies

In both colluvium and alluvium the form of the anomalous patterns depends on the origin and mode of transport of the anomalous material. Colluvial deposits below residual soil anomalies commonly show a broadly spreading, fan-shaped pattern of anomalous values. Fanglomerate deposits built up by streams draining mineralized outcrops commonly show a somewhat similar spreading pattern. The patterns in the alluvial deposits of confined valleys take the form of alluvial trains. These processes have been discussed and illustrated in more detail in Chapter 10.

The mode of occurrence of the metal may vary greatly according to the element concerned and the mechanism of dispersion. In colluvial anomalies the syngenetic material occurs in essentially the same form as in the parent soil anomaly. In alluvial anomalies the metal may be derived by the erosion of residual soil anomalies or of upstream anomalous seepage areas. Alluvial anomalies of this type are discussed in Chapter 15.

Sampling of fines from talus has been successfully utilized to detect small to moderate-sized mineralized zones in mountainous areas in Chile (Maranzara, 1972) and western Canada (Hoffman, 1977). Samples of fine particles were collected at the base of talus cones at a spacing of tens to hundreds of meters, sieved, and analyzed. In arid areas, dispersion is entirely clastic, derived from mineralized zones higher on the hillside, but in temperate or arctic regions, some hydromoprhic dispersion may be involved. The technique allows sampling in arid regions lacking streams, and the following up of large stream-sediment anomalies.

B. Epigenetic Anomalies

Hydromorphic haloes may develop in barren colluvium and alluvium overlying buried ore by the upward movement of metal-bearing solutions. They are subject to the same conditions and controls that are common to similar hydromorphic patterns developed in other materials.

The anomalies detected by Fulton (1950) in Tertiary alluvium overlying mineralized bedrock in the Appalachian Zn district are almost certainly of hydromorphic origin. A small but well-developed halo over Pb–Zn ore in Nigeria was observed in an alluvial deposit that was dated archaeologically as 400 years old (Fig. 12.19). The vertical, pipelike shape of the Pb pattern in the Nigerian example contrasts strongly with the broader halo of decreasing Zn values in the same material. It is possible that the Pb anomaly is biogenic and formed by the growth and decay of grass roots in which the Pb had been concentrated.

In arid and semi-arid environments it is reasonable to expect a substantial upward movement of ground water to replenish the moisture lost by evaporation and transpiration from the fringe zone above the water table as well as lateral transport of dissolved metals in ground water.

At the Mission and Pima porphyry-copper mines in the Basin and Range of Arizona, conglomerate at the base of 60 m of Pleistocene alluvium overlying ore contains up to 4000 ppm Cu (Fig. 12.18). Most of the Cu is contained in the calcite cement forming the matrix of the conglomerate. Anomalies related to the covered ore are not detectable in the surface alluvium. Acid Cu-bearing ground waters emerging from the orebody evidently flowed in the conglomerate just above bedrock, and precipitated Cu on neutralization. Present-day well waters several miles downslope are anomalous in Mo derived from the deposits, indicating that hydromorphic processes are still active. At Chuquicamata and in south-west U.S.A. oxidized Cu deposits are found in alluvium adjacent to porphyry-copper ore bodies. Near Milford, Utah, coatings of caliche on pebbles in near-surface gravel overlying Cu ore at a depth of several tens of meters were distinctly anomalous in Cu (Fig.

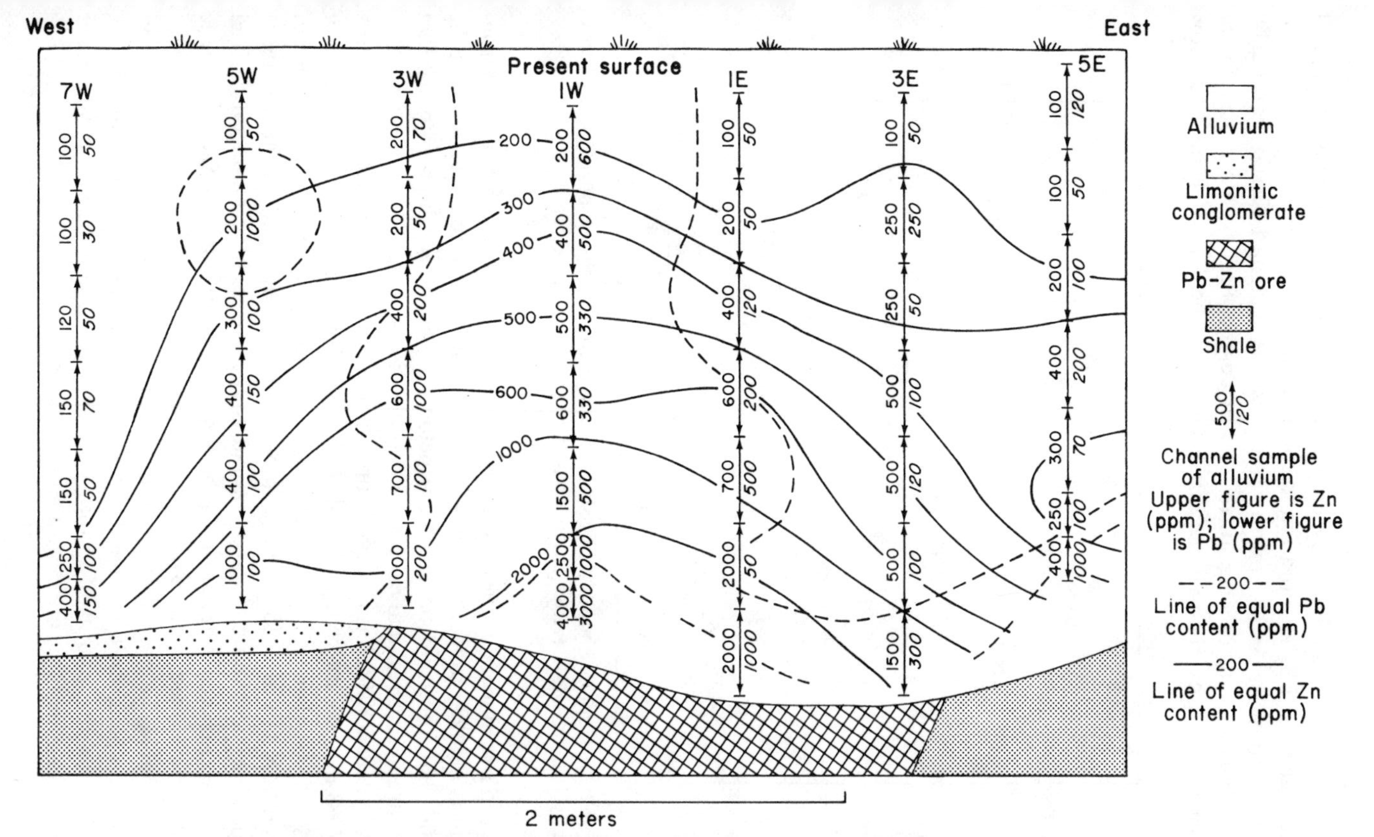

Fig. 12.19. Distribution of Pb and Zn in alluvium overlying Pb–Zn ore in bedrock, Nyeba district, Nigeria. Data on −80-mesh fraction. (After Hawkes, 1954.)

12.20). The Cu was evidently emplaced by upward movement of ground waters and fixed by evaporation in forming the caliche.

In South West Africa, on the fringe of the Kalahari Desert, erratically anomalous content of Zn and, to a lesser degree, Pb have been detected in massive caliche overlying mineralized dolomite. The data of Table 12.1 show

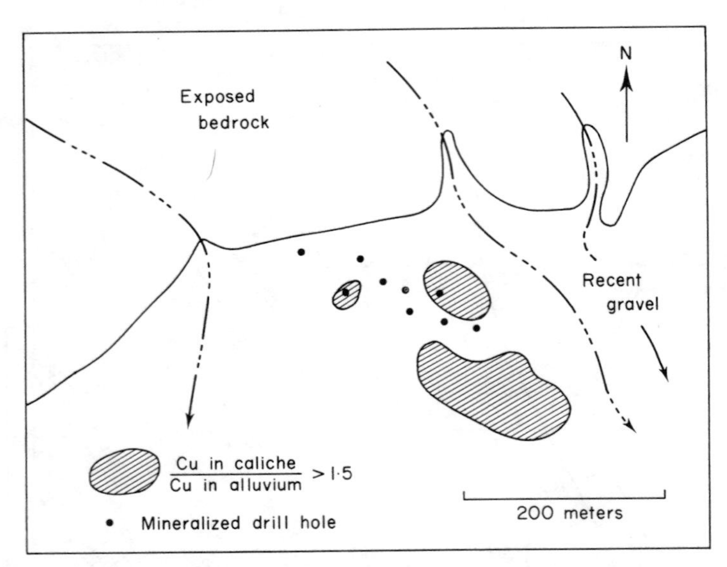

Fig. 12.20. Copper anomaly in caliche from gravel overlying buried Cu deposit, Rocky Range, Utah. (After Erickson and Marranzino, 1960.)

a sharp rise in metal values toward the base of the caliche layer. The environment is inimical to leaching, and the pattern of distribution suggests that the metals have been introduced from below. Nevertheless, the possibility that they are in fact residual cannot be entirely discounted.

Lotspeich (1958) has recorded enrichment of Cr, V, P, and Zn in colluvium 0·5–1·3 m thick overlying phosphate beds containing abnormal amounts of these metals (Table 12.2). Under semi-arid conditions and in the absence of vegetation, upward movement of moisture is considered to be the dominant dispersion mechanism. In a subhumid environment, biogenic metal derived from deep-rooted vegetation is believed to contribute to the epigenetic pattern developed in the overburden.

In Zambia, lateral superimposed patterns for Cu have been observed in river flood plains below mineralized ground, on the surrounding valley slopes, and around the mouths of tributary streams draining mineralized catchment areas. These features are described in detail in Chapter 15.

Table 12.1
Metal distribution in caliche overlying mineralized dolomite, South West Africa

Description	Depth (m)	Zn (ppm)	Pb (ppm)
Brown topsoil	0–0·3	75	70
	0·3–0·6	125	120
Massive caliche	0·6–0·9	115	20
	0·9–1·2	240	50
	1·2–1·5	150	20
	1·5–1·8	95	15
	1·8–2·1	300	35
	2·1–2·4	240	55
	2·4–2·7	850	120
Mineralized dolomite	2·7–3·0	6400	290
Background for caliche over unmineralized dolomite	—	20–100	5–30

Table 12.2
Enrichment of metals in colluvium overlying phosphatic bedrock[a]

Element	Content in phosphatic bedrock (ppm)	Content in overlying colluvium (ppm) Arid environment	Content in overlying colluvium (ppm) Subhumid environment	Content in barren colluvium (ppm)
V	5000	800	150	65
Cr	5000	500	200	35
Zn	3500	500	500	100
P	10 000	7000	12 000	1000

[a] Mean values computed from data by Lotspeich (1958).

In Africa, Cu in samples of the interior of termite mounds was strongly anomalous and could be used to outline favorable areas for Cu ore in an area of extensive talus and transported soil (d'Orey, 1975).

V. OTHER TYPES OF TRANSPORTED OVERBURDEN

Relatively little information is available on dispersion patterns in aeolian deposits. Some evidence for syngenetic dispersion in wind-blown sand has been reported from the Sahara. Here, the pattern of values in the sand surrounding a mineralized hill is distorted asymmetrically in the direction of the prevailing wind (I. B. Lambert, personal communication). Hydromorphic anomalies can also be formed in aeolian deposits, given a shallow water table. For example, in sands of the Kalahari Desert in southern Africa, Watson (1970) detected distinct Zn anomalies in the C horizon of soils developed on sands over weakly mineralized bedrock. Termite mounds at the same locality were less anomalous than the C horizon and contained no recognizable fragments of the bedrock, indicating that the termites were not effective in dispersion from the deeper layers. Tooms and Webb (1961, p. 840) reached a similar conclusion for an area in Zambia. In residual soils containing an admixture of wind-blown dust, the coarser fraction may show the greatest contrast. In Wisconsin, Kennedy (1956) found no pattern of Zn developed in a 3 m layer of Pleistocene loess over known Zn mineralization in the underlying bedrock.

Thin layers of volcanic ash do not inhibit the development of anomalies in overlying soil, as evidenced by well-developed anomalies recorded above 5–10 cm of ash at a prospect in Yukon Territory (Doyle and Fletcher, 1975), but in the same region, ash extending to 0·7 m depth obscured strong anomalies in the underlying residual overburden (Coope, 1975). Post-ore volcanic cover presents a major prospecting problem in south-western U.S.A. and in other areas, but to date little study has been given to this problem.

Chapter 13

Geochemical Soil Surveys

The preceding chapters have described principles of dispersion in the surficial environment and processes of forming anomalies in soils and overburden. This chapter presents the techniques and methods of planning, carrying out, interpreting, and following up surveys of residual soil and transported overburden.

Sampling and analysis of residual soils is one of the most widely used of the geochemical methods described in this book. The popularity of residual-soil surveying as an exploration method is a simple reflection of the reliability of soil anomalies as ore guides. General experience in many climates and in many types of geological environments has shown that where the parent rock is mineralized, some kind of chemical pattern almost always can be found in the residual soil that results from the weathering of that rock. Where residual-soil anomalies are not found over known ore in the bedrock, further examination usually shows either that the material sampled was not truly residual or that an unsuitable horizon or size fraction of the soil was sampled, or possibly that an inadequate extraction method was used. In other words, when properly used, the method is exceptionally reliable in comparison with most other exploration methods.

Residual-soil surveys have been found particularly applicable in areas of deep residual cover and sparse outcrops, where other exploration methods either are too expensive or are technically ineffective. Thus, the areas of deep weathering in the southern Appalachians of the U.S.A. and in east and central Africa have been found to be well suited for geochemical soil surveying. Although any of the common ore metals may form pronounced anomalous patterns in residual soil, the strongest contrast is developed for the relatively immobile elements such as Pb.

More specialized types of geochemical surveys are usually necessary in areas of transported overburden, such as glacial deposits and alluvium. This greater specialization is not surprising in view of the more complicated dispersion processes involved in the development of anomalies in transported overburden. Sampling at depth may be needed in glaciated areas, and partial extractions are more widely applied in order to emphasize anomalies due to clastic, hydromoprhic, or biogenic processes. With the help of such methods, anomalies may be detected from deposits covered by several tens of meters of transported overburden, though not with the complete assurance of surveys in residual soils. Geochemical exploration techniques for glacial overburden are well developed; for other types of transported overburden, much scope for improvement remains.

I. ORIENTATION SURVEYS

An orientation survey is a preliminary investigation designed to obtain information for planning an optimum routine survey. The intent is to develop procedures that maximize the contrast of significant anomalies and minimize the number of false anomalies, all for the lowest price. Orientation surveys are absolutely essential in terrain where there has been little or no prior experience with geochemical exploration methods. In terrain where considerable previous geochemical work has already been done and effective routines have already been established, elaborate orientation surveys are not so vital.

An orientation survey normally consists of a series of preliminary experiments aimed at determining the existence and characteristics of anomalies associated with mineralization of the type to be sought. This information may then be used in selecting adequate prospecting techniques and in determining the factors and criteria that have a bearing on interpretation of the geochemical data.

Although the orientation study will provide the necessary technical information upon which to base operational procedures, the final choice of methods to be used must also take into account other factors, such as cost of operation and availability of personnel. These considerations are discussed in Chapter 20, and only the technical aspects of orientation are considered here.

If possible, these preliminary experiments should be undertaken in the vicinity of known deposits that have not been disturbed or contaminated by human activity, so that the natural geochemical pattern can be observed. Where deposits of economic size are not available for study, it may be necessary to depend on experiments with vein extensions or minor subeconomic deposits. It is important, however, that orientation should be conducted in areas where the geological and geomorphological characteristics are representative of those likely to be encountered during prospecting.

Determination of the distribution of metal values in unmineralized terrain is of equal importance. Background studies must be carried out well away from the possible influence of known mineralization. They should also cover the full range of environmental conditions that exist in the exploration area.

The nature of the overburden, whether it is residual or is of glacial, alluvial, or wind-borne origin, is the first question that must be answered by the orientation survey. Sometimes it is surprisingly difficult to discriminate between residual and transported soil. The safest method, therefore, is to make critical and careful examinations of complete sections of the overburden at the start of every new field survey. If road-cut exposures are not available, the soil profile should be examined by pitting or augering.

Although the basic procedure in conducting an orientation survey is the same both for residual and for many kinds of transported soil, there are some differences in detail to which attention is drawn below.

A. Residual Soil

A completely thorough orientation survey in a new area starts with the collection of a series of vertical sections through the soil profile, arranged as a traverse across the suboutcrop of the mineralized ground. Comparable profiles from background areas should be sampled at the same time (Table 13.1). The initial traverse over the deposit should extend for a substantial distance on either side of the suboutcrop. Along the line of traverse, pits are dug at close intervals; if the deposit is narrow and the ore metals are relatively immobile, a complete trench section may be required. Every attempt should be made to sink the orientation pits to bedrock, particularly if any doubt remains as to whether the overburden is of residual or transported origin. If bedrock cannot be reached by pitting, deep samples may be collected with the aid of specialized augering equipment.

The next step is to log the pits in detail and establish the residual or transported origin of the overburden at all depths. The complete profile should then be channel sampled. No sample should represent a section of more than 30–60 cm, or should include material from more than one visibly distinguishable horizon. Background soil profiles should be sampled in a similar manner at selected points over representative types of unmineralized rock in otherwise comparable environments. A typical example of an orientation pit log, including the results of subsequent analysis, is shown in Fig. 13.1.

Preliminary analysis of the −80-mesh or −100-mesh fraction for the predominant ore metal or associated pathfinder elements will usually suffice to

Table 13.1
Major factors to be evaluated by an orientation survey in residual soil or transported overburden

(i) Optimum contrast between samples at a mineralized zone representative of that being sought, as compared to a range of background conditions in the survey area, considering the factors below.
(ii) Determination of the most suitable indicator element or elements, either ore elements or pathfinder elements, or both.
(iii) Nature of overburden.
   (a) Residual vs transported, and transport mechanism and direction.
   (b) Soil profile development.
   (c) Depth variation of indicator elements.
   (d) Effects of topography, drainage, vegetation, rock types.
(iv) Optimum depth of sampling.
(v) Optimum size or density fraction (clays, silts, heavy minerals, etc.).
(vi) Most suitable analytical procedure.
   (a) Extraction method (total, hot-acid extractable, cold-extractable, etc.).
   (b) Determination method (detection limit, interferences, cost).
(vii) Range of background and intensity of anomaly near ore.
(viii) Shape, extent, and homogeneity of anomaly, using preferred method and one or two traverses across ore.
(ix) Reproducibility of sampling and analysis.
(x) Possibility of contamination.

show whether an anomaly is present or not. Analytical determinations should include the total metal content (Me) and, in the case of relatively mobile elements, the readily extractable metal content (cxMe) by one or more methods as well; the choice of extractants is based on the possible modes of occurrence of the metals in question. Selected anomalous and background samples should then be subjected to a series of experiments to determine the range of concentration of the key elements, as well as the size fraction and analytical method that shows the greatest contrast between anomaly and background. When dealing with the elements that characteristically occur in the soil as components of readily identifiable clastic minerals, the heavy-mineral fraction of the soil should be examined mineralogically. On the basis of these experiments, procedures of preparation and analysis that show the maximum effective contrast (eqn 13.1) of anomaly over background are selected. Further pitting may be necessary to cover fully the width of the anomalous pattern. Analysis of the complete suite of samples by the selected procedures provides the basic information upon which to choose the most practicable horizon for sampling, i.e. the minimum depth at which an adequate anomaly contrast is obtained over the greatest width.

Pit traverses are now supplemented by traverses along which closely spaced samples are collected from the selected optimum depth, in order to

AREA: Kiswani River Location: Trav 2 (see map) Pit no. 2/6

Sampler: L. M. James Date: 20 Aug 59

Analyst: R. Hickman

Date: 2 Sep 59

Methods: Minus 80 mesh

Pb, Zn } $KHSO_4$ fusion, dithizone

cxHM, citrate, dithizone

Description of site: Residual overburden on granite bedrock on 5-degree slope, 7 m below presumed suboutcrop. Grass and shrub vegetation (secondary growth? Ground may have been cultivated). No possibility of contamination from old trench spoil 65 m east.

Soil horizon	Depth in meters	Pictorial log	Sample No. Depth (cm)	ppm Pb	ppm Zn	ppm cx HM	
A1. Gray, humic.			7642	800	400	40	
			25				
A2. Buff, sandy.			7641	500	350	15	
			60				
			7640	400	180	5	
B. Red-brown, concretionary, sandy loam matrix.	1		92				
			7639	400	750	20	
			125				
B/C. Transition.			7638	250	600	30	
			155				
C. Mottled clay-sand, mottling decreasing in depth. Limonitic qtz vein.			7637	110	400	40	
			190				
Transition to decomposed granite.	2		7636	30	250	30	
			230				
			7635 244	20	200	30	
Limonitic qtz vein at 182 cm			7643	5000	1000	>80	
	3						

Fig. 13.1. Typical example of an orientation pit log.

determine the shape, spread, and homogeneity of the anomaly (Fig. 13.2). Replicate samples at selected points may be collected to determine sampling error. In addition to examination of the superjacent anomalies, tests should also be carried out on the base-of-slope colluvium and in seepage areas down-drainage from mineralization, to determine the characteristics of the lateral anomalies, if any. Determination of the ratio of readily extractable to total metal (cxMe/Me) in the various kinds of material is particularly important at this stage.

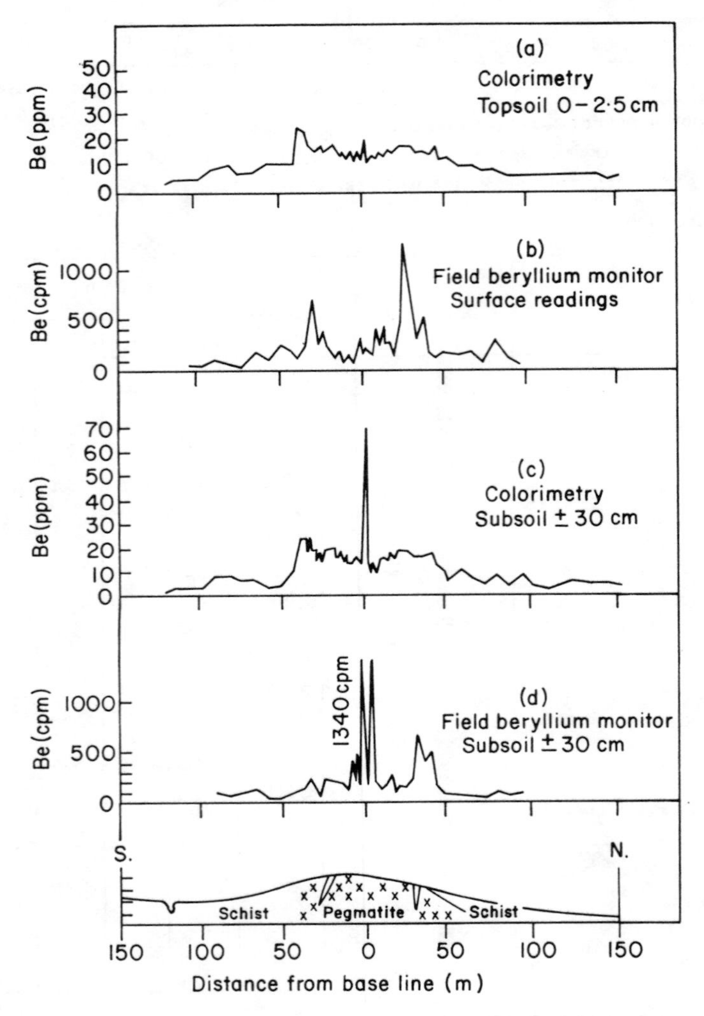

Fig. 13.2. Example of orientation traverse to determine spread and homogeneity of a residual soil anomaly at two depths and by two analytical methods. Chemical data on −80-mesh fraction. (After Debnam and Webb (1960); Be-monitor readings by K. C. Burke.)

Similar experiments should be made in each type of environment present in the exploration area. Particular attention should be given to ascertaining, if possible, the effect on metal content of variations in the grade and oxidation of the deposit, depth of overburden, soil type, and bedrock lithology. The metal content of mineralized and non-mineralized gossans, if present, should also be investigated.

While the foregoing is generally applicable to most soil surveys in a new area, each problem will usually call for some special modifications and additions in the design of the orientation experiments. Pitting may not always be practicable, or even necessary, if adequate information is available from a closely comparable area. In such cases, the orientation study may be restricted to determining the metal content only of the near-surface soil horizons, preferably including the B horizon.

B. Transported Overburden

In transported overburden, the nature of the overburden and the effective mechanisms of dispersion are the most important factors to be determined, but all aspects described under residual soils must be considered, with special attention to the optimum sampling depth. Deep overburden, either at a constant depth or just above bedrock, is commonly the most effective but is an expensive method. Seepage anomalies and biogenic concentrations are particularly useful in some areas because of the ease of sample collection. Heavy mineral fractions or sulfide-selective leaches may be helpful in deep unoxidized till, and readily extractable metal can be helpful in detecting hydromorphically dispersed metal. If deep sampling is contemplated, the drill or sampling device should be tested to determine the time and effort required to sample typical profiles to various depths.

Organic overburden (bogs, peats, etc.) should, if possible, be test sampled in both anomalous and background areas in the same way as described above for residual soils. The distribution of metals at different depths in the profile and the cxMe/Me ratios of the mobile elements should be determined. Tests using chelating agents (EDTA) and strong acids are desirable, in addition to analysis of the total metal content (Maynard and Fletcher, 1973). Careful attention should be directed to possible relations of metal content in the organic samples to the chemical character of water entering the bog (Eh, pH, source) and to the content of organic matter in the samples.

The degree of decomposition of the organic matter may be an important factor in this connection.

The influence of internal ground-water drainage on the pattern of anomalies should be studied in three dimensions if possible. Anomalies should be sought not only immediately above the suboutcrop of the mineralized ground but also down-drainage. Where mineralization occurs on the higher ground surrounding an area of organic soil, all likely seepage areas should, of course, be tested for lateral anomalies.

C. Contamination

A thorough orientation study of a new area should also include appropriate sampling to determine the extent of possible contamination arising from human activity.

As a rule, contamination from trash, fertilizers, road metaling, and farming or industrial installations is very localized and is generally restricted to the surface horizons. Leaching from old, oxidized spoil and mine dumps can disperse the metal deeper and may even contaminate seepage areas some distance downslope. Orientation studies of the effect of these sources of contamination are a simple matter of systematic sampling at progressively greater distances from a known source and noting any systematic decay in metal concentration. Regional or global contamination from industrial activities or radioactive fallout undoubtedly exists, as suggested by McNeal and Rose (1974) for Hg, but is a very minor effect.

Contamination by condensation from smelter fumes can be a more extensive feature and can sometimes be difficult to recognize and overcome. One key to smelter contamination is the fact that the effect is usually confined to the top few centimeters of the soil. For example, within a kilometer of the copper smelter at Superior, Arizona, the Cu content of surface soil is 5000 ppm but decreases rapidly to the normal background of about 20 ppm at a depth of 15 cm. Canney (1959) found that evidence of contamination existed in the upper 5 cm of the soil 29 km downwind from the Kellogg smelter in the Coeur d'Alene district, Idaho. Nevertheless, subsoil at a depth of 15 cm, even in the immediate vicinity of the smelter, was not sufficiently contaminated to interfere seriously with geochemical prospecting operations. Rather deeper penetration has been recorded for As (Table 13.2) where the effect is still noticeable at a depth of 30 cm at a point 1 km downwind from a smelter in Rhodesia.

Ancient smelting sites can present a far more serious problem. In the Bawdwin Mine area in Burma, for example, the soil has been extremely contaminated with Pb, often to depths of a meter or more (Table 13.3);

Table 13.2
Contamination of soil profile one-half mile downwind from smelter, Rhodesia[a]

Description	Depth (cm)	As (ppm)
Sandy layer transported by sheetwash	2–5	560
Brown loam subsoil	5–20	200
	20–33	120
Red loam with rubble	33–45	50
Soft, mottled, decomposed greenstone	45–70	45

[a] Source: Webb (1958a); data for −80-mesh fraction.

Table 13.3
Intense contamination of soil beneath ancient smelting sites[a]

(a) Bawdwin Mine, Burma

Sample	Depth (m)	Pb (ppm)
Residual overburden	0–0·3	4700
	0·3–0·6	1400
	0·6–0·9	1300
	0·9–1·2	1100
	1·2–1·5	900
	1·5–1·8	450
	1·8–2·1	400
	2·1–2·4	300
Sandstone bedrock	2·4–2·8	50

(b) Bushman Mine, Botswana

Description	pH	Depth (cm)	Cu (ppm)
Brown–black topsoil	8·0	0–15	7500
		15–30	1600
Brown–black clay subsoil	8·4	30–75	200
Brown sandy clay, some $CaCO_3$ spots	9·0	75–130	50
Main zone of calcification with bedrock fragments	9·0	130–160	40
Decomposed granite gneiss with patches of $CaCO_3$	8·2	160–210	40

[a] Source: (a) Ledward (1960), (b) Webb (1958a); data for both (a) and (b) for −80-mesh fraction.

here the proportion of cxPb is more than twice as much as in natural anomalous soils. Strong contamination to a depth of about 75 cm has also been noted at the Bushman Mine in Botswana in an area where the evidence of ancient smelting has been almost completely eradicated; much deeper contamination would have occurred but for the high soil pH (Table 13.3b).

Orientation work is almost always needed in the vicinity of modern or ancient smelters for the purpose of establishing a sampling technique that avoids the collection of contaminated samples. Inasmuch as the intensity of contamination falls off with both horizontal distance from the smelter site and vertical distance below the surface, the experimental samples should be collected as a series of vertical profiles of equivalent soils taken at progressively increasing distances from the source. The results should then indicate the minimum depth for uncontaminated samples at any given distance from the smelter.

D. Choice of Procedures

In choosing the optimum size fraction, extractant, sampling depth, or other parameter, both the contrast of the anomaly and the variability of background and anomaly values are important, in addition to the absolute value of the anomaly. Thus, in Fig. 13.3, method (a) gives the highest absolute concentration values near ore, but background values are also very high and overlap

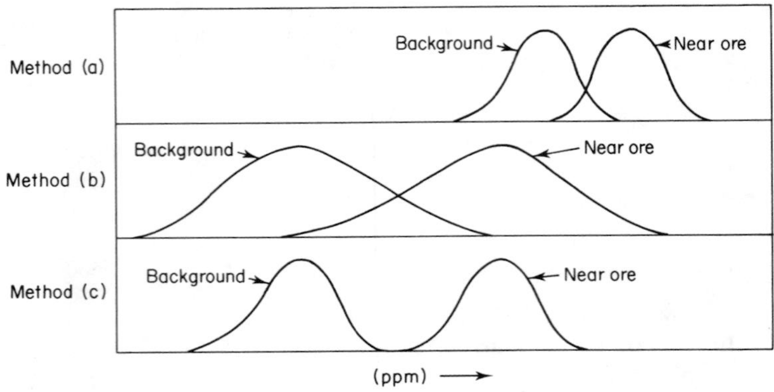

Fig. 13.3. Hypothetical frequency distributions for populations of background and anomalous values for three methods or types of samples. Despite the high absolute values of method (a), and the equal contrast of (b) and (c), method (c) gives the best resolution.

appreciably with those near ore, so that the contrast of method (a) is much lower than for methods (b) and (c). The latter two methods have equal contrast (calculated as mean anomaly/mean background), but the variability of background and anomalous values is much larger for method (b), so that a large overlap is observed. Other factors being equal, method (c) is preferred.

A statistical comparison of two procedures can be made if the background and anomalous populations have approximately log-normal frequency distributions and equal standard deviations (Rose and Keith, 1976). After conversion of the data to logarithms, the statistic t is calculated (Dixon and Massey 1951, p. 102):

$$t = \frac{(X_a - X_b)}{S_p\sqrt{[(1/n_a)+(1/n_b)]}} \qquad (13.1$$

where X_a and X_b are the means for groups of anomalous and background samples, n_a and n_b are the numbers of anomalous and background samples, and S_p is the pooled standard deviation:

$$S_p = \left(\frac{(n_a - 1)S_a^2 + (n_b - 1)S_b^2}{n_a + n_b - 2}\right)^{\frac{1}{2}} \tag{13.2}$$

for which S_a and S_b are the standard deviation of the anomalous and background samples, respectively. In general, the method with the highest value of t should be chosen.

II. FIELD OPERATIONS

An operational prospecting survey to be successful must have two qualifications: it must be based on valid principles, and it must be efficiently carried out. Establishing and confirming the validity of the principles is the function of the orientation survey, as already discussed. Selection of the field procedure for the operational survey is then only a matter of determining the cheapest and most efficient means of gathering the required data. Although to a large degree the systems used in sampling, analysis, and preparing the geochemical data maps are simply a matter of common sense, a summary of the systems most widely used might be instructive.

A. Sampling Pattern

The selection of the best sample pattern is determined primarily by the size and shape of the target. In looking for superjacent anomalies, the most suitable pattern is a simple rectilinear grid of samples taken at equal intervals along evenly spaced lines. Rectilinear grids are preferred because of the ease in laying out the field work and in plotting the data.

If the strike direction is known, the traverse lines should be laid out at right angles to the ore structure, at intervals such that every anomaly of interest will be intersected by at least two lines. This means that the traverse-line interval should be not more than one-third of the minimum economic strike length. Sample points along the line should then be spaced at intervals that ensure that at least two points fall within every important anomaly. Thus the interval is fixed by the probable minimum width of the expected anomalies. If the anomalies are not homogeneous, a corresponding closer sampling interval will be required, so that statistically at least two samples and preferably more will exceed threshold within the anomalous area.

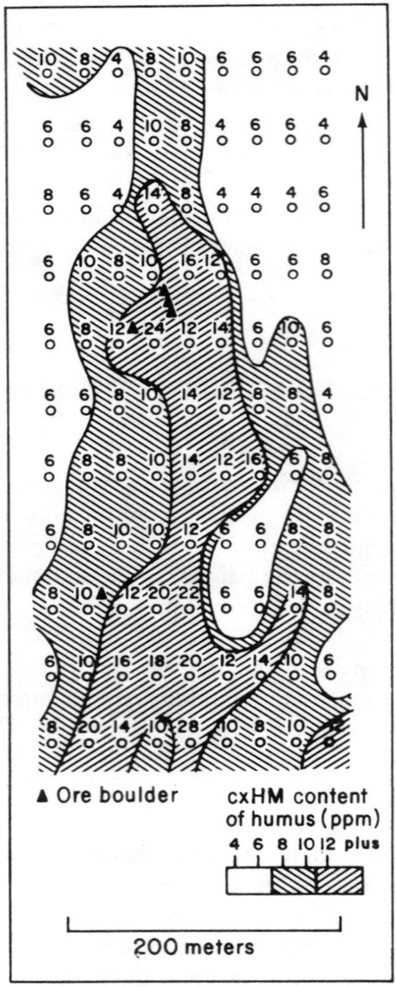

Fig. 13.4. Rectilinear sampling grid, Petolahti, Finland. (From Kauranne, persona communication, 1959.)

Linear, elongated, or fan-shaped dispersion patterns should be sought by systems of traverses laid out at right angles to the long dimension of the pattern. If the strike direction of elongated targets is not known, or if the anomalies are expected to be equidimensional or irregular, the most practica pattern is a square grid. Here the dimensions of the grid and the sample interval should be chosen so that at least four samples will fall within the limits of the smallest expected anomaly. Typical examples of rectilinear patterns are illustrated in Figs 13.4–13.7.

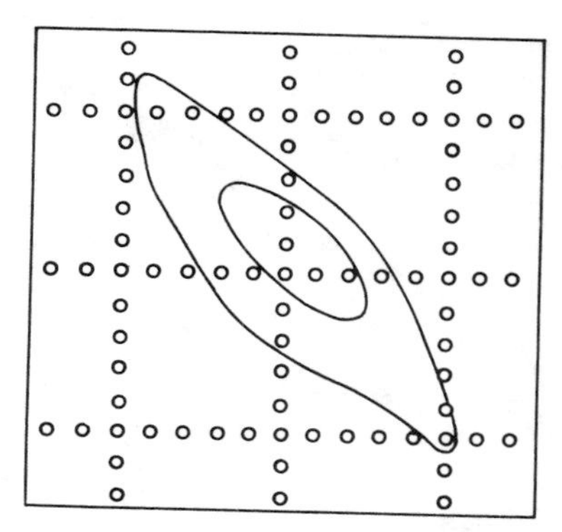

Fig. 13.5. Square grid used where strike direction of elongated targets is not predictable.

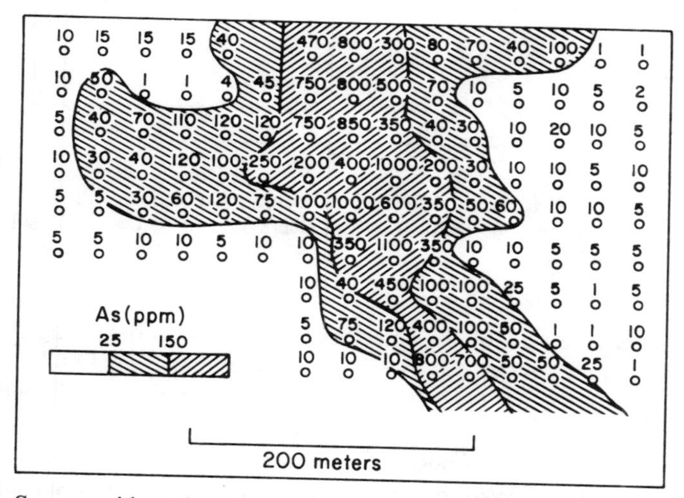

Fig. 13.6. Square grid used to detect irregular patterns of As in residual soil overlying arsenical Au mineralization, Sierra Leone. Data on −80-mesh fraction. (After Mather, 1959.)

Where the terrain is very steep it may be desirable to lay out a sample pattern to conform to the topography. Under these circusmtances, soil sample traverses following the crests and spurs of ridges have been used successfully on a number of occasions (Fig. 13.7). In order to obtain adequate coverage, it may be necessary to supplement ridge-crest traversing with contour or base-of-slope traverses. If downslope distortion of soil anomalies is

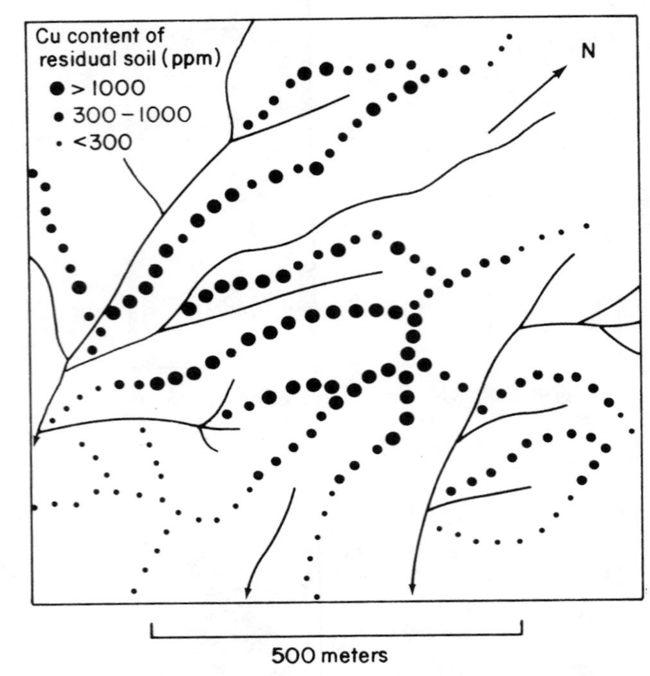

Fig. 13.7. Example of ridge-and-spur soil-sampling pattern, Cebu Project, Republic of Philippines. Data on −80-mesh fraction. (Reproduced by courtesy of Newmont Mining Company.)

extreme, it may be possible to detect any important anomaly on the higher ground by sampling only the base-of-slope colluvium (Fig. 13.8.)

Sampling patterns designed for mapping lateral hydromorphic anomalies are necessarily determined by the distribution of seepage areas. These are most commonly found in topographic depressions, at the base of slopes, and along the margins of swamps and flood plains. Spring and seepage areas may also occur, of course, at any point along the outcrop of aquifers, such as faults and permeable horizons, that may be carrying water under hydrostatic pressure. Sampling patterns for deep overburden will depend on the cost of sampling, the amount of geological and other information on the area, and the type of orebody sought. In relatively deep overburden, where costs are high, sample spacing may be designed to detect important anomalies by only one or two samples, to be followed by more closely spaced sampling around anomalies. Available geological and geophysical data should be used to select sample sites or orientation of sample traverses. Because of the higher costs of deep sampling, the statistical techniques suggested by Sinclair

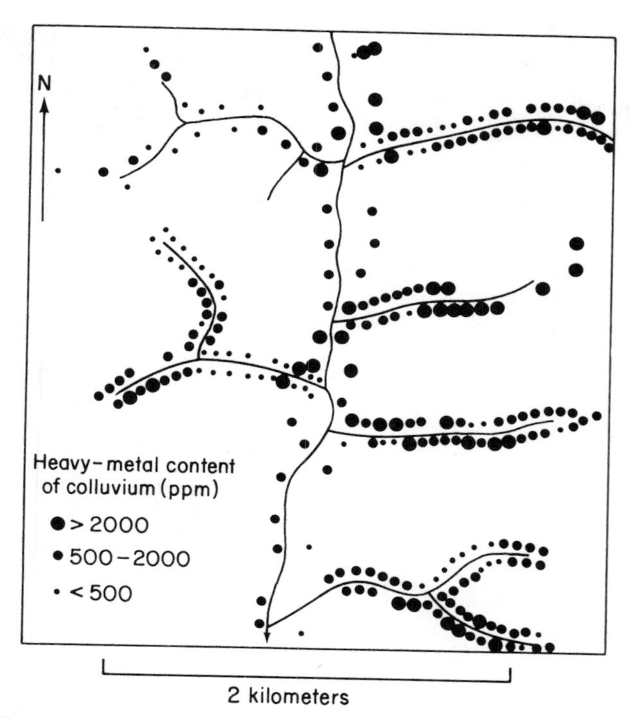

Fig. 13.8. Example of base-of-slope sampling pattern, Lemieux District, Quebec. Data on −1-cm fraction. (After Riddell, 1954.)

(1975) and Hodgson (1972) may help in selecting a sampling pattern. However, these methods generally require more knowledge of the size, orientation, and other characteristics of anomalies than is usually available in a covered area.

B. Sampling Procedure

As a rule, 20–50 g of sample provides enough material after sieving for analysis. Probably the best all-purpose containers are 3″ × 5″ water-resistant kraft paper envelopes that may be secured by folding over a non-contaminating stiffened tab attached to the open end. The advantages of such envelopes are low cost, convenience in handling, and the fact that wet samples may be dried without removing them from their containers. With very wet samples, it is advisable to pack the envelopes into a narrow, lightweight cardboard box, in order to prevent chafing during transport.

Soil samples at depths up to 30 to 60 cm may easily be collected from small pits. Deeper samples of loamy soils can usually be collected more economically by means of a soil auger. The simple type shown in Fig. 13.9. has been found satisfactory up to depths of 1–2 m. Augering to a predetermined depth

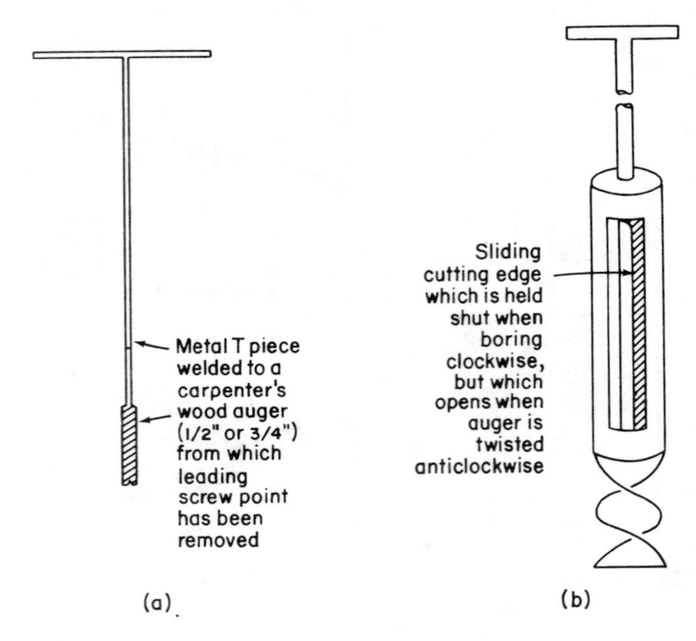

Fig. 13.9. Simple tools for soil and peat sampling. (a) Soil auger; (b) Hiller peat borer.

is greatly facilitated if a hole is first made with a crowbar. Curtin and King (1972) describe a modified auger for stony soils. When deep holes are required, light power augers may prove economical and have been used for taking samples as deep as 10m. Special augers, such as the Hillier peat borer (Fig. 13.9), are often more convenient than conventional soil augers when sampling organic overburden at depths greater than 50 cm below the surface.

Larsson (1976) describes a technique of sampling frozen and underlying unfrozen peat during winter along the margins of bogs (Fig. 13.10).

Three types of sampling equipment are in use for deep sampling of glacial overburden. For overburden with relatively sparse cobbles and boulders, a portable gasoline-powered percussion drill may be used, under favorable conditions (Gleeson and Cormier, 1971; Wennervirta, 1973; Tilsley, 1975; Wilbur and Royall, 1975). The drill and rods can be handled by two men. An initial hole is made with a solid point to near the desired sampling depth

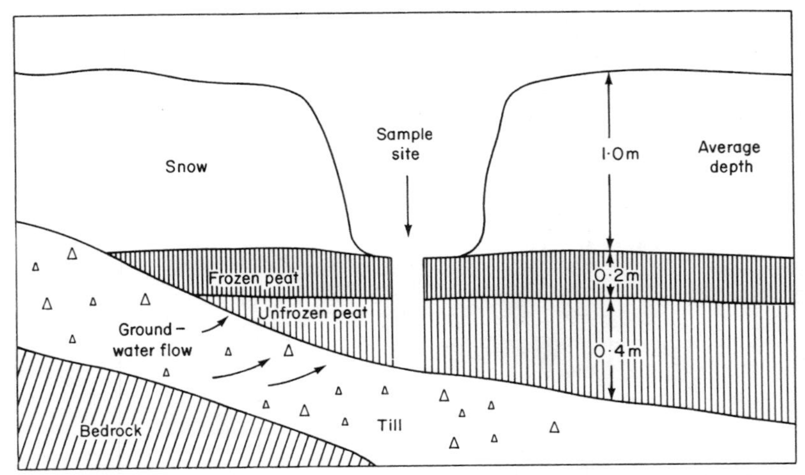

Fig. 13.10. Sampling of peat at the margin of bogs during winter. (After Larsson, 1976.)

after which an opened sampling tube is driven into the overburden (Fig. 13.11). Rods and sampler are removed with a hydraulic jack, if necessary. If till just above bedrock is to be sampled, a preliminary hole is driven to determine the depth to bedrock, and a second adjacent hole is sampled at the appropriate depth. Up to 30 holes per day can be drilled to 2 m depth with this equipment. In stony or bouldery till, deep sampling may require three or more holes per site in order to reach bedrock, or sampling at depth may not be possible at all.

A second type of sampling system utilizes a larger rotary or pneumatic drill, usually mounted on a tracked vehicle (Wennervirta, 1973; Van Tassell, 1969; Tilsley, 1975; Kokkola, 1976). Samples of bedrock or till just above bedrock can be taken at a rate of six to nine per day under favorable conditions. In systems using water as the drilling fluid, special precautions must be used to avoid loss of the silt and clay fraction which may contain hydromorphically transported metal.

A third type of overburden sampling utilizes a backhoe, bulldozer, or similar equipment to dig pits or trenches (Kauranne, 1976). This system is most useful in understanding complex layered overburden, or in very stony or bouldery till.

For reconnaissance sampling in remote areas, Sainsbury *et al.* (1973) describe methods of collecting samples of soil and other material from small airplanes and helicopters in flight, and A. R. Barringer (personal communication) describes an airborne "vacuum cleaner" for collecting fine surficial particles.

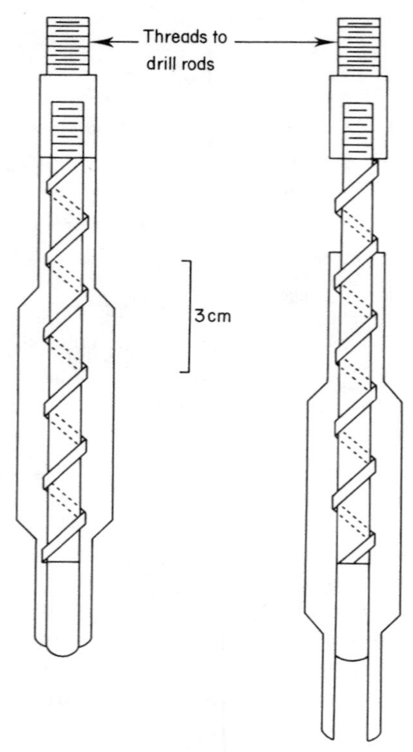

Fig. 13.11. Sampler for till used with percussion drill. (a) Sampler closed; (b) sampler partly opened. (After C. F. Gleeson (personal communication) and Hornbrook *et al.* (1975).)

C. Locating and Identifying Samples

Location of sample sites or sample traverses need only be accurate enough to enable any anomalous sites to be revisited in the field and to define the position relative to other sites and geological features. Inasmuch as precise surveying is not necessary, pace-and-compass traverses, without the cutting of lines beyond those necessary for access, are normally quite adequate. Samples collected on iregular patterns, such as along ridge crests and contours, or in seepage areas, may be located directly on aerial photographs or topographic maps if available.

Knowledge of the local topography is usually essential when interpreting soil surveys. If published contour maps are not available, the necessary information may be obtained by taking aneroid barometer readings along the traverse lines, or by measuring and recording the direction and angle of slope at each sample point. Providing the traverses are close enough together, approximate form lines can then be drawn on the geochemical map.

It is almost always desirable to mark the sample site in such a way that it can be found again when revisiting the area. Along traverse lines it is rarely necessary to identify more than ever third or fourth site. Stakes or blazes, marked with weatherproof crayon, are usually adequate to ensure legibility for the period of the survey. Baselines and other critical points may be identified by more permanent metal tags or cairns.

Two systems of numbering are in common use. On geometric grids, samples are usually identified by the grid coordinates with reference to the baseline. When the sampling net is irregular, the best procedure is to number the samples consecutively in the order in which they are collected. Duplication of samples can be avoided by prenumbering the containers, which can readily be done with a standard office-type automatic enumerator. A similar system can usefully be employed when grid sampling, to serve as a safeguard against errors in assigning grid coordinates.

The field notebook should record the traverse number, sample number, sample interval, and depth, together with a description of the material collected and any special information such as whether the ground is freely or poorly drained, topographic situation, and so on. Careful consideration of the types of data that will be useful, based on the orientation survey, will minimize wasted time in recording data that are never used.

D. Sample Preparation and Analysis

Preparation of samples for analysis is usually carried out at base camp or in the laboratory. If the overburden is particularly stony, however, it may be desirable to sieve the samples to -2 mm on the spot and discard the coarse fraction. Generally speaking, samples are too moist to pass through a fine mesh without being dried. In some climates samples will dry out by evaporation through the paper walls of their containers if laid out in the sun or stored in a dry room for a few days. Otherwise it is necessary to use some form of low-temperature oven, which may be improvised in the field.

The optimum system for sample preparation should be determined as part of the orientation experiments. For most problems, it is usually necessary only to sieve the sample, discard the coarse fraction, and retain the fines for analysis. Crushing at any stage of preparation of soil samples here is generally avoided. Rarely, the element sought may occur preferentially as clastic grains, such as Be in the mineral beryl; in this case, it may be found preferable to retain and crush one of the coarse fractions for analysis. Non-contaminating sieves should always be used. The principal danger of contamination is from Cu and Zn in standard brass sieves, where samples are to be analyzed for these metals. A sieve made of stainless steel, nylon, or bolting silk set in a plastic or wooden frame is satisfactory.

The elements to be determined, as well as the analytical procedure, should be decided from the orientation survey. The choice of elements is generally based on considerations of relative mobility, together with suitability of available analytical techniques. Thus, because Zn tends to form broader dispersion patterns than Pb, it is usual to analyze for Zn in intial stages of exploration surveys for Pb–Zn ores, and for Pb when a more precise location of the bedrock metal source is desired. Determination of both total and readily extractable metal is helpful in differentiating between syngenetic and hydromorphic anomalies. The precision of the analytical procedure should be good enough that the analytical errors will be small compared with the natural range of background fluctuation. Some of the chemical procedures most commonly employed in soil analysis are listed in Table 3.2; factors in the organization of an analytical program are discussed in Chapter 20.

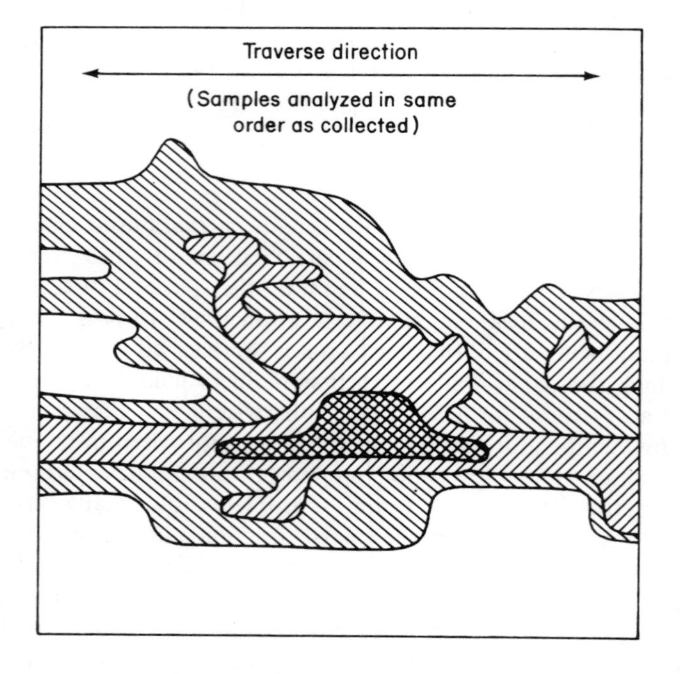

Fig. 13.12. Typical example of a false anomaly (or a false trend in a real pattern) due to variation in analytical bias.

The order in which samples are analyzed can be important because of bias in the analytical procedure from one day to the next, or over longer periods. Two approaches have been used to minimize this problem. If the entire group of samples from the project is available at one time, the samples may be submitted for analysis in random order, along with replicates to check the reproducibility of analysis. Any bias for one batch of samples is then spread randomly throughout the map area, and essentially appears as an increase in the analytical error. Alternatively, the samples may be analyzed in batches correlating with traverse location in the area. Bias can be recognized by abrupt changes in the background level from one row of samples to the next (Fig. 13.12). If this bias is significant, the samples should be reanalyzed after correcting the cause of the bias. This approach is generally preferable to randomization provided the time is taken to check the results for abrupt changes in background. The possibility of spurious anomalous values is further reduced if anomalous samples are automatically reanalyzed as a matter of routine. Randomization is time-consuming and adds to the possibility of human error, as well as allowing unnecessarily large errors to remain in the data.

E. Preparation of Geochemical Maps

Two different kinds of maps are commonly prepared in the course of recording and interpreting the data of geochemical prospecting surveys. One of these, the data map, is simply a vehicle for the chemical data. The other is an interpretation map that embodies some degree of graphical generalization of the data.

The purpose of the data map is to record the actual observations and to show their relationship to observed features of geology, topography, drainage, and possible sources of contamination (Table 13.4). Where this cannot be done on a single sheet without causing congestion, then a series of transparent overlay maps should be prepared. Inasmuch as the map should present the data objectively and in full, all sampling points and the actual analytical values obtained should always be given, along with the other data in Table 13.4.

Data may be plotted as profile curves where it is desired to emphasize the distribution of metal along separate lines of samples. As a rule, curves are used on maps when the traverse lines are too far apart to justify joining anomalies from one line to another. Figure 13.13. illustrates some common conventions for plotting data in the form of profile curves.

Interpretation maps usually involve the grouping of the data in ranges of concentration. These ranges then may be represented by simple graphical

Table 13.4
Check list of information to be presented on maps of geochemical soil surveys

Essential

(i) Element analyzed and units (ppm, %).
(ii) Analytical method, especially extraction or decomposition method.
(iii) Type of samples (depth, horizon, grain-size fraction, other identifying information).
(iv) Sampling locations, as dots or other symbols.
(v) Analytical results, plotted numerically next to each sampling location. Alternatively, the data may be presented as profiles (Fig. 13.13) where traverses are spaced widely or irregularly.
(vi) Scale bar, north arrow. The scale of the map should be chosen so that minimum spacing between sampling locations is about 0·5 cm.
(vii) Location of survey (state, province, locality, latitude, and longtiude), and reference points to local topographic or other identifiable features.
(viii) Date, reference to report giving other information on the survey.

Desirable (may be on separate sheets at same scale)

(i) Topography or form lines to facilitate interpretation.
(ii) Drainage, springs, lakes, swamps, flood plains.
(iii) Geology of bedrock.
(iv) Possible sources of contamination.
(v) Evidence of mineralization, including exposures, prospect pits, drill holes, and geophysical anomalies.
(vi) Contours of geochemical data (Figs 13.4–13.6) or symbols distinguishing different concentration ranges (Figs. 13.7 and 13.8), usually on a geometric scale.
(vii) Names of samplers, analysts, supervisor.
(viii) Direction of ice movement in glaciated areas or prevailing wind in arid climates.

symbols (Figs 13.7 and 13.8) or by contours. When selecting contour intervals or group concentration ranges for bringing out the pattern and relative significance of anomalies on the geochemical map, it is usual to select the intervals as factorial multiples of the threshold or background value, e.g. for a threshold of 100 ppm and a factor of 2, the intervals would be 100, 200, 400, 800, 1600 ppm, and so on. Generally speaking, contours should only be employed where anomalies are homogeneous and where there are adequate data to ensure validity of the contours. It is especially important to avoid closing contours at the edge of the sampled area, because important extensions of anomalies can be missed in this way.

In large surveys, maps can conveniently be prepared by computer methods, as described in Chapter 19.

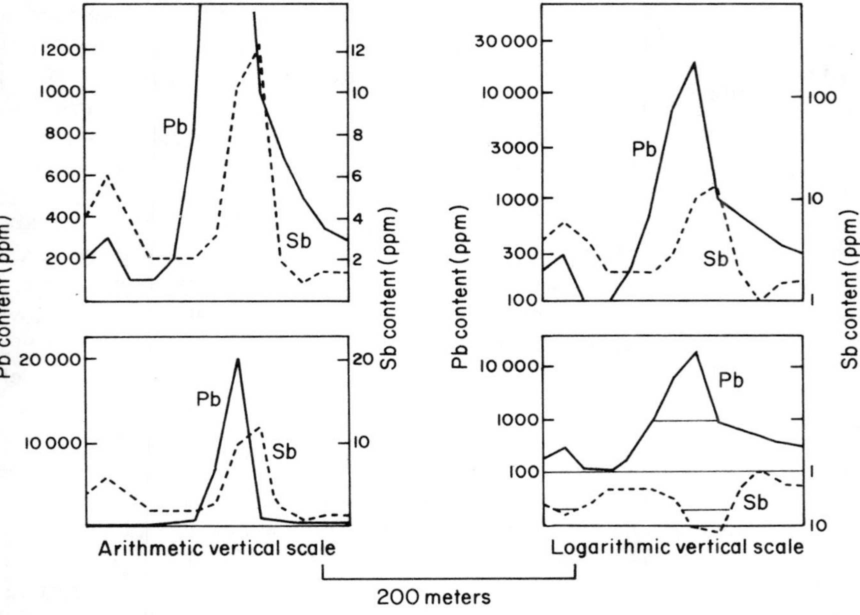

Fig. 13.13. Systems for plotting geochemical data as profile curves. Data on −80-mesh fraction. (After Hawkes, 1954.)

III. INTERPRETATION OF DATA

In general terms the interpretation of geochemical soil data involves four main problems: (i) estimation of background and threshold values; (ii) distinguishing between significant and non-significant anomalies; (iii) distinguishing between lateral and superjacent anomalies; and (iv) appraising the significance of anomalies in terms of possible ore, with a view to selecting those that merit further investigation.

It is naturally extremely difficult to generalize on interpretation, as each area presents its own problems. No attempt is therefore made in the following paragraphs to indicate more than a general philosophy, in the knowledge that any instructions as to interpretative techniques will often be found incomplete in detail and may well need to be modified in practice to suit the needs of particular problems.

A. Estimation of Background and Threshold Values

The general approach to this problem has been outlined in Chapter 2. In practice, the principal difficulty lies in recognizing the different background patterns that are related only to bedrock types or to variations in overburden and drainage, and that are not connected with mineralization. Geological or geomorphic features responsible for major variations in geochemical patterns may easily pass unnoticed at the time the samples are collected. This problem assumes increasing importance as anomaly contrast diminishes and peak anomalous values related to significant mineralization approach the threshold value.

The best practice is to study the overall geochemical pattern on the data map with a view to recognizing any correlations that exist between the "geochemical relief" and the observed and recorded features of geology and geomorphology. Some form of preliminary contouring or coding of high and low values may be helpful. It will be remembered that the units comprising the geochemical landscape are defined not only by differences in the mean level of the values but also by differences in scatter, or deviation about the mean. The boundaries of these geochemical units should be outlined on the data map or studied as an overlay to the geological, topographical, and such geophysical maps as may be available. In this way, not only is attention drawn to the more obvious anomalies but also to the broader features of the background pattern. Provisional correlations between geochemical features and the geology or topography provide a basis for determining the background and threshold values that should be assigned to each component geochemical unit of the overall background pattern. A computer may be helpful in handling the data at this stage (Chapter 19). A review of the geochemical pattern within each unit may then disclose anomalies that would not have been apparent had the data been assessed against the yardstick of a uniform threshold value. Other advantages of this approach are that it can provide additional information on the general geology of the area, as well as facilitating the subsequent interpretation of the anomalies.

B. Recognition of Non-significant Anomalies

Recapitulating from Chapter 10, the principal types of anomalies that are not related to mineral deposits include those resulting from (i) barren rock types characterized by a relatively high background metal content; (ii) human contamination; and (iii) sampling and analytical errors.

Anomalies resulting from the weathering of high-background rocks may be suspected wherever the patterns coincide with all or a major part of a mapped rock unit, or cover very large areas compared with the expected dimensions of an ore deposit. Such anomalies are particularly suspect where these broad patterns lie parallel to the known trend of the regional geological structure. Associations of elements that are characteristic of certain specific rocks, such as the Ni–Co–Cr association in ultramafic rocks, may also be useful as clues to the origin.

Contamination as a cause of anomalous patterns is immediately suspect where the anomalies are located near mine dumps and smelters or are related to some recorded pattern of agriculture, roads, or railways. Difficulties in intepretation can occur, however, in old mining areas where visible evidence of past activities may be lacking or inconspicuous. In these circumstances it is necessary to rely on the geometry of the anomalies, the fact that there is a general tendency for values to decrease with depth and that, for mobile elements in freely drained residual overburden, the cxMe/Me ratio is commonly higher than would be expected if the anomalous metal were truly residual. As mentioned earlier, however, biogenic anomalies also commonly decrease with

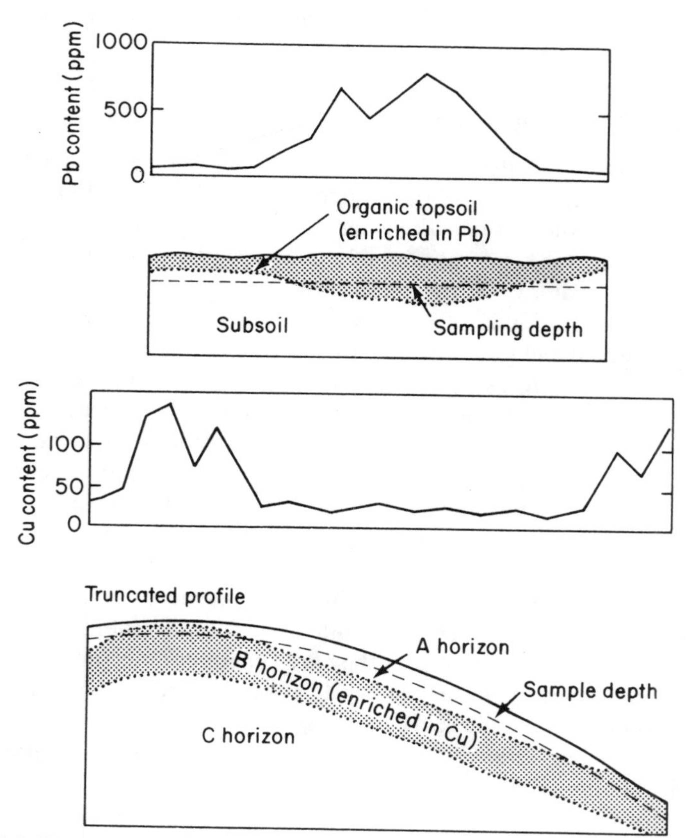

Fig. 13.14. Hypothetical examples of false anomalies due to inadvertent collection of samples from more than one soil horizon.

depth, and all epigenetic anomalies, both biogenic and hydromorphic, are characterized by a high cxMe/Me ratio.

False anomalies due to the inadvertent collection of samples that contain a natural enrichment of metal may be difficult to recognize unless the field area can be revisited. They most commonly reflect unnoticed variations in (i) the nature of the sample material, particularly with regard to the content of organic matter, clays, or hydrous Fe–Mn-oxides; (ii) the depth of the different soil horizons in relation to the sampling depth (Fig. 13.14); (iii) the nature of the vegetation and its influence on the accumulation of biogenic metal in the soil; and (iv) local drainage conditions. False anomalies due to analytical errors are more readily spotted and can easily be corrected by repeat analyses. Erratic high values are obviously suspect, as are patterns that are related to particular batches of samples. If samples were analyzed in the order in which they are collected, batch variation shows immediately as a pattern that trends parallel to the traverse direction (Fig. 13.12).

C. Distinction Between Lateral and Superjacent Anomalies

Many costly errors in interpreting geochemical data have occurred because the assumption was made that the source of metal lay immediately beneath the anomaly. In other words, the anomaly was assumed to be superjacent when in reality it was a lateral anomaly developed by horizontal movement either of ground water or of the soil itself.

The two principal kinds of superjacent dispersion patterns are the relic patterns in residual soil due to simple weathering in place and hydromorphic or biogenic patterns in transported cover resulting from upward movement of metal-rich solutions. Residual patterns are of course syngenetic, and may be recognized by the presence of diagnostic primary minerals and by a low cxMe/Me ratio in freely drained overburden; only under rare circumstances is the cxMe/Me ratio high, such as when the concealing overburden is poorly drained or organic-rich, or where the anomalous metal occurs in the superjacent pattern as relatively soluble secondary minerals.

Superjacent patterns of hydromorphic or biogenic origin, on the other hand, are characterized by the predominance of more mobile metals and a high cxMe/Me ratio. In this respect they are generally indistinguishable from lateral patterns, since the latter are nearly always hydromorphic in origin. A more certain criterion is given by the location and geometry of the pattern. Thus, after allowing for any distortion that may have taken place during dispersion, a soil anomaly whose shape and trend shows little or no correlation with land forms yet is consistent with a probable geological trend is more likely to be a superjacent anomaly. Lateral anomalies, on the other hand, will usually show a close relationship with local land forms.

Fossil anomalies related to past topography and drainage conditions may)ffer considerable difficulty in interpretation. Thus, one-time seepage areas :an become freely drained, leading to a decrease in the original ratio of readily :xtractable to total metal. Occurrences are known (Mather, 1959) where lownslope extensions of residual anomalies have become isolated from the 'est of the anomaly by erosion of the intervening ground, giving the im-)ression of two anomalies related to two bedrock sources. Relic patterns of nomalous terrace alluvium on valley slopes can be another source of diffi- :ulty. Past changes in the climate, ground-water level, and other features of he environment can leave their mark on the present-day dispersion pattern. 5uch effects have been noted in the central African peneplain where, in the)ast, there have been at least five alternating arid and pluvial periods.

). Appraisal of Anomalies

The principal considerations in assessing the possible economic significance of ;eochemical anomalies are: (i) the magnitude of the values, often expressed .s the contrast between the peak values and background; (ii) the size and hape of the anomalous area; (iii) the geological setting; and (iv) the extent o which the local environment may have influenced the metal content and he pattern of the anomaly.

It must be emphasized that the contrast between anomalous values and ackground constitutes no more than a provisional guide to help in the first tages of interpretation. Ratios of sample values to background values can be .seful for comparing soils developed from different parent materials and aving different background values (Govett and Galanos, 1974). Classifica- ion of anomalies solely on the basis of metal content tacitly assumes a eliable correlation between anomaly contrast and primary tenor of the bed- ock source—an assumption that is rarely justified in practice.

As a general rule, in residual soil, all anomalies whose dimensions are onsistent with the possibility of a sizable deposit should be listed for further ıvestigation. Priorities will generally depend on the favorability or other- 'ise of the geological setting. It will be appreciated, however, that shallow verburden, restricted mobility, and many other factors may lead to local nhancement of metal values and to the development of a relatively strong nomaly from a low-grade bedrock source. At the same time, a local variation ı the environment, such as a deeper overburden, more intensive leaching, nd so on, can result in relatively feeble values that are in fact related to a igh-grade source. Consequently, providing that the geological setting is not ntirely unfavorable, even weak anomalies of adequate extent should never e discounted without first considering the possibilities of local suppression f the values.

IV. GENERAL PROCEDURE FOR FOLLOW-UP

The object of the follow-up investigation is to provide further information as to the possible significance of the selected anomalies and to pinpoint targets for drilling or other means of direct subsurface exploration. The first obvious step is to revisit the area with a view to (i) confirming the cause and seeing whether there is any evidence for local enhancement or suppression of anomalous values and (ii) planning the follow-up work. If practicable, it is useful to recheck the anomalous pattern on the spot with a field chemical test. In the case of hydromorphic patterns, this can easily be done with a pocket cold-extraction kit. With residual or other types of anomaly, where the cxMe content is low, this test can give very unreliable results.

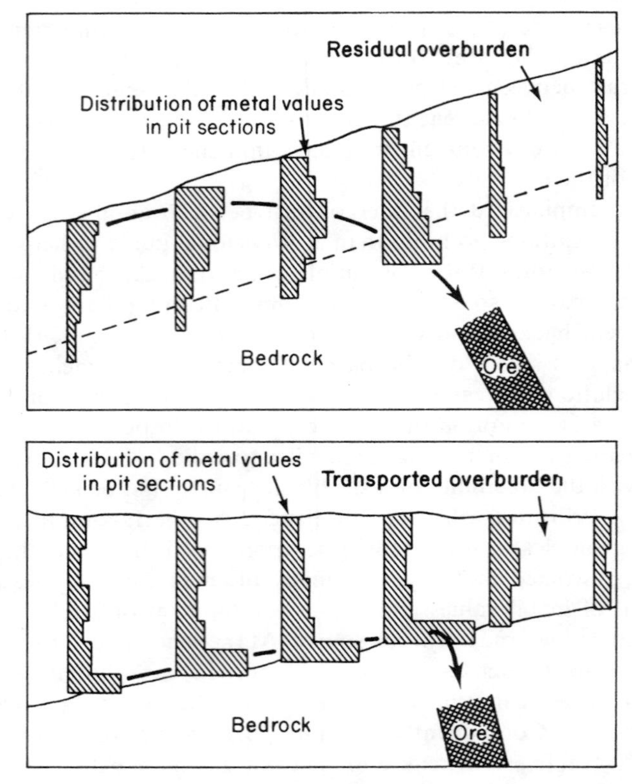

Fig. 13.15. Idealized diagrams illustrating the manner in which distribution of metal with depth may assist in locating bedrock source.

The source of lateral anomalies will be sought upslope, up-glacier, or up-drainage, according to the mechanism of dispersion. Lateral geochemical anomalies almost never provide drill targets without further work. Usually an elaborate system of follow-up involving geochemical soil surveys, geophysical surveys, conventional prospecting, or deep sampling will be necessary before a decision can be made regarding the chances of ore and the desirability of drilling.

Superjacent patterns may be followed up by close-spaced sampling aimed at delineating the axis of the anomaly. This may be facilitated by taking deep samples, insofar as the intensity of such patterns often increases with depth. Where deposits contain associated metals of contrasting mobility, a sharper pattern may be obtained by analyzing the samples for the relatively immobile constituents. The more promising sections of the anomaly may be opened up by pitting, trenching, or stripping, preferably to bedrock if possible. Pits should be logged and channel-sampled vertically, in the same manner as described for the orientation survey. The distribution of metal values down the pits may then aid in pinpointing the location of the bedrock source (Fig. 13.15). Trench exposures may be horizontally channel-sampled with the same end in view.

Where the nature of the deposit responsible for a superjacent geochemical anomaly cannot be readily determined, it may be helpful to carry out an appropriate geophysical survey in order to come to decisions regarding the desirability of drilling and the selection of drill sites. Inasmuch as drilling is usually the most expensive stage in the follow-up sequence, every possible means should be taken to localize suitable targets before drilling is undertaken.

Chapter 14

Anomalies in Natural Waters

Anomalous patterns of elements contained in ground and surface water are known as hydrogeochemical anomalies. The elements most likely to travel in solution in natural waters and hence to be useful in exploration are, by definition, the relatively mobile elements. Perhaps the most successful application of water analysis to reconnaissance exploration has been the determination of U in underground and surface waters. Although the first practical demonstration of this method was by Ostle (1954) many years ago, it was not until very recently that it has really come into its own as a major reconnaissance method applicable under almost all climatic conditions. The content of Mo, Zn, Cu, and SO_4 in natural water has also been found useful in special environments, and test surveys have been conducted for a wide range of other elements, including He, Rn, F, Cl, I, Se, As, Sb, Bi, Ge, Sn, Pb, Ag, Au, Cd, Hg, Ni, Co, Cr, W, V, Nb, B, Be, K, Rb, and Cs, especially in the U.S.S.R. (Shvartsev *et al.*, 1975). The development and use of hydrogeochemical methods have been slowed by the difficulties in analyzing the very low concentrations of many trace elements in natural waters, commonly in the order of parts per billion (ppb, μg/l, 10^{-9} g/ml). However, because ground waters furnish a sample from depths not easily sampled by other means, an increased use of hydrogeochemical techniques is likely.

I. MODE OF OCCURRENCE OF ELEMENTS

Inorganic material can occur in natural water in a variety of forms. If an element is to move with the water, it must occur either in soluble form or as a component of a stable suspension. The form is also important in deciding how to collect, treat, and analyze the water. The following are the most important of the mobile phases.

(i) Cations. Many of the water-soluble ore metals occur in natural water as simple cations. Examples are Zn^{2+}, Cu^{2+}, and Co^{2+}. Cationic complexes with OH^- are also common, as for example, $CaOH^+$. A few elements are mobile as oxy-cations, such as UO_2^{2+}, one of the major forms of U in solution.

(ii) Anions. A number of elements typically travel as anions. For example, S and Mo are stable in mildly oxidizing waters as the complex oxy-anions SO_4^{2-} and MoO_4^{2-}; As and Se in natural waters are also probably present as analogous oxy-anions. In alkaline waters, U is normally transported as $UO_2(CO_3)_2^{2-}$ and $UO_2(CO_3)_3^{4-}$ anions, or as complexes with phosphate (Langmuir, 1978).

(iii) Uncharged atoms, molecules, and ion pairs. A few elements occur in aqueous solution as simple uncharged atoms or molecules; examples are Rn, He, and O_2. The major form of Si in water is undissociated $H_4SiO_4^0$. Other elements also can occur as ion pairs or uncharged complexes, such as $PbCO_3^0$ (Fig. 8.8).

(iv) Organic complexes. Metals may be complexed or chelated by organic molecules, as described in Section VI of Chapter 8, and transported in natural waters. If the metal–organic complex is relatively small, the metal is effectively dissolved. If the complex is large and uncharged, it may separate gradually by gravity, or be filtered or centrifuged. If the complex is large and charged, it is effectively colloidal, and may be flocculated or dispersed by changes in the solution.

(v) Suspended colloidal particles. In addition to the above organic colloidal particles, elements that are extremely insoluble under some but not all natural conditions commonly occur in stable colloidal compounds in natural waters. Examples are colloidal Fe-, Al-, and Mn-oxides and hydroxides in oxidizing near-neutral solutions. Particle size ranges from large molecules (5 nm, 50 Å) to coarse clay sizes (2 μm, 20 000 Å). Trace elements may occur as analogous colloidal oxides, or they may be coprecipitated in Fe- or Mn-oxide particles. In addition, an inorganic colloidal particle of one type may be coated by another material which stabilizes it as a colloid. For example, Ong and Swanson (1969) have suggested that colloidal Au may be covered with an organic layer which changes its properties and allows a stable colloidal suspension. Floating algae, bacteria, and other living organisms also incorporate trace components.

(vi) Ions adsorbed on suspended matter (as contrasted to elements within particles). The small particle size and resulting large surface area of suspended matter give it a high exchange capacity. An active equilibrium is maintained between ions in solution and ions adsorbed to the surfaces of suspended particles.

The partition of minor elements between these six mobile phases is dependent on the chemical properties of the element, the chemical and physical parameters of the environment, and the history of the solution. Cations, anions, uncharged ion pairs, and adsorbed ions readily exchange from one form to another in seconds in response to changes in the chemistry of the solution. The other forms react more slowly, and once formed in an upstream environment, may persist for long distances. Suspended particles tend to remain in suspension only in relatively turbulent waters.

In collecting, treating, and analyzing water samples, care must be taken to identify the form of the desired element, to retain it in the sample, and if possible to exclude complicating effects of other forms. Precipitation from solution, collection on ion-exchange resins, or extraction into an immiscible phase may extract only the ionic forms. Filtration through membrane filters and similar devices can remove the coarser suspended material, but in practice, particles less than 0.1 μm are difficult to separate. Acidification may dissolve particles and change the form of others. Oxidation by ashing the evaporated residue or by treatment with an oxidizing acid may be required to liberate metal strongly bound by organic matter.

Ground water usually carries most of its load of metal in one of the ionic forms, with lesser proportions traveling as stabilized colloidal sols and organic complexes.

Surface water, on the other hand, because of the effect of both sunlight and increased aeration and turbulence, carries a large, variable, and unpredictable fraction of its metal content in non-ionic form, principally as constituents of soluble organic matter and of both inorganic and organic suspensoids (Perhac and Whelan, 1972). Sunlight and warmth promote the growth of both rooted plants and floating organisms, which may absorb ionic constituents from the water and which serve as the source material for both dissolved and suspended organic matter. The organic content of the water is also liable to vary according to the amount of organic-rich soil in the spring area or along the banks of the stream; seasonal variation in the decay of vegetation is another important factor in this connection. The amount of suspended inorganic matter in natural water also depends on the environment and tends to increase with increased turbulence of the water and availability of a source of fine-grained material. Figure 14.1 shows an example of

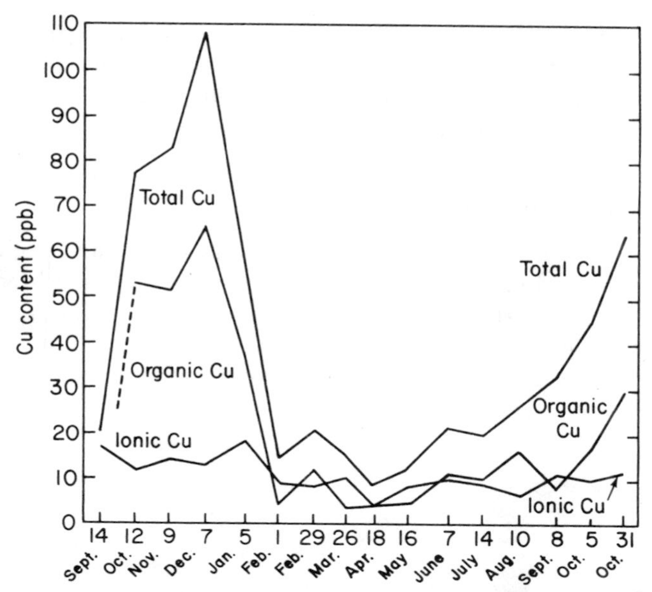

Fig. 14.1. Seasonal variation in the various fractions of Cu in surface water from Lake Quonnapaug, Connecticut. (After Riley, 1937.)

variations with time of year in the fraction of the total Cu content of lake water that occurs respectively in ionic and soluble organic forms.

II. FACTORS AFFECTING COMPOSITION OF NATURAL WATERS

The content of trace elements in waters unrelated to ore deposits depends on the processes that have affected the water since it fell as rain. Among the factors determining the level of trace and major elements are the content of solutes in the initial rain or snow, the extent of reaction with rock and soil, loss of constituents by precipitation or adsorption, and loss of water because of evaporation, transpiration, or reaction with minerals (Hem, 1970).

A. Composition of Rain Water

Rain and snow normally contain small amounts of Cl, SO_4, K, Mg, and other elements derived from sea spray, dust, volcanic emanations, and industrial pollution. Typical concentrations are 0·2–0·8 ppm Cl, 1–3 ppm SO_4, and 4–10 ppm total dissolved solids, but much higher values are observed near the sea coast (Hem, 1970; Junge and Werby, 1958). Rain water is normally slightly acid from dissolved CO_2, and contains other dissolved gases including N_2, O_2, Ar, SO_2, H_2S, HCl, NH_3, NO_2, and Hg. Metals are normally low in concentration except in areas of unusual pollution.

B. Reaction with Soil and Rock

On encountering soil and rock, water incorporates major and trace elements by decomposing and dissolving the rock minerals (Garrels and MacKenzie, 1967). For the most part this is accomplished by reaction of the H^+ ion with the aluminosilicate minerals and carbonates, to liberate cations and silica into solution and leave behind clay minerals, as reviewed in Chapter 6.

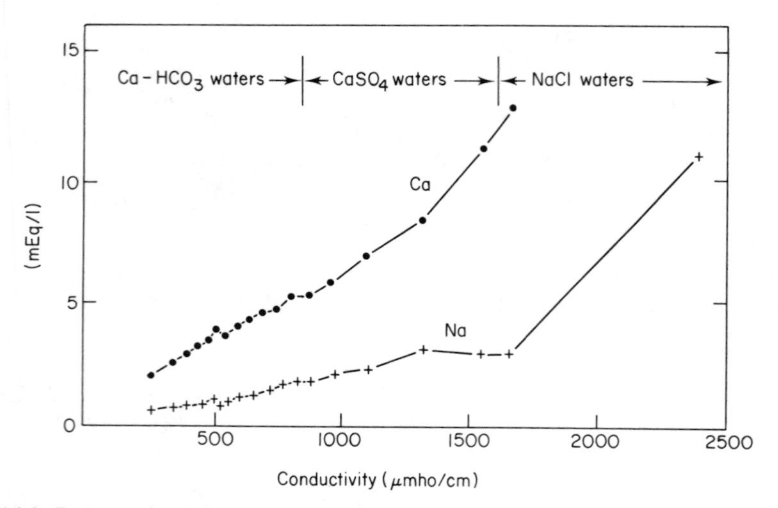

Fig. 14.2. Increase in Ca and Na with increase in conductivity, for stream water in Italy. Each point represents an average for 50 samples. Up to about 800 μmho/cm, Ca and Na increase because of reaction with silicates; above this level, solution of evaporites is important. (After Dall'Aglio and Tonani, 1973.)

H_2CO_3 converted to HCO_3^- by this process is at least partly replenished by absorption of CO_2 from the soil air, supplied by decomposition of organic matter. The amount of rock or soil "dissolved" by the water is dependent on the amount of time available for reaction, and on the amount of CO_2 and other acids that are available. As the minerals are dissolved by this process, major and trace metals in the water tend to increase proportionally to the total dissolved solids in the water (Fig. 14.2). The background contents of many elements in water are best expressed by the ratio of the elements to total dissolved solids. Essentially equivalent expressions are the ratio to conductivity, or the content in the evaporated solids.

The background amounts of trace elements in waters are also related in a general way to the composition of the rocks of the drainage, but the form of the trace element in the rock must be considered. If the element occurs as a trace substitution in one or more major minerals, it will be liberated into solution at essentially the same rate as the major elements are liberated. In this situation, the trace element in water will show a correlation with the total dissolved solids and the content in rock (Figs 14.3 and 14.4). However,

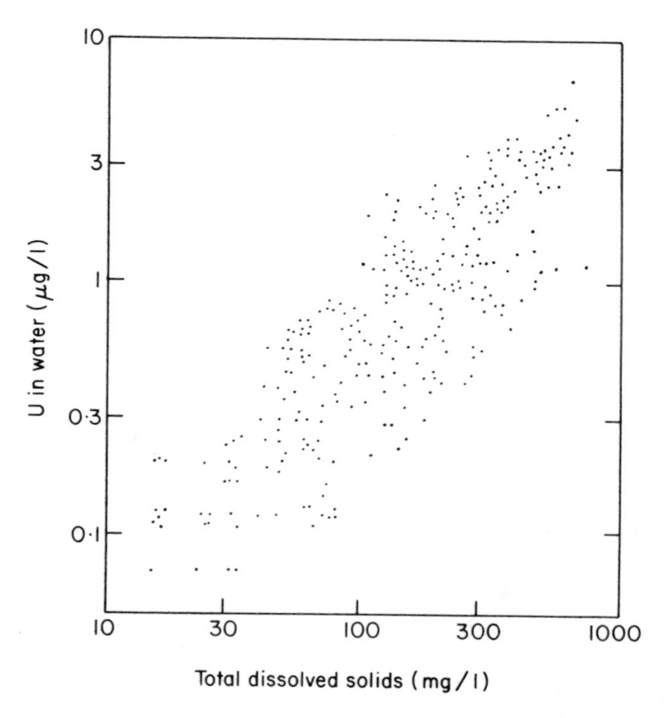

Fig. 14.3. Relation of background U content in waters to total dissolved solids for streams and river in the U.S.S.R. (After Lopatkina, 1964.)

if the element occurs as a major constituent of a trace mineral, it may be either more or less easily leached than the major elements, or may be preferentially leached under certain conditions, as is the case for sulfides exposed to oxidation. For example, Dall'Aglio (1971) found that within one drainage in the Alpine Range of Italy, very high U was observed in stream waters and sediments, but only small pegmatites with traces of U minerals could be identified as a source.

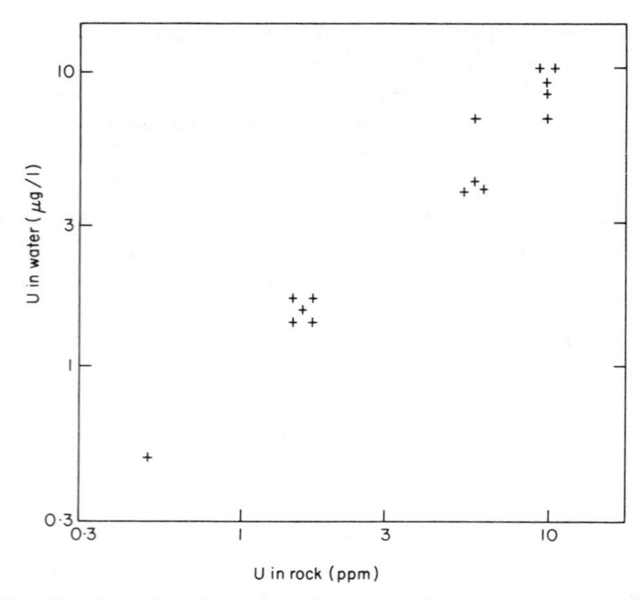

Fig. 14.4. Relation between U content in rock and in water with salinity of about 50 mg/l. (After Lopatkina, 1964.)

C. Regional and Climatic Factors

Distinct regional and climatic patterns in the element content of waters are observed. Tardy (1971) and Shvartsev (1975) illustrate the differences in major element content of ground and surface waters as a function of climate, abundance of CO_2, rate of water circulation, and extent of previous leaching of the weathered zone. In general, tropical regions of high rainfall and abundant organic matter have distinctly acid ground waters with relatively low dissolved solids contents because of previous leaching, rapid circulation of water through the soil and rock, and abundance of organic acids and dissolved CO_2. Dissolved solids and pH progressively increase with passage from forested lands in temperate and arctic regions, to plains and steppes, to semi-arid and arid regions.

In arid and semi-arid regions, two additional factors assume importance: precipitation of minerals and evapotranspiration of water. Evaporation of water from the soil and transpiration by plants may concentrate the dissolved solids to a significant degree. The ratio of trace-metal content to total dissolved solids remains essentially constant in this process. In the absence of significant additional sources of Cl, the extent of evaporation and transpiration may sometimes be estimated from the increase in Cl content of the water. If the content of Ca and HCO_3 rises enough to precipitate calcite, the ratio of Ca to other elements and to total dissolved solids is changed. Similar relations exist for other components of precipitated minerals. Solution of evaporites and consumption of water by reactions with minerals also tend to increase the content of dissolved solids.

D. Age of Ground Water

A similar trend of increasing dissolved solids and pH is observed along the path of ground-water flow through a single aquifer (Olmsted *et al.*, 1973). In the recharge areas near the tops of hills or headwaters of the area, the water is low in solutes and relatively acid, at least in tropical and humid areas, and dissolved solids increase along the flow path. Background levels of trace elements increase with major elements, in the absence of complicating factors. The progressive increase in major elements has been suggested as a means of determining the time spent by the water in an aquifer, and from this the maximum possible distance to the source of an anomaly (Hoag and Webber, 1976a).

E. Oxidation–Reduction

In rocks containing organic matter, which includes essentially all sedimentary rocks, an important vertical zoning in oxidation state and other parameters of ground water is usually found (Germanov *et al.*, 1958), and the levels of trace elements may be affected. Near the surface, waters are relatively dilute and oxidizing. With increasing depth or distance along the aquifer, oxygen is consumed by reaction with organic matter, and CO_2 becomes the major dissolved gas. Still deeper, H_2S, CH_4, and other hydrocarbons predominate and produce strongly reducing conditions. Precipitation of sulfides and low-valence oxides, and solution of Fe, Mn, and other constituents may affect the levels of trace elements.

F. Adsorption

A major factor affecting the abundance of trace elements in waters is adsorption onto solids and ion exchange with clays. This process is especially important for ground waters. Weathering of normal rocks rarely leads to saturation of ground waters with minerals of the trace metals, except very close to orebodies, but cations and anions may be considerably depleted by adsorption onto clays, Fe-oxides, and other minerals encountered by the water. Because most of the transition-series trace metals of interest in geochemical exploration are strongly sorbed compared to the major elements, the final levels in water, as controlled by adsorption, may be relatively low.

Table 14.1
Content of elements in ground and surface waters[a]

Major elements		
Element	Surface water (mg/l)	Ground water (mg/l)
C (as HCO_3^-)	58	200
Ca	15	50
Cl	7·8	20
K	2·3	3
Mg	4·1	7
Na	6·3	30
S (as SO_4)	3·7	30
Si (as SiO_2)	14	16
pH	—	7·4
Total dissolved solids	120	350

Trace elements

Element	Median content (μg/l)	Element	Median content (μg/l)
Al	10	La	0·2
Ag	0·3	Li	3
As	2	Mn	15
Au	0·002	Mo	1·5
B	10	Nb	1
Ba	20	Ni	1·5
Be	5	P	20
Bi	0·005	Pb	3
Br	20	Rb	1
Cd	0·03	Sb	2
Co	0·1	Se	0·4
Cr	1	Sn	0·1
Cs	0·02	Sr	400
Cu	3	Th	0·1
F	100	Ti	3
Fe	100	W	0·03
Hg	0·07	U	0·5
I	7	V	2
		Zn	20

[a] Source: Turekian (1977), Fig. 14.5, Davis and DeWeist (1966, Fig. 4.1), and other sources indicated in the Appendix.

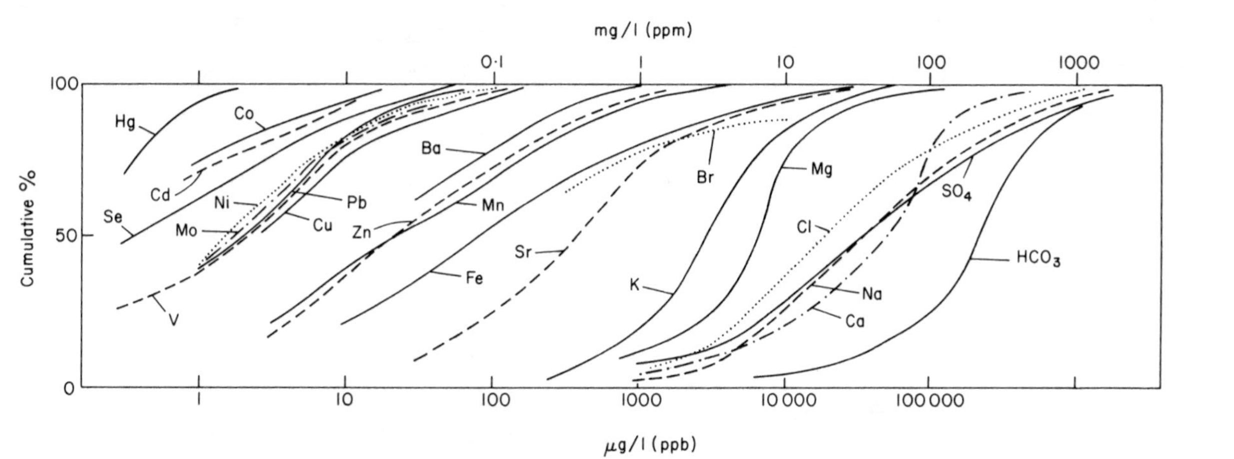

Fig. 14.5. Cumulative percentages of some major and trace elements in ground waters. Number of analyses: 13 000–18 000 for major elements; 750–8000 for trace elements. (Data compiled by D. Langmuir and H. L. Barnes from data in U.S. Geological Survey data bank.)

G. Mixing of Waters

At various points along their path of migration, ground and surface waters may mix with more dilute waters from near the surface, more concentrated waters from depth, or chemically distinct waters from a different environment. Obviously, the levels of trace elements are changed in this process, either by simple mixing or by precipitation of some constituents.

H. Background Content of Natural Waters

As indicated in the preceding discussion, the background content of a given trace element in natural waters varies over a considerable range, depending on the rock types traversed, the history of the water, and the levels of major elements. Table 14.1 and Fig. 14.5 illustrate typical levels of trace elements in background waters, and the frequency distribution of some major and minor elements in common ground and surface waters. Data on levels of many trace metals in water are sparse, and these values should be regarded as only a general guide to the levels expected in a given area. Only analyses in the survey area can establish the background for that area.

III. PERSISTENCE OF ANOMALIES

The usefulness of hydrogeochemical anomalies in exploration depends to a large extent on how far they extend down-drainage from the source before they are lost in the normal range of background variations. The length, or persistence, of a hydrogeochemical anomaly is conditioned largely by (i) contrast at the source, (ii) dilution, and (iii) precipitation or adsorption.

A. Contrast at Source

Both the background metal content and anomaly contrast can vary widely according to local conditions (Table 14.2). Rapid solution of the ore minerals favors a high contrast at the source. The extent of solution of most primary ore minerals is largely dependent on the stability of the minerals under weathering conditions, the accessibility of the ore to percolating solutions, and the solubility of the secondary products.

Under many conditions, the rate of solution of primary ore minerals depends roughly on the difference between their solubility in the surrounding water and the actual metal content of that water. Most oxides (cassiterite, magnetite, chromite) have low solubility and dissolve slowly. In contrast,

Table 14.2
Metal contents of natural waters in mineralized and unmineralized areas, U.S.S.R.[a]

District	Average metal content (ppb) of ground water			
	Zn at Pb–Zn deposits		Cu at Cu deposits	
	Background	Anomalous	Background	Anomalous
Transcaucasia	200	200–500	—	—
Altai	10	50–300	4	20
The Sayans and Kuznetzk Altau	1	40–90	1	10–100
Central Kazakhstan	80–200	300	30	130
Central Asia	—	—	20	50
Ural	—	—	30	80

[a] Source: Brodsky, quoted by Ginzburg (1960, p. 202). Reprinted with permission from "Principles of Geochemical Prospecting", © Pergamon Press Inc., 1960.

evaporite minerals are relatively soluble and dissolve rapidly. Sulfide minerals, in spite of their relative insolubility in pure water, are vulnerable to attack in the oxygen-rich environment of weathering because of the high solubility of the products of their oxidation. The free-oxygen content of the water affects the rate of oxidation. Also, the rate of decomposition of sulfides may be accelerated if pyrite or marcasite is present, as these minerals on oxidation liberate extremely corrosive solutions of free sulfuric acid that greatly increase the solubility of metals and cause rapid attack on the other primary ore minerals. Thus, other factors being equal, an ore high in pyrite will release metal into solution at a very much higher rate than an ore of equivalent grade that is low in pyrite. Similarly, strongly pyritized rock containing only minor quantities of the ore metals may release as much or more metal into the ground water as high-grade ore contianing no pyrite, at least during an initial period of oxidation. Rapid discharge of metals into the drainage is inhibited by calcareous country rock that either quickly neutralizes the acids before they have a chance to attack the ore minerals, or causes the precipitation of the metals in secondary minerals.

Fracturing promotes the liberation of soluble metals by increasing the reactive surface between the ore and the oxidizing solutions. A similar effect is seen in the case of fine-grained sulfides disseminated in a permeable host rock. Water draining highly sheared or disseminated permeable ores will therefore tend to show stronger anomalies than water from compact, massive ores of equivalent grade and size. The high contrast at the source illustrated in Fig. 14.6 results largely from the occurrence of sulfides exposed to oxidation and leaching in mine workings or in fissure zones.

The rate of flow of water through an oxidizing deposit also can have a pronounced effect on the intensity of a hydrogeochemical anomaly. A low rate of flow will provide more time for the weathering products of the ore minerals to go into solution and hence will result in a higher metal content of the water.

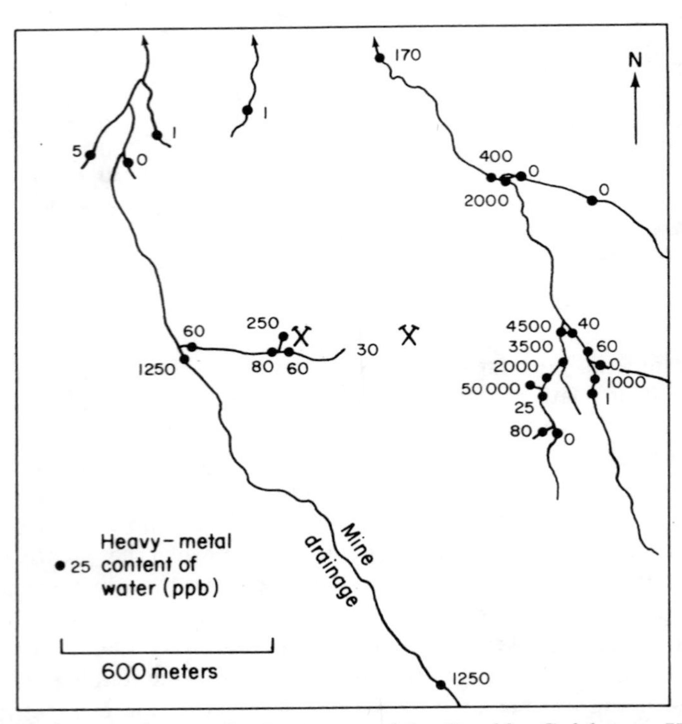

Fig. 14.6. Heavy-metal anomalies in streams of the Gambler Gulch area, Keno Hill, Yukon Territory. (After Boyle *et al.*, 1955.)

The solubility of an element may also be governed by the presence of other ions in solution. For example, the concentration of U in a solution is dependent not only on the availability of a source of the metal but also on the concentration of HCO_3^-, with which it reacts to form the soluble uranyl-carbonate complex (Langmuir, 1978). In these circumstances, it may be necessary to determine not only the metal content but also the concentration of relevant major constituents in order to obtain a better guide to the distribution of the metal in the bedrock of the catchment area.

The contrast of anomalous values over background in any given area will reflect the balance of these bedrock factors as conditioned by the local geomorphic environment. Topographic relief and climate are particularly important in this connection. Other things being equal, maximum contrast will normally be expected in areas of moderate rainfall and moderate relief. High rainfall and strong relief both tend to decrease contrast by increasing the flow of water and hence the chances that the anomaly will be swamped by water of background composition. In areas of strong relief, the rate of erosion may be so rapid as to keep pace with or even outstrip the rate of oxidation, so that the amount of metal available for leaching is restricted. Much depends, however, on the depth of effective oxidation. In areas of weak relief, the increased contrast resulting from a slow rate of flow of ground water is often offset by the fact that the water table and hence the zone of active oxidation lies close to the surface. Nevertheless, even in areas of low relief, appreciable solution of metal may still take place from ores that have been deeply oxidized during an earlier climatic cycle when the water table stood at a lower level. Climate can have an effect also on the pH and hence indirectly on the corrosive power of the percolating solutions. Thus, in heavily vegetated tropical and temperate regions, near-surface waters tend to be more acid than in desert areas.

B. Decay by Dilution

The pattern of decay of anomalous metal contained in waters draining an oxidizing ore deposit is in part the effect of simple dilution by water from unmineralized areas. Where this dilution by gradual small increments of barren water is the only factor in the decay of a hydrogeochemical anomaly, the anomalous pattern may be traced for a considerable distance, until it is lost in normal, background fluctuations (Fig. 14.7). Anomalous water may, however, be diluted abruptly by a very great volume of background water at a point relatively close to the source. Catastrophic dilution of this kind may obliterate anomalous patterns before they reach a point where they would normally be sampled, and thus make them difficult to detect without excessively detailed sampling patterns.

Figure 14.8, from the classic work of Sergeyev, shows the decay of the heavy-metal content of an anomalous stream as the result of dilution by relatively small increments from side tributaries and stream-bed seepage derived from unmineralized ground. The lowest sampling point is from a large river and illustrates the destruction of the anomalous pattern by catastrophic dilution.

The effects of dilution or of addition of metal-rich waters to a stream water may be evaluated from the areal metal load (L), equal to the mass of the element carried past a given point on a stream divided by the area of drainage

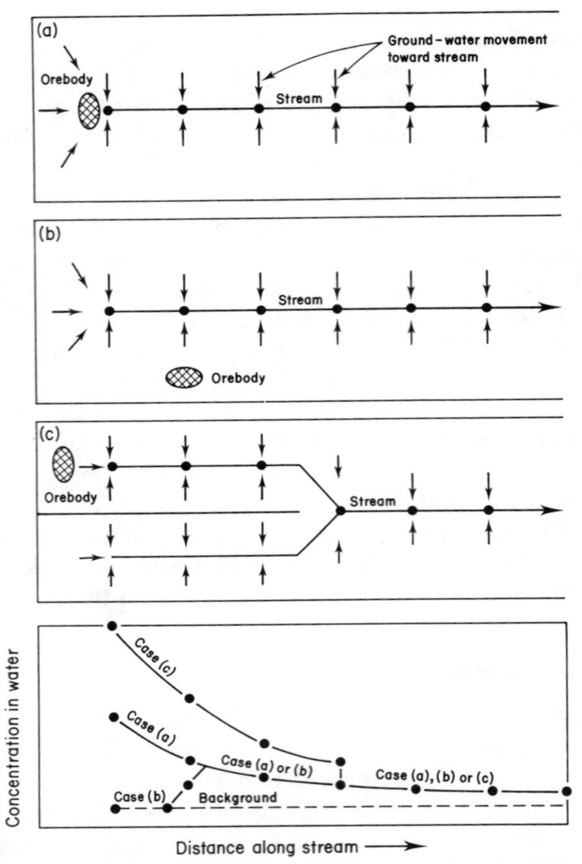

Fig. 14.7. Downstream decay of stream water anomalies in schematic drainage basins, as a result of ground-water seepage into streams, and dilution by background streams.

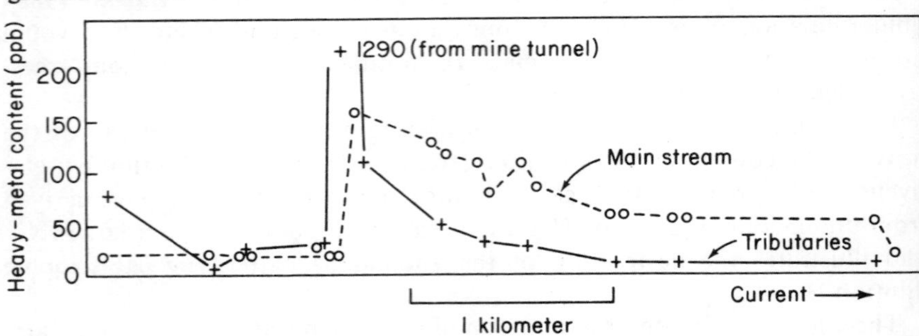

Fig. 14.8. Heavy-metal content of water, Berezov Brook, U.S.S.R. (After Sergeyev, 1946.)

upstream (Boberg and Runnells, 1971). It is calculated as

$$L = \frac{Me.D}{A}$$

where *Me* is the metal concentration in the water (mg/l), *D* is the discharge of the stream (l/s), and *A* is the upstream drainage area. In an unmineralized area, the parameter *L* will remain low and relatively constant downstream. Local increases or high values suggest an unusual source of the element, possibly an orebody.

C. Decay at Precipitation Barriers

Hydrogeochemical anomalies normally decay more rapidly than can be accounted for by simple dilution. Metal-rich water, as it moves away from the source of the metal, ordinarily soon comes into an environment where changing conditions of some kind cause precipitation or adsorption of part or all of the metal from the water. The change in chemical conditions that causes precipitation or unusual adsorption from flowing surface or ground waters is termed a precipitation barrier.

Precipitation barriers commonly arise because of changes in pH, oxidation potential, or concentration of precipitating substances. Causes for precipitation barriers include interaction of waters with solids (enclosing rocks, stream sediments), mixing with other waters, and loss or addition of gases on emergence of ground waters at the surface. Major types of precipitation barriers and elements affected include (Perel'man, 1967):

(i) Oxidation type. Iron- and Mn-oxides or native sulfur precipitated by oxidation of reducing solutions; usually caused by emergence at the surface, or flow of reducing water out of a swamp.
(ii) Reducing type. Uranium, V, Cu, Se, and Ag precipitated as metals or lower-valent oxides by reduction of oxidizing water; usually caused by encountering organic matter or mixing with reducing waters or gases.
(iii) Reducing sulfide type. Iron, Cu, Ag, Zn, Pb, Hg, Ni, Co, As, and Mo are precipitated as sulfides by reduction of oxidizing sulfate waters, usually by action of sulfate-reducing bacteria; U, V, and Se may also be precipitated; causes are the same as for reducing type, but require the presence of dissolved sulfate.
(iv) Sulfate–carbonate type. Barium, Sr, and Ca precipitated by increased sulfate or carbonate, as a result of mixing waters, oxidation of sulfide, or passage into carbonate rock.
(v) Alkaline type. Calcium, Mg, Sr, Mn, Fe, Cu, Zn, Pb, Cd, and other elements precipitated by increased pH; usually caused by interaction

of acid waters with carbonates or silicate rocks, or mixing with alkaline waters.

(vi) Adsorptive type. Adsorption or coprecipitation of ions on accumulations of Fe–Mn-oxides, clays, and organic materials; cations of transition metals and those with high valence tend to be more strongly adsorbed than anions and low-valent cations.

Precipitation barriers characteristically occur in spring and seepage areas where ground waters coming to the surface encounter an environment of increased availability of oxygen, sunlight, and organic activity. Soluble ferrous iron derived from the oxidation and leaching of Fe-sulfides tends to be oxidized to ferric iron and precipitated as limonite. Other ore metals may be coprecipitated either with the limonite or with the organic matter tha accumulates in the swampy soil of spring areas. This effect is illustrated by the distribution of Zn around a Zn-rich spring in the south-western Wisconsi Zn district, as shown in Fig. 14.9. Here, in a distance of less than 65 m the Zn content of the water is reduced by a factor of 20.

A precipitation barrier may also occur where acid waters emerge from th reducing environment of an organic swamp into the oxidizing environmen of an open stream channel. This change in environment commonly cause the precipitation of hydrous Fe- and Mn-oxides in the stream bed and th consequent removal of dissolved metals from the solution by coprecipitation

The mixing of waters of two confluent streams of contrasting chemica composition may result in a chemical reaction and a resulting precipitatio of material in the stream bed. Theobald *et al.* (1963), observing the point o mixing of a stream water of pH 3·5 with a stream of pH 8·0, found heav precipitates of hydrous Fe- and Mn-oxides together with large concentration of other metals immediately below the confluence. Aluminous precipitate were observed in the stream bed for several miles below the confluence. Th pH of the mixed water was 6·5.

A similar decay in metal content was observed in the highly contaminate drainage below the mines at Butte, Montana, where acid, metal-rich min water is progressively diluted by background water of normal pH. Figur 14.10 shows the sudden decay in the Cu content of the water as the pH i increased beyond 5·3, the point at which Cu normally begins to precipitat as insoluble basic salts. Presumably the more regular decline in the Zn con tent of the same series of samples is the result of simple dilution, probabl combined with some adsorption.

A similar pattern of decay was found by Cameron (1977b) for Zn, Cu, an Pb in lake waters in a permafrost area of northern Canada (Fig. 14.11 The anomaly for Zn has the greatest extent of the three metals in water, an Pb falls off the most rapidly. Silver and As show very short anomalies. Iro oxide coats the lake bottom in the localities of steeply declining metal co

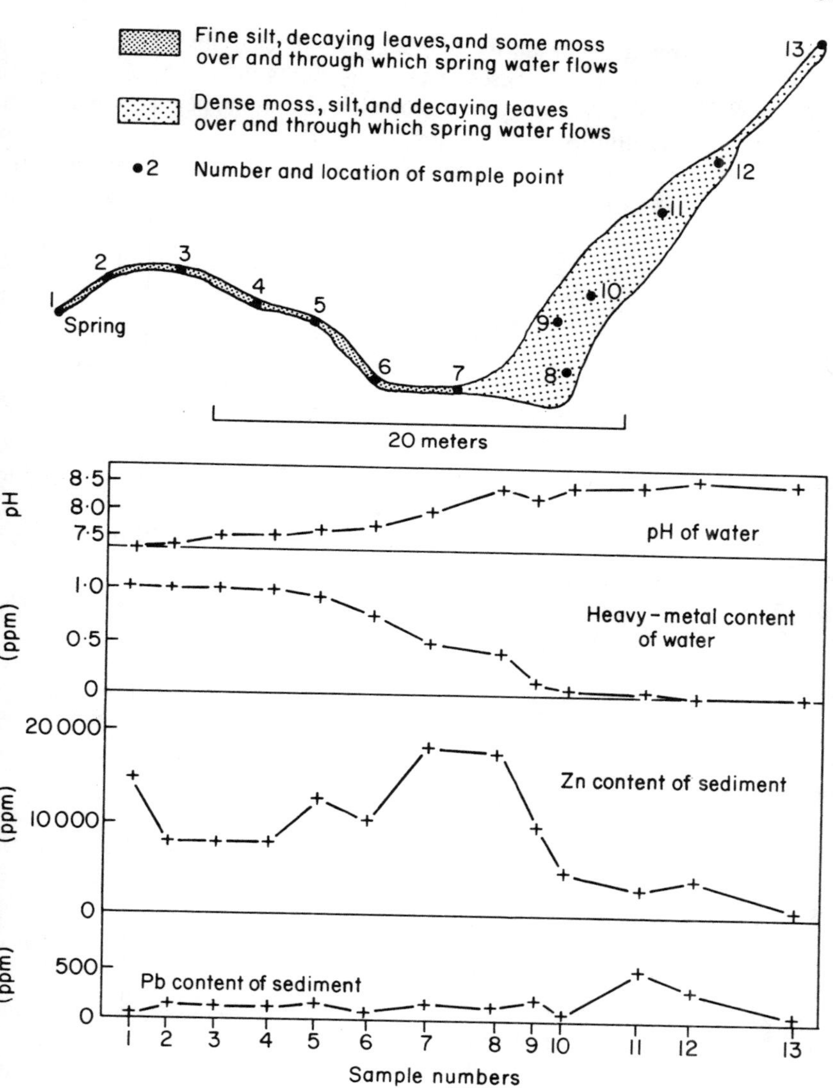

ig. 14.9. Precipitation of heavy metals from spring water, Potosi area, south-estern Wisconsin Zn district. Data on bulk samples. (After Kennedy, 1956.)

nt. An increase of Zn in lake sediment complements the decrease of Zn in e water.

Adsorption of cations from surface waters is an important mode of decay anomalies, as illustrated by Polikarpochkin *et al.* (1965b) for streams drain-g base-metal prospects in eastern Transbaikalia. Copper, Pb, and Zn anomalies in stream water decay to background levels in 1–3 km, accompanied by corresponding increases of these elements in stream sediments. In contrast, anomalies of As, Mo, and S, traveling as anions, persist downstream and decay only by dilution from major tributaries.

Ground water, as a result of the lower content of free oxygen and the absence of sunlight and biological activity that goes with it, is not as much at

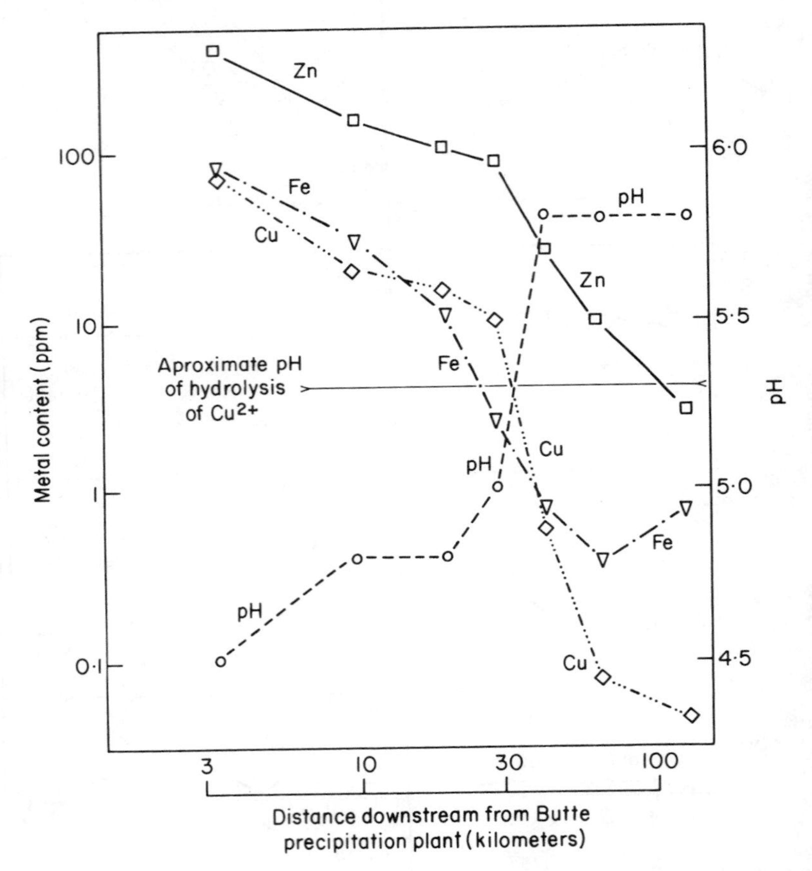

Fig. 14.10. Copper, Zn and Fe content and pH of drainage water below Cu mine at Butte, Montana. (Data collected by L. C. Huff.)

the mercy of precipitation barriers as surface waters. Even so, conspicuou examples of precipitation of metals from ground water as it moves int chemically reactive environments, such as the zones of secondary enrich ment in Cu deposits, and the concentrations of U at oxidation–reductio boundaries in roll-type U deposits, are well known.

The most mobile elements are the ones that are the least likely to be affected by precipitation barriers as they are carried away from their source in the bedrock. However, even the extremely mobile elements may be immobilized on occasion. Sulfur, for example, may be precipitated as sulfide in a strongly reducing environment. Similarly, molybdate may be precipitated as the insoluble ferric molybdate in Fe-rich solutions. It is important to appreciate

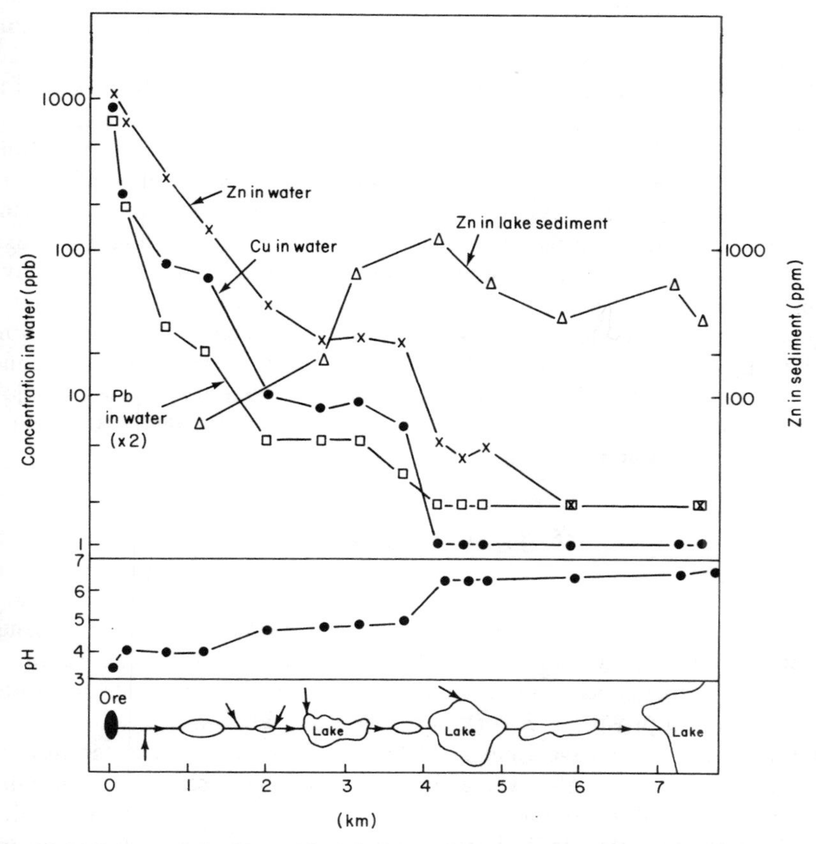

Fig. 14.11 Decay of Cu, Zn, and Pb in lake and stream water, Northwest Territories, Canada, and corresponding increase of pH and Zn in lake sediment. (After Cameron, 1977b.)

that precipitation of a particular metal will depend on the mode of occurrence of the element. Thus, the fraction of the total content of a metal that is traveling in an ionic form may be precipitated by adsorption or coprecipitation, while the colloidal fraction may be unaffected.

D. Re-solution at Precipitation Barriers

The chemical environment at any given spot may change with time. A site that at one time was a precipitation barrier may, as a result of changing conditions, become a site of re-solution of precipitates of anomalous material originally derived from the leaching of an ore deposit. Precipitated material thus subjected to later solution may consist of relatively insoluble secondary ore minerals, anomalous residual soil or material precipitated in hydromorphic and biogenic anomalies. Hydrogeochemical anomalies, therefore, can be of complex origin insofar as they may comprise soluble materials derived both directly from the leaching of the primary ore and indirectly by way of a number of intermediate materials, as illustrated diagrammatically in Fig. 10.2.

The secondary release of Cu in soluble form appears to be an important factor in the development of the drainage anomalies of the Zambian Copperbelt. Here, as a result of extremely deep oxidation during an earlier epoch, no primary sulfide minerals now occur within 100 m of the present surface. At the same time, the present-day ground-water level and zone of active chemical leaching is relatively shallow. Thus, the Cu moving in the ground waters draining mineralized ground in these areas must come from re-solution of metal from the zone of past oxidation rather than from the underlying zone of Cu-bearing sulfides (Tooms and Webb, 1961). Secondary release of major quantities of soluble metals can also result from the natural decay or burning of metal-rich organic matter.

IV. TIME VARIATIONS

Hydrogeochemical anomalies, especially in surface water, show a very strong tendency to change in intensity with changes in weather. The resulting instability of geochemical anomalies in water is one of the most serious problems in practical prospecting work based on water analysis.

Rainfall and runoff have a marked effect on the intensity of water anomalies, particularly in climates where the rainfall is highly seasonal. Normally the composition of ground and surface water is relatively stable during dry periods. In streams, this stable flow is termed the base flow and is derived from gradual discharge of relatively deep ground waters. During and following the first rain immediately after a period of dry weather, two factors come into play. One is dilution of the base flow by relatively direct runoff that has had little interaction with soil or rock, The second is flushing of accumulated soluble minerals. The mechanism of the flushing process is illustrated in Fig. 14.12. Oxidation of the ore during dry weather produces a large amount of readily soluble material. This material is flushed into the ground water by

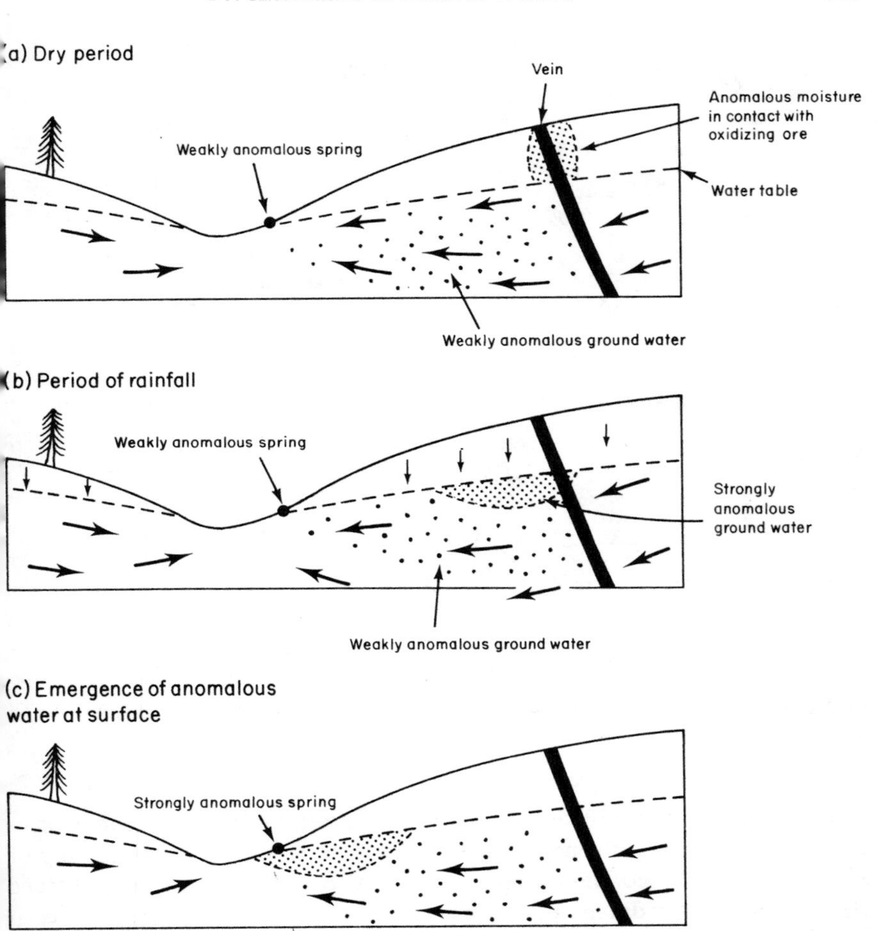

Fig. 14.12. Diagrammatic illustration of flushing of U caused by a period of rainfall following a dry period during which readily soluble oxidation products accumulated. (After Germanov *et al.*, 1958.)

the rainfall and emerges at the surface at a time dependent on the hydrology of the area and the location of ore relative to the discharge point.

The effects of the two processes are illustrated in Fig. 14.13. If dilution is the predominant process, a decrease in metal content of the stream water is noted shortly after the rain. The ratio of metal content to total dissolved solids (or to conductivity) is maintained during dilution. If flushing dominates, a period of high metal content in the stream water is observed; dilution may be evident before and after the period of flushing. The ratio of metal to total dissolved solids increases during the period of flushing. In large streams, very

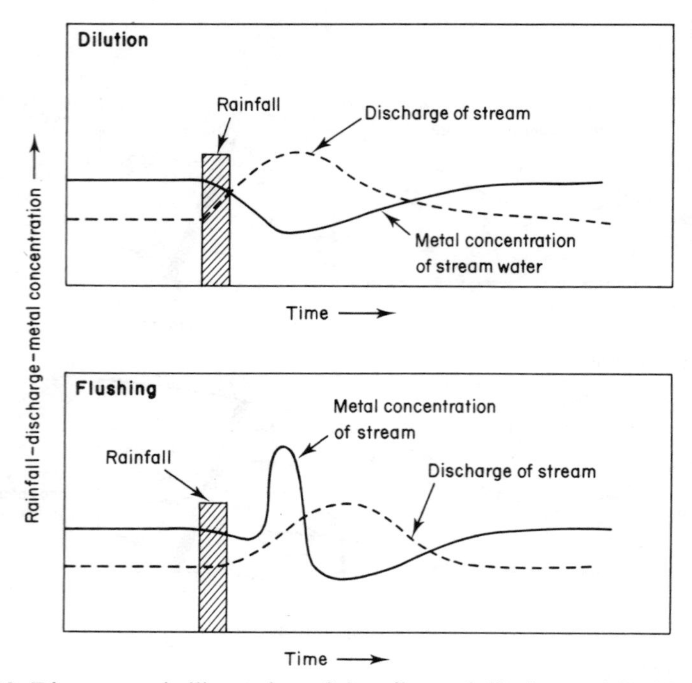

Fig. 14.13. Diagrammatic illustration of the effects of dilution and flushing on the metal content of stream water.

complicated effects are obviously possible as a result of differing behavior in different tributaries. In general, values tend to be erratic at the beginning of a period of increased rainfall and runoff. Also, different major and trace elements may show different patterns because of differing sources. With continuing rain the metal content may settle down to a value less than the peak at the start of the rains, but still somewhat higher than that during the previous dry season.

This rather complex correlation between the metal content of water and the incidence of rainfall and runoff has been noted in many parts of the world. Webb and Millman (1950) in Nigeria were the first to note the characteristic pattern of erratic variation in the anomalous heavy-metal content of stream waters at the onset of the rainy season as compared with other times of the year. Later work in the same areas showed that the contrast between anomaly and background is sharpest in the latter half of the rainy season (Webb, 1958a). The distribution of Cu in stream water shows the same seasonal trends in Angola (Atkinson, 1957). Wodzicki (1959) in New Zealand measured a marked variation in the U content of surface water with rainfall over period of 55 days (Fig. 14.14). In the Transbaikal region of Siberia, Polikarpochkin *et al.* (1958) found a stable heavy-metal content of river water at periods of low discharge and a high and erratic content at periods of high discharge (Fig. 14.15). Rose and Keith (1976) found variations in U content of more than 50-fold in a small stream near a U prospect in eastern U.S.A. In permafrost areas, the initial waters from thaw of frozen surface layers and basal snow may carry elevated metal contents derived from the oxidation of minerals and upward movement of saline solutions during the winter (Jonasson and Allan, 1973).

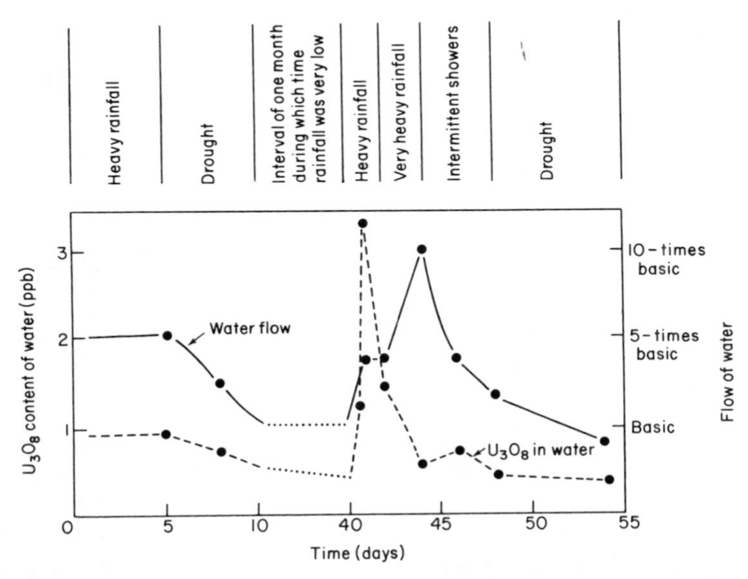

Fig. 14.14 Variation of U content of stream water with changing weather conditions. (After Wodzicki 1959). Reprinted with permission from *N.Z.J. Geol. Geophys.*)

Other factors in addition to the combined flushing and diluting effect of rain water may play a part in the variability of water anomalies. Melting snow in the springtime may cause dilution. In arid climates, evaporation during dry weather results in an increase in the content of metals as well as other salts dissolved in the water (Fix, 1956). According to Ginzburg (1960, p. 211), the contents of sulfate and chloride tend to vary sympathetically with rate of flow. An unusually high SO_4/Cl ratio in ground or surface water, therefore, may provide a better indication of a sulfur-rich bedrock source than the absolute content of sulfate alone. Changes in weather may be reflected in the biological activity, with resulting changes in the partition of metal between the various ionic and non-ionic aqueous phases (Fig. 14.1).

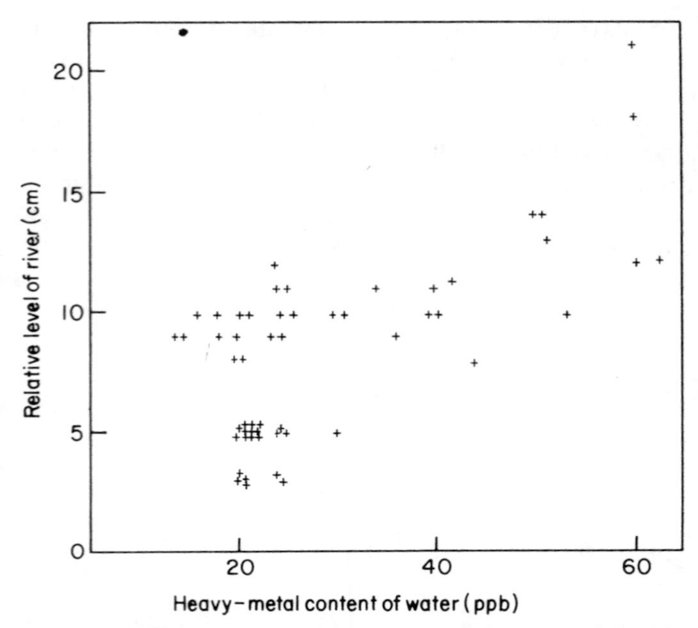

Fig. 14.15. Correlation of the heavy-metal content of the water of the Altachi River with water level. (After Polikarpochkin *et al.*, 1958.)

The metal contents of all but the smallest lakes are relatively homogeneous and stable with time, compared to streams. Macdonald (1969) found that the U content in a given lake in the Beaverlodge district of Canada was essentially constant between May and September, except for the brief period during the spring breakup. During this period, some surface water samples had anomalously low U contents, apparently because they consisted largely of newly melted snow and ice, and others had unusually high U content, possibly reflecting colloidal material mobilized from the bottom by the spring overturn. Surface-water samples from small lakes ($<2{\cdot}5\ km^2$) were essentially identical in U content to samples at "mid-depth" in the same lake. A later study by Dyck *et al.* (1971) in the same area confirmed the relative stability of U contents of surface lake water during the summer. In contrast, Meyer (1969) found that samples from smaller lakes ($<0{\cdot}25\ km^2$) in Labrador had distinctly higher contents of U in September than the following summer in July and August. Copper and Zn contents of lake water of permafrost areas in Canada were quite uniform from one part of the lake to another, even in lakes several kilometers in diameter, and were essentially constant over the summer (Cameron and Ballantyne, 1975). However, Jonasson (1976) found appreciable differences within lakes in the southern Canadian Shield, and other workers have reported similar differences in large lakes (Levinson, 1974).

V. GROUND-WATER ANOMALIES

Ground water may be obtained from wells, springs, or seepages. Spring and seepage water tends to be relatively shallow ground water, unless the spring is localized by a major fault or fissure zone, or is the discharge of a major aquifer. Wells yield water from locations limited by the well depth, casing, and any other means of inducing permeability that have been applied. Because of their vertical penetration, wells provide a sample from depths not commonly attained by other sources.

The form of ground-water anomalies depends on the flow pattern of the water, which is determined by the permability and geometry of the aquifers through which the water moves. Three general types of aquifers may be distinguished: fractured crystalline rocks, layered sedimentary rocks, and surficial materials.

In crystalline materials (mainly igneous and metamorphic rocks) the major permeability is furnished by fractures. If the fractures are closely spaced and interconnected, fan-shaped anomalies generally result; as illustrated by Fig. 14.16 for Mo in springs issuing from biotite granite downslope from a series of contact metasomatic molybdenite deposits in the mountainous terrain of the Caucasus. A somewhat similar pattern exists downslope from a disseminated molybdenite deposit in the desert climate of Central Asia (Fig. 14.17). If the fractures are widely spaced and incompletely interconnected, flow patterns are irregular and anomalies have a correspondingly complex form. However, detailed investigations show that water descending through well-defined cracks and fracture zones generally reflects the chemical character of the wallrock of the fractures. Lovering (1952) has shown that the metal content of descending water in the relatively dry mines of the Tintic district, Utah, reflects the presence or absence of ore in the channels through which the water has passed. He concluded that the composition of this water has been modified by the chemical character of the rocks adjoining the channels and by the distance traveled by the solutions. Work in Japan showed a correlation between the content of Fe, Zn, and SO_4 in fracture-zone waters and the occurrence of known ore above the underground workings in which the samples were collected (Kimura *et al.*, 1951). Figure 14.18a illustrates the use of this method in a mine or drill hole to find undiscovered ore.

For some elements at least, the relative mobility is the same in ground water as in surface water. Figure 14.19, for example, shows a more rapid decay

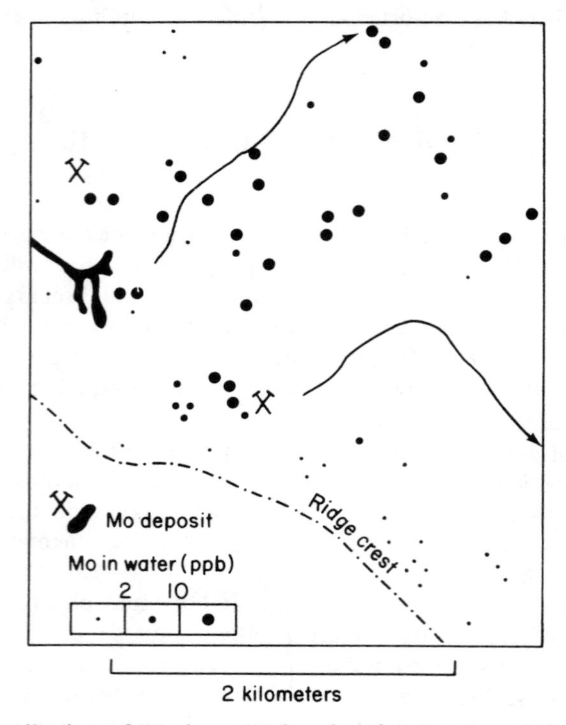

Fig. 14.16. Distribution of Mo in ground and spring water near the Lyangar Mo deposit, U.S.S.R. (After Vinogradov, 1957.)

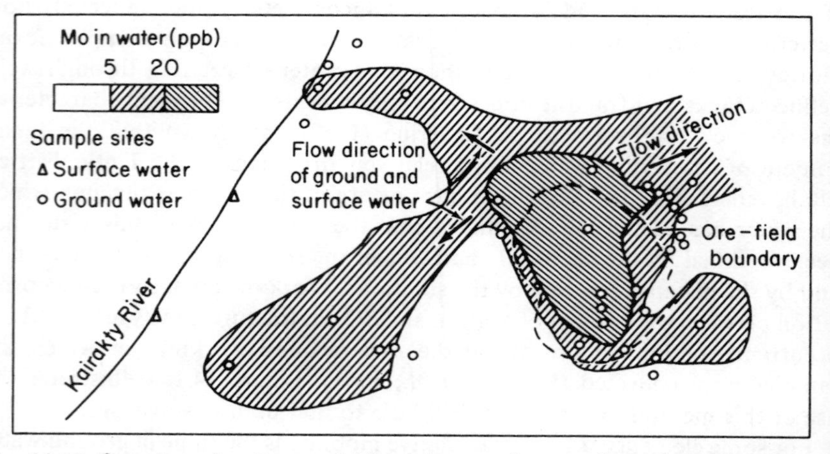

Fig. 14.17. Fan-shaped patterns of Mo-rich ground water downslope from Upper Kairakty Mo deposit, Kazakh, S.S.R. (After Belyakova, 1958.)

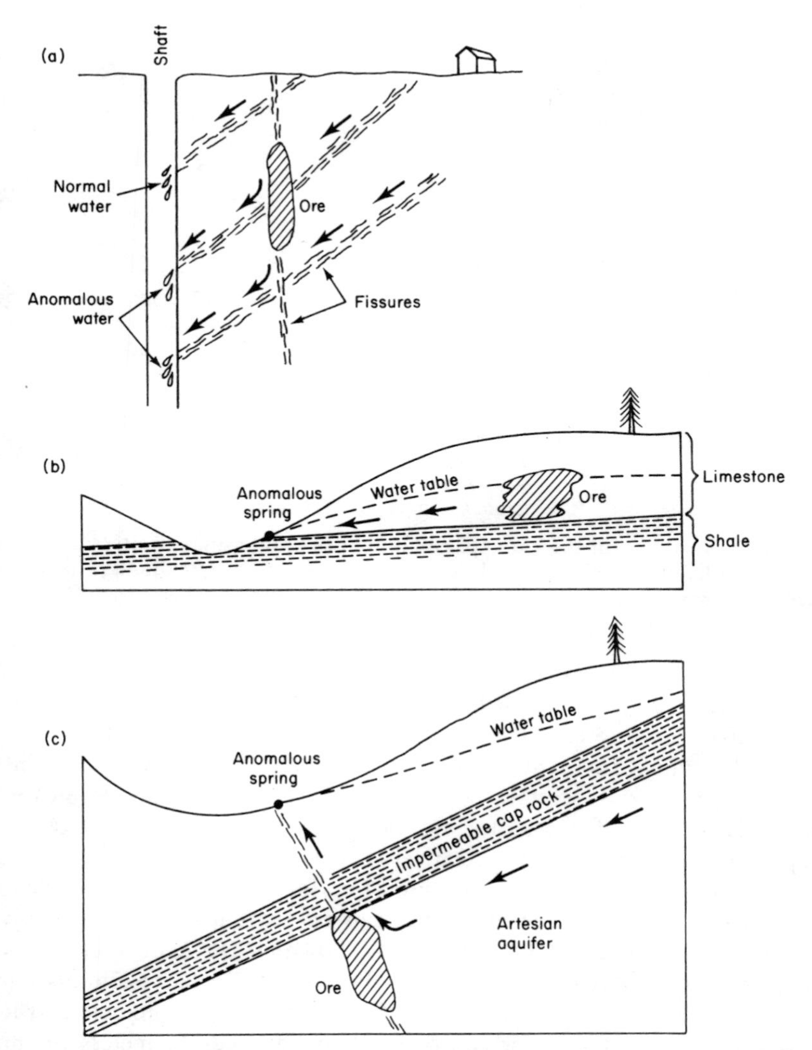

Fig. 14.18. Diagrammatic illustrations of anomalous ground water localized by permeable fissures and by impermeable layers. (a) Anomalies in water flowing into underground workings. (After Boyle *et al*., 1971.) (b) Restriction of flow by shale in gently dipping sediments. (c) Flow from artesian aquifer controlled by impermeable cap rock and fissure zone.

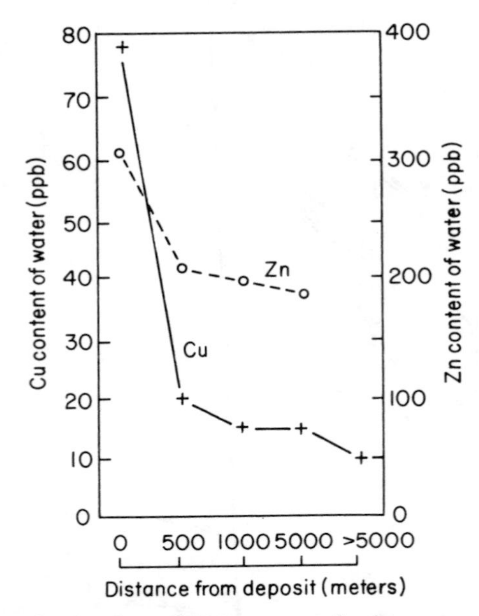

Fig. 14.19. Variation in the Cu and Zn content of subterranean waters in relatio to distance from deposit (Central Asia). (After Ginzburg (1960). Reprinted wit permission from "Principles of Geochemical Prospecting", © Pergamon Press Inc 1960.)

with distance from the source in the Cu content of ground water than in th Zn content. This trend parallels that found for surface water, as shown i Fig. 14.10.

In sedimentary rocks and some volcanic piles, where abundant inte connected pores exist, certain beds usually have much greater permeabilit than others, so that anomalies are elongated along these aquifers. Fa shaped anomalies are found in relatively flat-lying sediment, and more linea patterns in dipping beds. The location of springs tends to be controlled b impermeable beds, and the anomalies can be attributed to sources in th adjacent permeable units (Fig. 14.18b), as shown by DeGeoffroy *et al.* (196 for springs in the Wisconsin–Illinois Zn–Pb district. Recharge and discharg points in areas of dipping beds may be separated by many kilometers or te of kilometers, and the water can descend to hundreds of meters or mo (Williams, 1970). Drastic changes in Eh, pH, and other chemical characteri tics may also occur. Surface drainage divides do not necessarily mark th limits of ground-water flow systems in such environments.

In surficial materials overlying less permeable bedrock, flow is main lateral along the surficial material and may be concentrated in channe along the bedrock surface. Fan-shaped anomalies are expected in sheet-lik layers of alluvium or till, especially on long gentle slopes (Fig. 14.17) and linear anomalies are found in valley alluvium, as illustrated by Fig. 10.8. Mehrtens and Tooms (1973) have shown that ground-water underflow in alluvium along tributary streams can cross beneath trunk streams and cut across low drainage divides.

The chemical characteristics of ground water can serve as a valuable guide to the source of anomalies (or background values). Hoag and Webber (1976a) have used measurements of Na content in spring and drill-hole waters combined with empirical estimates of the rate of reaction of water with feldspar to determine the time that water has spent underground. They then used ground-water flow equations to estimate the distance to the source of the anomaly. Tilsley (1975) has attempted a similar type of analysis to evaluate the amount of fluorite encountered by a water sample, using the F content of the water and the rate of fluorite dissolution per unit surface area of fluorite.

VI. STREAM-WATER ANOMALIES

Stream water and the load of solid material that is dissolved or suspended in it come primarily from three sources: direct surface runoff, springs, and ground-water seepage. The soluble salts carried by the streams include all the constituents originally dissolved in the ground water that have escaped being precipitated, together with any soluble constituents picked up from the stream itself. Anomalous metal in any of the materials through which the water has flowed may dissolve in the water and produce an anomalous hydrogeochemical pattern in the stream water.

The homogeneity of stream water is one of its outstanding characteristics. As a rule, only a minor amount of turbulence is needed to mix waters of contrasting chemical or physical characteristics. Where turbulence is absent, however, or where the stream is wide in proportion to its depth, lateral variations in the composition of stream water may persist for considerable distances downstream. A lack of homogeneity of this kind is particularly common in the broad, shallow channels of the larger rivers (Fig. 14.20).

Anomalous metal may enter a stream at the headwater spring (Fig. 14.21) or at some point along its course (Fig. 14.8). In the latter event, the transition from anomalous to background values in the upstream direction is referred to as the cutoff. Where there is a strong influx of metal at a single point, the anomaly will tend to show a progressive increase in values to a well-developed peak and a sharp cutoff. More commonly, the metal enters at a number of points along a section of the stream course, in which case the anomalous pattern tends to be more erratic, and no single, well-defined cutoff point should be expected (Figs 14.6, 14.22, and 14.23).

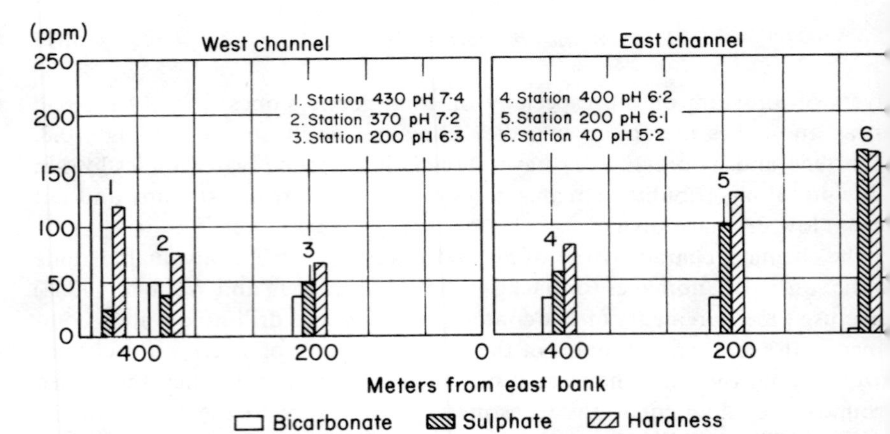

Fig. 14.20. Bicarbonate, sulfate, hardness, and pH of water samples collected in cross-section of Susquehanna River at Harrisburg, Pennsylvania. (From Hem 1970.)

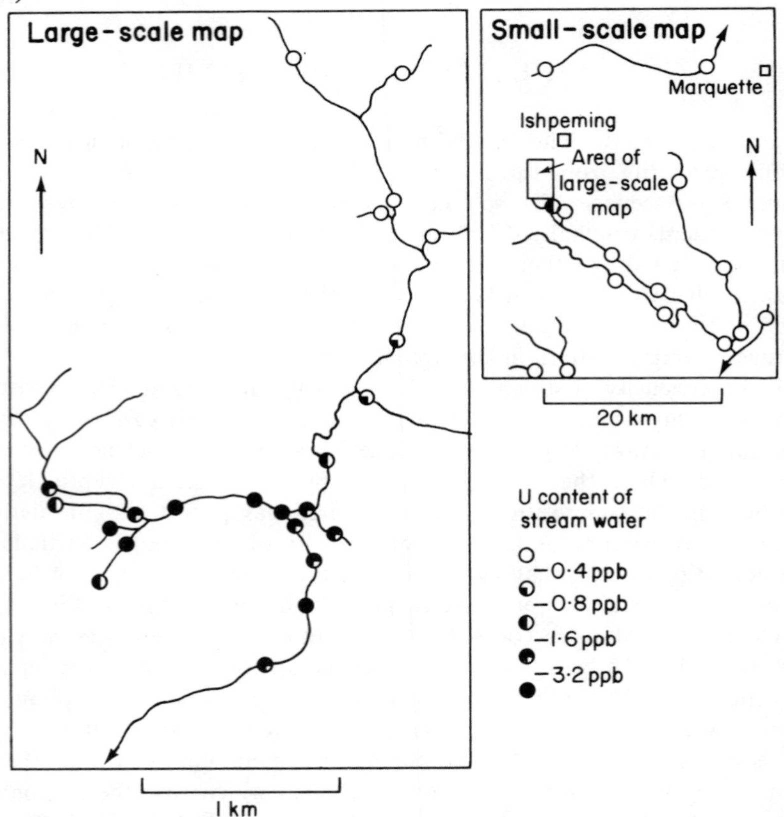

Fig. 14.21. Uranium content of stream water, Ely Creek Area, Michigan. (After Illsley *et al.*, 1958.)

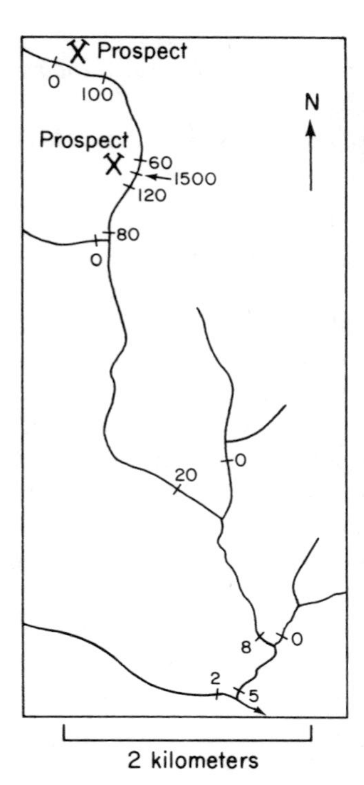

Fig. 14.22. Heavy-metal content of Missouri Creek, Colorado. Numbers are heavy-metal content of stream water in ppb. (After Huff, 1948.)

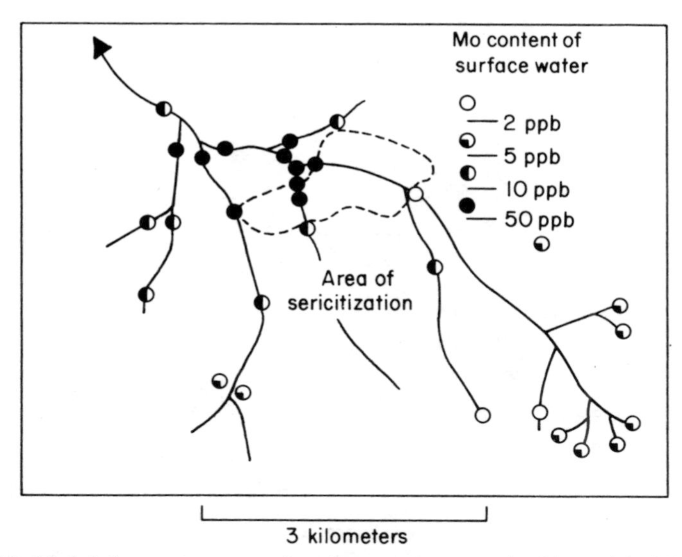

Fig. 14.23. Molybdenum content of surface water near the Yangokly Mo deposit, U.S.S.R. (After Vinogradov, 1957.)

VII. LAKE-WATER ANOMALIES

Lake-water anomalies can be derived from anomalous tributary streams,from seepage of ground water around the shores, and from ground water entering by way of springs in the lake bottom. The chemical composition of lake water is complicated by thermal stratification of the water and by the resulting heterogeneities in oxygen supply, organic activity, pH, and Eh, not only laterally and vertically within the lake, but seasonally. However, the studies discussed below indicate that these effects are usually small in suitably conducted surveys. An important cause of variations in the chemical composition of lake water is the scavenging effect of floating organisms and their decay products. This activity is most pronounced in the surface layer of lakes with moderate to high contents of dissolved nutrients in forested regions.

Chisholm (1950), working in Ontario, was the first to report that the metal content of lake water could be correlated with mineralization in surrounding terrain. A systematic survey of lakes in northern Maine by Kleinkopf (1960)

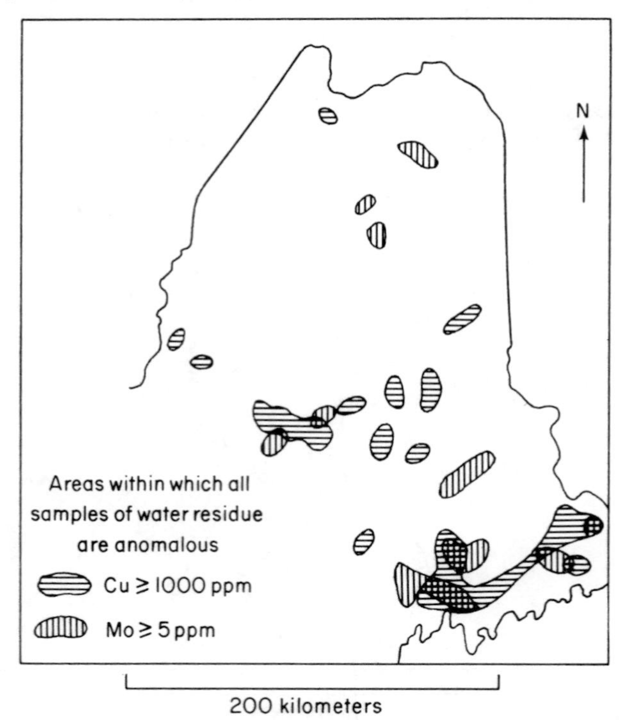

Fig. 14.24. Lake-water anomalies for Cu and Mo in northern Maine. (After Kleinkopf, 1960.)

showed that high Cu and Mo content of the lake waters correlated with known occurrence of Cu- and Mo-sulfides in the south-western corner of his map area (Fig. 14.24).

Lake waters have been most commonly utilized in the search for U. Regional surveys in the Beaverlodge area of Saskatchewan show clusters of lakes with U contents of about twice background, defining anomalous areas several kilometers in extent, with stronger anomalies within these areas apparently related to local sources (Macdonald, 1969; Dyck *et al.*, 1971). Surveys by the Geological Survey of Canada indicate that anomalies for U in lake water have about the same extent "downstream" as do anomalies in lake sediments, probably because of extensive primary dispersion to form anomalous districts and provinces in bedrock (Cameron, 1976).

Surveys of base metals in lake water show that Zn and Cu are distinctly anomalous near mineralized areas, but the anomalies disappear within a few kilometers down-drainage, apparently by adsorption onto particulate materials. The metal content of center-lake sediments increases to compensate for the decreases in the metal content of lake water (Fig. 14.11). High total dissolved solids are characteristic of many anomalous areas, apparently reflecting the unusual amounts of materials mobilized by oxidation of pyrite in and near the ore, and the resulting acid attack on minerals.

VIII. ANOMALIES NOT RELATED TO MINERAL DEPOSITS

Variations in the bulk composition of normal unmineralized rocks may cause variations in the composition of the drainage waters similar to those resulting from the weathering of ore deposits. Sulfate-rich waters identical in many respects to water draining actively oxidizing sulfide ores, for example, have been observed draining areas of disseminated pyrite and of gypsum. Although few reliable data are available, it is probable that the minor-element content will be similarly affected and that anomalous patterns can develop in waters in the vicinity of any rock characterized by unusually high background metal values.

Any condition promoting a local increase in the rate of decomposition of normal rocks can lead to the release of trace constituents in higher than normal amounts. Marmo (1953) has described well waters with as much as 500 ppb Cu and 5000 ppb Zn resulting from intense leaching by acid solutions derived from the weathering of a pyritiferous but otherwise unmineralized granite. The metal content of natural waters may be enhanced under arid climatic conditions by evaporation and the resulting concentration of all the dissolved solids. It may also vary in response to variations in the content of soluble organic matter. Finally, anomalous patterns have been found in areas of thermal activity that are apparently the effect of solutions of deep-seated origin.

Chapter 15

Anomalies in Drainage Sediments

Drainage sediments include spring and seepage sediments, active stream sediments, flood plain sediments, and lake sediments. These types differ in character of sediment, processes introducing metal, and application. Many drainage systems start in seepages and springs, where important interchanges between water and sediment may occur, as described previously (Chapter 12). Sediments in seepage areas and near springs tend to furnish strong anomalies and are useful in the follow-up and detailed stages of surveys. Active stream sediments include clastic and hydromorphic material from seepages, clastic material eroded from banks of streams, and hydromorphic material adsorbed or precipitated from stream water. Anomalies in active sediments may extend tens of kilometers from their source and are the most widely used medium for reconnaissance exploration. Sediments on flood plains are similar to active sediments in many respects, but may be distinctive in sorting, weathering, or content of organic matter. If lakes occur in a drainage pattern, sediments collect in these both as clastic particles and as adsorbed or precipitated material. In areas of abundant lakes, such as the glaciated shield area of Canada, lake sediments may furnish the most economical and effective means of reconnaissance and follow-up exploration.

I. SPRING AND SEEPAGE AREAS

As discussed in Chapters 10 to 12, consideration of hydromorphic anomalies in seepage areas is necessarily an integral part of the study of anomalies developed in residual and transported overburden. These same anomalies, however, are an equally integral component of dispersion patterns in drainage sediments, inasmuch as they serve as a source of a large fraction of the anomalous clastic material that goes to make up stream-sediment anomalies. It is from this point of view that relevant features of seepage anomalies are recapitulated and further described in this chapter.

Precipitation of insoluble compounds in areas where natural waters pass from one chemical environment into another tends to remove metal from solution, and thus cause an accelerated decay of hydrogeochemical anomalies. Up-drainage from a precipitation barrier of this kind, hydromorphic anomalies show up more clearly in the water. Below the barrier, they are stronger in the clastic material with which the water is in contact. Within the area where precipitation is actively taking place, anomalous values arising from material precipitated in the clastic matrix will be at a maximum.

The immediate source of the anomalous concentrations of metal in the soil and muck of seepage areas is, of course, the ground water that comes to the surface at these points. The ultimate source of the metal, however, must be sought somewhere along the route traversed by the ground water. Where the ground water flows through generally pervious, unconsolidated overburden, the route is determined by the slope of the water table and the topography of the bedrock surface. Where the ground water issues directly from the bedrock, the route is determined by the pattern of fissures, fractures, caverns, and high-pressure artesian channels in the bedrock.

The mode of occurrence of anomalous metal in seepage anomalies is commonly as components of precipitated hydrous oxides and as readily soluble components of the fine-grained fraction of the matrix. A close correlation of extractable metal with the organic content of many seepage soils suggests precipitation of an organic complex of some kind. The high base-exchange capacity of organic matter is undoubtedly a factor in this connection.

The location and form of seepage anomalies is controlled by the local relationships between the relief and water table and by the flow of water within the seepage area. Thus, seepage anomalies commonly occur in hollows and low places in sloping terrain, where the ground water emerges or comes relatively near the surface. Figures 10.6 and 10.7 illustrate some of the characteristic relations of spring and seepage areas to land forms.

Chemical precipitates of strictly inorganic origin are characteristic of many spring areas. Hydrous Fe-oxides are commonly precipitated from Fe-rich spring water. If these waters are also rich in ore metals, the limonitic material will be correspondingly enriched in these metals. Lead in amounts as high as 1% has been found occurring in an unidentified white, chalky mineral coating pebbles near the discharge of an acid spring in the Judith Mountains of central Montana. Accumulations of Pb up to 2% and probably more have been observed in organic-rich spring areas in the United Kingdom.

The edges of organic swamps are especially favorable sites for the development of seepage anomalies (Fig. 15.1). This distribution results from the fact that the organic matter of swamps tends to precipitate many of the ore metals out of ground-water solutions at the points where they first encounter

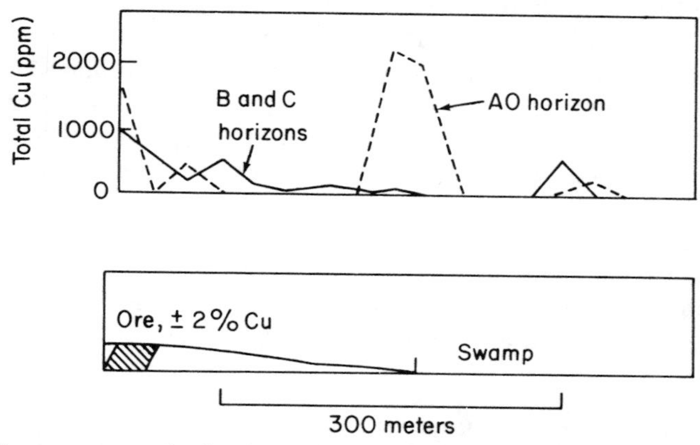

Fig. 15.1. Seepage anomaly developed in humus horizon at edge of swamp, downslope from source of metal, Campbell–Merrill ore zone, Chibougamau district, Quebec. Data on −80-mesh fraction. (After Ermengen, 1957.)

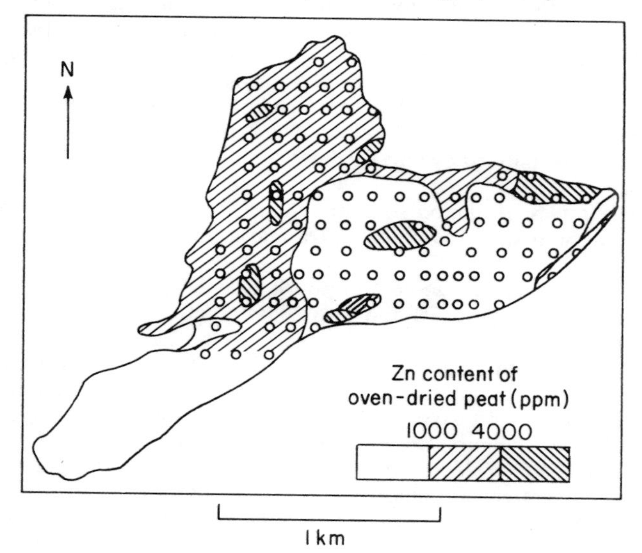

Fig. 15.2. Distribution of Zn in peat bog near Manning, New York. (After Staker and Cummings, 1941.)

the swamp environment. As the ground water moves toward the center of the swamp, it becomes progressively impoverished in metal, with the consequence that the anomalies are commonly restricted to the edges of the bogs. Figure 15.2 shows Zn anomalies of this kind resulting from precipitation of metal where Zn-rich ground and surface water enters the immobilizing organic environment. The Zn in the water here has been derived from the leaching of Zn-rich beds in Silurian dolomites. The oval patterns in the middle of the bog are related to springs rising from the Zn-rich bedrock. A deposit of native Cu of somewhat similar origin has been observed in a peat bog where Cu-rich water from neighboring mines enters the reducing environment of a swamp (Forrester, 1942). Here the precipitation of the Cu appears to be related to the low redox potential of the swamp rather than to chemical reactions with organic matter.

Although seepage anomalies have been used extensively in prospecting, it is only in Zambia that they have received intensive study. The following description of the distribution of Cu in the seepage areas, or dambos, of this area is therefore quoted at length from Webb and Tooms (1959).

> The sub-tropical climate is markedly seasonal, with an annual rainfall of 50 to 55 inches. For the most part, the vegetation is relatively thin forest savannah, except for small scattered areas of grassland known as "dambos", the majority of which are seasonal swamps.
>
> The country is drained by an open dendritic pattern of well-graded streams and tributaries which commonly rise in irregular pan-shaped "head-water" dambos. During the wet season these dambos are waterlogged and the outflow is considerable. Although some of the swamps may dry out completely at the surface as the ground-water level falls during the dry season, there is often a continual, though diminished, flow of water from the outlet of the dambo into the conventional stream channel below. The latter are generally moderately incised and may be flanked by narrow dambos more or less continuously along their course.
>
> Over 500 samples of dambo soil, stream sediment and stream banks were collected from the Baluba River and its tributaries in the area shown in Figure 15.3.
>
> In barren areas the metal content of poorly drained organic-rich dambo soil ranges from 30 to 200 ppm Cu and 4 to 20 ppm cxCu. These wide ranges in values are probably due to the sporadic distribution of very minor mineralization such as may occur in most of the rocks of the area. In dambo I, which overlies the Baluba ore-horizon, the metal content of the soil within 18 inches from surface exceeds 4000 ppm Cu and 1000 ppm cxCu. Strongly anomalous values up to about half these figures are recorded in the adjoining dambo II, which at its nearest point is over 2000 feet from the suboutcrop of the deposit. Even dambo III is moderately anomalous, despite the fact that only a small section of the mineralized horizon projects beyond the watershed into this particular catchment area.
>
> The peak values immediately over the ore-horizon in dambo I (Fig. 15.4) are no doubt largely of residual origin, being derived by weathering of the deposit

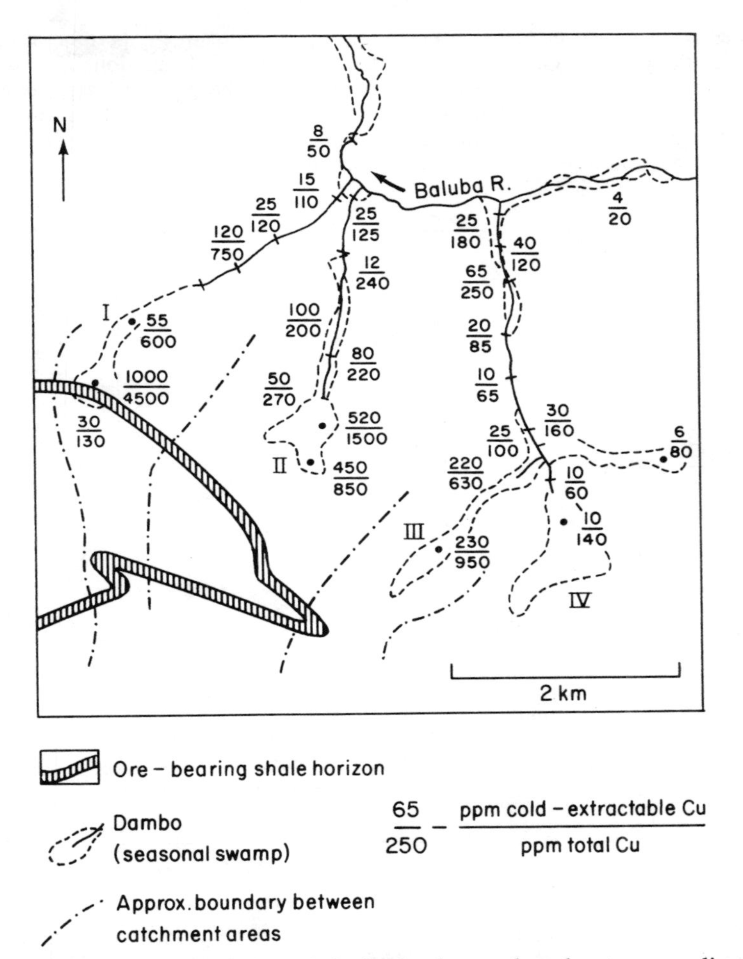

Fig. 15.3. Map showing relation of Cu in dambos and outlet-stream sediments to location of ore horizon, Baluba area, Zambia. Data on −80-mesh fraction. (After Webb and Tooms, 1959.)

in situ, which was almost certainly subject to oxidation before the advent of swamp conditions. The prolongation of peak values down-drainage from the ore suboutcrop is probably due to additional copper precipitated from ground waters, enriched in metal leached from the deposit rising in the main zone of seepage towards the centre of the dambo. Much less equivocal evidence of metal deposition from ground waters draining mineralized ground is shown in Figure 15.5 where a moderate copper anomaly in freely drained soil extends down-slope from the ore-horizon towards dambo II. On entering the dambo, the total copper content rises from about 250 ppm to 1500 ppm Cu, and the cold-extractable content from <5 ppm to over 800 ppm cxCu. Maximum values occur in the principal zone of seepage and not in the peripheral areas nearest the deposit. Occasionally small saucer-shaped depressions may occur in the broad head of the dambo, through which the ground waters emerge preferentially as the water-table rises during the wet season. In anomalous dambos, peak copper values are concentrated around these incipient springs—and also around more obvious springs, should any exist.

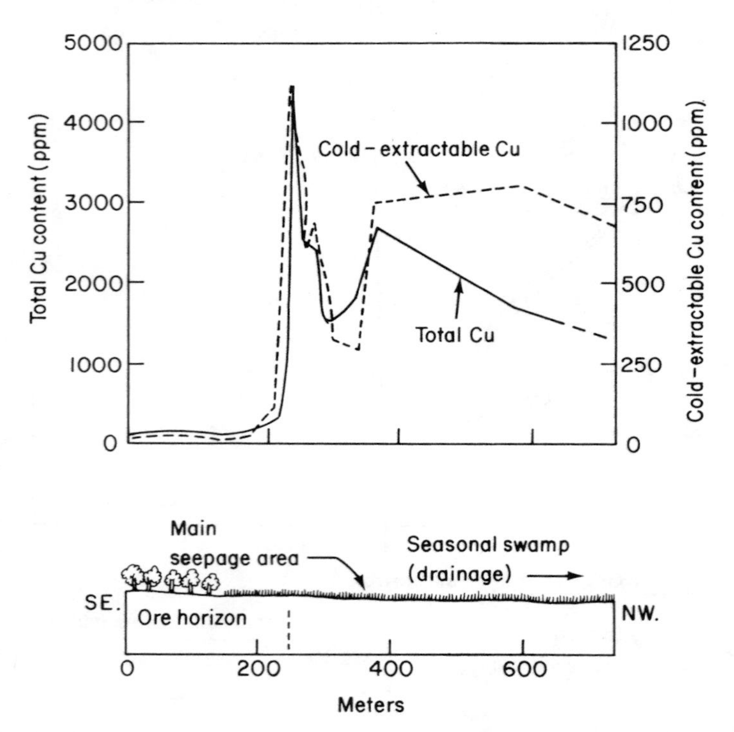

Fig. 15.4. Section showing relation of Cu in dambo I (Fig. 15.3) to underlying ore horizon. Data on −80-mesh fraction. (After Webb and Tooms, 1959.)

Although subject to considerable variation, the ratio of cold-extractable to total copper increases from 5 to 10 percent in background samples to 20 to 50 percent (occasionally up to 80 percent) in anomalous dambos. It follows that, although there is a degree of proportionality between the total and cold-extractable metal contents, the latter gives appreciably the greater contrast between anomaly and background. This observation does not apply to nearly the same extent in the areas of freely drained soils surrounding the dambos, where the cxCu : Cu ratio is consistently very much lower in samples collected at a comparable depth.

As a rule, the Cu content of anomalous soils in Zambia tends to decrease with depth from surface except in the immediate vicinity of the main points

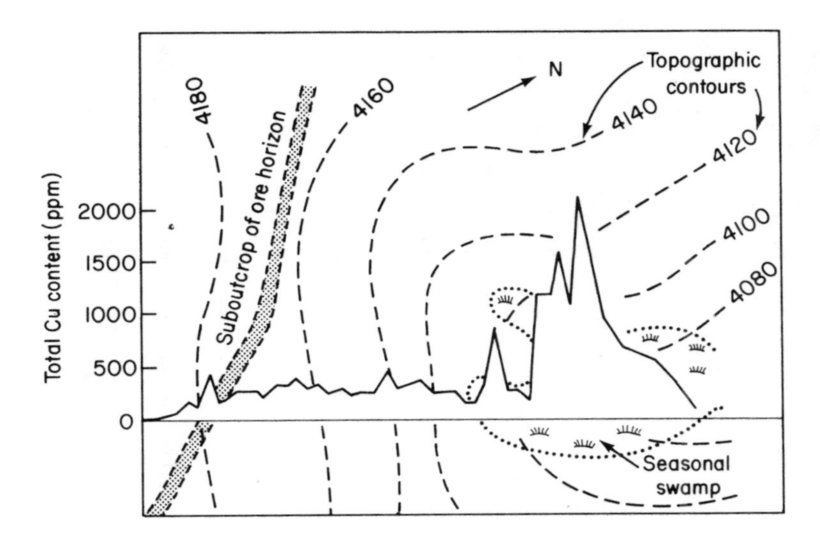

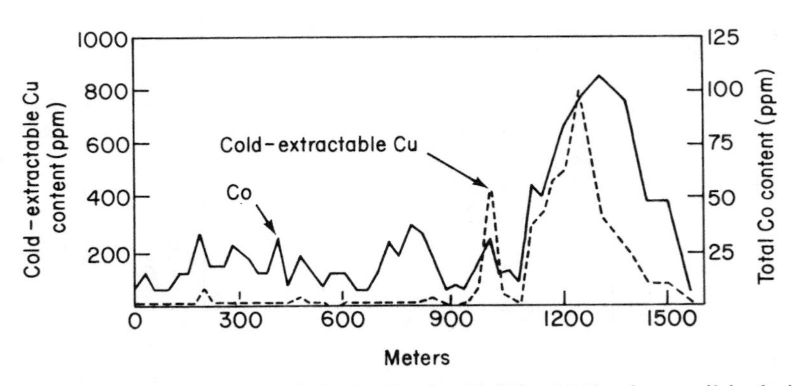

Fig. 15.5. Plan showing Cu and Co in dambo II (Fig. 15.3) where soil is derived from ore horizon located outside of swamp area. Data on −80-mesh fraction. (After Webb and Tooms, 1959.)

of seepage from suboutcropping mineralization (Table 15.1). Dispersion of Co follows a broadly similar pattern to that of Cu (Jay, 1959). According to Govett (1958), Cu values may also increase with depth away from the main seepage areas in the direction from which the Cu-rich waters are entering the spring area (Fig. 15.6). The general validity of this trend has not been established. In some dambos, the seepage anomalies are developed only at the break of slope along the dambo margin (F. H. Cornwall, personal communication).

Table 15.1
Metal distribution in typical soil profiles from background and anomalous dambos[a]

Soil horizon	General description	Average depth (m)	Background profile[b] (1)			Anomalous profiles[b] (2)			(3)		
			Cu (ppm)	cxCu (ppm)	cxCu/Cu (%)	Cu (ppm)	cxCu (ppm)	cxCu/Cu (%)	Cu (ppm)	cxCu (ppm)	cxCu/Cu (%)
A	Black organic-rich topsoil containing 2–5% organic carbon; pH 4·7–5·4	0–0·5	90	4	4·4	550	110	20·0	960	220	23·0
G	Blue–gray, orange-spotted sand/clay subsoil containing 0·07–0·4% organic carbon; pH 5·8–6·4	0·5–2+	80	4	5·0	750	170	22·6	480	45	9·4

[a] Source: Webb and Tooms (1959); data for −80-mesh fraction.
[b] Profiles: (1) Draining unmineralized basement granite; (2) in main seepage area of dambo II draining mineralized ground; (3) 1500 m down-drainage from mineralized ground.

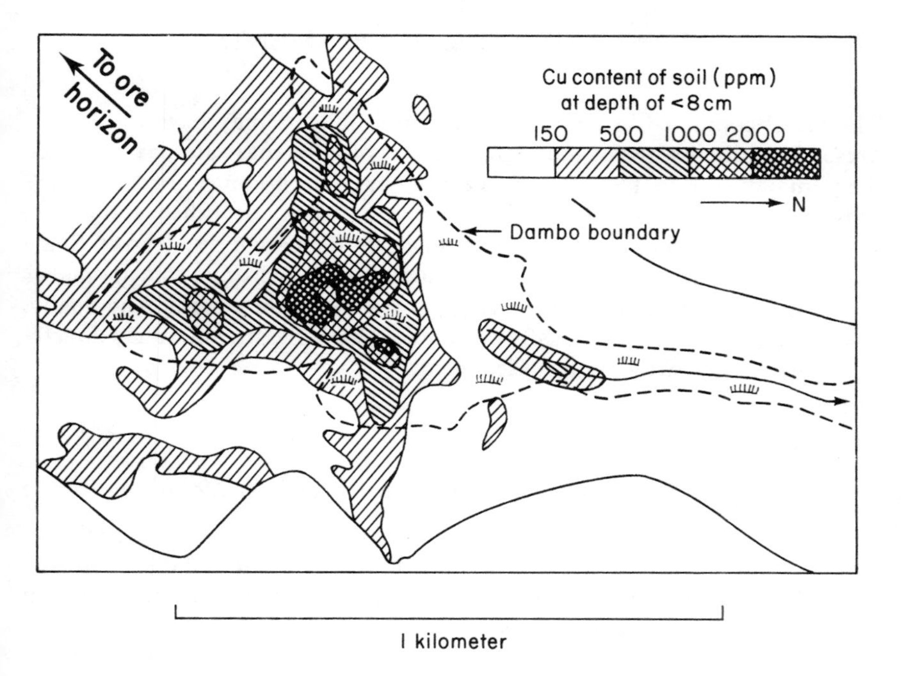

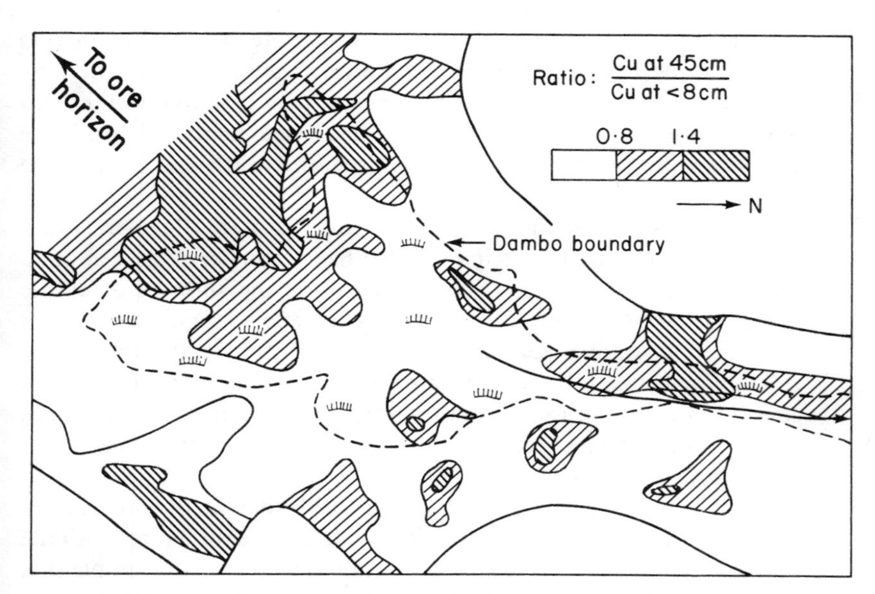

Fig. 15.6. Plans showing relative increase in Cu content of dambo soil with depth on side of dambo nearest source. Data on −80-mesh fraction. (After Govett, 1958.)

II. ACTIVE STREAM SEDIMENTS

The metal contained in anomalous stream sediments may have reached its present position by one or more of a number of different routes, as suggested diagrammatically in Fig. 10.2. Some may have been contributed from the erosion and transport of metal-rich soils, gossans, or other anomalous weathering products that originate very near to the parent ore deposit. Some may come from the erosion of clastic material from hydromorphic anomalies in spring and seepage areas or in the upper reaches of the stream. And finally, some metal may have been locally precipitated from the stream waters directly on to the clastic particles of the sediment.

A. Mode of Occurrence

Anomalous elements in stream sediment may occur in the following forms:

(i) Primary ore minerals, such as wolframite, columbite, pyrochlore, cassiterite, or Au. Most such minerals are resistant and dense, so that they travel with the heavy mineral fraction of the stream sediment. An outstanding exception is beryl, which has the same density as quartz and thus tends to travel with the light fraction of stream sediment.

(ii) Eroded secondary ore minerals, such as malachite, anglesite, or carnotite, and fragments of anomalous gossan, limonite, or other weathering products. Many of these materials are relatively friable and become finer in grain size downstream. As a result, they tend to be winnowed out and lost into the water. Although most sediments contain some clay-size material, the proportion is generally much less than in the parent soil that is eroded.

(iii) Precipitates from stream water, including both trace minerals and Fe–Mn-oxides or other precipitates that have incorporated the anomalous element. This material commonly coats the surfaces of clastic grains, or occurs as very fine particles that tend to remain in suspension, as noted above. Perhac and Whelan (1972) have shown that the colloidal particles filtered from stream water have much higher metal concentrations than the bottom sediment, although their abundance is very low and hence their contribution to the total metal transported by the stream is very small.

(iv) Exchangeable elements adsorbed on Fe- or Mn-oxides, organic matter, clay, or other major phases. This fraction readily equilibrates with the stream water and reflects the concentrations in water at the sample site.

(v) Organic material with incorporated ore elements. This fraction includes debris from plants growing on the ore or in seepage areas, degraded organic material from soils near the ore, and organic matter formed by biological activity in the stream. Organic matter may be a major component of stream sediment in swampy areas of poor drainage, or in streams of steep gradient and very little fine inorganic sediment.

The content of an anomalous element in the above forms is superimposed on the normal background levels and forms of the element. A high proportion of the background content usually occurs as a trace substitution in the lattices of silicates, oxides, and other major minerals, but some occurs in the Fe-oxides, organic matter, and in some localities in the other forms listed above.

In any given anomaly, the relative proportions of these different materials will depend on a number of factors, including the nature of the primary bedrock source, the previous dispersion history of the anomalous constituents, the origin of the stream sediment, and the degree of sorting of material within the stream channel. Many of the Cu sediment anomalies in Zambia are thought to be due mainly to the erosion of anomalous seepage-area soils. In

Table 15.2
Partition of Cu in stream sediments, Kilembe, Uganda[a]

Location of sample	Presumed mode of occurrence: Exchangeable[b]	Incorporated in secondary Fe-oxides[c]	Lattice-held[d]
		Cu content (ppm)	
Anomalous stream			
Near mineralization	15	120	135
340 m downstream	12	50	68
540 m downstream	5	15	65
Background stream	1	3	26
	2	3	30
		Presumed partition (%)	
Anomalous stream			
Near mineralization	6	44	50
340 m downstream	9	39	52
540 m downstream	6	18	76
Background stream	3	10	87
	3	9	85

[a] Source: J. S. Webb and R. E. Stanton (unpublished data); data for −80-mesh fraction.
[b] Copper by cold-citrate extraction at pH 7·3.
[c] Additional Cu released by dithionite extraction by method of White (1957).
[d] Additional Cu released by $KHSO_4$ fusion, after citrate and dithionite extraction.

the Kilembe area of Uganda, the sediment anomalies are apparently largely derived from the erosion of anomalous residual soils. In this case, nearly 50% of the anomalous metal is incorporated in precipitates of secondary Fe-oxides; about the same amount is "lattice-held", presumably in clay minerals, while only a small fraction, usually less than 5%, may be present as adsorbed ions (Table 15.2). Similarly, in sediment draining from a U prospect in Pennsylvania, a high proportion of the U is incorporated in organic material and Fe-oxide, whereas in a background sediment most U is in the sand and silt fractions (Fig. 15.7). A high proportion of Zn in both anomalous and background residual soils in Tennessee was found to occur in Fe-oxides and clays (White, 1957).

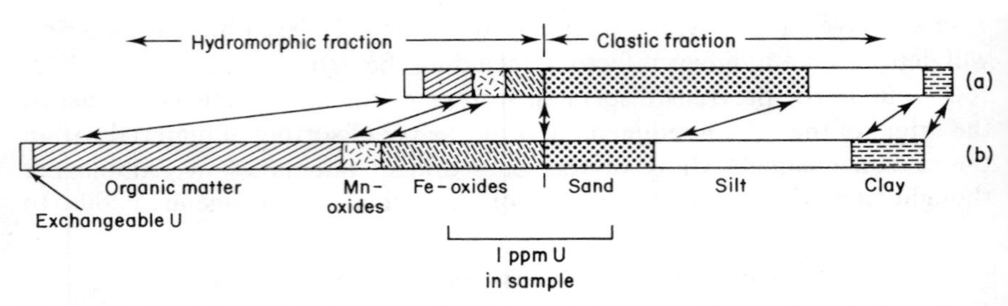

Fig. 15.7. Partition of U between fractions of stream sediment from mineralized and background drainages, determined by selective extraction methods. (a) Background sediment; (b) anomalous sediment. (From Rose *et al.*, 1977.)

In predominantly clastic anomalies derived by erosion of anomalous soil, the cxMe/Me ratio is typically low, generally less than 5%. In hydromorphic anomalies derived from precipitation or adsorption of the anomalous element, the cxMe/Me ratio is generally greater than 5%.

In other areas, the major proportion of the anomalous metal has probably been precipitated directly from metal-rich stream waters. Clear-cut examples of this may be seen where contaminated water from mine workings discharges into an otherwise normal drainage channel. Here, stream sediments showing cxMe/Me ratios greater than 50% are not uncommon. With time, the readily extractable metal tends to become fixed, with a resulting decline in the cxMe/Me ratio.

B. Contrast

A major factor controlling the contrast of anomalies in stream sediments is the primary contrast at the bedrock source and the extent to which this is diluted with barren material from the surroundings. This aspect of contrast is taken up in the following section on decay of anomalies.

In a more direct fashion, contrast in stream sediment is a function of the contrast at the point where the anomalous material is being fed into the stream channel. Where sediment anomalies are derived by the erosion of anomalous residual or seepage soil, the contrast in the sediment will simply reflect the contrast in the parent soil anomaly. Similarly, where the anomalous pattern in sediments results from precipitation from metal-rich stream water, the sediment anomaly contrast will depend most directly on the contrast between anomalous and background waters. In either case, the contrast in a stream-sediment anomaly will decrease with increasing distance between the bedrock source and the point of entry into the stream channel, as a result of dilution with barren material.

Contrast also varies with the mode of occurrence of the anomalous metal in the sample, and thus with the methods used to detect it. Depending on the mineralogy of the primary ore, the anomalous metal of stream sediments may be preferentially concentrated in the fine-grained fraction, the coarse-grained light fraction, or the coarse-grained heavy fraction.

Most of the anomalous metal in stream sediments is usually concentrated in the finer size fractions. Thus an inverse relationship between metal content and grain size is observed under a wide variety of climates, as illustrated by the data of Tables 15.3 to 15.5. Where comparable figures for anomalous

Table 15.3
Content of Zn and exchangeable heavy metal of different size fractions of stream sediments, New Brunswick[a]

Size fraction (mesh size)	Weight percent	Total Zn (ppm)	cxHM[b] (ppm)
−8+32	65	500	6
−32+80	35	500	15
−80+115	4	800	25
−115+200	3	800	45
−200	3	1000	75

[a] Source: Hawkes and Bloom (1956).
[b] Undifferentiated Zn, Pb, and Cu.

and background samples were reported, it is seen that maximum contrast is brought out by analysis of the intermediate grain sizes, between 80 and 200 mesh, even though these sizes do not show the maximum absolute values. These experiments also show that the contrast may be markedly greater for cxMe than total Me.

Primary resistant minerals often tend to be concentrated in the coarser size fractions of the sediment. An example of this relationship is illustrated

Table 15.4
Copper content and particle size in stream sediments, Zambia[a]

Size fraction (B.S.S. mesh)	Weight percent of −2-mm fraction		Total Cu (ppm)			cxCu (ppm)		
	Sample B	Sample A	B Background sediment	A Anomalous sediment	Contrast A/B	B Background sediment	A Anomalous sediment	Contrast A/B
−20+ 35	5·4	24·1	80	180	2·3	8	80	10·0
−35+ 80	64·4	43·5	40	160	4.0	2	35	17·5
−80+135	21·1	21·4	40	210	5·2	3	70	23·3
−135+200	5·4	4·6	80	250	3·1	12	110	9·2
−200	1·7	0·6	110	360	3·3	22	170	7·7
−80	—	—	50	220	4·4	4	80	20·0

[a] Source: Webb (1958a).

Table 15.5

Copper content of different size fractions of stream sediments, Ruwenzori Mountains, Uganda[a]

Size fraction (B.S.S. mesh)	Weight percent of −2-mm fraction		Total Cu (ppm)			cxCu (ppm)		
	Sample B	Sample A	B Background sediment	A Anomalous sediment	Contrast A/B	B Background sediment	A Anomalous sediment	Contrast A/B
−20+ 35	27·7	23·0	30	100	3·3	0·2	5	25
−35+ 80	26·6	13·8	30	150	5·0	0·4	9	22
−80+135	9·8	7·8	30	210	7·0	0·4	16	40
−135+200	4·8	5·0	40	260	6·5	0·6	30	50
−200	3·0	2·8	90	500	5·5	2·5	55	22
−80	—	—	45	280	6·2	0·8	28	35

[a] Source: Webb (1958a).

in Fig. 15.8. Here the higher contrast for Be in the coarser fractions is due to beryl, whereas the values in the finer fractions are probably due largely to the presence of Be in clay minerals. Most of the primary resistant ore minerals are relatively heavy and tend to be enriched in the heavy-mineral fraction of the sediment. Thus, in prospecting for Sn and Nb, much higher relative values may be obtained by analyzing only the heavy-mineral fraction in which the detrital cassiterite and columbite have been concentrated. The

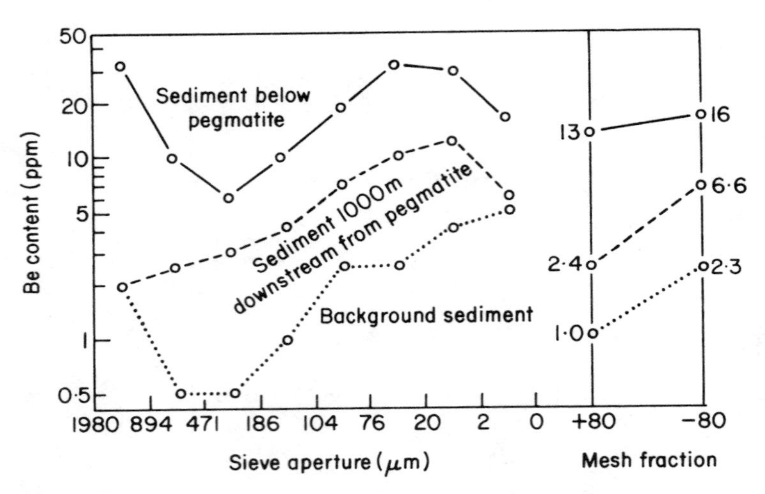

Fig. 15.8. Beryllium content of different size fractions of stream sediments, Ishasha Claims, Uganda. 80 mesh = 186 μm aperture. (From Debnam and Webb, 1960.)

extent to which contrast will be enhanced in this way will depend on whether the same detrital mineral also occurs as a normal accessory of the host rocks. Thus, the background level for Sn in detrital cassiterite, which often occurs in the bedrock only as an ore mineral, may be relatively low, whereas the background for Cr in chromite, a mineral that occurs as a minor accessory in ultramafic rocks, may be fairly high.

In glaciated regions and other terrains with poorly developed drainage, the organic matter in the stream bed can furnish the best contrast and the most reliable guide to mineralized areas (Brundin and Nairis, 1972; Larsson, 1976). The organic material scavenges metals from solution so that it is especially effective in detecting hydromorphic anomalies. The organic-rich sediments are also less sensitive than normal stream sediment to the distorting effects of Fe- and Mn-oxide precipitation from the water, and may be more abundant and easy to collect.

C. Decay Patterns

The general form of the downstream decay pattern of sediment anomalies is more or less the same whether the anomaly occurs as residual detrital grains or as precipitates of hydromorphic origin. The principal factors affecting the persistence of these anomalies are (i) the contrast at the source, (ii) the input of metal along the stream course, and (iii) dilution by erosion of the bank material and by confluence with barren tributaries.

The effects of dilution and contrast at the source may be expressed in the following idealized model, as proposed by Hawkes (1976c). Assume a mineralized zone of Area A_m and surface metal content Me_m, in a drainage basin with a total area A_t down to point S and background metal content Me_b contributed from barren tributaries and erosion of barren bank material (Fig. 15.9a). If erosion operates so as to remove material from all parts of the drainage basin at an equal rate, so that the stream sediment is a representative sample of all surface material in the basin, then the metal content of the sediment sample at point S (Me_S) can be found by weighting the metal contents of the mineralized and background areas according to their area:

$$Me_S A_t = Me_m A_m + Me_b(A_t - A_m). \quad (15.1)$$

This equation may be rearranged to give

$$Me_m A_m = (Me_S - Me_b)A_t + Me_b A_m. \quad (15.2)$$

If the drainage area is large relative to the mineralized area, then the term $Me_b A_m$ can be neglected, so that

$$Me_m A_m = (Me_S - Me_b)A_t \quad (15.3)$$

The term $Me_m A_m$ is a measure of the size (A_m) and grade (Me_m) of the deposit or source, at least to the extent that it is exposed at the surface. The quantity $(Me_S - Me_b)$ is a measure of the strength of the anomaly above background. The exposed size and grade of the source ($Me_m A_m$) can be estimated from an observed value for anomaly strength $(Me_S - Me_b)$, multiplied by the measured area of the drainage (A_t). The product $(Me_S - Me_b)A_t$ has been termed the productivity (P) of the sediment anomaly by Solovov and Kunin (1961) and Polikarpochkin (1971, 1976). The expected downstream decay pattern of stream sediment anomalies in two simplified drainage basins, and the behavior of the productivity for these ideal cases, is illustrated in Fig. 15.9.

Field data show that the above equation does indeed closely describe the contrast and decay of some real sediment anomalies. Hawkes (1976c) showed that the content of Cu in stream sediments as far as 45 km downstream from four porphyry-copper deposits was consistent with the model and could be used to estimate the average Cu content of soil at the prospects. Rose *et al.*

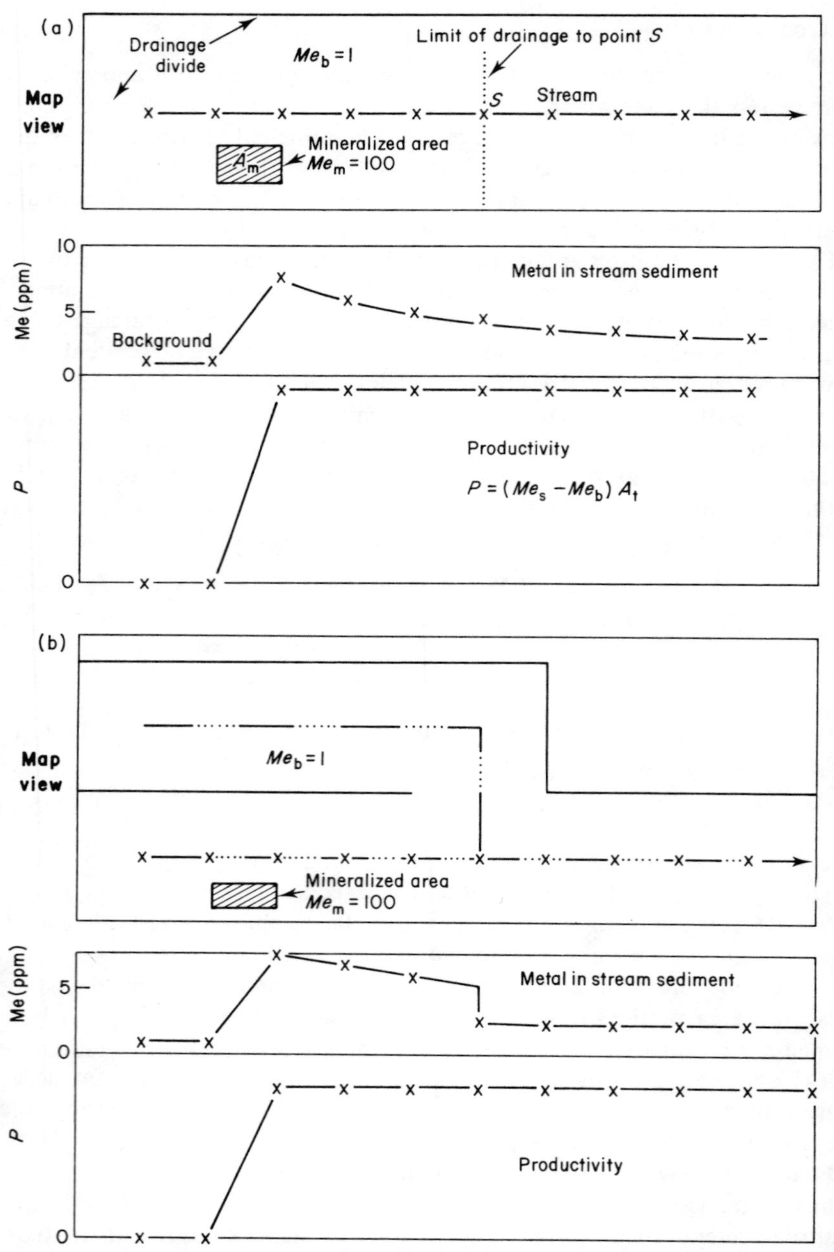

Fig. 15.9. Diagrammatic illustrations of metal in stream sediment and productivity for two simplified drainage basins. See text for explanation of symbols.

(1977) found that stream sediment from a 2·5 km^2 drainage basin with several U prospects in Pennsylvania could be used to estimate the average content of U in soils of the drainage basin. MacKenzie (1977) demonstrated a similar relation for Cu in stream sediment of several mineralized areas in the tropics.

Many factors can complicate the simple model described above. Probably the most important factor is interchange of metal with the stream water. If appreciable amounts of the element are carried past the sample site in dissolved or suspended form, the relations are distorted. This behavior is especially important for the more mobile elements and for acid streams near oxidizing sulfide deposits. Hawkes (1976c) pointed out anomalously low Cu content in several sediment samples from acid streams near the porphyry-copper deposits he studied. Further examples of this effect are discussed in the following section on precipitation barriers.

Unusually rapid downstream decay of anomalies can also be produced by washing out of metal-rich fines from the sediment (Govett, 1961) and slow decay by accumulation of fines (Rose *et al.*, 1970). Polikarpochkin (1971, 1976) observed that anomalies decreased downstream at a more rapid rate than could be accounted for by eqn 15.3, and was able to correct for this by adding an exponential loss term to the equations (Fig. 15.10). Loss of fines is most important when the anomalous metal is concentrated in the clay-size fraction. Effects attributable to washing out of fines are not observed in the data of Hawkes (1976c), and are observable only for Al and Si in the data of Rose *et al.* (1977).

The assumption of equal erosion throughout the drainage basin is also untrue for many areas. In large drainages, the erosion rate may differ markedly from one part of the drainage to another. Some rocks, including many ores, are either more or less resistant than their surroundings, and are eroded at a different rate.

The metal content of the banks alongside the active stream channel is also important in determining the persistence of sediment anomalies, and the degree to which eqn 15.3 can be applied. In most cases this factor is related to the alluvial or colluvial origin of bank material. At Nash Creek, New Brunswick, the streams are bordered by flood-plain alluvium containing about the same concentration of anomalous metal as the active sediment. In this and similar areas, the persistence of an anomaly is determined in large measure simply by the distance between confluences with barren streams and the size of these streams relative to the anomalous stream. Where the banks are composed of colluvium derived from the adjoining valley slopes, on the other hand, continuous progressive dilution of the sediment anomaly begins as soon as the anomalous stream leaves the source area. Under these circumstances, drainage trains may be drastically curtailed. Figure 15.11 shows an example of dilution of anomalous patterns by erosion of barren bank material.

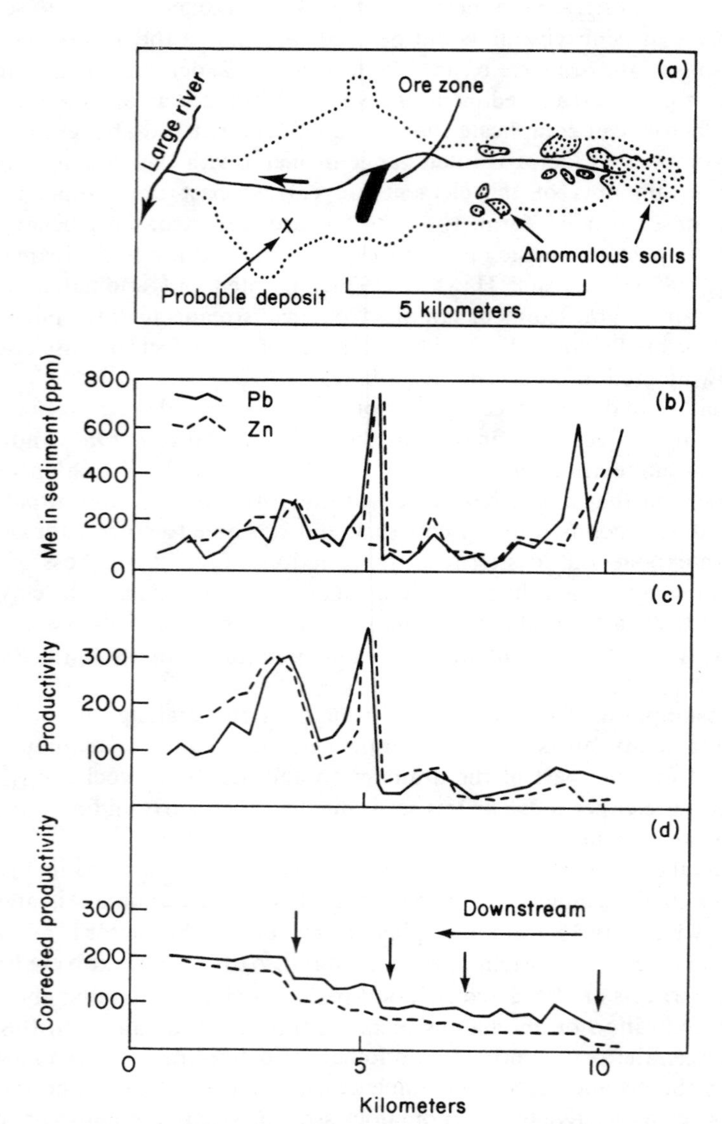

Fig. 15.10. Downstream dispersion in a valley in East Transbaikalia. (a) Map of drainage showing ore-bearing zones; (b) observed metal content of sediment; (c) productivity calculated from eqn 15.3; (d) productivity calculated by eqn 15.3 modified to account for exponential loss of metal content with distance along stream. Arrows indicate input areas from deposits and anomalous soils. (After Polikarpochkin, 1971.)

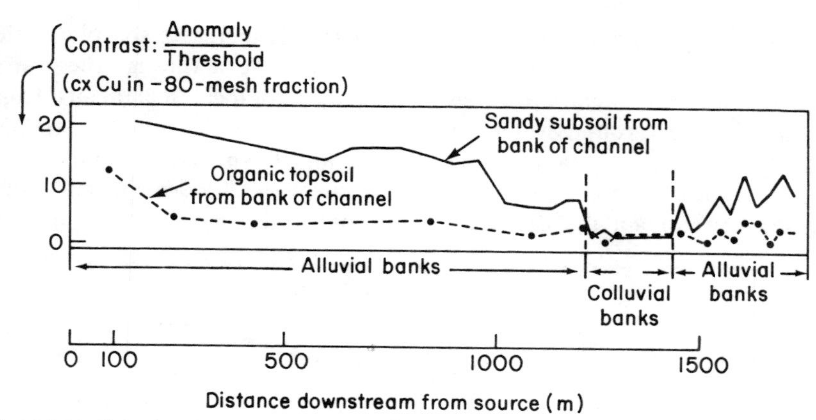

Fig. 15.11. Longitudinal diagram of drainage anomaly in alluvial sediments from bank of active channel, Baluba, Zambia. Data on −80-mesh fraction. (After Govett, 1960.)

If the mineralized area is covered by transported overburden, no correlation of productivity with area and grade of the weathered ore can be made. Fractionation processes in the soil and bedrock of the mineralized area, such as supergene enrichment of Cu, or mobilization of Fe, Mn, and other elements from gley-type soils and bogs, followed by precipitation in the stream, also limit the application of eqn 15.3. Even in these situations, although the size and grade of the source cannot be predicted, the decay pattern of any anomaly should follow the predicted pattern.

Other complications pointed out by Hawkes (1976c) are the effects of sampling and analytical errors (which increase in importance with size of drainage), variable background in the drainage, multiple sources of the anomalous element, and contamination by mine dumps or industrial waste. Variable background and multiple sources can be treated, at least in principle, by adding terms to the equations. It should be noted that the source term must account for not only the weathered ore but also the primary and secondary haloes around it. At many relatively small orebodies, the weakly mineralized halo zone may contribute more metal to the drainage than the ore itself. Sampling and analytical errors probably account for most of the erratic departures from the downstream decay patterns predicted by the model, and are discussed further in Chapter 16.

In general, eqn 15.3 appears to be most applicable to immobile elements in areas of vigorous physical erosion, but it is a good approximation in many other situations.

A type case is illustrated by the contrast and decay of a sediment anomaly at Nash Creek, New Brunswick, as shown in Figs 15.12 and 15.13. At site 1, upstream from the mineralized area, the metal content of the sediments is near the background value. The content of metals increases sharply as the stream enters the source area (site 2). As the stream flows through the source area, the metal content fluctuates in response to local increments from lateral drainage, ground water, and springs. Metal-rich tributary streams are known to enter the main channel at sites 3 and 4, where they can be correlated with local increases in the metal content of the flood-plain sediments. Further downstream the decay in metal content is largely a reflection of

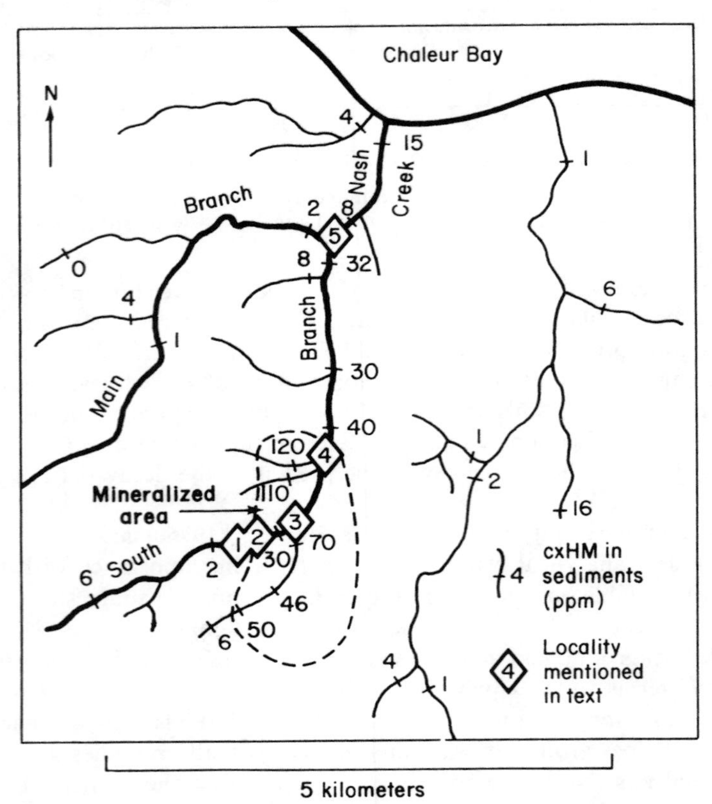

Fig. 15.12. Plan of drainage-sediment anomaly at Nash Creek, New Brunswick. Data on −80-mesh fraction. (After Hawkes and Bloom, 1956.)

simple dilution. At site 5 the anomalous south branch of Nash Creek is joined by the main branch, with about four times the drainage area. The ratio of the discharge of the two streams almost exactly accounts for the drop in metal content of the sediments. Very commonly, the dilution effect where a small anomalous tributary enters a large stream is enough to obliterate the anomalous pattern completely, as illustrated in Fig. 15.14.

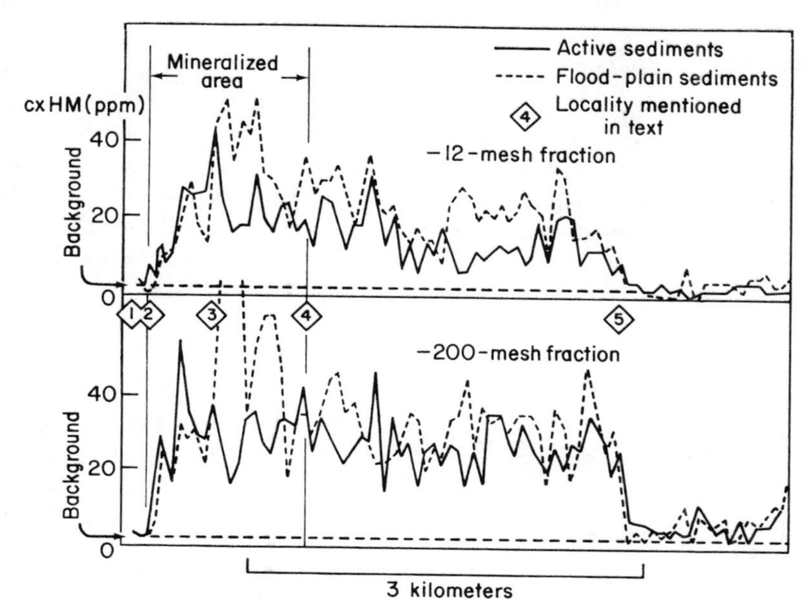

Fig. 15.13. Longitudinal diagram of drainage-sediment anomaly at Nash Creek, New Brunswick. (After Hawkes and Bloom, 1956.)

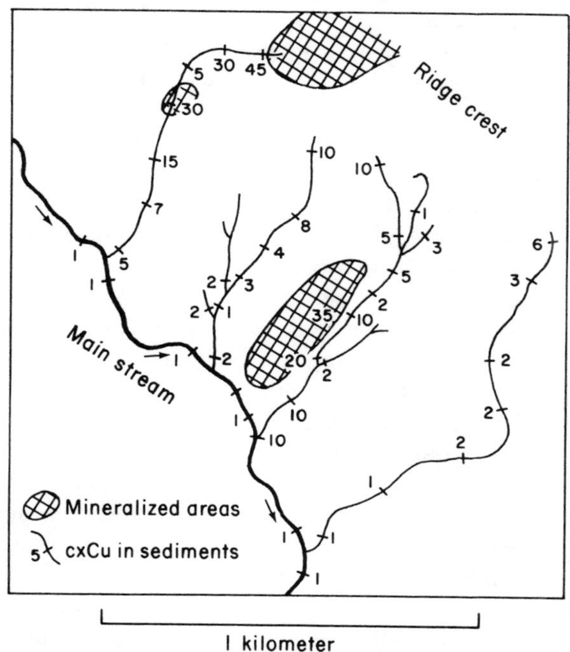

Fig. 15.14 Dispersion of Cu (ppm) in stream sediments draining Cu mineralization, Uganda. Data on −80-mesh fraction (After Research Report for 1954–57, Royal School of Mines, Imperial College, London.)

D. Anomaly Enhancement

According to the mode of occurrence of the metal, the effect of dilution may sometimes be reduced by treatment of the samples to separate the barren diluents from the anomalous components of the sediment. This process may be termed anomaly enhancement. For example, very persistent W anomalies were obtained in streams draining huebnerite deposits by analyzing the heavy-mineral fraction (Fig. 15.15). Other descriptions of the use of heavy minerals

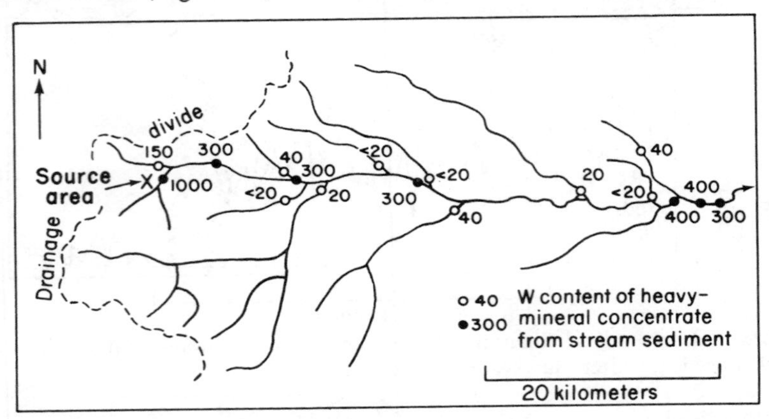

Fig. 15.15 Tungsten content (ppm) of heavy-mineral concentrates of stream sediments, Clear Creek, Colorado. (After Theobald and Thompson, 1959.)

are by Leake and Smith (1975), Overstreet (1962), and Bell (1976). Heavy-mineral anomalies are liable to decay more rapidly in areas where the local barren rocks contain abundant heavy accessory minerals, such as magnetite, ilmenite, garnet, zircon, and monazite. Where this happens, it may sometimes be possible to improve the persistence of anomalies by magnetic or electrostatic separations or by procedures involving selective chemical attack. Some anomalies are unrecognizable in the total sediment, but easily detected in panned heavy-mineral concentrates, as reported for Sn by Sainsbury and Reed (1973). Some detrital minerals are preferentially enriched in the fine-grained clastic fraction of stream sediments. Figure 15.16 shows the greater contrast and persistence of anomalous Nb, which occurs as pyrochlore, in the fine-size fractions of sediments downstream from a carbonatite deposit in Zambia. Analysis of the −80-mesh fraction of the local drainage disclosed a sediment anomaly extending for 20 km to the confluence with a major river (Fig. 15.17). The persistence of this anomaly is due in large measure to the fact that the anomalous stream flows for much of its course through flood-plain alluvium. Light detrital minerals, such as beryl, cannot be so readily separated from the quartz which serves as the principal diluent mineral in most sediments. Even so, under appropriate conditions, the anomalous

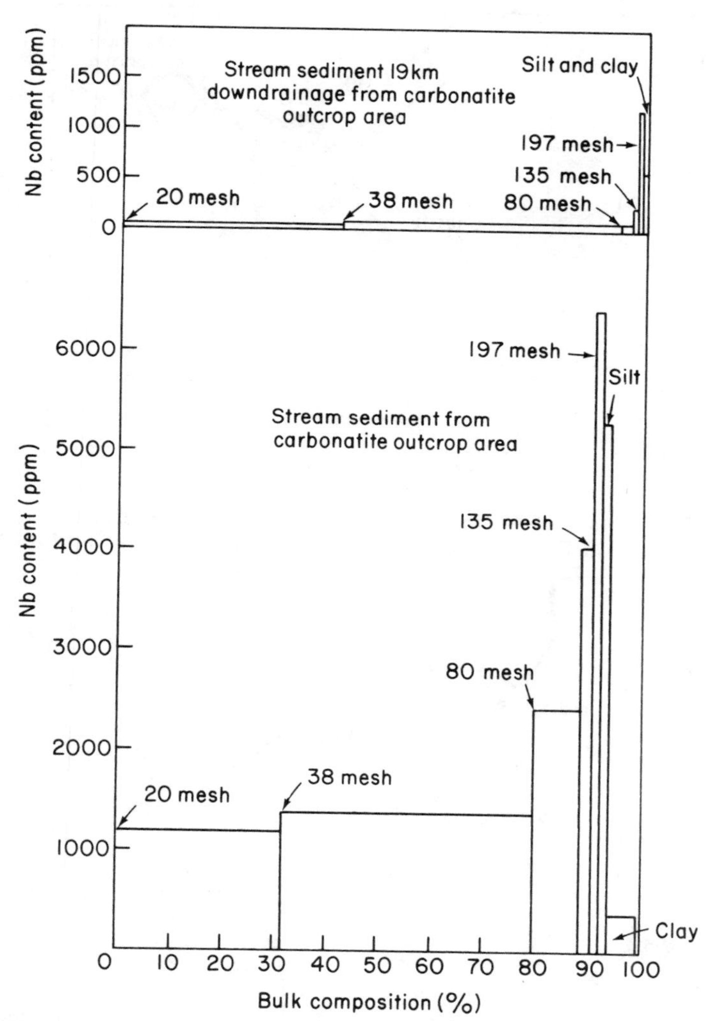

Fig. 15.16. Diagrams showing changes in Nb content of individual size fractions of stream sediments with distance from source, Kaluwe pyrochlore carbonatite, Zambia. (After Watts *et al.*, 1963.)

metal content may be detected for appreciable distances downstream (Fig. 15.18).

Another method of anomaly enhancement is selective extraction of Fe–Mn-oxides. Anomalies in Cu and Zn extracted by hydroxylamine treatment of stream sediments from south-eastern U.S.A. extended 5 km or more downstream from their source, compared to only 1·5 km for total Cu and

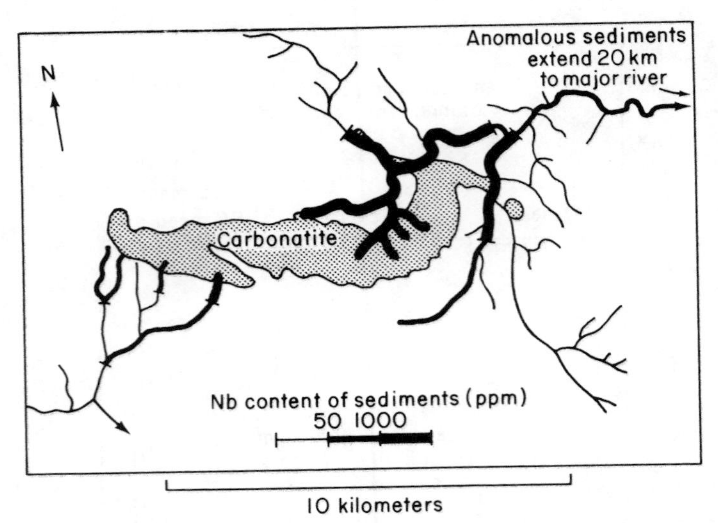

Fig. 15.17. Distribution of Nb in stream sediments draining pyrochlore carbonatite, Kaluwe, Zambia. Data on −80-mesh fraction. (After Watts *et al.*, 1963.)

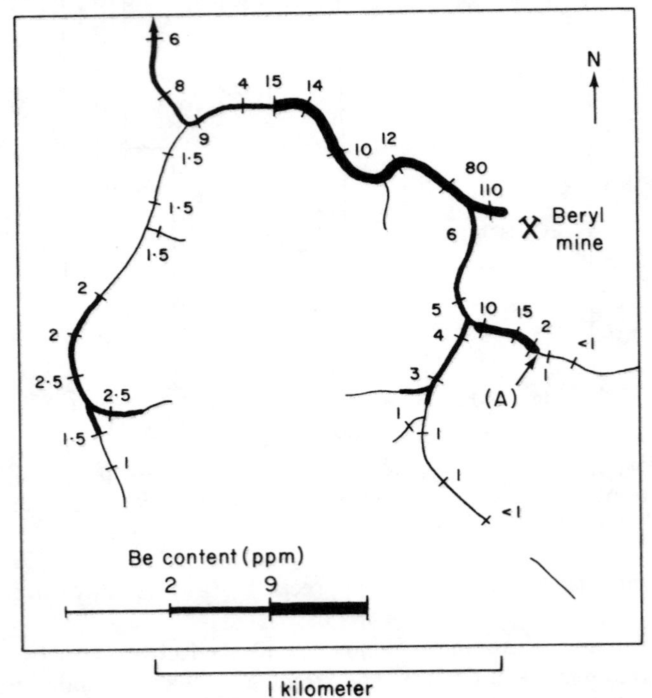

Fig. 15.18. Distribution of Be in stream sediment, Ishasha area, south-western Uganda. (A) Indicates position of undisturbed pegmatite discovered by sediment sampling. (After Debnam and Webb, 1960.)

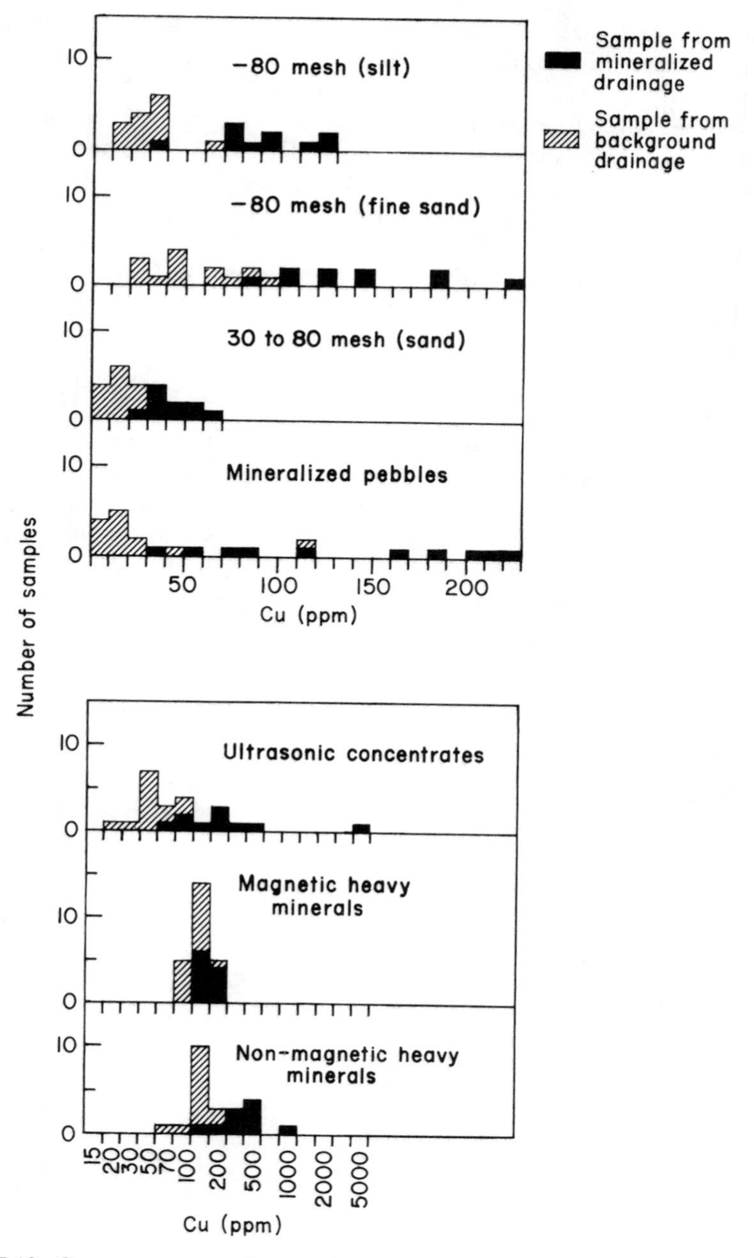

Fig. 15.19. Copper content of seven fractions of stream sediment from Arizona. (After Huff, 1971.)

Zn in −80-mesh stream sediment (Carpenter *et al.*, 1975). For Pb, a less mobile element, no anomaly was observed in the Fe–Mn-oxides. Similar patterns were obtained by Whitney (1975) for sediments in New York.

Other specialized fractions or types of stream sediments which can increase contrast are organic sediments (Brundin and Nairis, 1972), and suspended particulate matter (Perhac and Whelan, 1972). Selective extraction of organic matter and Fe-oxides improved contrast for U in stream sediments in Pennsylvania (Rose and Keith, 1976).

Results from a comprehensive study of sediments near porphyry-copper deposits in Arizona are illustrated in Fig. 15.19. The types of samples studied in this investigation were as follows (Huff, 1971):

(i) −80-mesh silt selectively sampled from silt-rich areas of the dry stream-beds.
(ii) −80-mesh sand sieved from normal sediment of the stream bed.
(iii) 30- to 80-mesh sand sieved from normal sediment of the stream bed.
(iv) Mineralized pebbles, consisting of the most stained, veined, and altered pebbles that could be found in sediment at the site.

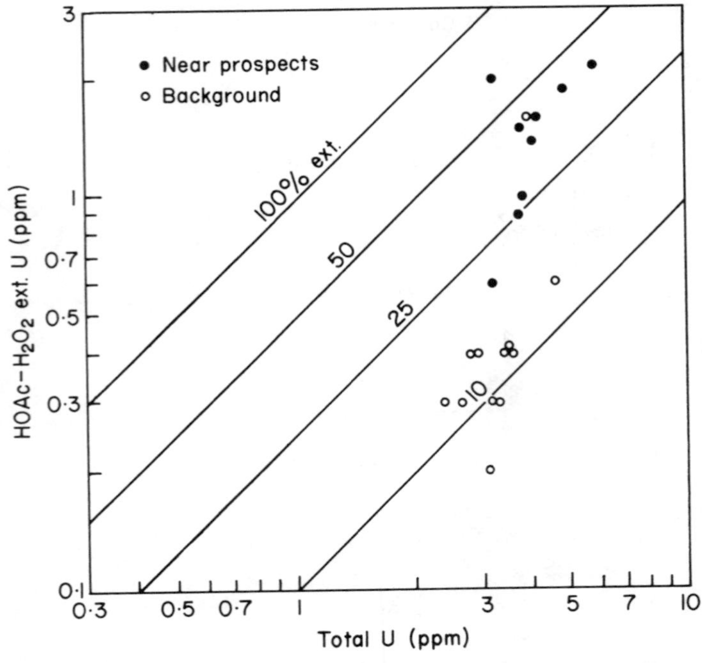

Fig. 15.20. Total U and acetic acid/H_2O_2-extractable U for stream sediment samples in Pennsylvania, and percentage extraction. (After Rose and Keith, 1976.)

(v) Concentrates of coatings on +80-mesh grains, liberated with an ultrasonic cleaning bath.
(vi) Magnetic heavy minerals.
(vii) Non-magnetic heavy minerals.

All types except the magnetic heavy minerals gave useful distinctions between the mineralized and background drainages, but the range of values for 30- to 80-mesh sands was small, approaching sampling and analytical error. The ultrasonic concentrates and the non-magnetic minerals required considerable effort in sample processing. The pebbles and ultrasonic concentrates have higher contrast than the normal −80-mesh sediment, and might be used to extend anomalies and allow wider-spaced sampling. Erickson *et al.* (1966a) found anomalies in altered and mineralized pebbles where no anomaly was evident in stream sediment.

Uranium anomalies in stream sediment can be enhanced by extracting the U contained in organic matter and Fe-oxides (Fig. 15.7) and leaving silicates in the residue. Figure 15.20 illustrates the enhancement produced by an oxidizing weakly acid leach (hydrogen peroxide and acetic acid) as compared to total U.

E. Effect of Precipitation Barriers

Most ore metals tend to be soluble in the acid environment of an oxidizing sulfide deposit. As the acidity of the water draining a deposit is progressively reduced with distance, the metals tend to be precipitated as trace minerals or incorporated in newly precipitated Fe–Mn-oxides. The distribution of metal in the stream sediments thus will naturally be strongly conditioned by the pH of the stream water in contact with it. The effect in the sediments will be precisely the reciprocal of that in the drainage water. An example is provided by the partition of Cu in the stream sediment below the disseminated Cu deposits of Cebu in the Philippines (Coope and Webb, 1963). Where the pH is below 4·0 as a result of the strong acid coming into the surface drainage from the oxidation of pyrite, the ratio of cxCu to total Cu is extremely low (Fig. 15.21). As the pH increases past 5·0, this ratio also increases to a value of about 10%.

Reference has already been made to the precipitation of metal below the confluence of two streams of contrasting composition (Theobald *et al.*, 1963). Similar effects can be expected at any point along a stream course when abrupt changes in the chemical environment contrive to produce a precipitation barrier, as discussed in Chapter 14. For example, in an area of Zambia the sediment of certain streams draining mineralized ground is not anomalous. Where these streams feed into the main river, anomalous cxCu patterns

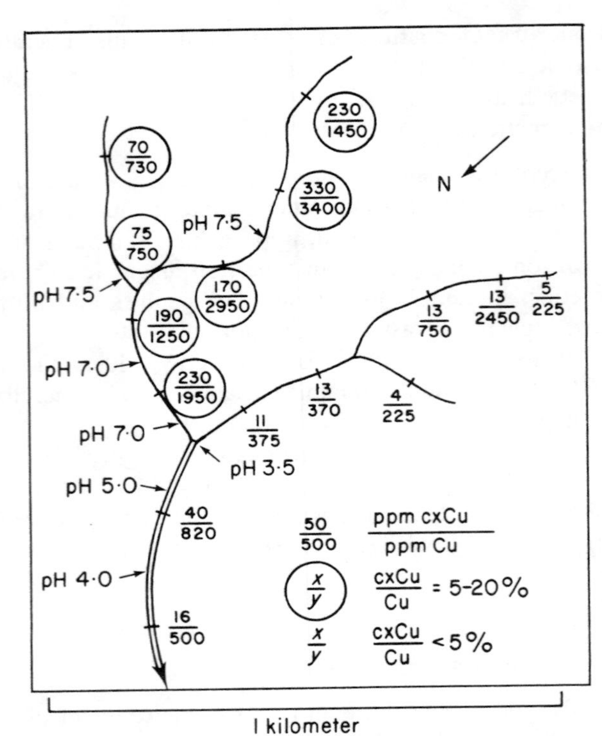

Fig. 15.21. Plan of drainage anomaly showing relation of pH of water to ratio of cold-extractable Cu (cxCu) to total Cu in stream sediments, Cebu Project, Republic of Philippines. (After Coope and Webb, 1963.)

are developed in the swampy ground bordering the river. It is presumed that Cu traveling in the tributary stream water in either suspension or solution is deposited in the sluggish organic-rich environment alongside the main river (D. R. Clews, personal communication).

Another cause for unusually high metal values in stream sediment is incorporation in Fe–Mn-oxides mobilized by reducing ground waters and precipitated in the stream. This process is common in glaciated areas and other regions of poor drainage, moderate to high rainfall, and cool climate where waterlogged soils and bogs are present. Horsnail *et al.* (1969) report enrichment by a factor of 30 for Mn, 13 for Co, and 5 for As in sediments in Wales affected by Fe–Mn-oxide precipitation. Nickel and Zn are also strongly concentrated in the precipitates. The phenomenon has been reported in eastern Canada (Boyle *et al.*, 1966), Maine (Canney, 1967), Sweden (Ljunggren, 1955), and Ireland (Horsnail, 1975). Effects of Fe–Mn-oxide precipitation are generally at a maximum in the smallest streams, and are diluted and averaged out in larger streams (Plant, 1971).

At a precipitation barrier, local very high concentrations of the elements being precipitated may be expected in favorable environments on the stream bottom. With transport downstream, these enrichments are expected to be homogenized by mixing with the rest of the sediment to approach a uniform enrichment at a particular sample site. Values higher than the level predicted from the contrast and decay model (eqn 15.3) may be obtained if the element is preferentially leached from rocks and soils of the drainage. Selective leaching is most likely in areas of youthful topography. Corresponding deficiencies may be expected in more mature, well-leached terrain.

F. Lateral Homogeneity

Homogeneity may be considered in terms of variability of sediments at a single sample site (area of a few meters of stream bed) or over longer distances where significant tributaries are lacking. Variability at a site depends largely on local differences in the grain size and mineralogy of the sediment, and on the mode of occurrence of the anomalous element. Homogeneity over longer distances involves, in addition, the distribution of points of entry of metal into the stream channel.

A single point of entry of anomalous metal into a drainage system results in a progressive building up of values to a well-defined and sharp cutoff at the upstream end of the anomaly. More commonly, metal-rich material is fed into the stream at a number of points, with the result that the anomalous pattern in the source area may be fairly complicated, as illustrated in Figs 15.10, 15.13, and 15.22.

The occurrence of anomalous metal as a principal constituent of one or more specific minerals, such as cassiterite or beryl, results in a lack of homogeneity both in samples and in anomalous patterns. Samples tend to be more homogeneous where the metal is pervasively impregnated through the finer fractions, as in hydromorphic patterns.

Variations from one sample to another in the relative proportions of coarse and fine material and of organic matter are the major cause for local inhomogeneity. These variations occur both laterally, because of differences in current velocity, and vertically in the sediment profile, reflecting differences with time. The differences in Fig. 15.19 between −80-mesh silt (selected from areas of fine sediment) and −80-mesh sand (sieved from coarser sediment) illustrate the kind of effects that can occur. Where the metal is concentrated in the silt or clay fraction, sieving to remove the coarser barren sand will often suppress some of the apparent irregularities in the anomalous pattern (Table 15.6), if these result from dilution with barren sand. However, if the

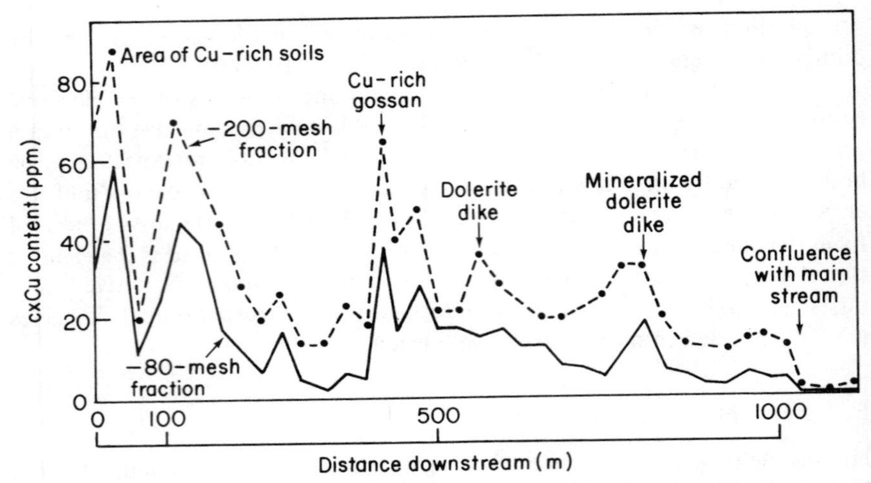

Fig. 15.22. Longitudinal diagram of sediment anomaly showing distribution of readily soluble Cu in −80- and −200-mesh fractions, Uganda. (After Webb, 1958a.)

Table 15.6
Variation in metal content with sediment type, Nash Creek, New Brunswick[a]

Sediment type	cxHM[b] (ppm) −12 mesh	−200 mesh
Site A		
Silt	74	60
Sand	30	50
Gravel	30	55
Site B		
Organic ooze	47	50
Silt	37	40
Gravel	13	40

[a] Source: Hawkes and Bloom (1956).
[b] Exchangeable heavy metal (undifferentiated Zn, Pb, and Cu).

anomalous material is concentrated in the fines, sieving out of the fine fraction may furnish a useful improvement in the consistency of anomalies, although at the expense of contrast (Rose *et al.*, 1970).

Based on replicate sampling studies of Bølviken and Sinding-Larsen (1973), Chork (1977), Howarth and Lowenstein (1971), and Leake and Smith (1975), the normal coefficient of variation for replicate sampling of stream sediments using standard procedures averages about 20%, with a range of 10–50%. To this must be added any analytical error. In typical areas, then, 95% of analytical results will fall within ±50% of the true value if a relatively precise analytical method is used. However, by collecting a large sample of stream gravel and sieving at the sample site, a lower coefficient of variation has been obtained in sampling in Scotland by the British Institute of Geological Sciences.

G. Vertical Homogeneity

Vertical variations at a sample site may result from deposition of differing material due to seasonal changes in stream velocity, supply of organic material or Fe–Mn-oxide precipitation, or from post-depositional chemical changes in the sediment. Coarse-grained heavy minerals tend to be sorted and concentrated at depth in stream sediments, commonly directly on the surface of bedrock. Such concentrations may be patchily distributed along the stream bed according to variations in the currents around bends in the stream course or by the "riffle" action of irregularities in the stream-bed topography. In some situations, an anomaly may be lacking in surface sediment, but easily detected in the heavy minerals at depth. When the metal occurs in the fine fractions, the apparent homogeneity may be reduced where increased stream velocity prevents deposition of all but the coarser barren material. Stream-bed sediments are sometimes stratified, so that the metal content may vary with depth in the sediment profile. In Zambia, for example, it is not uncommon to find alternating layers of coarse sandy material and fine organic sediment in flood-plain sediment adjacent to streams. In this case, the metal content of the fine material can be appreciably higher than in the coarse layer (Govett, 1958).

Post-depositional loss of Mn, Fe, Co, Zn, and U from sediment at depths of 0·1–0·3 m by mobilization out of reducing zones and incorporation into Fe–Mn-oxides at the stream bed has been demonstrated by Plant (1971).

H. Time Variations

Although variations with time in the metal content of stream sediments are clearly documented, the evidence indicates that they are normally much smaller than time-variations in waters. Three reasons for variations with time have been suggested: (i) washing out of fines during high water, or accumulation during low water, (ii) changes in the organic content of the sediment, and (iii) changes sympathetic to the changes in water, as a result of adsorption or coprecipitation.

The largest variations are reported in a study of the cxCu content of anomalous sediments collected from active stream-channels in Zambia (Govett, 1961), where seasonal variations up to a factor of about 3 occur near the source and increase to more than 5 as the pattern decays downstream (Fig. 15.23). This variation thus reduces the effective length of the dispersion train during the wet season. The reason for this seasonal variation is thought to be the accumulation of fine-grained metal-rich material during the period of minimum, extremely sluggish flow in the dry season.

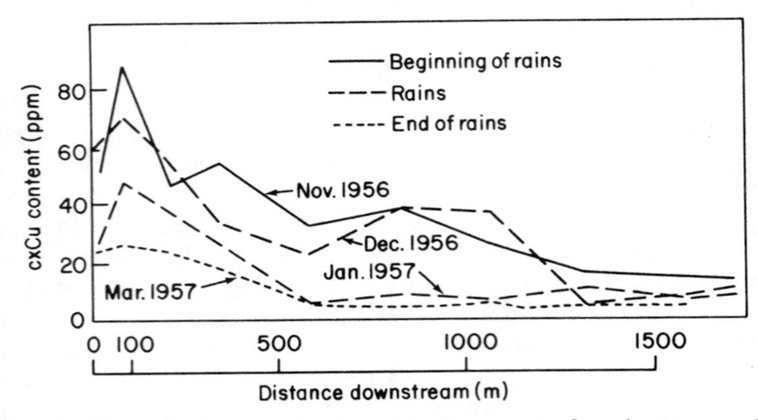

Fig. 15.23. Effect of rain on cold-extractable Cu content of sandy stream sediment, Baluba area, Zambia. Data on −80-mesh fraction. (After Govett, 1958.)

Other reported seasonal variations, all in regions with a cooler climate and more uniform distribution of precipitation during the year, are much smaller than in Zambia. Stream sediments in Chile varied by a factor of 2–3 for Cu, and 1·5 for Zn (deGrys, 1962). The high values were attributed to an increased hydromorphic contribution during the late summer. In Pennsylvania, U in stream sediment varied by a factor of 2 during the year, compared to a factor of 50 for the water. The U at this locality is concentrated in the organic matter, which was apparently washed out of the sediment during periods of high flow and required several months to accumulate its initial higher U content. In one study in British Columbia, both cxCu and total Cu in anomalous and background sediments showed no significant variation, in spite of a decline in runoff during the period of the experiment by a factor of more than 20 (Barr and Hawkes, 1963).

III. FLOOD-PLAIN SEDIMENTS

The pattern of anomalous metals in flood-plain sediments will reflect their distribution in the abandoned channels previously followed by the stream. As a rule, therefore, the part of the flood-plain anomaly that is due to detrital minerals will tend to be most pronounced toward the base of the alluvium. Where the anomalous metal is dispersed in the finer-sized fractions, however, other factors intervene. Flood-plain sediments normally are characterized by a higher proportion of fine material, and thus may carry a greater content of anomalous metal than active sediments from an equivalent site. This effect may be offset by the fact that flood-plain sediments are more subject to leaching by rain water or by barren ground-water entering the channel from the side, whereas the metal content of active sediments may be maintained by chemical exchange with the anomalous water in contact with it.

The lateral distribution of anomalous metals across the flood plain varies somewhat depending on local conditions. For example, in eastern Canada, anomalous patterns of exchangeable heavy metal in the −200-mesh fraction of sediment collected immediately on the bank of the active channel is roughly the same as that of samples of equivalent organic content from the channel itself, although the values normally decay with distance from the active channel (Figs 15.13 and 15.24). Observations in Zambia indicate a clear correspondence between the metal content of flood-plain sediments and the

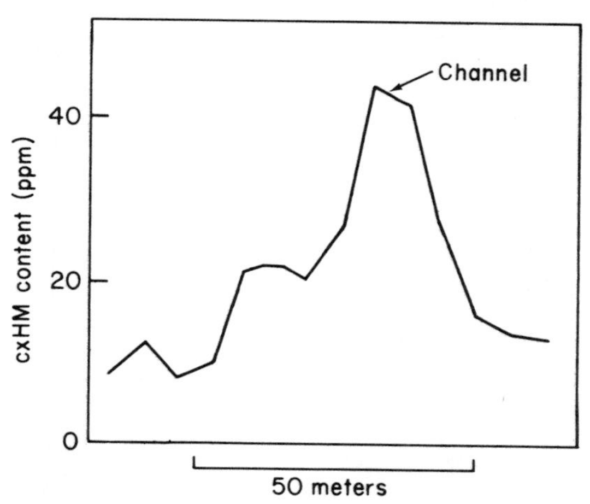

Fig. 15.24. Cold-extractable heavy-metal content of sediments from traverse across flood plain, Nash Creek, New Brunswick. Data on −80-mesh fraction. (After Hawkes and Bloom, 1956.)

location of abandoned channels. In the levee separating the flood plain from the active channel, the values are relatively low compared with the finer-grained, organic-rich flood-plain sediments beyond the levee (Fig. 15.25). Experience in Borneo has shown anomalous sediments in the active channel incised into flood-plain sediments, the near-surface horizons of which are completely negative (Fitch and Webb, 1958). Here the flood-plain alluvium was derived from the barren headwater areas at times of excessive floods and now stands well above the normal water table. As a result, Cu-bearing ground waters draining from a deposit on the valley slopes feed into the active channel without passing through the upper parts of the alluvium (P. Walker, personal communication).

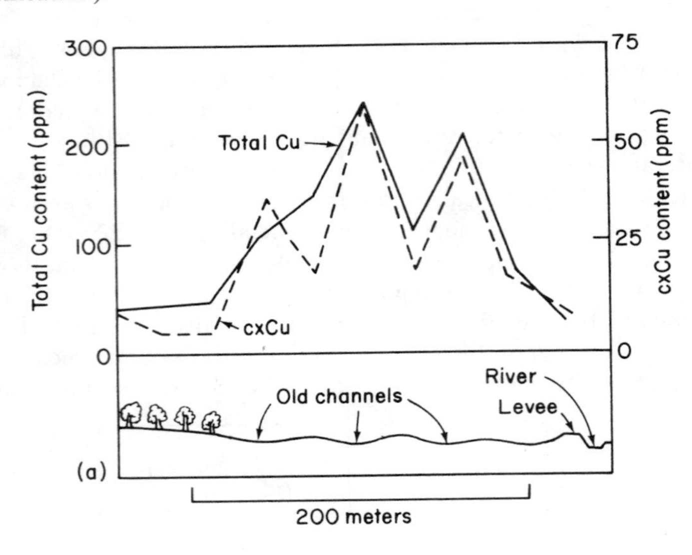

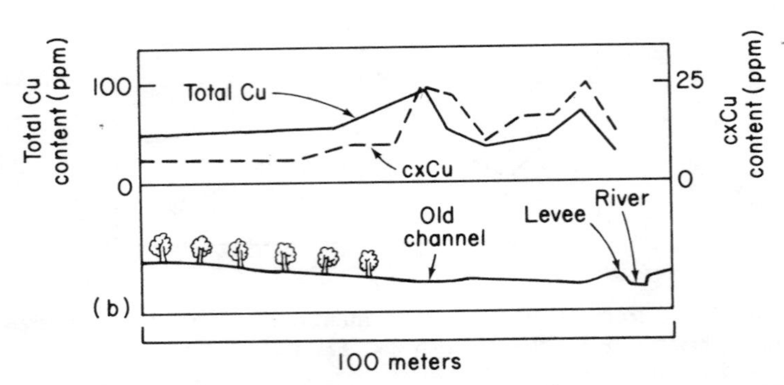

Fig. 15.25. Cold-extractable Cu content of sediments from traverses across flood plain, Baluba River, Zambia. Data on −80-mesh fraction. (After Webb and Tooms, 1959.)

The profile distribution of metals in flood-plain sediments is liable to be affected by soil-forming processes. In Zambia, for example, the Cu content tends to be higher in the organic-rich topsoil and is not so much influenced by seasonal leaching as the sandy subsoil (Govett, 1961).

IV. LAKE SEDIMENTS

Where lakes are abundant, the sediment trapped in these lakes can furnish a convenient sampling medium for the reconnaissance and follow-up stages of exploration (Allan *et al.*, 1973a). Some basic characteristics of lakes are described in Section I of Chapter 9.

Metals may reach the lake sediment by a number of routes:

(i) In clastic particles washed in by streams or eroded from the margins of the lake.
(ii) As elements adsorbed to or incorporated within colloidal organic or inorganic materials entering in streams.
(iii) As dissolved materials entering the lake in streams or in ground water, and later precipitated, adsorbed to suspended particles, or incorporated in living matter, and finally settling to form bottom sediment.

The character of the sediment and the occurrence of elements in it are complex functions of the climate, vegetation, depth, area, inflow and outflow of the lake, and the geology. Thermal stratification of the lake water and oxidation–reduction potentials in the water and sediment have important impacts on the fixation of elements in the lake sediment. Elements derived from weathering and erosion of orebodies may enter a lake in any of the forms described above, but because strong currents are lacking in lakes, the coarser clastic particles usually do not migrate far from their point of entrance. In contrast, dissolved constituents and colloidal particles may move considerable distances before being fixed in the lake sediment.

In most lakes larger than about $0{\cdot}5\ km^2$, distinction may be drawn between near-shore sediments and center-lake sediments. The near-shore materials range in size from clays to boulders, although the clays tend to be winnowed out from these sediments. Near-shore sediment originates by input from streams, erosion of lake shores, or drowning of older surficial materials. The near-shore sediments reflect the geology and characteristics of the shore and streams, with only minimal sorting and reworking by the lake water. In contrast, the center-lake sediments in all but the smallest and shallowest lakes are predominantly silt- and clay-size particles, and may contain a high proportion of precipitated or flocculated organic matter. The characteristics of these sediments depend strongly on the nature of the lake and the chemistry of its water. The details of geology and shoreline processes become blurred and homogenized.

Hydromorphic anomalies overlying ore can develop wherever there is upward movement of metal-bearing ground water into the lake. Although no well-defined example of this phenomenon has been reported in the literature, the authors have observed dispersion patterns of Pb and Zn in lake-bottom sediments in New Brunswick that are almost certainly the result of deposition from metal-rich spring water entering from the bottom of the lake.

A. Mode of Occurrence

Elements occur in lake sediments in the same general form as in stream sediments, but the importance of the various forms is markedly different, especially for center-lake sediments. The following major forms of occurrence may be distinguished:

(i) As substitutions for major constituents in minerals of clastic particles. This fraction is important in near-shore sediment and in the fine rock flour of many inorganic lake sediments, and is present in minor amounts even in organic gels.

(ii) As constituents incorporated within colloidal and fine particles of organic material formed by flocculation in the lake, tissues of plants and animals grown in the lake, or coarser organic material washed into the lake. This fraction is most important in stratified lakes with reducing bottom water, but is present in nearly all lake sediments to at least a minor extent.

(iii) As elements adsorbed to the surfaces of particles, including flocculated organic material, Fe–Mn-oxides, clays, and other small grains. This type of occurrence rarely amounts to an appreciable fraction of the total metal, but may be a helpful guide to the amount of mobile metal in the system.

(iv) As inorganic precipitates formed in the lake. The most widespread and abundant of these are Fe–Mn-oxides, sulfides, and carbonates, but a variety of trace minerals of heavy metals and other rare elements (heavy-metal phosphates, carbonates, hydroxides, and oxides) may be important in anomalous lakes. The Fe–Mn-oxides may be washed in by streams, precipitated in oxidizing lake water as a result of neutralization of acid stream water or oxidation of reducing ground water, or grow as nodules on oxidizing bottoms underlain by reducing sediments. In the latter case, a significant movement of Mn, Fe, and other elements out of the sediment and into the nodule may take place, distinctly changing the composition of the deeper bottom sediment. Copper sulfide is known to be stable in the reducing environment of organic gels, and might also precipitate from reducing waters. Zinc, Pb, Fe, and other metals may occur as sulfides, but it is uncertain whether their solubilities are low enough to compete with complexing by the organic material. Carbonates form a major proportion of the sediment in some lakes in northern U.S.A. and southern Canada, and incorporate trace metals such as Zn in small amounts.

B. Homogeneity and Contrast

Pronounced differences in composition are usually observed between near-shore lake sediment and center-lake sediment (Fig. 15.26). In near-shore sediment, local clastic components are dominant, and differences of a factor

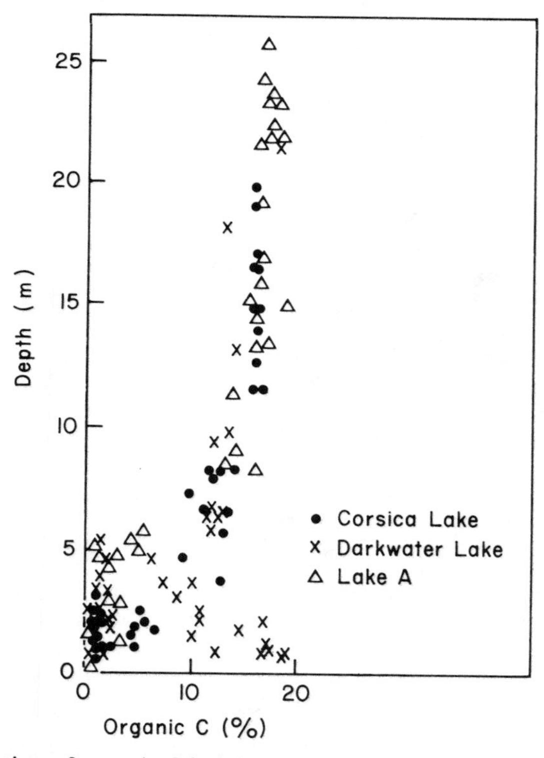

Fig. 15.26. Relation of organic C in lake sediment to water depth for three lakes in Ontario, showing uniform organic content of deep lake sediment. (After Coker and Nichol, 1975.)

of 100 or more may be observed within a lake of a few square kilometers area. For example, in samples of near-shore lake sediment only a few hundred meters apart, Allan *et al.* (1972) observed differences of a factor of 10 or more in the Coppermine district of Northwest Territories, including some differences by a factor of 100. Figure 15.27 illustrates spotty anomalies in near-shore lake sediment apparently attributable to a combination of discharge of ground water on the lake bottom and erosion of the ores. Even more local inhomogeneities are suggested by the observation of frost-boil-like structures on the bottom of some shallow lakes in permafrost areas (Allan *et al.*, 1973b). By analogy with the differences observed within frost boils in soils, considerable local differences are expectable in such lake

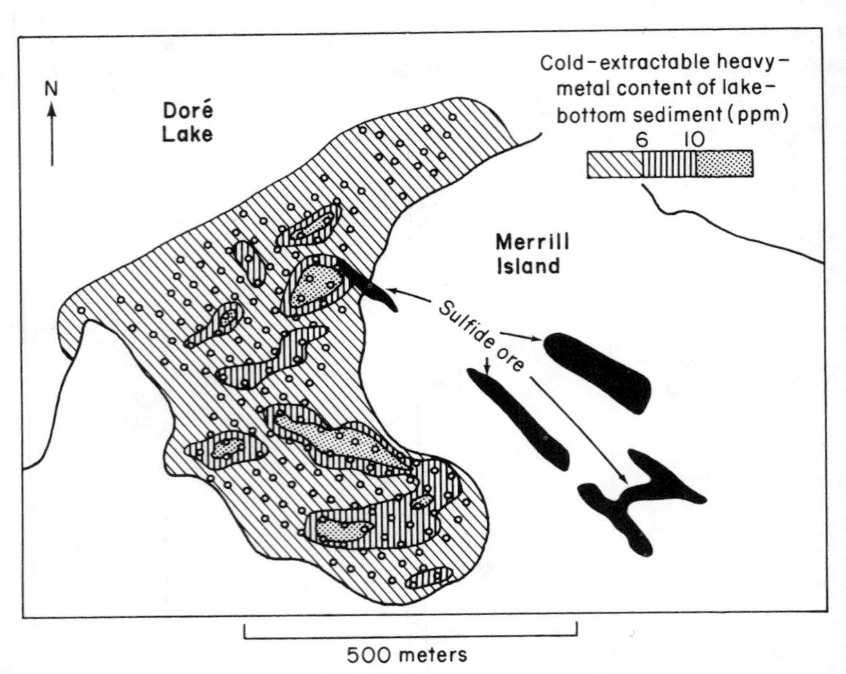

Fig. 15.27. Lake-sediment anomaly, Doré Lake, Chibougamau district, Quebec. Data on −80-mesh fraction. (After Schmidt, 1956.)

sediments. Other factors that promote the inhomogeneities in near-shore sediments include fluctuations in lake level, accumulation of organic material on the down-wind side of lakes, and local differences in pH, Eh, temperature, or dissolved gases, as related to water depth, local sediment sources, or location relative to inlets and outlets.

Center-lake sediments are relatively homogeneous within themselves in lakes of a few square kilometers or less. Hornbrook and Garrett (1976) found sampling variance for pairs of samples collected up to 20 m apart was negligible compared to analytical variance, for which they found coefficients of variation of 5–25%. Jonasson (1976) determined coefficients of variation of about 30% for replicate samples of deep organic gels from two background akes in Ontario. Coker and Nichol (1975) list coefficients of variation of 5–35% for Cu and Zn within groups of sediments collected from the centers of four lakes, except for one much higher value for Zn.

The contrast of anomalies in lake sediments is partly dependent on the relative areas of the drainage basin and the orebody. as discussed under active stream sediment, but because of the importance of hydromorphic dispersion into deep sediment and the local nature of most near-shore sediments, wide variability in contrast is found from place to place. If the orebody is covered by thick till, glacial clay, or other overburden, as is the case in many areas of glacial lakes, little or no anomaly may be observed. Adsorption of metal from ground waters along their path of flow may similarly decrease the amount of metal reaching the lake and incorporated in sediment. Lakes with very little inflow or outflow of water are most dominated by lake characteristics, and hydromorphic anomalies may be obscured by unusual features of the sediment. Strong anomalies may be produced near exposed or shallow ore in areas of moderate relief, where the ore is being actively weathered, as at Agricola Lake, Northwest Territories (Cameron, 1977a). An abundance of weathering fragments of ore in the till of the area will contribute to strong anomalies. Careful study of the local geochemical environment is necessary to evaluate the possible effects of these variables in a specific area. As with many exploration methods, an anomaly in lake sediment is a positive indication for ore, but lack of an anomaly is not necessarily negative. The lack of anomalies near several significant orebodies is evidence for this philosophy (Coker and Nichol, 1975; Nichol *et al.*, 1975).

The relative contrast of different fractions of lake sediments has been very little studied. Deep organic-rich sediments tend to be enriched in metals, although background values are also enhanced in these materials. The contrast is expected to be improved for the more mobile metals.

The organic content of lake sediments exerts a distinct but complex control over the content of mobile metals in lake sediments (Fig. 15.28). In a large set of background lake sediments, the content of Zn increases with organic material up to about 12%, and then is relatively constant or slowly decreasing at higher values. Similar relationships are observed for U (Cameron and Hornbrook, 1976). For Zn, the initial increase is attributed to adsorption and chelation of an increasing proportion of the dissolved Zn as more organic sorbent is available. In this range, adsorption is apparently not able to seriously deplete the Zn content of the water. Above 12% organic matter most of the Zn is contained in organic matter, and only very small amounts are left in the water. An inorganic component with an average of about 0 ppm Zn is indicated by the intercept of the curve. The Zn in the inorganic fraction is increasingly diluted with organic matter, and the fixed amount of mobile Zn is contained in an increasing mass of organic matter, resulting in

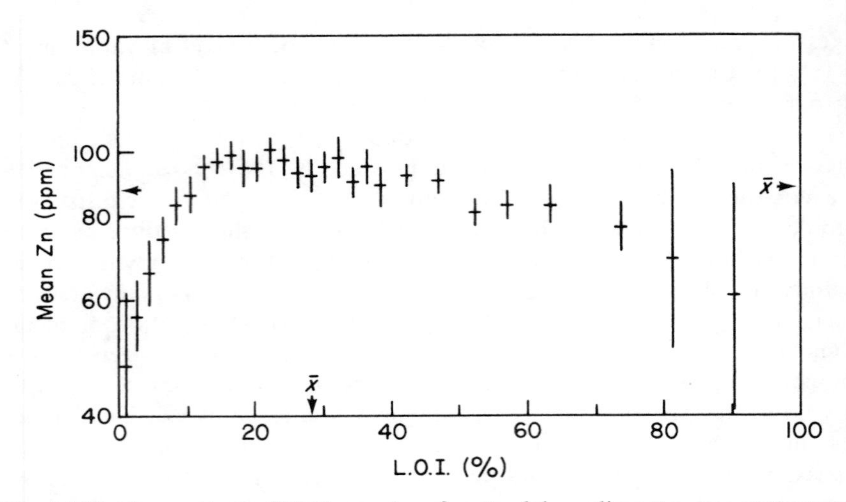

Fig. 15.28. Zinc content of 3844 samples of center-lake sediment vs per cent organi matter, as indicated by loss on ignition (LOI). Bars indicate 95% confidence limit and tick marks the mean for samples in a 2–10% range of LOI. (After Garrett an Hornbrook, 1976)

the negative slope of the upper parts of the curve. The details of the relation ship for high contents of organic matter are not completely understood, bu the relatively constant values of background Zn for levels of 10–60% organi matter mean that variations of organic matter within this range are a second ary concern. Presumably, in an anomalous lake with a higher than norma input of dissolved Zn the organic matter would be able to extract much of th additional Zn and the sediment would be anomalous.

C. Decay Patterns

Anomalies that are primarily clastic in origin decay down-drainage mainly b dilution, in a manner similar to that discussed for stream sediments. Typically such anomalies can be detected for distances of tens to hundreds of meter from the point that the anomalous stream enters the lake, as illustrated i Fig. 15.29. The length of this dispersion train depends on the size of th tributary furnishing the anomalous clastic sediment compared to the size o other drainages and sources of sediment along the shoreline, and on th bottom topography of the lake.

Hydromorphic anomalies in center-lake sediments can be much mor extensive. A hydromorphic anomaly derived from weathering of a base metal massive-sulfide deposit in the Northwest Territories can be detecte in sediments at least 7 km down-drainage from the deposit through a chai

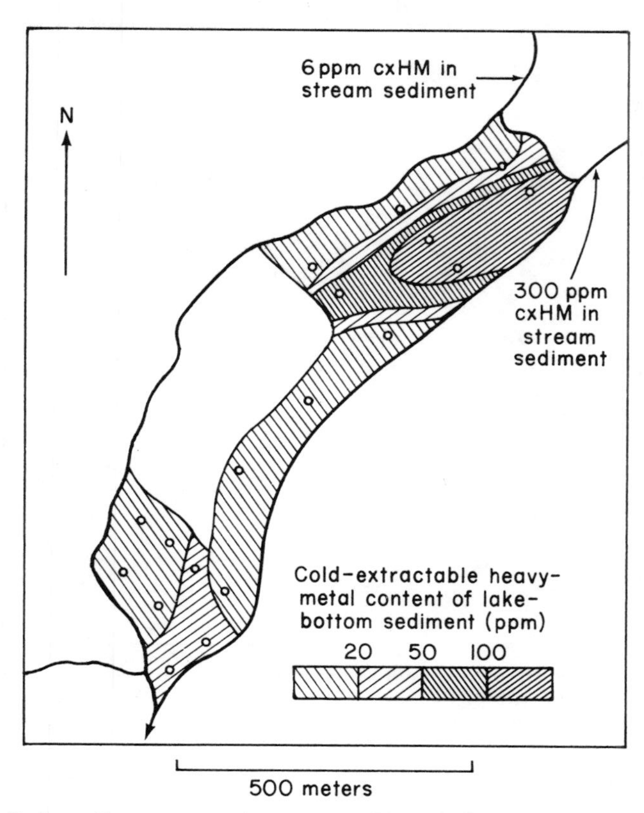

ig. 15.29. Lake-sediment anomaly, Second Portage Lake, New Brunswick. Data n −80-mesh fraction. (After Schmidt, 1956.)

f seven lakes (Fig. 14.11). The values in sediment increase down-drainage the first few kilometers of the anomaly because of precipitation from the riginally acid water. In Newfoundland, Zn was detectable in lake sediments r about 3 km from Zn deposits in limestones near Daniels Harbor, and in ream sediment a further 3 km downstream (Hornbrook *et al.*, 1975).

The U anomaly in lake sediments near the Rabbit Lake U deposit in askatchewan extends downstream about 5 km through several lakes before mptying into a larger lake (Fig. 15.30). The anomaly derived from the eposit is surrounded by a much larger area of weakly anomalous U in lake diment. This large anomaly is believed to reflect a U-rich province containg numerous small deposits and anomalous rocks. Similar extensive anomaes for U are reported by Cameron and Allan (1973) in Northwest Territories nd by Hornbrook and Garrett (1976) in north-eastern Saskatchewan.

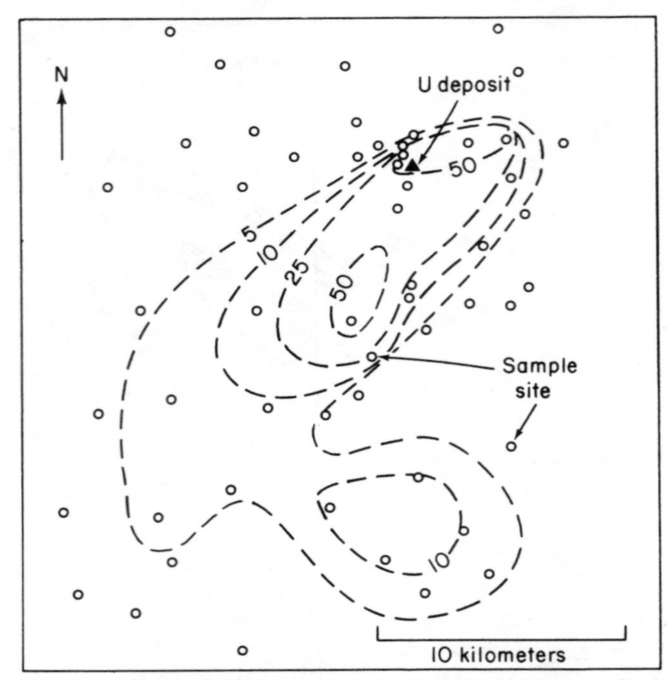

Fig. 15.30. Uranium (ppm) in lake sediments, Rabbit Lake area, Saskatchewar
(After Cameron and Hornbrook, 1976.)

V. DRAINAGE ANOMALIES NOT RELATED TO MINERAL DEPOSITS

Insofar as barren rocks characterized by a high background metal conten may give rise to distinctive anomalous patterns in either the soil or groun water, so may they affect the metal content of stream sediments that ar derived by erosion of the metal-rich soil or that come in contact with th metal-rich water. In prospecting surveys in Uganda, for instance, dolerit dikes containing trace amounts of disseminated sulfides can give rise to C anomalies in the sediments similar in many respects to those associated wit the Cu mineralization (Fig. 15.22). Sediment anomalies in Ni and Cu ar commonly developed from the erosion of ultramafic rocks with their charac teristically high contents of these metals (Fig. 15.31). Metal-rich black shale and extensive pyritic areas with only slightly anomalous base-metal value can give rise to distinct anomalies in drainage sediments.

The relationship between the metal content of drainage sediments and th geology of an area has been utilized in regional geochemical mapping t

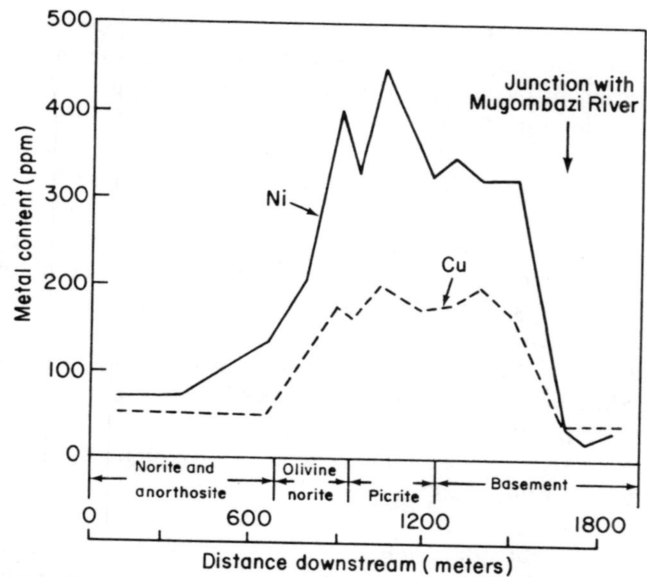

Fig. 15.31. Longitudinal diagram of sediment anomaly resulting from erosion of ultramafic rocks, Kungwe Bay, Tanzania. Data on −80-mesh fraction. (After Coope, 1958.)

detect geochemical provinces (Chapter 4), and has proven useful in identifying and outlining areas of agricultural and health problems induced by trace metals (Webb, 1964; Webb *et al.*, 1968; Webb, 1971). Webb (1970) and other workers have suggested that the method may be useful in reconnaissance geological mapping.

VI. MARINE SEDIMENTS

Recent years have seen an increasing interest in exploration and exploitation of mineral resources from the continental shelf and ocean floor. Major efforts have been directed at oil and gas, but a wide variety of solid minerals have also attracted attention (Siegel, 1974; Tooms, 1972). These include Mn nodules on the deep sea floor, muds and brines rich in Mn, Fe, Cu, Zn, Ag, and other metals, phosphorites, heavy minerals (cassiterite, Au, diamonds, and others), and sand and gravel. The great extent of the deposits is attractive, but the costs of exploration and production of materials covered by tens to thousands of meters of water are a deterrent to development. Inexpensive techniques for locating and outlining the deposits are obviously needed.

Several types of geochemical surveys have been developed to search for marine deposits. Sampling and analysis of bottom sediments appears to have wide application in recognizing regional favorability and gradients toward some types of deposits. Similarly, sampling and analysis of waters and suspended sediments can indicate favorability for some deposits. Special instruments for *in situ* analysis of the chemistry of bottom sediment have potential in exploration and production.

Sediments, waters, and suspended particles have proven useful in detection of metalliferous sediments of the type found in the Red Sea and elsewhere near mid-ocean ridges. In the Red Sea, metal-rich muds form from warm to hot brines that emerge on the sea floor in certain areas of deep water (Degens and Ross, 1969). The brines are dense and remain trapped beneath normal sea water. Slow mixing and diffusion at the interface between brines and overlying water results in oxidation and dilution of the brine, and transfer of metals to overlying water. Sulfides tend to precipitate at an early stage from this type of brine, and are generally localized within the sea floor or very close to the orifice of the submarine spring. Iron oxides are precipitated over a larger area, but still mainly in the deeps in the Red Sea area. Manganese oxides, which require more oxidizing conditions for precipitation, spread more widely, along with the trace metals, as illustrated by Fig. 15.32 (Bignell *et al.*, 1976; Cronan, 1976). Anomalous values of Mn and base metals extend many kilometers from the source of the brines.

The composition of bottom sediments responds not only to the currently active springs but also to past areas of hot spring activity on the sea floor. The haloes of Mn enrichment detected around the Tynagh Pb–Zn deposit in Ireland (Chapter 5; Russell, 1974) and the enrichments of As, Au, and other metals in Fe formations associated with some volcanogenic massive-sulfide deposits are fossil examples of this same phenomenon. Extensive areas of low-grade metalliferous sediment appear to occur in relatively recent sediments near mid-ocean ridges (Cronan, 1976; Field *et al.*, 1976; Bonatti *et al.*, 1976) and probably contain local higher grade zones that can be discovered by sampling of bottom sediments. Similarly, analysis of beach and offshore sediments has been used to detect favorable areas for placer deposits.

Chemical analysis of sea-floor sediments by instruments towed near the sea floor would allow scanning of large area at relatively low expense. Phosphorites tend to have high values of U that can be detected by measurement of gamma radiation near the sea floor (Summerhayes *et al.*, 1970; Noakes *et al.*, 1974). Some heavy mineral placers and Mn nodules have high concentrations of U or Th that can be detected by similar means (Kunzendorf, 1976). Analysis of sediments by neutron activation using a ^{252}Cf source of neutrons (Noakes and Harding, 1971) holds potential for geochemical surveys of a variety of elements in bottom sediments.

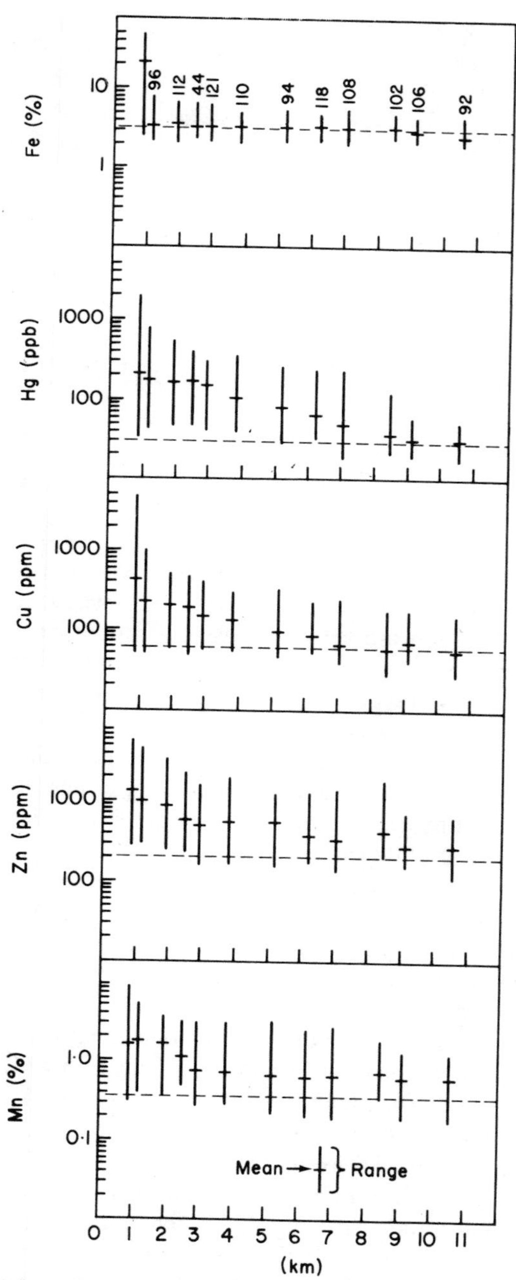

Fig. 15.32. Anomalous Cu, Zn, Hg, Fe, and Mn in sediment of the Atlantis Deep of the Red Sea, as a function of distance from major occurrences of metalliferous sediment. Dashed line indicates background. (After Bignell *et al.*, 1976.)

Chapter 16

Geochemical Drainage Surveys

The two outstanding areas of applicability of geochemical drainage surveys are (i) primary reconnaissance, as a method of locating geochemical provinces, mineralized districts, and individual deposits of moderate to large size, and (ii) appraisal of the location, extent, and contained metals at prospects, geophysical anomalies and favorable geological features. Drainage surveys are now the most widely used geochemical reconnaissance technique.

Very large areas can be scanned for their mineral potentialities by drainage surveys because of the great distances over which the weathering products of orebodies are carried by ground and surface waters. At least one representative of the group of mobile (S, Mo, U) and semi-mobile (Zn, Cu, Ni, Co) elements occurs characteristically either as a principal component or a minor constituent of many types of ore. If to these we add the ore metals that occur in the heavy-mineral fraction of stream sediments (Au, Sn, W), then almost every type of ore should give some kind of a recognizable drainage anomaly.

Appraisal of mining prospects or of anomalous areas found by other prospecting methods has become an important application of drainage studies. Under favorable conditions, field tests of sediments from seepage areas and small drainage channels can provide an on-the-spot indication of the mineral possibilities of the area under examination.

This chapter is devoted primarily to a review of the techniques of planning, conducting, and interpreting the data of geochemical drainage surveys that have been found effective under a wide variety of conditions throughout the world. These discussions represent the conclusions of a considerable number of experimental and operational surveys. Many or most of the conclusions will be pertinent in any new area that is being considered for a geochemical drainage survey. However, it is not always safe to assume that the applicability of techniques will necessarily carry over from one area to another. For this reason an experimental orientation survey in an area known to be mineralized can usually be recommended as a prelude to routine exploration work in a geologically similar unknown area.

I. ORIENTATION SURVEYS

An orientation survey preparatory to undertaking a geochemical drainage survey is directed toward choice of the optimum sample medium (water, stream sediment, seepage sediment) and procedures for collecting and analyzing samples. Evaluation of different media and procedures requires collection of both anomalous and background samples from localities representative of

Table 16.1
Check list of factors to be optimized and evaluated by an orientation survey preparatory to drainage sampling

	Sediment	Water
(i)	Best indicator elements, including both major and minor constituents of ore	
(ii)	Optimum material (sediment from seepages, stream channels, flood plain, center-lake, near-shore lake)	Optimum material (ground water vs surface water)
(iii)	Optimum fraction (size, heavy minerals, organic fraction)	For ground water, the relation to recharge areas, difference between aquifers, controls on water flow, and availability of points where ground water can be sampled. For lake waters, possible variations with depth and type of lake
(iv)	Most effective extractant or method of anomaly enhancement	
(v)	Magnitude of contrast of anomaly at source	
(vi)	Length of downstream decay pattern; controls on decay pattern	
	Metal content of bank material	pH, Eh, precipitants, adsorbents
(vii)	Background values of indicator elements, range, relation to rock types	
	Correlation with Fe–Mn-oxides, organic matter	Relation to total dissolved solids, and major elements
(viii)	Analytical methods (detection limits, precision, accuracy, form of element)	
(ix)	—	Seasonal or temporal variations
(x)	Cost of sampling and analytical procedures, and elapsed time for reporting results	

both the deposit being sought and the range of background conditions within the survey area. The known deposits that provide the source of anomalous metal in the orientation surveys need not be of economic grade. In fact, non-economic deposits may be preferable because of the usual absence of serious contamination resulting from mining operations.

The major factors to be considered are summarized in Table 16.1.

A. Water

Opportunities may exist in known mineralized districts for sampling ground waters in mine workings, bore holes, and wells, and for sampling surface waters in springs, streams, and lakes. Providing contamination can be avoided, none of these opportunities should be neglected. In general terms, samples should be taken at the following locations: (i) up-drainage from a known, preferably undisturbed metal source, (ii) at or near the deposit, (iii) at every point down-drainage where a tributary stream or a change of environment may modify the composition of waters draining the known deposit, and (iv) at background locations in barren areas representing all variations in bedrock type and surface environment in the survey area. Down-drainage from the deposit, intermediate samples should be taken where the distance between critical points exceeds, say, 200 m. If the deposit has been opened up by underground workings, it is necessary to bear in mind that the resulting accelerated oxidation may lead to enhancement of anomaly contrast relative to what might be expected had the deposit been undisturbed.

Analysis should include the determination of pH, total salinity, and appropriate major constituents in addition to the assemblage of ore metals. The processing of the sample should be planned with due regard for the fact that the content of ionic constituents may change during storage. Tests suitable for analysis directly in the field are available for some metals; usually, however, it is preferable to concentrate the ionic metal at the sample site by solvent extraction, coprecipitation, or treatment with ion-exchange resins, followed later by comprehensive analysis in the laboratory. The water, after the deionizing treatment, should be retained for determination of the non-ionic fraction. Total analysis of an acidified sample of the untreated water can be used as a check. All determinations should be carried out with maximum precision and sensitivity. This is particularly important in comparing ratios of the different constituents at the extremely low concentration levels commonly encountered in natural waters. In view of the concentration level and the number of different constituents that may need to be determined, water samples for an orientation survey will usually need to be quite large, say 1–5 l. Where filtration is necessary, the fine suspended matter may be removed by using micropore filters and a simple hand-operated vacuum pump (Oldershaw, 1969) or a small tank of compressed gas.

The downstream decay pattern of metals may be the effect of either dilution or precipitation. As a first approximation, the dilution effect may be estimated by analyzing and determining the discharge rate of all sources of water from unmineralized areas entering the anomalous drainage pattern. For a more nearly complete study it is necessary to take into account ground-water seepage via the stream bed by carrying out similar determinations at intervals along the main stream. The possibility of precipitation may be investigated by sampling stream sediment, swamp soil, or other solid material in contact with the waters. Particular attention should be given to possible relationships between the metal content of the water and the presence of organic matter, ferruginous scums, and other precipitates that may have scavenged ionic metals from the water.

The effect of seasonal variations commonly related to rainfall or melting ice and snow should be investigated by periodic resampling of critical sites at different seasons of the year. Particularly in lakes, variations in metal content and anomaly contrast due to seasonal changes in organic activity and thermal overturn should be monitored. Analysis of replicate samples taken at the same sites may provide a correction factor to apply to operational data collected during different weather conditions.

B. Drainage Sediment

In orientation studies of metal in drainage sediments, experimental samples should be collected from both the flood plain and the active channel at intervals no more than 50 m for the first 300 m or so downstream from sources of anomalous metal (Fig. 16.1). The interval may be progressively expanded with increasing distance from the metal source. In this phase of the program, samples should be taken much closer together than would be necessary for routine exploration; on the basis of these results, the optimum sample spacing may then be selected. Sediment from all tributary streams and seepage areas must be sampled. At each location, the various types of sediment represented, e.g. sand, silt, clay, and ooze, should be sampled separately. Sediment samples from the active stream channel should be collected well away from the banks to avoid dilution from collapsed bank colluvium of local origin. At each sediment sampling point, bank samples should be taken on either side of the stream and a record kept of the alluvial or colluvial origin of the bank material at that point. The metal content of flood-plain soils should be investigated by transverse soil-sample traverses.

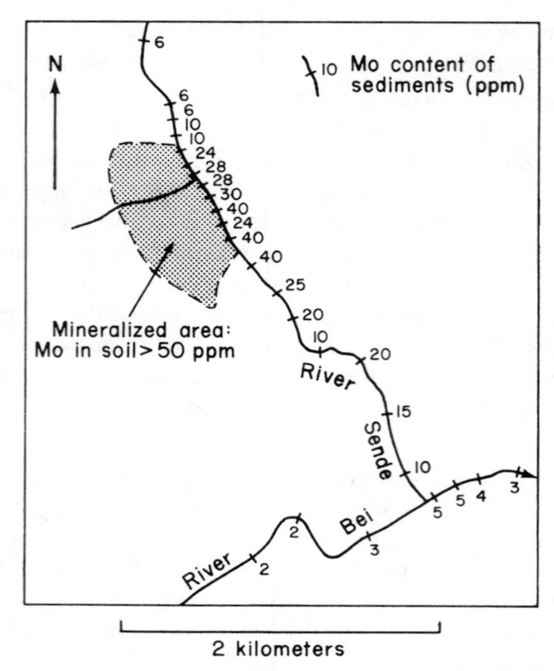

Fig. 16.1. Reconnaissance stream sediment survey of River Sende, Sierra Leone. Data on −80-mesh fraction. (After Mather, 1959.)

The variation of metal content with depth in all kinds of sediments, whether from seepages, active channels, or flood plains, should be checked by collecting samples from a number of experimental vertical profiles.

When dealing with metals that may occur in readily extractable form, both cxMe and the total metal content should be determined and the ratio computed for purpose of interpretation. The optimum sieve size should be determined by analyzing different size fractions of representative material from both anomalous and background areas and observing which fraction gives the strongest contrast of anomly over background, as well as the most homogeneous and the most persistent pattern. Mineralogical and chemical examinations should also be carried out on the heavy minerals, particularly when dealing with metals such as Sn and W.

C. Contamination

Surface drainage samples, particularly water, are far more susceptible to contamination than soil samples, and careful attention must be given to this problem during the orientation survey.

Contamination may come from trash, metal-rich drainage from factories and mechanized farms, metalliferous insecticides and algicides, roads and railway beds graded with mine waste, and condensates from smog and industrial fumes. The most common industrial and domestic contaminants are Zn and Cu. In wells where pumping machinery has been installed, it may be impossible to obtain water samples that are not severely contaminated with these metals. Spring water is almost always free of contamination; the only exceptions reported are very rare cases of springs that drain old mine workings.

Mining and smelting activity constitutes one of the most serious sources of contamination of surface water. Extremely high metal concentrations normally occur in water draining old workings and ore piles where the primary sulfide minerals have been artificially exposed to the air and the oxidation process thereby greatly accelerated. It has been estimated that the metal content of such waters may be 10 or even 100 times higher than if the deposits had not been opened up.

Contaminated stream water often contaminates the sediments with which it is in contact. Except where the metal content of the water is extremely high, such as in areas of past mining activity, the contamination passed on to the sediment is often not nearly as severe as in the water. Sediment surveys have been conducted in well-populated areas where satisfactory water sampling would have been impossible because of contamination.

II. CHOICE OF MATERIAL TO BE SAMPLED

The choice of material to be sampled from streams, whether water, seepage soils, fine-grained sediments, or heavy-mineral concentrates from the sediments, must naturally be made on the basis of which medium gives the strongest and most reliable pattern that at the same time can be readily detected by the techniques at hand. The principal factors in this connection are (i) the mobility of the constituents of the ore being sought, (ii) the influence of local conditions on the dispersion pattern, (iii) opportunities for sampling, and (iv) availability of a suitable analytical method. It is often possible to carry out a reliable drainage survey on the basis of a single sample medium; for complete and reliable coverage it may be found profitable under some conditions to collect samples of more than one kind of material.

A. Water

Water sampling is naturally most effective where the elements sought are the relatively mobile ones that would not normally be carried with detrital material. The elements that have been used with greatest success in hydrogeochemical prospecting are the mobile Mo, U, and SO_4. Alkaline, calcareous, ferruginous, and organic-rich environments tend to reduce the general effectiveness of water sampling. A humid climate coupled with moderate to strong relief and a relatively low pH provide the most favorable conditions.

The choice between stream water, lake water, and ground water is dependent on the availability of these media in the area, and on the interactions of geology with the flow patterns of ground water. In most humid regions, streams are more easily sampled than ground water and represent a well-defined drainage basin. Lakes are preferred in glaciated shield areas because of the abundance and easy access by airplane or helicopter, especially in forested terrain. In arid regions, where surface water is scarce, ground water is the obvious alternative. Ground water may also be preferred if the desired ore is restricted to a certain stratigraphic unit, but the ground water either does not rise to the surface in significant quantities, or is too diluted at the surface to permit anomalies to be detected readily.

The principal advantages of water over other kinds of geochemical samples are its physical homogeneity and its more direct dispersion path from some buried mineral deposits. Difficulties are the time variations and other factors affecting the composition of the water, combined with practical problems in transporting, analyzing, and storing water samples. In scale, water is most applicable in detailed surveys and in the more detailed stages of reconnaissance work, except for the most mobile elements.

B. Seepage Soils

The soils of swamps, seepages, and spring areas are commonly enriched in the same elements as the ground water that comes to the surface in these areas. By virtue of the hydromorphic origin of the anomalies, seepage-soil surveys are limited to the semi-mobile metals. By far the greatest amount of work has been concerned with prospecting for Cu and Zn. Sampling soils from spring and seepage areas is usually preferable to sampling the spring water at the same localities because (i) soils can be sampled when water is difficult to collect or is even completely absent in the seepage areas, (ii) the composition of the soil in the seepage anomaly is not as strongly affected by changes in weather as the composition of the water, (iii) in some areas seepage soils show a stronger contrast of anomaly over background than water samples, (iv) in most cases trace analysis of soil presents fewer technical problems than does water, and (v) soil samples may be easily shipped or stored for future reference. Under rare conditions, spring water will be preferable to seepage soils, particularly where the element sought does not tend to precipitate from the water or where the composition of the soil has been shown to be excessively unreliable or lacking in homogeneity.

The anomalous patterns in seepage soils will be somewhat different depending on whether they are analyzed for total or for readily extractable metals (cxMe). Experience in the dambo areas of Zambia, for example, favors cxCu over total Cu. Here the principal factor in the choice is the substantially greater contrast of anomaly over background for cxCu as compared with total Cu. In other projects, the principal points in favor of cxMe may be greater speed and economy of the determination and the possibility of making the analyses either at the base camp or directly at the sample site. Patterns defined by the distribution of total metal, however, tend to be more homogeneous. They also are not as susceptible to changes with the seasons as the cxMe content. A determination of total metal in addition to cxMe may be desirable as a confirmatory check. Furthermore, the cxMe/Me ratio may be very helpful in determining the origin of an anomalous pattern. With certain exceptions, a high cxMe/Me ratio suggests a hydromorphic anomaly, whereas a low ratio suggests a residual pattern.

C. Stream Sediments

The semi-mobile and immobile elements are the ones best suited for stream-sediment surveying. Under some conditions, however, sediment sampling may also be used for elements that are normally highly mobile, such as Mo. Anomalous stream sediments are not necessarily accompanied by an anomaly in the water with which they are in contact (Fig. 16.2). Even where there is a corresponding hydrogeochemical anomaly, stream sediment sampling may be preferable because of the scarcity of water in the drainage channels, excessive seasonal variations in the composition of anomalous water, and the greater ease of collecting, analyzing, shipping, and storing sediment samples. However, where anomalies occur in both sediment and water, both media should be sampled concurrently if maximum information from each sample site is desired, particularly if the partition of the metal between water and sediment is variable. Data on the dispersion of semi-mobile and immobile elements in the sediment anomalies will also complement the data on the mobile elements from the water anomalies.

In some areas, equally strong anomalies occur in both the active sediment and the flanking flood plain. If the banks are made up either wholly or in part of colluvium, then the choice is limited to the active sediment. Flood-plain sediments have the advantage of containing a higher proportion of fine material and are usually easier to collect. Generally speaking, however, they are liable to give more erratic results than the active sediment, particularly in the case of semi-mobile metals. In spite of its shortcomings, flood-plain

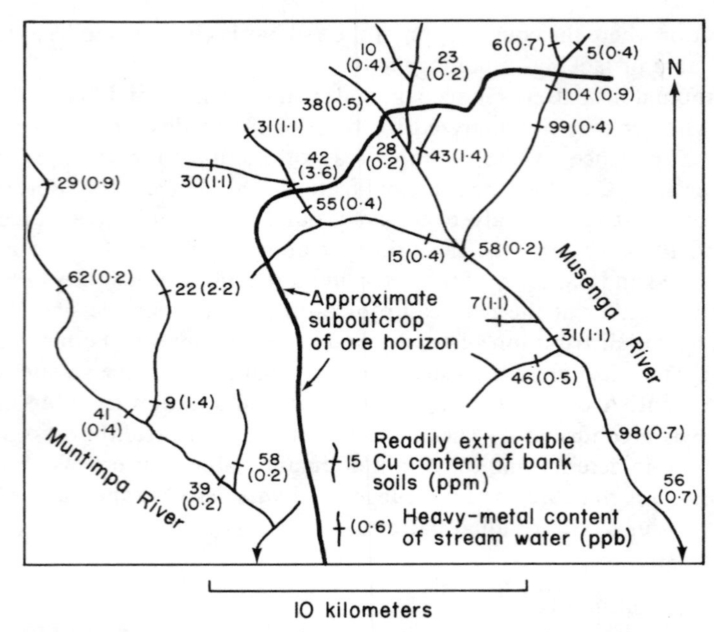

Fig. 16.2. Reconnaissance geochemical survey of the Musenga and Muntimpa drainage systems, Zambia. (After Govett, 1958.)

sampling has been used extensively in Canada with satisfactory results. In seasonal swamps there may not always be a well-defined surface channel, and in this case samples of the soil should be collected along the preferred drainage lines within the swamp.

Immobile elements usually must be determined by total-metal analysis. For mobile and semi-mobile elements, the choice between total and cold-extractable metal as the most useful ore guide is much the same for stream sediments as for seepage soils. With sediments, a high cxMe/Me ratio suggests chemical precipitation either directly in the stream bed or in a seepage area undergoing erosion further upstream; a low cxMe/Me ratio indicates that the anomaly may be derived by the mechanical erosion and dispersion of metal-rich residual soil.

D. Heavy Minerals

Heavy-mineral surveys are mostly restricted to the immobile elements, Au, Sn, W, Hg, and Ta–Nb, although in exceptional circumstances some of the more mobile ore elements may also occur in the heavy fraction. In some areas, anomalies are completely lacking in normal stream sediment but are easily detected in heavy mineral concentrate panned from the deeper layers of sediment (Sainsbury and Reed, 1973).

Heavy-mineral patterns often have the advantage of giving a longer dispersion train than that obtained by chemical analysis of sediment samples. Sometimes, however, the heavy-mineral grains are too small to be readily separated from the light fraction.

The choice between chemical analysis and mineralogical examination of heavy-mineral concentrates depends on the ability to identify the diagnostic constituent and on the technical facilities available. In some ways, mineral studies are simpler, but they are subject to errors both in identifying the mineral species and in estimating the relative abundance of the different minerals. Chemical analysis, particularly of fine-grained fractions that cannot be easily examined optically, tends to reduce such errors. Chemical analysis also may bring out patterns that cannot be appraised by optical examination, such as the Cu content of magnetite from magnetite-bearing Cu deposits.

E. Organic Sediment

Organic sediment is a preferred sampling medium in poorly drained areas with sparse inorganic sediment but abundant organic matter along the stream bottom. It has been extensively used in Sweden for detailed reconnaissance surveys (Larsson, 1976). The technique is most applicable for the mobile and semi-mobile Cu, Zn, Ni, Co, U, and Mo, which are adsorbed and incorporated into the partly decayed organic matter.

F. Lake Sediment and Water

These media are chosen in regions of abundant lakes, especially where access to streams is difficult. In general, center-lake sediments generally provide anomalies at least as extensive as lake waters, even for U, and are easier to analyze, so that center-lake sediments are currently preferred over waters for wide-spaced reconnaissance surveys (Cameron and Hornbrook, 1976). However, for follow-up and detailed surveys, lake waters can be very useful because anomalies are stronger and more easily detected, and are less subject to the complications in sediments caused by adsorption capacity and sedimentation rate. Near-shore lake sediment also has potential applications for follow-up. If the sampling and analytical problems of lake waters can be solved, the simplicity and speed of sampling lake water can be an advantage. Sampling of both lake water and lake sediment should be considered because of the complementary nature of the media and the high cost of access as compared to collection and analysis of a second sample.

III. SAMPLE LAYOUT

The main objectives of a sampling plan are to maximize the probability of detecting any economic orebodies that occur in the surveyed area and to minimize the cost and time for the survey. Approaches to balancing these two conflicting objectives are discussed in Chapter 20.

The main bases for deciding on sample spacing are the contrast and down-drainage decay patterns as determined in an orientation survey. As a rule of thumb, reconnaissance surveys should be laid out so that an important orebody will be indicated by two distinctly anomalous samples. The spacing, defined in terms of drainage areas or stream lengths, can be determined directly from the orientation data, by calculations using eqn 15.3, from data in the literature on similar surveys, or by combinations of these.

An alternative method of selecting a sample spacing is to use statistically determined variability to estimate the best sample spacing. A simple form of this approach was used by DeGeoffroy *et al.* (1968) for a spring water survey in Wisconsin. They first computed the mean Zn content of several spring samples in an anomalous area of 0·04 km^2. They then progressively increased the size of the area for which the mean was calculated, and plotted these

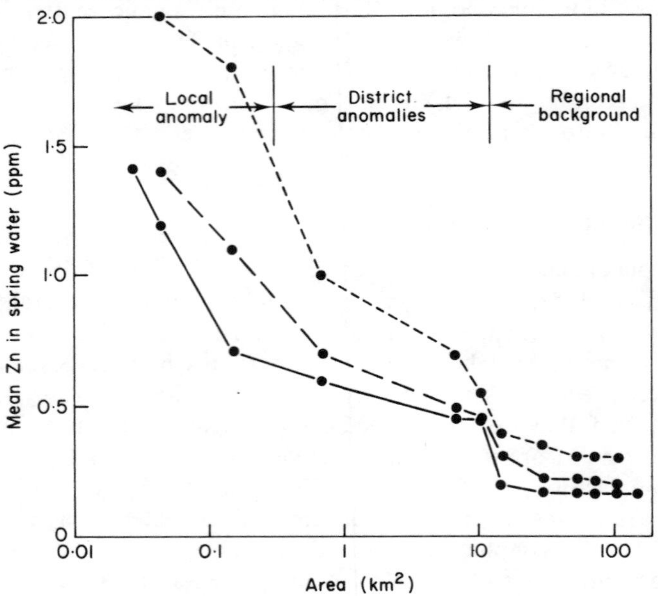

Fig. 16.3. Relation of mean Zn content of spring waters to area averaged for three areas centered on known anomalies in south-west Wisconsin. (After DeGeoffroy *et al.*, 1968.)

means against the area (Fig. 16.3). The plot shows breaks at about 0·65 km^2 and at about 10 km^2. The smaller dimension is suggested as the average extent of local anomalies around orebodies, and the larger the extent of weaker regional anomalies around sub-districts. For reconnaissance sampling, the spacing would be adjusted so that one or two samples were collected per 10 km^2. A similar approach is the use of "variograms" plotting variability of samples against increasing size of area considered (Matheron, 1963; David and Dagbert, 1975).

A. Ground-water Patterns

The layout of sampling points in ground-water surveys will depend not only on the dimensions of the patterns that are being sought but also on the distribution of available wells, springs, and seepage areas. Where these are densely distributed, surveys can be run by systematic cross-country traversing to collect samples at all wells, springs, and seepage areas that can be seen from the traverse lines. In other areas, it may be preferable to traverse along certain land forms that are especially likely to represent a line of effluence of ground water, as for example at the break in slope where the terrain begins to flatten or along an impermeable layer that forces water to the surface, or along the edges of swamps, lakes, and the flood plains of streams. In reconnaissance ground-water surveys, in areas with closely spaced wells, collection from wells closest to grid points laid out at the desired sample spacing is satisfactory.

Figure 16.4 shows the results of a survey based in part on the sampling of a series of springs located along topographic lines. In this survey, mineralized areas were found to lie within a threshold contour of 10 ppb over a background of 0·2 ppb U. Anomalous values up to 250 ppb led to the discovery of ore deposits in areas not otherwise thought to be favorable.

Springs may also be sought along geological features such as faults and the contacts between permeable and impermeable rocks. The sample interval along traverses of this kind must necessarily depend on the natural spacing of the spring and seepage areas, as well as on the probable size, shape, and geological control of the ore that is being sought.

Where the seepage areas are very large, for example the seasonal swamps of Zambia, it is necessary to sample the internal drainage at the main points of entry of ground water and along the lines of preferential flow within the seepage area as a whole. Where the internal drainage is not evident, it may be necessary to resort to systematic soil sampling within or around the periphery of the seepage area, depending on the most likely points of inflow.

In sampling well water, attention should be directed to the aquifer tapped by the well. Depending on the geological character of the survey area and

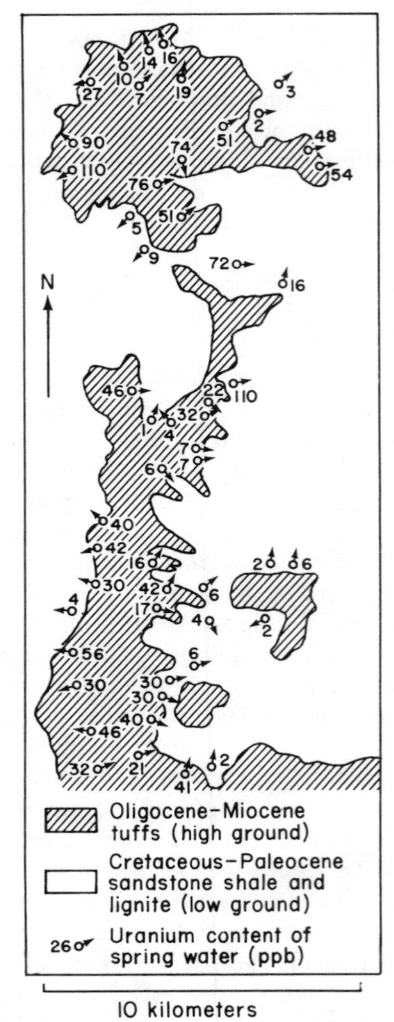

Fig. 16.4. Distribution of U in spring water, Slim Buttes, South Dakota. (After Denson, 1956.)

the ore being sought, sampling of wells in alluvium along drainages or deep wells penetrating certain stratigraphic units may be desirable.

B. Drainage Channel Patterns

The sample layout for drainage surveys depends on the maximum area of drainage basin which can be expected to show an anomaly, as defined by orientation surveys, supplemented by estimates of downstream dilution from eqn 15.3 and published data. These areal considerations must be modified, especially for water samples, by the effects of drainage length, because of adsorption and precipitation of dissolved elements and washing out of fines from sediment samples. For deposits of moderate size, the decay distance, determined in orientation surveys, is commonly found to lie between 300 m and 3000 m, and the limiting catchment area between 10 km^2 and 50 km^2 For large deposits, such as porphyry-copper deposits, anomalies may be detectable tens of kilometers downstream in drainage basins of several hundred square kilometers. For complete coverage of areas containing large rivers, sampling of the larger tributaries selected according to the above concepts must be supplemented by sampling of small tributaries flowing directly into the river.

In regional stream-sediment surveys in which the objective is to detect metallogenic and geochemical provinces with dimensions of tens to hundreds of kilometers, a spacing of one sample per 25 km^2 has proven satisfactory, although the spacing will clearly depend on the amount of previous knowledge of the region, access to sample sites, and the purposes of the surveys. Regional surveys described in the literature have used drainages smaller than the sample spacing (that is, drainages of 10 km^2 have been sampled in surveys with an average spacing of one sample per 25 km^2). Limited data suggest that alluvium along the larger drainages tends to dilute or distort the sediment composition so that samples are no more representative of their drainages than smaller tributaries. Regional samples are laid out as evenly as possible taking into consideration the problems of access and the distribution of the selected size of drainage.

In reconnaissance lake-water and sediment sampling, a single sample of water or sediment from the center of lakes up to about 3 km^2 is usually considered to be sufficient (Coker and Nichol, 1975). However, data presented by Tenhola (1976) indicate that such a spacing may not be adequate to detect many small to medium-sized mineralized zones. Generalizations on lake sediments and waters await further research and case histories.

For reconnaissance lake surveys, lakes are chosen to be larger than about 0·5 km^2, more than 3 m deep, and preferably at the focus of several drainages. Stagnant lakes are avoided. In detailed lake surveys, these constraints are relaxed.

Complete coverage is not always possible or even desirable in areas where access is difficult or where the geology is not particularly favorable. Inasmuch as most of the cost of a geochemical drainage survey is the cost of personnel time and expenses traveling to the sample site, optimizing the ratio of cost to completeness of coverage may justify confining traverses to roads, trails, navigable streams, or coastlines (Fig. 16.5).

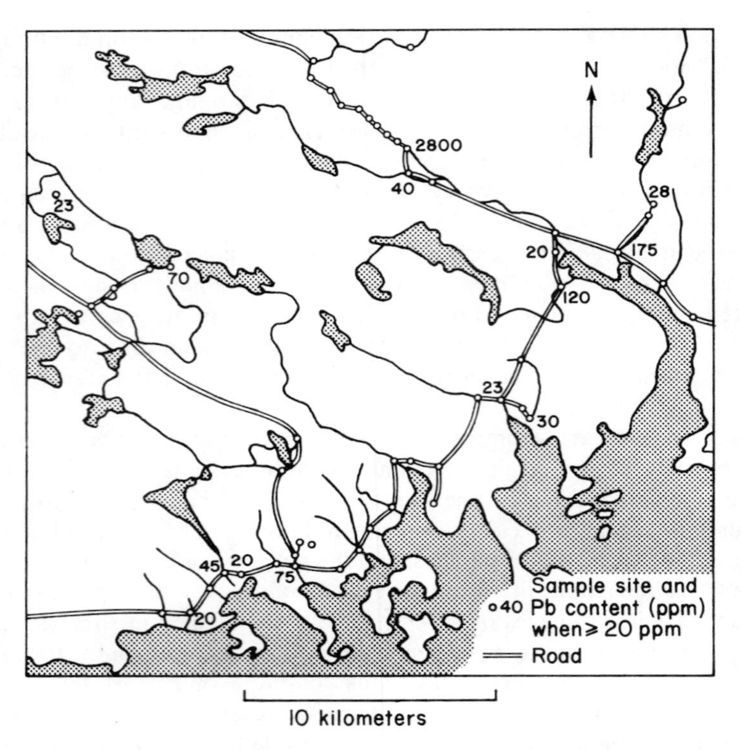

Fig. 16.5. Example of stream sediment survey based on road traverses, Nova Scotia. Data on −80-mesh fraction. (After Holman, 1959.)

IV. COLLECTION AND PROCESSING OF SAMPLES

The falsification of field observations by dishonest samplers presents a very special kind of problem. The temptation is always there for the sampler to fill all his containers with material from a single convenient spot, and then relax for the rest of the day. For large programs where many samplers of unknown qualifications must be hired, precautions are essential to ensure valid samples. The most widely used system for making sure that samples were actually taken at the points marked on the sampler's map is to have the sampler leave a mark of some kind either at every site, or at a certain proportion of sites. Spray paint on trees or rocks, or plastic ribbon tied to bushes and marked with the sample number are useful for this purpose. These may then be spot checked by a supervisor at a later date.

A. Water

Polyethylene or other types of plastic bottles are usually employed for water sampling because of their durability, light weight and the decreased likelihood of contamination by the bottle, as compared to glass bottles. Plastic bottle liners have been used to advantage in samples that were evaporated to dryness for analysis (Tinney, 1977).

Samples of stream water should be collected from the active flow by first rinsing the container several times in the water. The container should have been thoroughly cleaned by washing and rinsing in acid and distilled water before coming to the field. Most new bottles furnished by manufacturers are clean when received, but each new batch should be tested to confirm this, because the consequences of contaminated bottles are serious (Cameron and Durham, 1975). Bottles used for storing chemical reagents should not be used, because despite vigorous cleaning, contaminants may remain on the walls.

If the waters contain significant amounts of particulate matter, as indicated by the orientation survey, filtering of the sample may be necessary. Micropore filters with pores of 0·1–1 μm are commonly used because these are the finest that can be used in practical field surveys; pressure filtration using a hand pump, a small tank of compressed N_2 or Ar, or gas in pressure cans is usually necessary to speed the filtration process. Filtering through fast filter paper removes coarse suspended particles but allows colloidal material to pass, and is of questionable value. Care must be taken to avoid contamination from the filter or holder during the filtration process.

Trace elements may be lost to the wall of the container in several ways. In very dilute samples, adsorption of ions to the walls can be appreciable (Maienthal and Becker, 1976), but studies by Macdonald (1969), Wenrich-Verbeek (1977), and Applin and Langmuir (1978) indicate that non-acidified waters with U contents above 1 ppb do not lose U when stored in plastic bottles. Precipitation of Fe-oxides, organic material, or other substances as a result of oxidation, loss of CO_2, growth of algae or bacteria, or other changes can remove metal from solution. The CO_2 produced by organic activity can also be responsible for complicating reactions. Acidification of the sample with metal-free acid to about pH 2 at the sample site will prevent the precipitation of Fe-oxide and at the same time inhibit the growth of microorganisms. Samples should be essentially free of suspended sediment when acidified, because the acid will tend to dissolve the particles. The addition of a few milliliters of chloroform to a water sample at the time of collection will inhibit organic activity.

Collection of surface water from a lake can be accomplished by methods similar to those used for stream-water samples, but collection of deep samples from a lake (or open well) may require special equipment. A device easily improvised is an empty polyethylene bottle fitted with a friction stopper. The bottle is weighted and lowered to the desired depth on a length of line, and the sample collected by pulling out the stopper which is attached to a separate line. Commercial water samplers which can be opened and closed at depth are also available. Equipment for semi-automatic sampling of lake water from a helicopter has been described by Cameron and Durham (1975).

Collection of water from a pumped well is complicated by the storage of water in the system and by contamination, precipitation, adsorption, or other changes induced by the pipes, pump, and tanks. Water should be collected from as close to the well as possible, and before it passes through a water softener or other treating apparatus. Ideally, the water should be run until the temperature or conductivity stabilizes, or until the tank has been refilled by the pump several times. In practice, this process takes too long (five minutes to several hours (Back and Barnes 1964), so the significance of changes with pumping should be evaluated by the orientation survey. Changes are most severe for dissolved gases (CO_2, Rn, He) and elements added from the pipes and pump.

Electrical conductivity, which can be quickly measured at the sample site with a simple meter, gives considerable useful information on the character of the water. Measurements of pH, alkalinity (bicarbonate), and dissolved oxygen or Eh should be done at the sample site as these parameters are changed by loss of dissolved gases or acidification. Meaningful Eh measurements are very time-consuming and difficult (Back and Barnes, 1964; Morris and Stumm, 1967), and are less valid than measurement of dissolved oxygen using a simple meter. Most surface samples are saturated with dissolved oxygen, but in ground waters the dissolved oxygen content can be very helpful in interpretation.

Analysis of waters directly at the collection site is possible for a number of elements (Chapter 3). These include SO_4, Mo, Zn, and undifferentiated heavy metals. When they are present in concentrations exceeding 10 ppb, Cu and Pb may also be determined in the field; values as high as this, however, are rarely found in background areas. The advantage of analysis in the field is that the sample does not have a chance to change composition, either by contamination or by precipitation of trace metals in the course of the time lapse between collection and analysis. Furthermore, the cost and inconvenience of shipping bulky water samples is eliminated. The disadvantages of sample site analysis are that in general these methods are less precise; they measure only the ionic content of the water, and they can be time-consuming.

Pre-enrichment of the material dissolved in the water has been widely practiced in recent years. Pre-enrichment may be carried out at the sample site by adsorption on ion-exchange resins, by solvent extraction, or by filtering after coprecipitation with some suitable collector such as cadmium sulfide or aluminum phosphate. These methods of pre-enrichment recover only the ionic constituents of the water sample. Where a determination of the total content of a trace element in both dissolved and suspended forms is desired, evaporation is necessary. This operation can be carried out conveniently only in the laboratory or base camp. Freeze-drying is another possible method but is slow and expensive. The concentrates prepared in the field or in base camp are then shipped to a central laboratory for analysis. As with chemical analysis at the sample site, the advantages of pre-enrichment are that the sample does not have a chance to change composition on shipment and that the cost of shipment of the sample to the laboratory is greatly reduced. An added advantage is the considerably greater precision and economy in having the analysis made under controlled conditions in a well-organized laboratory. The disadvantages are the technical difficulties and the time involved in carrying out the pre-enrichment procedures in the field.

B. Stream and Seepage Sediment

Sediment from streams and seepage areas may vary in composition from place to place because of size-sorting and differences in content of organic matter or precipitated oxides. In order to obtain a representative sample of sediment at a site, small amounts of fine sediment should be collected at three to six localities along 5–10 m of stream bed and composited into a single container. If possible, material of about the same fineness and organic content should be collected at different sample sites so that values will be comparable. Results for cold-extractable metal are particularly sensitive to the content of fines and organic matter in the sample. Where samples are collected along a longitudinal stream traverse rather than at isolated sites at junctions, the same effect may be achieved by taking single samples at a reduced sampling interval. Samples so collected may be either analyzed individually or later mixed and analyzed as composite samples.

At all sample points, care must be taken to avoid sampling collapsed bank material of local origin, particularly when the banks are composed of colluvium derived from the adjoining slopes.

No special tools are necessary for collecting samples of seepage soils or stream sediments, although various types of non-contaminating scoops and trowels have been used. In flowing streams, waterproof boots are helpful in order to get into the active stream away from the banks. Small samples of 10–50 g normally furnish enough fine material, inasmuch as most chemical procedures require less than a gram. For relatively coarse-grained samples from fast-flowing streams, however, it may be necessary to collect 100–200 g or more to obtain sufficient fine material for analysis. Wet sieving in the field is helpful in reducing the volume of large samples poor in fines (Bølviken *et al.*, 1976). A simple design for a wet sieve is illustrated in Fig. 16.6. The effect of wet sieving should be tested during the orientation survey to ensure that loss of the ultrafine fraction does not degrade contrast or anomalies. A partial recovery of colloidal particles can be accomplished by flocculation with alum (Skinner, 1969). In some streams where fine sediment is scarce, it may be possible to shake out fines trapped in moss growing on rocks in the stream.

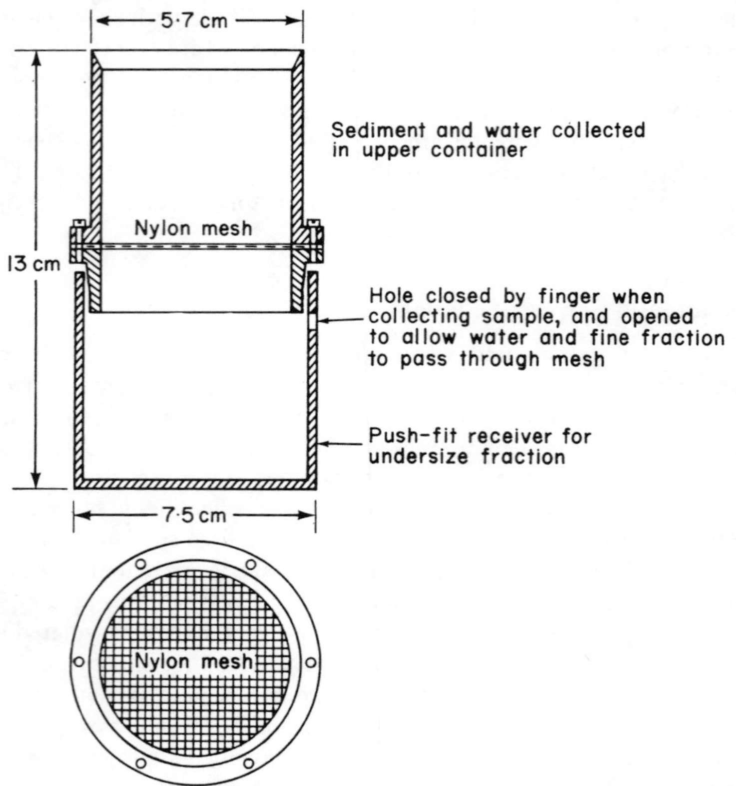

Fig. 16.6. Combination sediment sampler and wet sieve.

Sediment samples, even when very wet, can conveniently be collected in envelopes made of heavy kraft paper, waterproof glue, and a stiffened fold-down tab. A light, rigid, carrying box will prevent chafing of the paper in transit. Plastic, aluminum, or steel containers have been used on some surveys with varying degrees of success.

Notes taken at the sample site should include location, any unusual sediment or stream conditions, and any indication of contamination. Data on rock types in the stream are helpful in surveys of poorly mapped regions.

Most of the problems in processing sediment samples are the same as lready discussed in connection with soil samples and need not be repeated n detail. If possible, samples should be dried in the field using a low-tem-erature oven or simply strong sunlight, followed by sieving. If this is not racticable, the moist samples should be placed in zip-lock plastic containers nd subsequently dried in the laboratory. Sieving to about -200 µm and etaining only the fines for analysis is desirable for the purpose of homogeniz-ng the samples and enhancing the anomalous values. Sieves should always e made of non-contaminating materials.

:. Lake Sediment

'hree general types of sampling devices are used for lake sediments: augers, orers, and dredges. In shallow water, a conventional soil or peat auger may e used. For deeper water, a simple type of coring sampler is suitable (Fig. 6.7). A modification of this corer, with fins added to stabilize it and an nnular lead weight to assist penetration of the bottom, is described by

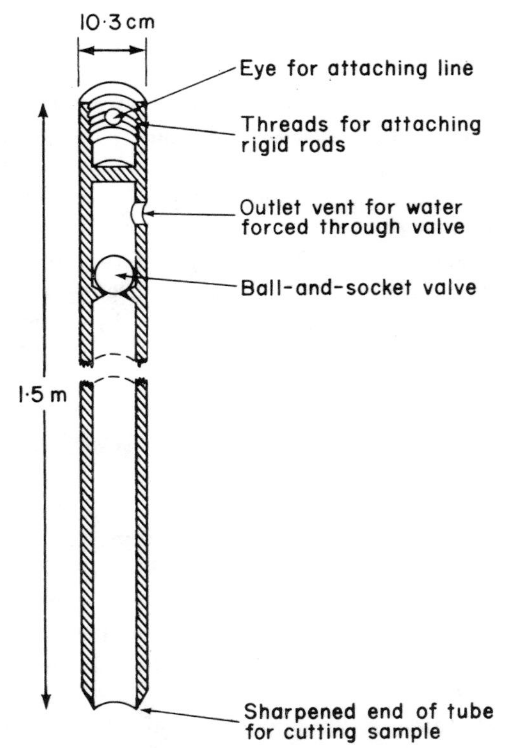

Fig. 16.7. Cut-away section of sample bailer for lake-sediment sampling.

Hornbrook and Garrett (1976). Dredges scrape sample from the surface layer and thereby tend to accumulate Mn nodules and other surficial material that may not be representative of the sediment. The corer captures deeper material that is generally a more satisfactory sample because it is more representative and homogeneous. In addition, if contamination from mining or other human activities is present, the layers below about 10 cm are generally uncontaminated (Allan and Timperley, 1975). The organization of a survey using helicopters is described by Allan *et al.* (1973b). Detailed surveys of lake sediment may be conducted through the ice during the winter, when locations can be precisely surveyed.

Data recorded at the sample site should include location, the character of the sediment (organic, inorganic, Fe–Mn-oxides, odor, H_2S), and depth of water, plus any other features suggested from the orientation survey.

Samples can be stored in paper envelopes like those described for stream sediments and dried in the air or an oven. Samples rich in organic materials dry to a very hard cake which can be disaggregated with a hammer or broken up with a food blender before sieving to about −80-mesh.

D. Heavy Minerals

Heavy detrital minerals characteristically occur in greatest abundance near the base of a sequence of alluvial sediments, usually just above the surface of bedrock. Collecting samples from the surface of bedrock involves either digging pits or sinking holes with augers or post-hole diggers. The heavy-mineral fraction then may be either separated in the field by panning, or shipped to the laboratory for whatever system of mineral separation or chemical analysis is called for. Techniques of collecting and panning heavy-mineral concentrates from sediments are discussed by Raeburn and Milner (1927), Mertie (1954), Theobald (1957), Griffith (1960), and Overstreet (1962).

V. PREPARATION OF MAPS

Most of the comments made in Chapter 13 regarding the preparation of geochemical soil maps also apply to geochemical drainage maps. The principal difference is the considerably greater complexity of much of the information that must be presented on a drainage map. Computer methods are equally helpful in recording the data, producing maps, and examining the character of subsets of the data (Chapter 19).

Maps of drainage surveys, like those for soil surveys, can present either factual data or interpretations. During the interpretation process, a set of maps showing only one element or type of information per map, with complete data on sample location, is essential in order to have a clear picture available of each individual element. At later stages of interpretation and report preparation, several types of data may be combined on one map. The symbols on a small-scale reconnaissance map must be highly simplified if all the necessary information is to be plotted on a single map. Examples of series of symbols for progressively increasing concentration are shown on Fig. 16.8. Normally a scale such that sample points are separated by not less than 0·5 cm results in an attractive and legible map.

A wide variety of different symbols has been used in representing the data of water and sediment analysis. The direction of flow of the water, if it is known, should be indicated by arrows, unless it is already apparent from the drainage pattern. Appropriate symbols should be used to distinguish the data

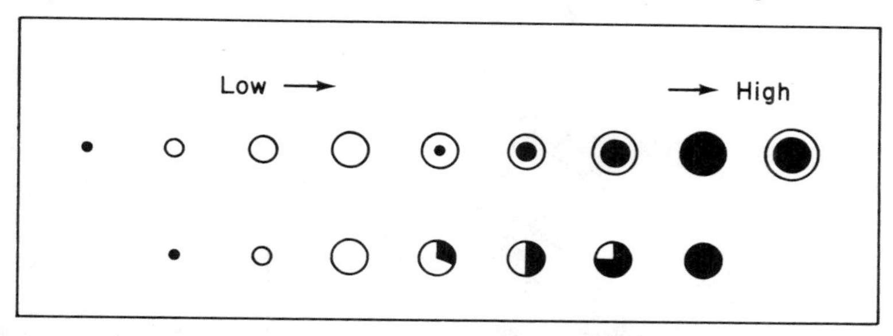

Fig. 16.8. Series of symbols commonly used to denote increasing concentration or intensity of anomalies.

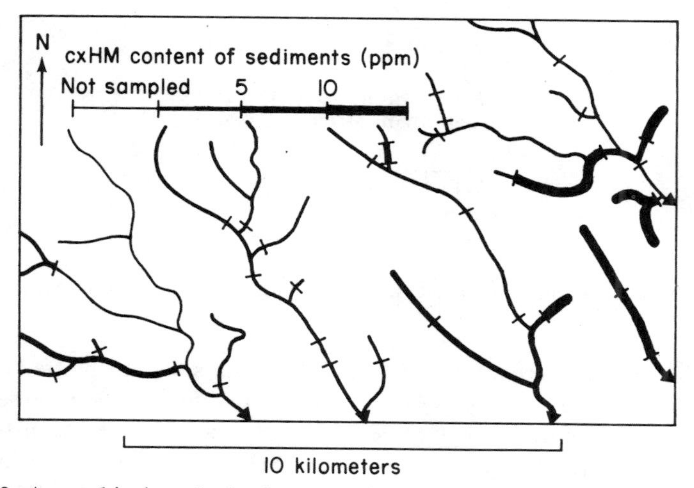

Fig. 16.9. A graphical method of representing drainage anomalies. Data on −80-mesh fraction of sediment samples. (After Hawkes *et al.*, 1960.)

of different kinds of samples, such as active or flood-plain sediments, swamps and seepage soils, water, plants, and residual soils, where they are all shown on the same map. Separate symbols may also be used to distinguish anomalous from background values. Examples of different methods of plotting geochemical drainage data are shown in Figs 16.1, 16.2, 16.4, 16.5, 16.9, and 16.10.

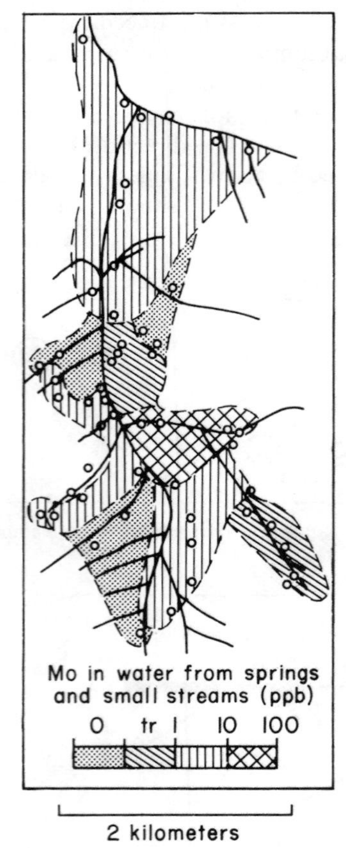

Fig. 16.10. A graphical method of representing drainage anomalies. (After Dolukhanova, 1957).

With particular reference to water data, the results may be plotted in terms of the metal content of the water or of the total dissolved salts (total salinity). At other times it may be preferable to plot the ratio of the metal to some major constituent, such as U/HCO_3^-, and SO_4^{2-}/Cl^-. The choice is made according to which method gives the most significant and reliable guide to ore.

VI. INTERPRETATION OF DATA

The general approach to interpretation is common to all kinds of geochemical anomalies and has already been discussed in Chapter 2. With drainage data, however, it is especially important to determine the range of background fluctuations related not only to bedrock geology and sample type but also to changes in environmental factors, such as pH, bulk composition of water, and varying amounts of organic matter and Fe–Mn-oxides in sediment.

Distinguishing between significant and non-significant anomalies is always one of the more difficult tasks in interpretation. Before any time is spent in follow-up work, every anomaly should be carefully examined to see whether it may be considered as a possible bona fide indicator of ore or dismissed as the result either of natural accumulations of metal unrelated to ore, of artificial sources of contamination, or of errors in technique. First, it is advisable to reanalyze all the critical samples and to check the field description against the sample material as a guard against simple mistakes. At the same time, it may be possible to recognize assemblages of associated elements that may help to indicate the nature of the parent source. Available maps and air photos should be examined with a view to the possibility both of contamination and of correlation of the anomaly with recorded topographic and geologic features. In many cases, it may be necessary to revisit the sample site before it is possible to decide whether an anomaly is significant or not.

Appraisal of significant anomalies, once they have been identified, calls for a critical review of the intensity and form of the anomaly taken in the context of the general favorability of the geological environment, together with consideration of all those environmental factors that may enhance or suppress anomalous patterns. A simple study of the geochemical data alone should never warrant a prediction of the grade and tonnage of the bedrock deposit responsible for the anomaly. Although relations of the type expressed by eqn 15.3 may apply, the complexities influencing these are such that a careful study of the geology and surficial environment should be carried out before even a qualitative assessment is possible.

The intensity of a drainage anomaly is a function of the amount of material eroded or leached from the catchment area, minus what has been precipitated from ground water below the surface, accumulated in overburden, or carried past the sample site. A strong anomaly, therefore, may mean (i) a very large area of low-grade mineralization, (ii) swarms of very small deposits of high-grade metalliferous material, (iii) small deposits of weakly mineralized but highly fractured rock that is unusually accessible to the leaching action of circulating ground water, or (iv) one or more large deposits of ore grade. If the anomaly is clastic in origin, possibility (iii) is unlikely, but the other possibilities cannot be distinguished; conversely, the absence of a strong anomaly does not necessarily mean the absence of an orebody. It may be only the effect of a low rate of mechanical or chemical attack on the bedrock. Just as commonly, the absence of an anomaly may be caused by either dilution or precipitation of the metal somewhere along the drainage system between the source and the sample site.

VII. FOLLOW-UP TECHNIQUES

The technique of following up anomalies disclosed during geochemical drainage reconnaissance will of course depend very much on local conditions. Opportunities for detailed sampling aimed at further delimiting ground-water patterns are often lacking, and the follow-up then passes directly to geological examinations, coupled with geophysical investigations and detailed geochemical soil surveys when conditions are appropriate. However, careful study of the ground-water pattern in relation to the local topography and other factors that may be influencing the subsurface drainage may help to define the area of maximum interest wherein to concentrate the detailed follow-up work.

With regard to anomalies detected in the surface drainage, the following outline presents many of the steps that have come into common use under a variety of different operating conditions.

After confirming the original reconnaissance indication, traverse upstream from the discovery site to find the cutoff. This may be done either by field analysis with a portable test kit or by sampling at close intervals and submitting the samples for laboratory analysis. The choice between these two methods will depend on local conditions of accessibility, personnel, analytical methods used, etc. Figure 14.6, for example, shows the data of a follow-up survey based on field analysis of water samples.

When using the field-test kit, it is often found that values are erratic owing to variable proportions of coarse and fine material in the unsieved samples. In very gravelly streams, it may even be difficult to confirm laboratory analyses which are carried out on the fine fraction. In the field, reproducible determinations giving a better-defined picture of the anomaly may often be obtained by sieving the sample on the spot.

At the cutoff, check carefully for possible contamination, natural enrichment, and any visible evidence of mineralization. Determine the principal points at which anomalous metal is entering the drainage system. According to the nature of the problem, this may best be done by collecting samples at close intervals along both banks of the active channel, along the zones of eepage and spring areas on opposite edges of the flood plain, or along the ase-of-slope colluvium. The most highly anomalous samples will normally ıdicate the side of the valley from which most of the metal is coming. Where pring or seepage anomalies are caused by near-surface water, the source will ormally occur directly up the slope of the valley side. Where the anomaly ; the result of water flowing from bedrock channels, the source will lie omewhere along the upslope projection of the water-bearing fracture or other quifer. Such a source may be deeply buried and not come to the surface of edrock.

As a result of these studies, it is often possible to delimit, quickly and more r less precisely, the area of interest wherein to concentrate detailed exploraon by more intensive geological, geophysical, and geochemical surveys.

During interpretation and follow-up it should always be kept in mind that n anomaly may be caused by multiple sources. A well-mineralized district ypically has many small mineralized zones and extensive weak mineralizaon, as well as one or more orebodies. The discovery of one source should be ollowed by careful consideration of whether it explains all the anomalous amples. The penalty for stopping too soon may be that the best ore of the rea is overlooked.

Chapter 17

Vegetation

Chemical analysis of systematically sampled trees and shrubs for traces o
ore metals was one of the first geochemical methods to be investigated. I
the early 1930s, V. M. Goldschmidt, pioneer in geochemistry, made th
observation that the humus of forest soils was very much enriched in mos
of the minor elements. From this he deduced that the same trace element
must be correspondingly enriched in the plants from which the humus wa
derived. He made the first suggestion that analysis of plant material migh
be an effective method of prospecting. In later years this method of explora
tion came to be known as the biogeochemical method, following the termi
nology of the Russian geochemist Vernadsky.

Visual observation of plants, when used as a guide to buried ore, is know
as geobotanical prospecting. Whereas biogeochemical methods requir
chemical analysis of plant organs, the geobotanical methods depend on direc
observations of plant morphology and the distribution of plant species
Where applicable, therefore, geobotanical methods have very great advan
tages over geochemical methods of prospecting in that the results of the surve
are immediately available without further treatment of the samples.

Detailed discussions of biogeochemical and geobotanical exploratio
methods are available in the books by Brooks (1972) and Malyuga (1964).

Recently, research on remote sensing methods has included studies of th
optical differences between vegetation growing on ore and that in backgroun
areas. These studies, if successful in developing practical systems of explora
tion, can be considered as extensions of geobotanical prospecting, and ar
briefly discussed in this chapter. The chemistry of airborne particulat
material shed by plants is essentially a part of biogeochemical prospecting

ut is included in Chapter 18 because of its relationship with other methods epending on vapor geochemistry.

I. UPTAKE OF MINERAL MATTER BY PLANTS

he principles underlying both biogeochemical and geobotanical studies of egetation as methods of locating buried ore deposits are basically simple. he root systems of trees act as powerful sampling mechanisms, collecting queous solutions from a large volume of moist ground below the surface. hese solutions then serve as a source of inorganic salts that may be deposited the upper parts of the plant, or that may stimulate, inhibit, or otherwise odify the growth habits of the plant.

In detail, however, the internal vascular systems of terrestrial plants are y no means simple. A vast and complex set of equilibrium relationships is ctive from the time the soil solution comes within the influence of the root p until the water is discharged into the atmosphere. Whether or not a metal-ch nutrient solution will result in an easily measured variation in the com-osition of the plant, or in a diagnostic variation in plant morphology or cology, depends on the balance of these many relationships.

More specifically, the factors that must be considered in the development f plant anomalies are (i) the nutritive requirements of plants, (ii) the avail-bility of elements in the soil in a form that can be ingested by plants, (iii) actions that take place in the root tips of plants, and (iv) the mechanisms f movement and storage of elements within the plant.

. Plant Nutrition

ach species of plant has its own individual nutritive requirements that iffer somewhat from those of every other species. Thus different species of lants all supported by identical nutrient solutions will contain widely varying oncentrations of many of the minor elements.

For whatever reasons the elements are needed by plants, it has been efinitely established that in addition to the common nutritive elements (N, , P, S, Ca, and Mg) most plants require small quantities of many minor ements, including principally Cu, Zn, Fe, Mo, Mn, and B (Brooks, 1972, 92). If the soil solutions do not contain adequate quantities of these ele-ents, the plant will be unhealthy or may not survive. Some workers have ven postulated that plants require a certain small amount of every element the periodic table, but perhaps in quantities so small that the need would e extremely difficult to demonstrate in greenhouse experiments.

At the other extreme, an excess of a particular element above a critic
level in the nutrient solution will impair the health of the plant or may eve
kill it. In this case, the element is present in such quantities that it interfere
with rather than assists the normal metabolism of the plant and thus has a
overall toxic rather than a nutritive effect.

B. Availability of Elements in Soil

An element may occur in the pore spaces of a moist soil in a variety
forms, including free ions, complex ions, ions adsorbed on clay minerals
other colloidal particles, organic complexes, and either primary or secondar
minerals (Mortvedt *et al.*, 1972). The free ions, weakly adsorbed ions, an
some of the inorganic and organic complexes can be adsorbed by the pla
directly. These, therefore, are the forms of an element that, strictly speakin
can be considered as available. At the same time, all the forms mentione
above will tend toward equilibrium with each other, so that the chemic
activities or concentrations of the different forms are all interrelated. Th
as the concentration of an available phase is reduced by ingestion in th
plant, it is replenished by release from the less available components of th
system.

Availability is a concept closely related to mobility in the surface enviro
ment. Like mobility, it is clearly influenced by pH, oxidation potential, an
the presence of complexing agents, precipitants, and adsorbents, all facto
discussed at length in Chapter 8. For example, Fe is very insoluble in oxidi
ing soils near neutral pH, but is relatively soluble as Fe^{2+} under reducin
conditions. Copper and Zn are more available (mobile) in acid soils than
neutral soils. Phosphorus, one of the important plant nutrients, is strong
influenced in its availability by the presence of Fe, which tends to fix it
highly insoluble Fe-phosphate compounds.

Availability of an element to plants can be measured empirically by dete
ming the amount of the element that can be removed from the soil by leachin
with chemical reagents, or by growing a plant in the soil and determining th
amount of the element taken up by the plant. Another method is to equilibra
the soil with an ion-exchange device and measure the amount on the e
changer (D. E. Baker, personal communication).

C. Reactions in the Root Tips of Plants

The root tips of a growing plant can absorb the salts dissolved in the so
moisture. Under certain conditions, they also have the capacity to mobili
and absorb elements that are more or less firmly bonded on the surfaces
clastic soil particles. According to currently held concepts, the surface of th

oot tip of a plant and the immediately surrounding solutions are character- zed by a relatively high acidity. This effect is so local that it is not generally pparent from a measurement of the pH of the soil as a whole. The cause of he acid condition here is probably the hydrolysis of CO_2, which escapes hrough the roots in substantial quantities. The effect of the abundance of ydrogen ions is to set up active cation-exchange reactions between the clay ninerals of the soil and the surface of the roots, which have been shown to ave a high exchange capacity (Fig. 17.1). On the surface of clay minerals or olloidal organic matter, hydrogen ions exchange for metal ions, which are

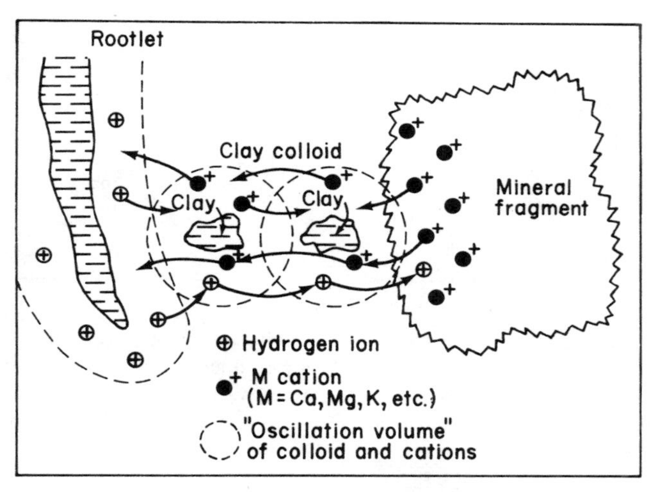

ig. 17.1. Diagram illustrating cation exchange reactions near root tips. (After eller and Frederickson, 1952.)

ien free to diffuse through the soil moisture to the roots. When the metal ons reach the outer surface of the root tip, they exchange for hydrogen ions, hich are released to repeat the process (Jenny and Overstreet, 1939; Keller nd Frederickson, 1952; Mehrlich and Drake, 1955).

Empirical observations have shown that the local but extremely corrosive nvironment near the root tips of plants can extract mineral matter well in xcess of what is present in readily exchangeable form. Even the primary licate minerals can be broken down and their components made available the plant (Lovering, 1959). A spectacular example of the dissolving effect of oots is the converter plant, which takes up Se from stable and relatively nsoluble Se compounds in the soil. With the death and decay of the plant, e Se returns to the soil in soluble form and is then available for uptake by ther plants that lack the power to dissolve Se (Beath *et al.*, 1939, p. 266).

Movement of inorganic constituents into a plant is selectively controlle in such a way that some elements are freely admitted while others are in peded to greater or lesser degree. The general relations of element uptake fro soil by plants are indicated by Fig. 17.2. At low levels of an element in th soil, increased availability normally leads to a higher content in the abov ground portions of the plant (segment A). Plants may exhibit deficienc

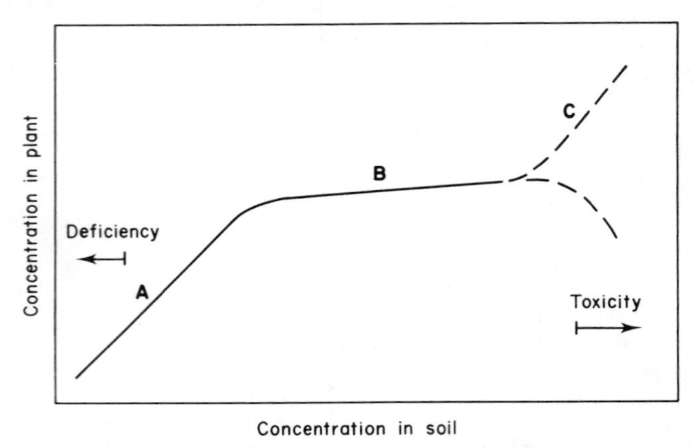

Fig. 17.2. Generalized relations between activity (concentration) of an element i soil, and the concentration in a plant. Along segment A, the content in the pla reflects variability of the underlying soil. Along segment B, an exclusion or stora mechanism is operating. Along segment C, toxicity may develop. (After Kov levskiy, 1977.)

symptoms in part or all of this range. For some elements, particularly fo the elements that are essential for plant growth or are toxic to plants, a lev is reached at which absorption from the soil and/or translocation to the upp parts of the plant are considerably reduced. In this range, the amount in th plant is unrelated to the availability in the soil (segment B). At still higher level the exclusion or root-storage mechanism may break down, and addition amounts may reach the plant, and cause toxicity so that the plant is u healthy, deformed, or dies (segment C). The relative lengths and absolu contents along segments A, B, and C vary widely among plants. Relatio for most elements are normally along segment A, as illustrated by Zn i Fig. 17.3. Other elements, such as Pb in Fig. 17.3, are along segments B–C and exclusion is nearly complete until very high levels are reached in the soi Plants under conditions along segment B are obviously of limited use i biogeochemical exploration, but those along segment C may find applicatio in geobotanical methods because of deformities and other observable sym toms of toxicity. Plants of arid regions that grow during short rainy seaso tend to be more susceptible to toxicity than those in which rain is concentrated in winter before the growing season (Cole, 1971).

The exclusion mechanism may operate at the soil–root interface, within the roots, or at a higher level in the plant (Peterson, 1971). Lead, for example, is an element that is apparently largely immobilized by precipitation in the root tissues of some plants (Hammett, 1928). Thus, toxic excesses of Pb may not reach the active centers of growth in the upper parts of the plant. If the quantity of Pb in the soil solution is too great, the precipitated Pb salt apparently impedes the flow of solutions, and the plant does not grow normally. Uranium and V, elements which are also toxic, are apparently

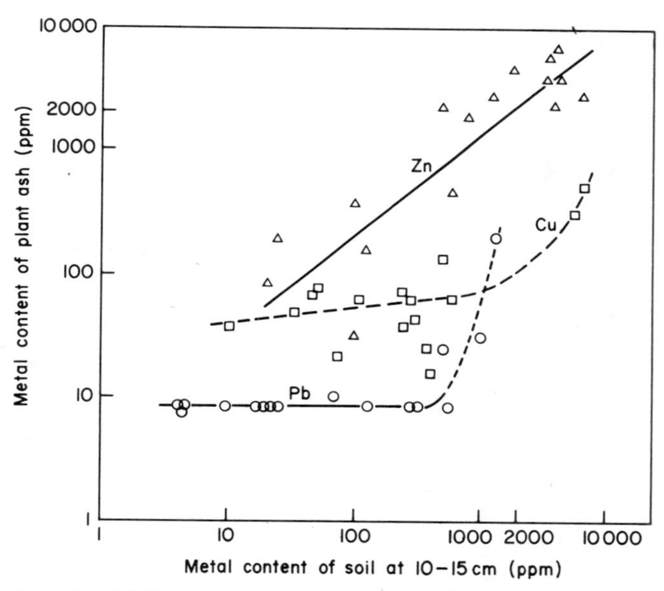

Fig. 17.3. Relationship between the Cu, Pb, and Zn contents of plant ash of *Triodia pungens* and associated soils. (□) Cu; (○) Pb; (△) Zn. (After Nicolls *et al.*, 1965.)

precipitated in the roots in the same way, as suggested by the data of Table 17.1.

Although the toxic elements, such as U and Pb, are largely precipitated in the root cells, enough reaches the upper parts of most plants to be readily detected. They normally occur in plant ash in lower concentrations than in the supporting soil (Fig. 17.4). In spite of this impoverishment, however, the geochemical patterns formed by the toxic elements may be a more faithful reflection of the composition of the soil moisture than the patterns of some of the more readily accepted nutrient elements.

Table 17.1
Uranium and V content of roots compared to tops of vegetation in the Thompson district, Utah[a]

Plant species	U (ppm) in ash			V (ppm V_2O_5) in ash		
	Tops	Roots	Roots/tops	Tops	Roots	Roots/tops
Deep roots collected from mine workings						
Juniperus monosperma	7·8	1600	200	20	3000	150
Juniperus monosperma	·2	140	70	50	4000	80
Quercus gambeli	10	190	19	90	1700	19
Near-surface roots						
Juniperus monosperma (average of 40 samples)	1·2	7	5·6	54	110	2
Atriplex confertifolia	3	5	1·6	10	90	9
Oryzopsis hymenoides	30	40	1·3	70	1600	23
Artemisia spinescens	3	5	1·6	70	100	1·4
Artemisia bigelovi	2	2	1	50	5	0·1
Astragalus preussi	70	70	1	3000	2600	0·8
Aplopappus armerioides	40	20	0·5	260	180	0·7

[a] Source: Cannon (1960a).

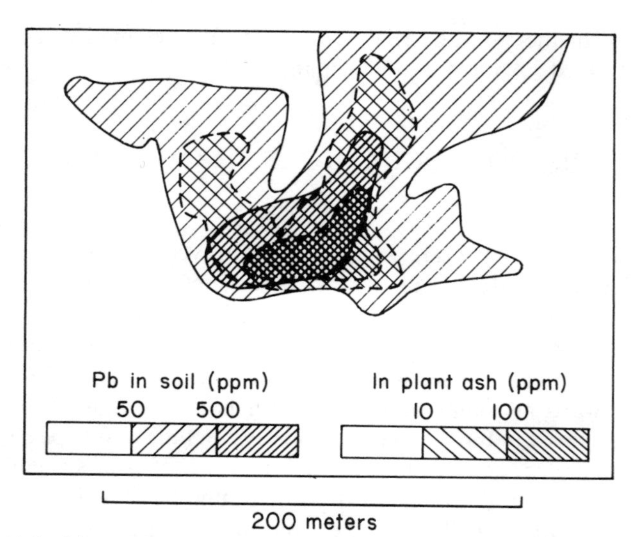

Fig. 17.4. Coincident Pb anomalies in plants and soil over an ore deposit in the Rudny Altay region, U.S.S.R. (After Nesvetaylova, 1955a.)

D. Movement and Storage Within Plants

Once the mineral matter has entered the vascular system of a plant its movement and storage within the plant is controlled by many factors. These include free and restrained diffusion, movement of the solvent, electrical and thermal effects, exchange reactions and, perhaps most important, the accumulation of mineral nutrients in metallo-organic molecules (Broyer, 1947). Of the last type of reaction, the entrance of Mg into the chlorophyll molecule is a familiar example. Many other elements play a similar dominant role as components of enzymes and other catalysts that speed the many vital reactions in a growing plant (Brooks, 1972).

Stated in a different way, the movement of inorganic matter throughout the structure of a plant is largely a response to the plant's nutritional requirements. In the upper part of the plant, the nutritive elements are commonly enriched in the actively growing cells, particularly in the seed structures and growing tips. As the cell matures, the concentration of these elements declines. There is abundant evidence that the distribution of nutrient elements changes with changing requirements and that the vascular systems of plants maintain a dynamic balance whereby inorganic material can be supplied to fulfill the nutritive needs and also removed when it is no longer needed. The specific nutritive or toxic effect of any one element may depend on the overall composition of the nutrient solution. The uptake of one element in a plant may thus be suppressed or increased by the presence of other elements in the solution (Evans *et al.*, 1951; Ahmed and Twyman, 1953).

II. BIOGEOCHEMICAL ANOMALIES

Biogeochemical anomalies are areas where the vegetation contains an abnormally high concentration of metals. If the metal content of a sample of plant material is to be useful in prospecting, it should bear a fairly simple relationship to the metal content of the bedrock. Consideration of the factors discussed in the preceding section indicates that the relationship may not be as straightforward as might be desired. Other factors that affect the background and contrast in biogeochemical surveys include plant species sampled, plant organ sampled, age of plant or organ, depth of root system, health of plant, and aspect (amount and direction of sunlight), plus the factors of soil pH, Eh, temperature, soil moisture, and interferences created by other elements in the soil, as discussed in the preceding section (Brooks, 1972, p. 99 *et seq.*).

A. Variation Between Plant Species

Different species of plants take up different amounts of inorganic material from the soil. The closest correlation between the composition of the plant with that of the supporting medium is not necessarily found in the plant that is the most highly enriched in a given element. Each plant has its own peculiar habits that must be determined empirically. An example of contrasting uptake of certain minor elements by different types of plants all living in essentially the same climate and all growing in an unmineralized environment is shown in Table 17.2. An extreme example of selective enrichment is the Arctic dwarf birch, in which the background Zn content of the ash of mature twigs is commonly about 1% (Warren *et al.*, 1952). It would obviously be extremely misleading, therefore, to attempt geologic interpretations of the minor-element content of plant samples taken without regard to species. In a given environment, a group of species may have a similar response to certain elements, and where this can be established it is permissible to include two or perhaps more species in one set of comparable samples.

B. Variation Between Plant Parts

The part of the plant selected for analysis is an especially critical factor. Two considerations are important here; the differences in trace element level

Table 17.2
Average metal content (ppm) in the ash of five types of vegetation growing in unmineralized ground[a]

Element	Grasses (above ground)	Other herbs (above ground)	Shrubs (leaves)	Deciduous trees (leaves)	Conifers (needles)	Averages and totals
Cr	19 (30)	10 (139)	14 (67)	5 (100)	8 (120)	9 (462)
Co	10 (30)	11 (192)	10 (70)	5 (101)	<7 (119)	9 (512)
Ni	54 (28)	33 (226)	91 (182)	87 (209)	57 (213)	65 (858)
Cu	119 (102)	118 (429)	223 (853)	249 (293)	133 (370)	183 (2047)
Zn	850 (62)	666 (355)	1585 (735)	2303 (278)	1127 (333)	1400 (1763)
Mo	34 (32)	19 (217)	15 (104)	7 (118)	5 (145)	13 (616)
Pb	33 (29)	44 (311)	85 (877)	54 (339)	75 (352)	70 (1908)
V	25 (4)	23·5 (39)	25 (36)	16 (14)	21 (77)	22 (180)

[a] Source: Cannon (1960b); reprinted from *Science* by permission. Figures in parentheses show the number of analyses used in the calculations.

from one organ to another, and the stability of element concentration within one organ as a function of time and location on a single plant. The concentration of trace elements is frequently found to decrease in the order: leaves, twigs, cones, wood, roots, and bark (Carlisle and Cleveland, 1958; Brooks, 1972, p. 101). However, the concentration in leaves, green tips, and seed structures that are actively growing tends to be less consistent with time than that in twigs (Warren and Delavault, 1949). Guha (1961) found differences of a factor of 3–4 in leaves collected in different seasons. Dead branches of piñon and juniper have been reported as more satisfactory than live branches in geochemical sampling for U (Cannon and Starrett, 1956).

The branches on a given side of a tree are connected most directly with the roots on the same side, unless the trunk has twisted. Thus the metal content may vary greatly even from one side of the tree to the other.

C. Depth of Root Penetration

Deep-rooted plants that habitually obtain their water from the zone of saturation below the water table are defined as phreatophytes, in contrast to

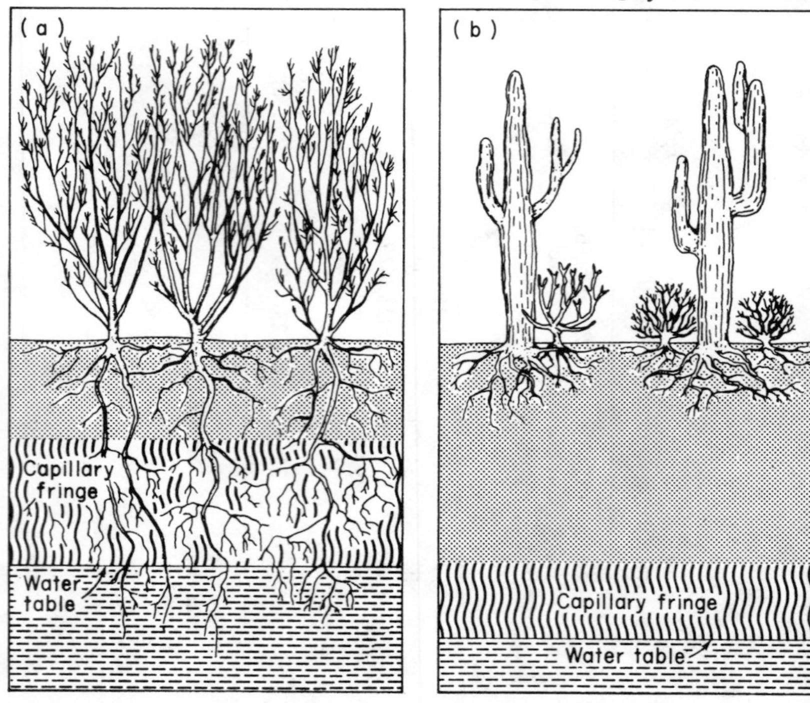

Fig. 17.5. Distinction between (a) phreatophytes and (b) xerophytes shown by their occurrence in relation to the water table. (After Robinson, 1958.)

the shallow-rooted xerophytes that can survive exclusively on vadose water derived from rainfall (Fig. 17.5). Observations in areas of transported overburden under a variety of climatic conditions have clearly demonstrated that many species of plants not uncommonly take up anomalous metal from orebodies buried at depths of 10–15 m. In the mine workings of the Shasta copper district of northern California, a live root was observed in a mine tunnel at a depth of 50 m below the surface. In the arid climate of the Colorado Plateau area, live juniper roots have been observed in U mines at depths on the order of 100 m (Cannon and Starrett, 1956). In Arizona, live roots have been observed at depths as great as 55 m in unconsolidated overburden overlying porphyry-copper deposits (Phillips, 1963). A field survey in the desert of New Mexico showed a positive correlation between the U content of the branches of piñon and juniper trees and the location of U mineralization in a flat-lying horizon 20 m below the surface (Fig. 17.6). A few scattered observations of similar responses at depths of 50 m or more have been reported by Russian observers (Vinogradov, 1954).

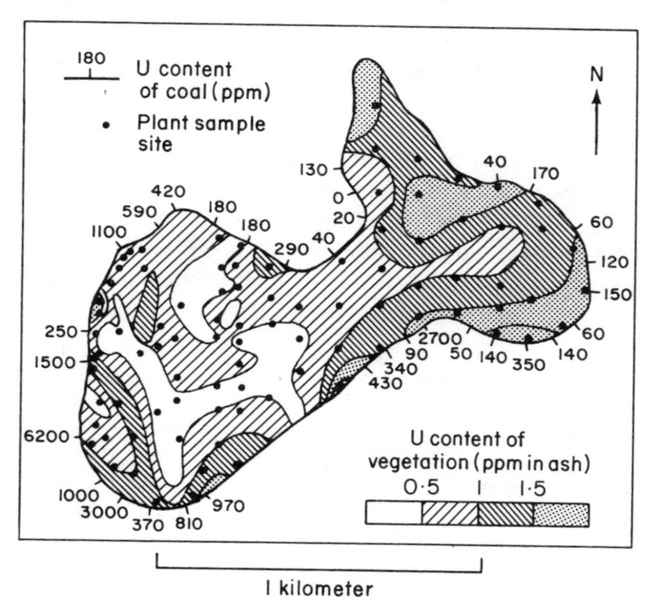

Fig. 17.6. Uranium content of dead piñon and juniper branches from top of La Ventana Mesa, 20 m above U-rich coal horizon. (After Cannon and Starrett, 1956.)

D. Variation with Other Factors

In well-drained soils, the roots of plants must go deeper for their water. Drainage conditions also affect the relative acidity of the soil. Variations in soil pH will cause variations in the relative solubility of elements in the soil and hence the availability of those elements for uptake by the plants. The movement of mineral nutrients into plants varies with the amount of sunlight received. Thus, the composition of plants on a sunny slope will be somewhat different from that of the same plants growing on the same soil on a shady slope. Plants may also change in composition with time of year, the mineral content often increasing in the spring during the active growing period, followed by a gradual decline in mineral content with maturity.

E. Contrast

The contrast of anomalous biogeochemical values against the normal background content appears to be related to the mobility of the element in soil solutions. Of the metals for which reliable data are available, Mo shows a fairly consistently high contrast. Expressed as the ratio of anomaly to average background, the contrast in Mo anomalies ranges from 10/1 up to as much as 100/1 (Warren *et al.*, 1953; Baranova, 1957; Malyuga, 1958). For Co, Pb, Fe, and U the contrast is normally in excess of 5/1 (Webb and Millman, 1951; Nesvetaylova, 1955a; Cannon and Kleinhampl, 1956; Goldsztein, 1957; Warren and Delavault, 1957), and Cu, Zn, and Ag usually show contrasts of 2/1 or 3/1 (Webb and Millman, 1951; Warren *et al.*, 1952). The tendency for low contrast of Cu and Zn anomalies leads, at the extreme, to a lack of recognizable Cu and Zn anomalies at some deposits of these metals (Brooks, 1972, p. 146). This lack of anomalies may be partly related to the fact that these elements are essential for plant growth and are limited by metabolic processes. Table 17.3 shows the relatively greater contrast of Pb as compared with Ag in plants from an area of Pb–Zn mineralization in Nigeria.

Table 17.3
Contrast of anomaly to background shown in Pb and Ag content of oven-dried twigs from trees from Nyeba Pb–Zn District, Nigeria[a]

	Pb (ppm)			Ag (ppm)		
Species of tree	Background	Anomaly peak	Ratio peak/bg	Background	Anomaly peak	Ratio, peak/bg
Afzelia africana	0·8	140	175	0·05	0·31	6
Baphia nitida	0·4	16	40	<0·03	0·06	>2
Albizzia zygia	0·4	23	57	0·04	0·07	2
Vitex cuneata	0·6	6	10	0·04	0·08	2
Parkia oliveri	0·4	13	32	<0·03	0·05	>2
Millettia sp.	0·3	7	23	<0·03	0·04	>1

[a] Source: Webb and Millman (1951).

Figure 17.7 illustrates the same relationship graphically. For some elements, the contrast in soil anomalies is equal to or greater than that in the corresponding biogeochemical anomaly, whereas the reverse has been reported for some other elements. These two relationships are illustrated in Figs 17.8 and 17.9, respectively.

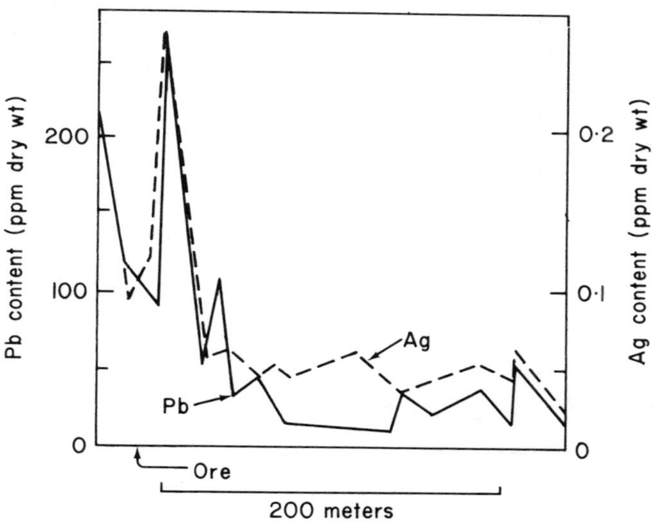

Fig. 17.7. Lead and Ag content of oven-dried twigs of *Rubiaceae* sp. over ore, Nyeba Pb–Zn district, Nigeria. (After Webb and Millman, 1951.)

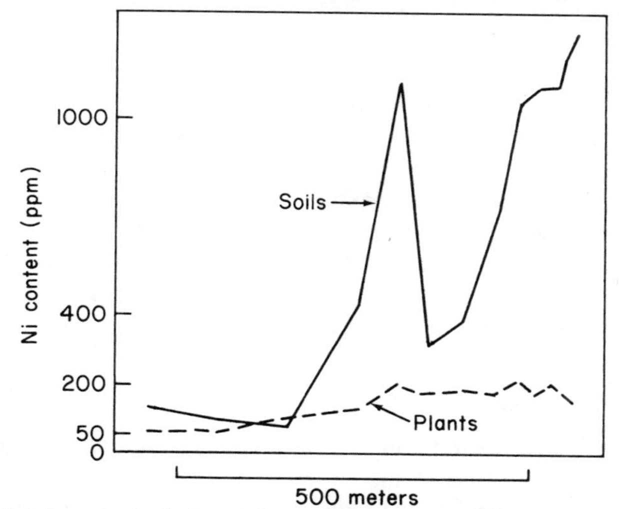

Fig. 17.8. Nickel content of plant ash as compared with corresponding soils at the Novo–Tayketken deposit, U.S.S.R. (After Vinogradov and Malyuga, 1957.)

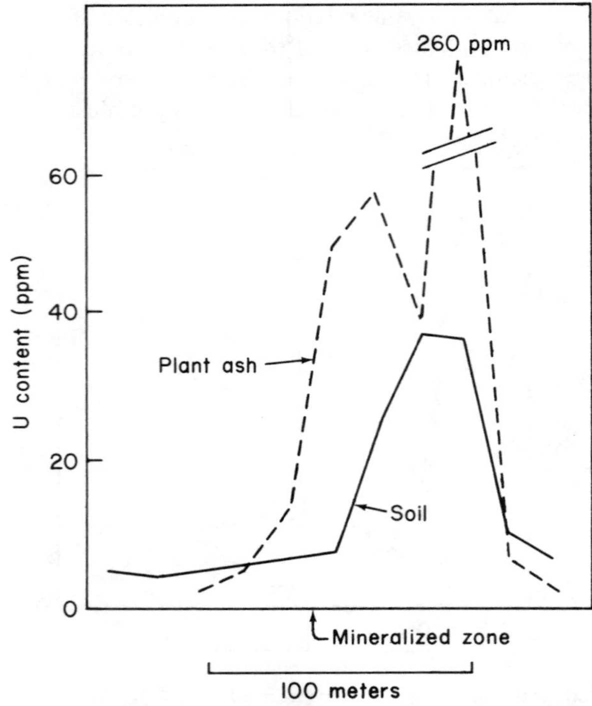

Fig. 17.9. Uranium content of ash of pine needles as compared with corresponding soils, Estérel region, France. (After Goldsztein, 1957.)

F. Homogeneity

The mineral content of plants is the combined effect of a great many unpredictable variables, of which only a very few are related even indirectly to the composition of the underlying rock. It is therefore not surprising to find that biogeochemical anomalies, at least for the mobile elements, are generally more irregular than the corresponding residual soil anomalies. An example showing the relative homogeneity of soil and plant samples collected from the same traverse is presented in Fig. 17.10.

G. Form of Anomalies

Subject to the several modifying factors previously discussed, variations in the chemical composition of the upper parts of the plant correspond with variations in the composition of the solutions tapped by the root system. Thus the form of biogeochemical anomalies is merely a composite of the

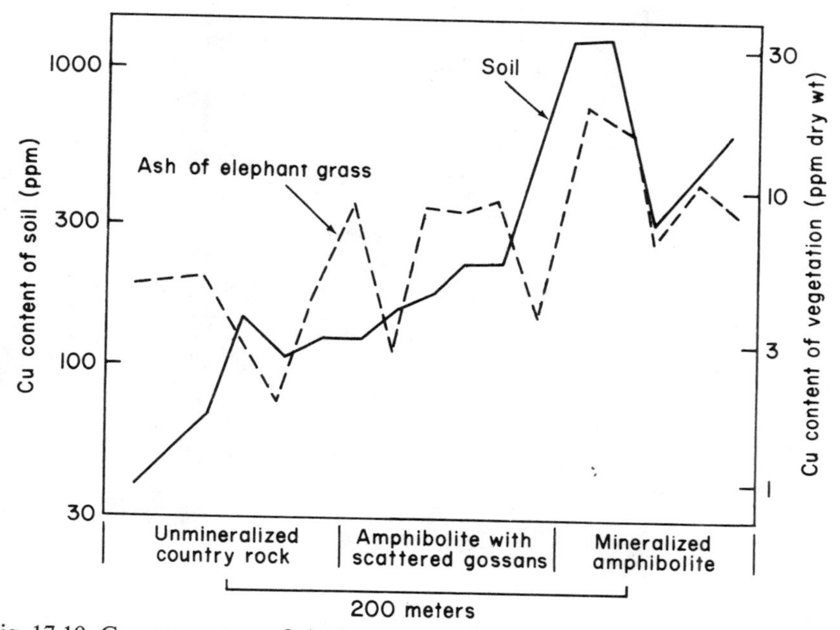

Fig. 17.10. Copper content of elephant grass as compared with corresponding soils, suggesting greater homogeneity of soil pattern, Kilembe area, Uganda. Data on −80-mesh fraction. (After Jacobson, 1956.)

form of the combined syngenetic, hydromorphic, and biogenic anomalies in the underlying soil together with whatever ground-water anomalies may be present. Where the plant is either rooted directly in ore or in a superjacent soil anomaly, the biogeochemical anomaly will also be superjacent. Where the parent anomaly is a lateral ground-water or hydromorphic soil anomaly, the resulting biogeochemical anomaly will show a corresponding displacement with respect to the ore. An example of the characteristic coincidence of plant and soil anomalies is shown in Fig. 17.11.

III. BIOGEOCHEMICAL SURVEYING TECHNIQUES

Operating conditions vary from area to area so much that a universally applicable system of biogeochemical surveying cannot be recommended. The general rule of starting with a review of previous experience in similar problems and following with a carefully planned orientation survey is, however, a safe rule to follow.

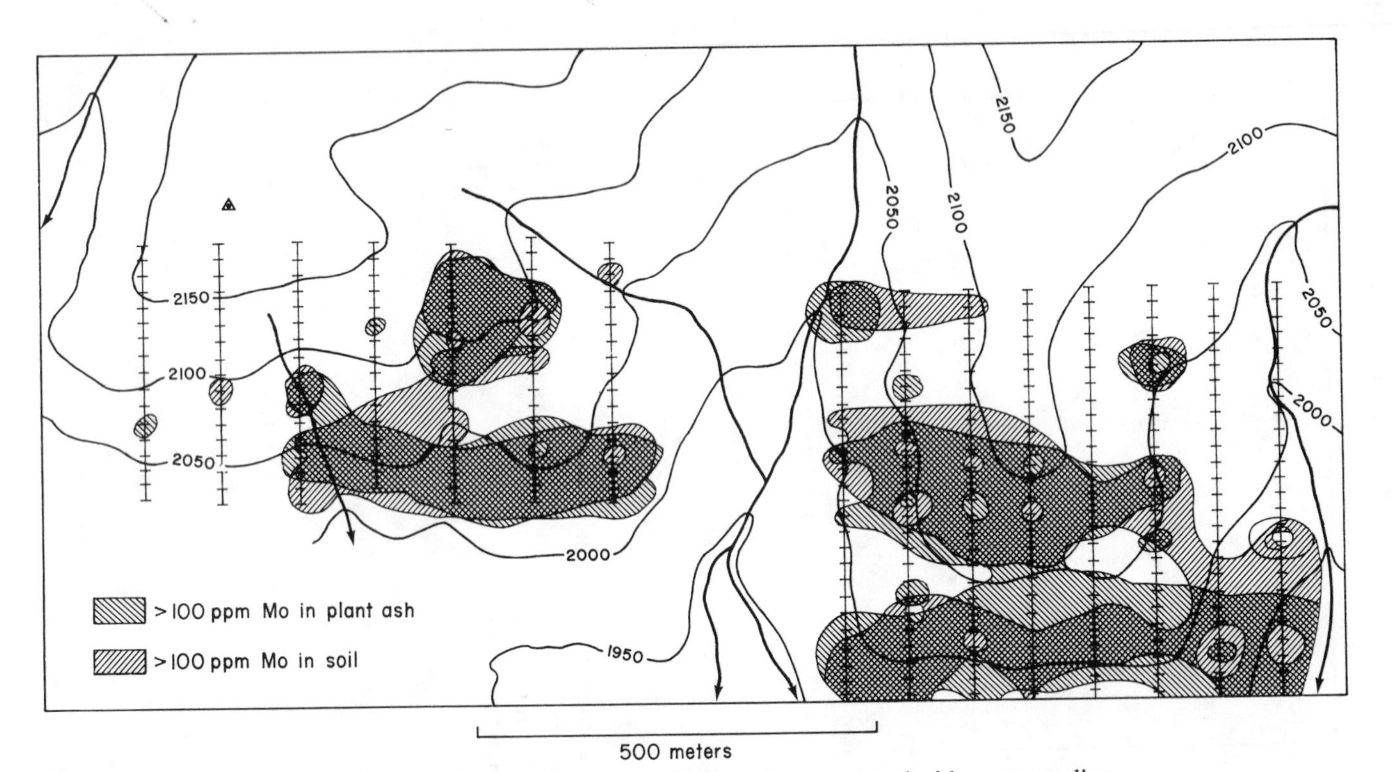

Fig. 17.11. Plan showing Mo content of plant ash as compared with corresponding soils, Okhchi River, Armenian Republic, U.S.S.R. (After Malyuga, 1958.)

A. Orientation Survey

The system of preliminary orientation experiments preparatory to biogeochemical survey work for U in the Colorado Plateau follows a pattern that could profitably be copied in other areas. According to Cannon and Kleinhampl (1956), the orientation study

> should include observations on the extent, depth and inclination of the ore-bearing strata, the size and habits of the ore bodies, the probable grade of the ore, the presence or absence of a chemical halo in the surrounding barren rock, and the relation of the ore-bearing bed to the water table and the plant roots. Botanical studies should be made of the growth habits of species available for sampling. Preliminary samples should be collected on both mineralized and barren ground and then analyzed to determine the amount of uranium absorbed by trees in the area under study. Finally, from this geological and botanical information, the sampling medium, the sampling pattern, and sampling interval may be determined.

Major factors to be evaluated during orientation surveys are listed in Table 17.4.

Table 17.4
Check list of factors to be determined during biogeochemical orientation surveys

(i) Optimum species, based on distribution (must be widespread), contrast of anomalies, homogeneity of anomalies, ease of recognition, ease of sampling, and depth of root system.
(ii) Part of plant to be sampled (twigs, leaves, fruits, bark, wood).
(iii) Best indicator element or elements.
(iv) Effects of aspect (sunlight), drainage, shading, antagonistic effects of other elements.
(v) Amount of vegetation needed to give adequate ash.
(vi) Contamination from dust or other sources.
(vii) Sampling pattern and interval.

The possibility of contamination should receive special attention in the course of the orientation study. Contamination of plant samples most commonly results from airborne material such as automobile exhaust, industrial gases, smelter fumes, fungicides, fertilizers, and dust from ore-processing mills. An example of the latter is shown in Table 17.5. Biogeochemical surveying in general should not be attempted in seriously contaminated areas.

B. Choice of Sampling Medium

Deep-rooted plants are generally preferable to shallow-rooted species. For adequate coverage, a common and uniformly distributed species is desirable. If it can be demonstrated that under the same conditions, two or more species

Table 17.5
Contamination of junipers near a uranium mill[a]

Distance from mill	Number of samples	U content (ppm ash)
600–1200 m	6	40 (av.)
250–460 m	4	150 (av.)
Adjacent to the mill	2	700 and 1100

[a] Source: Cannon (1952).

take up proportional amounts of metal, then they may be included in the same set of samples. The first or second year's growth of twigs from high shrubs or trees has been used successfully in many different climatic environments. The metal content of mature twigs does not vary appreciably through the growing season, in contrast to the fruit, leaves, needles, or other actively growing parts; furthermore, twig samples are easier to collect and process than samples of bark or wood. The taller shrubs and trees are less likely to be contaminated by rain splatter than low shrubs and non-woody plants. In bog areas, however, where rain splatter is not a danger, mosses and low shrubs may be successfully used (Salmi, 1956; Hawkes and Salmon, 1960).

C. Collecting and Processing of Samples

In areas densely populated by the species to be sampled, sampling points may be laid out on geometrical grids of profile lines. In areas of widely scattered individuals, it may be necessary to sample plants wherever they can be found and to survey the location of each sample separately. Standard pruning tools are quite satisfactory for collecting twig samples. A sample of 20 g is large enough to yield the 1 g of ash normally needed for analysis. The most commonly used containers are small envelopes or paper bags. Maps should be prepared to show the location of sample sites, species sampled, and pertinent aspects of the bedrock geology and of land forms. Standard procedures call for the drying, pulverizing, and homogenizing of plant samples before analysis. In relatively dry climates, samples collected in paper containers will normally dry out sufficiently in the course of the two weeks or so that usually elapse between collection and analysis. In moist climates, oven-drying of samples before shipment may be necessary to prevent the development of mold and the disintegration of the paper containers.

D. Choice of Analytical Method

A problem in the chemical analysis of plant material is the possible loss of trace elements on burning the sample to its ash. Ashing should usually be conducted with a relatively low oxygen supply so that the sample does not actively burst into flame, but with enough oxygen so volatilization of unoxidized organic matter is minimized. For most elements, loss is negligible if vegetation is ashed at about 450 °C, and increases at higher temperatures (Brooks, 1972, pp. 117 and 118). If losses during ashing are suspected, it may be preferable to use wet-ashing methods, such as the use of a 1/5 mixture of perchloric and nitric acids or sulfuric and nitric acids. The ash of plants is commonly about 2–5% by weight. Other than this, the choice of the analytical method depends only on economy of method as balanced against precision and sensitivity. Normally, plants cannot conveniently be analyzed for trace elements in field laboratories, but must be sent to a central laboratory set up with adequate equipment of pulverizing and ashing.

The analytical results may be expressed in terms of the metal content of the ash or of the dry plant material (dried at either room temperature or 110 °C) according to which method gives the most significant pattern.

E. Interpretation of Data

The first problem in interpretation of biogeochemical data, as with other geochemical prospecting data, is distinguishing between significant anomalies that are related to bedrock mineralization and that can be used as exploration guides, and similar patterns that arise from irrelevant natural or artificial factors.

Non-significant anomalies may appear as the result of normal variations in soil pH, drainage conditions, or exposure to sunlight. Under some conditions, variations in the metal content of plants due to such factors may be partially eliminated by determining the ratio of two elements that have a similar response to the various non-geologic factors (Brooks, 1972, p. 125). Warren *et al.* (1952) report that, although the absolute content of Cu and Zn in plants in an unmineralized area may vary through a wide range due to variations in local conditions, the ratio will remain fairly constant. This otherwise constant ratio would be modified by the presence of either Cu or Zn ore within the reach of the plant roots. They consider that a Cu/Zn ratio above 0·23 would indicate Cu ore in the bedrock, and a ratio below 0·07 would indicate Zn ore.

F. Advantages and Disadvantages of Biogeochemical Surveys

Disadvantages of biogeochemical methods are the lower degree of confidence in the results caused by the many complications of pH, Eh, drainage, age of organ, plant metabolism, plant exclusion mechanism, amount of sunlight, and other variables. In addition, biogeochemical surveys require a high degree of skill and experience in identifying and selecting plant species, orientation surveys are more local in application than for soils, and species may have an erratic occurrence which causes gaps in coverage.

Biogeochemical prospecting methods have advantages over other geochemical techniques in several respects. The most important are the potential for accumulation of elements from considerable depths below the surface, and the easy access of vegetation for sampling. In areas of transported overburden and immature soil, analyses of trees may allow recognition of anomalies not detectable in near-surface soils. Ease of collection may be an important factor in areas where soil is absent, frozen, cemented, or covered with thick humus or snow.

IV. GEOBOTANICAL INDICATORS

All living plants respond in one way or another to the chemical, physical, and biological environment in which they find themselves. This response normally takes the form of a characteristic habit of growth. For example plants in a warm, moist, and nutritive environment may grow much more luxuriantly than the same species in a more rigorous and less fertile environment. Where conditions are too harsh, the plant cannot develop at all. In the same area, other species will be more tolerant of, or may even prefer, the more rigorous environment. Thus by a process of natural selection, the distribution pattern of plant species becomes adjusted to local variations in the environment.

Geobotany is the study of plants as related specifically to their geologic environment. Many factors that have little or no relation to the geology can have a major influence on the health and distribution of plants. Chief among these are sunlight, length of growing season, elevation, forest fires, blights and insect pests. Many important factors in the natural development of plants do, however, arise either directly or indirectly from their geologic and geochemical environment.

Four aspects of plant appearance and distribution are useful in mineral exploration:

(i) The distribution of individual indicator species.
(ii) The distribution of groups or communities of plants.
(iii) Morphological features of plants, such as unusual size, deformities or colors.
(iv) The gross effects of one or more of the above aspects detectable by aerial photography or other remote sensing techniques.

Individual indicator species were recognized long ago. These indicator species replace normal vegetation of the region either because their growth is promoted by the unusual chemical conditions over ore, or because they are tolerant of conditions in which normal species cannot survive. The distribution of the indicator species may be determined by a major metal of the orebody being sought, a pathfinder element associated with the ore, or some condition associated with ore, such as pH or availability of water. The use of the Se-accumulating species *Astragalus* for detecting Se-bearing U ores on the Colorado Plateau of western U.S.A. is an example of the use of a species sensitive to a pathfinder element. In a majority of cases, the reason for the association of an indicator species with ore is not known.

Two types of indicator plants may be distinguished: universal indicators and local indicators. The universal indicators are found only on mineralized soils and do not grow elsewhere. An example is the calamine violet, which grows only on soils with anomalous Zn content. Universal indicators usually have limited range, but are extremely valuable within limited areas of similar climate, geology, and topography.

Local indicators are species that grow preferentially on mineralized ground within limited areas but grow over non-mineralized ground in other regions.

Plant communities, or associations of plants, can extend the usefulness of individual indicators by allowing recognition of mineralized ground over a wider range of conditions or by allowing recognition of more subtle differences. For example, Se floras include a number of species that are tolerant of Se and of relatively saline soils, and grow preferentially in Se-rich areas (Cannon, 1960a). Serpentine areas are usually characterized by stunted and thinly developed vegetation and different species than adjacent areas, perhaps because of the unusually high contents of Ni, Co, Cr, and other elements, deficiencies of K and other essential elements, or the poor drainage often associated with clay-rich overburden over ultramafic rocks.

Morphological changes induced by anomalous concentrations of trace elements include dwarfism, gigantism, and other changes of form, mottling and yellowing of leaves (chlorosis), and other color changes, abnormal fruits, changes of flower color, and disturbances of the rhythm of growth and flowering. The abnormal colors and morphological features of a plant caused by a poisonous element in the nutrient solution are collectively referred to as toxicity symptoms. Between the level where a plant can tolerate the concentration of a given element in the soil solutions and the level where it cannot survive is a fringe zone where the plant can live but where it shows visible injury as a result of the poisoning effect of the toxic element. Most commonly, the toxemia takes the form of a simple stunting of growth that cannot be uniquely associated with any specific cause. A few plants, however, develop peculiarly diagnostic symptoms that can be interpreted directly in terms of probable excesses of a particular element in the soil (Table 17.6).

A variety of mechanisms for these changes have been proposed (Brooks, 1972, p. 39). Enzymes may be poisoned by Cu, Ag, and Hg. Essential nutrients such as P may be replaced by As with deleterious effects on the plant. Essential nutrients may be precipitated or immobilized in the soil or plant; for

Table 17.6
Physiological and morphological changes in plants due to metal toxicities[a]

Element	Effect
Aluminum	Stubby roots; leaf scorch; mottling.
Boron	Dark foliage; marginal scorch of older leaves at high concentrations; stunted, deformed, shortened internodes; creeping forms; heavy pubescence; increased gall production.
Chromium	Yellow leaves with green veins.
Cobalt	White dead patches on leaves.
Copper	Dead patches on lower leaves from tips; purple stems; chlorotic leaves with green veins; stunted roots; creeping sterile forms in some species.
Iron	Stunted tops; thickened roots; cell division disturbed in algae, resulting in greatly enlarged cells.
Manganese	Chlorotic leaves; stem and petiole lesions; curling and dead areas on leaf margins; distortion of laminae.
Molybdenum	Stunting; yellow-orange coloration.
Nickel	White dead patches on leaves; apetalous sterile forms.
Uranium	Abnormal number of chromosomes in nuclei; unusually shaped fruits; sterile apetalous forms; stalked leaf rosette.
Zinc	Chlorotic leaves with green veins; white dwarfed forms; dead areas on leaf tips; roots stunted.

[a] Source: Cannon (1960b); reprinted from *Science* by permission.

example, phosphate is susceptible to immobilization by a number of elements and minerals. Other proposed mechanisms include catalytic decomposition of metabolites, and effects on cell walls and other structures.

Geobotanical indicators have been used in locating and mapping ground water, saline deposits, hydrocarbons, and rock types, as well as ores.

A. Indicators of Ground Water

Desert plants respond in a very spectacular manner to the availability of water. Phreatophytes, the plants that habitually obtain their water supply from the zone of saturation, indicate a water table within reach of their roo systems. Representatives of this group in the western U.S.A. are alfalfa, mesquite, greasewood, and paloverde (Meinzer, 1927; Robinson, 1958). In water-supply work, phreatophytes are important because of their effective wastage of ground water by transpiration. In contrast to phreatophytes are the xerophytes, which depend on occasional rain water and have only shallow roots (Fig. 17.5). Most other desert plants are indiscriminate in their choice of water and are grouped as mesophytes.

Many plants cannot grow in a soil that is saturated for any length of time, while other plants may thrive under these conditions. In eastern Canada, for example, alders, willows, and some ferns indicate waterlogged soil or ground-water seepage areas. Most other areas of the world also have their water-indicating plant assemblages. In prospecting, these botanical indications are important in that they point to areas where hydromorphic anomalies may have developed as a result of precipitation of ore metals from shallow ground-water.

B. Indicators of Saline Deposits

Many desert plants tolerate or even prefer a nutrient solution with a high content of dissolved salts. Specific indicators of high salinity, or halophytes, are useful to stockmen in recognizing areas unsuitable for grazing cattle. Borate-rich saline deposits in Central Asia are indicated by a series of plant species that are highly tolerant to B. Where the B content exceeds 100 ppm, many plants show deformities (Fig. 17.12) or are subject to diseases such as rotting of the root, increased gall formation, and chlorosis (Buyalov and Shvyryayeva, 1955).

C. Indicators of Hydrocarbons

Experimental investigations have shown that plants rooted in bituminous soils tend to have peculiar forms, distinguished by gigantism and deformity (Viktorov, 1955). Some of these plants show a tendency for abnormal repeated blooming (Kartsev *et al.*, 1959). These responses are not often seen under field conditions, and no immediate applications of indicator plants to petroleum exploration seem likely.

The population of microscopic plant indicators in surface materials may under favorable circumstances be useful as a guide in petroleum exploration (Kartsev *et al.*, 1959). The most successful microbiological methods are based on the identification of bacteria that can obtain their vital energy only from the oxidation of certain hydrocarbons, principally propane. According to Russian workers, the light hydrocarbon gases dissolved in the connate

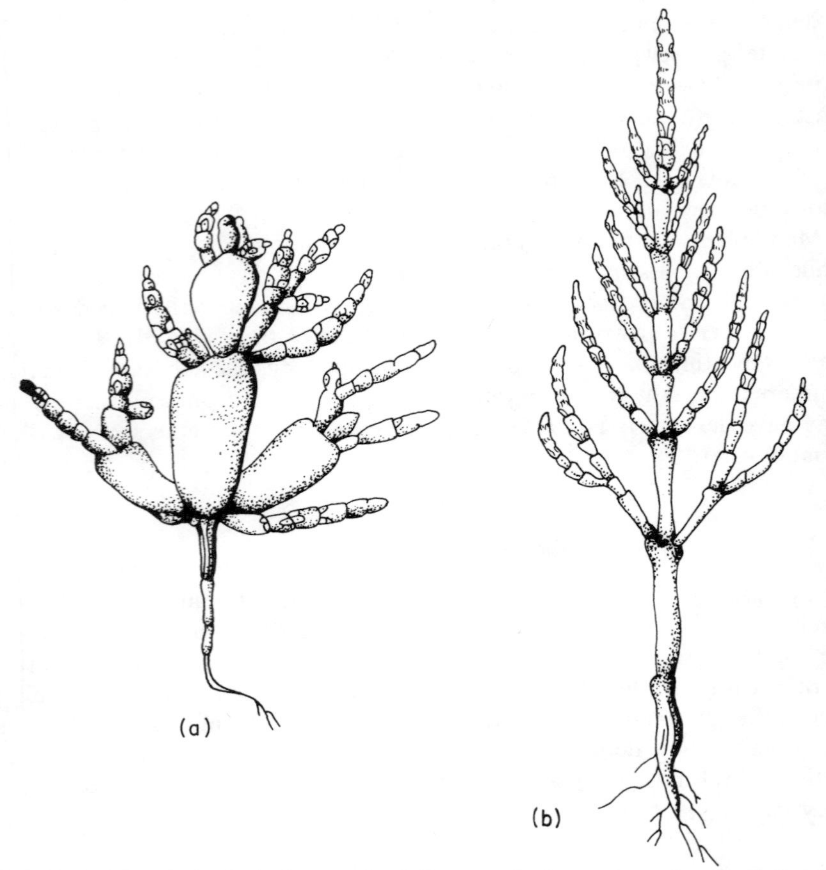

Fig. 17.12. Morphological changes in *Salicornea herbacea* L. under the influence of B. (a) *Salicornea* grown on soil with an increased B content; (b) grown on control sections which contained practically no B. (After Buyalov and Shvyryayeva, 1955.)

water of sedimentary rocks can diffuse upward from their source in deep-lying oil pools. Under the same conditions, the heavier hydrocarbons and the water itself are held firmly in a static condition. The Russian authors claim further that propane is a specific indicator of liquid or gaseous hydrocarbons and cannot be derived from bituminous sediments or black shales. They have reported the observation of a dynamic flow of light hydrocarbons extending continuously from an oil pool buried at depths of a thousand meters directly to the surface of the earth. Where hydrocarbons reach the zone of oxidation, they are met by oxidizing bacteria, among which are the specific propane-oxidizing bacteria. These bacteria are determined by incubating a culture of soil or soil water in an atmosphere of propane mixed with sterile air. After a period of 10 days, the cultures are examined and appraised qualitatively. The development of bacteria of any kind, as evidenced by a scum on the surface of the cultured samples, is an indication of propane bacteria, inasmuch as no other bacteria could survive or multiply in this environment. A success ratio of 65% in obtaining petroleum from wells drilled on the basis of microbiological anomalies is reported. Of these discoveries, a large number are reported to have developed into commercially producing oil fields (Mogilevsky, 1959).

D. Indicators of Rock Types

Limestone soils commonly support a diagnostic assemblage of indicator species. It is not always clear whether this response is due to soil pH, or to the Ca content of the soil solutions. Recognition of limestone indicators may be helpful both in geologic mapping and in prospecting for ores that occur preferentially in limestone country rock.

Areas of rock alteration and pyritization may support a characteristic plant assemblage (Billings, 1950). Here again it is not clear whether the geobotanical response is from the acid environment of the oxidizing pyrite, from the abundance of some minor element, or from the texture of the material.

Ultramafic rocks in virtually all climates have a very profound influence on the ecological assemblages of plants growing on them. In general, the vegetation in areas underlain by ultramafic rocks is conspicuously stunted and thinly developed. This relationship is ordinarily so clearly defined that the contacts of ultramafic rocks can be sketched accurately on aerial photographs.

Indicator assemblages have been noted over many other rock types, such as granite, quartzite, shale, or basalt. In many of these associations, the cause seems to be a matter of relative drainage and availability of water.

Plant communities are sensitive to types of overburden, as well as to types of bedrock. For example, in the Dugald River area of Australia, different floras occur over alluvial overburden along streams, mesas with fossil laterite development, siliceous rocks, calcareous rocks, and zones of Pb–Cu–Zn mineralization (Nicolls *et al.*, 1965). Similarly, in South West Africa, the distribution of plant species could be used to distinguish shallow bedrock, relatively deep sand cover, laterite, calcrete, and normal soils over siliceous and calcareous rocks (Cole, 1971).

E. Indicators of Ore

Long lists of indicator plants for specific elements have been compiled by various authors. These lists include specific indicators for virtually all the ore metals (Lidgey, 1897; Dorn, 1937; Vinogradov, 1954; Nesvetaylova, 1955b; Cannon, 1960b). Unfortunately, only a very few of all the plants listed are now, or ever have been, of any real help in prospecting. The reason is that most of these species were recorded from the unnaturally acidified and contaminated soils in the vicinity of old mine workings rather than from virgin areas. Indicator plants that are definitely known to have been used in prospecting are listed by Brooks (1972, Table 4.1).

The first of the indicator plants to be used in prospecting was the calamine violet, which thrives only on Zn-rich soils in the Zn districts of central and western Europe (Schwickerath, 1931). In the early days of development of these districts, the distinctive yellow blossoms of this plant are reported as having led to the discovery of many Zn deposits buried under shallow cover.

Selenium indicators were first described by Beath *et al.* (1939). These are a group of plants, chiefly species of *Astragalus*, in the Rocky Mountain states that will grow only on soils containing an excess of Se. This relationship is so consistent that it is possible to make accurate maps of seleniferous geologic formations simply by outlining the distribution of Se indicators. Inasmuch as Se-rich forage is highly toxic to sheep and other livestock, the recognition of seleniferous areas by means of the Se-indicator plants is of considerable practical importance. In prospecting, the Se indicators are valuable because they point to the location of the Se-rich U ores of the Colorado Plateau. Here a substantial ecological assemblage of Fe-indicators and also S-indicators has been described (Cannon, 1957). Deep-rooted members of this association of indicator plants have been found over U ore buried at depths up to 20 m.

The "copper flower", *Becium homblei* de Wild (formerly *Ocimum homblei*) discovered in Zambia in 1949 by G. Woodward, has been perhaps the most successful of all the geobotanical indicators (Horizon, 1959). This plant, a herb of the Labiatae family, grows to a height of 45–60 cm. Although it has a characteristic flower and leaf form that can be recognized on careful examination, it is very similar in appearance to the species *B. obovatum*, which is widespread away from mineralized areas. *Becium homblei* was originally thought to grow only in soil with a Cu content greater than 100 ppm (Horizon, 1959). However, more recent stuuies show that *B. homblei* is relatively widespread in northern Zambia, and occurs in Rhodesia in both Cu-rich and Cu-poor areas (Howard-Williams, 1970). The preference of *B. homblei* for Cu-rich soils appears to result from its competition with *B. obovatum*. The latter species grows more rapidly on soils of normal pH and Ca content, but is unable to grow in soils of low pH, low Ca, or high Cu, and perhaps high Ni. In the latter areas, *B. homblei* takes its place. Exchangeable Cu may be more important than total Cu. General experience in the Zambian Copperbelt shows that outlines of copper-flower distribution are almost identical with the outlines of either underlying Cu deposits or lateral seepage anomalies (G. Woodward, personal communication; Horizon, 1959).

Another copper flower, *Haumaniastrum robertii* (formerly *Acrocephalus*), has been described from the Shaba area in Zaire, immediately north of the Zambian Copperbelt (Duvigneaud, 1958). This is reported as a small annual mint whose resistance to toxicity appears to be infinite. This also is a herb of the Labiatae family of the Ocimoidae tribe.

Still another copper flower, *Gypsophila patrini*, or "kachim", is associated with Cu in the Rudny Altay deposits of central Asia (Nesvetaylova, 1955b). Kachim grows so selectively on Cu-bearing rocks that even small Cu-bearing dikes may be marked by a strong growth of this Cu indicator.

A local indicator of Cu in Arizona is the California poppy, observed over the outcrop of the San Manuel Cu deposit and at other deposits (Chaffee and Gale, 1976). The distribution of this species is locally confined to Cu-rich soil, and its population density is closely proportional to the Cu content of the soil. In neighboring areas of slightly different climate, the poppy can grow almost anywhere, whereas in other areas not too far distant it cannot grow at all, even where Cu is abundant (Lovering *et al.*, 1950).

Chlorosis, or the yellowing of the leaves of plants, has been a useful though non-specific guide to ore. Nickel, Cu, Co, Cr, Zn, and Mn are all antagonistic to Fe in the plant metabolism. Excesses of these elements produce a deficiency of Fe necessary in the formation of chlorophyll, with a resulting decoloration of the leaves (Cannon, 1960b). Chlorosis can result from a variety of other causes, including plant infections, improper drainage, and excess acidity in the soil. At the same time, abnormally colored leaves always deserve the attention of the prospector, inasmuch as chlorotic symptoms, particularly in the grasses and small flowering plants growing in seepage areas, are common in many mineralized areas.

The absence of normal vegetative cover rather than the presence of specific geobotanical indicators may, under some conditions, be useful as an ore guide. Acid from the oxidizing pyrite of sulfide deposits or toxic excesses of soluble metals may prevent the development of a normal plant ecology. In Zambia some of the big Cu deposits were originally discovered because of lack of tree growth over the suboutcrop. Five areas of naturally occurring Pb-poisoned soil have been found in Norway by Låg and Bølviken (1974). Only *Deschampsia flexnosa* grows in the bare areas. Similarly, Grimes and Earhart (1975) have found *Eriogonum ovalifolium* to be confined to bare areas in otherwise forested mountains of Montana. Extensive chemical analysis has shown a strong correlation only with Cu in the soil.

In tundra and mountain regions, mosses and lichens are widespread and are distinctly affected by the nature of the substrate on which they grow. A number of copper mosses preferentially growing on substrates high in Cu have been reported (Brooks, 1972, p. 33), but require expert training to identify. Czehura (1977) found that the color of a lichen growing on rock of a Cu prospect in California was determined by the Cu content of the underlying rock.

V. GEOBOTANICAL SURVEYING TECHNIQUES

The first step is to carry out a preliminary orientation study involving the preparation of population maps of all plant species growing both in the vicinity of known but undisturbed mineralization and in barren areas of otherwise similar characteristics. The abundance and distribution of individual species and plant communities may be mapped and compared from area to area by counting the individuals of all species within small areas called quadrats, which may range from a meter to a few tens of meters on a side. Alternatively, the abundance of species within a specified distance of a traverse line may be recorded.

Cannon (1957, p. 408) describes her experimental technique as follows:

> The information on indicator plants was established by marking off 10 × 5 foot areas over known ore bodies in a number of districts, and similar areas over unmineralized parts of the same bed with similar exposure and slope. Complete lists of plants in each plot were made, and final lists of indicator and tolerant plants were derived from them. When any of these plants are observed in a new area, a careful study of their distribution should be made to determine whether and how they can be used in prospecting.

Indicator plants are best found and mapped when they are in bloom. Mapping plants at other times of year can, of course, be done, but the work progresses at a very much slower rate and with very much greater danger of overlooking an occurrence of useful species. Once a system of identifying geobotanical indicators has been established, a geobotanical survey consists merely in plotting their occurrence on a map and then sketching the outlines of areas where they occur.

Several workers have discovered new prospects by searching libraries and herbariums for additional localities of known indicator plants, and visiting these areas in the field. Several Cu deposits are said to have been found by this method (Cannon, 1960b).

Major advantages of geobotanical exploration are that immediate results are available from visual observations, and that costs are low once an orientation survey has been carried out. Surveys can obtain considerable information on the distribution of bedrock types and overburden, as well as metal anomalies in poorly exposed areas. Geobotanical anomalies may be detectable in areas of transported overburden, which would furnish little or no clue to mineralization with conventional soil surveys.

Major disadvantages of the technique are the high degree of botanical skill required and the relatively local applicability of orientation surveys. The variety of conditions influencing the growth of plants leads to complex relationships between the geology, geochemistry, overburden, topography, and drainage which must be evaluated in each new area. Timing of surveys and orientation work may be crucial in the case of some plants that are recognizable only during the seasons of flowering or growth. In general, these disadvantages have inhibited geobotanical work, but the cost and time advantages of aerial, photographic, and other methods of remote sensing are focusing renewed attention on this method.

VI. REMOTE SENSING OF GEOBOTANICAL ANOMALIES

Airborne or satellite-borne methods are an obvious means of speedily conducting a geobotanical survey if reliable methods of interpretation can be developed. In principle, remote methods could detect changes in the distribution of species and in the size, spacing, or color of plants affected by anomalous metal concentrations. Possible remote-sensing methods, using different sensing methods and different portions of the electromagnetic spectrum, include visual observation from an aircraft or satellite, aerial photography (using black-and-white, color, or color-infrared film), instrumental sensing of reflectance in the visible and near infrared or of emission in the far infrared, and measurement of microwave and radar reflectance (Fig. 17.13).

Visual observations from a helicopter have been used by Cole (1971) to examine the distribution of tree species in jungle terrain. This method is applicable to detection of readily identifiable indicator species and plant communities, and to some types of toxic effects. In addition, many geological features, such as rock types, structures, and areas of shallow ground-water or seepage, can be identified in this manner. Similar results may be obtainable by low-altitude color air photography, or in favorable cases, by black-and-white photography.

The Fraunhofer line discriminator is a specialized method for measuring luminescence of natural materials. The discriminator operates mainly in the visual region of sunlight and measures the light emitted in the narrow wavelength regions of the dark bands created by absorption in the sun's outer atmosphere. Tests indicate that vegetation stressed by metal toxicity or other

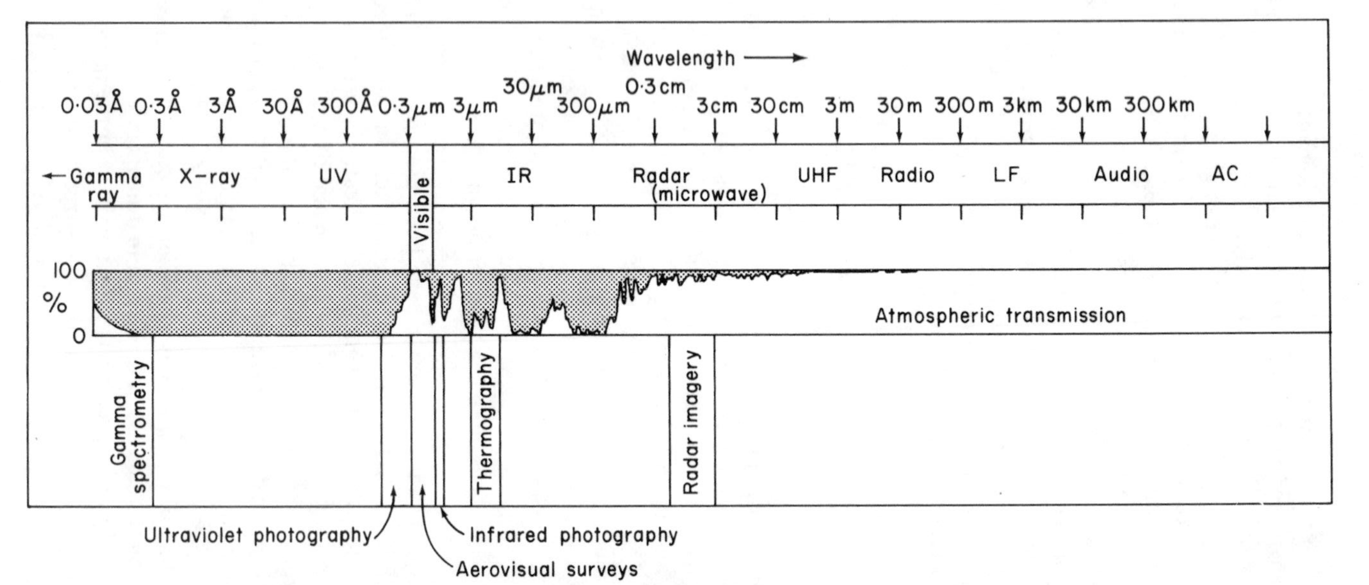

Fig. 17.13. Schematic representation of the electromagnetic spectrum and its significance in remote sensing. (After Colwell *et al.* (1963), from Brooks (1972). Reprinted with permission, © American Society of Photogrammetry, 1963.)

processes interfering with chlorophyll metabolism has anomalous luminescence (Hemphill *et al.*, 1977). Juniper trees growing on soils anomalous in Mo in the Pinenut Mountains of Nevada emit higher than normal luminescence.

Photography of reflected light in the near infrared adds information not visible to the unaided eye. Infrared can be recorded on special color film that is sensitive to visible and infrared light (Brooks, 1972, p. 53), by camera and film combinations that detect only infrared, or by electronic sensors of the type used in the Landsat satellite which record radiation in the 0·7–0·8 μm and 0·8–1·1 μm bands of the near infrared, as well as the 0·5–0·6 μm and 0·6–0·7 μm bands of the visible spectrum. Several investigators have shown that chlorotic vegetation resulting from metal toxicity has a different reflectance spectrum than normal vegetation (Fig. 17.14). However, the chlorotic vegetation has a higher reflectance for some species and a lower reflectance in others. Lyon (1977) and Bølviken *et al.* (1977) have detected an anomaly in

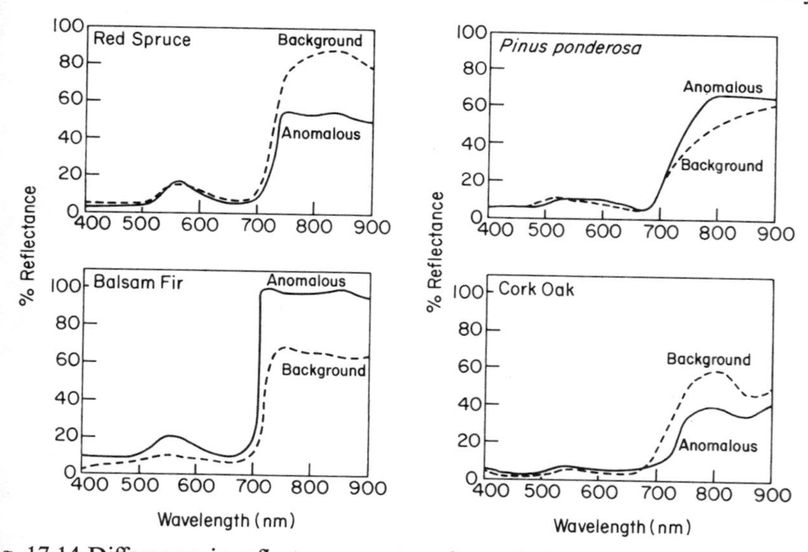

Fig. 17.14 Differences in reflectance spectra of trees in background and mineralized areas. (Top and bottom left after Yost and Wenderoth (1971), top right after Howard *et al.* (1971), bottom right after Press (1973); all cited in, and composite illustration after, Press (1974).)

the ratios of reflected light in the Landsat channels over Mo-rich vegetation in the Pinenut Mountains in Nevada and in Pb-poisoned areas in Norway. Encouraging results have been obtained with film techniques in arid regions, where metal toxicity symptoms are most pronounced (Press, 1974). Research continues into methods of distinguishing anomalies caused by metal toxicity from the effects of changes in leaf area, plant density, season, moisture, an many other effects. The limited resolution of the satellite equipment (abou 70 m) has also been a problem, because many geobotanical anomalies ar small in size.

To date, no application of thermal infrared techniques to geobotanica surveying have been suggested. At wavelengths of radar which are about th same as the dimensions of shrubs and trees, research on detection of veget tion has been undertaken, but no useful techniques have been developed.

Research on detection and interpretation of geobotanical anomalies b remote methods offers the potential of rapid scanning of large areas at lo cost. However, the numerous problems related to species, plant spacing, pla age, rock types, depth and type of overburden, moisture, sunlight, and oth variables will require a good deal of research to develop methods that can b used with confidence. The recognition of chlorotic or incipiently chlorot vegetation seems to offer the best possibilities.

Chapter 18

Volatiles and Airborne Particulates

he gases emanating from ore deposits or oil fields have recently attracted ncreased attention as possible exploration guides. The characteristic smell of mine dump containing sulfides is known to every prospector and mining eologist. In Scandinavia, dogs have been trained to smell out buried glacial oulders containing oxidizing sulfide minerals (Nilsson, 1973; Ekdahl, 1976). f these odors are real, it is logical to assume that the gases responsible can e sampled and analyzed quantitatively for the same constituents that come o us physiologically in the form of smells.

Until recently, however, very little has been done to develop the use of apor-phase exudations and emanations from the earth as indicators of arious geological phenomena. The reasons are not hard to find. Naturally ccurring gases are difficult to sample and even more difficult to analyze hemically for their trace constituents. Further, the trace constituents of a as can suddenly disappear by reaction with the matrix and with other gases, y oxidation, by dilution with atmospheric air, by biological reactions, and y sorption to the walls of the sample container. Thus time variations in the omposition of natural vapors can be extreme, often making acceptable eplication nearly impossible. On the other hand, gases have an undeniable ttraction as geochemical indicators because of their extreme mobility ompared with other naturally occurring fluids.

Three main environments may be identified in which gases can be studied: he open atmosphere, the pore space of soil and overburden, and surface and round waters with dissolved gases. Most aspects of dispersion in natural aters are discussed in Chapter 14, so only specialized aspects involving gases re covered in this chapter.

I. PROPERTIES OF GASES AND AEROSOLS

A gas is distinguished from a liquid in that it has a spontaneous tendency tc become distributed uniformly throughout any container. For our purposes the term "gas" is used to include any atomic or molecular species that car occur in significant concentrations as a vapor under ambient conditions without regard for whether it actually does occur primarily in the vapo phase, the aqueous phase, or otherwise dispersed through geological mate rials. By this definition, gases include all substances that are significantl volatile.

At pressures less than about 10 atm and at the temperatures of the surficia environment, gases closely follow the ideal gas law (Garrels and Christ 1965):

$$PV = nRT \tag{18.1}$$

where P is pressure of the gas, V is the volume of n moles (gram-molecula weights) of the gas, T is the absolute temperature (°K = °C + 273°) and R is the gas constant (82·06 cm^3-atm/deg-mole). This relationship leads to th important characteristic that equal volumes of any gas, held at identica pressures and temperatures, contain equal numbers of molecules and moles Specifically, at 0 °C and 1 atm (standard temperature and pressure, or STP), 22 400 cm^3 or 22·4 l of any gas contain one mole. In addition, at surficia temperatures and pressures, gases tend to mix ideally (unless they reac chemically) so that in a mixture of gases, the effective pressure (P_i) or fugacity (f_i) of a gas i is proportional to its mole fraction (X_i) in the mixture:

$$P_i = f_i \cong X_i P_t \tag{18.2}$$

where P_t is the total pressure, usually about 1 atm in the surficial environment

Gases tend to equilibrate with liquid phases with which they are in contact Specifically, all gases dissolve in water to a measurable degree. At and above the water table, the gas in soil air tends to equilibrate with the gas dissolved in the ground water and the adsorbed moisture on mineral surfaces.

In addition to gases, the atmosphere also contains fine particulate material This fine material, mostly less than about 10 μm in diameter, includes both liquid and solid particles which are small enough to be kept suspended by the normal random movements of the air. This mixture of gas and particles form an aerosol with most of the bulk properties of a gas, but with a chemica composition strongly dependent on the nature of the particulate material especially for trace constituents.

The discussion in this chapter deals mainly with substances that occur in the gas phase at the present time. Vapor-phase constituents of ore-formin fluids now occurring as solid-phase aureoles surrounding mineral deposits are considered separately in Chapters 4 and 5.

II. UNITS OF MEASUREMENT

With non-radioactive vapor species, the common unit of measurement is the volume ratio, which is a dimensionless unit. Thus 3 ppm v/v of He means that a million volume units (cm^3, m^3, or whatever) of the sampled gas contain three of the same units of He. This is the most satisfactory unit, as it is independent of temperature and pressure, at least under normal working conditions. By virtue of the ideal gas and ideal mixing behavior mentioned above, volume concentrations are equivalent to molecular concentrations. Thus 3 ppm v/v is also 3 ppm of He atoms out of the total number of gas atoms and molecules in the sample, and has a partial pressure of 3×10^{-6} atm where the total pressure is 1 atm.

A less commonly used unit is the ratio of weight to volume. In this case, the weight and volume units and the temperature must be specified, for example micrograms per cubic meter at 25 °C. Inasmuch as equal volumes of a gas contain equal numbers of molecules, a conversion of one of these units to the other involves the molecular weight of the constituent and the standard volume (22·4 l) containing one gram-molecular weight of the constituent at STP (Table 18.1). For other temperatures and pressures, accurate conversion involves a further adjustment using the ideal gas law (eqn 18.1).

Table 18.1
Units of measurement for gases

$$1\ \text{v/v unit} = \frac{22{\cdot}4}{\text{Mol. wt}} \times \text{wt/v unit (at 0 °C and 1 atm)}$$

$$1\ \text{wt/v unit} = \frac{\text{Mol. wt}}{22{\cdot}4} \times \text{v/v unit (at 0 °C and 1 atm)}$$

$$\begin{aligned}\text{Example: } 3\ \text{ng/m}^3\ \text{Hg} &= 3 \times 10^{-9}\ \text{g Hg/m}^3 \\ &= 3 \times 10^{-12}\ \text{g Hg/l} \times \frac{22{\cdot}4\ \text{l/mole}}{200{\cdot}6\ \text{g/mole}} \\ &= 3{\cdot}35 \times 10^{-13}\ \text{vol. Hg/vol. gas} \\ &= 3{\cdot}35 \times 10^{-7}\ \text{ppm Hg v/v}\end{aligned}$$

Radon and other radioactive elements
1 curie (Ci) = $3{\cdot}7 \times 10^{10}$ disintegrations/s
1 picocurie (pCi) = 10^{-12} Ci
1 pCi Rn = $6{\cdot}5 \times 10^{-18}$ g
1 pCi U = 3×10^{-6} g

Radon, or more specifically ^{222}Rn, is the only important radioactive gas. The unit of measurement commonly used for radioactive elements is the curie, usually abbreviated Ci. One curie is the quantity of any radioactive element (U, Th, Ra, Rn, K) that produces $3{\cdot}7 \times 10^{10}$ distintegrations per second. In the case of Ra, this is approximately 1 g. At radioactive equilibrium, approached after many half-lives of the daughter element, each member of a radioactive series produces the same number of disintegrations per second, and hence will contain the same number of curies. The most practical unit for most natural radioactivity is the picocurie (10^{-12} Ci), as summarized in Table 18.1.

Inasmuch as aqueous solutions of gases can be reported in a variety of units, the units used in each case must be clearly stated. One common unit is the volume ratio of the gas in question to the volume of total dissolved gases. To convert this ratio to content of the gas in the sample of water, multiply by the content of total gas in the water. Thus if the dissolved gases in a water sample contain 39 ppm v/v He and the water contains $29{\cdot}6\ cm^3$/l dissolved gas, then the water contains $39 \times 29{\cdot}6 = 1150 \times 10^{-6}\ cm^3$ He/l H_2O (liquid), or $1{\cdot}15 \times 10^{-6}\ cm^3$ He/cm^3 H_2O. These ratios may be converted to weight of gas per weight or volume of water using the conversion factors of Table 18.1 and the density of water.

III. SOURCE OF NATURALLY OCCURRING GASES

The gases contained in the pore spaces of rocks and soils represent mixtures of atmospheric gases, gases of deep-seated origin, and gases of shallow origin, with their vapor-phase reaction products with each other and with the rocks through which they migrate. Figures 18.1a and 18.1b illustrate diagrammatically the behavior of some common gases in the outer crust of the earth.

A. Atmospheric Gases

Except for water vapor, the composition of the atmosphere is remarkably stable, with average values shown in Table 18.2. Of these, oxygen is the dominant reactive constituent. Surface water containing dissolved O_2 can circulate to a depth of tens or sometimes hundreds of meters below the surface, where it sets up an oxidizing environment intermediate between the earth's surface and subjacent crustal rocks. Atmospheric CO_2 is maintained at a level of about 0·035% by equilibrium with CO_2 dissolved in sea water. The content of He in the atmosphere represents a balance between the amount

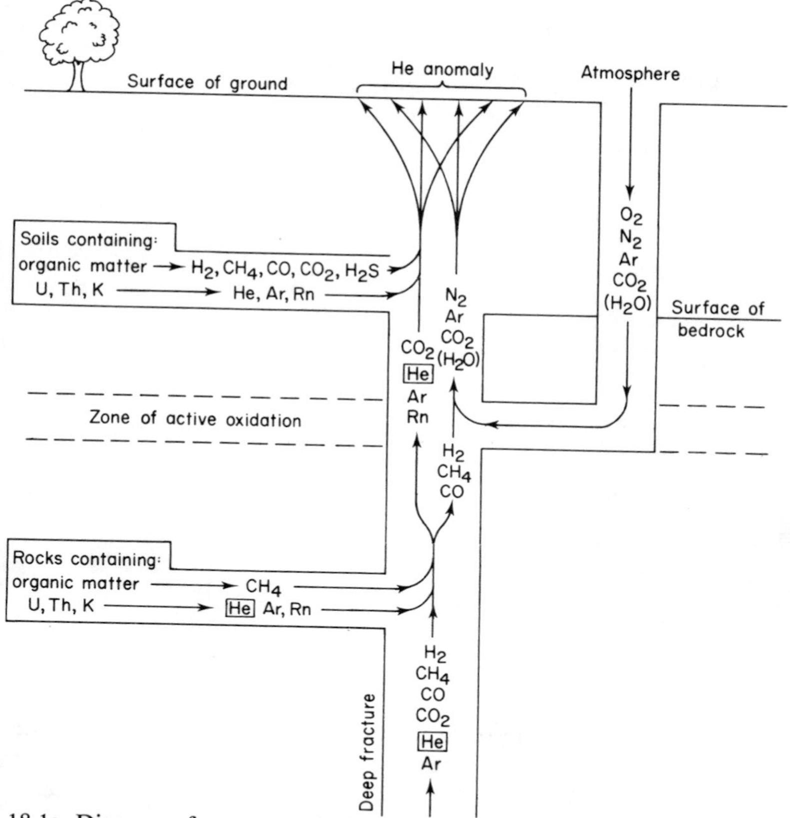

Fig. 18.1a. Diagram of sources and migration of some gases in a deep fracture zone.

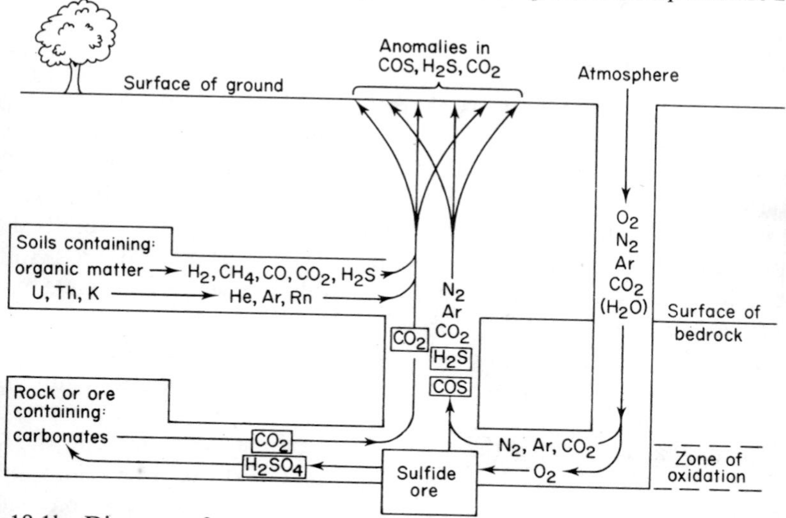

Fig. 18.1b. Diagram of sources and migration of some gases over an oxidizing sulfide deposit.

Table 18.2
Composition of the atmosphere[a]

	% v/v	ppm v/v
Nitrogen	78·088	—
Oxygen	20·949	—
Water	0·1–4	—
Argon	0·93	9300
Carbon dioxide	—	350
Neon	—	18
Helium	—	5·24
Methane	—	1·4
Krypton	—	1·14
Hydrogen	—	0·5
Nitrous oxide	—	0·2–0·4 (rural areas)
Xenon	—	0·09
Carbon monoxide	—	0·08–0·5 (rural areas)
Sulfur dioxide	—	0·007 (rural areas)
Mercury	—	$0{\cdot}2\text{–}0{\cdot}8 \times 10^{-6}$
Radon	—	0·01–0·45 pCi/l

[a] Source: Fairbridge (1967), Butcher and Charlson (1972), and Dyck (1973).

being fed to the atmosphere from the depths of the earth and that escaping into outer space from the upper atmosphere. The content of H_2 represents a similar balance, but with an additional source by dissociation of water vapor and other reactions in the upper atmosphere.

B. Deep-seated Gases

The deep-seated environment is characterized by an extremely low oxygen fugacity (partial pressure or effective concentration of elemental oxygen), and relatively high fugacities of reducing gases, such as H_2 and CH_4. The reduced state of the gases is determined by the occurrence of Fe in the ferrous rather than in the ferric state. For example, water at a pressure of 1000 atm (a pressure reached about 4 km below the surface) and 700 °C, and in equilibrium with quartz, magnetite, and fayalite (a ferrous silicate), has an H_2 partial pressure or fugacity (f_{H_2}) of about 20 atm, compared to an oxygen fugacity of $10^{-17.5}$ atm (Wones and Gilbert, 1969; Robie *et al.*, 1978). If 1% of the fluid is composed of carbon species, then at equilibrium $f_{CO_2} = 7{\cdot}5$ atm, $f_{CH_4} = 2{\cdot}5$ atm, and $f_{CO} = 0{\cdot}07$ atm. Any sulfur present will occur mainly as H_2S rather than SO_2. These estimates assume that the fluid contains only H, O, C, and S; in a more complex mixture, other species might be present, but the fugacity ratios among carbon gases and among sulfur gases would remain the same. Ratios of gases at lower temperatures are discussed by Holland (1965).

Under some conditions, sea water or rocks containing sea salts may be introduced into the deep-seated environment with the generation of a family of volatile species that are well known at volcanic fumaroles (HCl, H_2S, NH_3, etc.). The escape of gases representing the primordial composition of the earth has been postulated, but authorities differ as to whether they make up a significant fraction of the total complex of deep-seated gases.

C. Radiogenic Gases

Helium and Rn are generated by the radioactive decay of members of the U and Th series, and Ar by the decay of ^{40}K. These may be of either deep-seated or shallow origin. A large fraction of the radiogenic gases may be trapped in the parent rock and remain there until the rock is fractured during tectonic activity. Hydrogen is also reported as a product of breakdown of water by energetic radiation (Kravtsov and Fridman, 1965).

D. Biogenic Gases

Light hydrocarbons, of which methane (CH_4) is the most abundant member, can be generated by the disintegration of organic matter. Even in some fumarolic gases, the isotope ratios in CH_4 suggest an origin from sedimentary organic matter (Gunter and Musgrave, 1971). Where this process takes place within an impermeable sedimentary sequence, it can lead to accumulations of natural gas. In fractured rocks, however, CH_4 can escape to the surface. At and near the surface, CH_4, H_2, H_2S, CO, and H_2O can be generated by bacteria acting on plant debris (Schlegel, 1974). Organic sulfur gases such as dimethyl sulfide ($S(CH_3)_2$), dimethyl disulfide ($S_2(CH_3)_2$), and methyl mercaptan (CH_3SH) are also generated in relatively oxidizing zones of soils and decaying organic matter (Rasmussen, 1974). Appreciable amounts of CO are found in the ocean and are probably generated biogenically (Swinnerton *et al.*, 1970; Seiler, 1974). There is some evidence that living plants above the surface of the ground exude metals into the atmosphere in the form of liquid- or vapor-phase organometallic complexes (Curtin *et al.*, 1974).

E. Gases Generated in Sulfide Deposits

The abundance of free oxygen in buried sulfide deposits undergoing oxidation falls midway between that in the atmosphere and that at greater depth. Depending on conditions, any of a considerable number of sulfur-containing gases could be generated by reactions between the sulfide minerals and O_2, CO_2, and CH_4. Of these, H_2S, COS, and SO_2 have been reported in the literature (Shipulin *et al.*, 1973; Kahma *et al*, 1975; Hinkle and Turner, 1976), and several organic sulfur gases have been suspected. The stability of these gases depends on oxidation potential, pH, pressures or concentration of CO_2, CH_4, sulfur species, and other variables. Their abundance also depends on reaction rates forming and destroying each species, which, in turn, depend on the activity of biological and inorganic catalysts of the reactions. Figure 18.2 illustrates the equilibrium stability of some of the sulfur gases under conditions approximating many oxidizing orebodies. Hydrogen sulphide is calculated to be the most abundant sulfur gas, and COS the next most abundant. Except for H_2S, the sulfur gases shown are significant only

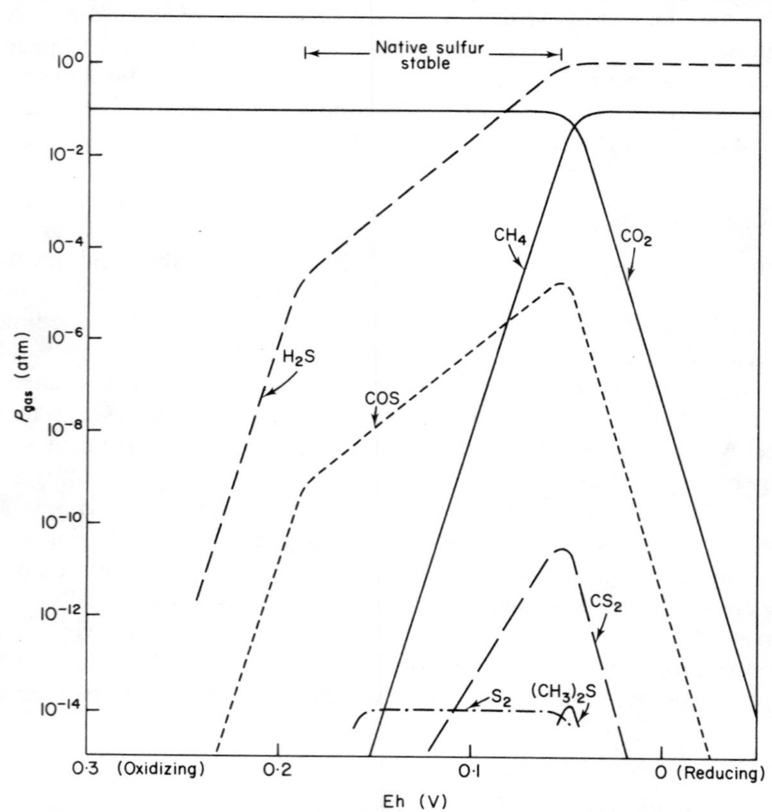

Fig. 18.2. Theoretical stability of some sulfur gases as a function of oxidation potential in equilibrium with a solution of pH 2, activity of SO_4^{2-} or H_2S = 10^{-1} M, at $P_{CO_2} + P_{CH_4} = 0{\cdot}1$ atm at 25 °C. SO_2 is less stable than any other species shown.

at intermediate oxidation potentials. Sulfur dioxide is much less stable than any of the gases shown, because it tends to be converted to SO_3 and SO_4^{2-} in the presence of water. Changes of pH, P_{CO_2}, and other variables can change the equilibrium ratios of these gases, but changes large enough to affect the relative order of stability (H_2S–COS–other species) are unlikely in the natural environment. However, generation and destruction of the gases in non-equilibrium proportions is likely, as is seen in soils where a preponderance of organic sulfur gases is formed rather than the more stable H_2S (Rasmussen, 1974). In the open air, COS is relatively stable, with a probable life of more than 600 days, compared to 1 day for H_2S, 3 days for $(CH_3)_2S$, 40 days for CS_2, and 9 days for SO_2 (Graedel, 1977).

The sulfur in sulfides can react with water and O_2 to form sulfuric acid which may attack whatever carbonates are present with the release of CO_2. During oxidation of sulfide minerals, any contained Hg may be released in vapor form to escape to the surface with the other gases.

F. Atmospheric Particulates

Water droplets, in the form of clouds and fog, are the most common particulate material in the atmosphere, but are not directly significant in exploration, so will not be considered further. Solid particles range in size from large molecules in the nanometer (nm $= 10^{-9}$ m $= 10$ Å) range to tens of micrometers (μm $= 10^{-6}$ m), but the mass of particulate material is generally most concentrated between 0·1 μm and 100 μm (Butcher and Charlson, 1972). At the smaller end of this range, normal turbulence in the air keeps particles suspended almost indefinitely. The higher levels of the atmosphere contain a relatively uniform population of particles resulting from mixing on a hemispheric scale (Peirson *et al.*, 1974).

One important source of solid particles is dust on land which is entrained by wind (Weiss, 1971). This component has the composition of rock and soil in the source area, somewhat fractionated according to size. A second major source arises by the formation of sea spray and the evaporation of the resulting droplets to produce salt crystals. A third source is vegetation, which sheds particles mainly from the surfaces of leaves. Beauford *et al.* (1975, 1977) observed that in gentle air currents most particles are less than 0·2 μm in size, but with normal surface breezes, particles of 1 μm are common. This component of particles can reflect the composition of soils in which the plants are growing. Suggested origins of these particles are the waxy material broken off from the surfaces of leaves, condensed volatile organic material exuded by plants, and loss of salt crusts formed during evapotranspiration by plants. Other types of particles that have been identified in the atmosphere include spores, pollen, microorganisms (Barringer, 1976, 1977), and a wide variety of material resulting from man-made pollution. Agglomeration of small particles is not uncommon, and reactions can take place within particles, especially those including a liquid component.

IV. REACTIVITY OF GASES

The high mobility of gases is the feature that makes them of outstanding value as geochemical indicators. For them to be of maximum usefulness, therefore, they should not be subject to chemical reactions that tend to restrict their mobility. Thus the relative reactivity of various naturally occurring gases is an important factor in assessing their usefulness.

Some reactions convert the gas species to a completely different chemical form having very different properties. Examples are oxidation of sulfur gases to dissolved SO_4^{2-} or of H_2 to H_2O, and formation of solid sulfides from H_2S. In this type of reaction the gas is permanently destroyed.

Another very important type of reaction is absorption in water. The solubility of some gases in water, expressed as cubic centimeters of the gas per 100 cm^3 of the liquid, in equilibrium with 1 atm of the gas, is listed in Table 18.3. In general, the concentration of a gas dissolved in water is proportional to the pressure of the gas, so that

$$m_i = kP_i \tag{18.3}$$

where m_i is the concentration of gas i in the water (expressed in any convenient units), P_i is the effective pressure of the gas, and k is an empirical constant which may be derived from the data in Table 18.3. Thus, the amount of gas dissolved in water decreases proportionately at pressures lower than 1 atm. Note that k is not necessarily constant from 1 atm down to natural levels; for example k for CO increases by a factor of eight between 1 atm and natural CO levels of air (Meadows and Spedding, 1974).

Table 18.3 shows that in terms of their solubility in water, gases fall into two well-defined groups. Most of them show relatively low solubility, less than 6 $cm^3/100\ cm^3$. These gases do not react with water to any significant extent, and tend to react slowly with most other natural substances. The group includes N_2 and the noble gases He, Ar, Ne, and Rn. Although CH_4, H_2, and CO are not stable in the presence of abundant free oxygen, a substantial fraction can survive for a considerable time because their rates of decomposition are relatively slow. Oxygen itself falls into the low-solubility group, and is known to react relatively slowly in performing its function of oxidation. However, the rate of decomposition of all these gases can be greatly increased by catalysts, which are usually substances containing an element of variable oxidation state, such as Fe or Mn, combined with a large surface area.

In contrast to the low-solubility gases are the species that react instantaneously with water to form ionized and acid radicals. Carbon dioxide dissolves to form H_2CO_3, HCO_3^-, and CO_3^{2-}; H_2S goes to HS^- and S^{2-}; and SO_2 goes to HSO_3^-, SO_3^{2-}, and (by oxidation) to SO_4^{2-}. The products of hydrolysis of COS are uncertain but probably include HS^- and S^{2-}.

Table 18.3
Solubility of gases in pure water in the presence of 1 atm of the gas

Gas	Solubility[a]	*T* (°C)	Reference[c]
Argon (Ar)	5·6	0	(1)
	3·01	50	(1)
Carbon dioxide[b] (CO_2)	171·3	0	(1)
	90·1	20	(1)
Carbon monoxide (CO)	3·5	0	(1), (5)
	2·32	20	(1)
Carbon disulfide (CS_2)	70	22	(1)
Carbonyl sulfide (COS)	80	13·5	(2)
	54	20	(1)
Helium (He)	0·94	0	(1)
	0·94	25	(1)
Hydrogen (H_2)	2·14	0	(1)
	1·91	25	(1)
Hydrogen sulfide[b] (H_2S)	437	0	(1)
	186	40	(1)
Methane (CH_4)	3·3	20	(3)
Neon (Ne)	1·47	20	(1)
Nitrogen (N_2)	2·33	0	(1)
	1·42	40	(1)
Oxygen (O_2)	4·89	0	(1)
	3·16	25	(1)
Sulfur dioxide[b] (SO_2)	3937	20	(4)

[a] Solublity in cm^3 of gas (STP)/100 cm^3 of liquid water.
[b] Solubility strongly dependent on pH—increases at higher pH.
[c] (1) Weast (1977, Table of physical constants of inorganic compounds); (2) Mellor (1924, Vol. 5, p. 974); (3) Dean (1973, Table 7-4); (4) Dean (1973, Table 4-1); (5) Meadows and Spedding (1974).

The proportions of the above ionized species are strongly dependent on pH (Garrels and Christ, 1965), with resulting effects on the solubility of the gases. The high solubility of this group of gases suggests that only a small fraction of them would be likely to escape in the vapor phase from an environment containing abundant water.

The biogenic gases can be removed from soil gases by microbial action. H_2, H_2S, CH_4, CO, CO_2, N_2O, and probably others are subject to this process of removal, and the net composition of soil gas depends on the rates of competing production and removal reactions (Schlegel, 1974; Schmidt, 1974).

All gases have a tendency to be adsorbed to mineral particles with a large surface area, particularly clay minerals and Fe-oxides. This effect is especially marked for large-diameter atoms such as Hg and Rn. In addition, Hg vapor tends to be adsorbed by organic matter with which it apparently reacts to form immobile organometallic complexes.

V. MOVEMENT OF GASES

The principles of diffusion and infiltration governing the movement of gases near the earth's surface are entirely analogous to those already discussed in Chapter 5 in connection with epigenetic anomalies. The only differences are the conditions of pressure and temperature under which the processes operate.

The pore spaces in soils and rocks below the water table are for practical purposes completely filled with water, except for relatively unusual phenomena such as natural gas traps. Thus the movement of gases at depth primarily involves movement in an aqueous solution. This can take place either by diffusion through the water or by mass transport of the water itself. Above the water table, movement of gases takes place by a combination of diffusion and mass transport. At the water table, the movement of gases across the water–air interface is by evaporation or condensation.

A. Diffusion

Diffusion may be defined as the flow of a molecular species from a region of high concentration to one of low concentration as a result of random thermal motion of the molecules. It can take place in any medium, solid, liquid, or vapor. The rate of flow is controlled by the freedom of movement provided by the medium though which the molecules are diffusing. Diffusion through a very viscous liquid, or through the very restricted pore spaces of a rock, will be slower than in open, unrestricted systems.

Rates of diffusion are expressed in terms of the diffusion coefficient D, of Fick's laws (eqns 5.1 and 5.2). Measured values of the diffusion coefficient of various gases through dry or moist soil fall in the range 10^{-1}–10^{-2} cm^2/s (Dyck, 1975; Schroeder *et al.*, 1965), whereas in water and water-saturated sediments D values in the range 10^{-5}–10^{-6} cm^2/s have been measured (Kroepelin, 1967). This is to say that a gas will migrate by diffusion about

10 000 times faster in soil air than in ground water. If flow is by diffusion only, Fick's laws make it possible to compute the expected concentration and rate of dispersion of a gas at any point, provided the concentration at some other point below the surface is known. Antonov, as quoted by Klimenko (1976), computed the rate of migration through a material (rock or water) with a D value of $3{\cdot}3 \times 10^{-6}$ cm^2/s, with results shown in Fig. 18.3. Schroeder *et al.* (1965) did the same thing for Rn in a desert soil and compared the results with Rn actually measured (Fig. 18.4).

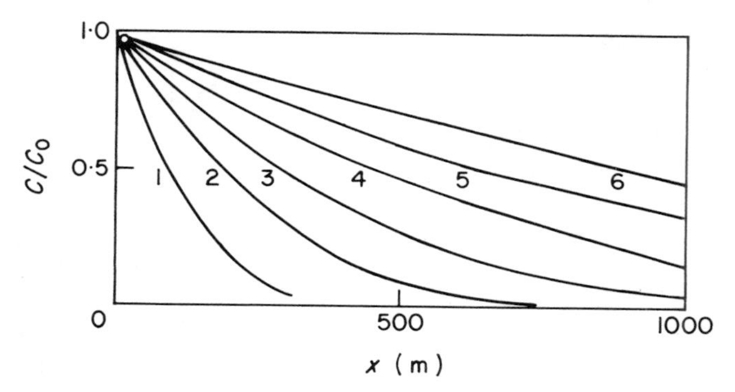

Fig. 18.3. Concentration (C) of gas resulting from pure diffusion, plotted against distance (x) from source (concentration $= C_0$). $D = 3{\cdot}3 \times 10^{-6}$ cm^2/s. Duration of diffusion: curve 1, 1 m. yr; curve 2, 4 m. yr; curve 3, 10 m. yr; curve 4, 25 m. yr; curve 5, 50 m. yr; curve 6, 100 m. yr. (After Klimenko, 1976.)

Rocks have a finite though often extremely small diffusion coefficient, the size of which is a function of the degree of interconnection of the water-filled pore spaces, a feature that is closely related to electrical conductivity. It is not necessarily related to permability, which is the openness of a rock to fluid flow. Thus shale, which can form an "impermeable" cap rock for oil pools, may have a high enough coefficient of diffusion for certain light hydrocarbons (in the order of 10^{-6} cm^2/s) that gaseous haloes may form by diffusion in the rocks and soil above the oil field. This is the basis of certain methods of geochemical exploration for petroleum.

Other factors being equal, the diffusion rate for light molecules of small diameter is greater than that of heavier, larger molecules. Thus free H_2 and He diffuse many times faster than Ar or CH_4 which, in turn, diffuse somewhat more rapidly than Hg vapor or Rn. The differences arise partly because, at a given temperature, the molecular velocity of light molecules is greater than heavy ones, and partly because their small size makes them less liable to blockage by molecular-scale obstructions. For example, radiogenic He trapped in crystal lattices slowly escapes, while radiogenic Ar in the same situation may be tightly held throughout geologic time.

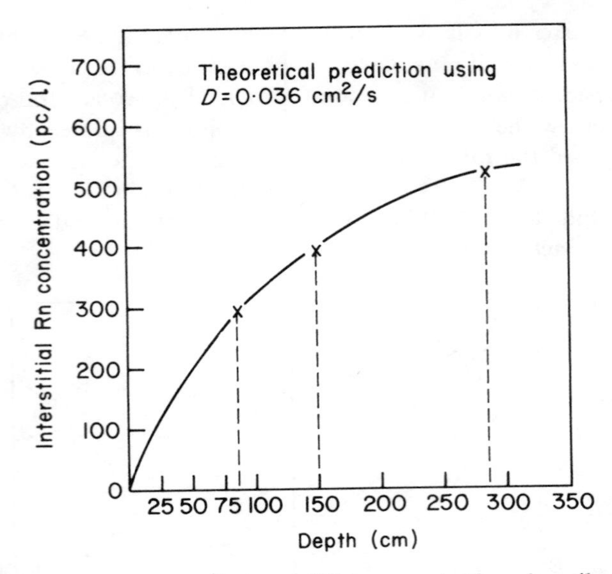

Fig. 18.4. Diffusion theory predictions of Rn concentrations in soil gas as a function of depth compared with observed concentration values at three depths. (×) Represents average concentration of 60 samples obtained between 26 October and 28 November, 1961, at U3ah, Yucca Flat, NTS. Soil type: alluvium, virgin. (After Schroeder *et al.* (1965). Reprinted with permission from *J. Geophys. Res.*, © American Geophysical Union, 1965.)

B. Water Transport

Mass movement, or infiltration, of aqueous solutions of volatile species through permeable rocks and through open fissures can be several orders of magnitude faster than movements by diffusion, as shown in Fig. 18.5. Such movement is restricted, however, to aquifers. For that reason, the distribution of volatile species that travel with the water is governed almost entirely by hydrologic factors. Thus, an understanding of the local hydrological conditions is necessary before the distribution of gases emanating from depths below the water table can be completely understood. Movement is expected to be along permeable stratigraphic units and along fractured zones. Russian workers have used gases in soils, ground water, and rock to trace fault zones of both local and regional extent (Ovchinnikov *et al.*, 1973; Eremeev *et al.*, 1973; Fridman and Petrov, 1976; Fridman and Makhlova, 1972).

In recently deposited sediments, or those undergoing diagenesis, some upward transport is expected, driven by the water released during compaction and diagenetic reactions.

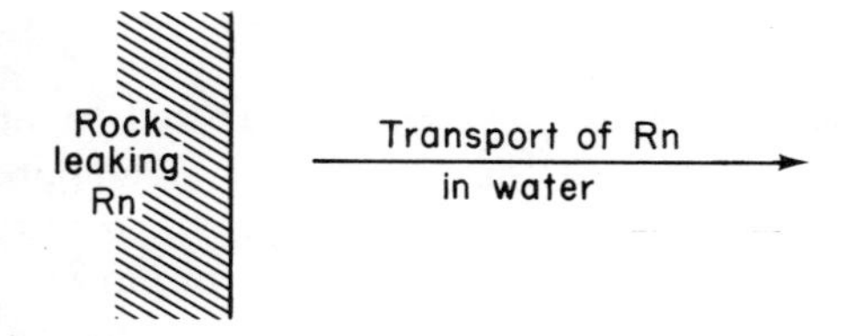

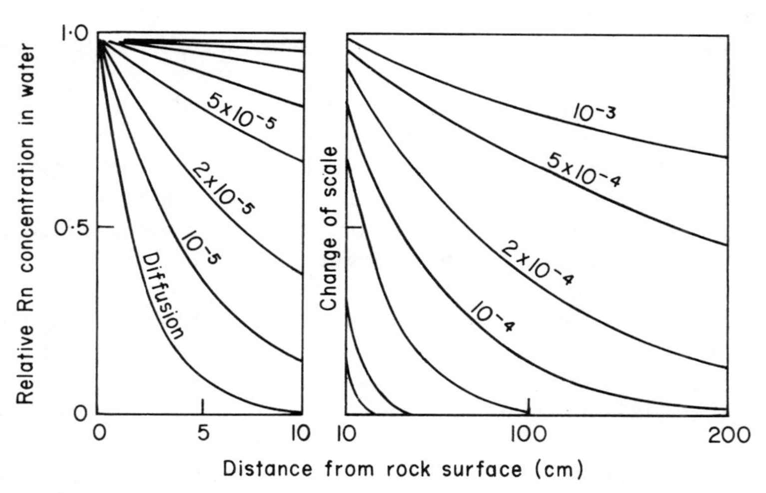

Fig. 18.5. Relative Rn concentrations in water adjacent to rock surface due to Rn movement by diffusion ($D = 10^{-5}$ cm^2/s) and by flow of water at various velocities (10^{-5}–10^{-3} cm/s). (After Andrews and Wood, 1972, p. 205.)

C. Transfer between Water and Vapor Phases

Evaporation and condensation of gases are continually taking place at any interface between an aqueous phase and a vapor phase. If this process is at equilibrium, no net transfer of gas occurs from one phase to the other. For gases generated at depth, however, a net upward movement of gas from the water to the overlying vapor will generally occur because the gas is continually replenished in the water, and continually tends to leak away towards the atmosphere. The actual concentration of a given species in each phase may be computed from eqn 18.3 and the data in Table 18.3. For example, the He content of water in contact with 1 atm of He is 0·94 cm^3/100 cm^3 of water. The earth's atmosphere contains 5 ppm He by volume, equivalent to a pressure of 5×10^{-6} atm He. Therefore water in equilibrium with th sphere contains $5 \times 10^{-6} \times 0{\cdot}94$ cm^3 He/100 cm^3 of water.

D. Vapor Transport

Except in volcanic systems, mass movement of vapor is restricted to the aerated zone above the water table. The only factors that can have an appreciable effect on the flow of air in the zone of aeration are changes in pressure, temperature, and moisture, all of which are related either directly or indirectly to changes in the weather. An increase in barometric pressure tends to push atmospheric air into the ground, while a decrease lets the soil air escape (Kovach, 1945; Tanner, 1964). The possible magnitude of this barometric effect can be illustrated by the following example: if soil and rock to 100 m depth have a uniform percentage of air-filled pore space, then a decrease of 1% in atmospheric pressure (a decrease often observed) will move soil air from 1 m depth to the surface. This motion will occur to a proportionally lesser extent throughout the thickness of the unsaturated pore space, down to the level of the water table, and will tend to mix and disperse constituents both upward and (during increases of pressure) downward. An increase in barometric pressure tends to trap the accumulating gases and increase their concentration at depth, whereas a decrease has the opposite effect.

An increase of surface temperature not only causes soil air to expand and escape, but also tends to release vapor species adsorbed to the surface of soil particles. If the heating of surface soils is uneven, lateral flow of air may occur. Similarly, convection of soil air may occur in the pore space within hills, because of temperature gradients created in horizontal planes by heating the hillsides. Strong winds are another possible cause for flow of soil air. Rain and increased moisture in the soil may result in adsorption of the vapor species from the soil air. Rain also introduces atmospheric gases into the soil and eventually into the ground water. Over a short period of time, these

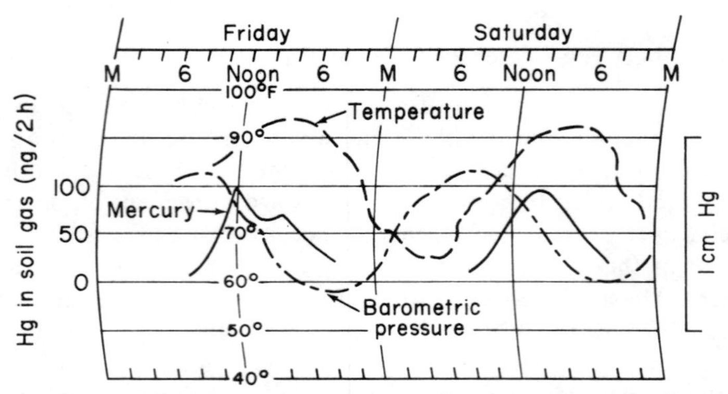

Fig. 18.6. Daily variation of Hg in air at the ground surface, temperature and barometric pressure, Silver Cloud mine near Battle Mountain, Nevada. (After McCarthy *et al.*, 1970, p. 38.)

meteorological factors acting together can result in a kind of pumping or breathing action, as illustrated in Fig. 18.6. The fluctuations in pressure and temperature operate on a daily cycle (Turk, 1975) and over longer periods. The net effect of this process is an outward movement of the gases generated at depth at rates much faster than diffusion, and a corresponding inward movement of atmospheric air. Experiments on CO_2 by Glebovskaya and Glebovskii (1960) suggest that short-period variations in the composition of soil air are not significant below a depth of about 1·5 m, but this depth obviously depends on the size of the underlying vapor reservoir, the movement of water, and other conditions.

Rain, frost, and low temperatures tend to block the pumping action, with a resulting decrease in the rate of escape of the upward-moving gases and an accompanying increase in their concentration in both soil air and ground water (Dyck *et al.*, 1976a; Kovach, 1945). Seasonal changes in temperature and rainfall can thus give rise to a seasonal, or at least a long-term variation in the gas content of near-surface materials, as illustrated in Fig. 18.7.

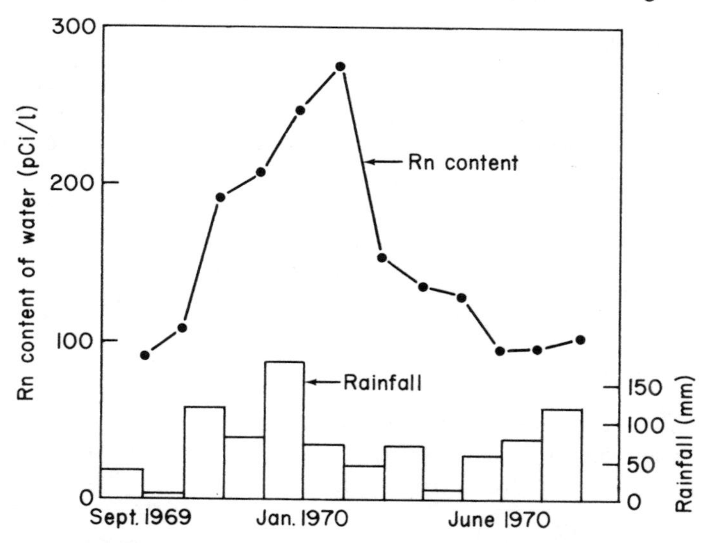

Fig. 18.7. Seasonal variations in rainfall and Rn content of spring water from Rickford Rising, U.K. Rainfall totals are for 30 days prior to Rn measurement. (After Andrews and Wood, 1972, p. 198.)

E. Relative Effects of Infiltration versus Diffusion

Diffusion is certainly the principal factor in the movement of gases through impermeable rocks such as shale where there is no water movement at all. Transport of volatile species by infiltration of aqueous solutions, however, must dominate in aquifers, both in permeable sedimentary rocks and in fractured igneous and metamorphic rocks. This relationship has been confirmed by studies of He in deep sedimentary sections (Golubev *et al.*, 1974). Diffusion through soil air is probably the principal factor in movement through soil and rock formations above the water table at levels that are not significantly affected by variable atmospheric effects, as shown in Fig. 18.4. Mass movement of the soil air becomes dominant toward the land surface where the effects of variations in the weather can be felt.

VI. GEOCHEMICAL BEHAVIOR OF SELECTED GASES

The genesis of the principal gases that are found near the earth's surface has already been reviewed. The following discussion reviews the geochemical characteristics of those gases that are or may be of particular importance as geochemical indicators. The arrangement is alphabetical.

1. Carbon dioxide (CO_2)

Anomalous concentrations of CO_2 in ground water have often been found associated with buried sulfides, a relationship that is explained by the action of acid waters releasing CO_2 from carbonates in the ore or country rock. Care should be taken not to confuse this effect with anomalous concentrations of CO_2 coming from deep-seated sources through deep fissure zones. The content of CO_2 in soil air commonly falls in the range from 0·2% to 4% v/v, as compared with 0·035% for atmospheric air (Brady, 1974, p. 256). Most of this excess of CO_2 in soil air probably comes from the oxidation of decaying organic matter.

2. Carbon disulfide (CS_2)

Anomalous concentrations of CS_2 up to 300 ppb of the total gases adsorbed to a molecular-sieve trap have been observed in soil gas overlying a geothermal area at Roosevelt Hot Springs, Utah (Hinkle *et al.*, 1978).

3. Carbon monoxide (CO)

The presence of CO suggests a source at depth. It is usually present in very small amounts in gases over deep fissures, and may enter into reactions at the site of oxidation of sulfide deposits.

4. Carbonyl sulfide (COS)

Carbonyl sulfide, sometimes referred to as carbon oxysulfide, has been identified in gases over oxidizing sulfide deposits. The reaction $COS + H_2O = CO_2 + H_2S$ is very rapid at ambient temperature and pressure. In particular, in the presence of moisture and sunlight, COS breaks down into CO_2 and H_2S. Its solubility and reactivity with water is lower than either H_2S or SO_2, so that it might be expected to travel further from the site of generation than the other sulfur gases.

5. Helium (He)

Helium-4 at depth is being continually generated by radioactive decay of U, Th, and their daughter products, and, hence, must be continually escaping at the earth's surface; the average He flux through the ocean floor has been measured at $1\text{–}2 \times 10^6$ atoms/cm^2/s (Barnes and Bieri, 1976). Helium-3 is formed in the atmosphere and at depth by decay of tritium (^3H). Unlike the other radiogenic gases, He can escape by diffusion out of the crystal lattice of source minerals. The flow rate of He is strongly affected by the permeability of the matrix. Thus, He generated at depth tends to be channeled into fracture zones (Eremeev *et al.*, 1973). The geochemistry of He has been summarized at length by Dyck (1976), and a bibliography has been compiled by Adkisson and Reimer (1976).

6. Hydrogen (H_2)

At high temperatures and pressures, elemental H_2 is a reaction product of water in contact with minerals containing Fe^{2+}. Thus, it is probable that free H_2 is continually filtering upwards from depth from what amounts to an unlimited source (Hawkes, 1972b). Hydrogen may also be formed by decomposition of water caused by energy released in radioactive decay (Kravtsov and Fridman, 1965). At ambient temperatures in the zone of weathering, H_2 is metastable and normally does not enter chemical reactions except where certain specialized bacteria are active. One of the technical problems in studying the geochemistry of H_2 is that is can be generated artificially by the electrochemical reaction of naturally occurring water on steel tools or drill-hole casing (Dyck, 1974). Earlier workers taking gas samples from drill holes and mine workings for analysis were not fully aware of this source of contamination, with the result that many published reports of free H_2 in the pore spaces or rocks are open to suspicion.

7. Hydrogen sulfide (H_2S)

Hydrogen sulfide, when dissolved in water, tends to dissociate into H^+, HS^-, and S^{2-} ions. Its solubility in water is therefore relatively high, so that it would not normally be expected in soil gases in a humid climate. Also, it is rarely observed in water containing free O_2 (Dyck, 1974). One commonly used analytical method involves trapping all sulfur gases, analyzing for total sulfur, and reporting the results as H_2S plus SO_2 without positively identifying the species. Thus, some published reports of H_2S in soil gas may be open to reinterpretation.

8. Mercury (Hg)

Background contents of Hg in air are in the range 2–8 ng/m^3, or 0·2–0·8 $\times 10^{-12}$ v/v (Williston, 1968). Saturated air in contact with liquid Hg contains 1·58 ppm v/v at 20 °C; saturated water in contact with liquid Hg at the same temperature contains about 25 ppb by weight (Hem, 1970) or 2·8 $\times 10^{-6}$ cm^3 Hg/cm^3 H_2O. Thus at 20 °C, a given volume of water saturated with Hg contains slightly more Hg than the same volume of air.

Mercury occurs in rocks and soils in a number of different phases, of which the principal ones are as a major or minor constituent of sulfides, as halides, oxides, organic complexes, and the native metal. During weathering, most minerals containing Hg as either a major or minor constituent are unstable and slowly break down with the release of elemental Hg. Native Hg itself can occur sorbed to active mineral surfaces, dissolved in water, or as liquid or gaseous Hg. On heating, the different naturally occurring Hg compounds release elemental Hg vapor at different temperatures (Fig. 18.8). Thus by observing the temperature and amplitude of the Hg vapor peaks recorded

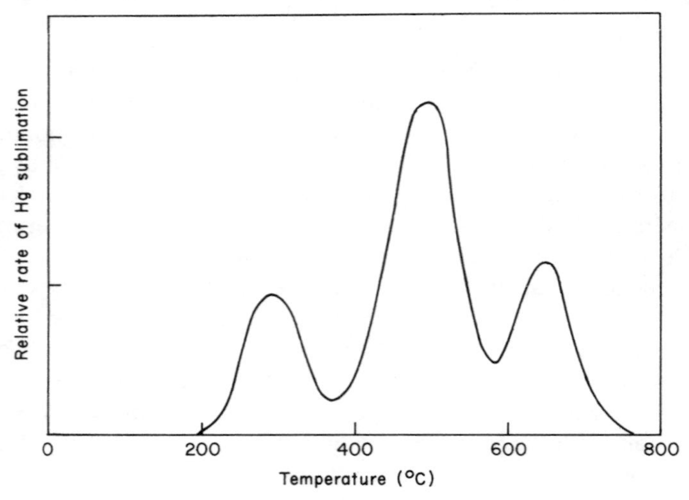

Fig. 18.8. Example of progressive sublimation of Hg, Kansay skarn Pb–Zn deposit, U.S.S.R. (After Fursov, 1973, p. 184.)

during heating of a rock or soil sample, the identity and concentration of the various Hg phases can be determined. In addition to the peaks representing specific mercury compounds, natural samples show a very pronounced peak at about 105 °C, probably representing Hg dissolved in moisture adsorbed to the mineral grains. The sorption of Hg vapor to the surfaces of fine-grained minerals and particularly organic matter strongly inhibits the escape of Hg vapor at the surface of the ground.

9. Dimethyl mercury ($Hg(CH_3)_2$)

Mercury may be acted upon by bacteria in the presence of organic matter to form dimethyl mercury (Jonasson, 1970), a gas which has received considerable attention because of its extreme toxicity. It has been identified in organic lake and river sediments (Jacobs, 1974; Olson and Cooper, 1974). However, it is thermodynamically stable only in an extremely reducing environment, so that it probably does not play a major part in the geochemical cycle.

10. Methane (CH_4)

Methane, together with other gaseous hydrocarbons that often accompany it, is slightly soluble in water. Although it is thermodynamically unstable in the presence of free O_2, it does not normally react except with the help of certain bacteria. It may react with metals in the environment of an oxidizing sulfide deposit to form volatile compounds, such as methyl compounds with sulfur and some metals.

11. Neon (*Ne*)

The only significant source of Ne in gas samples is atmospheric air, which contains 18 ppm v/v. The Ne content of a sample, therefore, can be used as a measure of atmospheric contamination (Clarke and Kugler, 1973).

12. Oxygen (O_2)

The free O_2 of the atmosphere is the primary agent and driving force behind almost all the geochemical reactions that take place near the surface of the earth. Its content in atmospheric air is 21 %. Water in equilibrium with the air under ambient conditions contains 0·66 cm^3 O_2/100 cm^3 H_2O, or 9·5 ppm by weight. Circulating underground water brings O_2 down from the atmosphere to the site of active oxidation of deep-seated minerals. As the water continues to circulate, the O_2 is progressively depleted, and at the same time the content of non-atmospheric gases tends to build up (Dyck, 1974). Thus, it is very common to find an antithetic relation between O_2, on the one hand, and He, CH_4, CO, and H_2S, on the other.

13. Radon (*Rn*)

Radon-222, with a half-life of 3·8 days, results from the radioactive decay of ^{238}U; thus, it is a unique indicator of U. The direct parent of ^{222}Rn is ^{226}Ra. Other isotopes of Rn which come from the disintegration of ^{232}Th and ^{235}U have much shorter half-lives (see Appendix) and are of little geochemical importance as indicators. The background content of Rn in the atmosphere is estimated to be on the order of 0·01–0·45 pCi/l (Dyck, 1973). In certain Canadian lakes, the background for Rn in water ranged from

1 pCi/l to 7 pCi/l (Dyck and Cameron, 1975). Background in stream water may be much higher, in the order of 50 pCi/l (Dyck and Smith, 1969), but the level depends strongly on turbulence and the rate of loss to the atmosphere (Rose and Korner, 1979).

Radon anomalies in soil gas decrease away from their source because of radioactive decay, as well as dilution and other processes. Tanner (1964) estimates that 7 m is the maximum depth at which diffusion through stagnant air in pore spaces can produce a detectable anomaly (1/1000 its initial concentration) above a buried orebody. Field observations, however, have shown that Rn haloes extend for as much as 100 m above parent U deposits (Gingrich and Fisher, 1976; Tanner, 1964; Beck and Gingrich, 1976). At least part of the explanation for this phenomenon may be found in the breathing action mentioned earlier, but a major explanation must be found in the geochemical behavior of the earlier members of the Rn series (U, Th, Pa, and Ra). Uranium (^{238}U and ^{234}U), if it dissolves in water, is likely to stay in solution and be dispersed by circulating ground water. Thorium (^{234}Th and ^{230}Th) and protactinium are among the most immobile elements in the periodic table, and are unlikely to contribute anything to the dispersion pattern. Radium (^{226}Ra) is another matter, as it may dissolve from the ore and be reprecipitated with secondary Ca, Fe, or Mn minerals, or be adsorbed to Fe–Mn-oxides and similar adsorbents (Rose and Korner, 1979). The Rn would then have to diffuse only from the site of reprecipitation of its immediate parent Ra. In this connection, it is common to find that the Rn/Ra ratio of natural waters greatly exceeds the radioactive equilibrium ratio (Scott and Barker, 1962).

14. Sulfur dioxide (SO_2)

Sulfur dioxide dissolves in water to form a weak acid, H_2SO_3, and its dissociation products (HSO_3^-, SO_3^{2-}) which in turn are readily oxidized to sulfuric acid, H_2SO_4. Thus it seems unlikely that any very large amount of SO_2 could escape from the site of oxidizing sulfides except under very arid conditions. SO_2 has been identified in gases desorbed by heating of molecular sieves buried over oxidizing sulfide deposits (Hinkle and Kantor, 1978). As with H_2S, some other reports of SO_2 may have come from a misinterpretation of the results of a determination of total sulfur in a gas sample.

15. Other gases

The literature contains scattered references to other vapor species emanating from depth. Clews (1966) reports readily extractible Ni in soils over oil fields, which he ascribes to precipitates of volatile Ni-porphyrins. Frederickson *et al.* (1971) determined free F and Cl by mass spectrometry in atmospheric air over precious metal deposits, which is remarkable as these species are both unstable and extremely reactive in the presence of water. Iodine, nitrogen oxides, and the hydrides and methylated compounds of As, Se, Te, and Sb have also been mentioned (Barringer, 1969, 1971; Bristow and Jonasson, 1972; Frederickson *et al.*, 1971; Jonasson, 1970).

VII. APPLICATIONS OF VAPOR GEOCHEMISTRY

This section briefly reviews the performance of vapor geochemistry as applied in the solution of geological problems, together with a presentation of the principal references to published case histories and experimental surveys. For most fields of applied vapor geochemistry, it is still too early to make generalized comments on the effectiveness of the various methods.

A. Exploration for Sulfide Deposits

All of the investigators working with free or dissolved gas associated with oxidizing sulfide ores report strong anomalies in CO_2 (Bolotnikova, 1965; Dadashev *et al.*, 1971, 1974; Danilova, 1968; Elinson *et al.*, 1970; Fridman, 1974; Glebovskaya and Glebovskii, 1960; Khayretdinov, 1971; Kravtsov and Fridman, 1965; Netreba *et al.*, 1971; Nigrini, 1971; Ovchinnikov *et al.*, 1973; Shipulin *et al.*, 1973). Anomalies in soil air reported in these investigations fall in the range 1·5–4% v/v. Although background CO_2 in soil air has been reported as high as 4% v/v, the normal contents usually fall well below 1%. The CO_2 content of gases dissolved in ground water near oxidizing ore may be as high as 10% compared to a background of 0·2%. Increases in CO_2 are not uncommonly paralleled by decreases in O_2, as illustrated in Fig. 18.9. Anomalies in CO_2 and O_2 appear to be consistently present in

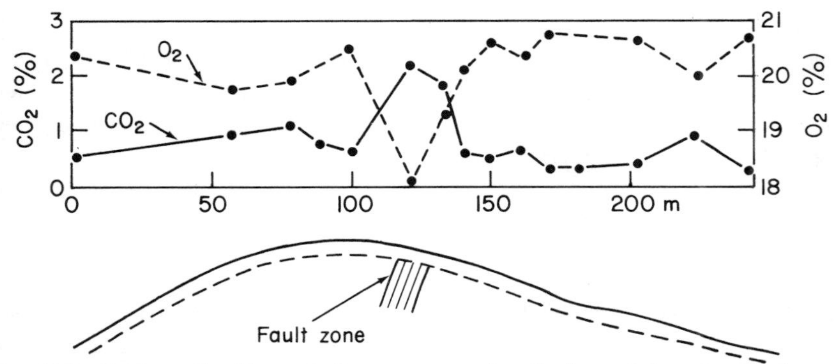

Fig. 18.9. Carbon dioxide and O_2 contents of soil air at 1·5 m depth above a pyritized fault zone. (After Glebovskaya and Glebovskii, 1960, p. 54.)

soil air or ground water adjacent to actively oxidizing sulfide bodies. The extent of dispersion through water-saturated overburden or country rock remains to be ascertained.

All the reported work with Hg vapor in soil air over Hg deposits has revealed well-defined anomalies (Fursov, 1970; Hawkes and Williston, 1962; Karasik and Bol'shakov, 1965; Khayretdinov, 1971; McCarthy, 1972; Meyer and Evans, 1973). Anomalies are in the range 100–5000 ng/m^3 compared to a background of 10 ng/m^3. Anomalous Hg was reported in soil air over polymetallic and precious metal deposits that presumably contain Hg concentrations (McNerney and Buseck, 1973; McCarthy *et al.*, 1970; Meyer and Evans, 1973; Hawkes and Williston, 1962; Stevens *et al.*, 1969), although it has been reported as background over porphyry-copper deposits (McNerney and Buseck, 1973; Robbins as quoted by McCarthy, 1972).

Anomalous He over sulfide deposits that are controlled by fracture zones was reported by Elinson *et al.* (1970), Elinson (1972), Eremeev *et al.* (1973) and Ovchinnikov *et al.* (1973). Anomalies in He and other deep-seated gases are characteristic of most deep permeable zones and are not restricted to ore deposits.

Anomalous contents of sulfur gases in soil air, in the open atmosphere over sulfide deposits, or in ground water near sulfides are reported by Kravtsov and Fridman (1965), Dyck (1974), Elinson *et al.* (1970), Elinson (1972), Rouse and Stevens (1971), Frederickson *et al.* (1971), Meyer and Peters (1973), and Shipulin *et al.* (1973, pp. 227–229). These experimental measurements are encouraging, but further tests and actual surveys are needed to understand the conditions under which anomalies are easily detectable and reliable. Ground water rich in H_2S has been suggested as a guide to native sulfur deposits (Popov and Shil'zhenko, 1972).

Anomalies in airborne particulate material have been described by Barringer (1977). The sampling of these materials requires size segregation of the particles and devices for collection only during updrafts in order to obtain local particulates rather than those from a considerable distance. More recent airborne equipment that sucks particulates directly off the vegetation may alleviate these sampling difficulties and lead to rapid ground and air surveys of particulates (A. R. Barringer, personal communication).

B. Exploration for Uranium

The use of Rn in soil gas and in natural waters in locating buried U deposits is becoming well established as a prospecting method (Smith *et al.*, 1976). Extensive programs of orientation work have been carried out by the Geological Survey of Canada (Dyck, 1973; Dyck *et al.*, 1976b; Dyck and Smith,

1969), British government agencies (Bowie *et al.*, 1971; Michie *et al.*, 1973; Miller and Ostle, 1973), and some independent investigators (Caneer and Saum, 1974; Gingrich, 1975; Stevens *et al.*, 1971; Rose and Korner, 1979). Anomalies for Rn in soil gas can have greater contrast than those for U in soil (Dyck, 1968b). In several areas, Rn anomalies have been detected above ores covered by up to 100 m of transported overburden. In other circumstances, strong control of Rn anomalies by faults or other permeable zones, or by U dispersed away from its primary source, has been demonstrated (Michie *et al.*, 1973).

With He, the association with U ores is not always as clear as with Rn, because its distribution is complicated by the large volumes of He that are brought to the surface through fracture zones and the relatively large amounts in the atmosphere. Under favorable conditions and in conjunction with other ore guides, He can be useful in U exploration (Clarke and Kugler, 1973; Clarke *et al.*, 1977; Dyck, 1975, 1976; Goldak, 1973). The He in the natural gas of the Texas Panhandle was, in fact, the clue that eventually led to the discovery of the U-bearing asphaltites that form the caprocks of that field (Pierce *et al.*, 1964). However, large He anomalies are commonly controlled by structures rather than U deposits (Dikun *et al.*, 1976).

C. Exploration for Petroleum

The principal gases associated with oil pools are the light hydrocarbons and the radiogenic gases He and Rn. Methane, the dominant member of the hydrocarbon group, tends to be most concentrated in the highest part of the oil structure, where it may accumulate as a separate vapor phase in the form of natural gas. The radiogenic gases are generated by U which tends to be precipitated at the water–oil interface at the base of the oil accumulation. In theory, He and CH_4 can diffuse directly upward through the impermeable cap rocks to form projected images near the surface that reflect their distribution at depth. While there is abundant evidence that this general concept is valid, the actual usefulness of near-surface gases in petroleum exploration has been hotly debated. Much of the disagreement comes from the claims of contractors who are more intent on promoting their services than on getting at the truth. Clearly, shallow petroleum reservoirs are much easier to detect than deep reservoirs (Devine and Sears, 1978). A statement of the status of the art throughout the world has been prepared by Siegel (1974, Chapter 9). Russian workers have been particularly active in gas analysis as a guide to oil fields, as summarized by Kartsev *et al.* (1959) and Sokolov (1971). Armstrong and Heemstra (1973), and Ball and Snowdon (1973) have written critical reviews of radiogenic gases as guides in petroleum exploration.

D. Location of Buried Faults

Russian workers have carried out extensive studies of He in soils and sedimentary cover as a guide to deep faults and basement fractures (Bulashevich, 1974; Eremeev *et al.*, 1973; Ivanov and Medovyi, 1975; Ovchinnikov *et al.*, 1973; Tugarinov and Osipov, 1974; Dikun *et al.*, 1976). Mercury has also shown promise as a guide to deep fracture zones (Fursov *et al.*, 1968; Oelsner *et al.*, 1973). Carbon dioxide and other gases show anomalies along near-surface faults and fracture zones (Fridman and Petrov, 1976: Rösler *et al.*, 1977).

E. Forecasting of Earthquakes

Recent studies have suggested that immediately prior to an earthquake, pervasive crackling and dilatancy occurs throughout the rocks adjoining the potential fault surface, resulting in the release of certain occluded vapor species. These gases are then free to move toward the surface where they may be measured as components of the soil air. The only vapor indicators of forthcoming earthquakes so far reported are the radiogenic gases Rn, Ar, and He (Gorbushina *et al.*, 1972; Kasimov, 1973; Noguchi and Wakiti, 1976; Ovchinnikov *et al.*, 1973; Plyusnin *et al.*, 1972; Sultankhodzhaev *et al.*, 1973; Tyminskii and Sultankhodzhaev, 1973).

F. Forecasting of Volcanic Eruptions

Stoiber and Rose (1974) found that the Cl/SO_4 ratio in fumarolic condensates, leachates, and encrustations decreases before a volcanic eruption, stays constant during the eruption, and then returns to its normal value afterwards. Chirkov (1976) has reported that the Rn content of volcanic fumaroles and hot springs in the U.S.S.R. increased before and during the eruption of a volcano.

G. Location of Geothermal Areas

Certain gases in soil air, particularly He, CO_2, CS_2, and Hg, may under favorable conditions indicate the presence of thermal water at depth (Koga and Noda, 1975; Reimer *et al.*, 1976; Tikhomirov and Tikhomirova, 1971; Hinkle, 1978). These constituents are apparently released by the heat and active water circulation of the thermal zone.

VIII. VAPOR SURVEY TECHNIQUES

The particular advantage of vapor as an indicator of geological phenomena that are hidden either in the ground or in the future lies in its high mobility.

This same quality also makes vapors probably the most difficult of all geochemical indicators to sample, analyze, and interpret. If the geochemical signatures that are being carried upward by vapors from the depths are to be deciphered at all, the techniques of sampling and analysis must be perfected so that even very low concentrations can be reproduced. More research and trial surveys are also needed to understand the natural processes whereby gases are transported and destroyed, and the limitations of vapor surveys.

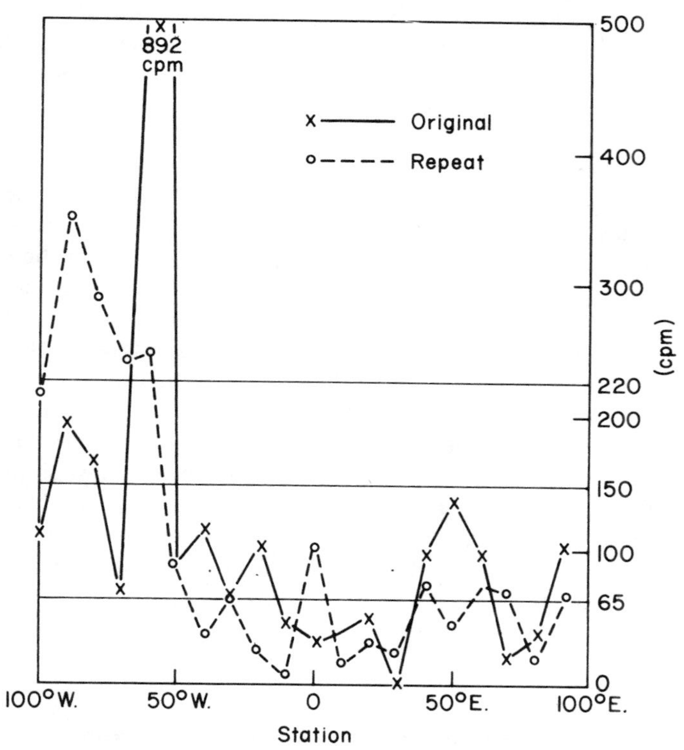

Fig. 18.10. Radon content of soil air, repeated after three days. (After Soonawala, 1974, p. 110.)

Variations in pressure, temperature, and moisture strongly affect the mobility and hence the relative concentration of the various gas species in soil air, as has been discussed earlier in Section V. (D). Because of these factors, repeated sampling may fail to replicate the original observations by a factor of ten or more, as illustrated by the data of Fig. 18.10. Well-defined anomalies in soil air have even been known to vanish entirely on resampling. Sampling soil air directly, therefore, can be an extremely frustrating experience. Fortunately, however, soil air is not the only medium that will give us a measure of the gases escaping from depth. The gases dissolved in ground water or sorbed to natural traps (soil, humus) or artificial traps (adsorbents), though less direct, may give values that tend to level out the time variations that are so dependent on the weather.

A. Soil Air

Air may be extracted from holes either bored with an auger (Caneer and Saum, 1974) or prepared by driving a hollow probe (Dyck and Meilleur, 1972; Miller and Ostle, 1973). Generally auger holes are simpler for shallow holes less than 1 m in depth, and probes are easier for deeper holes, up to 10 m or more. Air samples extracted by either method will be diluted with atmospheric air unless the top of the hole is tightly sealed after the sampling tube is inserted. Thick mud, water, and inflatable seals are commonly used to seal the hole. It has been reported that dilution may be minimized by discarding the first few liters of air, which contain atmospheric air left in the hole after augering or driving. Prolonged extraction of air results in dilution by atmospheric air drawn down from the surface in the vicinity of the hole. Any simple device such as bicycle pump is satisfactory for removing air from the hole. The sample should be stored pending analysis in a tight metal or glass container, with no plastic joints or gaskets, as most gases diffuse readily through plastic. Where H_2 or He is to be measured, the container should be all metal, not glass, as the light gases can diffuse to a certain extent through some kinds of glass (Barrer, 1951). Pogorski (1975) has designed a suitable all-metal sample container for H_2 and He. The problems of loss on storage of soil-air samples can be avoided if mobile analytical units are available (Friedman and Denton, 1976).

B. Water

Soil gases emanating from depth must pass through the water table on their way up. Therefore, a sample of gases dissolved in ground water should give data at least as significant as soil air. Collecting ground water in the absence of pre-existing wells and springs involves the drilling of deeper holes and hence somewhat greater effort and expense than is needed for soil-air samples. Stream and lake water samples do not suffer from this disadvantage. However, the dissolved gas originally present in the ground water quickly escapes into the atmosphere when it emerges at the surface, so that much of the information is lost. Where lakes are frozen during the northern winters, the loss of dissolved gas may not be enough to invalidate the method (Dyck *et al.*, 1976c).

One liter of water normally provides enough dissolved gas for analysis. The water should be collected with a minimum of agitation and turbulence to avoid degassing. Plastic containers are unsatisfactory, as dissolved gases can escape through the plastic walls. Dyck *et al.* (1976a) recommend glass beverage bottles with metal caps for preventing the escape of light gases. Whatever type of container is used, it should be filled completely with water, leaving no air space at all. Retention of dissolved gas may be improved by freezing water samples during winter sampling programs. For analysis, the dissolved gas is swept out, or outgassed, by bubbling through a carrier of any non-reactive gas such as Ar or N_2 that is not to be reported in the analysis.

C. Gas Traps

The variations of concentration of gases with time that make soil air sampling so unsatisfactory can be smoothed out by the use of gas traps of various kinds. These either collect all the gas with which they come in contact, or reach an equilibrium that is proportional to the average concentration in the soil air or soil moisture. Mercury vapor may be quantitatively removed from air or water by fine-textured Au or Ag in the form of wool or screening. This system was used by McCarthy (1972) who designed a plastic dome that funneled air escaping at the surface of the ground through an Au trap. More recently, the U.S. Geological Survey has been experimenting with artificial zeolites that are buried in the soil for periods of two weeks or more and that collect many vapor species with which they come in contact (Hinkle and Kantor, 1978). Activated charcoal has been used successfully in concentrating Rn (Megumi and Mamuro, 1972; Wennervirta, 1967). The gases collected by any of these traps may then be released by heating prior to analysis. Alpha-track film or similar devices that may be buried for substantial periods of time in the soil or in drill holes have been used successfully in detecting Rn in soil and rocks (Gingrich, 1975). An example of a naturally occurring gas trap is reported by Sutherland-Brown (1967) who found an easily recognizable anomaly of Hg sorbed to fine-grained glacial soils overlying Hg-bearing deposits.

D. Atmospheric Air

Whereas ground water and gas traps provide samples that are more reliable but more difficult to collect than soil air, the sampling of the air above the ground is less reliable but much easier to collect. Here the concentration of trace constituents coming from below varies with the weather and the time of day, just as with soil air. An added uncertainty comes from pollution of the atmosphere from various human activities. Atmospheric air does have one logistic advantage, however, in that it can be sampled from aircraft with a speed and unit economy that cannot be matched by any other geochemical method.

Geochemical anomalies in the atmosphere can take a variety of forms. They may be reflected in vapor-phase constituents, or as aerosols, with suspended particles that are organic or inorganic and solid or liquid. Mercury vapor in the air immediately above the ground surface has shown a distribution pattern that can be correlated with Hg-bearing deposits at depth (McCarthy, 1972). Sulfur gases in the atmosphere over sulfide deposits (Rouse and Stevens, 1971) and Rn near U deposits (Milly, 1971) have been reported. Metal-bearing aerosols discharged into the atmosphere by living plants and trees have been postulated by Curtin *et al.* (1974). Experimental plants fed with radioactive Zn were found to discharge organic particulates into the air carrying detectable concentrations of the Zn tracer (Beauford *et al.*, 1975). In the arid climate of southern Africa the metal content of suspended dust particles could be correlated with outcrops of known metalliferous deposits (Weiss, 1971). Various airborne methods for measuring anomalies in the atmosphere have been developed, some involving collection of a sample for later analysis (Barringer, 1977; Weiss, 1971), and some direct chemical analysis in the aircraft (Barringer, 1971; Turkin *et al.*, 1971). Kolotov *et al.* (1965) report anomalies in rain collected near ore.

Chapter 19

Computerized Data-handling and Statistical Interpretation

★★★★★★★★

The purpose of a geochemical exploration survey is the discovery of abnormal geochemical patterns, or geochemical anomalies, related to mineralization. The recognition of something abnormal, however, requires an understanding of what is normal. From this have come the concepts of the background, or typical range of an element in a naturally occurring material, together with the threshold, or lower limit of values that are worth considering as anomalous, and contrast, or ratio of anomalous to background values. When the field data are plotted on plans or sections, these parameters may become almost intuitively obvious, so that the recognition of anomalies worthy of follow-up becomes a relatively simple matter.

Unfortunately, the recognition of anomalies is not always so easy. Some geochemical surveys involve the collection of hundreds or thousands of samples, each one to be analyzed for ten or more elements. The sheer time and effort involved in tabulating and plotting such a mass of data, together with the interpretation of the plotted results, becomes a major cost of the survey. Errors easily creep in, and require additional time for checking. Furthermore, the anomalies are not always easy to recognize by simple inspection of the distribution of a single element. The presence of mineralization may be indicated by ratios of elements or groups of elements, or statistical treatment may be necessary before the anomalous patterns can be recognized.

Some statistical methods of interpretation can be carried out without any specialized equipment, provided the body of data to be handled is not too great. An outstanding example is the application of cumulative frequency curves in the recognition of anomalous sample populations, as discussed in Chapter 2. Another area where statistics plays an important role, but where the problems can generally be solved by conventional methods, is in the consideration of the accuracy and precision of geochemical analyses, as reviewed in Section VI of Chapter 3. A great many statistical methods of data analysis, however, would be essentially impossible with only a pocket calculator or slide rule.

The message is clear—that for both data-handling and a great deal of statistical interpretation of the results of large-scale geochemical surveys, computer-processing methods are virtually mandatory. Furthermore, the field is developing rapidly with the appearance of new statistical concepts and with computer services becoming progressively less expensive. In the following sections, the major types of computer processing and statistical methods are briefly reviewed.

I. DATA-HANDLING BY COMPUTER

In large surveys involving thousands of samples, computer data-handling is useful for organizing the data, recording data in tables, and plotting maps (Garrett, 1974b; Gustavsson, 1976). The major advantages of the computer are the ease and speed of handling large volumes of data, and the relative freedom from errors once the data have been entered into a proven system and verified.

A. Recording of Data

The initial stage in data-handling is to record the sample numbers and locations (coordinates), along with any relevant field data, such as sample type and depth, drainage, contamination, and geology. This information should be recorded in the field on forms from which it can be easily transferred to cards or other computer-readable media. Care must be taken in choosing symbols for non-numerical data so that the computer can easily manipulate the data (Chapman, 1975). Numerical codes for sample types and site characteristics are usually preferable to letters. If sample locations are recorded on a map, the coordinates may be easily obtained on punched cards or other computer-readable form using a digitizer.

When analytical data are obtained in the laboratory, these should also be recorded on standard forms and entered into the data bank. Entries of "not detected" should be clearly distinguished from "not analyzed", using different numbers (Chapman, 1975). Automated output delivered directly from the analytical instrument to cards or tape minimizes errors and saves work on large projects.

B. Production of Tables, Histograms, and Summaries

Once the above steps have been carried out, the data can be printed as tables for records and reference, and can be manipulated for a variety of purposes. One of the most helpful manipulations is the production of statistical and graphical summaries for subgroups of the data, such as samples from different rock types or drainage characteristics, if these groups are identified in the data bank by a coded designation for each sample. Comparisons of average values and standard deviations for subgroups can aid in deciding whether different background and threshold values are desirable for different subgroups. If the data within groups are distributed approximately normally or log-normally, statistical tests can aid in making such decisions, using a "t-test" for significant differences between the mean values for the groups (Koch and Link, 1970). Non-parametric statistical methods are applicable if the frequency distribution is distinctly non-normal (Siegel, 1956). Computer-generated histograms, showing numbers of samples at various concentration levels, allow visual comparison of subgroups and may suggest threshold values for the subgroup, as described in Section VII of Chapter 2.

C. Computer-generated Maps

Computer-generated maps are useful in large multi-element surveys, first as plots of the data, and second for smoothing out local "noise" and bringing out the broader patterns in the data. However, in all computer work, the original data should be frequently consulted in order to understand the origin of unusual features, because such an understanding may lead to improved interpretation.

The simplest type of computer-generated map is a plot of the sample location and element content for each sample. A cheap method of producing such a map utilizes the computer line printer (Fig. 19.1). Limitations of the line printer are the resolution, which is limited to the space occupied by one character (about 0·2 cm wide), and overlapping of numbers and characters from adjacent sample sites. An alternative to plotting actual analytical values is the use of a series of symbols of increasing darkness or distinctive character (Howarth, 1971; Gustavsson, 1976); here some experimentation with the levels for different symbols may be required. R. J. Howarth (personal communication) has found that a series based on percentiles, such as 10, 50, 75, 90, 98, 99, usually gives good results. Alternative to the line printer are pen-type plotters, photography of cathode ray tubes, and lasergraphic plotters, as

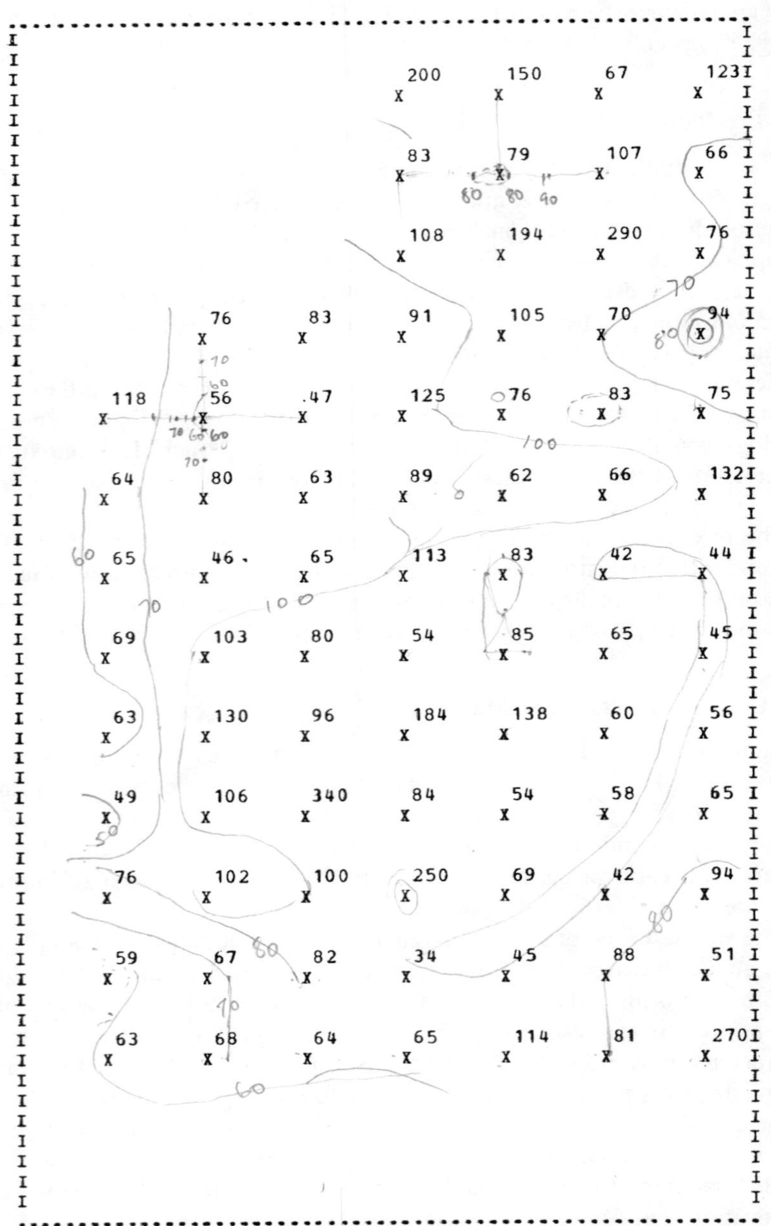

Fig. 19.1. A plot of data produced on a line-printer.

described by Howarth (1977). Systems for handling data using a computer are described by Howarth (1971) and Nichol *et al.* (1966).

D. Contouring of Concentrations

A number of methods for computer contouring of data are available, but because no method is unique, some thought must be given to the choice of contouring method (Crain, 1970). A commonly used program is SYMAP (Dudnik, 1971) which produces its output on a line printer and uses an inverse distance function of points within a preselected radius to assign a value to a particular location. The contour interval also requires some experimentation or previous thought, but a geometric or logarithmic interval is usually best. Comparisons of manual and computer contouring suggest that the computer may be more effective in the absence of auxiliary data, but a geologist or geochemist is better if geologic, geochemical, geomorphic, or other data are available and can be applied to choose trends or alternate contour patterns (Dahlberg, 1975).

E. Moving Averages and Trend Surfaces

Two general types of smoothing techniques are used for geochemical maps: moving averages and trend surfaces. The moving average is accomplished by averaging all determinations within a fixed area or window which is moved across the map to produce a uniformly spaced series of averages (Fig. 19.2). These average values are then contoured or plotted using the methods described above (Howarth, 1971; Applied Geochemistry Research Group, 1974; Webb *et al.*, 1978). Alternatively, the central point on a grid can be weighted more heavily than adjacent points in the window. The size of the window and the spacing between averaging points must be determined empirically. Adjacent positions of the window may be overlapped. Overlapping tends to broaden the resulting anomalies and reduce contrast, but the averaging process decreases the random fluctuations and clarifies the pattern (Govett *et al.*, 1975).

Smoothing by a trend surface involves approximation of the data by an equation in which Me, the concentration of an element, is given as a function of x and y, the coordinates of the point. Polynomials of the form

$$Me = a + bx + cy + dx^2 + exy + fy^2 + \dots \quad (19.1)$$

are the most commonly used type of equation, although many others are possible (Crain, 1970). The numerical values of a, b, c, etc. are chosen to minimize the squared differences between the observed concentrations and the calculated values derived from the equation. Use of only the first three

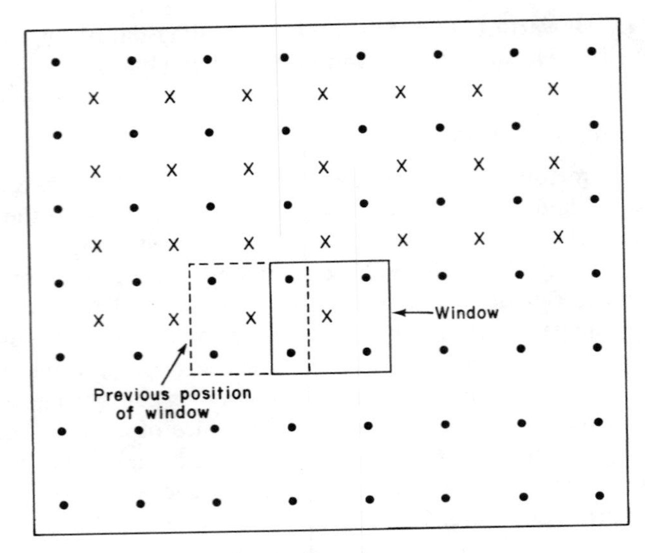

Fig. 19.2. Construction of a simple moving average of samples located on a grid.

terms of the above equation leads to a planar trend; increased numbers of terms lead to curved surfaces of increasing complexity (Fig. 19.3). If the anomaly occupies a large part of the map area, it is expressed in the trend. If the anomaly occupies only a small part of the map area, it is best expressed in the residuals (deviations from the trend).

In general, geochemical data do not fit any particular mathematical function, and have a complex pattern of highs and lows. The moving average can

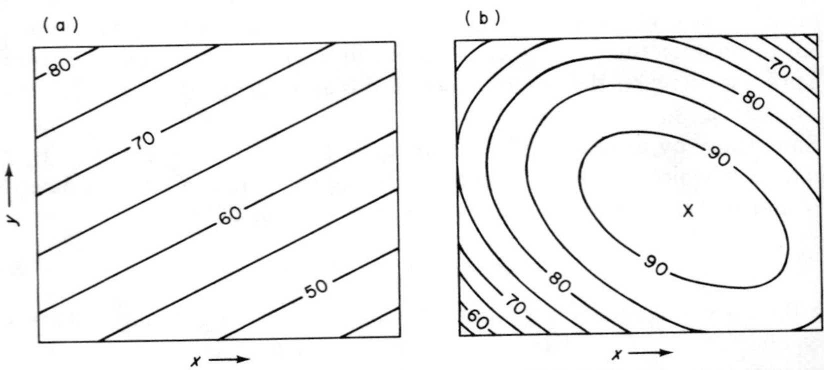

Fig. 19.3. Simple trend surfaces. (a) Planar trend surface from equation of the type $Me = a+bx+cy$; (b) quadratic trend formed by eqn 19.1 of text.

form any shape of surface inherent in the data, and the degree of smoothing can be varied widely by changes in the window size and degree of overlap. For these reasons, moving averages are generally to be preferred to trend surfaces.

II. MULTIVARIATE STATISTICAL METHODS

The relations among two or more elements in a group of samples may carry considerable information not expressed by any single element. For example, consider the data for Fe and Zn in a hypothetical set of stream sediments (Fig. 19.4). Some high values of Zn are correlated with high Fe, and may result from processes concentrating Fe. Other high Zn values occur independently of Fe, and may indicate an unusual source of Zn in the drainage.

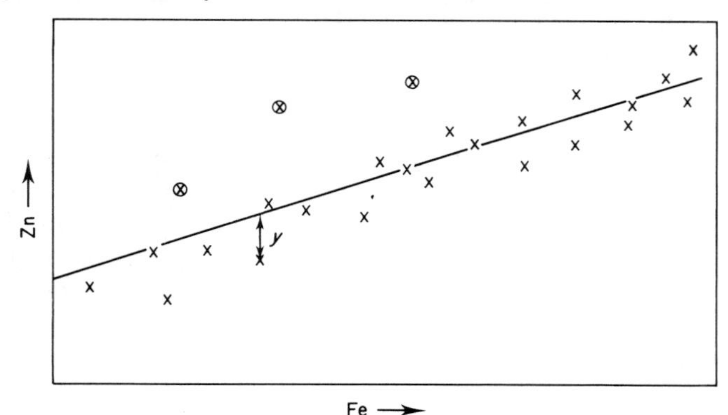

Fig. 19.4. Example of simple regression of Zn (dependent variable) against Fe (independent variable). The line minimizes the squared deviations (y) of Zn from the regression line. The circled points are anomalous based on the regression, but would not be evident in the raw Zn data.

The latter type of sample is not distinguishable by Zn alone, but by considering the relationship of Zn and Fe these samples are easily recognized. A variety of statistical methods are currently or potentially applicable for interpretation of relations involving multiple variables. These include regression analysis, discriminant analysis, factor analysis, and cluster analysis. Some general aspects of this topic are discussed by Howarth (1973) and Koch and Link (1971).

A. Regression Analysis

Natural situations in which a trace element shows distinct dependence on the concentration of a major element, mineral component, or other "master variable", such as pH or total dissolved solids, can be usefully treated by regression. Relations of trace elements to Fe-oxides, Mn-oxides, organic matter (loss on ignition), and clay content (Al) are common examples.

The simplest regression equations are of the form

$$Me = a_0 + a_1 X \tag{19.2}$$

where Me is the concentration of the dependent element (Zn in the example above), X is the concentration of the independent variable (Fe in the example), and a_0 and a_1 are, respectively, a constant and a coefficient computed to give a good fit to the data. The constant and coefficient are usually selected to minimize the squared deviations (y-values of Fig. 19.4) of the points from the regression equation, which in this simple case plots as a straight line. Statistical tests may be used to determine whether the regression line is a significant improvement over considering the two variables as unrelated (Draper and Smith, 1966; Koch and Link, 1971, p. 94).

In order to fit the data mathematically in a meaningful way, most of the samples must be "background" samples, that is, they must approximately fit the linear relationship of the two variables. If numerous anomalous samples are included, the regression line will be pulled away from the trend of the background samples into a compromise position, and some anomalous samples will not be recognizable in terms of their deviations from this trend. For simple sets of data, visual fitting of a regression line on a plot of the data may be adequate to express the relationship. Examination of scattergrams of the data is recommended before extensive regression tests in order to avoid misleading conclusions from a few aberrant analyses (Chapman, 1976).

In more complex situations, several possible independent variables may be measured, and an equation of the type

$$Me = a_0 + a_1 X_1 + a_2 X_2 + \ldots a_n X_n \tag{19.3}$$

is fitted, where $X_1 \ldots X_n$ are the independent variables. The entire set of coefficients may be calculated and tested for statistical significance in one step, or a stepwise procedure that adds (or subtracts) one variable at a time to the equation may be used (Koch and Link, 1971, pp. 89–90; Efroymson, 1965). The final stepwise equation contains only those independent variables that account for significant amounts of variability in the element being investigated.

Once the regression equation has been derived, the difference between values observed and values calculated from the regression is computed for each sample. Unusually large differences are considered anomalous.

As an example of regression, Rose *et al.* (1970) computed a stepwise regression equation for Zn in 267 drainage samples as a function of Fe, Mn, and the areal percentage of 20 rock types in each drainage basin. The equation accounted for 36% of the total variability in Zn content, and included terms for Fe, Mn, and four rock types. In the raw data, many of the high Zn values were in samples draining gneiss, one of the rock types included in the regression. Using the equation, previously unrecognizable anomalies near a known but abandoned Zn mine in a limestone area could be detected (Fig. 19.5), as well as several additional drainages of potential interest; most of the gneiss drainages were no longer anomalous.

B. Discriminant Analysis

In this method, several variables are combined into a single new variable which optimizes the distinction between two populations of samples. In mineral exploration, one population might be anomalous samples in the area under study, and the other population background samples. A representative set of known samples (the training set) from each population is necessary to form the discriminant equation. Using these two sets of known samples, an equation of the form

$$D = a_1X_1 + a_2X_2 + \dots a_nX_n \tag{19.4}$$

is calculated, where D is the new single variable used in discrimination, the X_i are values of the several independent variables, and the a_i are coefficients or weighting factors calculated to give the optimum separation of the two sets of samples. Once the discriminant equation has been calculated from the known groups, then values of D may be calculated for unknown samples which then may be classified as anomalous or background.

Figure 19.6 shows a simple two-dimensional example. Suppose that Ni and Cu have been measured on a group of gossans over known Cu–Ni ore and a group of known barren gossan-like materials. Although both Ni and Cu individually can be used to subdivide the groups with some success, the value of D, which is equivalent to the projected location of the points onto the line, is more effective than either Cu or Ni individually. That is, more samples are classified correctly using the discriminant than using either variable individually. Although two-variable cases can be handled graphically, the discriminant procedure can be used with any number of variables.

Discriminant analysis is an extremely useful method if sufficient samples representative of the desired anomalies can be obtained. Unfortunately, in a great many areas of interest, uncontaminated samples near known orebodies with a representative combination of geology, weathering, cover, size, and other characteristics are difficult to obtain in sufficient numbers to form a reliable training set, and great care must be exercised in deciding whether a discriminant function can be properly employed.

(a)

Deviations from grand mean: 1·5–2·0s, >2·0s; Precambrian rocks

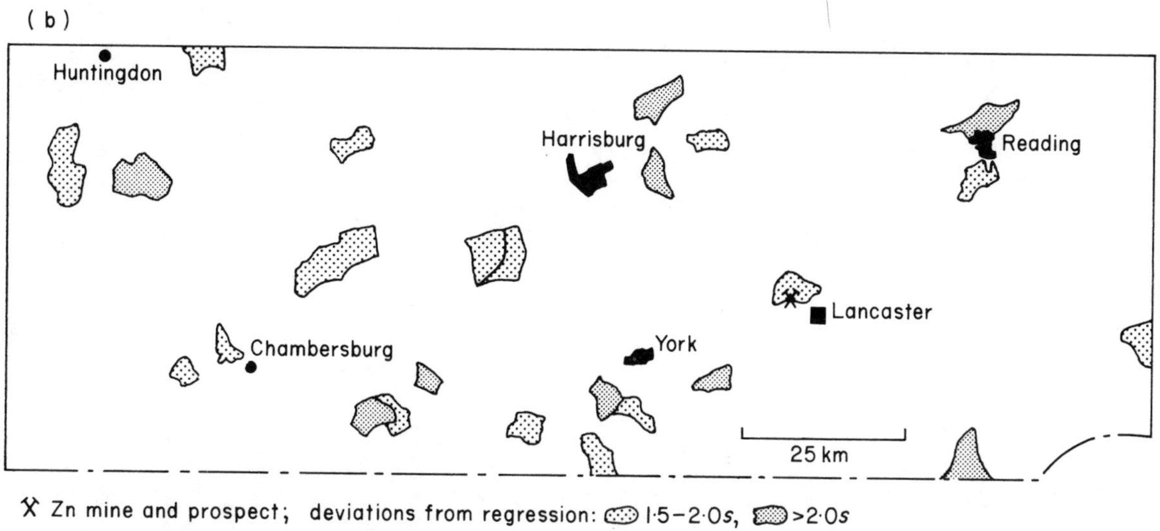

Fig. 19.5. Comparison of anomalies based on raw Zn values of stream sediment (a), and residuals from regression against Fe, Mn and rock types (b). Anomalies tend to reflect Precambrian gneisses in raw data, but residuals detect abandoned prospects near Lancaster, Pennsylvania. (After Rose *et al.*, 1970.)

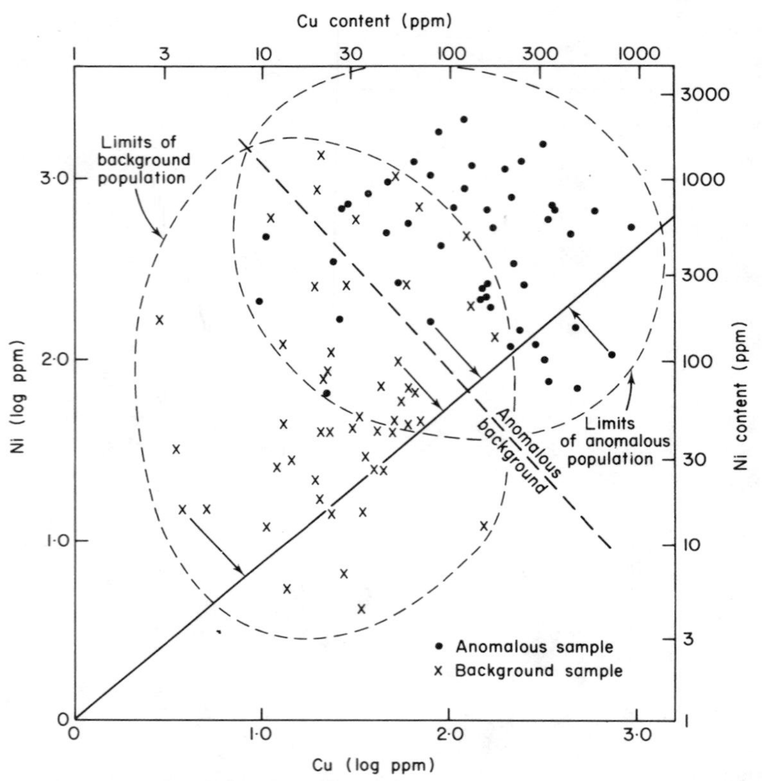

Fig. 19.6. Illustration of discriminant analysis for hypothetical anomalous and background samples analyzed for Cu and Ni. Projection to the line D (arrows) gives the single value which separates the two populations most effectively, using the dashed line as the dividing point. Equation is $D = 3{\cdot}08 \log \text{Cu} + 2{\cdot}70 \log \text{Ni}$; dividing value is $D = 11{\cdot}37$.

Most computer programs calculating discriminant functions assume a normal (or log-normal) distribution of the variables. Castillo-Munoz and Howarth (1976) demonstrate the use of empirical discriminant functions that do not require normally distributed data. Cameron *et al.* (1971) formed a discriminant function from the Cu, Ni, Co, and S contents of ultramafic rocks that improved discrimination of intrusions associated with Cu–Ni sulfide ore. Various graphical procedures can be used to construct simplified discriminating measures (Govett, 1972; Bull and Mazzucchelli, 1975; Joyce and Clema, 1974). Geological, geophysical, and geochemical variables were combined into a discriminant equation for evaluation of the exploration potential of small areas by Rose (1972) and Prelat (1977).

The method of characteristic analysis has a purpose similar to discriminant analysis. As used by Botbol *et al.* (1978), each elemental analysis in a multivariate set of data from an area is classified into one of two groups, according to whether it is higher or lower than adjacent samples. This procedure amounts to a separation of determinations into either positive anomalies or negative anomalies relative to immediately adjacent samples. Data from areas with known mineralization of the type sought are then used to select variables and assign weights in a way that most clearly distinguishes the known areas. These variables and weights are then used to evaluate the remaining samples. The method differs from discriminant analysis in the classification of all data into two categories dependent on magnitude relative to immediately adjacent samples, and in the lack of necessity for a background population.

C. Factor Analysis

Statistical correlations between elements are commonly found when a set of samples is analyzed for many elements. In many such cases, a group of elements correlate with each other and reflect the operation of a single process, geochemical characteristic, or master variable. The information contained in such sets of data can be expressed by a smaller number of variables that combine elements into groups with close correlation. R-mode factor analysis is a method of resolving a large number of elements into a smaller number of new combinations. These combinations may then be examined for significance in terms of process, types of samples, or other geological and geochemical information. Q-mode factor analysis is a method of recognizing different populations in a group of samples, based on multi-element similarity. Basic aspects of factor analysis are discussed by Harman (1967), Comrey (1973), Klovan (1975), and Joreskog *et al.* (1976).

As a simple two-variable example, the Zn–Fe data on Fig. 19.4 may be considered. A line approximately coincident with the regression line may be taken as one axis in a new set of rotated coordinates. A second new axis perpendicular to the regression line may also be constructed; these two coordinates are thus uncorrelated. Each sample point can be given coordinates on the new axes. For example, the value of the coordinate on the first axis may reflect the amount of fine-grained minerals, including Fe minerals, in the sample. Most of the variability in both Fe and Zn values is described by this single new axis. The value on the second axis may reflect availability of Zn, independent of grain size. R-mode factor analysis is a method for finding the best new axes in multivariate space, so that essentially the same information is expressed with fewer and/or more meaningful variables. The array of data is simplified in this process, and may be more easily understood.

Obviously factor analysis is closely related to regression, but does not require one variable to be dependent and others independent. In contrast to discriminant analysis, it does not require sets of known samples for calibration. Factor analysis allows an empirical analysis and simplification of the interrelations of many variables.

The relations between variables are commonly expressed in a table of factor loadings expressing the contribution of each variable to the new coordinates. In R-mode analysis, the loadings are in terms of elements, in Q-mode analysis each sample is expressed in terms of the contribution of a number of real (or hypothetical) "end member" samples of extreme composition. For example, 67% of the information in a set of data for 13 elements in stream sediments of Sierra Leone can be expressed in three R-mode factors (Table 19.1). Factor 1 has a high loading on Ti, V, Cr, Mn, Co, Ni, Cu, and Zn, and apparently represents the effect of mafic rocks on the composition of the stream sediments in question. Factor 2 (Ga, Sn, and Pb) and factor 3 (V, Ti, As, and Mo) are partly interpretable in terms of rock types (Nichol *et al.*, 1969). The communality for each element expresses the proportion of the total variability of that element that is contained in the factors.

Factor analysis is useful in larger sets (hundreds of samples) of multivariate data in recognizing relations among variables, and developing procedures for

Table 19.1
Factor loadings and communalities for stream sediments from Sierra Leone[a]

	Loadings[b]			
Element	Factor 1	Factor 2	Factor 3	Communality
Ti	*0·605*	*0·413*	0·047	0·539
V	*0·873*	—0·094	*0·308*	0·867
Cr	*0·729*	—0·135	*0·509*	0·808
Mn	*0·720*	—0·062	0·049	0·525
Co	*0·933*	−0·173	−0·075	0·906
Ni	*0·944*	−0·166	−0·058	0·923
Cu	*0·884*	−0·143	0·170	0·831
Zn	*0·854*	0·050	0·090	0·740
Ga	−0·117	*0·748*	0·161	0.599
As	0·311	−0·263	*0·663*	0·605
Mo	−0·027	0·200	*0·854*	0·770
Sn	0·070	*0·383*	−0·181	0·185
Pb	−0·217	*0·658*	0·012	0·480

[a] Source: Nichol *et al.* (1969).
[b] Italic indicates important loading.

removing the effects of unwanted or irrelevant processes on the composition of samples. In some sets of data, the presence of ore may be directly reflected in a recognizable factor which can be expressed quantitatively by calculating a factor score for each sample from the factor loadings. In other sets of data, the effects of ore are very small and are contained in the residual variability not explained by the factors. Use of this residual variability as a guide to ore (Nichol *et al.*, 1969) is probably preferable to extraction of weak factors of dubious statistical validity.

D. Cluster Analysis

This method attempts to separate items (samples) into groups or clusters on the basis of similarities in their measured attributes. Clustering refers to proximity in *n*-dimensional space where each of *n* elements or attributes is plotted on a separate axis. In geochemical exploration, the goal of cluster analysis might be recognition of separate ore and background clusters. For example, in a suite of 170 stream sediments from Derbyshire, England, analyzed for 19 elements, Obial and James (1973) detected three main clusters or groups of samples. Cluster 1 included sediments in limestone drainages, cluster 2 those in shale drainages, and cluster 3 sediments affected by Pb–Zn mineralization. In a ground water survey for U in Texas, Butz (1977) reports that clusters based on U, V, Mo, Se, As, SO_4, pH, and conductivity define the known U district and also a previously unrecognized area to the south-east. However, in both this example and that in Derbyshire, it appears that most or all of the anomalous samples could have been recognized by selection of samples high in obvious indicator elements. To date, no comparisons of conventional single-element interpretation with cluster analysis seem to have been made.

Active research continues in the development and application of cluster analysis and related methods. Summaries and descriptions of the method are published by Sokal and Sneath (1973). For large multi-element sets of data, this method deserves consideration as a technique for recognizing major types of samples and their relations to geological and geochemical processes. Best results with this and other statistical methods will be achieved if they are used as part of a comprehensive geochemical and geological assessment of data, rather than as a rote application of the statistical methods.

Chapter 20

Geochemical Methods in Mineral Exploration

★★★★★★★★

To help him in the increasingly difficult task of locating concealed mineral deposits, the present-day exploration geologist has at his disposal a wide selection of geological, geophysical, and geochemical techniques. At the same time he generally has better maps and mapping equipment than in the past, and vastly improved facilities for transport and communication. As a result of the multiplicity of new techniques and facilities, modern exploration has become a complex and often costly business. With the rising cost of field operations, maximum efficiency in the application of the modern technical aids is demanded in order to reduce the financial risk to reasonable proportions. This can be achieved only where adequately skilled personnel are using appropriate methods in carefully chosen areas at the most opportune time. To do this with legitimate economy requires that expenditure of funds must be correctly balanced against the chances of success and the possible financial return. Quite apart from technical considerations, it is important that the economic factor be given its full weight in each successive step in the evolution of a modern exploration program.

This chapter introduces some principles of planning exploration programs and attempts to put geochemical methods into perspective with other aids to exploration. Bailly and Still (1973) and Payne (1973) also provide summaries of current approaches to this subject. Suffice to emphasize at this point that the objective of all components of an exploration system is the same; namely, to detect and record any feature or property of the earth's crust that, when correctly interpreted, can serve as a guide to ore. The geological, geophysical, and geochemical surveys differ only with respect to technique. The basic philosophy of each is the same.

I. OPTIMIZATION OF EXPLORATION

Any exploration activity requires an immediate outlay of money and time, with the hope that discovery and exploitation of an orebody in the future will more than repay the initial cost. At the initiation of a typical exploration project, the probability of success on the project is very small, as indicated by records of exploration companies. For example, Roscoe (1971) estimated from data on Canadian exploration in the 1960s that an investment of \$30 000 in exploration in Canada has an average probability of success of 0·001. Brant (1968), Morgan (1969), and Peters (1978, p. 530) suggest probabilities for success of about 1 % for moderate-size deposits and 0·1 % for very large deposits. Nevertheless, the return for a discovery can be very large, in the range $\$10^7$–$\$10^9$, so that the outcome of a series of exploration projects, at least one of which is successful, can be very rewarding.

For an individual project, the expected financial return (E), taking into account the probability of success, is

$$E = pV \tag{20.1}$$

where p is the probability of discovering an economic mineral deposit and V is the net value of the discovery, after deduction of all post-discovery costs, including development, equipment, extraction, taxes, selling expenses, and overhead. The costs and selling prices must be adjusted to their present value, that is, the amount of money invested at present at a reasonable rate of compound interest that would return the future costs and proceeds anticipated for a successful mine (Megill, 1971).

The expected value must be balanced against the cost (C) of the exploration program. An exploration project is justified only if the expected return exceeds the cost of the program. In addition, an individual or organization investing in exploration normally wishes to maximize the ratio of expected return to cost (E/C) or some more complex function of these variables (MacKenzie, 1973; Brant, 1968; Newendorp, 1975). The ratio E/C has been termed the prospecting profit ratio (Slichter, 1955).

The probability of discovering an economic deposit may be broken down (Brant, 1968) into p_1, the probability of an economic orebody being present in the area being explored, and p_2, the probability of finding that orebody with the proposed exploration program, assuming that an orebody does exist in the area, so that

$$E = p_1 p_2 V. \tag{20.2}$$

Obviously, if no ore exists in the area ($p_1 = 0$), the best exploration technique cannot find it, and similarly use of an inappropriate technique ($p_2 = 0$) will also fail. The value of p_1 is set by nature (and by our definition of an economic deposit), but the value of p_2 can be controlled by the sample spacing, accuracy of analysis and interpretation, choice of method, and other factors involved in planning, execution, and interpretation. Note, however, that changes in sample spacing or other characteristics of the exploration plan affect the costs of the program. A good program optimizes the controllable variables in order to maximize E/C or some related variable.

The relations presented above (eqn 20.2) can at best be used semi-quantitatively, because p_1 is never known with any precision, p_2 is estimated from previous projects, orientation surveys or geometric considerations relating presumed target size to sample spacing, and V must be estimated for a hypothetical orebody. Only C can be closely estimated. Nevertheless, consideration of the interrelations among these variables is helpful in comparing methods, and in deciding which aspects of the program require the most care. Some comments on costs, effectiveness of methods, and size and value of ore are summarized in Section III below.

If exploration is conducted as a series of steps, in which favorable areas detected at one stage are further evaluated in a following stage, the overall probability of success is the product of the probabilities for the stages. For example, if a reconnaissance technique with a probability of detection of 0·8 is followed by detailed surveys with a probability of detection of 0·9, the total probability of detecting an orebody by the combined survey is 0·72. The costs of such a combination must include not only those for the reconnaissance survey and any successful detailed surveys, but also the costs for follow-up surveys of all anomalies from the reconnaissance that do not lead to ore. Such follow-up surveys can cost far more than the initial reconnaissance survey.

II. PLANNING OF EXPLORATION

Exploration normally involves a sequence of steps, both in the planning stage and in the execution stage. Figure 20.1 summarizes in diagrammatic form the sequence of administrative decisions normally followed in the planning and early operational stages of a mineral exploration program. The following sections discuss some considerations involved in those decisions.

A. Selection of Professional Leadership

Many exploration surveys have fallen short of expectations through lack of experience and technical training in the professional personnel responsible for

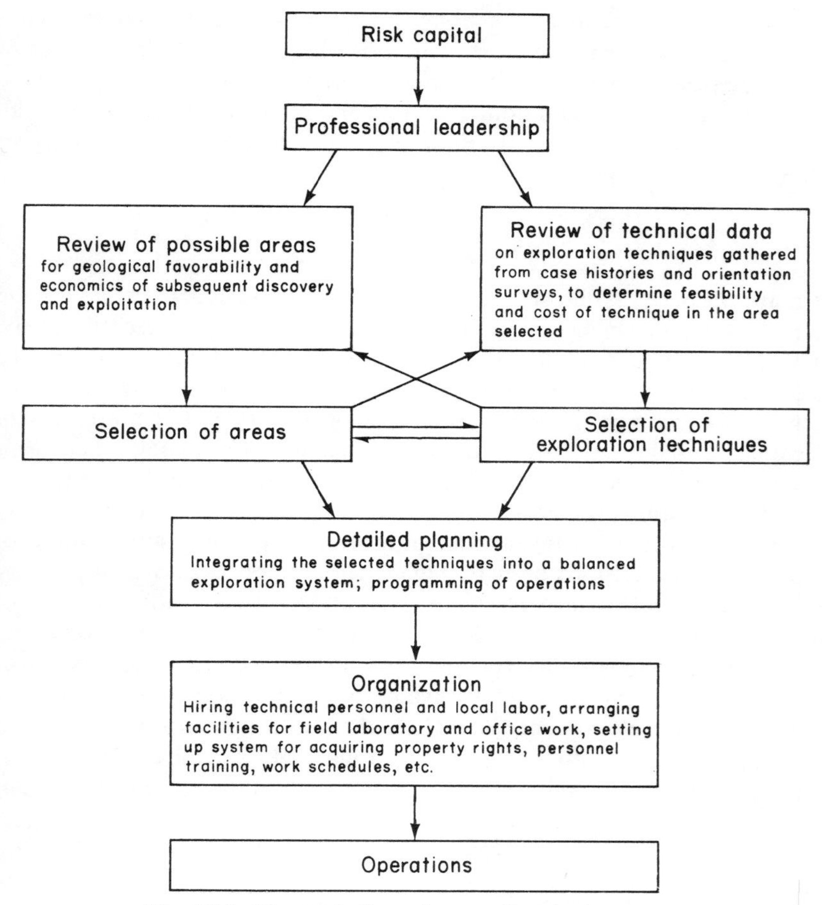

Fig. 20.1. The evolution of an exploration program.

planning and supervising the program. The essential requirements are a sound background in exploration geology, specialized skill in the different exploration techniques, and a natural flair for ore finding. Rarely are these to be found in any one man. More often it is necessary to build up a team, the members of which cooperate to provide the necessary virtues. The inability to obtain the services of a supervising geologist or of a specialist with the requisite experience and training may well be sufficient reason for ruling out a particular technique, no matter how much it may otherwise be appropriate for the problem on hand.

Not all techniques make the same demands on professional personnel. Full-time professional attention is certainly needed for most geological studies, but geochemical and geophysical surveys usually involve a considerable amount of routine work which can be done by trained teams of artisans and local labor, working under part-time professional supervision. The critical phases of orientation, planning, and interpretation, however, can be done effectively only by an appropriately qualified and experienced professional exploration geologist or geochemist.

B. Selection of Areas

The main purpose of this step is to select areas or regions that have good mineral potential (a high value of p_1) and that can be prospected effectively.

Initial selection of areas should be based on a most thorough review of the known geology and records of past mining and prospecting activity. This review should give particular attention to the possible types of deposit present in the areas, based on recorded mineralization and on the geological environment. In addition, careful attention should be given to the distribution of favorable rocks and structures, the nature of the overburden and weathering conditions, and other circumstances that may mask surface manifestations of bedrock mineralization. Examination of air photographs and satellite imagery can be an invaluable source of information on structural features, extent of rock units, and types of overburden at this stage. Possibilities for types and sizes of deposits compatible with the economic and commercial objectives of the investors should be carefully evaluated. In addition to technical considerations, other relevant aspects include the political environment, ground ownership, markets, taxation, communications, and labor. Preliminary visits to likely field areas may well be necessary before a well-founded selection can be made.

C. The Exploration Sequence

A large exploration program is commonly organized as a logical sequence of operations. Each stage in this sequence involves the study of an area by whatever exploration method or combination of methods is most effective for the purpose of delimiting smaller areas for more intensive study in the next stage (Fig. 20.2). At the end of each stage, the available information is evaluated and used to select smaller areas for detailed study, or to terminate the program if the initial premises have been disproved. The sequence may be entered at a relatively detailed stage if information on a specific area becomes available, as for example from a government agency or a prospector.

In a typical complete program, the first stage may be broad-scale reconnaissance using regional geological, geochemical, or geophysical criteria of favorability, to help decide which parts of a large area of interest have the best mineral potential and which parts can be eliminated as relatively unfavorable. The most promising regions constitute local areas of interest that are then followed up by more detailed surveys in order to determine whether an intensive exploration survey of some kind is desirable. This process of elimination of unfavorable areas and of increasingly detailed study of the

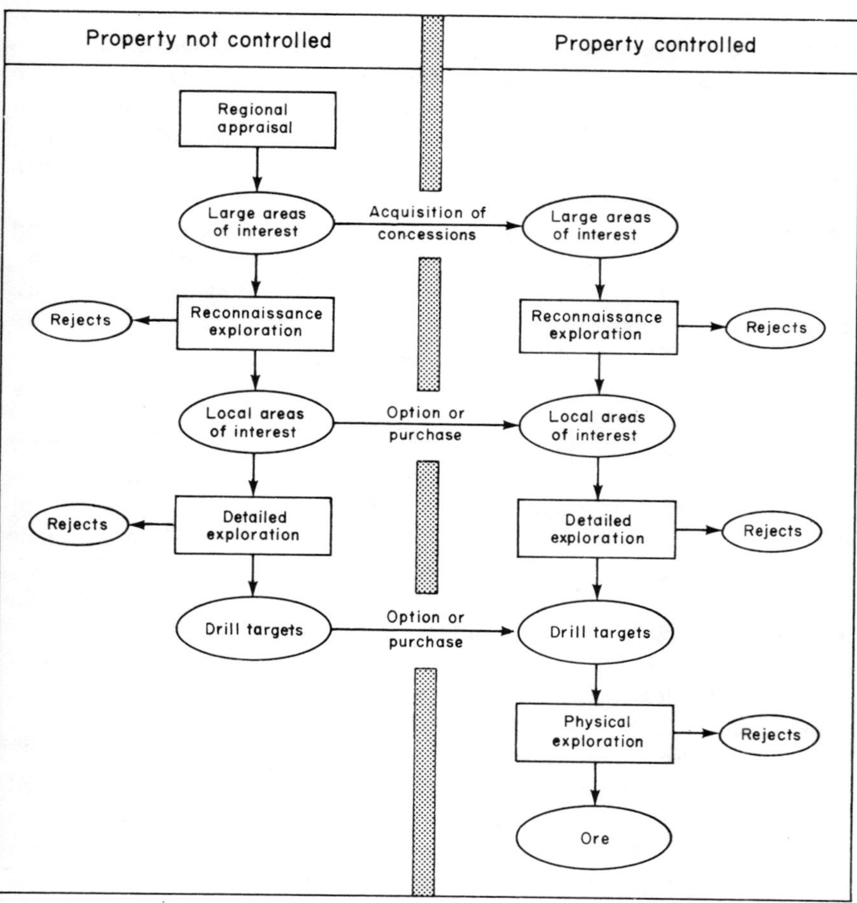

Fig. 20.2. Sequence of operations in an exploration program.

favorable areas is continued step by step up to the ultimate proving of a deposit by drilling and underground exploration. It is a process of progressively reducing the size of the target, where at each step the target is the area having the highest probability of containing ore. The ultimate purpose of a technical exploration program is the selection of drill sites. Almost invariably, the cost of drilling is by far the largest part of the exploration budget. The price of one hole saved will pay for a great deal of preliminary survey work.

III. CHOICE OF EXPLORATION METHODS

Normally, several ore guides with their corresponding methods of detection will be found to be more or less applicable for any given problem. The degree of usefulness of a particular exploration method will, of course, be different for each method and for each successive stage in the exploration sequence. For greatest overall effectiveness, an optimum balance must be struck between the various geological, geochemical, and geophysical methods that are known to be applicable. After the initial choice of area, the choice of methods and of the plan of operations that will result in this optimum balance is by far the most critical decision to be made by the geologist in charge. This choice must of necessity be based on relevant previous experience and case-history data, supported by the results of preliminary orientation surveys in the areas under consideration, and guided by the principles of optimization discussed in Section I of this chapter.

The journal literature carries frequent articles describing field experience gained in production geochemical surveys. Case-history studies and previous personal experience are useful only as suggestive guides in the choice of methods in a new area. The more nearly the past experience matches the new situation, the more reliable will be the conclusions that are drawn. However, the diversity of factors controlling metal dispersion is such that no two field problems are exactly alike. Quite subtle variations in the local environment can so radically change the course and pattern of metal dispersion as to demand more or less drastic modifications in technique and, particularly, in interpretation.

Preliminary orientation surveys or access to data on previous surveys in the area are therefore mandatory if any degree of success is to be expected from a geochemical survey. As an important corollary, the effectiveness of a geochemical survey must rest initially on the thoroughness and competence with which the orientation has been carried out. In addition to the vital preliminary studies, supplementary orientation may also be required from time to time as new conditions are encountered in the course of field operations.

As expressed in eqn 20.2, major factors bearing on the selection of exploration methods are the value of the ore deposit being sought (V), the reliability of the exploration method applied under the local conditions (p_2), and the costs of the exploration work (C). Selection of methods is also strongly

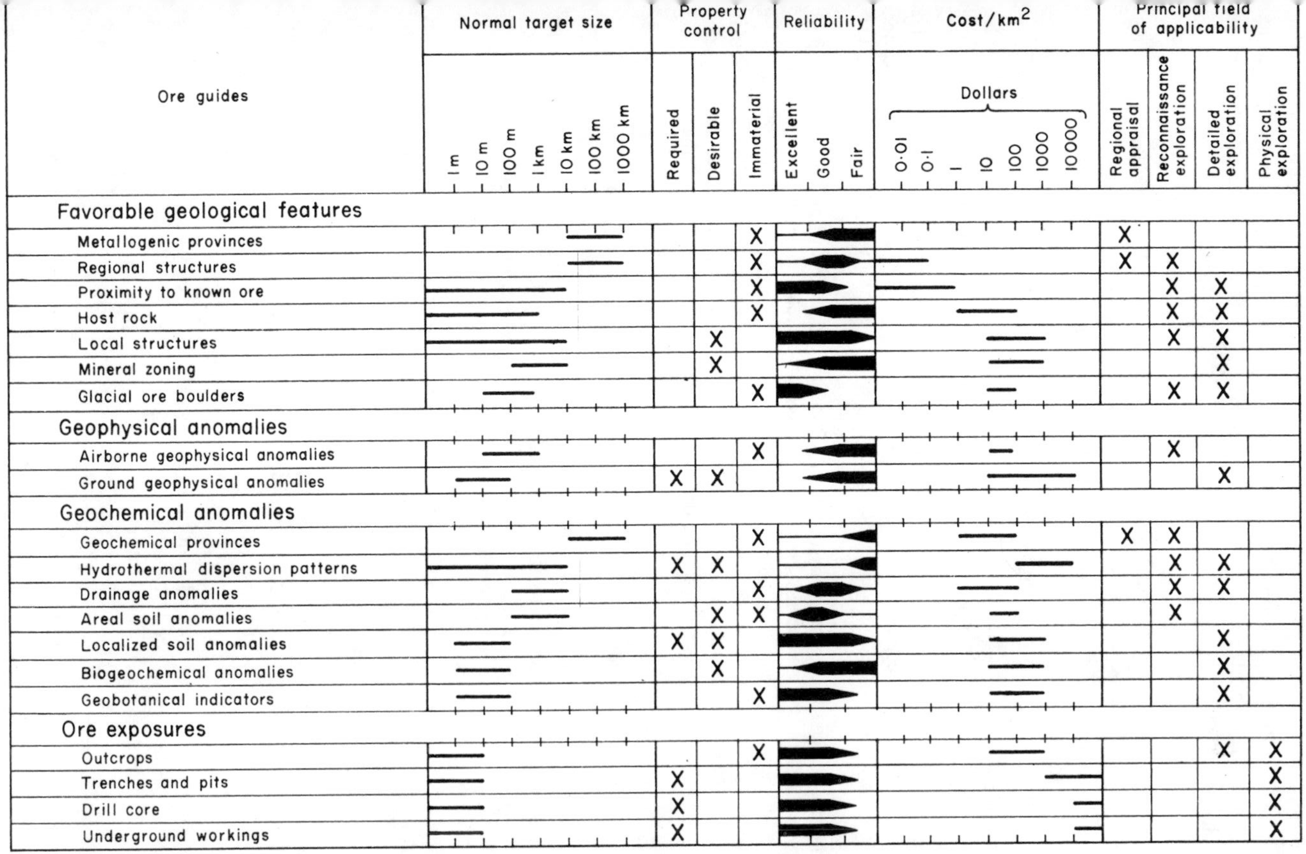

Ore guides	Normal target size (1 m, 10 m, 100 m, 1 km, 10 km, 100 km, 1000 km)	Property control: Required	Property control: Desirable	Property control: Immaterial	Reliability (Excellent, Good, Fair)	Cost/km² (Dollars: 0·01, 0·1, 1, 10, 100, 1000, 10000)	Principal field of applicability: Regional appraisal	Principal field of applicability: Reconnaissance exploration	Principal field of applicability: Detailed exploration	Principal field of applicability: Physical exploration
Favorable geological features										
Metallogenic provinces				X			X			
Regional structures				X			X	X		
Proximity to known ore				X				X	X	
Host rock				X				X	X	
Local structures			X					X	X	
Mineral zoning			X						X	
Glacial ore boulders				X				X	X	
Geophysical anomalies										
Airborne geophysical anomalies				X				X		
Ground geophysical anomalies		X	X						X	
Geochemical anomalies										
Geochemical provinces				X			X	X		
Hydrothermal dispersion patterns		X	X					X	X	
Drainage anomalies				X				X	X	
Areal soil anomalies			X	X				X		
Localized soil anomalies		X	X						X	
Biogeochemical anomalies			X						X	
Geobotanical indicators				X					X	
Ore exposures										
Outcrops				X					X	X
Trenches and pits		X								X
Drill core		X								X
Underground workings		X								X

Fig. 20.3. Factors in the selection of exploration methods.

influenced by the size of the target and by property control. Table 20.1, Fig. 20.3, and the following text summarize the nature of various ore guides and their reliability, cost, and applicability to various types and stages of exploration.

Table 20.1
Principal geological, geophysical, and geochemical ore guides

Ore guide	Explanation	Example
	Favorable Geological Features	
Metallogenic provinces	Preferential occurrence of some kinds of ore deposits in certain well-defined areas	Porphyry-copper deposits of the south-western U.S.A.
Regional structures	Ores preferentially located with respect to structural features of continental scale	Epigenetic precious metal deposits in the Basin and Range structural province, western U.S.A.
Igneous rock associations	Preferential association of some kinds of ores with certain kinds of igneous rocks	Nickel sulfide ores associated with gabbros and norites
Proximity to known ore	Better chances of finding additional ore near a known deposit than at a distance from it	Discovery of Mission deposit near Pima and Banner deposits, Twin Buttes district, Arizona
Host rock	Ores preferentially localized in specific rock types or in specific stratigraphic horizons	Zambian Copperbelt ores localized in the Mines Series
Local structures	Ores preferentially located with respect to the crests, troughs, or flanks of folds, or to certain features of fracture systems	Proximity of ore to Osborne Fault in the Coeur d'Alene district, Idaho
Mineral zoning	Changes in grade or type of ore either laterally or with depth as indicated by changes in mineralogy	Zoning of Pb ore at surface to Zn ore at depth in south-western Wisconsin Zn district
Glacial ore boulders	Ore buried by glacial debris indicated by glacially transported ore boulders (see Chapter 10)	Ore boulders south of Steep Rock Lake Fe deposit, western Ontario
	Geophysical Anomalies	
Airborne geophysical anomalies	Geophysical anomalies related to ore that can be detected by airborne equipment	Discovery of magnetite deposits by airborne magnetic surveys in south-eastern Pennsylvania
Ground geophysical anomalies	Geophysical anomalies related to ore that are mapped instrumentally at the surface of the ground	Discovery of New Brunswick massive-sulfide deposits with electromagnetic surveys

Table 20.1—continued

Geochemical Anomalies	
Geochemical provinces	See Chapter 4
Hydrothermal dispersion patterns	See Chapter 5
Drainage anomalies	See Chapters 14–16
Lateral soil anomalies	See Chapters 11–13
Superjacent soil anomalies	See Chapters 11–13
Biogeochemical anomalies	See Chapter 17
Geobotanical indicators	See Chapter 17

A. Target Size

Just as the center of the marksman's target is surrounded by concentric rings, so most ore deposits have associated with them, more or less closely, a series of larger target areas, each characterized by some geological, physical, or chemical feature that is diagnostic of the ore environment. The relationship between these features and the ore may be either direct or indirect. Thus, some features of the ore environment, such as favorable host rocks, geologic structures, or geochemical provinces, are related to the genesis and localization of the ore but are not in themselves necessarily indicative of mineralization. Other features, such as most geophysical anomalies, are a direct response to some unusual physical property of the orebody itself. Still others, as for example gossans, leached outcrops, and secondary geochemical anomalies, result from the weathering and dispersion of the primary components of the ore.

Each geological, geophysical, and geochemical feature of the ore environment defines a target area of characteristic size, shape, and relationship to the ore. Some extend over large areas, forming targets that may be detected by widely spaced observations. Others are more restricted in their extent and require more detailed examinations for their detection.

The figures for "target size" given in Fig. 20.3 refer to the longest dimension of the favorable area or target. Examination of regional structural trends on a geologic map of the U.S.A., for example, may show that the Appalachian belt is a favorable ore province. This presents a large target. Airborne magnetic surveys, on the other hand, will not indicate a magnetic anomaly much more than 300 m from the magnetic body responsible for the anomaly, whereas ground magnetic observations must usually be made within 30 m of the deposit in order to detect it. These are small targets. The normal sequence of operations in a large exploration program consists of surveys for progressively smaller targets by techniques capable of progressively greater resolving power.

B. Property Control

The acquisition of property rights, in most countries, is an extremely tedious and expensive phase of mineral exploration. Where this situation exists, the general pattern of exploration may be modified or even dominated by property considerations. Exploration methods that can be applied to appraising a tract of land without acquisition of property or of trespass rights are at a very definite premium. Subject to the local mining laws, airborne geophysical surveys and most geological and geochemical reconnaissance methods fall into this category. For detailed ground surveys of any type, and all physical exploration except for simple outcrop examination, property or trespass rights are usually desirable or required by law.

C. Reliability of Method

The reliability of a method refers to the probability of obtaining and recognizing indications of an orebody by the method being used. Reliability depends not only on how effective the exploration method is in locating the target, but also the extent to which an anomaly is specifically related to ore and the abundance of non-significant anomalies that may confuse interpretation and require follow-up surveys. As a simple illustration, a resistivity survey would not normally detect Zn deposits where the non-conducting sphalerite is the only ore mineral. At the same time, the resistivity survey might record strong anomalies due to graphite horizons or water-filled fractures that are of little or no prospecting significance and that would confuse the interpretation. Resistivity surveying, therefore, would be an unreliable method of prospecting for this kind of Zn deposit. Geochemical soil anomalies, however, are present over virtually all base-metal deposits that occur in the bedrock immediately beneath a cover of residual soil. Here the chances of finding a target are much greater, and the interpretation of the anomalies is likely to be fairly straightforward. In other words, residual soil sampling under these conditions would be a relatively reliable ore guide for Zn. Even in seeking conductive pyrite–chalcopyrite bodies for which resistivity would have a high probability of detection, the survey might lead to so many anomalies in an area of graphitic schists that the total costs and time for follow-up would be very large, and the method would only be of moderate reliability.

D. Cost

The cost of an exploration survey is the only one of the critical factors that can be estimated with any degree of accuracy. Costs should be compared on a unit-area basis. High total cost and high cost of capital equipment do not necessarily imply a high unit-area cost. Airborne geophysical surveys, for example, although having a high minimum cost and a high hourly cost, are often relatively inexpensive when computed in terms of cost per unit area, providing, of course, that the area surveyed is sufficiently large. Data on costs are listed by Peters (1978), Cazalet (1973), and Leyshon and Cazalet (1976).

E. Value of Expected Ore

The expected value of the orebody being sought and the chances of success may have an appreciable bearing on the applicability of a given method. Thus, a high-cost survey such as detailed geochemical soil sampling is justified if large orebodies are expected and there is relatively little chance of overlooking a deposit. Conversely, a low-cost survey, such as airborne magnetics, may be all that is justified in areas of small targets, poor chances of finding a target, or low probability for the occurrence of large orebodies.

F. Comparison of Programs

As an example of the interaction of the above factors, two simple exploration programs are compared, using eqn 20.2. Program 1 is a detailed survey of an area of 1000 km^2 using soil sampling; program 2 is a combination of a reconnaissance survey of the same area followed by detailed surveys of anomalous subareas. If the detailed method costs $200/$km^2$ and has a 90% probability of detecting an orebody, then the total cost is $1000 \times \$200 = \$200\,000$, and $E/C = 0{\cdot}9p_1V/200\,000 = 4{\cdot}5 \times 10^{-6}\,p_1V$. If the reconnaissance method costs $10/$km^2$, has a probability of detection of 50%, and leads to anomalies requiring that 5% of the area be covered by detailed follow-up surveys, then the total cost is $(1000 \times \$10) + (1000 \times 0{\cdot}05 \times \$200) = \$20\,000$, and $E/C = (0{\cdot}5 \times 0{\cdot}9)p_1V/20\,000 = 2{\cdot}25 \times 10^{-5}p_1V$. The sequential procedure has improved the profit ratio by a factor of 20 and decreased the total investment by a factor of 10, although some orebodies will clearly be missed using the reconnaissance. If E/C is to be greater than unity and p_1 is at levels of 10^{-2} to 10^{-3}, as suggested by Roscoe (1971), then V must be very high.

Although calculations of this type should not be accepted quantitatively, experimentation with different combinations of methods generally helps to refine one's thinking about the merits of various methods and combinations, and to indicate the kinds of information and activities that are most crucial to success.

G. The Role of Geochemistry in an Exploration System

It should be apparent from the foregoing discussion that geochemical methods have no unique claim to general applicability in mineral exploration. For any given problem, the pros and cons of the appropriate geochemical methods must be weighed against those of the other available prospecting methods, and a proper place assigned to each in the schedule of reducing the target size. The areas of particular applicability of geochemical methods, together with those of other exploration methods, are summarized in the right-hand column of Fig. 20.3.

IV. ORGANIZATION AND OPERATIONS

The best kind of organization for a geochemical exploration survey depends not so much on the nature of the problem as on the scale of operations. At one extreme is the quick reconnaissance appraisal of a prospect involving only a limited number of samples and simple analytical equipment. At the other extreme, large-scale operations involving comprehensive exploration over an extensive area may well require a complex organization comprising numerous field teams, centralized and mobile laboratories, and the services of specialized personnel. An enterprise of this kind presents a major problem in coordination.

Every geochemical survey organization, irrespective of its scale, is based on three main functional units: (i) the field party, engaged primarily in sampling, (ii) the laboratory, and (iii) the technical direction responsible for decisions on personnel, technique, operation, and interpretation of the results. The detail of organization within the functional units is largely a matter of circumstance and common sense. In the following paragraphs, however, attention is drawn to a few of the most critical factors in the organization of geochemical exploration.

A. Field Operations

In a sampling program, maximum efficiency requires that lower-grade personnel be used in preference to professional geologists wherever possible. Most operations in a systematic sampling program can be carried out by teams of local workers under the supervision of trained field assistants. In some surveys, however, proper selection of the material to be sampled and adequate recording of information needed in later interpretation may demand the experience and training of a member of the junior professional staff. The most rigid systematization of the sampling procedure is, of course, essential in order to ensure adequate reliability. Fullest efficiency also requires that all non-essential operations be eliminated and that the time required for each remaining operation be reduced to a minimum. Under no circumstances should time be wasted in unnecessarily precise surveying. The exact location of negative samples is of no interest. All that is required is enough information that any site can be revisited if it eventually proves to be anomalous. Productivity depends very largely on the time taken in traveling between sample points. Before a decision is made on the mode of operation of a sampling survey, careful estimates should be made of the relative combined cost of transport plus personnel time for travel on foot, by jeep, by helicopter, and by any other available means. Not infrequently, such studies show that an apparently expensive system of transport may actually be the cheapest in terms of cost per sample site.

B. Laboratory

In the analytical phases of a geochemical survey, the first question is whether to use a commercial laboratory or an internally run laboratory. A commercial laboratory is generally preferable for surveys involving up to a few hundred samples, and for surveys in areas close to good commercial laboratories. An internally run laboratory may be preferable for large surveys, for exploration groups conducting numerous surveys over a period of years, and for surveys in remote regions where shipping of samples would be difficult or time-consuming.

One advantage of a commercial laboratory is that the cost of analyses is generally as low or lower than those done in an internal laboratory. Another advantage of the commercial laboratory is that project personnel are relieved of the responsibility for organizing the lab and hiring, training, and supervising its personnel. A large commercial laboratory may have specialized equipment and procedures that are not economic for a smaller laboratory. On the other hand, considerable care is needed to select a commercial lab that does accurate and precise work, and has the capacity to complete analyses as rapidly as needed. A program of routine checks of the precision and accuracy of the commercial lab must be set up and conducted throughout the duration of the survey, as described in Section VI of Chapter 3.

If an internally run laboratory is to be used, the principal problems are: (i) location of the laboratory, (ii) the nature and number of the personnel employed, and (iii) systematization of the laboratory routine.

The promptness with which analytical data are required varies with the nature of the operation. In an orientation survey preparatory to laying out a program, data are often needed within a day or so from the time of collecting the sample, and can best be provided by a mobile analytical unit at the field camp. When following up anomalies already detected by drainage reconnaissance, it is also very desirable to be able to carry out determinations on the spot. Most of the colorimetric procedures listed in Table 3.3 can, if necessary, be set up under crude field conditions, and a few of the procedures are even adaptable for use at the sample site. Mobile spectrographic laboratories have also been designed for field use.

For large routine surveys, however, where there is no call for immediate day-to-day decisions, it is usually best to ship the samples to a central laboratory. In general, the most efficient, reliable, and economical analytical work is done in an established laboratory. The centralized laboratory may, if desirable, be supplemented by a forward field laboratory, where samples can be scanned before being sent back for more reliable or comprehensive analysis.

For most of the analytical techniques used in geochemical prospecting the bulk of the cost is salaries and wages. Experience has shown that when non-professional operators are properly trained and organized, they can be trusted with the entire laboratory routine. In many laboratories, for example, sample preparation and analysis is performed by locally recruited labor, working under trained non-professional supervisors. The services of a professional chemist are, of course, necessary for analytical development work and for any technical difficulties that may arise in the course of operations. The laboratory should be organized to handle about 30% productivity over and above the anticipated sampling rate, in order to cope with check samples, repeat determinations, and contingencies.

C. Supervision

The interpretation of the data of a geochemical prospecting survey in terms of possible mineral deposits is fundamentally a geological rather than a chemical problem. For this reason it is essential that the direction of geochemical surveys should always be the responsibility of one who is first an experienced exploration geologist, and second a man with sound training in geochemical techniques. His responsibilities will include (i) orientation to establish both field and laboratory technique, (ii) maintenance of technical efficiency and coordination throughout the operation, (iii) interpretation of data, and (iv) prompt follow-up of anomalies and preparation of recommendations for further work.

Larger organizations may have their own research and development sections; more usually, however, commercial exploration relies on the published literature and the specialized services of consultants for the development and application of new concepts and techniques. It goes without saying that in every case there must be the closest continuing liaison between the geochemical staff and all other technical groups that may be concerned in the exploration.

Appendix

Geochemical Characteristics of the Elements

In the following sections, the geochemical characteristics of the individual elements that are significant in geochemical exploration are summarized.

Except for water and air, all concentrations are expressed in ppm (parts per million). Wherever possible, the median, or middle, value of a series of figures is cited rather than the average; the advantage of the median value of a series is that it largely eliminates the influence of spurious high and low extremes.

Unless specifically noted otherwise, the sources of the numerical data are as follows:

Atomic weights: Clark (1966), on the carbon-12 scale.

Igneous rocks: (1) = median of figures cited in Wedepohl's "Handbook of Geochemistry" (1969–1978) for ultramafic rocks (Umaf), basalt and gabbro (Maf), and granite (Gran); (2) = average abundances cited by Turekian (1977).

Sedimentary rocks: (1) = median of figures cited in Wedepohl's Handbook for carbonate rocks (Ls), sandstones and quartzites (Ss), and normal shale (Sh); (2) = average abundances cited by Turekian (1977).

Soils: (3) = median of abundances in uncultivated soils cited by Connor and Shacklette (1975); (4) = average abundance cited by Brooks (1972).

Plants (*in ppm dry weight or ash, as indicated*)*:* (3) = median of abundances in native plant species cited by Connor and Shacklette (1975); (4) = average abundance cited by Brooks (1972). These values are computed without consideration for species, organ sampled or time of year.

Fresh water (in ppb): (2) = average cited by Turekian (1977). These values are estimated without consideration for changes with season, amount of rainfall or content of total solids; (6) = median value from Fig. 14.5, for ground waters.

Water-supply ceiling: (5) = upper limit for domestic water supplies stipulated by the U.S. Environmental Protection Agency (1977).

Associations: elements are classified as lithophile (affinity for silicates), chalcophile (affinity for sulfur), and siderophile (affinity for metallic Fe) following the system outlined by Goldschmidt (1954).

ANTIMONY (Sb)

Atomic number: 51; *atomic weight:* 121·75.

Igneous rocks (med): Umaf 0·1; Maf 0·1; Gran 0·2 (1).

Sedimentary rocks (av): Ls 0·3; Ss 1·0; Sh 1–2 (1).

Soils (av): 2.[f]

Plant ash (av): 1 (4).

Fresh water (av): 2 ppb (2).

Associations: chalcophile; Au, Ag, Hg, and As in complex precious-metal deposits, and with Pb and Zn in some base-metal deposits.

Industrial sources: stibnite (Sb_2S_3), tetrahedrite ($Cu_{12}Sb_4S_{13}$), sulfantimonides.

Aqueous species: SbO_2^-, $HSbO_2^0$, and anion complexes with Cl^-, SO_4^{2-}, HCO_3^-.[a]

Mobility: high average Sb content of fresh water relative to rock suggests high mobility in non-ore environment.

Geochemical prospecting applications: the Sb content of surficial materials has been used as a guide to Sb-bearing deposits,[b,c] Au deposits,[d,e] Ag deposits,[c,f] and polymetallic deposits.[g,h]

References: [a]Shvartseva (1972), [b]Sainsbury (1957); [c]Gott and Botbol (1973), [d]Webb (1958a), [e]Gustavson and Neathery (1976), [f]Boyle (1975), [g]Polikarpochkin *et al.* (1958), [h]Scott and Taylor (1977).

ARSENIC (As)

Atomic number: 33; *atomic weight:* 74·92.

Igneous rocks (med): Umaf 1·0; Maf 1·5; Gran 2·1 (1).

Sedimentary rocks (med): Ls 1·1; Ss 1·2; Sh 12 (1).

Soils (med): 7·5 (3).

Dry plants: <0·25 (3).

Fresh water (av): 2 ppb (2); *water-supply ceiling:* 50 ppb (5).

Associations: chalcophile; also in some ores As is associated with U, Sn, Bi, Mo, P, and F; As shows an especially strong coherence with Au in practically all types of Au deposits, except the Witwatersrand.[a]
Rock minerals: accessory sulfides.
Industrial source: by-product of Cu smelting; minerals are principally arsenopyrite (FeAsS) and complex sulfarsenides of ore metals.
Weathering products: Fe-oxides,[c] organic matter.
Aqueous species: $H_2AsO_4^-$, $HAsO_4^{2-}$, $HAsO_2^0$.
Biological response: As may become enriched in living vegetation; the As content of coal is extremely high, so that the burning of coal releases As to the air with resulting pollution.
Mobility: in oxidizing environments, controlled by coprecipitation with Fe-oxides; in Fe-poor or partially reducing environments, As may be relatively mobile, as suggested by moderately high average content in water relative to rock.
Geochemical prospecting applications: As in soils and sediments is an especially powerful guide to arsenical Au deposits;[a,b,d,f] it is also useful as a guide to many other ore types;[a,b,c,g] under some conditions, the As content of vegetation may be more effective as an ore guide than As in soils.[h]
References: [a]Boyle and Jonasson (1973), [b]Scott and Taylor (1977), [c]Presant and Tupper (1966), [d]Miesch and Nolan (1958), [e]Webb (1958a), [f]Burbank *et al.* (1972), [g]Granier (1958), [h]Warren *et al.* (1964).

BARIUM (Ba)

Atomic number: 56; *atomic weight:* 137·34.
Igneous rocks (av): Umaf 0·4; Maf 330; Gran 840 (2).
Sedimentary rocks (med): Ls 92; Ss 170; Sh 550 (1).
Soils (med): 300 (3).
Plant ash (med): 2800 (native species); 140 (cultivated plants) (3).
Fresh water (av): 20 ppb (2); *water-supply ceiling:* 1000 ppb (5).
Associations: lithophile; K in felsic igneous rocks; accompanies many Pb–Zn sulfide ores as barite.
Rock mineral: K-feldspar.
Industrial source: barite ($BaSO_4$).
Weathering products: residual barite; otherwise not known.
Aqueous species: Ba^{2+}.
Mobility: intermediate, probably limited by solubility of barite.
Geochemical prospecting applications: Hg has been cited as a guide to Ba deposits;[a] Ba in surficial materials has been used as a guide to Ba deposits,

polymetallic deposits,[b,c] and carbonatites;[d] Ba/Sr ratios in rocks and soils define zones mineralized with Cu–Mo in British Columbia.[e]

References: [a]Friedrich and Pluger (1971), [b]Merefield (1974), [c]Leake and Smith (1975), [d]Bloomfield *et al.* (1971), [e]Olade *et al.* (1975).

BERYLLIUM (Be)

Atomic number: 4; *atomic weight:* 9·01.
Igneous rocks (av): Umaf 0·x; Maf 1; Gran 3 (2).
Sedimentary rocks (av): Ls 0·x; Ss 0·x; Sh 3 (2).
Soils (range): 0·5–4.[a]
Plant ash (av): 0·7 (4).
Fresh water (av at pH 6·6–7·0): 5·5 ppb.[b]
Associations: lithophile; Al and Si in silicate minerals; Li, B, Nb, Th, and U in pegmatite veins; W veins.
Rock minerals: mica; feldspar.
Industrial source: bertrandite ($BeSi_2O_7(OH)_2$), beryl ($Be_3Al_2Si_6O_{18}$).
Aqueous species: Be^{2+}, BeF^+, $BeOH^+$ where pH <6–7.[b]
Mobility: in ore environments, limited by extreme resistance to weathering of beryl; in non-ore environments may be relatively mobile, as indicated by high average content in fresh water relative to rock.
Geochemical prospecting applications: Be in soils and sediments has been used as a guide to Be-bearing deposits;[a,c] beryl pegmatites may be identified by muscovite containing higher than average Be.[d]
References: [a]Debnam and Webb (1960), [b]Kraynov (1968), [c]Hawley and Griffitts (1968), [d]Heinrich (1962).

BISMUTH (Bi)

Atomic number: 83; *atomic weight:* 208·98.
Igneous rocks (med): Umaf 1·2; Maf 0·05; Gran 0·3 (1).
Sedimentary rocks (med): Ss 0·3; Sh 1·0 (1).
Soils (av): 0·8 (4).
Plant ash (av): 0·7 (4).
Fresh water: 0·005 ppb.[d]
Industrial source: by-product of Pb smelting.
Associations: chalcophile; Sb and As in sulfide deposits.
Mobility: probably very low, follows hydrolyzates.
Geochemical prospecting applications: the Bi content of surficial materials in mineralized areas does not appear to be a more effective ore guide than more abundant elements with which it is associated, especially Pb.[a,b,c]
References: [a]Erickson and Marsh (1974), [b]Alminas *et al.* (1975), [c]Silberman *et al.* (1974), [d]Udodov and Parilov (1961).

BORON (B)

Atomic number: 5; *atomic weight:* 10·81.
Igneous rocks (av): Umaf 3; Maf 5; Gran 10 (2).
Sedimentary rocks (av): Ls 20; Ss 35; Sh 100 (2).
Soils (med): 29 (3).
Plant ash (med): 230 (3).
Fresh water (av): 10 ppb (2).
Associations: lithophile; Be, Li, Nb, Th, and U in pegmatite veins; as a result of the enrichment of B in sea water, the B content of marine shales is very much higher than that of freshwater shales and may be used as a criterion of origin.[a]
Industrial source: borates from saline lake deposits.
Rock minerals: tourmaline, feldspar.
Weathering products: clay minerals, tourmaline (complex borosilicate), soluble borates.
Aqueous species: soluble borates.
Biological response: at a low level of concentration, B is a necessary nutrient for most plants; at a somewhat higher level of availability in the soil it is generally toxic; B toxicity may take the form of recognizable symptoms that can be used as guides to areas enriched in B.[b]
Mobility: if present as borate, B is extremely mobile; otherwise its mobility is intermediate, as indicated by its content in normal fresh water compared to rock.
Geochemical prospecting applications: B in surficial materials can be used as a guide to borate-rich areas[c] and to hydrothermal deposits containing tourmaline, axinite or other B-silicates.[d]
References: [a]Keith and Degens (1959), [b]Buyalov and Shvyryayeva (1955), [c]Smith (1960), [d]Boyle (1971b).

CADMIUM (Cd)

Atomic number: 48; *atomic weight:* 112·40.
Igneous rocks (med): Maf 0·2; Gran 0·1 (1).
Sedimentary rocks (av): Ls 0·035; Ss 0·0x; Sh 0·3 (2).
Soils (range): 0·1–0·5 (1).
Plant ash (med): 4·3 (3).
Fresh water (av): 0·032 ppb;[c] *water-supply ceiling:* 10 ppb (5).
Associations: chalcophile; Cd shows an almost universal association with Zn, with a Zn/Cd ratio that does not deviate by much more than a factor of 2 from an average of 500 to 1 in igneous, metamorphic, and sedimentary rocks.
Rock minerals: same as for Zn.

Industrial source: sphalerite (ZnS); produced as by-product of Zn smelting.

Weathering products: greenockite (CdS) over Zn-sulfide ores; adsorbed to organic matter and Fe–Mn-oxides.[d]

Aqueous species: Cd^{2+}, $Cd(OH)_2^0$, $CdHCO_3^+$.[b]

Mobility: in ore environment, relatively immobile, controlled by CdS; in non-ore environment, relatively mobile, as indicated by relatively high average content in fresh water compared to rocks.

Geochemical prospecting applications: Zn is normally a much better guide to Zn ores than Cd; thus in spite of the close coherence of the two metals, Cd has found little practical application as a pathfinder for Zn. The Zn/Cd ratio was found to be zoned at a district scale in the Coeur d'Alene Pb–Zn–Ag district.[a]

References: [a]Gott and Botbol (1973), [b]Long and Angino (1977), [c]Udodov and Parilov (1961), [d]Gong *et al.* (1977).

CHROMIUM (Cr)

Atomic number: 24; *atomic weight:* 52·00.

Igneous rocks (av): Umaf 2980; Maf 170; Gran 4·1 (2).

Sedimentary rocks (av): Ls 11; Ss 35; Sh 90 (2).

Soils (med): 43 (3).

Plant ash (med): 6·3 (3).

Fresh water (av): 1 ppb (2); *water-supply ceiling:* 50 ppb (5).

Associations: lithophile; strong association with Ni and Mg in ultramafic rocks.

Rock mineral: chromite ($FeCr_2O_4$).

Industrial source: chromite.

Weathering products: chromite if present in parent rock, Fe-oxides in normal soils.

Aqueous species: CrO_4^{2-}, $Cr_2O_7^{2-}$.

Mobility: low.

Geochemical prospecting applications: under most conditions, geochemical anomalies associated with chromite deposits are defined by residual or detrital grains of chromite in soils and sediments, which are generally so conspicuous that chemical analysis is not necessary. In rare cases where the residual chromite cannot be readily identified, however, the Cr content of residual soil may be helpful.[a, b]

References: [a]Webb (1958a), [b]Suvanasingha (1963).

COBALT (Co)

Atomic number: 27; *atomic weight:* 58·93.

Igneous rocks (av): Umaf 110; Maf 48; Gran 1 (2).

Sedimentary rocks (av): Ls 0·1; Ss 0·33; Sh 19 (2).
Soils (med): 10 (3)
Plant ash (med): 5·0 (3).
Fresh water (av): 0·1 ppb (2).
Associations: mainly chalcophile; Mg and Ni in mafic and particularly ultramafic rocks.
Rock minerals: mafic minerals.
Industrial source: mainly complex sulfarsenides and sulfantimonides, and as a by-product of Ni- and Cu-sulfide ores.
Weathering products: $CoCO_3$, $Co(OH)_3$, Co adsorbed and coprecipitated with Mn-oxides, or to a lesser extent Fe-oxides.
Aqueous species: Co^{2+}, organic complexes.
Biological response: Co is a necessary element in animal nutrition.[a]
Mobility: intermediate, controlled mainly by adsorption and coprecipitation with Mn- and Fe-oxides.
Geochemical prospecting applications: the Co content of soils and sediments has been used successfully as a guide to Co-bearing ores.[b,c,d] However Cu, if it accompanies Co in the primary ore, is usually more effective than Co as an ore guide.[e] Experimental plant sampling for Co shows promise.[f]
References: [a]Thornton and Alloway (1974), [b]Koehler *et al.* (1954), [c]Hawkes (1959), [d]Canney and Wing (1966), [e]Webb (1958a), [f]Warren and Delavault (1957).

COPPER (Cu)

Atomic number: 29; *atomic weight:* 63·54.
Igneous rocks (med): Umaf 42; Maf 72; Gran 12 (1).
Sedimentary rocks (med): Ls 5; Ss 10; Sh 42 (1).
Soils (med): 15 (3).
Plant ash (med): 130 (3).
Fresh water (av): 3 ppb (6); *water-supply ceiling:* 1000 ppb (5).
Associations: chalcophile; Pb, Zn, Mo, Ag, Au, Sb, Se, Ni, Pt and As in sulfide deposits.
Rock minerals: mafic minerals, chalcopyrite ($CuFeS_2$).
Industrial sources: chalcopyrite, bornite (Cu_5FeS_4), chalcocite (Cu_2S), complex Cu–As–Sb–S minerals.
Weathering products: sulfides, oxides, basic carbonates, sulfates, and silicates over chalcopyrite ores; Mn-oxides, limonite, organic matter, oxides, and carbonates in soils.
Aqueous species: Cu^{2+}, $Cu(OH)_2^0$, $CuHCO_3^+$, $CuCl_3^{2-}$, $CuCl_2^-$.[m,n]
Biological response: Cu is one of the more important trace elements in plant nutrition; where the Cu content of soil falls below 10 ppm, deficiency symp-

toms may develop in the vegetation, whereas somewhat higher concentrations may be toxic. A number of very interesting indicator plants for Cu have been reported.[a]

Mobility: intermediate; controlled by adsorption to Fe- and Mn-oxides and organic matter;[h, j, k, l] and precipitation by hydrolysis at pH >5·0.

Geochemical prospecting applications: determination of Cu in soils and sediments has become one of the most widely used and successful geochemical methods of locating and delineating Cu deposits. Details of techniques with literature citations are reviewed in earlier sections of this book, and may be found by checking under "Copper" in the index. The Cu content of vegetation has been shown to be a fairly reliable guide to bedrock ore,[b] but has not been widely used. Cu dissolved in fresh water is rarely effective because of the limited solubility of Cu at normal pH values. Various pathfinders for porphyry-copper ores in addition to Cu itself have shown some promise, particularly Mo, K,[c] Rb and Sr,[d, e] Ba/Sr ratio,[f] and Au.[g]

References: [a]Brooks (1972), [b]Warren *et al.* (1952), [c]Davis and Guilbert (1973), [d]Olade and Fletcher (1975), [e]Oyarzun (1975), [f]Olade *et al.* (1975), [g]Learned and Boissen (1973), [h]Ong and Swanson (1966), [j]McKenzie (1972), [k]McLaren and Crawford (1973), [m]Long and Angino (1977), [n]Rose (1976), [o]Rose and Suhr (1971).

FLUORINE (F)

Atomic number: 9; *atomic weight:* 19·00.

Igneous rocks (med): Umaf 20; Maf 420; Gran 810 (1).

Sedimentary rocks (med): Ls 250; Ss 280; Sh 680 (1).

Soils (med): 300 (3).

Dry plants (med): 0·77 (3).

Fresh water (av): 100 ppb (2); *water-supply ceiling:* 1400–2400 ppb (5).

Associations: lithophile; follows phosphate in apatite; appears to be the key element in the "rare metal" association of Be, B, Cs, Li, Mo, Nb, Rb, Sn, Ta, W, and U.

Rock minerals: apatite ($Ca_5(PO_4)_3F$) or fluorite (CaF_2) if present; otherwise up to 90% in micas, particularly biotite.[a]

Industrial source: fluorite.

Weathering products: not known, but possibly clay minerals and phosphates.

Aqueous species: F^-, HF^0, and fluoride–cation complexes.

Biological response: none known in plants; in animal nutrition, F plays a part in the structure of teeth and bones.

Geochemical prospecting applications: Hg and Zn in soils[b] and F in surface waters[c, d, e] and soils[d, p] show promise as guides to fluorite deposits; the Y content of fluorite in the northern Pennines of England is highest near vein intersection where the greatest tonnage of commercial fluorite is concentrated;[f] although the F content of igneous biotite has been studied as an indication of productive plutons, the results are not definitive.[g,h] F has been cited as a guide to rare-metal carbonatites,[j] W-skarns,[k] U,[m] and barite deposits.-

References: [a]Bailey (1977), [b]Friedrich and Pluger (1971), [c]Friedrich (1970), [d]Schwartz and Friedrich (1973), [e]Graham *et al.* (1975), [f]Smith (1974), [g]Parry and Jacobs (1975), [h]Kesler *et al.* (1975a), [j]Komarov and Glagolev (1969), [k]Cachau-Herreillat and Prouhet (1971), [m]Jonasson and Gleeson (1976), [n]Lalonde (1976), [p]Mukherjee (1978).

GOLD (Au)

Atomic number: 79; *atomic weight:* 196·97.

Igneous rocks (*med*)*:* Umaf 0·0032; Maf 0·0032; Gran 0·0023 (1).

Sedimentary rocks (*med*)*:* Ls 0·005; Ss 0·005; Sh 0·004 (1).

Soils (*av*)*:* 0·002 (4).

Plant ash: $<0{\cdot}0007$.[a]

Fresh water (*med*)*:* 0·002 ppb (2).

Associations: siderophile and to a lesser extent chalcophile; Ag, As, Sb, Hg, Se, and Te in precious metal deposits; also Fe, Zn, and Cu in many sulfide deposits.

Industrial source: native Au; Au tellurides.

Weathering products: there is some evidence that native Au from Au deposits can be locally dissolved and reprecipitated with resulting growth of alluvial Au fragments in place.[b]

Soil phases: not known.

Aqueous species: anion complexes, colloids of metallic gold, particulate suspensoids, and as Au adsorbed or chelated to soluble organic matter.[c,d,e]

Biological response: Au can enter plants and be concentrated as a cyanide complex.[f]

Mobility: in ore environment, limited by refractoriness of native Au; in non-ore environment, an extremely high ratio of average Au in normal fresh water relative to rock suggests high mobility.

Geochemical prospecting applications: As in soils and sediments is by far the most effective geochemical guide to arsenical Au ore;[g] determination of Au in fresh water is not recommended as an effective prospecting method;[c] the Au content of tropical soils has been found useful as a guide to deeply-weathered Au-bearing porphyry-copper deposits;[h] placer Au commonly contains alloyed metals in ratios that are characteristic of the type of bedrock deposit from which it was derived.[i]

References: [a]Cannon (1960b), [b]Voronin and Gol'dberg (1972), [c]Gosling *et al.* (1971), [d]Ong and Swanson (1969), [e]Boyle *et al.* (1975), [f]Lakin *et al.* (1974), [g]Boyle and Jonasson (1973), [h]Learned and Boissen (1973), [i]Antweiler and Campbell (1977).

HELIUM (He)

(*see also radioactive Decay Products of Uranium and Thorium*)

Atomic number: 2; *atomic weight:* 4·00.

Atmosphere: 5 ppm v/v.

Fresh water (aerated): 0·01 ppb (computed from data of Table 18.2).

Associations: as a noble gas, He is a component of virtually all naturally occurring gases; it is one of the stable end products of the radioactive decay of U and Th.

Mobility: extremely high.

Geochemical prospecting applications: He extracted from the pore spaces of soils and rocks has been used successfully as a guide to U ores[a] and buried faults.[b] The practical applications of He to solving geologic problems are discussed in greater length in Chapter 18.

References: [a]Dyck (1976), [b]Bulashevich *et al.* (1974).

IODINE (I)

Atomic number: 53; *atomic weight:* 126·90.

Igneous rocks (med): Umaf 0·12; Maf 0·11; Gran 0·17 (1).

Sedimentary rocks (med): Ls 4; Ss 0·5; Sh 1·7 (1).

Soils: transient and extremely variable.[a]

Dry plants (med): 4·6 (3).

Fresh water (av): 7 ppb (2).

Air (range): 5–7000 ng/m^3[a,b]

Associations: volcanic emanations, sea salt.

Industrial source: I-rich brines and saline deposits.

Biological response: I is necessary for animal nutrition; the I cycle is very similar to the N cycle and is primarily biogeochemical.[a]

Mobility: extremely high.

Geochemical prospecting applications: the I content of normal soil does not reflect the underlying geology;[a] iodides of Ag and Cu have been found in the zone of weathering directly overlying chalcopyrite ore.[c]

References: [a]Shacklette and Cuthbert (1967), [b]Moyers and Duce (1972), [c]Chitayeva *et al.* (1971).

IRON (Fe)

Atomic number: 26; *atomic weight:* 55·84.
Igneous rocks (av): Umaf 94 300; Maf 86 500; Gran 14 200 (2).
Sedimentary rocks (av): Ls 3800; Ss 9800; Sh 47 000 (2).
Soils (med): 21 000 (3).
Plant ash (med): 1600 (3).
Fresh water (av): 100 ppb (2); *water-supply ceiling:* 300 ppb (5).
Associations: siderophile; enriched in mafic igneous rocks.
Rock minerals: mafic minerals; pyrite (FeS_2); magnetite (Fe_3O_4).
Industrial sources: hematite (Fe_2O_3); magnetite.
Weathering products: Fe-oxides.
Aqueous species: Fe^{2+}, Fe^{3+}.
Biological response: Fe is necessary for the enzymatic synthesis of chlorophyll in plants; in animals it is essential as a constituent of hemoglobin in the blood.
Mobility: for Fe^{2+} moderate; for Fe^{3+} very low, as Fe^{3+} is precipitated as hydrous Fe-oxides where pH >3; under some conditions colloidal suspensions of undissociated hydrous Fe-oxides or Fe-bearing organic complexes may be stable.[a]
Geochemical prospecting applications: some of the ore metals are coprecipitated with Fe-oxides in soils overlying sulfide ores and in the beds of streams draining them; thus the metal content of the resulting limonitic material may be a useful ore guide.[b, c, d] The Fe content of organic matter in peat bogs has been cited as a promising guide to buried sulfide ores.[e, f]
References: [a]Hem (1960), [b]Dyck (1971), [c]Nowlan (1976), [d]Chao and Theobald (1976), [e]Salmi (1955), [f]Hawkes and Salmon (1960.)

LEAD (Pb)

Atomic number: 82; *atomic weight:* 207·19.
Igneous rocks (med): Umaf 1 (1); Maf 4; Gran 18.[n]
Sedimentary rocks (med): Ls 5; Ss 10; Sh 25 (1).
Soils (med): 17 (3).
Plant ash (med): 30.[n]
Fresh water (av): 3 ppb (2); *water-supply ceiling:* 50 ppb (5).
Associations: chalcophile; Ag in precious metal deposits; Fe, Zn, Cu, and Sb in many other sulfide deposits; F in rock-forming silicates.
Rock minerals: micas; K-feldspars.
Industrial source: galena (PbS).
Weathering products: cerussite ($PbCO_3$), anglesite ($PbSO_4$), pyromorphite ($Pb_5(PO_4)_3Cl$), plumbojarosite ($PbFe_2(SO_4)_4(OH)_{12}$) over Pb-sulfide ores; Mn- and Fe-oxides, pyromorphite in soils.[a]

Aqueous species: Pb^{2+}, $PbCO_3^0$, $Pb(OH)^+$, $Pb(OH)_2^0$,[b,m] soluble organic matter;[c] complexes with Cl^- and HCO_3^-.[d]

Biological response: Pb is generally toxic to vegetation and animal life when present in ionic form.

Mobility: relatively low; restricted by tendency for adsorption to Mn–Fe-oxides and insoluble organic matter,[a] but assisted by formation of soluble organic complexes[c] and anion complexes.[d]

Geochemical prospecting applications: the Pb content of residual soil and stream sediment is a well established and successful guide to Pb-rich deposits;[e,f] Pb is particularly useful as a guide to deposits containing argentiferous galena;[g] Pb-isotope ratios may help in identifying ore types,[j] in indicating the most favorable parts of mineralized districts,[h] and in distinguishing productive host rocks.[k]

References: [a]Antropova (1969), [b]Arakelyan and Kyuregyan (1969), [c]Bondarenko (1968), [d]Goleva *et al.* (1970), [e]Huff (1952), [f]Webb (1958a, b), [g]Boyle and Cragg (1957), [h]Cannon *et al.* (1971), [j]Doe and Stacey (1974), [k]Gulson (1977), [m]Long and Angino (1977), [n]Lovering (1976).

LITHIUM (Li)

Atomic number: 3; *atomic weight:* 6·939.

Igneous rocks (av): Umaf 0·x; Maf 17; Gran 40 (2).

Sedimentary rocks (av): Ls 5; Ss 15; Sh 66 (2).

Soils (med): 22 (3).

Plant ash (med): 6·2 (3).

Fresh water (av): 3 ppb, extremely variable (2).

Associations: lithophile; Mg in silicate rocks; Be, B, La, Nb, Th, and U in pegmatite veins.

Rock minerals: mafic minerals.

Industrial sources: spodumene ($LiAl(SiO_3)_2$, Li-rich brines lepidalite $4(K(Li, Al)_3(Si, Al)_4O_{10}(F, OH)_2)$.

Weathering products: montmorillonite.[a]

Aqueous species: Li^+.

Biological response: there are no known toxic effects of Li.

Mobility: intermediate.

Geochemical prospecting applications: Li content of accumulator plants has been used in locating Li-rich brines;[a] Li in surface fresh water and water-soluble Li in residual soil may serve as a guide to pegmatite deposits.[b]

References: [a]Cannon *et al.* (1975), [b]Miller and Danilov (1957).

MANGANESE (Mn)

Atomic number: 25; *atomic weight:* 54·94.

Igneous rocks (av): Umaf 1040; Maf 1500; Gran 390 (2).

Sedimentary rocks (av): Ls 1100; Ss x0; Sh 850 (2).
Soils (med): 320 (3).
Plant ash (med): 6700 (3).
Fresh water: 15 ppb (6); *water-supply ceiling:* 50 ppb (5).
Associations: lithophile; Mg and Fe in silicates.
Rock minerals: most mafic minerals.
Industrial sources: Mn-oxides.
Weathering products: Mn-oxides.
Aqueous species: Mn^{2+} in an acid reducing environment.
Biological response: Mn is an essential nutrient for most plants.
Mobility: intermediate to low, except in the acid, reducing environment of organic swamps and bogs where Mn can move very readily.
Geochemical prospecting applications: Mn is extremely important in geochemical exploration because of the very large number of trace elements that may be coprecipitated with or adsorbed on Mn-oxide minerals.[a, b, d] Selective extraction and determination of the ore metals contained in Mn-oxides in stream sediments has become a very powerful method of geochemical reconnaissance. Under favorable conditions, the Mn content of soils and plants may reflect the presence of Mn ore beneath the cover.[c]
References: [a]Chao and Theobald (1976), [b]Nowlan (1976), [c]Bloss and Steiner (1960), [d]Jenne (1968).

MERCURY (Hg)

Atomic number: 80; *atomic weight:* 200·59.
Igneous rocks (av): Umaf 0·004; Maf 0·01; Gran 0·04 (2).
Sedimentary rocks (av): Ls 0·04; Ss 0·03 (2); Sh 0·02–0·4 (2).[j]
Soils (med): 0·056 (3).
Plant ash (av): 0·01 (4).
Fresh water (av): 0·07 ppb (2); *water-supply ceiling:* 2 ppb (5).
Air (av): 3 ng/m^3.[a]
Associations: chalcophile; Sb, Se, Ag, Zn, and Pb in sulfide deposits.
Industrial sources: cinnabar (HgS), by-product of Zn smelting.
Weathering products: Hg_2Cl_2, HgO, Hg over Hg ores;[b] probably organic matter, $HgCl_2$ in soils.
Aqueous species: Hg_2^{2+}, $Hg(OH)_2^0$, $HgCl_2^0$, Hg^0;[k] soluble organic matter.[c]
Biological response: organic mercury compounds are extremely toxic to animals.
Mobility: relatively high, limited by adsorption to solid organic matter.
Geochemical prospecting applications: the close association of Hg with precious metals and the elements of volcanogenic deposits makes it a useful pathfinder for Au,[d] Ag,[e] Sb,[f] massive sulfides,[g, m] as well as Hg content of wallrock and residual soil. Anomalous Hg in rocks and soils

has been observed near fault zones.[i] Considerable experimental work has been done on the Hg content of soil gases and atmospheric air, without results that are adequately reproducible.[d,h]

References: [a]Williston (1968), [b]Watling *et al.* (1973), [c]Andren and Harriss (1975), [d]McCarthy *et al.* (1970), [e]Lovering *et al.* (1966), [f]Koksoy and Bradshaw (1969), [g]Friedrich and Hawkes (1966), [h]McCarthy (1972), [i]Rösler *et al.* (1977), [j]McNeal and Rose (1974), [k]Hem (1970), [m]Wu and Mahaffey (1978).

MOLYBDENUM (Mo)

Atomic number: 42; *atomic weight:* 95·94.

Igneous rocks (av): Umaf 0·3; Maf 1·5; Gran 1·3 (2).

Sedimentary rocks (av): Ls 0·4; Ss 0·2; Sh 2·6 (2).

Soils (av): 2·5 (4).

Plant ash (av): >5 in arid alkaline environments;[a] <5 over neutral to acid soils of humid environments (3).

Fresh water: 1·5 ppb (6).

Associations: siderophile, to a lesser extent chalcophile; with Cu, Re, and base metals in porphyry-copper deposits; with W, Sn, etc. in porphyry-molybdenum and skarn deposits.

Rock mineral: molybdenite (MoS_2).

Industrial source: molybdenite.

Weathering products: ferrimolybdite ($Fe_2O_3.3MoO_3.8H_2O$), powellite ($Ca(Mo,W)O_4$) over Mo-sulfide ores; Fe-oxides in soils.

Aqueous species: MoO_4^{2-}, possibly soluble organic complexes.

Biological response: Mo is essential for the functioning of nitrogen-fixing bacteria in the root nodules of legumes; Mo in forage is extremely toxic to livestock.[b]

Mobility: relatively high, limited (i) by rate of solution of primary MoS_2, (ii) by sorption on limonite or reaction with Fe^{3+} to form ferrimolybdite at pH 2·5–7·0, and to a lesser extent sorption on clay minerals at pH 2–4,[c] and (iii) by precipitation in carbonate-rich environments. Otherwise, the mobility of Mo is independent of pH variations in the natural environment.

Geochemical prospecting applications: the almost universal association of Mo with porphyry-copper deposits makes it a useful guide to that type of deposit in both residual soil[d] and in stream water and sediment. The Mo content of ground water shows particular promise as a method of locating porphyry-copper deposits buried beneath post-ore cover in the southwestern U.S.A.[e] The apparent non-selectivity of vegetation with respect to Mo makes the analysis of deep-rooted plants a possible ore guide.[d,f]

References: [a]Cannon (1960b), [b]Thornton and Alloway (1974), [c]Jones (1957), [d]Chaffee (1976b), [e]Trost and Trautwein (1975), [f]Huff and Marranzino (1961).

NICKEL (Ni)

Atomic number: 28; *atomic weight:* 58·71.
Igneous rocks (av): Umaf 2000; Maf 130; Gran 4·5 (2).
Sedimentary rocks (av): Ls 20; Ss 2; Sh 68 (2).
Soils (med): 17 (3).
Plant ash (med): 18 (3).
Fresh water (av): 1·5 ppb (6).
Associations: siderophile, to a lesser degree chalcophile; with Mg and Co in ultramafic and mafic rocks; with Co, Cu, and Pt in sulfide deposits.
Rock minerals: mafic minerals.
Industrial sources: Ni–Fe-sulfides, Ni-rich laterites.
Weathering products: Fe-oxides and nickeliferous silicates.
Aqueous species: Ni^{2+}
Mobility: relatively immobile, limited by coprecipitation with limonite, and hydrolysis where pH $>6{\cdot}5$.
Geochemical prospecting applications: Ni in residual soil is a valuable guide to underlying Ni-sulfides;[a, b] Ni in stream sediments may be useful in geochemical reconnaissance, but in areas of ultramafic rocks the high Ni background may swamp the pattern coming from sulfides. Ni in soils and sediments will outline ultramafic laterites,[c] but it is doubtful whether it would help in pinpointing commercial-grade lateritic ore without sampling at depth. Although the Ni content of plants tends to parallel that of the supporting soil,[d, e] the direct determination of Ni in soil would normally be more practical. Ni in surficial materials has been cited as a guide to kimberlites.[f] Determinations of sulfide-Ni in ultramafic rocks have been used to indicate productive plutons in the Canadian Shield.[g]
References: [a]Cox (1975), [b]Wilmshurst (1975), [c]Ong and Sevillano (1975), [d]Warren and Delavault (1954), [e]Timperley *et al.* (1972), [f]Gregory and Tooms (1969), [g]Cameron *et al.* (1971).

NIOBIUM (Nb)

Atomic number: 41; *atomic weight:* 92·91.
Igneous rocks (av): Umaf 1; Maf 20; Gran 20.[g]
Sedimentary rocks (av): Sh 20.[g]
Soils (av): 15 (4).
Plant ash (av): 0·3 (4).
Fresh water (av pH 6·6–7·0): 1·0 ppb.[a]

Associations: lithophile; Nb shows an almost universal association with Ta; Nb occurs with Ti, rare earths, U, Th, and P in alkaline igneous rocks.[a]

Rock minerals: pyrochlore (complex niobate), Ti minerals.

Industrial sources: columbite ($(Fe,Mn)(Nb,Ta)_2O_6$), pyrochlore.

Aqueous species: soluble organic complexes most abundant, then F and CO_3 complexes, and finally niobate anions.[b]

Mobility: relatively low, controlled by pH and abundance of F and CO_3.

Geochemical prospecting applications: Nb in soils and sediments has been used as a guide to Nb-bearing carbonatites[c, d] and kimberlites.[e] In prospecting for columbite, it is usually more practical to test for the mineral itself in heavy concentrates. Some columbite-bearing pegmatites are distinguished by Nb-rich muscovite.[f]

References: [a]Parker and Fleischer (1968), [b]Kraynov (1968), [c]van Wambeke (1960), [d]Watts *et al.* (1963), [e]Gregory and Tooms (1969), [f]Heinrich (1962), [g]Vinogradov (1962).

PHOSPHORUS (P)

Atomic number: 15; *atomic weight:* 30·97.

Igneous rocks (av): Umaf 220; Maf 1100; Gran 600 (2).

Sedimentary rocks (av): Ls 400; Ss 170; Sh 700 (2).

Soils (med): 300 (3).

Plant ash (med): 16 000 (3).

Fresh water (av): 20 ppb (2).

Associations: apatite ($Ca_5(PO_4)_3F$) is associated with rare earths and Nb in some alkaline igneous rocks; sedimentary phosphorites are commonly enriched in U.[a, b]

Industrial sources: Ca-phosphate (sed. rocks), apatite (igneous rocks).

Weathering products: Ca-phosphates.

Aqueous species: PO_4^{3-}, HPO_4^{2-}, $H_2PO_4^-$, phosphate–cation complexes.

Biological response: P, with N and K, is one of the principal plant nutrients.

Mobility: intermediate to low, limited by solubility of the Ca-phosphates.

Geochemical prospecting applications: the association of U with Ca-phosphate makes aerial radiometric prospecting for phosphorites technically feasible.[c]

References: [a]Bliskovskii and Smirnov (1966), [b]Summerhayes *et al.* (1970), [c]Bollo and Jacquemin (1963).

PLATINUM METALS (Ru, Rh, Pd, Os, Ir, Pt)

Atomic numbers: Ru, 44; Rh, 45; Pd, 46; Os, 76; Ir, 77; Pt, 78; *atomic weights:* Ru, 101·07; Rh, 102·90; Pd, 106·4; Os, 190·2; Ir, 192·2; Pt, 195·09.

Igneous rocks (av): Umaf: Pt, 0·032; Pd, 0·013; Maf: Pt, 0·030; Pd, 0·021; Gran: Pt, 0·0082; Pd, 0·002 (1).

Associations: siderophile with Ni–Cu in Sudbury-type sulfide deposits and Ni–Cu–Cr in Bushveld-type deposits; with Alpine-type ultramafic rocks.

Industrial sources: by-product of Ni smelting, native Pt in placer deposits.

Soil phases: Pd follows clay minerals and organic matter; Pt may be same but evidence not clear.[a]

Aqueous species: Pd as Cl^- anion complex;[a] Ru as stable "anion" with fulvic acid.[b]

Biological response: Pd has been identified in trees over ore at Stillwater Complex, Montana.[a]

Mobility: Pd and Ru probably moderate; Pt low.

Geochemical prospecting applications: Pt metals in residual soil have been suggested as promising indicators of Ni–Cu sulfide-ore.[c]

References: [a]Fuchs and Rose (1974), [b]Varshal *et al.* (1972), [c]Wilmshurst (1975).

POTASSIUM (K)

Atomic number: 19; *atomic weight:* 39·10.

Igneous rocks (av): Umaf 34; Maf 8300; Gran 42 000 (2).

Sedimentary rocks (av): Ls 2700; Ss 10 700; Sh 26 600 (2).

Soils (med): 11 000 (3).

Plant ash (med): 120 000 (3).

Fresh water (av): 2300 ppb (2).

Associations: lithophile, major constituent of granitic igneous rocks; K alteration is commonly associated with porphyry-copper mineralization.

Rock minerals: K-feldspars, micas.

Industrial sources: saline deposits; K-rich brines.

Weathering products: sericite, hydromica, soluble K.

Aqueous species: K^+.

Biological response: K, with N and P, is one of the principal plant nutrients.

Mobility: moderately high; restricted by adsorption to clay minerals and uptake by living organisms.

Geochemical prospecting applications: the distribution of K minerals in the alteration zones of porphyry-copper deposits is an ore guide of long standing.[a] Because of the radioactivity of K, radiometric surveying has been suggested as a guide to porphyry-copper deposits.[b] In the absence of ore-stage albitization, the K/Na ratio may be helpful in judging the proximity to ore for a wide variety of ore types.[c]

References: [a]Lowell and Guilbert (1970), [b]Davis and Guilbert (1973), [c]Boyle (1974).

RADIOACTIVE DECAY PRODUCTS OF URANIUM AND THORIUM

Atomic numbers and weights: see Fig. A.1 (pp. 568–570).

Abundance: if the decay product (daughter) has accumulated from its parent for at least several of its half-lives, then it will be approximately in radioactive equilibrium with its parent, and the abundance can be estimated from

$$\lambda_d N_d = \lambda_p N_p$$

where λ is the decay constant (0·693/half-life), and N is the number of atoms of the daughter (d) and parent (p). Abundances in radioactivity units (curies, or decays per unit time) will then be equal for parent and daughter. After several half-lives of all intervening decay products, the abundances of daughters may be estimated from the abundances of U and Th in undisturbed materials. Disequilibrium is especially common for ^{222}Rn and its decay products.

Associations: lithophile; U and Th ores and minerals; many decay products are strongly sorbed by Fe-oxides and probably by other natural adsorbents.

Mobility: see Fig. A.2. He and Rn are among the most mobile elements.

Geochemical prospecting applications: ^{214}Bi, and ^{208}Tl, daughters of ^{238}U and ^{232}Th, respectively, furnish most of the gamma rays used to detect U and Th by radiometric methods. The ratio ^{234}U/^{238}U is anomalously high near some sandstone-type U deposits.[a] ^{230}Th and other long-lived immobile daughters of ^{238}U are retained in soils and outcrops from which U has been leached. The mobile daughters (He, Rn, ^{234}U, possibly Pb) can form extensive haloes and dispersion trains in water and air near U deposits (see He, Rn). Unusual abundances of Pb isotopes may indicate the former presence of U in leached outcrops.[b,c] Other decay products may have similar applications.

References: [a]Cowart and Osmond (1977), [b]Hills and Delevault (1977), [c]Rossman *et al.* (1971).

RADIUM (Ra)

(*see also Radioactive Decay Products of Uranium and Thorium*)

Atomic number: 88; no stable isotopes; most abundant isotope is ^{226}Ra.

Average igneous rocks: 9×10^{-7} ppm (3).

Soils (av): 8×10^{-7} ppm (4).

Plant ash (av): 2×10^{-8} ppm (4).

Associations: lithophile; a daughter product of U; Ra in equilibrium with ^{238}U may be determined by multiplying the U content (q.v.) by $3{\cdot}5 \times 10^{-7}$.

Aqueous species: Ra^{2+}.

Mobility: moderate to low; adsorbed by Fe-oxides[e] and organic matter,[b] and coprecipitated with $CaCO_3$, $CaSO_4$, and $BaSO_4$;[a] mobility high in some reducing brines.

Geochemical prospecting applications: because of its mobility, Ra in spring water and stream sediments has been suggested as a guide to U ores,[c,d] although analysis of its daughter Rn is usually simpler.

References: [a]Armstrong and Heemstra (1973), [b]Morse (1969), [c]Cadigan and Felmlee (1977), [d]Foy and Gingrich (1977), [e]Rose and Korner (1979).

RADON (Rn)

(*see also Radioactive Decay Products of Uranium and Thorium*)

Atomic number: 86; no stable isotopes; most abundant isotope is ^{222}Rn.

Lake water (range): 1–7 pCi/l (6 to 45×10^{-12} ppb).[a]

Stream water (av): 50 pCi/l ($3{\cdot}2 \times 10^{-10}$ ppb).[b]

Atmosphere (range): 0·01–0·45 pCi/l (6 to 290×10^{-21} v/v).[c]

Associations: a radioactive daughter product of Ra and U.

Mobility: very high, limited by its radioactive half-life of 3·8 days.

Geochemical prospecting applications: Rn is becoming well established as a guide to U ore and faults,[d] and in forecasting earthquakes[e] and volcanic eruptions.[f] See Chapter 18 for fuller discussion with references.

References: [a]Dyck and Cameron (1975), [b]Dyck and Smith (1969), [c]Dyck (1973), [d]Armstrong and Heemstra (1973), [e]Gorbushina *et al.* (1972), [f]Chirkov (1976).

RARE EARTHS (Y, and La through Lu)

Atomic numbers: Y, 39; La, 57; Lu, 71; *atomic weights:* Y, 88·91; La, 138·91; Lu, 174·97.

Igneous rocks (av): Umaf: ΣRE, 32; Y, 5; La, 4; Ce, 9; Maf: ΣRE, 182; Y, 25; La, 17; Ce, 66; Gran: ΣRE, 226; Y, 41; La, 55; Ce, 57 (2).

Sedimentary rocks (av): Ls: ΣRE, 24; Y, 4; La, 4; Ce, 8; Ss: ΣRE, 52; Y, 10; La, 7; Ce, 15; Sh: ΣRE, 228; Y, 35; La, 39; Ce, 76 (2). (RE includes Y, La, Ce, Pr, Nd, Pm, Sm, Eu, Gd, Tb, Dy, Ho, Er, Tm, Yb, and Lu.)

Soils (med): Y, 27; La, 33 (3).

Plant ash (med): Y, <5, La, 38; Ce, 0·06 ppb (3).

Fresh water (av): Y, 0·07 ppm; La, 0·2 ppb (2).

Associations: lithophile, occurring principally in accessory minerals of igneous rocks.

Rock minerals: monazite ((RE,Th)PO_4) in silicic igneous rocks, bastnaesite (RE(F,CO_3)) in some carbonatites.

Industrial sources: bastnaesite in carbonatite, monazite in placer deposits.

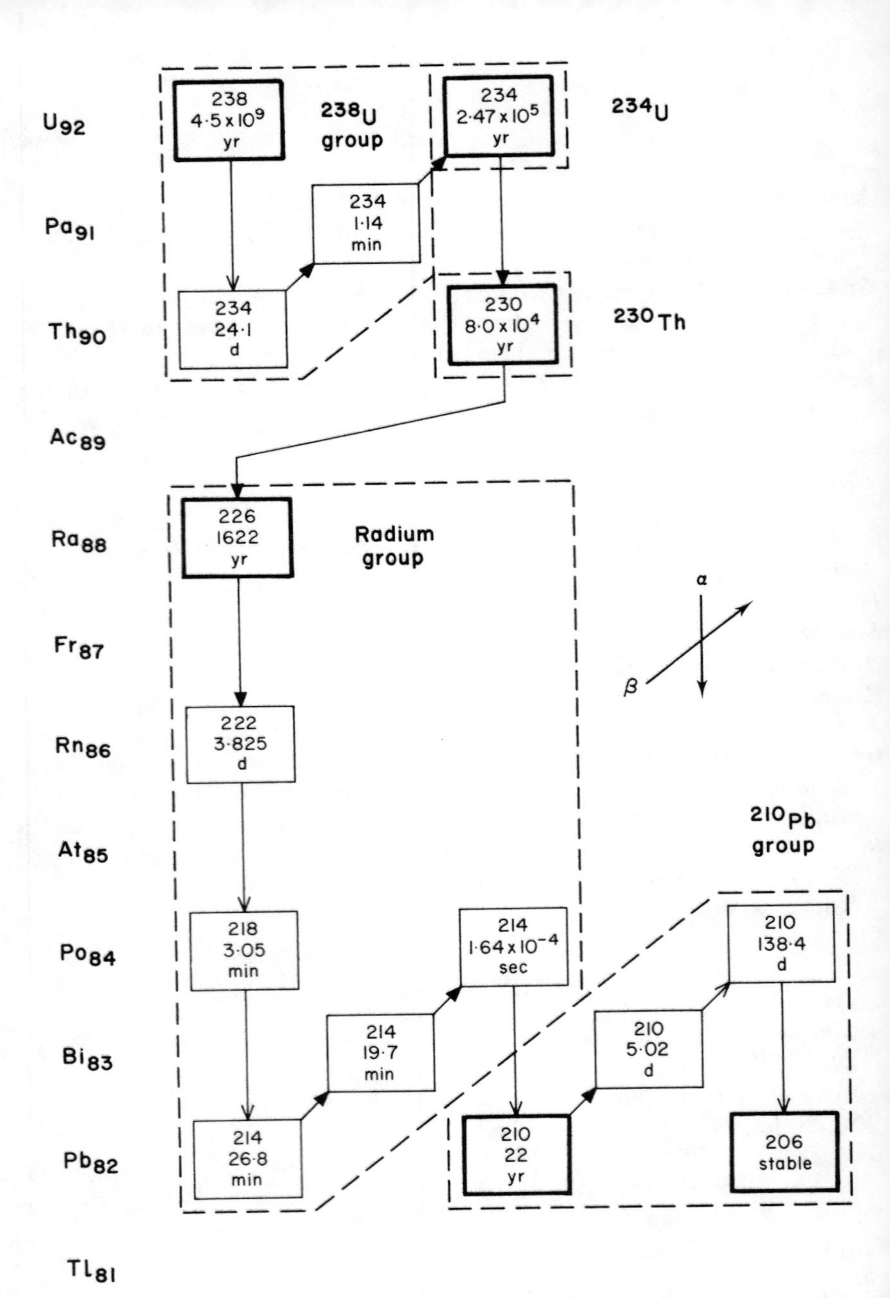

Fig. A.1a. Uranium-238 disintegration series.

Fig. A.1. Figures A. 1a–c (pp. 568–570) show the ^{238}U, ^{235}U, and ^{232}Th disintegration series, with half-lives and atomic weights (in small boxes), decay type and half life groupings; Arrows on Fig. A.1a indicate the direction for α and β decay; bold barbs on arrows indicate that >0·1% of decay events release γ radiation (Khattab 1970); broken-line boxes indicate groups of short-lived or immobile isotopes headed by a longer-lived isotope.

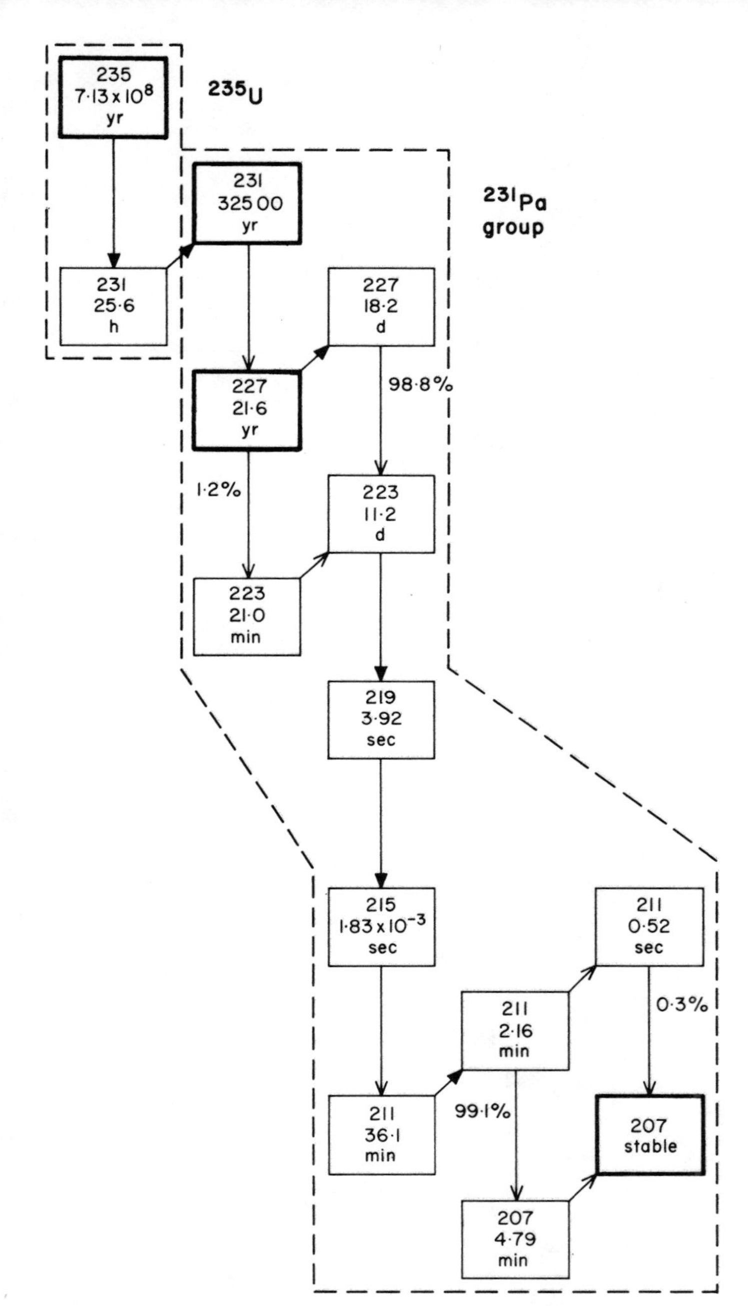

Fig. A.1b. Uranium-235 disintegration series.

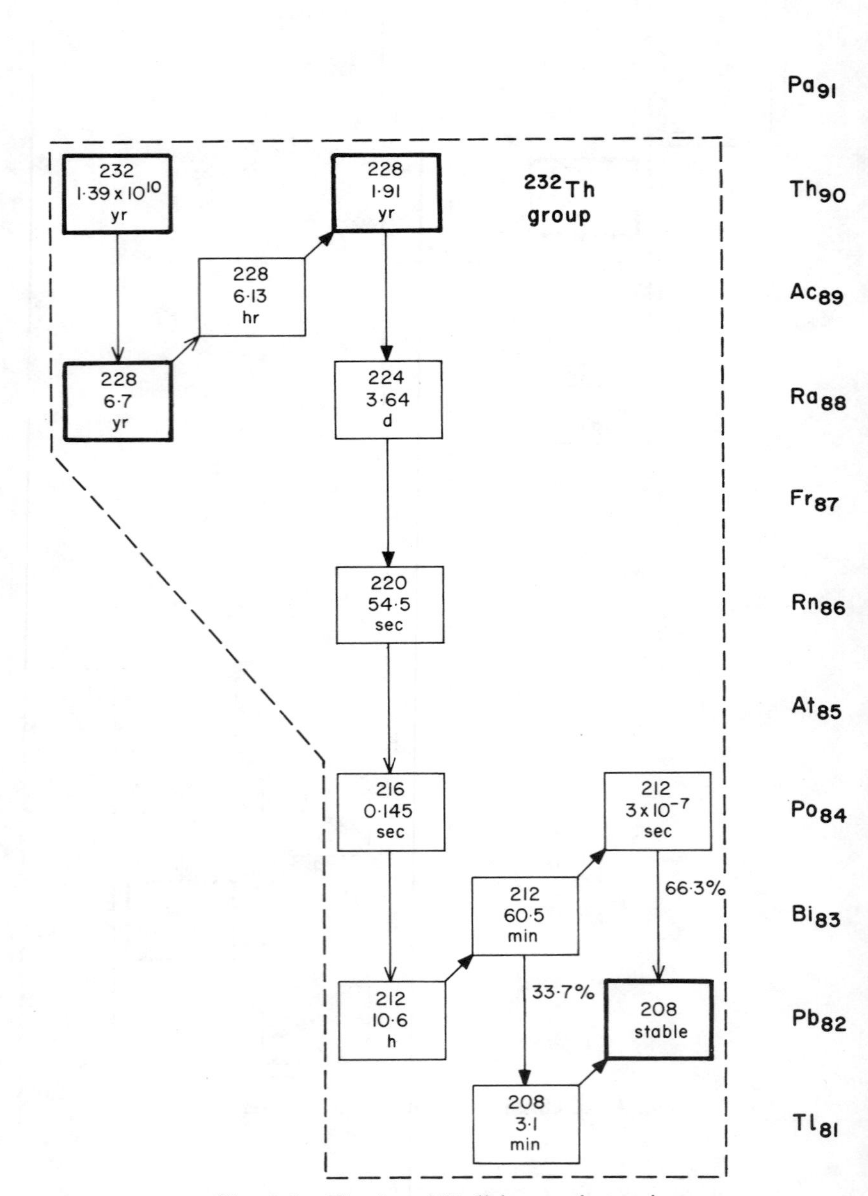

Fig. A.1c. Thorium-232 disintegration series.

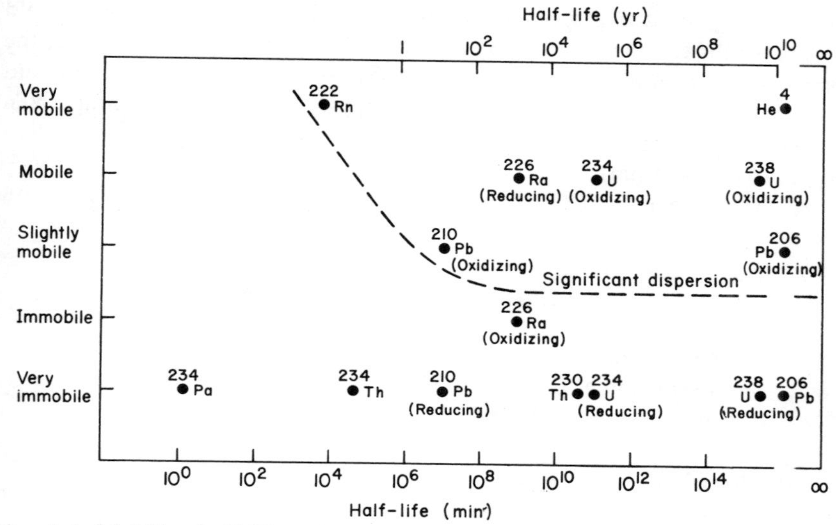

Fig. A.2. Mobility, half-life and expected degree of dispersion of some radioactive decay products of U. (After Rose and Korner, 1979.)

Mobility: relatively low, controlled by the refractory character of monazite, and by low solubility.

Geochemical prospecting applications: the content of rare earths in soils has been cited as a guide to carbonatite bodies in Uganda;[a] the Y content of fluorite in the northern Pennines of England is highest near vein intersections where the greatest tonnage of commercial fluorite is concentrated.[b]

References: [a]Bloomfield *et al.* (1971), [b]Smith (1974).

RHENIUM (Re)

Atomic number: 75; *atomic weight:* 186·2.

Igneous rocks (av): Maf 0·0006; Gran 0·0006 (2).

Sedimentary rocks (av): Ls 0·000x; Ss 0·0003; Sh 0·0005 (2).

Soils (av): 0·005 (4).

Plant ash (av): 0·005 (4).

Fresh water (range in Cu–Mo mine areas): 5–21 ppb.[a]

Associations: siderophile; Re follows Mo in molybdenite (MoS_2); also occurs highly enriched as ReS_2 in some cupriferous sandstone ores.[b]

Industrial sources: by-product of Mo smelting.

Aqueous species: ReO_4^- over wide pH range; in Cl-rich waters, ReO_3Cl^0.[b]

Mobility: probably very high, higher than Mo.[a]

Geochemical prospecting applications: higher mobility of Re over Mo could make Re in stream water a useful guide to Mo-bearing ores; however, the analytical difficulties make this impractical at the present time. Re could be used for ore typing, as porphyry-copper molybdenite contains ten times as much Re (mean 660 ppm) as molybdenite from most other ore types (mean 47 ppm).[c]

References: Klimenko (1965), [b]Mukanov and Aubakirova (1969), [c]Giles and Schilling (1972).

RUBIDIUM (Rb)

Atomic number: 37; *atomic weight:* 85·47.
Igneous rocks (av): Umaf 0·14; Maf 32; Gran 276 (1).
Sedimentary rocks (med): Ls 56; Ss 40; Sh 143 (1).
Soils (med): 35 (1).
Plant ash (med): 73 (3).
Fresh water (av): 1 ppb (2).
Associations: lithophile: follows K in igneous and sedimentary rocks.
Industrial source: by-product of K refining.
Aqueous species: Rb^{+}.
Mobility: apparently low, as indicated by very low content in average fresh water compared with rock.
Geochemical prospecting applications: high Rb and Rb/Sr in altered rocks may indicate associated base and precious-metal deposits;[a, b, c, d] Rb in surface fresh water and water-soluble Rb in residual soil may serve as a guide to pegmatite deposits.[e]
References: [a]Al-Atia and Barnes (1975), [b]Lawrence (1975), [c]Olade and Fletcher (1975), [d]Oyarzun (1975), [e]Miller and Danilov (1957).

SELENIUM (Se)

Atomic number: 34; *atomic weight:* 78·96.
Igneous rocks (med): Umaf 0·13; Maf 0·13; Gran 0·14 (1).
Sedimentary rocks (av): Ls 0·88; Ss 0·05; Sh 0·6 (2).
Soils (med): 0·31 (3).
Dry plants (med): 0·027 (3).
Fresh water (av): 0·4 ppb (6); *water-supply ceiling:* 10 ppb (5).
Associations: chalcophile; with As, Sb, Cu, Ag, and Au in sulfide deposits, especially precious-metal deposits; with U and V in Colorado Plateau ores; highly enriched in certain sedimentary beds, such as the Pierre Shale of the Dakotas.
Industrial source: by-product of Cu refining.

Weathering products: in acid soils as basic ferric selenite,[a] elemental Se, and as Se adsorbed to Fe-oxides, clay minerals, and organic matter,[b] in alkaline soils as soluble SeO_4^{2-}.

Aqueous species: SeO_4^{2-} under alkaline conditions.

Biological response: the biological cycle of Se is unusual in that certain plants will take up and concentrate extremely large quantities of Se from Se-rich soils. Se-rich plants often give off a strong and characteristic odor that has been identified as vapor-phase dimethyldiselenide.[a] Se in plants acts as a severe poison for most grazing animals. In addition to the concentrating effect, certain species of indicator plants will grow only on soils that are high in Se.[c,d]

Mobility: low in normal acid soil; very high under alkaline, oxidizing conditions where the insoluble SeO_3^{2-} is oxidized to the soluble SeO_4^{2-}.[a]

Geochemical prospecting applications: Se, because of its association with epigenetic sulfides, has been recommended as a guide to a wide variety of sulfide deposits.[e,f] Se has also been reported as a criterion for discriminating between gossans derived from sulfides and those derived from other ferruginous materials.[g] Indicator plants have received a great deal of study because of their usefulness in drawing attention to areas of poisoned grazing land. Studies of the distribution of Se indicator plants in the Colorado Plateau area of the western U.S.A. have led to the discovery of several blind Se-rich U deposits.[c]

References: [a]Lakin (1972), [b]Howard (1970), [c]Cannon (1957), [d]Cannon (1960a), [e]Goleva and Lushnikov (1967), [f]Koljonen (1976), [g]Sindeyeva (1955).

SILVER (Ag)

Atomic number: 47; *atomic weight:* 107·87.

Igneous rocks (av): Umaf 0·06; Maf 0·1; Gran 0·037 (1).

Sedimentary rocks (med): Ls 0·1; Ss 0·25; Sh 0·19 (1).

Plant ash (range): 0·1–1.[a]

Fresh water (av): 0·3 ppb (2); *water-supply ceiling:* 50 ppb (5).

Associations: chalcophile; Au, Sb, As, Pb, Zn, Cu, and Hg in sulfide deposits.

Rock minerals: trace constituent of sulfides and mafic minerals.[a]

Industrial sources: native Ag, argentite (Ag_2S); complex sulfo-salts; by-product of Cu, Pb, and Zn smelters.

Weathering product: cerargyrite (AgCl); Ag-jarosite ($AgFe_3(SO_4)_2(OH)_6$) over Ag ores; in soil probably associated with organic matter and Mn–Fe-oxides.

Aqueous species: in acid waters, Ag^+; in high-chloride water, $AgCl^0$, $AgCl_2^-$, $AgCl_3^{2-}$, $AgCl_4^{3-}$; in normal waters, $Na(AgCl_2)^0$.[a,b]

Mobility: controlled by sorption to Mn- and Fe-oxides[c] and organic matter;[a] very high average Ag content of fresh water relative to rock suggests high mobility in non-ore environment.

Geochemical prospecting applications: Ag deposits are most effectively prospected by analysis of soils and sediments for Pb, Cu, Zn, Hg, and Ag, and if present also As. Pb is the most useful guide to Ag deposits, as Ag itself is commonly below the detection limit of most analytical methods.

References: [a]Boyle (1968), [b]Singer (1973), [c]Chao and Anderson (1974).

STRONTIUM (Sr)

Atomic number: 38; *atomic weight:* 87·62.

Igneous rocks (av): Umaf 5·8; Maf 465; Gran 100 (2).

Sedimentary rocks (av): Ls 610; Ss 20; Sh 300 (2).

Soils (med): 67 (3).

Plant ash (med): 1800 (native species); 140 (cultivated plants) (3).

Fresh water (av): 400 ppb (6).

Aqueous species: Sr^{2+}.

Associations: lithophile; follows Ca, Ba; enriched in carbonatites.

Geochemical prospecting applications: Sr in soils and stream sediments shows promise as a guide to carbonatite deposits.[a] In the alteration haloes associated with porphyry-copper deposits, Sr is commonly leached where Rb is enriched; thus high Rb/Sr ratios may be useful in exploring porphyry-copper deposits.[b,c] The same relationship holds for some base-metal deposits.[d]

References: [a]Bloomfield *et al.* (1971), [b]Olade and Fletcher (1975), [c]Oyarzun (1975), [d]Lawrence (1975).

SULFUR (S)

Atomic number: 16; *atomic weight:* 32·06.

Igneous rocks (av): Umaf 300; Maf 300; Gran 300 (2).

Sedimentary rocks (av): Ls 1200; Ss 240; Sh 2400 (2).

Soils (range): 100–2000.[a]

Dry plants (av): 500 (4).

Fresh water (av): 3700 ppb (2).

Associations: chalcophile; as sulfate in saline deposits.

Rock minerals: sulfides; gypsum ($CaSO_4.2H_2O$), anhydrite ($CaSO_4$).

Industrial sources: native S deposits, by-product of natural gas and petroleum refining.

Weathering products: jarosite ($KFe_3(SO_4)_2(OH)_6$); Ca-sulfates and soluble S complexes.

Aqueous species: SO_4^{2-}, H_2S, HS^-.

Biological response: although S is an essential nutrient for plants and animals, the biological cycle does not materially affect the mobility of sulfate; under some conditions, bacteria can catalyze the oxidation of sulfides to sulfate, and vice versa. Indicator plants for S have been recognized on the Colorado Plateau.[c]

Mobility: as sulfate, extremely high, limited only by the Eh at which sulfate is reduced to sulfide, and the solubility of some saline minerals, principally gypsum.

Geochemical prospecting applications: sulfate in ground and surface water can be used as a pathfinder for sulfide deposits; its content may be expressed either as absolute concentrations or as ratios of sulfate either to chloride or to total dissolved solids.[b,d] Sulfate waters from the oxidation of sulfides can be distinguished from saline sulfates by their lower Ca and Cl content, and from geothermal sulfate waters by their lower B content.[e] In the Canadian Shield, the sulfide content is higher in ultramafic plutons hosting Cu–Ni deposits than in barren plutons.[f] S-isotope ratios can be used in ore-typing to distinguish sulfides of magmatic and hydrothermal origin from those of sedimentary origin.[g,h]

References: [a]Brady (1974, p. 24), [b]Hoag and Webber (1976b), [b]Cannon (1957), [d]Kraynov (1957), [e]Dall'Aglio and Tonani (1973), [f]Cameron *et al.* (1971), [g]Jensen (1971), [h]Seccombe (1977).

TANTALUM (Ta)

Atomic number: 73; *atomic weight:* 180·94.

Igneous rocks (av): Umaf 0·018; Maf 0·48; Gran 3·5.[a]

Sedimentary rocks (av): Sh 3·5.[a]

Industrial source: tantalite ($(Fe,Mn)(Ta,Nb)_2O_6$).

Associations: lithophile; close coherence with Nb except in alkaline igneous rocks; occurs in Li-rich pegmatites.

Industrial source: tantalite.

Mobility: low, follows hydrolyzates.

Geochemical prospecting applications: high Ta in cassiterite and tantalite may indicate Ta province;[b] ore-bearing granites have been distinguished from barren by comparing Ta and Nb in rock-forming biotite.[c]

References: [a]Vinogradov (1962), [b]Chetyrbotskaya *et al.* (1967), [c]Potap'yev *et al.* (1967).

TELLURIUM (Te)

Atomic number: 52; *atomic weight:* 127·60.

Lithosphere: 0·0018.[e]

Soils (range): 0·001–0·01.[a]
Associations: chalcophile; with As, Sb, Se, Ag, Au in epithermal precious-metal deposits.
Weathering products: as TeO_3^{2-} substituted for OH^- in Fe-oxides.[a]
Mobility: relatively low.
Geochemical prospecting applications: although Te is widely dispersed in hypogene mineralizing processes, it is relatively immobile in the supergene zone. Thus it shows promise of being a useful pathfinder element in soils for a wide variety of polymetallic ores.[a, b, c, d]
References: [a]Watterson *et al.* (1977), [b]Gott *et al.* (1969), [c]Gott and Botbol (1973), [d]Lovering *et al.* (1966), [e]Goldschmidt (1954, p. 74).

THORIUM (Th)

(*see also Radioactive Decay Products of Uranium and Thorium*)
Atomic number: 90; *atomic weight:* 232·04.
Igneous rocks (av): Umaf 0·004; Maf 2·7; Gran 20 (2).
Sedimentary rocks (av): Ls 1·7; Ss 5·5; Sh 12 (2).
Soils (av): 13 (4).
Plant ash (av): 20 ppb (4).
Fresh water (av): 0·1 ppb (2).
Associations: lithophile, occurring principally in accessory minerals of igneous rocks; in placer deposits, Th minerals occur with Au, magnetite and other heavy minerals; Th is the parent for a radioactive decay series ending in He and Pb.
Rock minerals: monazite ((RE,Th)PO_4), and as a minor constituent of allanite (hydrated silicate of Ca, Fe, Al), sphene ($CaTiSiO_5$), and zircon ($ZrSiO_4$).
Industrial source: monazite in placer deposits.
Mobility: very low, determined by the extremely refractory character of the principal primary Th minerals.
Geochemical prospecting applications: He has been suggested as a guide to Th deposits;[a] a low Th/U ratio in kimberlites has been reported as favorable for the occurrence of diamonds.[b]
References: [a]Clarke and Kugler (1973), [b]Aswathanarayana (1971).

TIN (Sn)

Atomic number: 50; *atomic weight:* 118·69.
Igneous rocks (med): Umaf 0·5; Maf 1·5; Gran 3·0 (1).
Sedimentary rocks (av): Ls 0·x; Ss 0·6; Sh 6 (2).
Soils (av): 10 (4).
Plant ash (med): 15 (3).

Fresh water (av): 0·09 ppb (6).
Associations: siderophile; to a lesser extent lithophile; B, F, Li, and Rb in late-stage differentiates of granites; Be, B, Li, Nb–Ta, W, and rare earths in pegmatite veins.
Rock minerals: biotite, cassiterite (SnO_2).
Industrial sources: cassiterite, stannite ($Cu_2(Fe,Zn)SnS_4$).
Weathering products: residual cassiterite, varlamoffite (hydrated SnO_2) over Sn ores.
Biological response: Sn is not known to be a plant nutrient, although it apparently does enter plants in amounts reflecting the Sn content of the supporting soil.[h]
Mobility: low.
Geochemical prospecting applications: the Sn content of residual soils and stream sediments[a, h] and of vegetation[c] has been used successfully as a guide to Sn in the bedrock. The Sn content of granitic rocks in the Malaysian alluvial Sn fields is in the order of 60 ppm[d] compared with a normal granitic background of 3 ppm. B, Li, F, and Rb as well as Sn are enriched in late-stage granitic differentiates, and have been cited as local but not universally applicable ore guides.[e, f, g]
References: Varlamoff (1969), [b]Udodov and Parilov (1961), [c]Millman (1957), [d]Aranyakanon (1972), [e]Hesp (1971), [f]Kesler *et al.* (1973), [g]Sheraton and Black (1973), [h]Groves (1972).

TUNGSTEN (W)

Atomic number: 74; *atomic weight:* 183·85.
Igneous rocks (med): Umaf 0·1; Maf 1·0; Gran 1·5 (1).
Sedimentary rocks (med): Ls 0·5; Ss 1·6; Sh 1·8 (1).
Soils (av): 1 (4).
Plant ash (av): 0·4 (4).
Fresh water (av): 0·03 ppb (2).
Associations: lithophile; Mo, Sn, and Nb in igneous differentiates.
Industrial sources: scheelite ($CaWO_4$), wolframite ($(Fe,Mn)WO_4$).
Weathering products: residual W minerals over W ores.
Aqueous species: probably HWO_4^-.
Mobility: intermediate to low at normal pH values; W may occur in aqueous solution in alkaline lakes.[a]
Geochemical prospecting applications: the W content of residual soils and stream sediments[b, c, d] and vegetation[e] has been cited as an ore guide to W-bearing deposits. Anomalies may be enhanced by determining W in heavy concentrates, where the W occurs as fine-grained detrital ore minerals.[f, g]

References: [a]Carpenter and Garrett (1959), [b]Holman and Webb (1957), [c]Granier (1958), [d]Cachau-Herreillat and Prouhet (1971), [e]Quin *et al.* (1974), [f]Theobald and Thompson (1959), [g]Lee Moreno and Caire (1975).

URANIUM (U)

(*see also Radioactive Decay Products of Uranium and Thorium*)

Atomic number: 92; *atomic weight:* 238·04.

Igneous rocks (med): Umaf 0·03; Maf 0·53; Gran 3·9 (1).

Sedimentary rocks (med): Ls 2·2; Ss 1·7; Sh 3·7 (1).

Soils (av): 1 (4).

Plant ash (av): 0·6.[b]

Fresh water (med): 0·5 ppb, but ranges from 0·05 in humid areas to 5 in arid regions.[cc, d]

Associations: lithophile; V, As, P, Mo, Se, Pb, Cu, etc. in Colorado Plateau ores; Co and Ag in some sulfide ores; Au in the South African Rand; P in phosphorite deposits;[e] C in black shales.[a]

Rock minerals: zircon ($ZrSiO_4$), apatite ($Ca_5(PO_4)_3F$), allanite (complex aluminosilicate).

Industrial sources: uraninite (UO_2), coffinite ($USiO_4$), uraniferous organic matter, carnotite ($K_2(UO_2)_2(VO_4)_2.3H_2O$).

Weathering products: complex carbonates, phosphates, vanadates, and silicates, including carnotite, over uraninite ores; organic matter, and to a lesser extent limonite in normal soils; soluble U complexes.

Aqueous species: UO_2^{2+}, $UO_2(CO_3)_3^{4-}$, $UO_2(CO_2)_2^{2-}$, $UO_2(HPO_4)_2^{2-}$.[z]

Biological response: in vegetation, most of the U in the nutrient solution is apparently precipitated in the root tips as autunite, $Ca(UO_2)_2PO_4$. Even so, some U gets through to the upper parts of some plant species, where it is useful in biogeochemical prospecting.[f, g] Because of the association of Se with U deposits of the Colorado Plateau, Se indicator plants have successfully been used in locating U-rich areas.[b, f]

Mobility: mobile under oxidizing conditions, especially in acid or carbonate-rich waters; immobile under reducing conditions: strongly sorbed to organic matter[ee] and Fe-oxides; may be mobile as organic complexes or colloidal particles.[h, z]

Geochemical prospecting applications: the content of U in normal soils[j, k, m] is not used extensively because of the greater effectiveness of radioactivity and Rn as indicators. U in organic soils and bogs[n, o, p] and particularly in organic lake sediments[q] has been shown to be highly effective under specialized conditions. U in stream sediment, especially extractable U, is a useful ore guide[m, p, r, s, aa], particularly in sediments containing some organic matter. Strong enrichments can occur in organic rich environments, however. U in surface waters is useful but is affected by very large temporal variations and by very low background values in humid regions.[g, r, u, v, bb] U in ground water seems to be a good guide if samples can be obtained.[c, d, dd, t] Analysis of plants for U has been used successfully in the Colorado Plateau, where it has been known to give an indication of ore through thicknesses of as much as 15 m of barren cover.[b, f] Experiments in the swamps of the Arctic taiga country of northern European Russia show that although both plants and soils show anomalies over U ore, soil sampling is generally more satisfactory.[w] U has been used as a guide to U bearing phosphorites.[x, y]

References: [a]Fix (1958), [b]Cannon (1960a), [c]Jonasson and Gleeson (1976), [d]Dyck *et al.* (1976d), [e]Altschuler *et al.* (1958), [f]Cannon (1957), [g]Ostle (1954), [h]Manskaia *et al.* (1956), [j]Grimbert (1963), [k]Michie *et al.* (1973), [m]Steenfelt *et al.* (1976), [n]Armands and Landergren (1960), [o]Armands (1967), [p]Bowie *et al.* (1971), [q]Cameron and Allen (1973), [r]Ferguson and Price (1976), [s]Little and Durham (1971), [t]Landis (1960), [u]Dall'Aglio (1971), [v]Asmund and Steenfelt (1976), [w]Moiseenko (1959), [x]Bollo and Jacquemin (1963), [y]Bliskovskii and Smirnov (1966), [z]Langmuir (1978), [aa]Rose and Keith (1976), [cc]Lopatkina (1964), [bb]Dyck *et al.* (1971), [dd]Denson *et al.* (1956), [ee]Szalay (1964).

VANADIUM (V)

Atomic number: 23; *atomic weight:* 50·94.
Igneous rocks (av): Umaf 40; Maf 250; Gran 44 (2).
Sedimentary rocks (av): Ls 20; Ss 20; Sh 130 (2).
Soils (med): 57 (3).
Plant ash (med): 5 (3).
Fresh water (av): 2 ppb (6).
Associations: lithophile; U in the secondary U minerals of the Colorado Plateau; sedimentary phosphorites; some Fe-ores; organic matter.
Rock minerals: mafic minerals.
Industrial source: by-product in production of Fe, P, and U.
Weathering products: carnotite ($K_2(UO_2)_2(VO_4)_2.3H_2O$) over Colorado Plateau U ores; Fe-oxides, organic matter in normal soils.
Aqueous species: $H_4VO_4^+$, $H_2VO_4^-$, HVO_4^{2-}, VO^{2+}.
Biological response: small amounts of V are stimulating to plants, but larger amounts (>10 ppm in nutrient solutions) are commonly toxic. Some legumes use V in the nitrogen-fixation process.[a]
Geochemical prospecting applications: None of any consequence reported to date.

Reference: [a]Cannon (1963).

ZINC (Zn)

Atomic number: 30; *atomic weight:* 65·37.

Igneous rocks (med): Umaf 58; Maf 94; Gran 51 (1).

Sedimentary rocks (med): Ls 21; Ss 40; Sh 100 (1).

Soils (med): 36 (3).

Plant ash (med): 570 (3).

Fresh water (av): 20 ppb (2); *water-supply ceiling:* 5000 ppb (5).

Associations: chalcophile; Cu, Pb, Ag, Au, Sb, As, and Se in base-metal and precious-metal deposits; Mg in some silicates.

Rock minerals: mafic minerals.

Industrial source: sphalerite (ZnS).

Weathering products: Zn sulfates, carbonates, hydrated silicates, Zn clay minerals over Zn-sulfide ores; Fe- and Mn-oxides, organic matter in normal soils.

Aqueous species: variable partition between Zn^{2+}, $Zn(OH)_2^0$, soluble organic complexes, and floating live organisms.

Biological response: Zn is an essential nutrient for almost all plants. For this reason algae growing in streams and lakes can absorb a large part of the Zn dissolved in the water. In addition to its nutritive effect, Zn is also toxic to most forms of plants if it is present in amounts exceeding certain limits.

Mobility: moderately high, limited by tendency to be adsorbed by MnO_2 and by insoluble organic matter.[a,b]

Geochemical prospecting applications: geochemical prospecting based on analysis of residual soils has been a very successful method.[c,d,e,f] Zn not uncommonly forms epigenetic patterns in transported overburden[g,h] and in peat swamps.[j,k] A considerable amount of experimental work has been done on Zn indicator plants and on analysis of vegetation for Zn as a prospecting method.[m] Soil sampling in general is preferable to plant sampling because of the greater ease of collecting and analyzing samples and the greater homogeneity of the anomalous patterns. Zn forms useful dispersion patterns in ground water,[n] stream water,[o,p,q,r] stream sediments,[f,r,s] and lake sediments.[t] Under some conditions, Hg in surficial materials shows promise as a guide to deeply buried Zn ore.[u,v].

References: [a]Garrett and Hornbrook (1976), [b]Nowlan (1976), [c]Hawkes and Lakin (1949), [d]Huff (1952), [e]Boyle and Cragg (1957), [f]Webb (1958a), [g]Hawkes (1954), [h]Cox and Hollister (1955), [j]Salmi (1950), [k]Cannon (1955), [m]Cannon (1960b), [n]Kennedy (1956), [o]Webb and Millman (1950), [p]Boyle *et al.* (1955), [q]Atkinson (1957), [r]Boyle *et al.* (1971), [s]Hawkes and Bloom (1956), [t]Allan *et al.* (1973a), [u]Friedrich and Hawkes (1966), [v]Gustavson (1976).

ZIRCONIUM (Zr)

Atomic number: 40; *atomic weight:* 91·22.
Igneous rocks (*av*)*:* Umaf 45; Maf 140; Gran 175 (2).
Sedimentary rocks (*av*)*:* Ls 19; Ss 220; Sh 160 (2).
Soil (*med*)*:* 270 (3).
Plant ash (*med*)*:* <20 (3).
Associations: lithophile, occurring principally in accessory zircon ($ZrSiO_4$) in igneous rocks.
Rock mineral: zircon.
Industrial source: zircon in placer deposits.
Mobility: relatively low, controlled by the refractory character of zircon.
Geochemical prospecting applications: none of importance reported to date.

References

Where *Golden Symposium Volume*, *Toronto Symposium Volume*, *London Symposium Volume*, *Vancouver Symposium Volume*, or *Sydney Symposium Volume* are cited in a reference, see under the appropriate volume for a fuller citation.

Adams, D. F. (1976). Sulfur compounds. *In* "Air Pollution" (A. C. Stern, ed.), Vol. 3, 213–257. Academic Press, New York and London.

Adkisson, C. W. and Reimer, G. M. (1976). Helium and radon-emanation bibliography, selected references of geologic interest to uranium exploration. *U.S. Geol. Survey* Open-file Rept 76-860, 44 pp.

Agnew, A. F. (1955). Application of geology to the discovery of zinc–lead ore in the Wisconsin–Illinois–Iowa District. *Mining Eng*. **7,** 781–795.

Ahmed, M. B. and Twyman, E. S. (1953). The relative toxicity of manganese and cobalt to the tomato plant. *J. Exp. Bot*. **4,** 164–172.

Ahrens, L. H. (1954). Lognormal distribution of the elements. *Geochim. Cosmochim. Acta* **5,** 49–73; **6,** 121–131.

Ahrens, L. H. (1957). Lognormal-type distributions. *Geochim. Cosmochim. Acta* **11,** 205–212.

Ahrens, L. H. and Liebenberg, W. R. (1950). Tin and indium in mica, as determined spectrochemically. *Am. Mineralogist* **35,** 571–578.

Ainsworth, B., Kowalchuk, J. M. and Rotherham, D. (1977). Howards Pass (Y.T./N.W.T.) synsedimentary deposit, the geochemical setting (abst.). *Geol. Assoc. Can. Program with Abstracts* **2,** 4.

Akright, R. L., Radtke, A. S. and Grimes, D. J. (1969). Minor elements as guides to gold in the Roberts Mountain Formation, Carlin Gold Mine, Eureka County, Nevada. *Colorado School Mines Quart*. **64**(1), 49–66.

Al-Atia, M. J. and Barnes, J. W. (1975). Rubidium, a primary dispersion pathfinder at Ogofau gold mine, southern Wales. *Vancouver Symposium Volume* 341–352.

Al-Hashami, A. R. K. and Brownlow, A. H. (1970). Copper content of biotites from the Boulder batholith. *Econ. Geol*. **65,** 985–992.

Allan, R. J. and Hornbrook, E. H. W. (1971). Exploration geochemistry evaluation studies in a region of continuous permafrost, Northwest Territories, Canada. *Toronto Symposium Volume* 53–66.

Allan, R. J. and Timperley, M. H. (1975). Prospecting by use of lake sediments in areas of industrial heavy metal contamination. *In* "Prospecting in Areas of Glaciated Terrain" (M. J. Jones, ed.), 87–111. Institution of Mining and Metallurgy, London.

Allan, R. J., Lynch, J. J. and Lund, N. G. (1972). Regional geochemical exploration in the Coppermine River area, District of McKenzie—a feasibility study in permafrost terrain. *Geol. Survey Can.* Paper 71-33, 52 pp.

Allan, R. J., Cameron, E. M. and Durham, C. C. (1973a). Lake geochemistry, a low sample density technique for reconnaissance geochemical exploration and mapping of the Canadian Shield. *London Symposium Volume* 131–160.

Allan, R. J., Cameron, E. M. and Durham, C. C. (1973b). Reconnaissance geochemistry using lake sediments of a 36 000 square mile area of the northwestern Canadian Shield (Bear–Slave Operation, 1972). *Geol. Survey Can.* Paper 72-50, 70 pp.

Allcott, G. H. and Lakin, H. W. (1975). The homogeneity of six geochemical exploration reference samples. *Vancouver Symposium Volume* 659–681.

Allcott, G. H. and Lakin, H. W. (1978). Tabulation of geochemical data furnished by 109 laboratories for six geochemical exploration reference samples. *U.S. Geol. Survey* Open-file Rept 78-163, 199 pp.

Allen, J. (1976). Development of a portable radon detector system. *U.S. Energy Res. Dev. Adm.* Open-file Rept GJBX-50(76), 47 pp.

Alminas, H. V., Watts, K. C., Griffiths, W. R. *et al.* (1975). Maps showing anomalous distribution of W, F, Ag, Pb, Sn, Bi, Mo, Cu, and Zn in stream sediment concentrates from the Sierra Cuchillo-Animas uplifts and adjacent areas, southwestern New Mexico. *U.S. Geol. Survey* Geol. Inv. Maps I-880, I-881, I-882.

Al-Shaieb, Z. and Bolter, E. (1973). Geochemical anomalies in the igneous wall rock at Mayflower Mine, Park City district, Utah (abst.). *Geol. Soc. Am.* Abst. with Programs **5,** 459.

Altschuler, Z. S., Clarke, R. S., Jr and Young, E. J. (1958). Geochemistry of uranium in apatite and phosphorite. *U.S. Geol. Survey* Prof. Paper 314-D, 45–90.

Altshuller, A. P. (1976). Hydrocarbons and carbon oxides. *In* "Air Pollution" (A. C. Stern, ed.), Vol. 3, 183–189. Academic Press, New York and London.

Andren, A. W. and Harriss, R. C. (1975). Observations on the association between mercury and organic matter dissolved in natural waters. *Geochim. Cosmochim. Acta* **39,** 1253–1257.

Andrews, J. N. and Wood, D. F. (1972). Mechanism of radon release in rock matrices and entry into groundwaters. *Inst. Mining Metall. Trans.* **81,** 198B–209B.

Andrews-Jones, D. A. (1968). The application of geochemical techniques to mineral exploration. *Colorado School Mines Bull.* **11**(6), 1–31.

Angoran, Y. and Madden, T. R. (1977). Induced polarization, a preliminary study of its chemical basis. *Geophysics* **42,** 788–803.

Antropova, L. V. (1969). Forms of occurrence of lead in dispersion trains (Estonia). *Int. Geol. Rev.* **11,** 24–30.

Antropova, L. V. (1975). "Forms of Occurrence of Elements in Dispersion Halos." Nedra, Leningrad, 144 pp.

Antweiler, J. C. and Campbell, W. L. (1977). Application of gold compositional analyses to mineral exploration in the United States. *J. Geochem. Explor.* **8,** 17–29.

Applied Geochemistry Research Group (1974). Data processing for the provisional geochemical atlas of Northern Ireland. *Appl. Geochem. Res. Group, London.* Tech. Comm. 61, 7 pp.

Applied Geochemistry Research Group (1975). Chelation/solvent extraction system for the determination of cadmium, cobalt, copper, iron, manganese, nickel, lead, and zinc in natural waters. *Appl. Geochem. Res. Group, London.* Tech. Comm. 62, 7 pp.

Applin, K. R. and Langmuir, D. (1978). Ground water geochemistry as a prospecting tool for uranium deposits in Pennsylvania. *U.S. Dept Energy* Open-file Rept GJBX-132(78), 94–96.

Arakelyan, G. B. and Kyuregyan, T. N. (1969). Forms of lead migration in mine waters (English abst.). *Chem. Abst.* **73,** item 100945.

Aranyakanon, P. (1972). Geochemical studies of tin in granite and basalt and of fine tin alluvium, Thailand. *In* "Proc. 2nd Seminar Geochem. Prosp. Methods and Techniques". U.N. Min. Res. Sev. Ser. No. 38, 199–202. United Nations, New York.

Archer, A. R. and Main, C. A. (1971). Casino, Yukon, a geochemical discovery of an unglaciated Arizona-type porphyry. *Toronto Symposium Volume* 67–77.

Arkhangel'sky, A. D. and Soloviev, N. V. (1938). Experimental investigation of the mechanism of accumulation of copper in sedimentary rocks (in Russian with English abst.). *Bull. Acad. Sci. U.S.S.R., Ser. Geol.* **1938(2),** 279–294.

Armands, G. (1967). Geochemical prospecting of a uraniferous bog deposit at Masugnsbyn, northern Sweden. *In* "Geochemical Prospecting in Fennoscandia" (A. Kvalheim, ed.), 127–154. Interscience, New York.

Armands, G. and Landergren, S. (1960). Geochemical prospecting for uranium—the enrichment of uranium in peat. *Proc. 21st Int. Geol. Cong., Copenhagen* 1960 51–66.

Armbrust, G. A., Oyarzun, J. and Arias, J. (1977). Rubidium as a guide to ore in Chilean porphyry copper deposits. *Econ. Geol.* **72**, 1086–1100.

Armour-Brown, A. and Nichol, I. (1970). Regional geochemical reconnaissance and the location of metallogenic provinces. *Econ. Geol.* **65,** 312–330.

Armstrong, F. E. and Heemstra, R. J. (1973). Radiation halos and hydrocarbon reservoirs—a review. *U.S. Bur. Mines* Inf. Circ. 8579, 52 pp.

Aruscavage, P. J. and Millard, H. T. (1972). A neutron activation analysis procedure for the determination of uranium, thorium, and potassium in geological samples. *J. Radioanal. Chem.* **11,** 67–84.

Aslin, G. E. M. (1976). The determination of As and Sb in geological materials by flameless atomic absorption spectrometry. *J. Geochem. Explor.* **6,** 321–330.

Asmund, G. and Steenfelt, A. (1976). Uranium analysis of stream water, east Greenland. *J. Geochem. Explor.* **5,** 374–380.

Aswathanarayana, U. (1971). Thorium/uranium ratio as an aid in prospecting for diamonds in the kimberlitic rocks of India. *Current Science (Bangalore)* **40,** 663–664.

Atkinson, D. J. (1957). Heavy metal concentration in streams in north Angola. *Econ. Geol.* **52**, 652–667.

Aubert, G. and Tavernier, R. (1972). Soil Survey. *In* "Soils of the Humid Tropics", 17–44. National Acad. Sci., Washington.

Back, W. and Barnes, I. L. (1961). Equipment for field measurement of electrochemical potentials. *U.S. Geol. Survey* Prof. Paper 424-C, 366–368.

Baes, C. F. and Mesmer, R. E. (1976). "The Hydrolysis of Cations." John Wiley, New York, 489 pp.

Bailey, J. C. (1977). Fluorine in granitic rocks and melts, a review. *Chem. Geol.* **19,** 1–42.

Bailly, P. A. and Still, A. R. (1973). Exploration for mineral deposits—purpose, procedure, methods, management. *In* "Mining Engineers Handbook" (A. B. Cummins and I. A. Given, eds), Vol. 1, 5-2–5-12. Soc. Mining Eng., New York.

Ball, N. L. and Snowdon, L. R. (1973). A preliminary evaluation of the applicability of the helium survey technique to prospecting for petroleum. *Geol. Survey Can.* Paper 73-1B, 199–202.

Banks, N. G. (1974). Distribution of copper in biotite and biotite alteration products in intrusive rocks near two Arizona porphyry copper deposits. *U.S. Geol. Survey J. Res.* **2,** 195–212.

Banwart, W. L. and Bremner, J. M. (1974). Gas chromatographic identification of sulfur gases in soil atmospheres. *Soil. Biol. Biochem.* **6,** 113–115.

Baranov, E. N., Zasukhin, G. N., Karpukhina, L. A. *et al.* (1972). Occurrence of copper, zinc, lead, and other elements in pyrite from aureoles of pyrite deposits. *Geochem. Int.* **9,** 834–844.

Baranova, V. V. (1957). Aureoles of molybdenum dissemination of the Tyrny–Auz mineral district. *Geochemistry* **1957,** 152–158.

Barnes, H. L. (1973). Chemical depositional controls of mineral deposits. *In* "Mining Engineers Handbook" (A. B. Cummins and I. A. Given, eds), Vol. 1, 4–28. Soc. Mining Eng., New York.

Barnes, H. L. and Czamanske, G. K. (1967). Solubilities and transport of ore minerals. *In* "Geochemistry of Hydrothermal Ore Deposits" (H. L. Barnes, ed.), 334–381. Holt, Rinehart and Winston, New York.

Barnes, H. L. and Lavery, N. G. (1977). Use of primary dispersion for exploration of Mississippi-Valley type deposits. *Sydney Symposium Volume* 105–115.

Barnes, R. O. and Bieri, R. H. (1976). Helium flux through marine sediments of northeast Pacific Ocean. *Earth Planet. Sci. Letters* **28,** 331–336.

Barnett, P. R., Huleatt, W. P., Rader, L. F. *et al.* (1955). Spectrographic determination of contamination of rock samples after grinding with alumina ceramic. *Am. J. Sci.* **253,** 121–124.

Barr, D. A. and Hawkes, H. E. (1963). Seasonal variations in copper content of stream sediments in British Columbia. *Am. Inst. Mining Eng. Trans.* **226,** 342–346.

Barrer, R. M. (1951). "Diffusion In and Through Solids." Cambridge University Press, Cambridge, 464 pp.

Barringer, A. R. (1966). Interference-free spectrometer for high-sensitivity mercury analyses of soils, rocks, and air. *Inst. Mining Metall. Trans.* **75**, B120–B124.

Barringer, A. R. (1969). Remote-sensing techniques for mineral discovery. *Proc. 9th Commonwealth Mining Metall. Cong., London* **2,** 649–690.

Barringer, A. R. (1971). Optical detection of geochemical anomalies in the atmosphere (abst.). *Toronto Symposium Volume* 474.

Barringer, A. R. (1976). Exploration method and apparatus utilizing atmospheric microorganic particulates. U.S. Patent 3 985 619.

Barringer, A. R. (1977). AIRTRACE—an airborne geochemical exploration technique. *U.S. Geol. Survey* Prof. Paper 1015, 231–251.

Barshad, I. (1966). The effect of a variation in precipitation on the nature of clay mineral formation in soils from acid and basic igneous rocks. *Proc. 6th Int. Clay Conf.* **1,** 167–173.

Barsukov, V. L. and Durasova, N. A. (1966). Metal content and metallogenic specialization of intrusive rocks in the regions of sulfide–cassiterite deposits. *Geochem. Int.* **3**, 97–107.

Barsukov, V. L. and Kurilchikova, G. J. (1966). On the forms in which tin is transported in hydrothermal solutions. *Geochem. Int.* **3**, 759–764.

Barsukov, V. L. and Volosov, A. G. (1967). A geochemical method for estimating depth of sulfide–cassiterite mineralization and discovery of blind ore bodies (abst.). *Geochem. Int.* **4**, 1105.

Bayrock, L. A. and Pawluk, S. (1967). Trace elements in tills of Alberta. *Can. J. Earth Sci.* **4**, 597–607.

Beath, O. A., Gilbert, C. S. and Eppson, H. F. (1939). The use of indicator plants in locating seleniferous areas in western United States—I. General. *Am. J. Bot.* **26**, 257–269.

Beauford, W., Barber, J. and Barringer, A. R. (1975). Heavy metal release from plants into the atmosphere. *Nature, Lond.* **256**(5512), 35–37.

Beauford, W., Barber, J. and Barringer, A. R. (1977). Release of particles containing metals from vegetation into the atmosphere. *Science* **195**, 571–573.

Beck, K. C., Reuter, J. H. and Perdue, E. M. (1974). Organic and inorganic geochemistry of some coastal plain rivers. *Geochim. Cosmochim. Acta* **38**, 341–364.

Beck, L. S. and Gingrich, J. E. (1976). Track-etch orientation survey in the Cluff Lake area, northern Saskatchewan. *Can. Bull. Mining Metall.* **69**(769), 104–109.

Bell, H. (1976). Geochemical reconnaissance using heavy minerals from small streams in central South Carolina. *U.S. Geol. Survey Bull.* **1404**, 23 pp.

Belyakova, Ye. Ye. (1958). Migration of elements in underground and surface waters of the Upper Kairakty District, Central Kazakhstan. *Geochemistry* **1958**, 176–188.

Belyakova, Ye. Ye., Reznikov, A. A., Kramarenko, L. E. *et al.* (1963). "Hydrogeochemical Method for Prospecting for Ore Deposits." English translation published by U.S. Dept Commerce, Office of Technical Services, Springfield, Virginia, 381 pp.

Bennett, C. A. and Franklin, N. L. (1954). "Statistical Analysis in Chemistry and the Chemical Industry." John Wiley, New York, 724 pp.

Bennett, R. A. and Rose, W. I. (1973). Some compositional changes in Archaean felsic volcanic rocks related to massive sulfide mineralization. *Econ. Geol.* **68**, 886–891.

Beus, A. A. (1969). Geochemical criteria for assessment of the mineral potential of igneous rock series during reconnaissance exploration. *Colorado School Mines Quart.* **64**(1), 67–74.

Beus, A. A. and Grigorian, S. V. (1977). "Geochemical Exploration Methods for Mineral Deposits." Applied Publishing Co., Wilmette, Illinois, 287 pp.

Bignell, R. D., Cronan, D. S. and Tooms, J. S. (1976). Metal dispersion in the Red Sea as an aid to marine geochemical exploration. *Inst. Mining Metall. Trans.* **85**, B274–B279.

Billings, W. D. (1950). Vegetation and plant growth as affected by chemically altered rock in the western Great Basin. *Ecology* **31**, 62–74.

Birkeland, P. W. (1974). "Pedology, Weathering, and Geomorphological Research." Oxford University Press, New York, 285 pp.

Blackwelder, E. (1927). Fire as an agent in rock weathering. *J. Geol.* **35**, 134–140.

Blanchard, R. (1968). Interpretation of leached outcrops, *Nevada Bur. Mines Bull.* **66**, 196 pp.

Bliskovskii, V. Z. and Smirnov, A. I. (1966). Radioactivity of phosphorites. *Geochem. Int.* **3,** 563–567.

Bloomfield, K., Reedman, J. H. and Tether, J. (1971). Geochemical exploration of carbonatite complexes in eastern Uganda. *Toronto Symposium Volume* 85–102.

Bloss, F. D. and Steiner, R. L. (1960). Biogeochemical prospecting for manganese in northeast Tennessee. *Geol. Soc. Am. Bull.* **71,** 1053–1066.

Boberg, W. W. and Runnells, D. D. (1971). Reconnaissance study of uranium in the South Platte River, Colorado. *Econ. Geol.* **66,** 435–450.

Bollo, R. and Jacquemin, M. (1963). Phosphate exploration by aerial radiometric surveying (in French). *Geophys. Prosp.* **11,** 550–560.

Bolotnikov, A. F. and Kravchenko, N. S. (1970). Criteria for recognizing tin-bearing granites. *Am. Geol. Inst. Dokl. Acad. Sci. U.S.S.R.* **191,** 186–187.

Bolotnikova, I. V. (1965). An experiment on using a gas survey in prospecting for deep-seated polymetallic deposits (in Russian). English abst. in *Chem. Abst.* **64,** item 7882h.

Bolter, E. and Al-Shaieb, Z. (1971). Trace element anomalies in igneous wall rocks of hydrothermal veins. *Toronto Symposium Volume* 289–290.

Bølviken, B. and Logn, Ø. (1975). An electrochemical model for element distribution around sulphide bodies. *Vancouver Symposium Volume* 631–650.

Bølviken, B. and Sinding-Larsen, R. (1973). Total error and other criteria in the interpretation of stream sediment data. *London Symposium Volume* 285–295.

Bølviken, B., Krog, J. R. and Naess, G. (1976). Sampling technique for stream sediments. *J. Geochem. Explor.* **5,** 382–383.

Bølviken, B., Honey, F., Levine, S. R. *et al.* (1977). Dectection of naturally heavy-metal-poisoned areas by Landsat-1 digital data. *J. Geochem. Explor.* **8,** 457–471.

Bonatti, E., Zerbi, M., Kay, R. *et al.* (1976). Metalliferous deposits from the Apennine ophiolites, Mesozoic equivalents of modern deposits from oceanic spreading centers. *Geol. Soc. Am. Bull.* **87,** 83–94.

Bond, R. G. and Straub, C. P. (1972). "Handbook of Environmental Control" Vol. 1, 82–88, 117. CRC Press, Cleveland, Ohio.

Bondarenko, G. P. (1968). An experimental study of the solubility of galena in the presence of fulvic acids. *Int. Geol. Rev.* **5,** 525–531.

Botbol, J. M., Sinding-Larsen, R., McCammon, R. B. *et al.* (1978). A regionalized multivariate approach to target selection in geochemical exploration. *Econ. Geol.* **73,** 534–546.

Bowen, N. L. (1922). The reaction principle in petrogenesis. *J. Geol.* **30,** 177–198.

Bowie, S. H. U., Darnley, A. G. and Rhodes, J. R. (1965). Portable radioisotope X-ray fluorescence analyzer. *Inst. Mining Metall. Trans.* **74,** 361–379.

Bowie, S. H. U., Ostle, D. and Ball, T. K. (1971). Geochemical methods in the detection of hidden uranium deposits. *Toronto Symposium Volume* 103–111.

Boyle, R. W. (1967). Geochemical prospecting—retrospect and prospect. *Geol. Survey Can.* Paper 66–54, 30–43.

Boyle, R. W. (1968). The geochemistry of silver and its deposits. *Geol. Survey Can. Bull.* **160,** 264 pp.

Boyle, R. W., ed. (1971a). "Geochemical Exploration", Spec. Vol. 11, Canadian Institution of Mining and Metallurgy, Toronto, 594 pp.

Boyle, R. W. (1971b). Boron and the boron minerals as indicators of mineral deposits (abst.). *Toronto Symposium Volume* 112.

Boyle, R. W. (1974). Elemental associations in mineral deposits and indicator elements of interest in geochemical prospecting (revised). *Geol. Survey Can.* Paper 74–45, 40 pp.

Boyle, R. W. (1975). The geochemistry of antimony, Keno Hill area, Yukon, Canada. *In* "Recent Contributions to Geochemistry and Analytical Chemistry" (A. I. Tugarinov, ed.), 354–370. John Wiley, New York.

Boyle, R. W. (1976). Report of retiring president to annual meeting of the Association of Exploration Geochemists. *J. Geochem. Explor.* **6,** 389–395.

Boyle, R. W. and Cragg, C. B. (1957). Soil analysis as a method of geochemical prospecting in Keno Hill–Galena Hill area, Yukon Territory. *Geol. Survey Can. Bull.* **39,** 27 pp.

Boyle, R. W. and Dass, A. S. (1967). Geochemical prospecting—use of the A horizon in soil surveys. *Econ. Geol.* **62,** 274–276.

Boyle, R. W. and Garrett, R. G. (1970). Geochemical prospecting—a review of its status and future. *Earth Sci. Rev.* **6,** 51–75.

Boyle, R. W. and Jonasson, I. R. (1973). The geochemistry of arsenic and its use as an indicator element in geochemical prospecting. *J. Geochem. Explor.* **2,** 251–296.

Boyle, R. W. and Smith, A. Y. (1968). The evolution of techniques and concepts in geochemical prospecting. *In* "The Earth Sciences in Canada" (E. R. W. Neale, ed.), Spec Pub. 11, 117–128. Royal Society of Canada.

Boyle, R. W., Illsley, C. T. and Green, R. N. (1955). Geochemical investigation of the heavy metal content of stream and spring waters in the Keno Hill–Galena Hill area, Yukon Territory. *Geol. Survey Can. Bull.* **32,** 34 pp.

Boyle, R. W., Koehler, G. F., Moxham, R. L. *et al.* (1958). Heavy metal content of water and sediment in the streams, rivers and lakes of southwestern Nova Scotia. *Geol. Survey Can.* Paper 58-1.

Boyle, R. W., Tupper, W. M., Lynch, J. *et al.* (1966). Geochemistry of Pb, Zn, Cu, As, Sb, Mo, Sn, W, Ag, Ni, Co, Cr, Ba, and Mn in the waters and stream sediments of the Bathurst–Jacquet River district, New Brunswick. *Geol. Survey Can.* Paper 65-42, 49 pp.

Boyle, R. W., Hornbrook, E. H. W., Allan, R. J. *et al.* (1971). Hydrogeochemical methods—application in the Canadian Shield. *Can. Inst. Mining Metall. Bull.* **64,** 60–71.

Boyle, R. W., Alexander, W. M. and Aslin, G. E. M. (1975). Solubility of gold. *Geol. Survey Can.* Paper 75-24, 6 pp.

Bradshaw, P. M. D., ed. (1975). Conceptual models in exploration geochemistry. *J. Geochem. Explor.* **4,** 1–213.

Bradshaw, P. M. D., Clews, D. R. and Walker, J. L. (1972). "Exploration Geochemistry." Barringer Research, Rexdale, Ontario, 49 pp.

Bradshaw, P. M. D., Thomson, I., Smee, B. W. *et al.* (1974). The application of different analytical extractions and soil profile sampling in exploration geochemistry. *J. Geochem. Explor.* **3,** 209–225.

Brady, N. C. (1974). "The Nature and Properties of Soils", 8th edn. Macmillan, New York, 639 pp.

Brant, A. A. (1968). The pre-evaluation of the possible profitability of exploration prospects. *Mineralium Deposita* **3,** 1–17.

Bristow, Q. (1972). An evaluation of the quartz crystal microbalance as a mercury vapor sensor for soil gases. *J. Geochem. Explor.* **1,** 55–76.

Bristow, Q. and Jonasson, I. R. (1972). Vapor sensing for mineral exploration. *Can. Mining J.* **93,** 39–44, 47, 85.

Brooks, R. R. (1972). "Geobotany and Biogeochemistry in Mineral Exploration." Harper and Row, New York, 290 pp.

Brotzen, O. (1967). Geochemical prospecting in northern Sweden. *In* "Geochemical Prospecting in Fennoscandia" (A. Kvalheim, ed.), 203–223. Interscience, New York.

Brown, E., Skougstad, M. W. and Fishman, M. J. (1970). "Methods for Collection and Analysis of Water Samples for Dissolved Minerals and Gases." U.S. Geol. Survey, Techniques of Water Resource Investigation, Book 5, Ch. A-1, 160 pp.

Brown, G., ed. (1961). "The X-ray Identification and Crystal Structure of Clay Minerals", 2nd edn. Mineralogical Society, London, 544 pp.

Broyer, T. C. (1947). The movement of material into plants. II. The nature of solute movement into plants. *Bot. Rev.* **13,** 125–167.

Brundin, N. H. and Nairis, B. (1972). Alternative sample types in geochemical prospecting. *J. Geochem. Explor.* **1,** 7–46.

Bulashevich, Yu. P. (1974). Helium and argon surveys in research on the tectonics of continental shelves and the ocean floor (in Russian). English translation in *Geochem. Int.* **11,** 255.

Bulashevich, Yu. P., Bashorin, V. N., Druzhinin, V. S. *et al.* (1974). Helium in ground water along the Sverdlovsk deep-seismic traverse. *Am. Geol. Inst. Dokl. Acad. Sci. U.S.S.R.* **208,** 15–17.

Bull, A. J. and Mazzucchelli, R. H. (1975). Application of discriminant analysis to the geochemical evaluation of gossans. *Vancouver Symposium Volume* 219–226.

Burbank, W. S., Luedke, R. G. and Ward, F. N. (1972). Arsenic as an indicator element for mineralized volcanic pipes in the Red Mountain area, western San Juan Mountains, Colorado. *U.S. Geol. Survey Bull.* **1364,** 31 pp.

Burnham, C. W. (1959). Metallogenic provinces of the southwestern United States and northern Mexico. *New Mexico Bur. Mines Mineral Res. Bull.* **65,** 1–76.

Bush, J. B. and Cook, D. R. (1960). The Chief Oxide–Burgin area discoveries, East Tintic district, Utah, a case history. Part II. Bear Creek Mining Company studies and exploration. *Econ. Geol.* **55,** 1507–1540.

Butcher, S. S. and Charlson, R. J. (1972). "An Introduction to Air Chemistry." Academic Press, New York and London, 241 pp.

Butt, C. R. M. and Sheppy, N. R. (1975). Geochemical exploration problems in Western Australia, exemplified by the Mt. Keith area. *Vancouver Symposium Volume* 391–415.

Butt, C. R. M. and Wilding, I. G. P. eds (1977). "Geochemical Exploration 1976." Elsevier, Amsterdam, 494 pp. Also in *J. Geochem. Explor.* **8,** 1–494.

Butz, T. R. (1977). Uranium geochemical survey of the Crystal City–Beeville quadrangles, Texas. *U.S. Dept Energy* Open-file Rept GJBX-(77)77, 99–132.

Buyalov, N. I. and Shvyryayeva, A. M. (1955). Geobotanical methods in prospecting for salts of boron. *Int. Geol. Rev.* **3,** 619–625.

Cachau-Herreillat, F. (1975). Towards a quantitative utilization of geochemical exploration—the threshold and its determination in soil surveys. *Vancouver Symposium Volume* 183–189.

Cachau-Herreillat, F. and Prouhet, J. P. (1971). The utilization of metalloids (arsenic, phosphorus, fluorine) as pathfinders for skarn tungsten deposits in the Pyrenees (France). *Toronto Symposium Volume* 116–120.

Cadigan, R. A. and Felmlee, J. K. (1977). Radioactive springs geochemical data related to uranium exploration. *J. Geochem. Explor.* **8,** 381–395.

Cameron, E. M., ed. (1967). Proceedings, Symposium on Geochemical Prospecting, Ottawa, April, 1966. *Geol. Survey Can.* Paper 66-54, 282 pp.

Cameron, E. M. (1975a). Integrated studies on mineral resource appraisal in the Beechey Lake belt of the northern Shield. *Geol. Survey Can.* Paper 75-1A, 189–192.

Cameron, E. M. (1975b). Geochemical methods of exploration for massive sulfide mineralization in the Canadian Shield. *Vancouver Symposium Volume* 21–49.

Cameron, E. M. (1976). Geochemical reconnaissance for uranium in Canada—notes on methodology and interpretation of data. *Geol. Survey Can.* Paper 76-1C, 229–236.

Cameron, E. M. (1977a). Geochemical dispersion in mineralized soils of a permafrost environment. *J. Geochem. Explor.* **7,** 301–326.

Cameron, E. M. (1977b). Geochemical dispersion in lake waters and sediments from massive sulfide mineralization, Agricola Lake area, Northwest Territories. *J. Geochem. Explor.* **7,** 327–348.

Cameron, E. M. and Allan, R. J. (1973). Distribution of uranium in the crust of the northwestern Canadian Shield as shown by lake sediment analysis. *J. Geochem. Explor.* **2,** 237–250.

Cameron, E. M. and Ballantyne, S. B. (1975). Experimental hydrogeochemical surveys of the High Lake and Hackett River areas, Northwest Territories. *Geol. Survey Can.* Paper 75-29, 19 pp.

Cameron, E. M. and Durham, C. C. (1975). Further studies of hydrogeochemistry applied to mineral exploration in the northern Canadian Shield. *Geol. Survey Can.* Paper 75-1C, 233–238.

Cameron, E. M. and Hornbrook, E. H. W. (1976). Current approaches to geochemical reconnaissance for uranium in the Canadian Shield. *In* "Exploration for Uranium Ore Deposits", 241–266. International Atomic Energy Agency, Vienna.

Cameron, E. M., Siddeley, G. and Durham, C. C. (1971). Distribution of ore elements in rocks for evaluating ore potential—nickel, copper, cobalt, and sulfur in ultramafic rocks of the Canadian Shield. *Toronto Symposium Volume* 298–313.

Caneer, W. T. and Saum, N. M. (1974). Radon emanometry in uranium exploration. *Mining Eng.* **26**(5), 26–29.

Canney, F. C. (1959). Effect of soil contamination on geochemical prospecting in the Coeur d'Alene district, Idaho. *Mining Eng.* **11,** 205–210.

Canney, F. C. (1967). Hydrous manganese–iron oxide scavenging—its effect on stream sediment surveys. *Geol. Survey Can.* Paper 66-54, 267.

Canney, F. C., ed. (1969). International Geochemical Exploration Symposium. *Colorado School Mines Quart.* **64**(1), 520 pp.

Canney, F. C. and Wing, L. A. (1966). Cobalt—useful but neglected in geochemical prospecting. *Econ. Geol.* **61,** 198–203.

Cannon, H. L. (1952). The effect of uranium–vanadium deposits on the vegetation of the Colorado Plateau. *Am. J. Sci.* **250,** 735–770.

Cannon, H. L. (1955). Geochemical relations of zinc-bearing peat to the Lockport Dolomite, Orleans County, New York. *U.S. Geol. Survey Bull.* **1000-D,** 119–185.

Cannon, H. L. (1957). Description of indicator plants and methods of botanical prospecting for uranium deposits on the Colorado Plateau. *U.S. Geol. Survey Bull.* **1030-M,** 399–516.

Cannon, H. L. (1960a). The development of botanical methods of prospecting for uranium on the Colorado Plateau. *U.S. Geol. Survey Bull.* **1085-A,** 1–50.

Cannon, H. L. (1960b). Botanical prospecting for ore deposits. *Science* **132**(3427), 591–598.

Cannon, H. L. (1963). The biogeochemistry of vanadium. *Soil Sci.* **96,** 196–204.

Cannon, H. L. (1964). Geochemistry of rocks and related soils and vegetation in the Yellow Cat area, Grand County, Utah. *U.S. Geol. Survey Bull.* **1176,** 127 pp.

Cannon, H. L. and Kleinhampl, F. J. (1956). Botanical methods of prospecting for uranium. *In* "Geology of Uranium and Thorium", Proc. 1st Int. Conf. Peaceful Uses Atomic Energy, Geneva, Aug. 8–20, 1955. **6,** 801–805.

Cannon, H. L. and Starrett, W. H. (1956). Botanical prospecting for uranium on La Ventura Mesa, Sandoval County, New Mexico. *U.S. Geol. Survey Bull.* **1009-M,** 391–407.

Cannon, H. L., Harms, T. F. and Hamilton, J. C. (1975). Lithium in unconsolidated sediments and plants of the Basin and Range Province, southern California and Nevada. *U.S. Geol. Survey* Prof. Paper 918, 23 pp.

Cannon, R. S., Pierce, A. P. and Antweiler, J. C. (1971). Suggested uses of lead isotopes in exploration. *Toronto Symposium Volume* 457–463.

Carlisle, D. and Cleveland, G. B. (1958). Plants as a guide to mineralization. *California Div. Mines* Spec. Rept 50, 31 pp.

Carpenter, L. G. and Garrett, D. E. (1959). Tungsten in Seales Lake, *Mining Eng.* **11,** 301–303.

Carpenter, R. H. (1975). Status of exploration geochemistry in U.S. and Canadian universities. *Assoc. Explor. Geochem. Newsl.* **15,** 4–8.

Carpenter, R. H., Pope, T. A. and Smith, R. L. (1975). Fe–Mn coatings in stream sediment surveys. *J. Geochem. Explor.* **4,** 349–364.

Castillo-Munoz, R. and Howarth, R. J. (1966). Application of the empirical discriminant function to regional geochemical data from the United Kingdom. *Geol. Soc. Am. Bull.* **87,** 1567–1581.

Cazalet, P. C. D. (1973). Notes on the interpretation of geochemical data in glaciated areas. *In* "Prospecting in Areas of Glacial Terrain" (M. J. Jones, ed.), 25–29. Institution of Mining and Metallurgy, London.

Chaffee, M. A. (1976a). The zonal distribution of selected elements above the Kalamazoo porphyry copper deposit, San Manuel district, Pinal County, Arizona, *J. Geochem. Explor.* **5,** 145–165.

Chaffee, M. A. (1976b). Geochemical exploration techniques based on distribution of selected elements in rocks, soils, and plants, Mineral Buttes copper deposit, Pinal County, Arizona. *U.S. Geol. Survey Bull.* **1278-D,** 55 pp.

Chaffee, M. A. (1977). Geochemical exploration techniques based on distribution of selected elements in rocks, soils, and plants, Vekol porphyry copper deposit area, Pinal County, Arizona. *U.S. Geol. Survey Bull.* **1278-E,** 78 pp.

Chaffee, M. A. and Gale, C. W., III (1976). The California poppy (*Eschsholtzia mexicana*) as a copper indicator plant—a new example. *J. Geochem. Explor.* **5,** 59–63.

Chamberlin, T. C. (1883). Terminal moraine of the second glacial epoch. *U.S. Geol. Survey 3rd Ann. Rept* 291–402.

Chan, S. S. M. (1969). Suggested guides for exploration from geochemical investigation of ore veins at the Galena Mine deposits, Shoshone County, Idaho. *Colorado School Mines Quart.* **64**(1), 139–168.

Chao, T. T. (1972). Selective dissolution of manganese oxides from soils and sediments with acidified hydroxylamine hydrochloride. *Soil Sci. Soc. Am.* **36,** 764–768.

Chao, T. T. and Anderson, B. J. (1974). The scavenging of silver by manganese and iron oxides in stream sediments collected from two drainage areas of Colorado. *Chem Geol.* **14,** 159–166.

Chao, T. T. and Sanzolone, R. F. (1973). Atomic absorption spectrophotometric determination of microgram levels of Co, Ni, Cu, Pb, and Zn in soil and sediment extract containing large amounts of Mn and Fe. *U.S. Geol. Survey J. Res.* **1,** 681–686.

Chao, T. T. and Sanzolone, R. F. (1977). Chemical dissolution of sulfide minerals. *U.S. Geol. Survey J. Res.* **5,** 409–412.

Chao, T. T. and Theobald, P. K,, Jr (1976). The significance of secondary iron and manganese oxides in geochemical exploration. *Econ. Geol.* **71,** 1560–1569.

Chao, T. T., Ball, J. W. and Nakagawa, H. M. (1971). Determination of silver in soils, sediments, and rocks by organic chelate extraction and atomic absorption spectrophotometry. *Anal. Chim. Acta* **54,** 77–82.

Chao, T. T., Sanzolone, R. F. and Hubert, A. E. (1978). Flame and flameless atomic absorption determination of tellurium in geological materials. *Anal. Chim. Acta* **96,** 251–258.

Chapman, F. W., Marvin, G. G. and Tyreem, S. Y. (1949). Volatilization of elements from perchloric and hydrofluoric acid solutions. *Anal. Chem.* **21,** 700–701.

Chapman, R. P. (1975). Data processing requirements and visual representation for stream sediment exploration geochemical surveys. *J. Geochem. Explor.* **4,** 409–423.

Chapman, R. P. (1976). Limitations of correlation and regression analysis in geochemical exploration. *Inst. Mining Metall. Trans.* **85,** B279–B283.

Chetyrbotskaya, I. I., Bykhovskii, L. Z., Getmanskii, I. I. *et al.* (1967). The tantalum content of wolframite and cassiterite, a criterion for tantalum prospecting (in Russian). English abst. in *Chem. Abst.* **68,** item 97454.

Chirkov, A. M. (1976). Radon as a possible criterion for predicting eruptions, as observed at Karymsky Volcano. *Bull. Volcanol.* **39,** 126–131.

Chisholm, E. O. (1950). A simple chemical method of tracing mineralization through light non-residual overburden. *Can. Inst. Mining Metall. Trans.* **53,** 44–48.

Chisholm, E. O. (1957). Geophysical exploration of a lead–zinc deposit in Yukon Territory. *Proc. 6th Commonwealth Mining Metall. Cong.* 269–277.

Chitayeva, N. A., Miller, A. D., Grosse, Yu. I. *et al.* (1971). Iodine distribution in the supergene zone of the Gay chalcopyrite deposit. *Geochem. Int.* **8,** 426–436.

Chork, C. Y. (1977). Seasonal, sampling, and analytical variations in stream sediment surveys. *J. Geochem. Explor.* **7,** 31–47.

Chowdhury, A. N., Bose, B. B. and Bose, S. K. (1972). Studies of the method of estimating of cold extractable copper in soil. *In* "Proceedings of the Second Seminar on Geochemical Prospecting Methods and Techniques", Mineral Res. Dev. Ser. No. 38, 263–266. United Nations, New York.

Clark, S. P. (1966). "Handbook of Physical Constants." Geol. Soc. America Memoir 97, 11–18.

Clarke, G. R. (1938). "The Study of Soil in the Field", 2nd edn. Clarendon Press, Oxford, 142 pp.

Clarke, W. B. and Kugler, G. (1973). Dissolved helium in groundwater—a possible method for uranium and thorium prospecting. *Econ. Geol.* **68,** 243–251.

Clarke, W. B., Top, Z., Beavan, A. I. *et al.* (1977). Dissolved helium in lakes, uranium prospecting in the Precambrian terrain of central Labrador. *Econ. Geol.* **72,** 233–242.

Clema, J. M. and Stevens-Hoare, N. P. (1973). A method of distinguishing nickel gossans from other ironstones in the Yilgarn Shield, Western Australia. *J. Geochem. Explor.* **2,** 393–402.

Clews, D. R. (1966). Geochemical method. U.S. Patent 3 285 698.

Coats, R. R. (1959). Uranium and certain other trace elements in felsic volcanic rocks of Cenozoic age in western United States. *U.S. Geol. Survey* Prof. Paper 300, 75–78.

Coker, W. B. and Nichol, I. (1975). The relation of lake sediment geochemistry to mineralization in the northwest Ontario region of the Canadian Shield. *Econ. Geol.* **70,** 202–218.

Cole, M. M. (1971). The importance of environment in biogeographical/geobotanical and biogeochemical investigations. *Toronto Symposium Volume* 414–425.

Colwell, R. N. (1963). Basic matter and energy relationships involved in remote reconnaissance. *Photogramm. Eng.* **29,** 761–799.

Comrey, A. L. (1973). "A First Course in Factor Analysis." Academic Press, New York and London, 316 pp.

Connor, J. J. and Shacklette, H. T. (1975). Background geochemistry of some soils, plants, and vegetables in the conterminous United States. *U.S. Geol. Survey* Prof. Paper 574-F, 164 pp.

Coope, J. A. (1958). Studies in geochemical prospecting for nickel in Bechuanaland and Tanganyika. D.I.C. Thesis, Imperial College, London.

Coope, J. A. (1975). Mount Nansen field area, Yukon Territory. *J. Geochem. Explor.* **4,** 89–92.

Coope, J. A. and Webb, J. S. (1963). Copper in stream sediments near disseminated copper mineralization, Cebu, Philippines Republic. *Inst. Mining Metall. Trans.* **72,** 397–406.

Cooper, J. R. and Huff, L. C. (1951). Geological investigations and geochemical prospecting experiment at Johnson, Arizona. *Econ. Geol.* **46,** 731–756.

Corn, R. M. (1975). Alteration-mineralization zoning, Red Mountain, Arizona. *Econ. Geol.* **70,** 1437–1447.

Cornwall, F. W. D. (1969). Discovery and exploration of the Fitula copper deposit, Nchanga area, Zambia. *Proc. 9th Commonwealth Mining Metall. Cong., London* **2,** 535–560.

Cowart, J. B. and Osmond, J. K. (1977). Uranium isotopes in groundwater, their use in prospecting for sandstone-type uranium deposits. *J. Geochem. Explor.* **8,** 365–379.

Cox, M. W. and Hollister, V. F. (1955). The Chollet project, Stevens County, Washington. *Mining Eng.* **7,** 937–940.

Cox, R. (1975). Geochemical soil surveys in exploration for nickel–copper sulfides at Pioneer, near Norseman, Western Australia. *Vancouver Symposium Volume* 437–460.

Cox, R. and Curtis, R. (1977). The discovery of the Lady Loretta zinc–lead–silver deposit, northwest Queensland—a geochemical exploration case history. *J. Geochem. Explor.* **8,** 189–202.

Crain, I. K. (1970). Computer interpolation contouring of two-dimensional data, a review. *Geoexploration* **8,** 71–86.

Cranwell, P. A. (1975). Environmental organic geochemistry of rivers and lakes, both water and sediment. *In* "Environmental Chemistry" (G. Eglinton, ed.), Vol. 1, 22–44. The Chemical Society, London.

Cremer, M. and Schlocker, J. (1976). Lithium borate decomposition of rocks, minerals, and ores. *Am. Mineralogist* **61,** 318–321.

Crenshaw, G. L. and Lakin, H. W. (1974). A sensitive and rapid method for the determination of trace amounts of selenium in geological materials *U.S. Geol. Survey J. Res.* **2,** 483–487.

Cronan, D. S. (1976). Implications of metal dispersion from submarine hydrothermal systems for mineral exploration on mid-ocean ridges and in island arcs. *Nature, Lond.* **262,** 567–569.

Crosby, G. M. (1969). A preliminary examination of trace mercury in rocks, Coeur d'Alene district, Wallace, Idaho. *Colorado School Mines Quart.* **64**(1), 169–194.

Curtin, G. C. and King, H. D. (1972). An auger-sleeve sampler for stony soils. *J. Geochem. Explor.* **1,** 203–206.

Curtin, G. C., King, H. D. and Mosier, E. L. (1974). Movement of elements into the atmosphere from coniferous trees in subalpine forests of Colorado and Idaho. *J. Geochem. Explor.* **3,** 245–263.

Cuturic, N., Kafol, N. and Karamata, S. (1968). Lead contents in K feldspars of young igneous rocks of the Dinarides and neighboring areas. *In* "Origin and Distribution of the Elements" (L. H. Ahrens, ed.), 739–747. Pergamon Press, London.

Czehura, S. J. (1977). A lichen indicator of copper mineralization, Lights Creek district, Plumas Co., California. *Econ. Geol.* **72,** 796–803.

Dadashev, A. M., Guliev, I. S. and Dadashev, F. G. (1974). Results of gas survey of pyrite–polymetallic deposits on the southern slope of the Great Caucasus (in Russian). English abst. in *Econ. Geol.* **71,** 1076.

Dadashev, F. G., Guseinov, R. A. and Zykov, Yu. S. (1971). Gas surveying in sulfide deposits (in Russian). English abst. in *Chem. Abst.* **76,** item 5615.

Dahlberg, E. C. (1975). Relative effectiveness of geologists and computers in mapping potential hydrocarbon exploration targets. *J. Int. Assoc. Math. Geol.* **7,** 373–394.

Dall'Aglio, M. (1971). A study of the circulation of uranium in the supergene environment in the Italian Alpine Range. *Geochim. Cosmochim. Acta* **35,** 47–60.

Dall'Aglio, M. and Tonani, F. (1973). Hydrogeochemical exploration for sulphide deposits—correlation between sulphate and other constituents. *London Symposium Volume* 305–314.

Danielson, A., Lundgren, F. and Sundqvist, G. (1959). The tape machine I. *Spectrochim. Acta* **15,** 122–125.

Danilova, T. R. (1968). Geology and geochemistry of natural gases in Talnakh deposit of copper–nickel ore. *Int. Geol. Rev.* **10,** 644–647.

Davenport, P. H. and Nichol, I. (1973). Bedrock geochemistry as a guide to areas of base metal potential in volcano-sedimentary belts of the Canadian Shield. *London Symposium Volume* 45–57.

David, M. and Dagbert, M. (1975). Lakeview revisited: variograms and correspondence analysis—new tools for the understanding of geochemical data. *Vancouver Symposium Volume* 163–181.

Davis, J. D. and Guilbert, J. M. (1973). Distribution of the radioelements potassium, uranium, and thorium in selected porphyry copper deposits. *Econ. Geol.* **68,** 145–160.

Davis, S. N. and DeWeist, R. J. M. (1966). "Hydrogeology." John Wiley, New York, 463 pp.

Dean, J. A., ed. (1973). "Lange's Handbook of Chemistry", 11th edn. McGraw-Hill, New York.

Dean, J. A. and Rains, T. C., eds (1975). "Flame Emission and Atomic Absorption Spectrometry", Vol. 3. Marcel Dekker, New York, 674 pp.

Dean, W. E. (1974). Determination of carbonate and organic matter in calcareous sediments and sedimentary rocks by loss on ignition: comparison with other methods. *J. Sed. Petrol.* **44,** 242–248.

Dean, W. E. and Gorham, E. (1976). Major chemical and mineral components of profundal surface sediments in Minnesota lakes. *Limnol. Oceanogr.* **21,** 259–284.

Debnam, A. H. and Webb, J. S. (1960). Some geochemical anomalies in soil and stream sediment related to beryl pegmatites in Rhodesia and Uganda. *Inst. Mining Metall. Trans.* **69**(7), 329–344.

Degens, E. T. and Ross, D. A. (1969). "Hot Brines and Recent Heavy Metal Deposits in the Red Sea." Springer-Verlag, New York, 600 pp.

DeGeoffroy, J., Wu, S. M. and Heins, R. W. (1967). Geochemical coverage by spring sampling method in the southwest Wisconsin zinc area. *Econ. Geol.* **62,** 679–697.

DeGeoffroy, J., Wu, S. M. and Heins, R. W. (1968). Selection of drilling targets from geochemical data in the southwest Wisconsin zinc area. *Econ. Geol.* **63,** 787–795.

De Grys, A. M. (1959). Factors affecting the secondary geochemical dispersion of metals associated with sulfide mineralization. D.I.C. Thesis, Imperial College, London.

De Grys, A. M. (1962). Seasonal variations in copper content of some Andean streams of central Chile. *Econ. Geol.* **57,** 1031–1044.

Denson, M. E., Jr (1956). Geophysical–geochemical prospectecting for uranium. *In* "Geology of Uranium and Thorium", Proc. 1st Int. Conf. Peaceful Uses Atomic Energy, Geneva, Aug. 8–20, 1955, **6,** 772–781.

Denson, N. M., Zeller, H. D. and Stephens, J. G. (1956). Water sampling as a guide in the search for uranium deposits and its use in evaluating widespread volcanic units as potential source beds for uranium. *U.S. Geol. Survey* Prof. Paper 300, 673–680.

Devine, S. B. and Sears, H. W. (1978). Soil hydrocarbon geochemistry, a potential petroleum exploration tool in the Cooper Basin, Australia. *J. Geochem. Explor.* **8,** 394–414.

Dikun, A. V., Korobeynik, V. M. and Yanitskiy, I. N. (1976). Some features of the development of helium surveying. *Int. Geol. Rev.* **18,** 98–100.

Dixon, W. J. and Massey, F. J. (1951). "Introduction to Statistical Analysis." McGraw-Hill, New York, 370 pp.

Doe, B. R. and Stacey, J. S. (1974). The application of lead isotopes to the problems of ore genesis and ore prospect evaluation: a review. *Econ. Geol.* **69,** 757–776.

Dolukhanova, N. I. (1957). The application of hydrochemical surveys to copper and molybdenum deposits in the Armenian S.S.R. *Int. Geol. Rev.* **2,** 20–42.

Donovan, P. R. and James, G. H. (1967). Geochemical dispersion in glacial overburden over the Tynagh (Northgate) base metal deposit, west-central Eire. *Geol. Survey Can.* Paper 66-54, 89–110.

d'Orey, F. L. C. (1975). Contribution of termite mounds to locating hidden copper deposits. *Inst. Mining Metall. Trans.* **84,** 150–153.

Dorn, P. (1937). Plants as indicators of ore deposits (in German). *Der Biologe, Munich* **6,** 11–13.

Dorrzapf, A. F. and Brown, F. W. (1970). Direct spectrographic analysis for platinum, palladium, and rhodium in gold beads from fire assay. *Appl. Spectrosc.* **24,** 415–418.

Doyle, P. J. and Fletcher, L. (1975). MacMillan Pass region, Yukon Territory. *J. Geochem. Explor.* **4,** 83.

Draper, N. R. and Smith, H. (1966). "Applied Regression Analysis." John Wiley, New York, 407 pp.

Dregne, H. E. (1976). "Soils of Arid Regions." Elsevier, Amsterdam, 237 pp.

Dreimanis, A. (1960). Geochemical prospecting for Cu, Pb, and Zn in glaciated areas, eastern Canada. *Proc. 21st Int. Geol. Cong., Copenhagen* Pt II, 7–19.

Dreimanis, A. and Vagners, U. J. (1971). Bimodal distribution of rock and mineral fragments in basal till. *In* "Till, a Symposium" (R. P. Goldthwaite, ed.), 237–250. Ohio State University Press, Columbus.

Drewes, H. (1967). A geochemical anomaly of base metals and silver in the southern Santa Rita Mts, Santa Cruz County, Ariz. *U.S. Geol. Survey* Prof. Paper 575-D, 176–182.

Drewes, H. (1973). Geochemical reconnaissance of the Santa Rita Mountains, southeast of Tucson, Arizona. *U.S. Geol. Survey Bull.* **1365,** 67 pp.

Dudal, R. (1968). "Definitions of Soil Units for the Soil Map of the World", F.A.O., World Soil Resources Rept 33, United Nations, New York.

Dudnik, E. E. (1971). "SYMAP Users Manual for Synagraphic Computer Mapping." Dept of Architecture, University of Illinois, Chicago, 114 pp.

Durst, R. A., ed. (1969). "Ion-selective Electrodes", National Bur. Standards Spec. Pub. 314, 452 pp.

Duvigneaud, P. (1958). The vegetation of Katanga and of its metalliferous soils (in French). *Bull. Soc. Roy. Bot. Belg.* **90,** 127–286.

Dyck, W. (1968a). Radon determination for geochemical prospecting for uranium. *Geol. Survey Can.* Paper 68-21, 30 pp.

Dyck, W. (1968b). Radon-222 emanations from a uranium deposit. *Econ. Geol.* **63,** 288–289.

Dyck, W. (1971). The adsorption and coprecipitation of silver on hydrous oxides of iron and manganese. *Geol. Survey Can.* Paper 70-64, 23 pp.

Dyck, W. (1973). An ionization chamber for continuous monitoring of atmospheric radon-222 levels in Ottawa and Gatineau Hills, Canada. *Geol. Survey Can.* Paper 73-28, 11 pp.

Dyck, W. (1974). Gases and their relevance to mineral exploration. *Geol. Survey Can.* Paper 74-1A, 61; Paper 74-1B, 57–59.

Dyck, W. (1975). Geochemistry applied to uranium exploration. *Geol. Survey Can.* Paper 75-26, 33–47.

Dyck, W. (1976). The use of helium in mineral exploration. *J. Geochem. Explor.* **5,** 3–20.

Dyck, W. and Cameron, E. M. (1975). Surface lake water uranium–radon survey of the Lineament Lake area, District of Mackenzie. *Geol. Survey Can.* Paper 75-1A, 209–212.

Dyck, W. and Meilleur, G. A. (1972). A soil gas sampler for difficult overburden. *J. Geochem. Explor.* **1,** 199–202.

Dyck, W. and Smith, A. Y. (1969). The use of radon-222 in surface waters in geochemical prospecting for uranium. *Colorado School Mines Quart.* **64**(1), 223–235.

Dyck, W., Dass, A. S., Durham, C. C. *et al.* (1971). Comparison of regional geochemical uranium exploration methods in the Beaverlodge area, Saskatchewan. *Toronto Symposium Volume* 132–150.

Dyck, W., Pelchat, J. C. and Meilleur, G. A. (1976a). Equipment and procedures for the collection and determination of dissolved gases in natural waters. *Geol. Survey Can.* Paper 75-34, 12 pp.

Dyck, W., Chatterjee, A. K., Gemmell, D. A. *et al.* (1976b). Well water trace element reconnaissance, eastern Maritime Canada. *J. Geochem. Explor.* **6,** 139–162.

Dyck, W., Jonasson, I. R. and Liard, R. F. (1976c). Uranium prospecting with ^{222}Rn in frozen terrain. *J. Geochem. Explor.* **5,** 115–128.

Dyck, W., Whittaker, S. J. and Campbell, R. A. (1976d). Well-water uranium reconnaissance, southwestern Saskatchewan. *Geol. Survey Can.* Paper 76-1C, 249–253.

Edwards, A. B. and Carlos, G. C. (1954). The selenium content of some Australian sulfide deposits. *Australas. Inst. Mining Metall. Proc.* **172,** 31–64.

Efroymson, M. A. (1965). Multiple regression analysis. *In* "Mathematical Methods for Digital Computers" (A. Ralston and H. S. Wilf, eds), 188–203. John Wiley, New York.

Ekdahl, E. (1976). Pielavesi—the use of dogs in prospecting. *J. Geochem. Explor.* **5,** 296–298.

Elinson, M. M. (1972). Gaseous surveys in prospecting for sulphide deposits (abst.). Handbook for 4th Int. Geochem. Explor. Symposium, London, 21.

Elinson, M. M., Pashkov, Yu. N., Agababov, G. M. *et al.* (1970). Gas haloes around copper–molybdenum ore bodies (in Russian). English translation available from the Geological Survey of Canada. English abst. in *Chem. Abst.* **74,** item 5470.

Elliott, I. L. and Fletcher, W. K., eds (1975). "Geochemical Exploration 1974." Elsevier, Amsterdam, 720 pp.

Ellis, A. J., Tooms, J. S., Webb, J. S. *et al.* (1967). Application of solution experiments in geochemical prospecting. *Inst. Mining Metall. Trans.* **76,** B25–B39, B216–B217; **77,** B136.

Ellis, M. W. and McGregor, J. A. (1967). The Kalengwa copper deposit in northwestern Zambia. *Econ. Geol.* **62,** 781–797.

Emmons, W. H. (1917). The enrichment of ore deposits. *U.S. Geol. Survey Bull.* **625,** 68–70.

Engel, A. E. J. and Engel, C. G. (1956). Distribution of copper, lead, and zinc in hydrothermal dolomites associated with sulfide ore in the Leadville limestone (abst.). *Geol. Soc. Am. Bull.* **67,** 1692.

Engels J. C. and Ingamells, C. O. (1970). Effect of sample inhomogeneity in K–Ar dating. *Geochim. Cosmochim. Acta* **34,** 1007–1018.

Eremeev, A. N., Sokolov, V. A., Solovov, A. P. *et al.* (1973). Application of helium surveying to structural mapping and ore deposit forecasting. *London Symposium Volume* 183–192.

Erickson, R. L. and Marranzino, A. P. (1960). Geochemical prospecting for copper in the Rocky Range, Beaver County, Utah. *U.S. Geol. Survey* Prof. Paper 400-B, 98–101.

Erickson, R. L. and Marsh, S. P. (1974). Geochemical aeromagnetic and generalized geologic maps showing distribution and abundance of Au, Ag, Cu, Mo, Sb, As, Hg, Pb, W, and Bi, Brooks Spring quadrangle, Humboldt County, Nevada. *U.S. Geol. Survey* Misc. Field Studies Maps MF-563–MF-567.

Erickson, R. L., Masursky, H., Marranzino, A. P. *et al.* (1961). Geochemical anomalies in the upper plate of the Roberts thrust near Cortez, Nevada. *U.S. Geol. Survey* Prof. Paper 424-D, 316–320.

Erickson, R. L., Masursky, H., Marranzino, A. P. *et al.* (1964a). Geochemical anomalies in the lower plate of the Roberts thrust near Cortez, Nevada. *U.S. Geol. Survey* Prof. Paper 501-B, 92–94.

Erickson, R. L., Marranzino, A. P., Oda, U. *et al.* (1964b). Geochemical exploration near the Getchell Mine, Humboldt Co., Nevada. *U.S. Geol. Survey Bull.* **1198-A,** 26 pp.

Erickson, R. L., Marranzino, A. P., Oda, U. *et al.* (1966a). Geochemical reconnaissance in the Pequop Mountains and Wood Hills, Elko County, Nevada. *U.S. Geol. Survey Bull.* **1198-E,** 20 pp.

Erickson, R. L., VanSickle, G. H., Nakagawa, H. M. *et al.* (1966b). Gold geochemical anomaly in the Cortez district, Nevada. *U.S. Geol. Survey* Circ. 534, 9 pp.

Erikson, J. E. (1957). Geochemical prospecting abstracts, July 1952 to December 1954. *U.S. Geol. Survey Bull.* **1000-G**, 395 pp.

Eriksson, K. (1973). Prospecting in an area of central Sweden. *In* "Prospecting in Areas of Glacial Terrain" (M. J. Jones, ed.), 83–86. Institution of Mining and Metallurgy, London.

Eriksson, K. (1976). Regional prospecting by the use of peat sampling. *J. Geochem. Explor.* **5,** 387–388.

Ermengen, S. V. (1957). Geochemical prospecting in Chibougamau. *Can. Mining J.* **78**(4), 99–104.

Evans, H. J., Purvis, E. R. and Bear, F. E. (1951). Effect of soil reaction on availability of molybdenum. *Soil Sci.* **71,** 117–124.

Everett, K. (1967). Handling perchloric acid and perchlorates. *In* "Handbook of Laboratory Safety" (N. V. Steere, ed.), 205–216. Chemical Rubber Co., Cleveland.

Ewers, G. R. and Keays, R. R. (1977). Volatile and precious metal zoning in the Broadlands geothermal field, New Zealand. *Econ. Geol.* **72,** 1337–1354.

Fairbridge, R. W., ed. (1967). "Encyclopedia of Atmospheric Sciences and Astrogeology", 67. Reinhold, New York.

F.A.O.–Unesco (1974). "F.A.O.–Unesco Soil Map of the World, 1:5 000 000, VI." Legend, Paris.

Ferguson, R. B. and Price, V., Jr (1976). National uranium resource evaluation (NURE) program—hydrogeochemical and stream sediment reconnaissance in the eastern United States. *J. Geochem. Explor.* **6,** 103–117.

Ficklin, W. H. (1970). A rapid method for the determination of fluoride in rocks and soils using an ion-sensitive electrode. *U.S. Geol. Survey* Prof. Paper 700-C, 186–188.

Ficklin, W. H. (1975). Ion-selective electrode determination of iodine in rocks and soils. *U.S. Geol. Survey J. Res.* **3,** 753–755.

Ficklin, W. H. and Ward, F. N. (1976). Flameless atomic absorption determination of bismuth in soils and rocks. *U.S. Geol. Survey J. Res.* **4,** 217–220.

Field, C. W., Dymond, J. R., Corless, J. B. *et al.* (1976). "Metallogenesis in southeast Pacific Ocean: Nazca Plate Project. In Circum-Pacific Energy and Mineral Resources", American Assoc. Petroleum Geologists Memoir 25, 539–549.

Fisher, F. S. (1972). Tertiary mineralization and hydrothermal alteration in the Stinkingwater mining region, Park Co., Wyo. *U.S. Geol. Survey Bull.* **1332-C,**

Fitch, F. H. and Webb, J. S. (1958). Contribution on geochemical aspects of the Sandakan Area. *In* "Copper Deposits of the Sandakan Area" (by F. H. Fitch), 125–152. Geol. Survey Dept, British Territories in Borneo, Memoir 9.

Fix, C. E. (1958). Selected annotated bibliography of the geology and occurrence of uranium-bearing marine black shales in the United States. *U.S. Geol. Survey Bull.* **1059-F,** 263–325.

Fix, P. F. (1956). Hydrogeochemical exploration for uranium. *In* "Geology of Uranium and Thorium", Proc. 1st Int. Conf. Peaceful Uses Atomic Energy, Geneva, Aug. 8–20, 1955, **6,** 788–791.

Flanagan, F. J. (1973). 1972 values for international reference samples. *Geochim. Cosmochim. Acta*, **37,** 1189–1200.

Flanagan, F. J. (1974). Reference samples for the earth sciences. *Geochim. Cosmochim. Acta* **38,** 1731–1744.

Flanagan, F. J. (1976). Descriptions and analyses of eight new U.S.G.S. rock standards. *U.S. Geol. Survey* Prof. Paper 840, 192 pp.

Fleischer, M. W. (1954). The abundance and distribution of the chemical elements in the earth's crust. *J. Chem. Educ.* **31,** 446–455.

Fleischer, M. W. (1955). Minor elements in some sulfide minerals. *In* "50th Anniversary Volume" (A. M. Bateman, ed.), 970–1024. Economic Geology Pub. Co., Lancaster, Pennsylvania.

Fleming, H. W. (1961). The Murray deposit, Restigouche County, N.B.: a geochemical–geophysical discovery. *Can. Inst. Mining Metall. Bull.* **54,** 230–235.

Flint, R. F. (1957). "Glacial and Pleistocene Geology." John Wiley, New York, 553 pp.

Flint, R. F. (1971). "Glacial and Quaternary Geology." John Wiley, New York, 892 pp.

Flinter, B. H., Hesp, W. R. and Rigby, D. (1972). Selected geochemical, mineralogical, and petrological features of granitoids of the New England complex, Australia, and their relation to Sn, W, Mo, and Cu mineralization. *Econ. Geol.* **67,** 1241–1262.

Forbes, E. A., Posner, A. M. and Quirk, J. P. (1976). The specific adsorption of divalent Cd, Co, Cu, Pb, and Zn on goethite. *J. Soil Sci.* **27,** 154–166.

Forrester, J. D. (1942). A native copper deposit near Jefferson City, Montana. *Econ. Geol.* **37,** 126–135.

Foster, J. R. (1971). The reduction of matrix effects in atomic absorption analysis and the efficiency of selected extractions on rock-forming minerals. *Toronto Symposium Volume* 554–560.

Foster, J. R. (1973). The efficiency of various digetion procedures on the extraction of metals from rocks and rock-forming minerals. *Can. Inst. Mining Metall. Bull.* **66**(736), 85–92.

Foy, M. F. and Gingrich, J. E. (1977). A stream sediment orientation programme for uranium in the Alligator River province, Northern Territory, Australia. *J. Geochem. Explor.* **8,** 357–364.

Franklin, J. M., Kasarda, J. and Poulsen, K. H. (1975). Petrology and chemistry of the alteration zone of the Mattabi massive sulfide deposit. *Econ. Geol.* **70,** 63–79.

Frederickson, A. F., Lehnertz, C. A. and Kellogg, H. E. (1971). Mobility, flexibility highlight a mass spectrometer–computer technique for regional exploration. *Eng. Mining J.* **172**(6), 116–118.

Fridman, A. I. (1974). Gas mapping during ore prospecting and geologic mapping in closed regions (in Russian). English abst. in *Chem. Abst.* **82,** item 114205.

Fridman, A. I. and Makhlova, N. K. (1972). Certain aspects of the origin and migration of CO_2 in Hg deposits. *Int. Geol. Rev.* **14,** 1345–1350.

Fridman, A. I. and Petrov, V. A. (1976). Principal results of gas surveying in ore deposits under permafrost conditions. *Int. Geol. Rev.* **18,** 545–550.

Friedman, I. and Denton, E. H. (1976). A portable helium sniffer. *U.S. Geol. Survey J. Res.* **4,** 35–40.

Friedrich, G. H. (1970). Dispersion of mercury in soils in the region of some fluorspar deposits in Nabburg–Woelsendorf (in German). *Erzmetall.* **23,** 482–486.

Friedrich, G. H. and Hawkes, H. E. (1966). Mercury as an ore guide in the Pachuca–Real del Monte district, Hidalgo, Mexico. *Econ. Geol.* **61,** 744–753.

Friedrich, G. H. and Plüger, W. L. (1971). Geochemical prospecting for barite and fluorite deposits. *Toronto Symposium Volume* 151–156.

Friedrich, G. H., Plüger, W. L., Hilmer, E. F. *et al.* (1973). Flameless atomic absorption and ion-sensitive electrodes as analytical tools in copper exploration. *London Symposium Volume* 435–443.

Friese, F. W. (1931). Study of the abrasion of minerals during transport in water (in German). *Mineralog. Petrolog. Mitt.* **41,** 1–7.

Fuchs, W. A. and Rose, A. W. (1974). The geochemical behavior of platinum and palladium in the weathering cycle in the Stillwater Complex, Montana. *Econ. Geol.* **69,** 332–346.

Fulton, R. B. (1950). Prospecting for zinc using semiquantitative analyses of soils. *Econ. Geol.* **45,** 654–670.

Fursov, V. Z. (1970). Mercurial atmosphere from mercury deposits. *Am. Geol. Inst. Dokl. Acad. Sci. U.S.S.R.* **194,** 209–211.

Fursov, V. Z. (1973). Mercury in rocks and ores and its sublimation temperatures. *Am. Geol. Inst. Dokl. Acad. Sci. U.S.S.R.* **204,** 184–187.

Fursov, V. Z., Vol'fson, B. N. and Khvalovskiy, A. G. (1968). Results of a study of mercury vapor in the Tashkent earthquake zone. *Am. Geol. Inst. Dokl. Acad. Sci. U.S.S.R.* **179,** 208–210.

Fyfe, W. S. and Kerrich, R. (1976). Geochemical prospecting: extensive vs intensive factors. *J. Geochem. Explor.* **6,** 177–192.

Gallagher, M. J. (1970). Portable X-ray spectrometers for rapid ore analysis. *In* "Mining and Petroleum Geology" (M. J. Jones, ed.), *Proc. 9th Commonwealth Mining Metall. Cong.* **2,** 691–730. Institute of Mining and Metallurgy, London.

Gamble, D. S. and Schnitzer, M. (1973). The chemistry of fulvic acid and its reactions with metal ions. *In* "Trace Metals and Metal–Organic Interactions in Natural Waters" (P. C. Singer, ed.), 265–302. Ann Arbor Sci. Pub., Ann Arbor, Michigan.

Garrels, R. M. (1960). "Mineral Equilibria at Low Temperature and Pressure." Harper and Row, New York, 254 pp.

Garrels, R. M. and Christ, C. L. (1965). "Solutions, Minerals and Equilibria." Harper and Row, New York, 450 pp.

Garrels, R. M. and MacKenzie, F. T. (1967). Origin of the chemical compositions of some springs and lakes. *In* "Equilibrium Concepts in Natural Water Systems" (W. Stumm, chairman), 222–242. American Chemical Society, Cleveland.

Garrett, R. G. (1969). The determination of sampling and analytical errors in exploration geochemistry. *Econ. Geol.* **64,** 568–569; **68,** 282–283.

Garrett, R. G. (1971a). Molybdenum, tungsten, and uranium in acid plutonic rocks as a guide to regional exploration. *Can. Mining J.* **92**(4), 37–40.

Garrett, R. G. (1971b). The dispersion of copper and zinc in glacial overburden at the Louvem deposit, Val d'Or, Quebec. *Toronto Symposium Volume* 157–158.

Garrett, R. G. (1973). Regional geochemical study of Cretaceous acidic rocks in the northern Canadian Cordillera as a tool for broad mineral exploration. *London Symposium Volume* 203–219.

Garrett, R. G. (1974a). Mercury in some granitoid rocks of the Yukon and its relation to gold–tungsten mineralization. *J. Geochem. Explor.* **3,** 277–289.

Garrett, R. G. (1974b). Computers in exploration geochemistry. *Geol. Survey Can.* Paper 74-60, 63–69.

Garrett, R. G. (1975). Copper and zinc in Proterozoic acid volcanics as a guide to exploration in the Bear province. *Vancouver Symposium Volume* 371–388.

Garrett, R. G. and Hornbrook, E. H. W. (1976). The relationship between zinc and organic content in centre-lake bottom sediments. *J. Geochem. Explor.* **5,** 31–38.

Garrett, R. G. and Nichol, I. (1967). Regional geochemical reconnaissance in eastern Sierra Leone. *Inst. Mining Metall. Trans.* **76,** B97–B112.

Germanov, A. I., Batulin, S. G., Volkov, G. A. *et al.* (1958). Some regularities of uranium distribution in underground waters. *Proc. 2nd U.N. Conf. Peaceful Uses Atomic Energy* **2,** 161–177.

Giles, D. L. and Schilling, J. H. (1972). Variations in the rhenium content of molybdenite. *Proc. 24th Int. Geol. Cong., Montreal 1972,* **10,** 145–152.

Gilluly, J. (1946). The Ajo mining district. *U.S. Geol. Survey* Prof. Paper 209, 112 pp.

Gingrich, J. E. (1975). Results from a new uranium exploration method. *Soc. Mining Eng. Trans.* **258,** 61–64.

Gingrich, J. E. and Fisher, J. C. (1976). Uranium exploration using the track-etch method. *In* "Exploration for Uranium Ore Deposits", 213–227. International Atomic Energy Agency, Vienna.

Ginzburg, I. I. (1960). "Principles of Geochemical Prospecting." Pergamon Press, New York, 311 pp.

Glebovskaya, V. S. and Glebovskii, S. S. (1960). The possibility of application of gas surveys in prospecting for sulfide deposits (in Russian). English translation announced in *Geol. Survey Can.* Paper 74-58, item 63.

Gleeson, C. F. and Coope, J. A. (1967). Some observations on the distribution of metals in swamps in eastern Canada. *Geol. Survey Can.* Paper 66-54, 145–166.

Gleeson, C. F. and Cormier, R. (1971). Evaluation by geochemistry of geophysical anomalies and geological targets using overburden sampling at depth. *Toronto Symposium Volume* 159–165.

Goldak, G. R. (1973). Helium-4 mass spectrometry in uranium exploration (abst.). *Mining Eng.* **25**(12), 47.

Golden Symposium Volume (1969). "International Geochemical Exploration Symposium", *Colorado School Mines Quart.* **64**(1), 520 pp.

Golden Symposium Volume (1979). "Geochemical Exploration 1978" (J. R. Watterson and P. K. Theobald, eds), 7th Int. Geochem. Explor. Symp., Golden. Assoc. Expl. Geochem., Rexdale, Ontario, 504 pp.

Goldich, S. S. (1938). A study in rock weathering. *J. Geol.* **46,** 17–58.

Goldschmidt, V. M. (1937). The principles of distribution of chemical elements in minerals and rocks. *J. Chem. Soc.* 1937, Pt 1, 655–673.

Goldschmidt, V. M. (1954). "Geochemistry." Clarendon Press, Oxford, 730 pp.

Goldsztein, M. (1957). Geobotanical prospecting for uranium in Esterel (in French). *Soc. Française Minéralog. Cristallographe Bull.* **80,** 318–324.

Goldthwaite, R. P. (1971). Introduction to till, today. *In* "Till, A Symposium" (R. P. Goldthwaite, ed.), 3–26. Ohio State University Press, Columbus.

Goleva, G. A. (1968). "Hydrogeochemical Prospecting of Hidden Deposits" (in Russian). Nedra, Moscow. English translation available from the Geological Survey of Canada.

Goleva, G. A. and Lushnikov, V. V. (1967). Selenium distribution in the subsurface waters of ore deposits and in certain types of mineralized waters. *Geochem. Int.* **4,** 378–385.

Goleva, G. A., Polyakov, V. A. and Nechaeva, T. P. (1970). Distribution and forms of migration of lead in subsurface waters. *Geochem. Int.* **7,** 256–268.

Golubev, V. S. and Beus, A. A. (1970). A theoretical model of the interaction between an ore-bearing solution and the country rocks. *Geochem. Int.* **6,** 836–844.

Golubev, V. S. and Garibyants, A. S. (1971). "Heterogeneous Processes of Geochemical Migration." Consultants Bureau, New York, 150 pp.

Golubev, V. S., Yurmeev, A. N. and Yanitskii, I. N. (1974). Analyses of some models of helium migration in the lithosphere. *Geochem. Int.* **11,** 734–742.

Gong, H. and Suhr, N. H. (1976). The determination of cadmium in geological materials by flameless atomic absorption spectrometry. *Anal. Chim. Acta* **81,** 297–303.

Gong, H., Rose, A. W. and Suhr, N. H. (1977). The geochemistry of cadmium in some sedimentary rocks. *Geochim. Cosmochim. Acta* **41,** 1687–1692.

Goodfellow, W. D. (1975). Major and minor element halos in volcanic rocks at Brunswick No. 12 sulfide deposit, N.B., Canada. *Vancouver Symposium Volume* 279–295.

Goodman, R. J. (1973). Rapid analysis of trace amounts of tin in stream sediments, soils, and rocks by X-ray fluorescence analysis. *Econ. Geol.* **68,** 275–277.

Gorbushina, L. V., Tyminskii, V. G. and Spiridonov, A. I. (1972). Formation mechanism of radiohydrogeological anomalies in seismically active areas and their significance in prognostication of earthquakes. *Int. Geol. Rev.* **15,** 380–383.

Gordiyenko, V. V. (1973). Cesium content of lepidolite as an indicator of the cesium potential of granite pegmatite. *Am. Geol. Inst. Dokl. Acad. Sci. U.S.S.R.* **209,** 191–194.

Gordon, G. E., Randle, K., Goles, G. G. *et al.* (1968). Instrumental neutron activation analysis of standard rocks with high resolution gamma-ray detectors. *Geochim. Cosmochim. Acta* **32,** 369–396.

Gorham, E. (1955). On the acidity and salinity of rain. *Geochim. Cosmochim. Acta* **7,** 231–239.

Gorsuch, T. T. (1959). Radiochemical investigations on the recovery for analysis of trace elements in organic and biological materials. *Analyst* **84,** 135–173.

Gosling, A. W., Jenne, E. A. and Chao, T. T. (1971). Gold content of natural waters in Colorado. *Econ. Geol.* **66,** 309–313.

Gott, G. B. and Botbol, J. M. (1973). Zoning of major and minor metals in the Coeur d'Alene mining district, Idaho, U.S.A. *London Symposium Volume* 1–12.

Gott, G. B. and McCarthy, J. H. (1966). Distribution of gold, silver, tellurium, and mercury in the Ely mining district, White Pine Co., Nevada. *US. Geol. Survey* Circ. 535, 5 pp.

Gott, G. B., McCarthy, J., Jr, VanSickle, G. H. *et al.* (1969). Distribution of gold and other metals in the Cripple Creek district, Colorado. *U.S. Geol. Survey* Prof. Paper 625-A, 17 pp.

Gottschalk, V. H. and Buehler, H. A. (1912). Oxidation of sulphides. *Econ. Geol.* **7,** 15–34.

Govett, G. J. S. (1958). Geochemical prospecting for copper in Northern Rhodesia. Ph.D. Thesis, Imperial College, London.

Govett, G. J. S. (1960). Geochemical prospecting for copper in Northern Rhodesia. *Proc. 21st Int. Geol. Cong., Copenhagen 1960* Pt II, 44–56.

Govett, G. J. S. (1961). Seasonal variation in the copper concentration in drainage systems in Northern Rhodesia. *Inst. Mining Metall. Trans.* **70,** 177–189.

Govett, G. J. S. (1972). Interpretation of a rock geochemical exploration survey in Cyprus—statistical and graphical techniques. *J. Geochem. Explor.* **1,** 77–102.

Govett, G. J. S. (1973a). Differential secondary dispersion in transported soils and post-mineralization rocks: an electrochemical interpretation. *London Symposium Volume* 81–91.

Govett, G. J. S. (1973b). Geochemical exploration studies in glaciated terrain, New Brunswick, Canada. *In* "Prospecting in Areas of Glacial Terrain" (M. J. Jones, ed.), 11–24. Institution of Mining and Metallurgy, London.

Govett, G. J. S. and Chork, C. Y. (1977). Detection of deeply buried sulfide deposits by measurement of organic carbon, hydrogen ion, and conductance in surface soils. *In* "Prospecting in Areas of Glaciated Terrain, 1977", 49–55. Institution of Mining and Metallurgy, London.

Govett, G. J. S. and Galanos, D. A. (1974). Drainage and soil surveys in Greece—the use of standardized data as an interpretative procedure. *Inst. Mining Metall. Trans.* **83,** B99–B111.

Govett, G. J. S., Goodfellow, W. D., Chapman, R. P. *et al.* (1975). Exploration geochemistry: distribution of elements and recognition of anomalies. *J. Int. Assoc. Math. Geol.* **7,** 415–446.

Graedel, T. E. (1977). The homogeneous chemistry of atmospheric sulfur. *Rev. Geophys. Space Phys.* **15,** 421–428.

Graham, G. S., Kesler, S. E. and Van Loon, J. C. (1975). Fluorine in ground water as a guide to Pb–Zn–Ba–F mineralization. *Econ. Geol.* **70,** 396–398.

Granier, C. (1958). Dispersion of tungsten and arsenic in residual soil (in French). *Soc. Française Minéralog. Cristallographie Bull.* **81,** 194–200.

Granier, C. (1973). "Introduction to Geochemical Prospecting for Mineral Deposits" (in French). Masson et Cie, Paris, 143 pp.

Graybeal, F. T. (1973). Copper, manganese, and zinc in coexisting mafic minerals from Laramide intrusive rocks in Arizona. *Econ. Geol.* **68,** 785–798.

Green, J. (1959). Geochemical table of the elements for 1959. *Geol. Soc. Am. Bull.* **70,** 1127–1184.

Greenland, L. P. and Campbell, E. Y. (1974). Spectrophotometric determination of niobium in rocks. *U.S. Geol. Survey J. Res.* **2,** 353–355.

Gregory, P. and Tooms, J. S. (1969). Geochemical prospecting for kimberlites. *Colorado School Mines Quart.* **64**(1), 265–306.

Griffith, S. V. (1960). "Alluvial Prospecting and Mining." Pergamon Press, New York, 245 pp.

Griggs, D. (1936). The factor of fatigue in rock exfoliation. *J. Geol.* **44,** 783–796.

Grigorian, S. V. (1974). Primary geochemical halos in prospecting and exploration of hydrothermal deposits. *Int. Geol. Rev.* **16,** 12–25.

Grim, R. E. (1953). "Clay Mineralogy." McGraw-Hill, New York, 396 pp.

Grimbert, A. (1963). The application of geochemical prospecting for uranium in forested zones in the tropics. *In* "Proceedings of the Seminar on Geochemical Prospecting Methods and Techniques", 81–94. Mineral Res. Dev. Ser. No. 21, United Nations, New York.

Grimes, D. J. and Earhart, R. L. (1975). Geochemical soil studies in the Cotter Basin area, Lewis and Clark County, Montana. *U.S. Geol. Survey* Open-file Rept 75-72, 25 pp.

Grip, E. (1953). Tracing of glacial boulders as an aid to ore prospecting in Sweden. *Econ. Geol.* **48,** 715–725.

Groves, D. I. (1972). The geochemical evolution of tin-bearing granites in the Blue Tier batholith, Tasmania. *Econ. Geol.* **67,** 445–457.

Guha, M. (1961). A study of the trace element uptake of deciduous trees. Ph.D. Thesis, University of Aberdeen, as quoted by Brooks (1972).

Gulson, B. L. (1977). Application of lead isotopes and trace elements to mapping black shales around a base metal sulfide deposit. *J. Geochem. Explor.* **8,** 85–104.

Gunter, B. D. and Musgrave, B. C. (1971). New evidence on the origin of methane in hydrothermal gases. *Geochim. Cosmochim. Acta* **35,** 113–118.

Gunton, J. E. and Nichol, I. (1975). Chemical zoning associated with the Ingerbelle–Copper Mountain mineralization, Princeton, British Columbia. *Vancouver Symposium Volume* 297–312.

Gustavson, J. B. (1976). Use of mercury in geochemical exploration for Mississippi Valley type of deposit in Tennessee. *J. Geochem. Explor.* **6,** 251–277.

Gustavson, J. B. and Neathery, T. L. (1976). Geochemical prospecting for gold in Alabama. *Soc. Mining Eng. Trans.* **260,** 177–184.

Gustavsson, N. (1976). Automatic data processing of regional geochemical data at the Geological Survey of Finland. *J. Geochem. Explor.* **5,** 389–393.

Gy, M. P. (1966). Sampling calculator (slide rule). *Revue de l'Industrie Minérale* **48,** 463–468.

Hamil, B. M. and Nackowski, M. P. (1971). Trace element distribution in accessory magnetite from quartz monzonite intrusives and its relation to sulfide mineralization in the Basin and Range province of Utah and Nevada—a preliminary report. *Toronto Symposium Volume* 331–333.

Hammett, F. S. (1928). Studies in the biology of metals. I. The localization of lead by growing roots. *Protoplasma* **4,** 183–186.

Harbaugh, J. W. (1953). Geochemical prospecting abstracts through June, 1952. *U.S. Geol. Survey Bull.* **1000-A,** 1–50.

Harden, G. and Tooms, J. S. (1964). Efficiency of the potassium bisulfate fusion in geochemical analyses. *Inst. Mining Metáll. Trans.* **74,** 129–141.

Hardon, H. J. (1936). Podzol profile in the tropics. *Natuurkundig Tijdschr.* **96,** 25–41.

Harman, H. H. (1967). "Modern Factor Analysis", 2nd edn. University of Chicago Press, Chicago, 474 pp.

Harms, T. F. and Ward, F. N. (1975). Determination of arsenic in vegetation. *U.S. Geol. Survey Bull.* **1408,** 13–20.

Hausen, D. M. and Kerr, P. F. (1971). X-ray diffraction methods of evaluating potassium silicate alteration in porphyry mineralization. *Toronto Symposium Volume* 334–340.

Hausen, D. M., Ahlrichs, J. W. and Odekirk, J. R. (1973). Application of sulfur and nickel analyses to geochemical prospecting. *London Symposium Volume* 13–24.

Hawkes, H. E. (1954). Geochemical prospecting investigations in the Nyeba lead–zinc district, Nigeria. *U.S. Geol. Survey Bull.* **1000-B,** 51–103.

Hawkes, H. E. (1957). Principles of geochemical prospecting. *U.S. Geol. Survey Bull.* **1000-F,** 225–355.

Hawkes, H. E. (1959). Geochemical prospecting. *In* "Researches in Geochemistry" (P. H. Abelson, ed.), 62–78. John Wiley, New York.

Hawkes, H. E. (1963). Dithizone field tests. *Econ. Geol.* **58,** 579–586.

Hawkes, H. E. (1972a). "Exploration Geochemistry Bibliography, 1965–1971", Spec. Vol. No. 1, Assoc. Expl. Geochem., Toronto, 118 pp.

Hawkes, H. E. (1972b). Free hydrogen in genesis of petroleum. *Am. Assoc. Pet. Geol. Bull.* **56,** 2268–2270.

Hawkes, H. E. (1976a). "Exploration Geochemistry Bibliography, 1972–1975", Spec. Vol. No. 5, Assoc. Expl. Geochem., Toronto, 195 pp.

Hawkes, H. E. (1976b). The early days of exploration geochemistry. *J. Geochem. Explor.* **6,** 1–13.

Hawkes, H. E. (1976c). The downstream dilution of stream sediment anomalies. *J. Geochem. Explor.* **6,** 345–358.

Hawkes, H. E. (1978). "Exploration Geochemistry Bibliography, 1976–1977", Assoc. Explor. Geochem., Toronto, 63 pp.

Hawkes, H. E. and Bloom, H. (1956). Heavy metals in stream sediment used as exploration guides. *Mining Eng.* **8,** 1121–1126.

Hawkes, H. E. and Lakin, H. W. (1949). Vestigial zinc in surface residuum associated with primary zinc ore in east Tennessee. *Econ. Geol.* **44,** 286–295.

Hawkes, H. E. and Salmon, M. L. (1960). Trace elements in organic soil as a guide to copper ore. *Proc.* **21***st Int. Geol. Cong., Copenhagen 1960* **2,** 38–43.

Hawkes, H. E. and Webb, J. S. (1962). "Geochemistry in Mineral Exploration", 1st edn. Harper and Row, New York, 415 pp.

Hawkes, H. E. and Williston, S. H. (1962). Mercury vapor as a guide to lead–zinc–silver deposits. *Mining Cong. J.* **48**(12), 30–32.

Hawkes, H. E., Bloom, H., Riddell, J. E. *et al.* (1960). Geochemical reconnaissance in eastern Canada. *In* "Symposium de Exploración Geoquímica", *Proc. 20th Geol. Cong., Mexico City 1956* **3,** 607–621.

Hawley, C. C. and Griffitts, W. R. (1968). Distribution of beryllium, tin, and tungsten in the Lake George area, Colorado. *U.S. Geol. Survey* Circ. 597, 18 pp.

Heinrich, E. W. (1962). Geochemical prospecting for beryl and columbite. *Econ. Geol.* **57,** 616–619.

Helgeson, H. C. (1969). Thermodynamics of hydrothermal systems at elevated temperatures and pressures. *Am. J. Sci.* **267,** 729–804.

Hem, J. D. (1959). Study and interpretation of the chemical characteristics of natural water. *U.S. Geol. Survey* Water-supply Paper 1473, 269 pp.

Hem, J. D. (1960). Complexes of ferrous iron with tannic acid. *U.S. Geol. Survey* Water-supply Paper 1459-D, 75–94.

Hem, J. D. (1970). Study and interpretation of the chemical characteristics of natural water. *U.S. Geol. Survey* Water-supply Paper 1473, 361 pp.

Hem, J. D. (1977). Reactions of metal ions at surfaces of hydrous iron oxide. *Geochim. Cosmochim. Acta* **41,** 527–538.

Hemphill, W. R., Watson, R. D., Bigelow, R. D. *et al.* (1977). Measurement of luminescence of geochemically stressed trees and other materials. *U.S. Geol. Survey* Prof. Paper 1015, 93–112.

Hesp, W. R. (1971). Correlations between the tin content of granitic rocks and their chemical and mineralogical composition. *Toronto Symposium Volume* 341–353.

Heydemann, A. (1959). Adsorption from very weak copper solutions on pure clay minerals (in German). *Geochim. Cosmochim. Acta* **15,** 305–329.

Hill, P. A. and Parker, A. (1970). Tin and zirconium in the sediments around the British Isles: a preliminary reconnaissance. *Econ. Geol.* **65,** 409–416.

Hills, F. A. and Delavault, M. H. (1977). Origin of uranium in the middle Precambrian Estes Conglomerate, eastern Black Hills, South Dakota: inferences from lead isotopes. *U.S. Geol. Survey* Circ. 753, 15–17.

Hinkle, M. E. (1978). Helium, mercury, sulfur compounds, and carbon dioxide in soil gases of the Puhimau thermal area, Hawaii Volcanoes National Park, Hawaii. *U.S. Geol. Survey* Open-file Rept 78-246, 15 pp.

Hinkle, M. E. and Kantor, J. E. (1978). Collection and analysis of soil gases emanating from buried sulfide mineralization, Johnson Camp, Cochise County, Arizona. *J. Geochem. Explor.* **9,** 209–216.

Hinkle, M. E. and Turner, R. L. (1976). Determination of soil gases. *U.S. Geol. Survey* Prof. Paper 1000, 19–20.

Hinkle, M. E., Denton, E. H., Bigelow, R. C. *et al.* (1978). Helium in soil gases of the Roosevelt Hot Springs known geothermal resource area, Beaver County, Utah. *U.S. Geol. Survey J. Res.* **6,** 563–569.

Hoag, R. B. and Webber, G. R. (1976a). Hydrogeochemical exploration and sources of anomalous waters. *J. Geochem. Explor.* **5,** 39–57.

Hoag, R. B. and Webber, G. R. (1976b). Significance for mineral exploration of sulfate concentrations in groundwaters. *Can. Inst. Mining Metall. Bull.* **69**(776), 86–91.

Hoagland, A. D. (1962). Distribution of zinc in soils overlying the Flat Gap zinc mine. *Mining Eng.* **14**(10), 56–58.

Hoagland, D. R., Chandler, W. H. and Stout, P. R. (1937). Little-leaf or rosette of fruit trees. VI. Further experiments bearing on the cause of the disease. *Am. Soc. Hort. Sci. Proc.* **34,** 210–212.

Hodgson, W. A. (1972). Optimum spacing for soil sample traverses. *In* "Proc. 10th Int. Symp. on Appl. of Computer Methods in Mineral Industries" (M. Salamon and F. H. Lancaster, eds), 75–78. South African Inst. Mining Metall.

Hoffman, J. I. and Lundell, G. E. F. (1939). Volatilization of metallic compounds from solutions in perchloric or sulfuric acid. *Nat Bur. Standards J. Res.* **22,** 465–470.

Hoffman, S. J. (1977). Talus fine sampling as a regional geochemical exploration technique in mountainous regions. *J. Geochem. Explor.* **7,** 349–360.

Holland, H. D. (1965). Some applications of thermochemical data to problems of ore deposits. II. Mineral assemblages and the composition of ore-forming fluids. *Econ. Geol.* **60,** 1101–1166.

Holland, H. D. (1972). Granites, solutions, and base metal deposits. *Econ. Geol.* **67,** 281–301.

Holman, R. H. C. (1959). Lead in stream sediments, northern mainland of Nova Scotia (map). *Geol. Survey Can.* Map 26-1959, Sheet 2.

Holman, R. H. C. and Webb, J. S. (1957). Exploratory geochemical soil survey at Ruhiza ferberite mine, Uganda. *In* "Methods and Case Histories in Mining Geophysics" (S. F. Kelly and E. W. Westrick, chairmen), 353–357. 6th Commonwealth Mining Metall. Cong., Montreal.

Holmes, R. and Tooms, J. S. (1973). Dispersion from a submarine exhalative orebody. *London Symposium Volume* 193–202.

Horizon (1959). A flower that led to a copper discovery. *Horizon* (*Salisbury, Southern Rhodesia*) **1**(1), 35–39.

Hornbrook, E. H. W. (1971). Mercury in permaforst regions; occurrence and distribution in the Kaminak Lake area, Northwest Territories. *Geol. Survey Can.* Paper 71-43, 13 pp.

Hornbrook, E. H. W. and Garrett, R. G. (1976). Regional geochemical lake sediment survey, east-central Saskatchewan. *Geol. Survey Can.* Paper 75-41, 20 pp.

Hornbrook, E. H. W., Davenport, P. H. and Grant, D. R. (1975). Regional and detailed geochemical exploration studies in glaciated terrain in Newfoundland. *Newfoundland Dept Mines Energy* Rept 75-2, 116 pp.

Horsnail, R. F. (1975). Strategic and tactical geochemical exploration in glaciated terrain: illustrations from Northern Ireland. *In* "Prospecting in Areas of Glaciated Terrain, 1975" (M. J. Jones, ed.), 16–31. Institution of Mining and Metallurgy, London.

Hcrsnail, R. F., Nichol, I. and Webb, J. S. (1969). Influence of variations in the secondary environment on the metal content of drainage sediments. *Colorado School Mines Quart.* **64**(1), 307–322.

Howard, J. H. (1970). Geochemical behavior of selenium in earth surface environment. *Diss. Abst. Int.* **30B,** 5554–5555.

Howard-Williams, C. 1970. The ecology of *Becium homblei* in central Africa, with special reference to metalliferous soils. *J. Ecol.* **58,** 745–763.

Howarth, R. J. (1971). Fortran I program for grey-level mapping of spatial data. *J. Int. Assoc. Math. Geol.* **3,** 95–121.

Howarth, R. J. (1973), The pattern recognition problem in applied geochemistry. *London Symposium Volume* 259–275.

Howarth, R. J. (1977). Cartogıaphy in geochemical exploration. *Sci. Terr. Ser. Inf. Geol.* **9,** 105–128.

Howarth, R. J. and Lowenstein, P. L. (1971). Sampling variability of stream sediments in broad scale regional geochemical reconnaissance. *Inst. Mining Metall. Trans.* **80,** B363–B372.

Huang, W. H. and Keller, W. D. (1971). Dissolution of clay minerals in dilute organic acids at room temperature. *Am. Mineralogist* **56,** 1082–1095.

Hubert, A. E. and Lakin, H. W. (1973). Atomic absorption determination of thallium and indium in geologic materials. *London Symposium Volume* 383–387.

Huff, L. C. (1948). A sensitive field test for heavy metals in water. *Econ. Geol.* **43,** 675–684.

Huff, L. C. (1952). Abnormal copper, lead, and zinc content of soil near metalliferous veins. *Econ. Geol.* **47,** 517–542.

Huff, L. C. (1970). A geochemical study of alluvium-covered copper deposits in Pima County, Arizona. *U.S. Geol. Survey Bull.* **1312-C,** 31 pp.

Huff, L. C. (1971). A comparison of alluvial exploration techniques for porphyry copper deposits. *Toronto Symposium Volume* 190–194.

Huff, L. C. and Marranzino, A. P. (1961). Geochemical prospecting for copper deposits hidden beneath alluvium in the Pima district, Arizona. *U.S. Geol. Survey* Prof. Paper 424-B, 308–310.

Hutchinson, G. E. (1957). "A Treatise on Limnology", Vol. 1, "Geography, Physics and Chemistry". John Wiley, New York, 1015 pp.

Hutchinson, G. E. (1967). "A Treatise on Limnology", Vol. 2, "Introduction to Lake Biology and Limnoplankton". John Wiley, New York.

Hvatum, O. (1965). Geochemical investigations in nine Norwegian bogs (abst.). *Norks. Geol. Tidssk.* **45**(1), 147.

Hyvarinen, L. (1958). Geochemical prospecting for lead ore in Korsnäs. *Geol. Tutkimuslaitos Geotekn. Julkaisuja.* **61,** 7–22.

Hyvarinen, L., Kauranne, K. and Yletyinen, V. (1973). Modern boulder tracing in prospecting. *In* "Prospecting in Areas of Glacial Terrain" (M. J. Jones, ed.), 87–95. Institution of Mining and Metallurgy, London.

Illsley, C. T., Bills, C. W. and Pollock, J. W. (1958). Geochemical methods in uranium exploration. *Proc. 2nd U.N. Conf. Peaceful Uses Atomic Energy* **2,** 126–130.

Ineson, P. R. (1970). Trace element aureoles in limestone wall rocks adjacent to fissure veins in the Eyam area of the Derbyshire ore field. *Inst. Mining Metall. Trans.* **79,** B238–B245.

Ingamells, C. O. (1970). Lithium metaborate flux in silicate analysis. *Anal. Chim. Acta* **52,** 323–334.

Ingersoll, L. R., Zobel, P. J. and Ingersoll, A. C. (1954). "Heat Conduction." University of Wisconsin Press, Madison, 325 pp.

Institution of Mining and Metallurgy (1977). "Prospecting in Areas of Glaciated Terrain 1977", 140 pp.

Intersociety Committee (1972). "Methods of Air Sampling and Analysis." American Public Health Assoc., Washington, 480 pp.

Ivanov, W. and Medovyi, V. I. (1975). Structure of the helium concentration field in the artesian basin of southern Mangyshlak (in Russian). English abst. in *Chem. Abst.* **84,** item 93128.

Ivanova, G. F. (1963). Content of tin, tungsten, and molybdenum in granites enclosing tin–tungsten deposits. *Geochemistry* **1963,** 492–500.

Jackson, M. L., Tyler, S. A., Willis, A. L. *et al.* (1948). Weathering sequence of clay-size minerals in soils and sediments. *J. Phys. Chem.* **52,** 1237–1260.

Jacobs, L. W. (1974). Methylation of mercury in lake and river sediments during field and laboratory investigations. *Diss. Abst. Int.* **34,** 4156.

Jacobson, J. D. (1956). Geochemical prospecting studies in the Kilembe area, Uganda. II. Dispersion of copper in the soil. Geochem. Prosp. Res. Centre, Imperial College, London, Tech. Comm. No. 6, 160 pp.

Jambor, J. L. (1974). Trace element variations in porphyry copper deposits, Babine Lake area, B.C. *Geol. Survey Can.* Paper 74-9, 30 pp.

James, C. H. (1957). The geochemical dispersion of arsenic and antimony related to gold mineralization in Southern Rhodesia. Geochem. Prosp, Res. Centre, Imperial College, London, Tech. Comm. No. 12.

James, C. H. (1967). The use of the terms "primary" and "secondary" dispersion in geochemical prospecting. *Econ. Geol.* **62,** 997–999.

James, R. O. and MacNaughton, M. G. (1977). The adsorption of aqueous heavy metals on inorganic minerals. *Geochim. Cosmochim. Acta* **41,** 1549–1558.

Jay, J. R. (1959). Geochemical prospecting studies for cobalt and uranium in Northern Rhodesia. Ph.D. Thesis, Imperial College, London.

Jenks, W. F. (1975). Origins of some massive pyrite ore deposits of Western Europe. *Econ. Geol.* **70,** 488–498.

Jenne, E. A. (1968). Controls on Mn, Fe, Co, Ni, Cu, and Zn concentrations in soils and water: the significant role of Mn and Fe oxides. *Am. Chem. Soc. Adv.* Chem. Ser. No. 73, 337–388.

Jenne, E. A. (1977). Trace element sorption by sediments and soils—sites and processes. *In* "Symposium on Molybdenum in the Environment" (W. Chappel and K. Petersen, eds), 425–533. Marcel Dekker, New York.

Jenny, H. (1941). "Factors of Soil Formation." McGraw-Hill, New York, 281 pp.

Jenny, H. (1950). Origin of soils. *In* "Applied Sedimentation" (P. D. Trask, ed.), 41–61. John Wiley, New York.

Jenny, H. and Overstreet, R. (1939). Surface migration of ions and contact exchange. *J. Phys. Chem.* **43,** 1185–1196.

Jensen, M. L. (1971). Stable isotopes in geochemical prospecting. *Toronto Symposium Volume* 464–468.

Joffe, J. S. (1949). "Pedology", 2nd edn. Rutgers University Press, New Brunswick, New Jersey, 662 pp.

Jonasson, I. R. (1970). Mercury in the natural environment—a review of recent work. *Geol. Survey Can.* Paper 70-57, 39 pp.

Jonasson, I. R. (1976). Detailed hydrogeochemistry of two small lakes in the Grenville geologic province. *Geol. Survey Can.* Paper 76-13, 37 pp.

Jonasson, I. R. and Allan, R. J. (1973). Snow: a sampling medium in hydrogeochemical prospecting in temperate and permaforst regions. *London Symposium Volume* 161–176.

Jonasson, I. R. and Gleeson, C. F. (1976). On the usefulness of water samples in reconnaissance surveys for uranium in the Yukon Territory. *Geol. Survey Can.* Paper 76-1C, 241–248.

Jonasson, I. R., Lynch, J. J. and Trip, L. J. (1973). Field and laboratory methods used by the Geological Survey of Canada in geochemical surveys. No. 12. Mercury in ores, rocks, soils, sediments and waters. *Geol. Survey Can.* Paper 73-21, 22 pp.

Jones, L. H. P. (1957). The solubility of molybdenum in simplified systems and aqueous soil suspension. *J. Soil Sci.* **8,** 313–327.

Jones, M. J., ed. (1973a). "Geochemical Exploration 1972." Institution of Mining and Metallurgy, London, 458 pp.

Jones, M. J., ed. (1973b). "Prospecting in Areas of Glacial Terrain." Institution of Mining and Metallurgy, London, 138 pp.

Jones, M. J., ed. (1975). "Prospecting in Areas of Glaciated Terrain." Institution of Mining and Metallurgy. London, 154 pp.

Jones, M. T., Reed, B. L., Doe, B. R. *et al.* (1977). Age of tin mineralization and plumbotectonics, Belitung, Indonesia. *Econ. Geol.* **72,** 745–753.

Joreskog, K. G., Klovan, J. E. and Reyment, R. A. (1976). "Geological Factor Analysis." Elsevier, Amsterdam, 178 pp.

Joyce, A. S. and Clema, J. M. (1974). Application of statistics to the chemical recognition of nickel gossans in the Yilgarn Block, Western Australia. *Australas. Inst. Mining Metall. Proc.* **252,** 21–24.

Junge, C. E. and Werby, R. T. (1958). The concentration of chloride, sodium, potassium, calcium, and sulfate in rainwater over the U.S. *J. Meteorol.* **15,** 417–425.

Kahma, A., Nurmi, A. and Mattson, P. (1975). On the composition of the gases generated by sulphide-bearing boulders during weathering and on the ability of prospecting dogs to detect samples treated with these gases in the terrain. *Geol. Survey Finland* Rept Inv. 6, 6 pp.

Karasik, M. A. and Bol'shakov, A. P. (1965). Mercury vapor in the Nikitovak ore field. *Am. Geol. Inst. Dokl. Acad. Sci. U.S.S.R.* **161,** 204–206.

Kartsev, A. A., Tabarasansky, Z. A., Subbota, M. I. *et al.* (1959). "Geochemical Methods of Prospecting and Exploration for Petroleum and Natural Gas." University of California Press, Berkeley, 349 pp.

Kasimov, Kh. K. (1973). Results of application of the radon method for predicting earthquakes in northern Fergana (in Russian). English abst. in *Chem. Abst.* **80,** item 39462.

Katz, M., ed. (1977). "Methods of Air Sampling and Analysis", 2nd edn. Am. Public Health Assoc., Washington D.C., 762 pp.

Kauranne, L. K. (1958). On prospecting for molybdenum on the basis of its dispersion in glacial till. *Soc. Geol. Finland* C. R. No. 30, 31–43.

Kauranne, L. K. (1959). Pedogeochemical prospecting in glaciated terrain. *Fin. Comm. Geol. Bull.* No. 184, 10 pp.

Kauranne, L. K. (1976). Conceptual models in exploration geochemistry, Norden 1975. *J. Geochem. Explor.* **5,** 173–420.

Keays, R. R. and Davison, R. M. (1976). Palladium, iridium, and gold in the ores and host rocks of nickel sulfide deposits in Western Australia. *Econ. Geol.* **71,** 1214–1228.

Keith, M. L. and Degens, E. T. (1959). Geochemical indicators of marine and fresh-water sediments. *In* "Researches in Geochemistry" (P. H. Abelson, ed.), 38–61. John Wiley, New York.

Keller, W. D. and Frederickson, A. F. (1952). Role of plants and colloidal acids in the mechanism of weathering. *Am. J. Sci.* **250,** 594–608.

Kennedy, V. C. (1956). Geochemical studies in the south-western Wisconsin lead–zinc district. *U.S. Geol. Survey Bull.* **1000-E,** 187–223.

Kesler, S. E. and Van Loon, J. C. (1973). Analysis of water-extractable chloride in rocks by use of a selective ion electrode. *London Symposium Volume* 431–436.

Kesler, S. E., Van Loon, J. C. and Bateson, J. H. (1973). Analysis of fluoride in rocks and an application to exploration. *J, Geochem. Explor.* **2,** 11–18.

Kesler, S. E., Issigonis, M. J. and Van Loon, J. C. (1975a). An evaluation of the use of halogen and water abundances in efforts to distinguish mineralized from barren intrusive rocks. *J. Geochem. Explor.* **4,** 235–246.

Kesler, S. E., Issigonis, M. J., Brownlow, A. H. *et al.* (1975b). Geochemistry of biotites from mineralized and barren intrusive systems. *Econ. Geol.* **70,** 559–567.

Khattab, K. M. (1970). Analysis of uranium and thorium and their daughter nuclides in uranium–thorium ores by high resolution gamma-ray spectrometry. Ph.D. Thesis, Pennsylvania State University, University Park, Pennsylvania, 112 pp.

Khayretdinov, I. A. (1971). Gas mercury aureoles. *Geochem. Int.* **8,** 412–452.

Khetagurov, G. V., Rekhviashvili, K. L. and Shchepetova, L. V. (1970). Zonal distribution of lead, zinc, copper, silver, cobalt, and molybdenum in endogenous haloes and ores of some complex ore deposits of the northern Caucasus. *Geochem. Int.* **7,** 764–774.

Kim, C. H., Owens, C. M. and Smythe, L. E. (1974). Determination of traces of Mo in soils by solvent extraction of the molybdenum-thiocyanate complex and atomic absorption. *Talanta* **21,** 445–454.

Kimura, K., Fujiwara, S. and Morinaga, K. (1951). Chemical prospecting in Hosokura Mine districts. *J. Chem. Soc. Japan, Pure Chem. Sect.* **72,** 398–402.

Kinniburgh, D. G., Jackson, M. L. and Syers, J. K. (1976). Adsorption of alkaline earth, transition and heavy metal cations by hydrous oxide gels of iron and aluminum. *Soil Sci. Soc. Am. J.* **40,** 796–799.

Kleinkopf, M. D. (1960). Spectrographic determination of trace elements in lake waters of northern Maine. *Geol. Soc. Am. Bull.* **71,** 1231–1242.

Klimenko, A. A. (1976). Diffusion of gases from hydrocarbon deposits. *Int. Geol. Rev.* **18,** 717–722.

Klimenko, I. A. (1965). Aqueous migration of rhenium (in Russian). English abst. in *Chem. Abst.* **64,** item 3245h.

Klovan, J. E. (1975). R- and Q-mode factor analysis. *In* "Concepts in Geostatistics" (R. B. McCammon, ed.), 21–29. Springer-Verlag, New York.

Koch, G. S. and Link, R. F. (1970). "Statistical Analysis of Geological Data", Vol. 1. John Wiley, New York, 375 pp.

Koch, G. S. and Link, R. F. (1971). "Statistical Analysis of Geologic Data", Vol. 2. John Wiley, New York, 438 pp.

Koehler, G. F., Hostetler, P. B. and Holland, H. D. (1954). Geochemical prospecting at Cobalt, Ontario. *Econ. Geol.* **49,** 378–388.

Koga, A. and Noda, T. (1975). New geochemical exploration methods for geothermal sources in vapor-dominated fields (in Japanese). English abst. in *Chem. Abst.* **85,** item 65816.

Kokkola, M. (1976). Geochemical sampling of deep moraine and the bedrock surface underlying overburden. *J. Geochem. Explor.* **5,** 395–400.

Kokkola, M. (1977). Application of humus to exploration. *In* "Propecting in Areas of Glaciated Terrain, 1977", 104–110. Institution of Mining and Metallurgy, London.

Köksoy, M. and Bradshaw, P. M. D. (1969). Secondary dispersion of mercury from cinnabar and stibnite deposits, west Turkey. *Colorado School Mines Quart.* **64**(1), 333–356.

Köksoy, M., Bradshaw, P. M. D. and Tooms, J. S. (1967). Notes on the determination of mercury in geological samples. *Inst. Mining Metall. Trans.* **76,** B121–B124.

Koljonen, T. (1976). Puikonlahti—selenium in the vicinity of copper ore. *J. Geochem. Explor.* **5,** 263–265.

Kolotov, B. A., Kiseleva, Ye. A. and Rubeykin, V. Z. (1965). On the secondary dispersion aureoles in the vicinity of ore deposits. *Geochem. Int.* **2,** 675–677.

Komarov, P. V. and Glagolev, A. A. (1969). Use of fluorimetric survey techniques during prospecting for mineral deposits affiliated with alkaline-ultrabasic and alkaline massifs (in Russian). English abst. in *Econ. Geol.* **69,** 582–583.

Kovach, E. M. (1945). Meteorological influence upon the radon content of soil gas. *Am. Geophys. Union Trans.* **26,** 241–248.

Kovalevskiy, A. L. (1974). "Biogeochemical Exploration for Mineral Deposits" (in Russian). English translation in press with Amerind Publishing Co., New Delhi.

Kovalevskiy, A. L. (1977). Biogeochemical prospecting for polymetallic deposits. *Int. Geol. Rev.* **18,** 1000–1011.

Kozlov, V. D., Sheremet, Ye. M. and Yanitskiy, V. M. (1975). Geochemical characterization of the Mesozoic plumasitic leucocratic granites of the Transbaykalia tin–tungsten belt. *Geochem. Int.* **11,** 997–1008.

Krauskopf, K. B. (1955). Sedimentary deposits of rare metals. *In* "Economic Geology 50th Anniversary Volume", 411–463. Economic Geology Pub. Co., Lancaster, Pennsylvania.

Krauskopf, K. B. (1967). "Introduction to Geochemistry." McGraw-Hill, New York, 721 pp.

Kravtsov, A. I. and Fridman, A. I. (1965). Gases in ore deposits. *Am. Geol. Inst. Dokl. Acad. Sci. U.S.S.R.* **165,** 192–193.

Kraynov, S. R. (1957). The feasibility of applying hydrochemical methods to the solution of certain problems in metallogenesis. *Geochemistry* 1957, 460–469.

Kraynov, S. R. (1968). Aspects of the occurrence and migration of niobium, beryllium, and rare earths in natural alkaline waters. *Geochem. Int.* **5,** 315–325.

Kroepelin, H. (1967). Geochemical prospecting. *Trans. 7th World Pet. Cong.* **1-B,** 37–57.

Kuenen, P. H. (1959). Experimental abrasion. Fluviatile action on sand. *Am. J. Sci.* **257,** 172–190.

Kunzendorf, H. (1973). Non-destructive determination of metals in rocks by radioisotope X-ray fluorescence instrumentation. *London Symposium Volume* 401–414.

Kunzendorf, H. (1976). Uranium and thorium in deep-sea manganese nodules from the central Pacific. *Inst. Mining Metall. Trans.* **85,** B284–B288.

Kvalheim, A., ed. (1967). "Geochemical Prospecting in Fennoscandia." Interscience, New York, 350 pp.

Låg, J. and Bølviken, B. (1974). Some naturally occurring heavy-metal poisoned areas of interest in prospecting. *Norges Geol. Undersok.* **304,** 73–96.

Lakin, H. W. (1972). Selenium accumulation in soils and its absorption by plants and animals. *Geol. Soc. Am. Bull.* **83,** 181–190.

Lakin, H. W. (1973). Geochemical analysis. *Assoc. Explor. Geochem. Newsl.* **9,** 6–9.

Lakin, H. W., Curtin, G. C. and Hubert, A. E. (1974). Geochemistry of gold in the weathering cycle. *U.S. Geol. Survey Bull.* **1330,** 80 pp.

Lalonde, J.-P. (1976). Fluorine—an indicator of mineral deposits. *Can. Inst. Mining Metall. Bull.* **69**(769), 110–122.

Lambert, I. B. and Sato, T. (1974). The Kuroko and associated ore deposits of Japan: a review of their features and metallogenesis. *Econ. Geol.* **69,** 1215–1236.

Lambert, I. B. and Scott, K. M. (1973). Implications of geochemical investigations of sedimentary rocks within and around the MacArthur zinc–lead–silver deposit, Northern Territory. *J. Geochem. Explor.* **2,** 307–330.

Landis, E. R. (1960). Uranium content of ground and surface waters in a part of the central Great Plains. *U.S. Geol. Survey Bull.* **1087-G,** 223–258.

Langmuir, D. (1971). Measurement of Eh and pH. *In* "Procedures of Sedimentary Petrology" (R. E. Carver, ed.), 597–634. John Wiley, New York.

Langmuir, D. (1978). Uranium solution–mineral equilibria at low temperatures with applications to sedimentary ore deposits. *Geochim. Cosmochim. Acta* **42,** 547–570.

Langmuir, D. and Whittemore, D. O. (1971). Variations in the stability of precipitated ferric oxyhydroxides. *In* "Non-equilibrium Systems in Natural Water Chemistry" (F. R. Gould, ed.), 209–234. Am. Chem. Soc. Adv. Chem. Ser. No. 106.

Larsson, J. O. (1976). Organic stream sediments in regional geochemical prospecting, Precambrian Pajala district, Sweden. *J. Geochem. Explor.* **6,** 233–249.

Larsson, J. O. and Nichol, I. (1971). Analysis of glacial material as an aid in geologic mapping. *Toronto Symposium Volume* 197–203.

Lavergne, P. J. (1965). Field and laboratory methods used by the Geological Survey of Canada. No. 8: Preparation of geologic materials for chemical and spectrographic analysis. *Geol. Survey Can.* Paper 65-18, 23 pp.

Lavery, N. G. and Barnes, H. L. (1971). Zinc dispersion in the Wisconsin zinc–lead district. *Econ. Geol.* **66,** 226–242.

Lawrence, G. (1975). The use of rubidium/strontium ratios as a guide to mineralization in the Galway granite, Ireland. *Vancouver Symposium Volume* 353–370.

Leake, B. E., Hendry, G. L., Kemp, A. *et al.* (1969). The chemical analysis of rock powders by automatic X-ray fluorescence. *Chem. Geol.* **5,** 7–86.

Leake, R. C. and Aucott, J. W. (1973). Geochemical mapping and prospecting by use of rapid automatic X-ray fluorescence analysis of panned concentrates. *London Symposium Volume* 389–400.

Leake, R. C. and Smith, R. T. (1975). A comparison of stream sediment sampling methods in parts of Great Britain. *Vancouver Symposium Volume* 579–594.

Learned, R. E. and Boissen, R. (1973). Gold—a useful element in the search for porphyry copper deposits in Puerto Rico. *London Symposium Volume* 93–103.

Leduc, C. and Boucetta, M. (1971). A study of the distribution and localization of lead and zinc in soil developed from the Fougeres granite (in French). *Toronto Symposium Volume* 204–213.

Ledward, R. A. (1960). Geochemical prospecting studies for base metals in Tanganyika and Burma. Ph.D. Thesis, Imperial College, London.

Lee, H. A. (1965). Investigation of eskers for mineral exploration. *Geol. Survey Can.* Paper 65-14, 17 pp.

Lee, H. A. (1968). An Ontario kimberlite occurrence discovered by application of the glaciofocus method to a study of the Munro esker. *Geol. Survey Can.* Paper 68-7, 3 pp.

Lee Moreno, J. L. and Caire, L. F. (1975). Results of geochemical investigations comparing samples of stream sediment, panned concentrate, and vegetation in the vicinity of the Caridad porphyry copper deposit, Sonora, Mexico (abst.). *Mining Eng.* **27**(12), 68c.

Lepeltier, C. (1969). A simplified statistical treatment of geochemical data by graphical representation. *Econ. Geol.* **64,** 538–550.

Lepeltier, C. (1971). Geochemical exploration in the United Nations Development Programme. *Toronto Symposium Volume* 24–27.

Lepeltier, C. (1974). Geochemical exploration in the United Nations Development Programme, 1970–73. *Assoc. Explor. Geochem. Newsl.* **14,** 6 pp.

Levinson, A. A. (1974). "Introduction to Exploration Geochemistry." Applied Pub. Co., Calgary, 612 pp.

Levinson, A. A. and dePablo, L. (1975). A rapid X-ray fluorescence procedure applicable in exploration geochemistry. *J. Geochem. Explor.* **4,** 399–408.

Leyshon, P. R. and Cazalet, P. C. D. (1976). Base metal exploration programme in Lower Paleozoic volcanic rocks, County Tyrone, Northern Ireland. *Inst. Mining Metall. Trans.* **85,** B91–B99.

Lidgey, E. (1897). Some indications of ore deposits. *Australas. Inst. Mining Eng. Trans.* **4,** 110–122.

Liptak, B. G. (1974). "Environmental Engineers Handbook", Vol. 2. Chilton Book Co., Radnor, Pennsylvania, 316–317.

Little, H. W. and Durham, C. C. (1971). Uranium in stream sediments in Carboniferous rocks of Nova Scotia. *Geol. Survey Can.* Paper 70-54, 17 pp.

Livingstone, D. A. (1963). Chemical composition of rivers and lakes: data of geochemistry. *U.S. Geol. Survey* Prof. Paper 440-G, 64 pp.

Ljunggren, P. (1955). Geochemistry and radioactivity of some Mn and Fe bog ores. *Geol. Foren. Forh.* **77,** 33–44.

Loftus-Hills, G. and Solomon, M. (1967). Cobalt, nickel, and selenium in sulfides as indicators of ore genesis. *Mineralium Deposita* **2,** 228–242.

London Symposium Volume (1973). "Geochemical Exploration 1972" (M. J. Jones, ed.), 4th Int. Geochem. Explor. Symp., London. Institution of Mining and Metallurgy, London, 458 pp.

Long, D. T. and Angino, E. E. (1977). Chemical speciation of Cd, Cu, Pb, and Zn in mixed freshwater, sea water, and brine solution. *Geochim. Cosmochim. Acta* **41,** 1183–1192.

Lopatkina, A. P. (1964). Characteristics of migration of uranium in the natural waters of humid regions and their use in the determination of the geochemical background for uranium. *Geochem. Int.* 1964, 788–795.

Lotspeich, F. B. (1958). Movement of metallic elements in shallow colluvium. *In* "Symposium de Exploración Geoquímica" (T. S. Lovering, ed.), Vol. 1, 125–142 20th Int. Geol. Cong., Mexico City.

Lovering, T. G., ed. (1976). Lead in the environment. *U.S. Geol. Survey* Prof Paper 957, 90 pp.

Lovering, T. G. and Hamilton, J. C. (1962). Criteria for the recognition of jasperoid associated with sulfide ores. *U.S. Geol. Survey* Prof. Paper 450-C, 9–11.

Lovering, T. G., Lakin, H. W. and McCarthy, J. H. (1966). Tellurium and mercury in jasperoid samples. *U.S. Geol. Survey* Prof. Paper 550-B, 138–141.

Lovering, T. G., Cooper, J. R., Drewes, H. *et al.* (1970). Copper in biotite from igneous rocks in southern Arizona as an ore indicator. *U.S. Geol. Survey* Prof. Paper 700-B, 1–8.

Lovering, T. S. (1927). Organic precipitation of metallic copper. *U.S. Geol. Survey Bull.* **795-C,** 45–52.

Lovering, T. S. (1952). Mobility of heavy metals in ground water. Part I of Supergene and hydrothermal dispersion of heavy metals in wall rocks near ore bodies, Tintic District, Utah. *Econ. Geol.* **47,** 685–698.

Lovering, T. S., ed. (1958–60). "Symposium de Exploración Geoquímica", Vols 1–3, 725 pp. 20th Int. Geol. Cong., Mexico City.

Lovering, T. S. (1959). Significance of accumulator plants in rock weathering. *Geol. Soc. Am. Bull.* **70,** 781–800.

Lovering, T. S., Sokoloff, V. P. and Morris, H. T. (1948). Heavy metals in altered rock over blind ore bodies, East Tintic District, Utah. *Econ. Geol.* **43,** 384–399.

Lovering, T. S., Huff, L. C. and Almond, H. (1950). Dispersion of copper from the San Manuel copper deposit, Pinal County, Arizona. *Econ. Geol.* **45,** 493–514.

Lowell, J. D. and Guilbert, J. M. (1970). Lateral and vertical alteration-mineralization zoning in porphyry copper deposits. *Econ. Geol.* **65,** 373–408.

Lur'ye, A. M. (1957). Certain regularities in the distribution of elements in sedimentary rocks of the northern Bayaldyr district in central Karatau. *Geochemistry* **5,** 470–479.

Lynch, J. J. (1971). The determination of copper, nickel, and cobalt in rocks by atomic absorption spectrophotometry using a cold leach. *Toronto Symposium Volume* 313–314.

Lyon, R. J. P. (1977). Mineral exploration applications of digitally processed Landsat imagery. *U.S. Geol. Survey* Prof. Paper 1015, 271–292.

Lyon, R. J. P. and Tuddenham, W. M. (1959). Quantitative mineralogy as a guide in exploration. *Mining Eng.* **11**(12), 1–5.

Macdonald, J. A. (1969). An orientation study of the uranium distribution in lake waters, Beaverlodge district, Saskatchewan. *Colorado School Mines Quart.* **64**(1), 357–376.

MacIntire, W. L. (1963). Trace element partition coefficients—a review of theory and applications to geology. *Geochim. Cosmochim. Acta* **27,** 1209–1264.

MacKenzie, B. W. (1973). Corporate exploration strategies. *In* "Application of Computer Methods in the Mineral Industry" (M. O. Salamon and F. H. Lancaster, eds), 1–8. South African Institution of Mining and Metallurgy, Johannesburg.

MacKenzie, D. H. (1977). Empirical assessment of anomalies in tropical terrain. *Assoc. Explor. Geochem. Newsl.* No. 21, 6–10.

Maes, A., Peigneur, P. and Cremers, A. (1975). Thermodynamics of transition metal ion exchange in montmorillonite. *Proc. Int. Clay Conf.*, **1975** 319–329.

Mahaffey, E. J. (1974). A spectrophotometric method for the determination of rhenium in geological materials. *J. Geochem. Explor.* **3,** 53–60.

Maienthal, E. J. and Becker, D. A. (1976). A survey on current literature on sampling, sample handling, for environmental materials and long-term storage. *Interface* **5,** 49–61.

Malyuga, D. P. (1958). An experiment of biogeochemical prospecting for molybdenum in Armenia. *Geochemistry* 1958, 314–337.
Malyuga, D. P. (1964). "Biogeochemical Methods of Prospecting." Consultants Bureau, New York, 205 pp.
Manskaia, S. M., Drozdova, T. V. and Emelianova, M. P. (1956). Binding of uranium in humine acids and by melanoidines. *Geochemistry* 1956, 339–356.
Maranzana, F. (1972). Application of talus sampling to geochemical exploration in arid areas: Los Pelambres hydrothermal alteration area, Chile. *Inst. Mining Metall. Trans.* **81,** B26–B33.
Markward, E. L. (1961). Geochemical prospecting abstracts, January 1955–June 1957. *U.S. Geol. Survey Bull.* **1098-B,** 160 pp.
Marmo, V. (1953). Biogeochemical investigations in Finland. *Econ. Geol.* **48,** 211–224.
Marmo, V. and Puranen, M., eds (1960). "Geological Results of Applied Geochemistry and Geophysics", 21st Int. Geol. Cong., Copenhagen, Vol. 2.
Marsh, S. P. and Erickson, R. L. (1975). Integrated geologic and geochemical studies, Edna Mtn, Nevada. *Vancouver Symposium Volume* 239–250.
Marshall, C. E. (1964). "The Physical Chemistry and Mineralogy of Soils." John Wiley, New York, 388 pp.
Marshall, T. J. (1959). Relations between water and soil. Commonwealth Bur. Soils, Harpenden, Tech. Comm. No. 50, 91 pp.
Martell, A. E. (1971). Principles of complex formation. *In* "Organic Compounds in Aquatic Environments" (S. J. Faust and J. V. Hunter, eds), 239–263. Marcel Dekker, New York.
Mason, B. (1958). "Principles of Geochemistry", 2nd edn. John Wiley, New York, 310 pp.
Mason, D. R. (1979). Chemical variations in ferromagnesian minerals: a new exploration tool to distinguish between mineralized and barren stocks in porphyry copper provinces. *Golden Symposium Volume* (*1979*) 243–250.
Mather, A. L. (1959). Geochemical prospecting studies in Sierra Leone. D.I.C. Thesis, Imperial College, London.
Matheron, G. (1963). Principles of geostatistics. *Econ. Geol.* **58,** 1246–1266.
Maynard, D. E. and Fletcher, W. K. (1973). Comparison of total and partial extractable copper in anomalous and background peat samples. *J. Geochem. Explor.* **2,** 19–24.
McCarthy, J. H. (1972). Mercury vapor and other volatile components in the air as guides to ore deposits. *J. Geochem. Explor.* **1,** 143–162.
McCarthy, J. H., Meuschke, J. L., Ficklin, W. H. *et al.* (1970). Mercury in the atmosphere. *U.S. Geol. Survey* Prof. Paper 713, 37–39.
McFarlane, M. J. (1976). "Laterite and Landscape." Academic Press, London and New York, 151 pp.
McKenzie, R. M. (1972). Sorption of some heavy metals by the lower oxides of manganese. *Geoderma* **8,** 29–35.
McKinstry, H. E. (1948). "Mining Geology." Prentice Hall, Englewood Cliffs, New Jersey, 680 pp.
McLaren, R. G. and Crawford, D. V. (1973). Soil copper—II. Specific adsorption of copper by soils. *J. Soil Sci.* **24,** 443–452.
McNeal, J. M. and Rose, A. W. (1974). The geochemistry of mercury in sedimentary rocks and soils in Pennsylvania. *Geochim. Cosmochim. Acta* **38,** 1759–1784.
McNerney, J. J. and Buseck, P. R. (1973). Geochemical exploration using mercury vapor. *Econ. Geol.* **68,** 1313–1320.

McNerney, J. J., Buseck, P. R. and Hanson, R. C. (1972). Mercury detection by means of thin gold films. *Science* **178,** 611–612.

Meadows, R. W. and Spedding, D. J. (1974). The solubility of very low concentrations of carbon monoxide in aqueous solution. *Tellus* **26,** 143–150.

Medlin, J. H., Suhr, N. H. and Bodkin, J. B. (1969). Atomic absorption analysis of silicates employing $LiBO_2$ fusion. *Atomic Absorption Newsl.* **8,** 25–28.

Megill, R. E. (1971). "An Introduction to Petroleum Economics." Petroleum Publishing Co., Tulsa, 159 pp.

Megumi, K. and Mamuro, T. (1972). A method for measuring radon and thoron exhalation from the ground. *J. Geophys. Res.* **77,** 3052–3056.

Mehrlich, A. and Drake, M. (1955). Soil chemistry and plant nutrition. *In* "Chemistry of the Soil" (F. E. Bear, ed.), 286–327. Reinhold, New York.

Mehrtens, M. B. and Tooms, J. S. (1973). Geochemichal drainage dispersion from sulfide mineralization in glaciated terrain, central Norway. *In* "Prospecting in Areas of Glaciated Terrain" (M. J. Jones, ed.),1–10. Institution of Mining and Metallurgy, London.

Mehrtens, M. B., Tooms, J. S. and Troup, A. G. (1973). Some aspects of geochemical dispersion from base metal mineralization within glaciated terrain in Norway, North Wales and British Columbia, Canada. *London Symposium Volume* 105–115.

Meinzer, O. E. (1927). Plants as indicators of ground water. *U.S. Geol. Survey* Water-supply Paper 577, 95 pp.

Mellor, J. W. (1924). "A Comprehensive Treatise on Inorganic and Theoretical Chemistry", Vol. 5. Longmans, Green and Co., New York.

Merefield, J. R. (1974). Major and trace element anomalies in stream sediments of the Teign Valley orefield. *J. Geochem. Explor.* **3,** 151–166.

Mertie, J. B., Jr (1954). The gold pan: a neglected geological tool. *Econ. Geol.* **49,** 108–129.

Meyer, W. T. (1969). Uranium in lake water from the Kaipokok region, Labrador. *Colorado School Mines Quart.* **64**(1), 377–394.

Meyer, W. T. and Evans, D. S. (1973). Dispersion of mercury and associated elements in a glacial drift environment at Keel, Eire. *In* "Prospecting in Areas of Glaciated Terrain" (M. J. Jones, ed.), 127–138. Institution of Mining and Metallurgy, London.

Meyer, W. T. and Leen, K. C. (1973). Microwave-induced argon plasma emission system for geochemical trace analysis. *London Symposium Volume* 325–335.

Meyer, W. T. and Peters, R. G. (1973). Evaluation of sulphur as a guide to buried sulphide deposits in the Notre Dame Bay area, Newfoundland. *In* "Prospecting in Areas of Glacial Terrain" (M. J. Jones, ed.), 55–66. Institution of Mining and Metallurgy, London.

Michie, U. M., Gallagher, M. J. and Simpson, A. (1973). Detection of concealed mineralization in northern Scotland. *London Symposium Volume* 117–130.

Miesch, A. T. (1967). Theory of error in geochemical data. *U.S. Geol. Survey* Prof. Paper 574-A, 1–17.

Miesch, A. T. (1977). Log transformations. *Math. Geol.* **9,** 191–198.

Miesch, A. T. and Nolan, T. B. (1958). Geochemical prospecting studies in the Bullwhacker Mine area, Eureka district, Nevada. *U.S. Geol. Survey Bull.* **1000-H,** 397–408.

Miesch, A. T., Shoemaker, E. M., Newman, W. L. *et al.* (1960). Chemical composition as a guide to the size of sandstone-type uranium deposits in the Morrison Formation on the Colorado Plateau. *U.S. Geol. Survey Bull.* **1112-B,** 17–61.

Millar, C. E., Turk, F. M. and Foth, H. D. (1958). "Fundamentals of Soil Science", 3rd edn. John Wiley and Sons, New York, 526 pp.
Miller, A. D. and Danilov, V. Ya. (1957). Salt dispersion halos of rare-metal pegmatites in the Kola Peninsula. *Geochemistry* 1957, 620–630.
Miller, J. M. and Ostle, D. (1973). Radon measurement in uranium prospecting. *In* "Uranium Exploration Methods", 237–247. International Atomic Energy Agency, Vienna.
Millman, A. P. (1957). Biogeochemical investigations in areas of copper–tin mineralization in southwest England. *Geochim. Cosmochim. Acta* **12,** 85–93.
Milly, G. H. (1971). Prospecting for mineral deposits having radioactive gaseous decay products. U.S. Patent 3 609 363.
Milne, G. (1936). A soil reconnaissance in Tanganyika Territory. East African Agr. Res. Stat. Rept, Amani, Tanganyika.
Mitchell, R. L. (1954). Trace elements in Scottish peats. *Int. Peat Symp., Dublin* Sect. B-3, 9 pp.
Mogilevsky, G. A. (1959). The role of bacteria in prospecting for petroleum. *5th World Pet. Cong., Gen. Pet. Geochem. Symp.* Preprint Vol. 111–115.
Mohr, E. C. J. and Van Baren, F. A. (1954). "Tropical Soils." Interscience, New York, 498 pp.
Mohr, E. C. J., Van Baren, F. A. and Van Schuylenborgh, J. (1972). "Tropical Soils." Mouton–Ichtiar Baru-van Hoeve, The Hague, 481 pp.
Moiseenko, U. I. (1959). Biogeochemical surveys in prospecting for uranium deposits in marshy areas. *Geochemistry* 1959, 117–121.
Moore, W. J., Curtin, G. C., Roberts, R. J. *et al.* (1966). Distribution of selected metals in the Stockton district, Utah. *U.S. Geol. Survey* Prof. Paper 550-C, 197–205.
Morgan, D. A. O. (1969). A look at the economics of the mineral industry. *Proc. 9th Commonwealth Mining Metall. Cong., London* **2,** 305–324.
Morissey, C. J. and Romer, D. M. (1973). Mineral exploration in glaciated regions of Ireland. *In* "Prospecting in Areas of Glaciated Terrain" (M. J. Jones, ed.), 45–53. Institution of Mining and Metallurgy, London.
Morris, H. T. (1952). Primary dispersion patterns of heavy metals in carbonate and quartz monzonite wall rocks. Part II of Supergene and hydrothermal dispersion of heavy metals in wall rocks near ore bodies, Tintic District, Utah. *Econ. Geol.* **47,** 698–716.
Morris, J. C. and Stumm, W. (1967). Redox equilibria and measurements of potentials in the aquatic environment. *In* "Equilibrium Concepts in Natural Water Systems" (F. R. Gould, ed.), Am. Chem. Soc. Adv. Chem. Ser. **65,** 270–285.
Morse, R. H. (1969). Radium geochemistry as applied to prospecting for uranium. *Can. Mining J.* **90**(5), 75–76.
Mortvedt, J. J., Giordano, P. M. and Lindsay, W. L., eds (1972). "Micronutrients in Agriculture." Soil Sci. Soc. America, Madison, Wisconsin, 666 pp.
Moxham, R. M. Senftle, F. E. and Boynton, G. R. (1972). Borehole activation analysis by delayed and capture gamma rays using a Cf-252 neutron source. *Econ. Geol.* **67,** 579–591.
Moyers, J. L. and Duce, R. A. (1972). Gaseous and particulate iodine in the marine atmosphere. *J. Geophys. Res.* **77,** 5229–5238.
Mukanov, K. M. and Aubakirova, R. V. (1969). Geochemistry of rhenium in oxidation zones of sulfide deposits. *Am. Geol. Inst. Dokl. Acad. Sci. U.S.S.R.* **184,** 176–178.

Mukherjee, K. K. (1978). Fluorine as a direct indicator in a local soil-geochemical survey in the humid tropics—a case history (abst.). *7th Int. Proc. Geochem. Explor. Symp., Denver, Program* 53–54.

Murray, J. W. (1975a). The interaction of metal ions at the manganese dioxide–solution interface. *Geochim. Cosmochim. Acta* **39,** 505–520.

Murray, J. W. (1975b). The interaction of cobalt with hydrous manganese dioxide. *Geochim. Cosmochim. Acta* **39,** 635–648.

Nakagawa, H. M. and Ward, F. N. (1975). Determination of molybdenum in natural waters and brines after separation using a chelating resin. *U.S. Geol. Survey Bull.* **1408,** 65–76.

Nakagawa, H. M., Watterson, J. R. and Ward, F. N. (1975). Atomic absorption determination of molybdenum in plant ash. *U.S. Geol. Survey Bull.* **1408,** 29–36.

Naumov, G. B., Ryzhenko, B. N. and Khodakovsky, I. L. (1974). "Handbook of Thermodynamic Data" (G. J. Soleimani, translator). Rept Pb 226722, Natl Tech. Inf. Serv., Springfield, Virginia, 328 pp.

Nedashkovsky, P. G. and Narnov, G. A. (1968). Tin distribution in tin-bearing granites, apogranites and replaced pegmatites in the Soviet Far East. *Geochem. Int.* 1968, 687–694.

Nesvetaylova, N. G. (1955a). Geobotanical investigations for prospecting for ore deposits. *Int. Geol. Rev.* **3,** 609–618.

Nesvetaylova, N. G. (1955b). Geobotanical method of prospecting for copper and polymetallic ores (in Russian). *Razvedka i Okhrana Nedr.* 1955(4), 17–20.

Netreba, A. V., Fridman, A. I., Plotnikov, I. A. *et al.* (1971). On the large-scale mapping of closed ore-bearing areas in the North Caucasus with the use of gas surveying as a geochemical method (abst.). *Geochem. Int.* **8,** 630.

Newcomb, G. S. and Millan, M. M. (1970). Theory, applications and results of the long-line correlation spectrometer. *I.E.E.E. Trans. Geosci. Electron.* **8,** 149–157.

Newendorp, P. D. (1975). "Decision Analysis for Petroleum Exploration." Petroleum Pub. Co., Tulsa, 668 pp.

Nichol, I. and Björklund, A. (1973). Glacial geology as a key to geochemical exploration in areas of glacial overburden with particular reference to Canada. *J. Geochem. Explor.* **2,** 133, 170.

Nichol, I., Garrett, R. G. and Webb, J. S. (1966). Automatic data plotting and mathematical and statistical interpretation of geochemical data. *Geol. Survey Can.* Paper 66-54, 195–210.

Nichol, I., Garrett, R. G. and Webb, J. S. (1969). The role of some statistical and mathematical methods in the interpretation of regional geochemical data. *Econ. Geol.* **64,** 204–220.

Nichol, I., Coker, W. B., Jackson, R. G. *et al.* (1975). Relation of lake sediment composition to mineralization in different limnological environments in Canada. *In* "Prospecting in Areas of Glaciated Terrain" (M. J. Jones, ed.), 112–125. Institution of Mining and Metallurgy, London.

Nicholls, G. D. (1971). Geochemical sampling problems in the analytical laboratory. *Inst. Mining Metall. Trans.* **80,** 299–304.

Nicolls, O. W., Provan, D. M. J., Cole, M. M. *et al.* (1965). Geobotany and mineral exploration in the Dugald River area, Cloncurry district, Australia. *Inst. Mining Metall. Trans.* **74,** 696–699.

Nigrini, A. (1971). Investigation into the transport and deposition of copper, lead and zinc in the surficial environment (abst.). *Toronto Symposium Volume* 235.

Nilsson, G. (1973). Nickel prospecting and the discovery of the Mjovattnet mineralization in northern Sweden: a case history of the use of combined techniques in drift-covered glaciated terrain. *In* "Prospecting in Areas of Glacial Terrain" (M. J. Jones, ed.), 97–109. Institution of Mining and Metallurgy, London.

Noakes, J. E. and Harding, J. L. (1971). New techniques in sea-floor mineral exploration. *J. Mar. Technol. Soc.* **5**(6), 41–44.

Noakes, J. E., Harding, J. L. and Spaulding, J. D. (1974). Locating offshore mineral deposits by natural radioactive measurements. *J. Mar. Technol. Soc.* **8**(5), 36–39.

Noble, J. A. (1970). Metal provinces of the United States. *Geol. Soc. Am. Bull.* **81,** 1607–1624.

Noble, J. A. (1974). Metal provinces and metal finding in the western United States. *Mineralium Deposita* **9,** 1–25.

Noguchi, M. and Wakiti, H. (1976). Radon—recent development in its measurement and its application to earthquake prediction (in Japanese). *Oyo Butsuri* **45,** 453–458.

Nowlan, G. A. (1976). Concretionary manganese–iron oxides in streams and their usefulness as a sample medium for geochemical prospecting. *J. Geochem. Explor.* **6,** 193–210.

Nurmi, A. (1976). Geochemistry of the till blanket of the Talluskanava Ni–Cu ore deposit, Tervo, central Finland. *Geol. Survey Finland* Rept Inv. 15, 84 pp.

Nuutilainen, J. and Peuraniemi, V. (1977). Application of humus analysis to geochemical prospecting: some case histories. *In* "Prospecting in Areas of Glacial Terrain, 1977", 1–5. Institution of Mining and Metallurgy, London.

Nyuppenen, T. I. (1966). A method for estimating potential nickel content in ultramafic massifs. *Geochem. Int.* **3,** 84–88.

Obial, R. C. and James, C. H. (1973). Use of cluster analysis in geochemical prospecting, with particular reference to southern Derbyshire, England. *London Symposium Volume* 237–257.

O'Brien, M. V. and Romer, D. M. (1971). Tara geologists describe Navan discovery. *World Mining* **24**(6), 38–39.

O'Connor, T. P. and Kester, D. R. (1975). Adsorption of copper and cobalt from fresh and marine systems. *Geochim. Cosmochim. Acta* **39,** 1531–1544.

Oelsner, C., Beuge, P., Peyer, K. *et al.* (1973). Complex detection of tectonic disturbances by geochemical and geophysical methods (in German). *Neue Bergbautech.* **3,** 482–489.

Oertel, A. C. (1969). Frequency distribution of element concentrations—II. Surface soils and ferromagnesian minerals. *Geochim. Cosmochim. Acta* **33,** 833–840.

Ohmoto, H. (1972). Systematics of sulfur and carbon isotopes in hydrothermal deposits. *Econ. Geol.* **67,** 551–578.

Olade, M. A. (1977). Nature of volatile element anomalies at porphyry copper deposits, Highland Valley, B.C., Canada. *Chem. Geol.* **20,** 235–252.

Olade, M. and Fletcher, K. (1974). Potassium chlorate–hydrochloric acid: a sulfide selective leach for bedrock geochemistry. *J. Geochem. Explor.* **3,** 337–344.

Olade, M. A. and Fletcher, W. K. (1975). Primary dispersion of rubidium and strontium around porphyry copper deposits, Highland Valley, British Columbia. *Econ. Geol.* **70,** 15–21.

Olade, M., Fletcher, K. and Warren, H. V. (1975). Barium–strontium relationships at the Highland Valley porphyry–copper deposits, British Columbia. *Western Miner.* **48**(3), 24.

Oldershaw, W. (1969). A portable vacuum pump and Buchner filter for stream sediment sampling. *Australas. Inst. Mining Metall. Proc.* **232,** 85.

Ollier, C. (1969). "Weathering." Elsevier, New York, 304 pp.

Olmsted, F. H., Loetz, O. J. and Irelan, B. (1973). Geohydrology of the Yuma area, Arizona and California. *U.S. Geol. Survey* Prof. Paper 486-H, 227 pp.

Olson, B. H. and Cooper, R. C. (1974). *In situ* methylation of mercury in estuarine sediment. *Nature, Lond.* **252,** 682–683.

Ong, H. L. and Swanson, V. E. (1966). Adsorption of copper by peat, lignite, and bituminous coal. *Econ. Geol.* **61,** 1214–1231.

Ong, H. L. and Swanson, V. E. (1969). Natural organic acids in the transportation, deposition and concentration of gold. *Colorado School Mines Quart.* **64**(1), 395–426.

Ong, H. L., Swanson, V. E. and Bisque, R. E. (1970). Natural organic acids as agents of chemical weathering. *U.S. Geol. Survey* Prof. Paper 700-C, 130–137.

Ong, P. M. and Sevillano, A. C. (1975). Geochemistry in the exploration of nickelliferous laterites. *Vancouver Symposium Volume* 461–478.

Ostle, D. (1954). Geochemical prospecting for uranium. *Mining Mag., London* **91,** 201–208.

Ostle, D., Coleman, R. F. and Ball, T. K. (1972). Neutron activation analysis as an aid to geochemical prospecting for uranium. *In* "Uranium Prospecting Handbook" (S. H. U. Bowie *et al.*, eds), 95–109. Institution of Mining and Metallurgy, London.

Ottley, D. J. (1966). Gy's sampling slide rule. *Mining Mineralog. Eng.* **2,** 390–395.

Ovchinnikov, L. N. and Baranov, E. N. (1972). Endogenic geochemical aureoles of massive sulfide deposits. *Int. Geol. Rev.* **14,** 419–429.

Ovchinnikov, L. N. and Grigorian, S. V. (1971). Primary halos in prospecting for sulfide deposits. *Toronto Symposium Volume* 375–380.

Ovchinnikov, L. N., Sokolov, V. A., Fridman, A. I. *et al.* (1973). Gaseous geochemical methods in structural mapping and prospecting for ore deposits. *London Symposium Volume* 177–182.

Overstreet, W. C. (1962). A review of regional heavy mineral reconnaissance and its application in the southeastern Piedmont. *Southeastern Geol.* **3,** 133–172.

Oyarzun, M. J. (1975). Rubidium and strontium as guides to copper mineralization emplaced in some Chilean andesitic rocks. *Vancouver Symposium Volume* 333–338.

Park, C. F. and MacDiarmid, R. A. (1975). "Ore Deposits", 3rd edn. W. H. Freeman, San Francisco, 529 pp.

Parker, R. L. and Fleischer, M. (1968). Geochemistry of niobium and tantalum. *U.S. Geol. Survey* Prof. Paper 612, 43 pp.

Parks, G. A. (1965). The isoelectric points of solid oxides, solid hydroxides and aqueous hydroxo-complex systems. *Chem. Rev.* **65,** 177–198.

Parks, G. A. (1972). Free energies of formation and aqueous solubilities of aluminum hydroxides and oxide hydroxides at 25 °C. *Am. Mineralogist* **57,** 1163–1189.

Parks, G. A. (1975). Adsorption in the marine environment. *In* "Chemical Oceanography" (J. P. Riley and G. Skirrow, eds), Vol. 1, 241–301. Academic Press, London and New York.

Parks, G. A. and de Bruyn, P. L. (1962). The zero point of charge of oxides. *J. Phys. Chem.* **66,** 967–973.

Parry, W. T. and Jacobs, D. C. (1975). Fluorine and chlorine in biotite from Basin and Range plutons. *Econ. Geol.* **70,** 554–558.

Parry, W. T. and Nackowski, M. P. (1963). Copper, lead, and zinc in biotites from Basin and Range quartz monzonites. *Econ. Geol.* **58,** 1126–1144.

Patten, L. E. and Ward, F. N. (1962). Geochemical field method for beryllium prospecting. *U.S. Geol. Survey* Prof. Paper 450-C, 103–104.

Payne, A. L. (1973). Exploration for mineral deposits. *In* "Mining Engineering Handbook" (A. B. Cummins and I. A. Given, eds), Vol. 1, 5-2–5-19. Soc. Mining Eng., New York.

Peachey, D. (1976). Extractions of copper from ignited soil samples. *J. Geochem. Explor.* **5,** 129–134.

Peachey, D., Roberts, J. L. and Scot-Baker, J. (1973). Rapid colorimetric determination of phosphorus in geochemical survey samples. *J. Geochem. Explor.* **2,** 115–120.

Peirson, D. H., Cawse, P. A. and Cambray, R. S. (1974). Chemical uniformity of airborne particulate material, and a maritime effect, *Nature, Lond.* **251,** 675–679.

Perel'man, A. I. (1967). "Geochemistry of Epigenesis." Plenum Press, New York, 266 pp.

Perhac, R. M. and Whelan, C. J. (1972). A comparison of water, suspended solid, and bottom sediment analyses for geochemical prospecting in a northeast Tennessee zinc district. *J. Geochem. Explor.* **1,** 47–53.

Perkin-Elmer, (1976). "Analytical Methods for Atomic Absorption Spectrophotometry." Perkin Elmer Co., Norwalk, Connecticut.

Peters, W. C. (1978). "Exploration and Mining Geology." John Wiley, New York, 696 pp.

Peterson, P. J. (1971). Unusual accumulations of elements by plants and animals. *Sci. Progress* **59,** 505–526.

Phillips, W. S. (1963). Depth of roots in soil. *Ecology* **44,** 424.

Pierce, A. P., Gott, G. B. and Mytton, J. W. (1964). Uranium and helium in the Panhandle gas field, Texas, and adjacent areas. *U.S. Geol. Survey* Prof. Paper 454-G, 57 pp.

Piper, C. S. (1950). "Soil and Plant Analysis." Interscience, New York, 368 pp.

Pitulko, V. M. (1968). Features of geochemical searches for rare metal deposits in permafrost areas. *Int. Geol. Rev.* **11,** 1239–1246.

Plant, J. (1971). Orientation studies in stream sediment sampling for a regional geochemical survey in northern Scotland. *Inst. Mining Metall. Trans.* **80,** B324–B345.

Plant, J., Jeffery, K., Gill, E. *et al.* (1975). The systematic determination of accuracy and precision in geochemical exploration data. *J. Geochem. Explor.* **4,** 467–486.

Plant, J., Goode, G. C. and Herrington, J. (1976). An instrumental neutron activation method for multi-element geochemical mapping. *J. Geochem. Explor.* **6,** 299–319.

Plüger, W. L. and Friedrich, G. H. (1973). Determination of total and cold-extractable fluoride in soils and stream sediments with an ion-sensitive fluoride electrode. *London Symposium Volume* 421–428.

Plyusnin, G. S., Volkhova, N. V., Godvinskii, G. P. *et al.* (1972). Trends in the distribution of argon and helium in contact regions and fault zones. (in Russian) English translation announced in *Geochem. Int.* **9,** 882.

Pogorski, L. A. (1975). Geochemical exploration method. U.S. Patents 3 835 710 and 3 862 576.

Pokorny, J. (1975). Geochemical prospecting for ores in the Bohemian massif, Czechoslovakia. *Vancouver Symposium Volume* 77–83.

Polikarpochkin, V. V. (1971). The quantitative estimation of ore-bearing areas from sample data of the drainage systems. *Toronto Symposium Volume* 585–586.

Polikarpochkin, V. V. (1976). "Secondary Aureoles and Trains of Dispersion" (in Russian). Nauka, Novosibirsk, 407 pp.

Polikarpochkin, V. V., Kasyanova, V. I., Utgov, A. A. *et al.* (1958). Geochemical prospecting for polymetallic ore deposits in the eastern Transbaikal by means of the muds and waters of the drainage system. *Int. Geol. Rev.* **2,** 237–253.

Polikarpochkin, V. V., Kitaev, V. A. and Sarapulova, V. N. (1965a). Structure and vertical zonation of the primary dispersion aureoles at the Baley gold deposits. *Geochem. Int.* 1965, 741–753.

Polikarpochkin, V. V., Korotayeva, I. Ya., Grechkina, A. A. *et al.* (1965b). The relation between the solid and liquid phases of dispersion aureoles. *Geokhim. Trans.* 1965, Pt I, 135–164. Am. Geol. Inst. (Abst. *Geochem. Int.* **2,** 140.)

Polynov, B. B. (1937). "The Cycle of Weathering" (A. Muir, translator). Murby, London, 220 pp.

Popov, V. S. and Shil'zhenko, V. N. (1972). Significance of hydrogen sulfide waters in the search for native sulfur deposits (in Russian). English abst. in *Chem. Abst.* **81,** item 155877.

Potap'yev, V. V., Malikova, I. N., Grebennikov, A. M. *et al.* (1967). Geochemical indications of tantalum mineralization in granitoids. *Am. Geol. Inst. Dokl. Acad. Sci. U.S.S.R.* **173,** 202–204.

Prelat, A. E. (1977). Discriminant analysis as a method of predicting mineral occurrence potentials in central Norway. *Math. Geol.* **9,** 343–367.

Presant, E. W. and Tupper, W. M. (1966). The distribution and nature of arsenic in the soils of the Bathurst, New Brunswick, district. *Econ. Geol.* **61,** 760–767.

Press, N. P. (1974). Remote sensing to detect the toxic effects of metals on exploration for mineral deposits. *Proc. 9th Int. Symp. Remote Sensing Environ.* **3,** 2027–2038.

Price, R. J. (1973). "Glacial and Fluvioglacial Landforms." Oliver and Boyd, Edinburgh, 242 pp.

Price, W. J. (1972). "Analytical Atomic Absorption Spectrometry." Heyden and Son, London, 239 pp.

Quin, B. F. and Brooks, R. R. (1972). The rapid determination of tungsten in soils, stream sediments, rocks and vegetation. *Anal. Chim. Acta* **58,** 301–310.

Quin, B. F., Brooks, R. R., Boswell, C. *et al.* (1974). Biogeochemical exploration for tungsten at Barrytown, New Zealand. *J. Geochem. Explor.* **3,** 43–51.

Raeburn, C. and Milner, H. B. (1927). "Alluvial Prospecting." Murby, London, 478 pp.

Rankama, K. and Sahama, T. G. (1950). "Geochemistry." University of Chicago Press, Chicago, 912 pp.

Rashid, M. A. and Leonard, J. D. (1973). Modification of the solubility and precipitation behavior of various metals as a result of their interaction with humic acid. *Chem. Geol.* **11,** 89–97.

Rasmussen, R. A. (1974). Emission of biogenic hydrogen sulfide. *Tellus* **26,** 254–260.

Reedman, J. H. (1974). Residual soil geochemistry in the discovery and evaluation of the Butiriku carbonatite, southeast Uganda. *Inst. Mining Metall. Trans.* **83,** B1–B12.

Rehrig, W. A. and McKinney, C. N. (1976). The distribution and origin of anomalous copper in biotite (abst.). *Mining Eng.* **27**(12), 68c.

Reiche, P. (1950). A survey of weathering process and products. *Univ. New Mexico Pub. in Geol.* No. 3, 95 pp.

Reimer, G. M. (1975). Uranium determination in natural water by the fission track technique. *J. Geochem. Explor.* **4,** 425–431.
Reimer, G. M., Roberts, A. A. and Denton, E. H. (1976). The use of helium detection to locate energy resources (abst.). *Geol. Soc. Am. Abst. with Programs* **8**(6), 1063–1064.
Reinking, R. L., Houghton, R. L. and Peterson, J. A. (1973). Trace metal dispersion at the Idarado Mine, Silverton, Colorado (abst.). *Geol. Soc. Am. Abst. with Programs* **5,** 506.
Reuter, J. H. and Perdue, E. M. (1977). Importance of heavy metal–organic matter interactions in natural waters. *Geochim. Cosmochim. Acta* **41,** 325–334.
Riddell, J. E. (1954). Geochemical soil and water surveys in Lemieux Township, Gaspe-North County, Quebec. *Quebec Dept Mines* Prelim. Rept 302, 23 pp.
Ridler, R. H. and Shilts, W. W. (1974). Exploration for Archaean polymetallic sulphide deposits in permafrost terrains: An integrated geological/geochemical technique, Kaminak Lake area, District of Keewatin. *Geol. Survey Can.* Paper 73-34, 33 pp.
Riley, G. A. (1937). The copper cycle in natural waters and its biological significance. Ph.D. Thesis, Yale University.
Robie, R. A., Hemingway, B. S. and Fisher, J. R. (1978). Thermodynamic properties of minerals and related substances at 298.15 K and 1 bar (10^5 pascals) pressure and at higher temperatures. *U.S. Geol. Survey Bull.* **1452,** 456 pp.
Robinson, T. W. (1958). Phreatophytes. *U.S. Geol. Survey* Water-supply Paper 1423, 84 pp.
Roedder, E. (1977). Fluid inclusions as tools in mineral exploration. *Econ. Geol.* **72,** 503–525.
Roscoe, W. (1971). Probability for an exploration discovery in Canada. *Can. Inst. Mining Metall. Bull.* **64**(107), 134–137.
Rose, A. W. (1970). Zonal relations of wall rock alteration and sulfide distribution at porphyry copper deposits. *Econ. Geol.* **65,** 920–936.
Rose, A. W. (1972). Favorability for Cornwall-type magnetite deposits in Pennsylvania, using geological, geochemical and geophysical data in a discriminant equation. *J. Geochem. Explor.* **1,** 181–194.
Rose, A. W. (1975). The mode of occurrence of trace elements in soils and stream sediments applied to geochemical exploration. *Vancouver Symposium Volume* 691–705.
Rose, A. W. (1976). The effect of cuprous chloride complexes in the origin of red-bed copper and related deposits. *Econ. Geol.* **71,** 1036–1048.
Rose, A. W. and Keith, M. L. (1976). Reconnaissance geochemical techniques for detecting uranium deposits in sandstones of northeastern Pennsylvania. *J. Geochem. Explor.* **6,** 119–137.
Rose, A. W. and Korner, L. A. (1979). Radon in natural waters and a guide to uranium deposits in Pennsylvania. *Golden Symposium Volume* (*1979*) 65–76.
Rose, A. W. and Suhr, N. H. (1971). Major element content as a means of allowing for background variation in stream-sediment geochemical exploration. *Toronto Symposium Volume* 587–593.
Rose, A. W., Dahlberg, E. C. and Keith, M. L. (1970). A multiple regression technique for adjusting background values in stream sediment geochemistry. *Econ. Geol.* **65,** 156–165.
Rose, A. W., Schmiermund, R. L. and Mahar, D. L. (1977). Geochemical dispersion of uranium near prospects in Pennsylvania. *U.S. Energy Res. Dev. Adm.* Open-file Rept GJBX-59(77), 87 pp.
Rösler, H. J., Beuge, P., Pilot, J. *et al.* (1977). Integrated geochemical exploration for deep-seated solid and gaseous mineral resources. *J. Geochem. Explor.* **8,** 415–429.

Rossman, G. I., Sychev, I. V., Tarkhanova, G. A. *et al.* (1972). Use of radiogenic Pb^{206} halos in prospecting for uranium in acid volcanic rocks. *Int. Geol. Rev.* **14,** 332–337.

Rouse, G. E. and Stevens, D. N. (1971). The use of sulfur dioxide gas in the detection of sulfide deposits (abst.). *Mining Eng.* **22**(12), 65.

Rudolfs, W. and Heilbronner, A. (1922). Oxidation of zinc sulfide by microorganisms. *Soil Sci.* **14,** 459–464.

Ruotsala, A. P., Nordeng, S. C. and Weege, R. J. (1969). Trace elements in accessory calcite—a potential exploration tool in the Michigan copper district. *Colorado School Mines Quart.* **64**(1), 451–455.

Russell, M. J. (1974). Manganese halo surrounding the Tynagh ore deposits, Ireland—a preliminary note. *Inst. Mining Metall. Trans.* **83,** B65–B66.

Russell, M. J. (1975). Lithogeochemical environment of the Tynagh base metal deposit, Ireland, and its bearing on ore deposition. *Inst. Mining Metall. Trans.* **84,** B128–B133.

Ruxton, B. P. (1958). Weathering and surface erosion in granite. *Geol. Mag., U.K.* **95,** 353–377.

Rye, R. O. and Ohmoto, H. (1974). Sulfur and carbon isotopes in ore genesis: A review. *Econ. Geol.* **69,** 826–842.

Rye, D. M. and Rye, R. O. (1974). Homestake gold mine, South Dakota: I. Stable isotope studies. *Econ. Geol.* **69,** 293–317.

Sainsbury, C. L. (1957). A geochemical exploration for antimony in southeastern Alaska. *U.S. Geol. Survey Bull.* **1024-H,** 163–1678.

Sainsbury, C. L. and Reed, B. L. (1973). Tin. *U.S. Geol. Survey* Prof. Paper 820, 637–651.

Sainsbury, C. L., Curry, K. J. and Hamilton, J. C. (1973). An integrated system of geologic mapping and geochemical sampling by light aircraft. *U.S. Geol. Survey Bull.* **1361,** 28 pp.

Salmi, M. (1950). Trace elements in peat (in Finnish). *Geotek. Julkaisuja* No. 51, 20 pp.

Salmi, M. (1955). Prospecting for bog-covered ore by means of peat investigations. *Fin. Comm. Geol. Bull.* **169,** 5–34.

Salmi, M. (1956). Peat and bog plants as indicators of ore minerals in Vihanti ore field in western Finland. *Finl. Comm. Geol. Bull.* **175,** 1–22.

Salmi, M. (1967). Peat in prospecting applications in Finland. *In* "Geochemical Prospecting in Fennoscandia" (A. Kvalheim, ed.), 113–126. Interscience, New York.

Sandell, E. B. (1959). "Colorimetric Determination of Traces of Metals", 3rd edn. Interscience, New York, 1032 pp.

Sato, M. (1960). Oxidation of sulfide orebodies. *Econ. Geol.* **55,** 928–961.

Sato, M. and Mooney, H. M. (1960). The electrochemical mechanism of sulfide self-potentials. *Geophysics* **25,** 226–249.

Schlegel, H. G. (1974). Production, modification, and consumption of atmospheric trace gases by micro-organisms. *Tellus* **26,** 11–20.

Schmidt, R. C. (1956). Adsorption of copper, lead, and zinc on some common rock-forming minerals and its effect on lake sediments. Ph.D. Thesis, McGill University, Montreal.

Schmidt, U. (1974). Molecular hydrogen in the atmosphere. *Tellus* **26,** 78–90.

Schmitt-Colerus, J. J. (1967, 1969). Investigations of the relationships between organic matter and uranium deposits. *Atomic Energy Comm.* Open-file Repts GJO-933-1 and GJO-933-2, 141 pp. and 192 pp.

Schnitzer, M. (1976). The chemistry of humic substances. *In* "Environmental Biogeochemistry" (J. O. Nraigu, ed.), Vol. 1, 89–108. Ann Arbor Sci. Pub., Ann Arbor, Michigan.

Schnitzer, M. and Hanson, E. H. (1970). Organometallic interactions in soils: 8. An evaluation of methods for the determination of stability constants for metal-fulvic acid complexes. *Soil Sci.* **109,** 333–340.

Schroeder, G. L., Kraner, H. W. and Evans, R. D. (1965). Diffusion of radon in several naturally occurring soil types. *J. Geophys. Res.* **70,** 471–474.

Schuiling, R. D. (1967). Tin belts on the continents around the Atlantic Ocean. *Econ. Geol.* **62,** 540–550.

Schwartz, M. O. and Friedrich, G. H. (1973). Secondary dispersion patterns of fluoride in the Osor area, Province of Gerona, Spain. *J. Geochem. Explor.* **2,** 103–114.

Schwickerath, M. (1931). *Viola calaminariae* in zinc soils near Aachen (in German). *Beitr. NatDenkmPflege* **14,** 463–503.

Scott, K. M. and Taylor, G. F. (1977). Geochemistry of the Mammoth copper deposit, northwest Queensland, Australia. *J. Geochem. Explor.* **8,** 153–168.

Scott, M. J. (1975). Case histories from a geochemical exploration programme—Windhoek district, South-West Africa. *Vancouver Symposium Volume* 481–492.

Scott, R. C. and Barker, F. B. (1962). Data on uranium and radium in ground water in the United States, 1954 to 1957. *U.S. Geol. Survey* Prof. Paper 426, 115 pp.

Searle, P. L. (1968). Determination of total sulphur in soil using high frequency induction furnace equipment. *Analyst* **93,** 540–545.

Seccombe, P. K. (1977). Sulphur isotopes and trace metal composition of stratiform sulphides as an ore guide in the Canadian Shield. *J. Geochem. Explor.* **8,** 117–137.

Seiler, W. (1974). The cycle of atmospheric CO. *Tellus* **26,** 116–135.

Sergeyev, Ye. A. (1946). Water analysis as a means of prospecting for metallic ore deposits (in Russian). English translation in "Selected Russian Papers on Geochemical Prospecting for Ores", 7–12. U.S. Geol. Survey, Washington.

Sesterenko, G. V. and Smirnova, N. P. (1964). The chromium–vanadium ratio as a search indicator of ore-bearing differentiated trap rocks of the Norilsk type. *Am. Geol. Inst. Dokl. Acad. Sci. U.S.S.R.* **154,** 162–165.

Shacklette, H. T. and Cuthbert, M. E. (1967). Iodine content of plant groups as influenced by variation in rock and soil type. *Geol. Soc. Am.* Spec. Paper 90, 30–46.

Shapiro, L. (1973). Rapid determination of sulfur in rocks. *U.S. Geol. Survey J. Res.* **1,** 81–84.

Shapiro, L. (1975). Rapid analysis of silicate, carbonate, and phosphate rocks—revised edition. *U.S. Geol. Survey Bull.* **1401,** 76 pp.

Sharpe, C. F. S. (1938). "Landslides and Related Phenomena." Columbia University Press, New York, 137 pp.

Shaw, D. M. (1961). Element distribution laws in geochemistry. *Geochim. Cosmochim. Acta* **23,** 116–134.

Sheraton, J. W. and Black, L. P. (1973). Geochemistry of mineralized granitic rocks of northeast Queensland. *J. Geochem. Explor.* **2,** 331–348.

Sheremet, Ye. M., Gormasheva, G. S. and Legeydo, V. A. (1973). Geochemical criteria for the productivity of potential ore-bearing granitoids in the Gudzhir intrusive complex in western Transbaikalia. *Geochem. Int.* **10,** 1125–1135.

Sherman, G. D. (1952). The genesis and morphology of the alumina-rich laterite clays. *In* "Problems of Clay and Laterite Genesis" (A. F. Frederickson, ed.), 154–161. Am. Inst. Mining Eng., New York.

Shilts, W. W. (1971). Till studies and their application to regional drift prospecting. *Can. Mining J.* **92,** 45–50.

Shilts, W. W. (1973). Till indicator train formed by glacial transport of nickel and other ultrabasic components: a model for drift prospecting. *Geol. Survey Can.* Paper 73-1A, 213–218.

Shipulin, F. K., Genkin, A. D., Distler, V. V. *et al.* (1973), Some aspects of the problem of geochemical methods of prospecting for concealed mineralization. *J. Geochem. Explor.* **2,** 193–235.

Shmakin, B. M. (1973). Geochemical specialization of the Indian Precambrian pegmatites in relation to alkali and ore element contents of the minerals. *Geochem. Int.* **10,** 890–899.

Shmakin, B. M., Makryghina, V. A., Glebov, M. P. *et al.* (1971). Use of the petrographical–geochemical prospecting method for the discovery of hidden muscovite deposits in different geological environments. *Toronto Symposium Volume* 391–393.

Shvartsev, S. L. (1972). Geochemical prospecting methods in regions of permanently frozen ground. *Proc. 24th Int. Geol. Cong., Montreal* **10,** 380–383.

Shvartsev, S. L. (1975). Volume and composition evolution for infiltration groundwater in alumino-silicate rocks. *Geochem. Int.* **12,** 184–194.

Shvartsev, S. L. (1976). Electrochemical dissolution of sulphide ores. *J. Geochem. Explor.* **5,** 71–72.

Shvartsev, S. L., Udodov, P. A. and Rasskazov, N. M. (1975). Some features of the migration of microcomponents in neutral waters of the supergene zone. *J. Geochem. Explor.* **4,** 433–439.

Shvartseva, N. M. (1972). Antimony in ground water of the Kadamdzhay deposit. *Am. Geol. Inst. Dokl. Acad. Sci. U.S.S.R.* **207,** 224–226.

Siegel, F. R. (1974). "Applied Geochemistry." Wiley-Interscience, New York, 353 pp.

Siegel, S. (1956). "Nonparametric Statistics for the Behavioral Sciences." McGraw-Hill, New York, 312 pp.

Silberman, M. L., Carten, R. B. and Armstrong, A. K. (1974). Geologic and geochemical maps showing distribution and abundance of Cu, Pb, Zn, Bi, Mo, W, and Ag, central Peloncillo Mountains, Hidalgo County, New Mexico. *U.S. Geol. Survey* Open-file Rept 74–112.

Sillen, L. G. and Martell, A. E. (1964, 1970). "Stability Constants of Metal-Ion Complexes." The Chemical Society, London, Spec. Pubs 17 and 25. 754 pp. and 865 pp.

Simonson, R. W. (1957). What soils are. *In* "1957 Yearbook of Agriculture", 17–31. U.S. Dept Agriculture, Washington.

Sinclair, A. J. (1974). Selection of thresholds in geochemical data using probability graphs. *J. Geochem. Explor.* **3,** 129–149.

Sinclair, A. J. (1975). Some considerations regarding grid orientation and sample spacing. *Vancouver Symposium Volume* 133–140.

Sinclair, A. J. (1976). "Probability Graphs." Assoc. Exploration Geochemists, Spec. Vol. No. 4, 95 pp.

Sindeyeva, N. D. (1955). A geochemical indicator of pyrite deposits (in Russian). *Dokl. Akad. Nauk S.S.S.R.* **104**(1), 114.

Singer, P. C., ed. (1973). "Trace Metals and Metal–Organic Interactions in Natural Waters." Ann Arbor Sci. Pub., Ann Arbor, Michigan, 380 pp.

Skinner, D. N. B. (1969). The use of flocculants in geochemical prospecting. *Australas. Inst. Mining Metall. Proc.* **230,** 49–50.

Skinner, R. G. (1972). Drift prospecting in the Abitibi clay belt. *Geol. Survey Can.* Open-file Rept 116.

Slawson, W. F. and Nackowski, M. P. (1959). Trace lead in potash felspar associated with ore deposits. *Econ. Geol.* **54,** 1543–1555.

Slichter, L. B. (1955). Geophysics applied to prospecting for ores. *In* "Economic Geology 50th Anniversary Volume", 885–969. Economic Geology Pub. Co., Lancaster, Pennsylvania.

Smith, A. Y. and Lynch, J. J. (1969). Field and laboratory methods used in geochemical prospecting by the Geological Survey of Canada, No. 11, Uranium in soil, stream sediment, and water. *Geol. Survey Can.* Paper 69–40.

Smith, A. Y., Barretto, P. M. C. and Pournis, S. (1976). Radon methods in uranium exploration. *In* "Exploration for Uranium Ore Deposits", 185–211. International Atomic Energy Agency, Vienna.

Smith, F. W. (1974). Yttrium content of fluorite as a guide to vein intersections in partially developed fluorspar ore bodies. *Soc. Mining Eng. Trans.* **256,** 95–96.

Smith, K. A. (1977). Gas chromatographic analysis of the soil atmosphere. *Adv. Chromatogr.* **15,** 197–231.

Smith, R. M. and Martell, A. E. (1976). "Critical Stability Constants", Vol. 4 "Inorganic Complexes". Plenum Press, New York, 257 pp.

Smith, W. C. (1960). Borax and borates. *In* "Industrial Minerals and Rocks", 103–122. Am. Inst. Mining Eng., New York.

Soil Survey Staff (1975). "Soil Taxonomy." U.S. Dept. Agriculture, Handbook 436, Washington D.C., 754 pp.

Sokal, R. R. and Sneath, P. H. (1973). "Principles of Numerical Taxonomy." W. H. Freeman, San Francisco, 573 pp.

Sokolov, V. A. (1971). The theoretical foundations of geochemical prospecting for petroleum and natural gas and the tendencies of its development. *Toronto Symposium Volume* 544–549.

Solovov, A. P. (1959). "The Theory and Practice of Metallometric Surveys" (in Russian). Akad. Nauk Kazakhskoy S.S.R., Alma-Ata.

Solovov, A. P. and Kunin, N. Ya. (1961). Metallometric surveying in mountainous areas based on alluvial deposits. *Int. Geol. Rev.* **3,** 998–1010.

Soonawala, N. M. (1974). Data processing techniques for the radon method of uranium exploration. *Can. Inst. Mining Metall. Bull.* **67**(744), 110–116.

Sotnikov, V. I. and Izyumova, L. G. (1965). Tungsten in the granite massifs of various ore districts, Gornyy Altay (in Russian). English translation announced in *Geochem. Int.* **2,** 132.

Stacey, J. S., Zartman, R. E. and Nkomo, I. T. (1968). A lead isotope study of galenas and selected feldspars from mining districts in Utah. *Econ. Geol.* **63,** 796–814.

Staker, E. V. and Cummings, R. W. (1941). The influence of zinc on the productivity of certain New York peat soils. *Soil Sci. Soc. Am. Proc.* **6,** 207.

Stanton, R. E. (1966). "Rapid Methods of Trace Analysis for Geochemical Application." Edward Arnold, London, 96 pp.

Stanton, R. E. (1976). "Analytical Methods for Use in Geochemical Exploration." Halsted Press, New York, 55 pp.

Stanton, R. E., Mockler, M. and Newton, S. (1973). The colorimetric determination of molybdenum in organic-rich soil. *J. Geochem. Explor.* **2,** 37–40.

Steenfelt, A., Kunzendorf, H. and Friedrich, G. H. W. (1976). Randbøldal—uranium in stream sediments, soils and water. *J. Geochem. Explor.* **5,** 300–305.

Stevens, D. N., Bloom, D. N. and Bisque, R. E. (1969). Evaluation of mercury vapor anomalies at Colorado Central Mines, Clear Creek County, Colorado (abst.). *Colorado School Mines Quart.* **64**(1), 513–514.

Stevens, D. N., Rouse, G. E. and Devoto, R. H. (1971). Radon-222 in soil gas: three uranium exploration case histories in the western United States. *Toronto Symposium Volume* 258–264.

Stevenson, F. J. and Ardakani, M. S. (1972). Organic matter reactions involving micronutrients in soils. *In* "Micronutrients in Agriculture" (J. J. Mortvedt *et al.*, eds), 79–114. Soil Sci. Soc. Am., Madison, Wisconsin.

Stoiber, R. E. and Rose, W. I., Jr (1974). Cl, F, and SO_2 in Central American volcanic gases. *Bull. Volcanol.* **37,** 454–460.

Stoll, W. C. (1945). The presence of beryllium and associated chemical elements in the wall rocks of some New England pegmatites. *Econ. Geol.* **40**, 136–141.

Stollery, G., Borcsik, M. and Holland, H. D. (1971). Chlorine in intrusives: A possible prospecting tool. *Econ. Geol.* **66,** 361–367.

Stumm, W. and Morgan, J. J. (1970). "Aquatic Chemistry." Wiley-Interscience, New York, 583 pp.

Sultankhodzhaev, A. N. (1974). Use of radon to forecast earthquakes (in Russian). Abst. in *Chem. Abst.* **82,** item 6157.

Summerhayes, C. P., Hazelhoff-Roelfzema, B. H., Tooms, J. S. *et al.* (1970). Phosphorite prospecting using a submersible scintillation counter. *Econ. Geol.* **65,** 718–723.

Sutherland-Brown, A. (1967). Investigation of mercury dispersion halos around mineral deposits in central British Columbia. *Geol. Survey Can.* Paper 66-54, 73–83.

Suvanasingha, A. (1963). Geochemical study of chromium in soil over the chromite deposit of the Narathivas Province, southern Thailand. *In* "Proc. the Seminar on Geochemical Prospecting Methods and Techniques", 164–168. Min. Res. Dev. Ser. No. 21, United Nations, New York.

Swain, F. M. (1970). "Non-marine Organic Geochemistry." Oxford University Press, Oxford, 445 pp.

Swinnerton, J. W., Linnenbom, V. J. and LaMontagne, R. A. (1970). The ocean, a natural source of carbon monoxide. *Science* **67,** 984–986.

Sydney Symposium Volume (1977). "Geochemical Exploration 1976" (C. R. M. Butt and I. G. P. Wilding, eds), 6th Int. Geochem. Explor. Symp., Sydney. Elsevier, Amsterdam, 494 pp. Also published in *J. Geochem. Explor.* **8,** 1–494.

Szalay, A. (1964). Cation exchange properties of humic acids and their importance in geochemical enrichment of UO_2^{2+} and other cations. *Geochim. Cosmochim. Acta* **28,** 1605–1614.

Tanner, A. B. (1964). Radon migration in the ground. *In* "Natural Radiation Environment", 161–190. Int. Symp., Houston, 1963. University of Chicago Press, Chicago.

Tanuskanen, H. (1976). Factors affecting the metal contents in peat profiles. *J. Geochem. Explor.* **5,** 412–414.

Tardy, Y. (1971). Characterization of the principal weathering types by the geochemistry of waters from some European and African massifs. *Chem. Geol.* **7,** 253–271.

Tauson, L. V. and Kozlov, V. D. (1973). Distribution functions and ratios of trace element concentrations as estimators of the ore-bearing potential of granites. *London Symposium Volume* 37–44.

Tauson, L. V. and Petrovskaya, S. G. (1971). Endogenic halo types of hydrothermal molybdenum deposits. *Toronto Symposium Volume* 394–396.

Taylor, H. P., Jr (1974). The application of oxygen and hydrogen isotope studies to problems of hydrothermal alteration and ore deposition. *Econ. Geol.* **69,** 843–883.

Taylor, S. R. (1964). Abundance of chemical elements in the continental crust: a new table. *Geochim. Cosmochim. Acta* **28,** 1273–1286.

Tedrow, J. C. F. (1977). "Soils of the Polar Landscapes." Rutgers University Press, New Brunswick, New Jersey, 638 pp.

Tenhola, M. (1976). Ilomantsi: distribution of uranium in lake sediments. *J. Geochem. Explor.* **5,** 235–239.

Tennant, C. B. and White, M. L. (1959). Study of the distribution of some geochemical data. *Econ. Geol.* **54,** 1281–1290.

Terashima, S. (1976). The determination of arsenic in rocks, sediments, and minerals by arsine generation and atomic absorption spectrometry. *Anal. Chim. Acta* **86,** 43–52.

Theobald, P. K. (1957). The gold pan as a quantitative geologic tool. *U.S. Geol. Survey Bull.* **1071-A,** 1–54.

Theobald, P. K. and Thompson, C. E. (1959). Geochemical prospecting with heavy-mineral concentrates used to locate a tungsten deposit. *U.S. Geol. Survey* Circ. 411, 13 pp.

Theobald, P. K. and Thompson, C. E. (1961). Relations of metals in lithosols to alteration and shearing at Red Mountain, Clear Creek County, Colorado. *U.S. Geol. Survey* Prof. Paper 424-B, 139–141.

Theobald, P. K., Lakin, H. W. and Hawkins, D. B. (1963). The precipitation of aluminum, iron, and manganese at the junction of Deer Creek with the Snake River in Summit County, Colo. *Geochim. Cosmochim Acta* **27,** 121–132.

Theodore, T. G. and Blake, D. W. (1975). Geology and geochemistry of the Copper Canyon porphyry copper deposit and surrounding area, Lander County, Nevada. *U.S. Geol. Survey* Prof. Paper 798-B, 86 pp.

Thompson, I. S. (1967). The discovery of the Gortdrum deposit, Co. Tipperary, Ireland. *Inst. Mining Metall. Trans.* **70,** 85–92.

Thompson, M. and Howarth, R. J. (1976). Duplicate analysis in geochemical practice. Part I. Theoretical approach and estimation of analytical reproducibility. *Analyst* **101,** 690–698.

Thornton, I. and Alloway, B. J. (1974). Geochemical aspects of the soil–plant–animal relationship in the development of trace element deficiency and excess. *Proc. Nut. Soc.* **33,** 257–266.

Thorp, J. (1931). The effect of vegetation and climate upon soil profiles in northern and northwestern Wyoming. *Soil Sci.* **32,** 283–301.

Thorp, J. and Baldwin, M. (1938). New nomenclature of the higher categories of soil classification as used in the Department of Agriculture. *Soil Sci. Soc. Am. Proc.* **3,** 260–268.

Thurlow, J. G., Swanson, E. A. and Strong, D. F. (1975). Geology and lithogeochemistry of the Buchans polymetallic sulfide deposits. Newfoundland. *Econ. Geol.* **70,** 130–144.

Tikhomirov, V. V. and Tikhomirova, V. G. (1971). Origin of nitrogen–helium anomalies in formation waters of the Dnieper–Donets Basin (in Russian). English abst. in *Chem Abst.* **78,** item 113928.

Tilsley, J. E. (1975). Application of geophysical geochemical, and deep-overburden sampling techniques to exploration for fluorspar in glaciated terrain. *London–Symposium Volume* 126–133.

Tilsley, J. E. (1977). Placosols: another problem in exploration geochemistry. *J. Geochem. Explor.* **7,** 21–30.

Timperley, M. H. and Allan, R. J. (1974). The formation and detection of metal dispersion halos in organic sediments. *J. Geochem. Explor.* **3,** 167–190.

Timperley, M. H., Brooks, R. R. and Peterson, P. J. (1972). Trend analysis as an aid to the comparison and interpretation of biogeochemical and geochemical data. *Econ. Geol.* **67,** 669–676.

Tinney, J. F. (1977). Hydrochemical and stream sediment reconnaissance program at Lawrence Livermore Laboratories. *U.S. Dept. Energy* Open-file Rept GJBX-77(77), 377–394.

Tischendorf, G. (1969). On the causal relationships between granitoids and endogenous tin deposits (in German). *Z. Angew. Geol.* **15,** 333–342.

Tischendorf, G. (1973). The metallogenetic basis of tin exploration in the Erzgebirge. *Inst. Mining Metall. Proc.* **82,** B9–B24.

Tolstoy, M. I., Ostafiychuk, I. M. and Gudimenko, L. M. (1965). The types of statistical distribution curves for chemical elements in rocks and the methods of computing their parameters. *Geochem. Int.* **2,** 993–1000.

Tooms, J. S. (1955). Geochemical dispersions related to copper mineralization in Northern Rhodesia. Ph.D. Thesis, Imperial College, London.

Tooms, J. S. (1972). Potentially exploitable marine minerals. *Endeavour* **31,** 113–117.

Tooms, J. S. and Webb, J. S. (1961). Geochemical prospecting investigations in the Northern Rhodesian Copperbelt. *Econ. Geol.* **56,** 815–846.

Toronto Symposium Volume (1971). "Geochemical Exploration" (R. W. Boyle, ed.), 3rd Int. Geochem. Explor. Symp., Toronto. Can. Inst. Mining Metall. Spec. Vol. No. 11, 594 pp.

Toth, J. (1963). A theoretical analysis of groundwater flow in small drainage basins. *J. Geophys. Res.* **68,** 4695–4812.

Trask, P. D., ed. (1950). "Applied Sedimentation." John Wiley, New York, 707 pp.

Travis, G. A., Keays, R. R. and Davison, R. M. (1976). Palladium and iridium in the evaluation of nickel gossans in Western Australia. *Econ. Geol.* **71,** 1229–1242.

Trost, P. B. and Trautwein, C. (1975). Interpretation of molybdenum content in groundwater utilizing Eh, pH and conductivity. *Vancouver Symposium Volume* 531–545.

Truesdell, A. H. and Jones, B. F. (1974). WATEQ, a computer program for calculating chemical equilibria of natural waters. *U.S. Geol. Survey J. Res.* **2,** 233–248.

Tugarinov, A. I. and Osipov, Yu. G. (1974). Flow of helium through granite massifs and zones containing radioactive ores. *Geochem. Int.* **11,** 981–986.

Turekian, K. K. (1977). Geochemical distribution of elements. *In* "Encyclopedia of Science and Technology", 4th edn, 627–630. McGraw-Hill, New York.

Turk, L. J. (1975). Diurnal fluctuations of water tables induced by atmospheric pressure changes. *J. Hydrol.* **26,** 1–16.

Turkin, Yu. I., Sveshnikov, G. B., Svistov, P. F. *et al.* (1971). Application of a remote distance method of absorption spectroscopy in prospecting geochemistry (in Russian). English abst. in *Chem. Abst.* **78,** item 126818.

Turneaure, F. S. (1955). Metallogenetic provinces and epochs. *In* "Economic Geology 50th Anniversary Volume", 38–89. Economic Geology Pub. Co., Lancaster, Pennsylvania.

Tyminskii, V. G. and Sultankhodzhaev, A. N. (1973). Radiohydrogeochemical anomalies as a reflection of the seismotectonic activity of a region (in Russian). English abst. in *Chem. Abst.* **79,** item 128210.

Udodov, P. A. and Parilov, Y. S. (1961). Certain regularities of migration of metals in natural waters. *Geochemistry* 1961, 762–771.

United Nations (1956). "Geology of Uranium and Thorium", Proc. 1st Int. Conf. Peaceful Uses Atomic Energy, Geneva, 8–20 Aug., 1955, **6,** 825 pp.

United Nations (1958). "Survey of Raw Material Resources", Proc. 2nd Int. Conf. Peaceful Uses Atomic Energy, Geneva, 1–13 Sept., 1958, **2,** 843 pp.

United States Environmental Protection Agency (1977), "Quality Criteria for Water, July 1976." U.S. Env. Prot. Agency, Washington, 256 pp.

Usik, L. (1969). Review of geochemical and geobotanical prospecting methods in peatlands. *Geol. Survey Canada* Paper 68-66, 43 pp.

Vancouver Symposium Volume (1975). "Geochemical Exploration 1974" (I. L. Elliott and W. K. Fletcher, eds), 5th Int. Geochem. Explor. Symp., Vancouver. Elsevier, Amsterdam, 720 pp.

Van Loon, J. C., Kesler, S. E. and Moore, C. M. (1973). Analyses of water-extractable chloride in rocks by use of a selective ion electrode. *London Symposium Volume* 429–434.

Van Tassell, R. E. (1969). Exploration by overburden drilling at Keno Hill Mines, Limited. *Colorado School Mines Quart.* **64**(1), 457–478.

van Wembeke, L. (1960). Geochemical prospecting and appraisal of niobium-bearing carbonatites by X-ray methods. *Econ. Geol.* **55,** 732–758.

Varlamoff, N. (1969). The bearing of tin minerals and ores in the weathering zone and the possibility of geochemical exploration for tin. *Colorado School Mines Quart.* **64**(1), 479–496.

Varshal, G. M., Koshcheeva, I. Ya. and Morozova, R. P. (1972). Possible forms of ruthenium migration in surface waters. *Geochem. Int.* **9,** 669–673.

Viktorov, S. V. (1955). "Applications of Geobotanical Method for Geological and Hydrogeological Investigations" (in Russian). Akad. Nauk S.S.S.R., Moscow, 200 pp.

Viljoen, R. P., Saager, R. and Viljoen, M. J. (1970). Some thoughts on the origin and processes responsible for the concentration of gold in the early Precambrian of southern Africa. *Mineralium Deposita* **5,** 164–180.

Vinogradov, A. P. (1954). Search for ore deposits by means of plants and soils (in Russian). *Akad. Nauk S.S.S.R., Trudy Biogeokhim. Lab.* **10,** 3–27.

Vinogradov, A. P. (1956). Regularity of distribution of chemical elements in the earth's crust. *Geochemistry* **1,** 1–43.

Vinogradov, A. P. (1959). "The Geochemistry of Rare and Dispersed Elements in Soils", 2nd edn. Consultants Bureau, New York, 209 pp.

Vinogradov, A. P. (1962). Average contents of chemical elements in the principal types of igneous rocks of the earth's crust. *Geochemistry* 1962, 641–664.

Vinogradov, A. P. and Malyuga, D. P. (1957). Biogeochemical methods of prospecting for ore deposits (in Russian). *In* "Geochemical Prospecting for Ore Deposits" (V. I. Krasnikov, ed.), 290–300. Gosgeoltekhizdat, Moscow.

Vinogradov, V. I. (1957). On the migration of molybdenum in the supergene zone. *Geochemistry* **2,** 120–126.

Virkkala, K. (1958). Stone counts in the esker of Hameenlinna, southern Finland. *Fin. Comm. Geol. B* **180,** 87–103.

von Olphen, H. (1963). "An Introduction to Clay Colloid Chemistry." Interscience, New York, 301 pp.

Voronin, D. V. and Gol'dberg, I. S. (1972). Electrochemical processes in placer deposits of native metals. *Am. Geol. Inst. Dokl. Acad. Aci. U.S.S.R.* **207,** 193–195.

Wagman, D. D., Evans, W. H., Parker, V. B. *et al.* (1968, 1969, 1971). "Selected Values of Chemical Thermodynamic Properties", U.S. Natl. Bur. Standards, Washington, Tech. Notes 270-3 (264 pp.), 270-4 (141 pp.), 270-5 (37 pp.).

Wahlberg, J. S., Baker, J. H., Vernon, R. W. *et al.* (1965). Exchange adsorption of strontium on clay minerals. *U.S. Geol. Survey Bill.* **1140-C,** 26 pp.

Wainerdi, R. E. and Uken, E. A., eds (1971). "Modern Methods of Geochemical Analysis." Plenum Press, New York, 397 pp.

Ward, F. N. (1970). Analytical methods for the determination of mercury in rocks and soils. *U.S. Geol. Survey* Prof. Paper 713, 46–49.

Ward, F. N. (1975). New and refined methods of trace analysis useful in geochemical exploration. *U.S. Geol. Survey Bull.* **1408,** 105 pp.

Ward, F. N., Lakin, H. W. and Canney, F. C. (1963). Analytical methods used in geochemical exploration by the U.S. Geological Survey. *U.S. Geol. Survey Bull.* **1152,** 100 pp.

Ward, F. N., Nakagawa, H. M., Harms, T. F. *et al.* (1969). Atomic absorption methods of analysis useful in geochemical exploration. *U.S. Geol. Survey Bull.* **1289,** 45 pp.

Wargo, J. G. (1964). Geology of a disseminated copper deposit near Cerrillos, New Mexico. *Econ. Geol.* **59,** 1525–1538.

Warren, H. V. and Delavault, R. E. (1949). Further studies in biogeochemistry. *Geol. Soc. Am. Bull.* **60,** 531–560.

Warren, H. V. and Delavault, R. E. (1954). Variations in the nickel content of some Canadian trees. *Roy. Soc. Can. Trans.* **48**(IV), 71–74.

Warren, H. V. and Delavault, R. E. (1957). Biogeochemical prospecting for cobalt. *Roy. Soc. Can. Trans.* **51**(IV), 33–37.

Warren, H. V., Delavault, R. E. and Irish, R. I. (1952). Biogeochemical investigations in the Pacific Northwest. *Geol. Soc. Am. Bull.* **63,** 435–484.

Warren, H. V., Delavault, R. E. and Routley, D. G. (1953). Preliminary studies of the biogeochemistry of molybdenum. *Roy. Soc. Can. Trans.* **47,** 71–75.

Warren, H. V., Delavault, R. E. and Barakso, J. (1964). The role of arsenic as a pathfinder in biogeochemical prospecting. *Econ. Geol.* **59,** 1381–1389.

Warth, H. (1905). Weathered dolerite of the Rowley Regis (South Staffordshire) compared with the laterite of the Western Ghats near Bombay. *Geol. Mag., U.K.* **V**(2), 21–23.

Watling, R. J., Davis, G. R. and Meyer, W. T. (1973). Trace identification of mercury compounds as a guide to sulfide mineralization at Keel, Eire. *London Symposium Volume* 59–70.

Watson, J. P. (1970). Contribution of termites to development of zinc anomaly in Kalahari sand. *Inst. Mining Metall. Trans.* **79,** B53–B59.

Watterson, J. R. and Neuerberg, G. J. (1975). Analysis for tellurium in rocks to 5 parts per billion. *U.S. Geol. Survey J. Res.* **3,** 191–195.

Watterson, J. R. and Theobald, P. K., eds (1979). "Geochemical Exploration 1978." Assoc. Expl. Geochem., Rexdale, Ontario, 504 pp.

Watterson, J. R., Gott, G. B., Neuerberg, G. J. *et al.* (1977). Tellurium, a guide to mineral deposits. *J. Geochem. Explor.* **8,** 31–48.

Watts, J. T., Tooms, J. S. and Webb, J. S. (1963). Geochemical dispersion of niobium from pyrochlore-bearing carbonatites in Northern Rhodesia. *Inst. Mining Metall. Trans.* **72,** 729–747.

Weast, R. C., ed. (1976). "Handbook of Chemistry and Physics", 57th edn. Chemical Rubber Co., Cleveland.

Webb, J. S. (1958a). Observations of geochemical exploration in tropical terrains. *In* "Symposium de Exploración Geoquímica", 20th Int. Geol. Cong., Mexico City 1956, **1,** 143–173.

Webb, J. S. (1958b). Notes on geochemical prospecting for lead–zinc deposits in the British Isles. *In* "Technical Aids to Exploration", 23–40. Paper 19 Symp. on the Future of Non-ferrous Mining in Great Britain and Ireland. Institution of Mining and Metallurgy, London.

Webb, J. S. (1964). Geochemistry and life. *New Scientist* **23,** 504–507.

Webb, J. S. (1970). Some geological applications of regional geochemical reconnaissance. *Geol. Assoc. Proc.* **81**(3), 585–594.

Webb, J. S. (1971). Regional geochemical reconnaissance in medical geography. *Geol. Soc. Am.* Memoir 123, 31–42.

Webb, J. S. (1973). Applied geochemistry and the community. *Inst. Mining Metall. Trans.* **82,** A23–A28.

Webb, J. S. and Millman, A. P. (1950). Heavy metals in natural waters as a guide to ore. *Inst. Mining Metall. Trans.* **59,** 323–336.

Webb, J. S. and Millman, A. P. (1951). Heavy metals in vegetation as a guide to ore. *Inst. Mining Metall. Trans.* **60,** 473–504.

Webb, J. S. and Tooms, J. S. (1959). Geochemical drainage reconnaissance for copper in Northern Rhodesia. *Inst. Mining Metall. Trans.* **68,** 125–144.

Webb, J. S., Nichol, I. and Thornton, I. (1968). The broadening scope of regional geochemical reconnaissance. *Proc. 23rd Int. Geol. Cong., Prague* **6,** 131–147.

Webb, J. S., Howarth, R. J., Thompson, M. *et al.* (1978). "Wolfson Geochemical Atlas of England and Wales." Clarendon Press, Oxford, 69 pp.

Webber, G. R. (1975). Efficacy of electrochemical mechanisms for ion transport in the formation of geological anomalies. *J. Geochem. Explor.* **4,** 231–234.

Wedepohl, K. H. (1956). Investigations on the geochemistry of lead (in German). *Geochim. Cosmochim. Acta* **10,** 69–148.

Wedepohl, K. H., ed. (1969). "Handbook of Geochemistry", Vol. 1. Springer-Verlag, Berlin, 442 pp.

Wedepohl, K. H., ed. (1969–1978). "Handbook of Geochemistry", Vols 2–4. Springer-Verlag, Berlin.

Wehrenberg, J. P. and Silverman, A. (1965). Studies of base metal diffusion in experimental and natural systems. *Econ. Geol.* **60,** 317–350.

Weiss, O. (1971). Airborne geochemical prospecting. *Toronto Symposium Volume* 502–514.

Welsch, E. P. and Chao, T. T. (1975). Determination of trace amounts of antimony in geological materials by atomic absorption spectrophotometry. *Anal. Chim. Acta* **76,** 65–69.

Welsch, E. P. and Chao, T. T. (1976). Determination of trace amounts of tin in geologic materials by atomic absorption spectrometry. *Anal. Chim. Acta* **82,** 337–342.

Wennervirta, H. (1967). Geochemical methods of uranium prospecting in Finland. *In* "Geochemical Prospecting in Fennoscandia" (A. Kvalheim, ed.), 155–170. Interscience, New York.

Wennervirta, H. (1973). Sampling of the bedrock–till interface in geochemical exploration. *In* "Prospecting in Areas of Glacial Terrain" (M. J. Jones, ed.), 67–71. Institution of Mining and Metallurgy, London.

Wenrich-Verbeek, K. J. (1977). Uranium and coexisting element behavior in surface waters and associated sediments with varied sampling techniques used for uranium exploration. *J. Geochem. Explor.* **8,** 337–355.

West, P. W. and Gaeke, G. C. (1956). Fixation of sulfur dioxide as disulfite mercurate (II) and subsequent colorimetric determination. *Anal. Chem.* **28,** 1818–1819.

White, D. E. and Bannock, W. W. (1950). The sources of heat and water supply of thermal springs with particular references to Steamboat Springs, Nevada. *Am. Geophys. Union Trans.* **31,** 566–574.

White, M. L. (1957). The occurrence of zinc in soil. *Econ. Geol.* **52,** 645–651.

Whitehead, R. E. S. and Govett, G. J. S. (1974). Exploration rock geochemistry—detection of trace metal halos at Heath Steele Mines by discriminant analysis. *J. Geochem. Explor.* **3,** 371–386.

Whitney, P. R. (1975). Relationship of manganese–iron oxides and associated heavy metals to grain size in stream sediments. *J. Geochem. Explor.* **4,** 251–263.

Whittaker, E. J. W. and Muntus, R. (1970). Ionic radii for use in geochemistry. *Geochim. Cosmochim. Acta* **34,** 945–956.

Wilbur, D. G. and Royall, J. J. (1975). Discovery of the Mallow copper–silver deposit, County Cork, Ireland. *In* "Prospecting in Areas of Glaciated Terrain" (M. J. Jones, ed.), 60–70. Institution of Mining and Metallurgy, London.

Williams, R. E. (1970). Applicability of mathematical models of ground water flow systems to hydrogeochemical exploration. *Idaho Bur. Mines Geol.* Pamphlet 144, 13 pp.

Williams, S. A. and Cesbron, F. P. (1977). Rutile and apatite: useful prospecting guides for porphyry copper deposits. *Mineralog. Mag.* **41,** 288–291.

Williston, S. H. (1968). Mercury in the atmosphere. *J. Geophys. Res.* **73,** 7051–7055.

Wilmshurst, J. R. (1975). The weathering products of nickeliferous sulphides and their associated rocks in Western Australia. *Vancouver Symposium Volume* 417–430.

Winters, E. and Simonson, R. W. (1951). The subsoil. *Adv. Agron.* **3,** 2–92.

Wodzicki, A. (1959). Geochemical prospecting for uranium in the Lower Buller Gorge, New Zealand. *N.Z. J. Geol. Geophys.* **2,** 602–612.

Wolfe, W. J. (1975). Zinc abundance in early Precambian volcanic rocks; its relationship to exploitable levels of zinc in sulphide deposits of volcanic exhalative origin. *Vancouver Symposium Volume* 261–278.

Wollenberg, H., Kunzendorf, H. and Rose-Hansen, J. (1971). Isotope-excited X-ray fluorescence analysis for Nb, Zr, and La+Ce on outcrops in the Ilimaussaq intrusion, south Greenland. *Econ. Geol.* **66,** 1048–1060.

Wones, D. R. and Gilbert, M. C. (1969). The fayalite–magnetite–quartz assemblage between 600°C and 800°C. *Am. J. Sci.* **267,** 480–488.

Wood, W. W. (1976). Guidelines for collection and field analysis of ground water samples for selected unstable constituents. *In* "Techniques of Water Resource Investigation", Ch. D-2. U.S. Geol. Survey, Washington.

Woodcock, J. R. and Carter, N. C. (1976). Geology and geochemistry of the Alice Arm molybdenum deposits. *In* "Porphyry Deposits of the Canadian Cordillera" (A. S. Brown, ed.), Can. Inst. Mining Metall. Spec. Vol. 15, 462–475.

Wu, I. J. and Mahaffey, E. J. (1978). Mercury-in-soils geochemistry over massive sulfide deposits in Arizona (abst.). *7th Int. Geochem. Explor. Symp., Denver, Program* 82–83.

Yoe, J. H. and Koch, H. J., Jr, eds (1957). "Trace Analysis." John Wiley, New York, 672 pp.

Young, A. (1966). "Tropical Soils and Soil Survey." Cambridge University Press, Cambridge, 468 pp.

Subject Index

A

Abundance of elements, *see* Background and under specific elements
Accuracy of data, 65–66
Acetic-acid digestion, 207–208, 410–411
Acid decomposition of samples, 51, 248–249
Active stream sediments, 392–418
Activity, chemical, 19–20, 182
Adsorbed ions, *see* under specific elements
Adsorbents, *see* Clay minerals, Iron-manganese oxides, and Organic matter
Adsorption,
  effect on mobility, 262, 263, 355, 364, 367
  general, 195–206
Aeolian deposits, 138, 233, 235, 243, 253, 317–318
Aerosols, 490, 518
Age of ground water, 355
Agricultural geochemistry, 11
Airborne pollution, 266, 512
Airborne survey techniques, 335, 446, 450, 485–487, 512, 517–518
Air photography, 485
Alberta, 235
Algae, 230, 238, 349
Alkali soils, 178
Alluvial fans, 253
Alluvium, anomalies in, 257, 311–317, 377
Alpha-track film, 517
Alteration haloes, 113, 114, 116, 119
Aluminosilicates, *see* Clay minerals
Analytical error, 61–66, 260, 339, 422–423
Analytical methods, 43–70, 338–339, 446, 474–475
Angola, 370
Animal activity, 133, 216, 235, 240, 260
Anion complexes, 263, 349
Anomalies, *see also* Non-significant anomalies
  in drainage sediment, 382–427
  in residual overburden, 261–287
  in transported overburden, 288–318
  in water, 348–382
Anomaly,
  contrast, *see* Contrast
  definition, 2
  enhancement, 406–411
Antimony,
  abundance, 30, 31
  associations, 27, 76
  epigenetic aureoles, 104, 106, 110, 118
  general, 550
  geochemistry, 25, 133
Appraisal of anomalies, *see* Interpretation of data
Appraisal of prospects, *see* Evaluation of prospects
Aquifers, *see* Ground water
Arctic soils, 166, 168, 176–177, 314
Argentina, 7

Arid climates,
effect on, exploration methods, 436
soils, 147, 156, 158, 159, 165, 174–178, 267, 284
vegetation, 157, 460, 467
water, 354, 371, 381
Arizona,
Ajo, 94, 265
Johnson Camp, 113–114
Kalamazoo–San Manuel, 116–117
Mission–Pima, 314
Patagonia, 94
Ray, 95
Red Mountain, 117–118
stream sediments, 409–411
vegetation, 467, 483
Arsenic,
abundance, 30–32
analytical methods, 55, 67, 69
anion complexes, 263, 349
associations, 28, 76
epigenetic aureoles, 26, 102, 104, 106, 108, 109, 118, 120
general, 550–551
geochemistry, 21, 24, 25, 133, 239
as guide to gold ore, 331
in iron–manganese oxides, 204, 239, 412
smelter contamination, 326
in soils, 267, 273, 304, 331
in water, 364, 366
Ashing of samples, 249, 350
Association of elements, 25–28, 76, 549–581
Association of Exploration Geochemists, 8, 9
Astragalus, 477
Atmosphere, 185, 187, 189, 489
Atmospheric gases, 492–494, 517–518
Atmospheric particulates, 490, 497–498, 512, 518
Atomic-absorption spectrometry, 55–57, 68, 69
Atomic weights, 549–581
Australia, 7, 75, 90, 120, 263, 481
Availability of elements to plants, 458, 460
Averages, moving, 523–524
Azonal soils, 163, 166

B

Background,
general, 30–34, 549–581
rocks, 30, 31
soils, 32
water, 132, 356
Bacteria,
decomposing organic matter, 155, 157, 208
oxidizing ferrous iron, 133
oxidizing hydrocarbons, 479–481, 495
oxidizing sulfur and sulfides, 238
precipitating metals, 238
Barium,
abundance, 30–32
analytical methods, 55, 60, 67
associations, 27, 76
epigenetic aureoles, 104–106, 117
general, 551–552
geochemistry, 21, 24, 25, 197
in iron–manganese oxides, 204
in productive plutons, 87
Barometric pressure,
effect on soil gases, 504
Basal till, 232, 292, 294–296, 301
Base-exchange capacity, 286
Base-metal deposits, *see* Copper, Lead–zinc, and Polymetallic deposits
Base metals, *see* Copper, Heavy metals, Lead and Zinc
Basin bogs, 307
Basin and Range province, 312–314
Beryl, 266, 406
Beryllium,
abundance, 30–32
analytical methods, 55, 67
associations, 27, 76
deposits, 87
epigenetic aureoles, 104, 106, 109
general, 552
geochemistry, 21, 24, 25, 250
monitor, 324
in productive plutons, 87
in sediments, 398, 408
in soils, 156, 268, 324
Bibliographies, 9
Bicarbonate, 181, 189, 190, 352, 446

Biogenic dispersion patterns, 236–238, 243–246, 250–252, 289, 304–305
Biogenic gases, 495, 500
Biogeochemical cycle, 237, 256
Biogeochemical exploration, 4, 456, 464–471; *see also* under specific elements, plant topics, and Vegetation
Biogeochemical survey techniques, 471–476
Biogeochemistry, 241
Biological agents in weathering and soil formation, 133–149, 157–158
Biological dispersion, *see* Biogenic dispersion patterns
Biotite, 134, 143, 265
Bismuth,
  abundance, 30
  analytical methods, 55, 67
  associations, 27, 76
  deposits, 106
  epigenetic aureoles, 104, 106, 110, 112, 113, 118, 121
  general, 552
  geochemistry, 25
Bisulfate, 188, 189
Black shale, 258, 426
Bog ore, 245
Bog soil, 167
Bog water, 188
Bogs, 232, 301, 306–312, 334–335, 385 386
Borneo, 264, 417
Boron,
  abundance, 30–32
  analytical methods, 55, 67
  associations, 27, 76
  epigenetic aureoles, 102, 113, 117
  general, 553
  geobotany, 478–480
  geochemistry, 22, 24, 25
  in soil, 480
Botswana, 218, 267, 279
Boulder clay, 240
Boulder fans, 291
Boxwork pseudomorphs, 148
British Columbia, 416, 552
Burma, 326, 327

C

Cadmium,
  abundance, 30
  analytical methods, 55, 67, 69
  associations, 27, 76
  epigenetic aureoles, 104, 110, 111
  general, 553–554
  geochemistry, 24, 25
Calamine violet, 477, 482
Calcareous rocks,
  effect on dispersion, 239, 258, 297, 307
  as host for sulfide ore, 267, 273, 280, 316, 317
  indicator plants for, 481
  weathering of, 131, 133, 148, 227, 231, 240, 274
Calcrete, *see* Caliche
Caliche, 131, 172, 175, 284, 314
California, 76, 144, 467, 484
California poppy, 483
Canada, *see also* specific provinces
  Appalachian Belt, 412, 479
  glaciated areas, 231, 293
  western mountains, 314
Capillarity, 155, 175, 222, 223, 236, 284, 307
Carbon dioxide,
  analytical methods, 69
  in atmosphere, 492–494
  general, 506
  geochemistry, 22, 352, 506
  in lake water, 230
  in soil air, 511
  in soil genesis, 152
  solubility in water, 499
  stability, 496
Carbon disulfide, 496, 497, 499, 506
Carbon monoxide, 494, 499, 506
Carbonaceous material, *see* Organic matter
Carbonate,
  analytical methods, 69, 248
  minerals, 136, 230, 261, 263
  precipitates, 155, 363
  rocks, *see* Calcareous rocks
  in soils, 158, 172–178
Carbonatite,
  pathfinders for, 552 (Ba), 557 (F), 564 (Nb), 571 (Th), 574 (Sr)

sediment anomalies, 406–408
weathering of, 270–271
Carbonic acid, *see* Carbon dioxide
Carbon-ion species, 190
Carbonyl sulfide, 493, 496, 497, 499, 506
Cassiterite, 139, 226, 266
Cation exchange, 459
Cationic complexes, 349
Cations in water, 349
Cesium, 87, 88
Characteristic analysis, 531
Chelation, 52, 170, 191, 207, 208, 210 211
Chemical analysis, *see* Analytical methods
Chemical weathering, 131–133
Chile, 314, 416
Chloride as an ore guide, 94, 371
Chlorophyll, 211
Chlorosis, 477–478, 483, 487–488
Chromatography, 59
Chromium,
abundance, 30–32
analytical methods, 55, 67
anomaly contrast, 251
associations, 27, 76
general, 554
geochemistry, 21, 23–25
in gossan, 264
ore, 73, 76, 81, 251
in peat, 307
in productive plutons, 89
in rocks, 258
in sediments, 398
in soils, 153, 268, 270, 286
Citrate extraction, 249
Clarke, *see* Background
Clastic dispersion patterns, 244, 252–254, 308–309
Clay content of soil, 152, 170–177
Clay minerals,
adsorbed ions, 45, 250, 255, 266–269, 273, 287, 295, 364
classification, 140–144
genesis, 131, 132, 135, 140–144, 156, 248
ion-exchange properties, 171–173, 196–197, 203–204, 263
stability of, 46, 48
Clay-sized material,
definition, 139
Climate, *see also* Arid, Cold, Humid, Semi-arid, Temperate, and Tropical climates
effect on soil formation, 154–157, 165, 168
effect on surficial dispersion, 238, 239, 345, 354
effect on weathering, 137–138
Cluster analysis, 533
Clustering of anomalies, 41
Cobalt,
abundance, 30–32
analytical methods, 55, 67, 69
anomaly contrast, 251
associations, 27, 76
deposits, 76, 81
epigenetic aureoles, 104–106, 110, 116, 117
general, 554–555
geobotany, 478
geochemistry, 21, 25, 206, 210
in gossans, 264
in iron–manganese oxides, 45, 204, 267, 412
ore, 251
in peat, 307
provinces, 81
regional zoning, 84
in rocks, 358
in seepages, 389
in soils, 156, 204, 205, 267, 268, 286, 287
in till, 301
in vegetation, 468
in water, 198–203
Coefficient of variation in samples, 63, 414–415
Cold climates, 157, 267, 306
Cold-extractable metal (cxMe), *see* Readily extractable metal and Selective extraction
Cold-zone soils, 166, 239
Colloidal suspensions, *see also* Clay minerals, Iron–manganese oxides, and Organic matter
gold, 212, 349
particles, 195–206, 448
silica, 152–155
Colluvium,

anomalies in, 219, 220, 265, 277, 311–317
Colorado, 107, 118, 277, 406
Colorado Plateau, 73, 75
Colorimetric analysis, 54, 55, 68–70
Complexing agents, 52, 206, 212, 215
Complex ions, 22, 191–192, 202, 229, 243, 349
Computerized data handling, 520–525
Conductivity,
soil, 303–304
water, 352–353
Connecticut, 238, 351
Contamination, *see also* Mine dumps, Mine water, and Smelter contamination drainage samples, 434, 435, 446, 450
general, 257–259
soil samples, 325–328, 337, 434
vegetation samples, 473–474, 482
Continental glaciation, 231–235
Contouring of data, 523
Contrast,
definition, 35
general, 250–252
sediment anomalies, 421–424
soil anomalies, 269–278, 328, 345
vegetation anomalies, 468–470
water anomalies, 358–361
Converter plants, 459
Copper,
abundance, 30
adsorption, 199, 200, 202–204, 262
in algae, 238
analytical methods, 55, 59, 68, 69
anomaly contrast, 251, 359
associations, 27, 76
in biotite, 45
in caliche, 314–316
carbonates, 267
cold-extractable (cxCu), *see* Readily extractable metal
epigenetic aureoles, 104, 106, 107, 113, 115, 121
equilibria in solution, 184–187
flower from Copperbelt, 482
general, 555–556
geobotany, 478, 481–484
geochemistry, 21–25, 86, 182, 210, 212, 239
in gossan, 264
in ground water, 376
in humus, 281
in iron–manganese oxides, 45, 206, 267, 282, 301
in lake sediments, 381, 423
in lake water, 238, 351, 364, 372, 380, 381
in marine sediments, 428–429
ore, 251, 263, 264, 274, 275, 296, 316
in peat, 307, 311, 386
in placer gold, 75
in productive plutons, 88, 89, 93
in productive volcanics, 96
provinces, 78, 79, 81, 82
smelter contamination, 326–327
in soil, 153, 211, 218, 268, 274–279, 282–287, 332
in stream sediments, 393, 396, 399, 405, 409, 416
in stream water, 364–367
sulfides, 238, 267
supergene enrichment, 74
in vegetation, 461, 468, 471, 475
Copper deposits, *see also* Nickel–copper, Polymetallic, Porphyry-copper, and Volcanogenic massive-sulfide deposits
epigenetic aureoles, 105
geobotany, 483
geochemical discoveries, 7
pathfinders for, 76
water anomalies, 359
Coprecipitation, 269
Crustal rocks,
composition, 30, 31
Cumulative frequency plots, 39
Cutoff of stream anomalies, 377, 413, 454
cxMe, cxHM, cxCu, etc., *see* Readily extractable metal
Cyprus, 119
Czechoslovakia, 79, 112

D

Dambos, 386–391, 437
Data handling, 520–525
Decay patterns in drainage anomalies,
lake sediment, 424–426,

stream sediment, 399–405
stream water, 361–363
Deep-sea sediments, 143
Deep-seated environment,
14, 20–23, 71–127, 494–495
Deficiency symptoms in vegetation, 460
Desert climate, *see* Arid and Semi-arid climates
Desert soils, 158, 161, 163, 175–176
Detailed surveys, 4, 439, 443
Diamonds, 576; *see also* Kimberlite
Differential thermal analysis, 139
Diffusion,
surficial water, 236
theory, 97–101
thermal water, 103, 105, 107–109
volatiles, 500–506
Discoveries, 6, 7
Discriminant analysis, 527–530
Dispersion, 16, 20–25, 238–260
Dispersion fans, 243–252
Dispersion haloes, 252
Dispersion trains, 243, 252
Dithionite leach, 265
Dogs, 291, 489
Dolomite, *see* Calcareous rocks
Downstream decay of anomalies, *see* Decay patterns
Drainage surveys, 430–455
Drill-hole water, 373, 377
Drilling, 334–336
Drying of samples, 474
Duricrust, 172

E

Earthquake forecasting, 514
Ecuador, 7
EDTA, 52, 207
Efflorescences, 236
Eh, *see* Oxidation potential
Eh–pH relationships, 46–47, 181–191, 238, 240, 255, 307
Electrochemical dispersion, 212
Electrochemical reactions, 136, 303
Electromagnetic spectrum, 486
Electrometric analysis, 59, 69; *see also* Oxidation potential and pH
Electron microscopy, 139
Electrostatic separations, 406
Eluviation, 151, 152, 175
Enhancement of anomalies, 406–411
Environmental geochemistry, 11
Epigenetic anomalies, 71, 97–127, 243, 245–246, 249, 289–290, 314–317
Equilibria,
chemical, 180–215
Equilibrium constant,
definition, 20
Errors,
analytical, 61–66, 257, 260
sampling, 257, 259–260
Eskers, 231, 232, 298–300
Evaluation of prospects, 74–76
Evaporites, 74, 243, 245, 355
Exchange capacity, 151, 171, 172, 175, 196–197, 286, 287
Extractants for soil samples, 328

F

Fans, dispersion,
clastic, 252–254, 313
hydromorphic, 255, 257, 373–376
Faults,
detection of, 227, 514
Feldspar,
dissolution of, 46
minor elements in, 45, 87, 93
stability of, 263
Fick's Laws, 98, 500
Field analysis, 346, 437, 446, 454
Field notes, 448, 450, 520
Field operations, 329–341, 546–547
Field techniques, *see* Survey techniques
Filtration of water samples,
350, 432, 445–446
Finland, 293, 295, 296, 298, 299, 307, 312, 330
Fissure water, 373–376, 384
Flood-plain sediments, 227, 240, 401, 416–418, 437
Fluid inclusions, 75
Fluorimetric analysis, 55, 58, 68, 69
Fluorine,
abundance, 30–32
analytical methods, 55, 59, 68, 69
associations, 27, 76
epigenetic aureoles, 102, 108
general, 556–557

geochemistry, 22, 25
as indicator of rock type, 89
in productive plutons, 91–94
in water, 377
Fluorite deposits, 76
Flushing, 368–371
Fluvial transport, 252
Fluxing of samples for analysis, 51, 248–249
Follow-up procedures, 346–347, 454
Fossil laterite, 173
Fossil-spring deposits, 265, 345
France, 470
Fraunhofer discriminator, 485
Frequency distribution diagrams, 33, 328, 329
Frost action, 156, 177, 218, 266
Frost boils, 177, 284, 296, 422
Fugacity,
definition, 490
Fulvic acid, 208, 210–212, 218
Fungi, 238
Fusion of samples for analysis, 51, 248–249

G

Gas,
analytical methods, 69
definition, 490
geochemistry, 5, 69, 181, 489–518
movement, 500–506
reactivity, 498–500
solubility in water, 499
traps, 517
in water, 502–503
Geobotanical surveys, 484–485
Geobotany, 4, 456, 476–488
Geochemical anomaly,
definition, 2, 34
Geochemical cycle, 15
Geochemical exploration,
definition, 2
Geochemical provinces, *see* Provinces, geochemical
Geochemical relief, 83
Geochemistry,
definition, 13
Geologic mapping by geochemical surveys, 426–427, 481
Geophysical surveys, 214, 310, 311, 332, 347
Geothermal activity, 382, 514
Geothermal brine, 427–429
Germany, 79, 113
Glacial boulders, 292–299, 234–235
Glacial deposits, 233, 234, 244, 253, 290–306
Glacial fans, 234–235, 253, 291–299
Glacial transport, 217, 231–234, 252
Glaciation, 227, 228, 231–234
Glaciofluvial deposits, 231, 233, 234, 243, 290
Glaciolacustrine deposits, 294, 298
Glaciology, 241
Gley soils, 151, 164, 165, 167, 174, 177, 197
Gold,
abundance, 30, 31
analytical methods, 55, 60, 67
anomaly contrast, 251
associations, 27, 76
colloidal, 212, 349
deposits, *see* Precious-metal deposits
epigenetic aureoles,
104, 106, 115–117
general, 557–558
geochemistry, 22, 25, 139
in gossan, 264
minor element content of, 75
native, 266, 298
ore, 251
in productive plutons, 93
remobilization of, 77, 81
in soil, 265
Gossan,
anomalies in, 203, 261–266
distinguished from laterite, 172, 173
general, 145–146
transported, 285
Grain size, 227
Granite,
classification of, 89
minor elements in, 93
ores associated with, 89
weathering products of, 153
Gravimetric analysis, 68
Gravity action, 243, 253, 266, 311
Great Britain, *see* United Kingdom
Ground moraine, 232

Ground water, *see also* Hydromorphic anomalies, Spring water, and Well water
  age estimation, 355
  anomalies, 276, 359, 373–377
  composition, 132, 147, 356, 357, 359
  geobotanical indicators, 478–479
  geochemistry, 188, 189
  movement, 220–224, 229, 254–257
Gyttja, 229–230, 423

H

Heavy metals, *see also* Copper, Lead, and Zinc
  analytical methods, 69
  in lake sediment, 422–425
  in rock, 113–114
  in soil, 305, 330, 333
  in stream sediments, 83, 395
  in water, 365, 370, 371, 372, 379
Heavy minerals,
  sampling techniques, 450
  from stream sediments, 398, 406–411, 438–439
  from till, 295, 298
Helicopter surveys, *see* Airborne survey techniques
Helium,
  analytical methods, 69
  distribution, 492–495
  general, 507, 558
  as indicator of uranium and hydrocarbons, 512–513
  mobility, 499–503
Hemoglobin, 207, 211
Histograms, 37, 521
Homogeneity of anomalies,
  definition, 252
  lake sediment, 421–424
  soil, 278–281
  stream sediment, 415–416
  stream water, 377
  vegetation, 470–471
Horizons, *see* Soil, horizons
Humic acid, 210; *see also* Organic acids and Organic matter
Humid climates, effect on,
  soils, 145, 155, 156, 163, 170–174, 267
  water, 223, 239, 436
  weathering processes, 218
Humus,
  definition, 151–152
  enrichment of metals in, 237, 269, 305, 456
  lack of, in dry climates, 157
  in soil profile, 170–174, 282
Hydration, 131, 132
Hydrocarbons, 206, 207
Hydrogen,
  molecular, 494, 499, 501, 507
Hydrogen electrode, 184
Hydrogen-ion concentration, *see* pH
Hydrogen-peroxide leach, 47, 53, 187, 410–411
Hydrogen sulfide, 230, 493–499, 507
Hydrogeochemical anomalies, 348–382
Hydrology, 241
Hydrolysis, 130–132, 178
Hydromorphic anomalies, *see also* Anomalies, in water, Ground-water, anomalies, and Stream water
  anomaly contrast, 250–252, 303
  form, 254–257, 394
  general, 243–247
  genesis, 220–224, 364
  mode of occurrence of metals, 269, 384, 394
  in residual soil, 383–391
  superjacent versus lateral, 344
  surveys for, 441, 447, 479
  in transported soil, 300–312, 314–317, 385
Hydromorphic soils, 173–174, 177
Hydroquinone leach, 47, 53
Hydrous oxides, 250
Hydroxylamine leach, 53, 407
Hypochlorite leach, 47, 53

I

Ice action, 243, 253, 254, 289
Idaho, 73, 111, 112, 326, 554
Ideal gas law, 490
Igneous rocks, 549–581; *see also* Granite, Productive plutons, and Ultramafic rocks
Illinois, 234

Illuviation, 151, 152, 171
India, 137, 197
Indicator plants, 477–485
Infiltration,
gases, 502, 505–506
general, 97
solutions, 101–105, 109–121
Infrared sensing, 487–488
Interpretation of data,
non-statistical, 126–127, 339–345, 450–454, 475
statistical, 32–42, 525–533
Intrazonal soils, 163, 166–167, 173, 178
Intrusive rocks, *see* Granite, Productive plutons, and Ultramafic rocks
Iodine, 558
Ion exchange, 102, 195–206, 355
Ion-exchange resins, 350, 446
Ionic charge, 23, 24
Ionic potential, 23, 24
Ionic radius, 23, 24
Ionic replacement index, 21
Ireland,
geochemical discoveries, 7
glacial erratics, 235
Mallow deposit, 297
peat bogs, 307
stream sediments, 412
Tynagh deposit, 120, 302, 304, 305, 428
Iron,
abundance, 30, 32
analytical methods, 55, 59, 68, 69
associations, 27, 76
carbonates, 261, 262
epigenetic aureoles, 117, 121
general, 559
geobotany, 478, 484
geochemistry, 21, 23–25, 47, 132, 133, 146, 172, 173
in marine sediments, 429
as ore-type indicator, 75
oxides, *see* Iron–manganese oxides
in peat, 307, 310, 312
in sediments, 204, 205
in vegetation, 468
in water, 192–193, 366, 373
Ironformation, 76, 77, 119, 120
Iron–manganese oxides,
colloidal suspensoids, 181
dissolution of, for analysis, 50, 53
erosion of, 225–226
ion-exchange properties, 196, 200–206
metal content, occluded, 45, 250, 266, 267, 393, 394
metal content, readily extractable, 45, 249, 295, 301, 364, 394, 395, 412
precipitation of, 131, 143, 145, 238, 248, 267, 384
in soil, 153, 171, 174, 223
solubility with changing Eh–pH, 146, 197, 229, 230
stability, 46, 47
Isotope ratios, 74–75, 90, 118
Italy, 352, 353

J

Japan, 73, 119, 120, 373
Jasperoid, 121
Journals, 10

K

Kaolinite, *see* Clay minerals
Kimberlite, 258, 286, 287, 298, 563, 564, 576
Kupferschiefer, 76
Kurbatov plots, 201
Kuroko deposits, 73, 119, 120

L

Laboratory organization, 547–548
Labrador, 372
Lakes, 227–234, 240
Lake sediment,
anomalies, 367, 419–426
surveys, 439, 443, 449–450
Lake water,
anomalies, 351, 364, 367, 372, 380, 381
surveys, 439, 443, 446
Landsliding, 217, 219, 227
Lateral dispersion patterns, 243, 252, 253, 255, 256, 344
Laterite deposits, 131, 137, 171–173, 264
Lateritic soil, 153, 163, 171, 197, 223, 263, 282, 283

Latosol, *see* Lateritic soil
Leached cavities, 262
Leached outcrops, 261–266
Leaching of soil, 273
Lead,
  abundance, 30–32
  adsorption, 263
  in algae, 238
  analytical methods, 55, 59, 68, 69
  anomaly contrast, 251
  associations, 27, 76
  in caliche, 316–317
  cold-extractable, 327
  complex ions, 191, 192, 194, 195, 202
  epigenetic aureoles, 104–121
  in feldspar, 45, 94
  general, 559–560
  geobotany, 483, 487
  geochemistry, 21, 22, 24, 25
  in gossan, 264–265
  in humus, 281–283
  in iron–manganese oxides, 267, 410
  isotopes, 74–75
  in lake sediments, 420
  ore, 251
  in peat, 307
  in productive plutons, 93, 94
  in sediments, 204, 205
  smelter contamination by, 326, 327
  in soils, 171, 272–274, 277
  in spring areas, 384, 385
  in vegetation, 461, 468, 469
  in water, 202, 364–367, 384
Lead–zinc deposits, *see also* Polymetallic and Volcanogenic massive-sulfide deposits
  epigenetic aureoles, 106, 107, 111, 120
  geochemical discoveries, 7
  hydromorphic anomalies, 314
  lake-sediment anomalies, 425
  pathfinders for, 76, 508
  vegetation, 469
  water anomalies, 359, 376
Leakage anomalies, 97, 101–104, 109–121
Lichens, 208, 483, 484
Limestone, *see* Calcareous rocks
Limonite, 144, 148, 261–266, 267, 364
Literature, 9
Lithium,
  abundance, 30–32
  analytical methods, 55
  associations, 27, 76
  deposits, 73
  epigenetic aureoles, 112, 113, 117
  general, 560
  geochemistry, 22, 24, 25, 197
  in productive plutons, 87, 89, 91
  in soils, 156
Loess, *see* Aeolian deposits
Log-normality, 32–33
Log probability plots, 39–41
Loss on ignition, 424
Luminescence of vegetation, 485–487

M

Mafic rocks,
  composition, 549–581
Magnetic separations, 406, 409
Magnetite deposits, 76, 310
Maine, 380, 412
Malaysia, 7, 268, 577
Manganese,
  abundance, 30, 31, 32
  analytical methods, 55, 68, 69
  anomaly contrast, 251
  associations, 28, 76
  carbonates, 267
  epigenetic aureoles, 108, 115–117, 121
  general, 560–561
  geobotany, 478
  geochemistry, 21, 25, 47, 132, 133, 146, 210
  in gossan, 264
  in marine sediments, 427–429
  nodules, 427, 450
  ore, 251
  as ore-type indicator, 75, 76
  oxides, *see* Iron–manganese oxides
  in peat, 307
Mangrove swamps, 307
Map preparation,
  computer plots, 521–522
  drainage data, 450–452
  plant distribution, 484
  rock data, 125–126
  soil data, 339–341
Marine sediments, 12, 427–429

Massive-sulfide deposits, *see* Volcanogenic massive-sulfide deposits
Mass spectrometry, 60
Mechanical composition of soils, 150, 155
Mechanical dispersion, 216–236, 289
Medical geochemistry, 11
Mercury,
 abundance, 30, 31
 analytical methods, 51, 55, 68, 69
 associations, 28, 76
 in atmospheric air, 494
 deposits, 76, 106, 512
 dimethyl, 509
 epigenetic aureoles, 102, 104, 106, 112, 117, 118, 120, 121
 general, 561–562
 geochemistry, 24, 25, 508–509
 in marine sediments, 429
 in organic compounds, 45, 500
 in productive plutons, 93
 in sediments, 266
 in soil air, 504, 512
 sublimation temperatures, 508
 in water, 183, 184
Metallogenic provinces, 78–84
Metalloorganic compounds, *see* Organometallic compounds
Methane, 60, 190, 207, 494–496, 499, 509, 513
Methyl compounds, 207, 495, 497, 509
Mexico, 73, 112
Michigan, 73, 121, 234, 378
Microboulders, 291–299
Microorganisms, *see* Algae, Bacteria, and Fungi
Mine dumps, 258, 326, 343
Mine water, 188, 189, 238, 364, 366, 373, 386, 394, 435
Mineral deposits, *see also* Copper, Lead–zinc, Molybdenum, Nickel–copper, Pegmatite, Polymetallic, Porphyry-copper, Precious-metal, Skarn, Tin, Uranium, and Volcanogenic massive-sulfide deposits
 classification, 73–74
Mineralized pebbles, 409–411
Minerals,
 resistance to weathering, 134–136
Missouri, 147
Mixing of surface waters, 364
Mobility,
 general, 23–25, 246–248
Mobility as affected by,
 chelation, 208
 climate, 284
 ion solubility, 23–25, 181–183, 192–195
 microorganisms, 238
 mineral solubility, 192–195
 organometallic complexes, solubility of, 206–210, 212
 pH, 284
 precipitation by solid organic matter, 211
 vegetation, 236, 251, 458, 468
Mode of occurrence of elements,
 general, 44–53
 in hydromorphic soil anomalies, 285
 in lake-sediment anomalies, 420–421
 in residual-soil anomalies, 348–351
 in stream-sediment anomalies, 392–395, 406–411
 in water anomalies, 348–351
Molybdenum,
 abundance, 30, 31, 33
 analytical methods, 55, 68, 69
 anion complexes, 267, 349, 367
 anomaly contrast, 251
 associations, 27, 28, 72, 76
 epigenetic aureoles, 104–106, 112, 113, 115–117, 121
 general, 562–563
 geobotany, 478, 487
 geochemistry, 21, 24, 25, 133, 239
 in ground water, 239, 314, 373, 374
 in iron–manganese oxides, 204, 239
 in leached outcrops, 265
 ore, 251
 in peat, 307
 in productive plutons, 93
 in soils, 265, 267, 268, 272, 273, 472, 487
 in surface water, 366, 379, 380
 in till, 294
 in vegetation, 239, 468, 472

Molybdenum deposits,
epigenetic aureoles, 106, 117
pathfinders for, 76
productive plutons for, 93
water anomalies, 373, 374, 379
Montana, 95, 96, 364, 366, 384, 385, 483
Montmorillonite, *see* Clay minerals
Moraine, *see* Till
Mosses, 483
Mountain areas, 231, 239, 314, 483
Mountain soils, 178, 219, 267
Moving averages, 523–524

N

Namibia, *see* South-West Africa
Negative geochemical anomalies, 34
Neutron-activation analysis, 58, 68, 69
Nevada,
Battle Mountain, 504
Ely, 117, 265
geochemical discoveries, 7
Pinenut Mountains, 487
Robinson, 94
New Brunswick,
Bathurst district, 303, 310
geochemical discoveries, 7
geochemical relief, 83
lake sediments, 420, 425
Nash Creek, 401–405, 417
stream sediments, 83, 395
Newfoundland, 425; *see also* Labrador
New Mexico, 73, 76, 467
New York, 385, 410
Nickel,
abundance, 30–32
analytical methods, 55, 68, 69
anomaly contrast, 251
associations, 27, 76
epigenetic aureoles, 104, 106
general, 563
geobotany, 478
geochemistry, 21, 24, 25, 206, 210, 211
in gossan, 264
in iron–manganese oxides, 204, 412
ore, 251, 263, 264, 296, 469
in peat, 307
in productive plutons, 88, 89
in rocks, 258
in soils, 29, 156, 268, 286, 469, 510
in till, 292, 293, 299
in ultramafic rocks, 258
in vegetation, 469
Nickel–copper deposits,
76, 88, 89, 292, 293, 296, 298, 299, 565
Nigeria, 73, 314, 316, 370, 468, 469
Niobium,
abundance, 30–32
analytical methods, 55, 68
associations, 27
deposits, 73, 76, 87, 89
general, 563–564
geochemistry, 21, 24, 25, 139
in productive plutons, 87, 95
in sediments, 398, 406–408
in soils, 268, 270, 271
Nitric acid digestion, 47, 53
Nitrogen fixation, 133
Non-significant anomalies,
definition, 34
general, 257, 260
in soils,
285–287, 290, 338–339, 342–344
in stream sediment and water,
381–382, 426–427, 453
in vegetation, 475
Non-specific adsorption, 199
North Carolina, 73, 272, 282
Northwest Territories,
Agricola Lake, 7, 423
Coppermine district, 421
geochemical discoveries, 7
lake sediments,
78–80, 82, 367, 421, 423, 425
uranium province, 78–80, 82
Norway, 300, 301, 483, 487

O

Ocean water, *see* Sea water
Ohio, 234
Ontario, 73, 76, 119
Ore deposits, *see* Mineral deposits
Ore guides, 540–542
Ore typing, 72–76
Organic acids, 152, 157, 170,
206–208; *see also* Fulvic acid and Humic acid
Organic matter, *see also* Fulvic acid,

Humic acid, Humus, Organic acids, and Organometallic compounds
as adsorbent for metal ions, 45, 266, 301, 364, 368
general, 206–212
in hydromorphic anomalies, 269, 301, 364, 384, 386–391
ion-exchange properties, 196, 197, 202
in lakes, 229–231, 421, 423
oxidation potential, 185, 187, 189
in soil genesis, 170–177
in soil profile, 150–159, 269
source, 246, 249
in stream sediment, 392, 398, 410
in weathering, 131
Organometallic compounds, *see also* Organic matter
bonding of metal, 207, 249
in vegetation, 463
in water, 349, 350, 382, 421
Orientation surveys,
rocks, 122–124
soil, 320–329
vegetation, 473, 484–485
water, 431–435
Overturn of lake waters, 228–230, 372
Oxalate leach, 47, 53, 191
Oxidation, 132, 133, 355, 363
Oxidation potential, *see also* Eh–pH relationships
definition, 184–185
determination of, 59, 69, 446
in electrochemical dispersion, 212–214
Oxidizing agents, 53
Oxygen,
in atmosphere, 185, 187, 492–494
dissolved, 59, 131, 446, 499
general, 509
isotopes, 118
in soil air, 212, 511
in water, 213, 230, 499

P

Panama, 7
Partial extration, *see* Selective extraction
Partial pressure of gases, 20
Particle size classes, 139, 224, 226, 231, 396
Particulates, *see* Atmospheric particulates
Partition of elements between phases, 85, 246–249, 350, 393, 394, 411
Pathfinders, 25–28, 72, 76, 477; *see also* under individual mineral-deposit types
Peat, 158, 174, 212, 477
Pebble coatings, 409–411
Pegmatite deposits, indicator elements in,
minerals, 80 (Sn), 87 (Be, Cs, Li, Nb, Rb, Tl), 88 (Cs)
sediment anomalies, 398 (Be), 408 (Be)
soil anomalies, 156 (Be, Li, Rb, Sr), 324 (Be)
wallrocks, 109 (Be)
water anomalies, 572 (Rb)
Pennsylvania, 378, 394, 401, 410, 411, 528, 529
Permafrost, 176, 177, 236, 284, 296, 334–335
Permeability of rocks, 136, 373–377
Permeability of soils, 225, 240
Persistence of water anomalies, 358–368
Personnel, 444, 536–538, 546–548
Peru, 78
Petroleum, 479–481, 513
pH, *see also* Eh–pH relationships
definition, 182
measurement of, 59, 69, 446
soils, 304
water, 263–266
pH, effect on,
exchange capacity of minerals, 197–204
solubility of elements, 181–184, 274
Phase diagrams, 47, 48
Philippines, 267, 332, 411–412
Phosphorite deposits, 89, 316, 317, 427, 428, 579
Phosphorus, 564
Phreatophytes, 466, 478, 479
Physical weathering, 130
Pit logs, 323
Placer deposits, 74, 75, 428

Placon soils, 174
Plankton, 229, 230, 238
Planning, 534–548
Plant activity, 266
Plant anomalies, 245–247, 251–256, 464–471
Plant exudates, 518
Plant nutrition, 457–458, 463–464
Plant sampling, 4, 471–476
Plant toxicity, 477–478
Plants, *see* Biogenic dispersion patterns and Vegetation
Plasma-emission analysis, 55, 56
Platinum electrode, 184
Platinum metals,
  abundance, 30, 31, 89
  analytical methods, 55, 60, 68
  associations, 27, 76
  deposits, 76
  general 564–565
  geochemistry, 25, 89, 133, 139
  native, 266
  in placer deposits, 74
Podzolic soils, 161–170, 282
Polarography, 59
Polymetallic deposits,
  epigenetic aureoles, 109–114
  geobotanical indicators, 481
  geochemical discoveries, 7
  pathfinders for, 28, 76, 550, 552, 572, 576
  soil anomalies, 272, 300
  volatile emanations, 512
Population maps of plant species, 484
Populations of data, 35–42
Pore spaces of soils, 220, 221, 263, 489
Porphyrins, 206, 211, 510
Porphyry-copper deposits,
  geochemical discoveries, 7
  hydromorphic anomalies, 314
  leached capping, 264–265
  pathfinders for, 27, 28, 72, 75, 552, 557, 562, 565, 572, 574
  productive plutons for, 77, 94–95
  provinces, 78
  secondary enrichment, 265
  sediment anomalies, 399, 411, 443
  vegetation, 467
  volatile emanations, 512
Potassium, 565
Potential-determining ions, 198
Precious-metal deposits (deposits of gold and silver),
  epigenetic aureoles, 106, 108, 118–119
  geochemical discoveries, 7
  pathfinders for, 76 (general), 118 (As), 331 (As), 550 (Sb), 551 (As), 560 (Pb), 561, (Hg), 572 (Rb)
Precipitates, 244, 392, 394, 420
Precipitation barriers, 103, 301, 363–368, 384, 411–413
Precision of data, 414–415
Pre-enrichment of water samples, 446–447
Primary dispersion, 17
Processing of samples, 444–450
Productive drainage basins, 399, 403
Productive environments, 77–96
Productive plutons, 84–96
Propane, 479–481
Provinces, geochemical, 2, 78–84, 381, 427, 443
Pseudogossans, 263–265
Puerto Rico, 265
Pyrite,
  acid from oxidation of, 258, 263, 265, 274, 307, 381, 411
  mineral stability, 47, 135, 214
  minor-element content, 75, 120, 135
  plant indicators for, 481, 483
  weathering, 136, 144, 262
Pyrochlore, 266, 270, 271, 406–408

Q

Quebec, 73, 76, 83, 119

R

Radioactive decay series, 566–570
Radiogenic gases, 495
Radiometric analysis, 58–59
Radium, 510, 566, 571
Radon,
  analytical methods, 69
  general, 567–571
  geochemistry, 492–495, 502–503, 509–510
  in ground water, 505

in soil air, 515
units of measurements, 491–492
in uranium exploration, 512–513
Rain water,
anomalies, 518
composition of,
131, 132, 188, 351–352
leaching by, 269, 284
movement of, 220–224
Rainfall, effect on,
anomaly contrast, 361, 368–371
soil air movement, 504–505
soil formation,
144, 145, 154–158, 170–176
Rare earths, 75, 567, 571
Rate of extraction during sample digestion, 50
Ratios of elements,
general, 104, 110
major elements, 87, 91, 120
ore metals, 75, 87, 89, 110, 111, 286
other elements, 87, 89, 91, 95
in vegetation, 475
in water, 369, 371, 452
Readily extractable metal,
contaminated soils, 343–344
general, 249–250
glacial soils, 306, 308–309
stream sediments, 394, 411
Zambian dambos,
269, 211, 386–391, 437, 438
Reconnaissance surveys,
2, 430, 439–441, 443
Red Sea, 120, 428–429
Reduction, 355, 363
Reflectance of vegetation, 485
Regression analysis, 525–527
Reliability of analytical data, 61–66
Relief, effect of,
in dispersion, 239, 274–278, 361
in soil formation, 158–161
in weathering, 138
Remote sensing, 456, 485–488
Research organizations, 8
Residual soil, 3, 138, 252–255, 261–287, 319, 325, 344
Rhenium,
abundance, 30
analytical methods, 68
associations, 76
general, 571–572
geochemistry, 25
ore-type indicator, 75
Rhodesia, 108, 109, 156, 268
Rock sampling, 5, 124–125
Rocks,
average composition, 30, 31, 549–581
Rock type,
effect on dispersion,
240, 270, 274, 276, 297
Root penetration, 466–467
Root-tips of plants, 458–462
Rubidium,
abundance, 30
analytical methods, 55, 68
associations, 27, 76
epigenetic aureoles, 102, 112, 116, 117
general, 572
geochemistry, 24, 25, 197
in productive plutons, 87, 89, 91
in soils, 156
Runoff, 224–225, 228, 368–371, 377

S

Saline deposits, 148, 181, 258
Saline soils, 158, 159, 165, 167, 477, 479
Saline water, 188
Salinity, 132, 147, 181
Sample containers for,
soil and sediment, 333, 448–450
vegetation, 474
volatiles, 516–517
water, 445
Sample decomposition,
49–53, 248–250
Sample layout,
drainage surveys, 2–4, 440–444
rock surveys, 125
soil surveys, 2–4, 329–333, 336–337
Sample numbering, 337, 444
Sample preparation,
46–49, 125, 444–450
Sample spacing, *see* Sample layout
Sample variance, 414–415, 422–423
Sampling procedures,
atmospheric air, 517–518
rocks, 125–126
sediments, 444–450
soil air, 516

soils, 173, 294, 321–325, 328–329, 333–336
vegetation, 474
water, 444–450, 516
Sampling tools, 447, 449–450, 474, 475
Saprolite, 148
Saskatchewan, 381, 425, 426, 372
Satellite surveys, 485–487
Sea water, 188, 247, 492
Seasonal climate, 223
Seasonal movement of soil moisture, 222, 241
Seasonal variations, 415–416, 433
Secondary dispersion haloes, 17
Secondary enrichment of copper, 265, 267, 274
Secondary minerals, 139–146, 266, 267, 268
Sedimentary rocks,
background composition, 549–581
Seepages and seepage soils, *see* Hydromorphic anomalies
Selection of areas for reconnaissance, 538
Selection of exploration methods, 540–546
Selective extraction, *see also* Readily extractable metal
analytical methods, 49–53, 68
applicability, 252
iron, 265, 267, 407
uranium, 394, 410, 411
Selenium,
abundance, 30–32
analytical methods, 55, 68, 69
anion complexes, 263, 349
associations, 27, 28, 76
epigenetic aureoles, 104, 106–108, 110, 111, 115–117
general, 572, 573
geobotany, 477, 482
geochemistry, 21, 23–25, 47, 133
in gossans, 265
in iron–manganese oxides, 204
in soils, 273
Semi-arid climates, effect on,
soils, 156, 158, 175–176, 178, 314, 316
surficial dispersion, 223, 235
water, 354
Sheetwash, 224–225, 243, 253, 254
Sierra Leone, 81, 264, 267, 268, 286, 434, 532
Sieving of samples, 49, 447, 449, 450
Signatures, trace-element, 89
Significant anomalies, 34; *see also* Anomalies and Non-significant anomalies
Silcrete, 172
Silicic acid, 132, 152, 181
Silver,
abundance, 30–32
adsorption, 262
analytical methods, 55, 60, 67, 69
anomaly contrast, 251
associations, 27, 28, 76
deposits, *see* Precious-metal deposits
epigenetic aureoles, 104, 106–108, 110, 111, 115–117
general, 573–574
geochemistry, 21, 23, 24, 25
in gossan, 264
ore, 251
as ore-type indicator, 75
in soils, 304
in vegetation, 468, 469
in water, 183, 184, 191, 194, 195, 364
Size fractions,
sediments, 395, 407, 409, 414
soils, 268, 328
Skarn deposits, 73, 76, 106, 109, 110, 113
Smelter contamination, 326–328, 343
Snow, 236, 371
Soil, *see also* Residual soil and Transported overburden
air, 504–506
analysis, 3, 4, 70
anomalies, 219
augers, 334
composition, 32, 549–581
creep, 217, 218, 225, 274, 278, 281, 311
definition, 149
effect on water composition, 352–353
formation, 149–179, 251, 266
gases, 493, 504–506
horizons, 150–151
moisture, 222, 223
parent material, 151–154
profile development, 150–153

profiles, 282, 283, 321–323, 327
survey technique, 319–347
Taxonomy, 162–169, 177
Solifluction, *see* Soil, creep
Solubility,
metal oxides, 183
minerals, 133, 146–147, 192–195, 368
Solvent extraction, 350, 446
South Africa, 73, 76, 81, 96, 551
South Dakota, 73, 77, 442
South-West Africa,
80, 274, 275, 316, 317, 318, 481
Soviet Union, *see* U.S.S.R.
Spain, 119
Specific adsorption, 199
Specific-ion electrodes, 55, 59
Spectrographic analysis, 55, 56, 68
Spring water,
364–365, 369, 373–377, 441–442
Springs, 220–224, 256
Stability diagrams, 47, 48
Stability of minerals, 46, 134–135, 226
Standard deviations, 328–329
Standard temperature and pressure,
definition, 490
Statistical interpretation of data,
32–42, 61–66, 525–533
Stratabound ore deposits, *see*
Ironformation and Volcanogenic
massive-sulfide deposits
Stream sediments, *see also* under
specific elements
general, 3, 244, 256, 392–418
survey techniques,
437-439, 442–444, 447–450
Stream water,
224, 229, 247, 256, 377–379, 442–444
Strontium,
abundance, 30–32
analytical methods, 55, 68
associations, 27
epigenetic aureoles, 117
general, 574
geochemistry, 22, 24, 25
isotope ratios, 90
in soils, 156
in water, 202
Subtropics, 132, 171, 267, 269, 386
Sulfate, *see also* Sulfur
analytical method, 69
bacterial production, 238
general, 574–575
ion stability, 190
as pathfinder for sulfides, 28
in soils, 178
in water, 132, 181, 349, 351, 371, 373
Sulfide minerals,
84, 145, 213, 238, 261, 310
Sulfide-selective extraction,
88, 295, 301
Sulfur,
abundance, 30, 31
analytical methods, 55, 68
associations, 76
epigenetic aureoles, 115–117
general, 574–575
geobotany, 482
geochemistry, 24, 25, 132, 133, 188
ion species, 188–191
isotope ratios, 75
as ore-type indicator, 75
oxidation of, 238
in productive plutons, 89
in productive volcanics, 96
Sulfur dioxide,
22, 60, 494–497, 499, 510
Sulfur gases,
22, 60, 69, 494–498, 506, 507, 512
Sulfuric acid, 262–263
Superjacent anomalies,
definition, 252
distinguished from lateral anomalies,
344
in lake sediments, 419
in rocks, 109, 111
in soils, 252–256, 305–306
Supra-ore anomalies, *see*
Superjacent anomalies
Surface water, 224–231, 243, 356, 368;
see also Hydromorphic anomalies,
Lake-water, anomalies, and
Stream water
Surficial dispersion patterns, 243–260
Surficial environment, 14, 23–24
Survey techniques,
rocks, 121–127
sediments and water, 430–455
soils, 319–347
vegetation, 471–476, 484–485
volatiles, 514–517

Suspended matter, 225, 349, 350, 410
Swamps, 167, 224, 227, 240, 256, 270, 276, 301, 364, 385, 386, 436, 437
Sweden, 110, 292, 310, 311, 412, 439
Syngenetic anomalies in,
- residual soils, 242–245
- rocks, 71, 77–96
- transported soils, 242–245, 289, 291, 313–314

T

Talus, 314, 317
Tannic acid, 208
Tantalum,
- abundance, 30, 31
- analytical methods, 55
- associations, 76
- deposits, 73, 76, 87
- general, 575–576
- geochemistry, 24, 25, 139
- in productive plutons, 87, 95

Tanzania, 29, 268, 277
Tellurium,
- abundance, 30
- analytical methods, 68
- associations, 76
- epigenetic aureoles, 105, 111, 117, 121
- general, 575–576
- geochemistry, 25
- in gossans, 264

Temperate climates, 218, 230, 239, 314, 354, 361
Temperate soils, 166, 170, 174
Temperate weathering, 137
Tennessee, 7, 73, 74, 267, 280, 394
Termites, 235, 317, 318
Texas, 513, 533
Textbooks, 9, 11
Thallium, 115–118
Thermal stratification of lakes, 228–230
Thermal water, 132, 428–429
Thorium, 118, 576, 570
Threshold, 34, 42, 341–342, 519; *see also* Background
Till, 231–235, 243, 290–309, 377
Time variations,
- sediment anomalies, 415–416
- water anomalies, 368–373

Tin,
- abundance, 30
- analytical methods, 55, 68
- anomaly contrast, 251, 406
- associations, 27, 76
- in biotite, 91–92
- epigenetic aureoles, 104
- general, 576–577
- geochemistry, 22–25, 133, 250
- in heavy-mineral concentrates, 406
- ore, 251
- in productive plutons, 87, 89–93

Tin deposits,
- epigenetic aureoles, 106, 112, 113
- geochemical discoveries, 7
- pathfinders for, 76
- productive plutons for, 89–93

Titanium in peat, 310, 312
Topography, 270
Total dissolved solids in water, 351, 352–353, 355, 381, 452
Total-metal content,
- definition, 49

Toxicity to vegetation, 458, 460, 485–487
Trains, dispersion, 252–256, 313
Transpiration, 177, 222–224, 307, 314, 354, 479
Transported gossans, 285
Transported overburden, 240, 254–255, 325, 403
Trend surfaces, 523–524
Tropical climates, 239, 354, 361
Tropical soils, 157, 171–173; *see also* Lateritic soil
Tropical weathering, 137, 145, 147, 306, 354
Truncated soil profiles, 151, 160
Tundra, 165, 166, 170, 177, 483
Tungsten,
- abundance, 30
- analytical methods, 55, 68
- associations, 27, 76
- epigenetic aureoles, 104, 106, 110, 118
- general, 577–578
- geochemistry, 21, 24, 25, 133
- in productive plutons, 87, 93
- in plants, 239
- in soils, 266

in water, 239
Tungsten deposits, 76, 89, 93, 106, 557

U

Uganda,
geochemical discoveries, 7
Ishasha, 398, 408
Kilembe, 285, 393, 394, 471
Ruwenzori, 275, 278, 397
soils, 275, 278, 285, 571
stream sediments,
393, 394, 397, 398, 405, 414, 426
vegetation, 471
Ultramafic rocks,
composition, 258, 549–581
geobotany, 477, 481
gossans over, 263
nickel–copper deposits in, 88–89
non-significant anomalies from,
426–427
Ultrasonic concentrates, 409–411
Underflow in drainage channels,
224, 257, 377
United Kingdom,
Derbyshire, lead–zinc deposits,
108, 111–112, 533
Isle of Man, lead–zinc deposits,
204–205
Pennines, fluorite deposits, 556–571
Scotland, uranium deposits, 302
Wales, arsenic in sediments, 412
United States, *see* under specific states
Units of measurement, 61, 491–492
Uptake by plants, 457–464
Uranium,
abundance, 30, 31
analytical methods, 55, 58, 68, 69
associations, 27, 76
carbonate complexes, 349, 360
in coal, 467
epigenetic aureoles, 106, 118
exploration, 512–513
general, 578–579
geobotany, 478
geochemistry, 21, 24, 25
in ground water, 441, 442, 533
ion species, 349, 360
in iron oxides, 411
in lake sediments, 79, 80, 381, 425, 426
in organic matter, 411, 416, 423
in peat, 212, 302, 307
in phosphorite, 428
in productive plutons, 87
provinces, 78–80, 82, 425
in sediments, 211, 401, 410, 411
in soils, 272, 470, 513
in surface water, 353, 370, 378, 381
in vegetation, 461, 462, 466, 467, 470
Uranium deposits,
geochemical discoveries, 7
ore-typing, 75
pathfinders for, 557, 558, 567, 573
productive plutons for, 89
provinces, 78–80, 82
sediment anomalies, 394, 425
vegetation, 467, 470, 477, 478
U.S.S.R.,
epigenetic aureoles,
88–90, 104–106, 109, 118, 120
geochemical exploration activity,
5–10
ground water, 373, 374, 376
stream sediments, 353, 401, 452
surface water, 348, 353, 359, 362, 365,
366, 371, 372, 379
vegetation, 469, 479, 483
volatiles, 508, 511
Utah, 7, 73, 76, 94, 107, 108, 113

V

Vadose water, 220
Vanadium,
abundance, 30, 32
analytical methods, 55, 68
anomaly contrast, 251
associations, 27, 76
epigenetic aureoles, 117
general, 579
geochemistry, 24, 25, 133, 211
ore, 251
in peat, 307, 310, 312
in productive plutons, 89, 95
in soils, 153, 316, 317
in vegetation, 461, 462
Vapor-phase ore guides, *see* Gas
Vegetation, 236–237, 456–488, 549–581;
see also Biogenic dispersion
patterns and plant topics

Volatiles, *see* Gas
Volatilization, 51, 52
Volcanic rocks, 230
Volcanism,
  forecasting of, 216, 227, 514
Volcanogenic massive-sulfide deposits,
  epigenetic aureoles, 119–120
  pathfinders for, 76, 428
  productive volcanic rocks, 95–96, 428
  soil anomalies, 305

W

Washington, 304
Water, *see also* Bog, Ground, Lake, Mine, Rain, Saline, Sea, Spring, Stream, Surface, Vadose, and Well water
  adsorbed, 199
  anomalies in, 348–382
  composition,
    180–181, 351–358, 549–581
  sampling, 435–436, 445–447
Water action, 243, 289
Water analysis, 69, 70
Water table, 220–224
Weather, *see also* Rainfall,
  effect on, 368–373, 504–505
Weathering,
  resistance of minerals to, 134–136
Weathering processes, 128–138
Weathering products, 137–148
Well water, 373, 381, 441, 446
Wet oxidation of plant samples, 249, 475
Wet sieving, 448
Wind action, 216, 228, 231, 235, 243, 253, 254, 289
Wind-borne material, *see* Aeolian deposits
Wisconsin–Illinois lead–zinc district, 107, 108, 120, 273, 364, 365, 376, 440
World soils, 169
Worms, 235
Wyoming, 7

X

Xerophytes, 466–467, 479
X-ray diffraction, 139
X-ray fluorescence analysis, 55, 56, 58, 68

Y

Yugoslavia, 93
Yukon Territory, 7, 305, 318, 360

Z

Zaire, 483
Zambia,
  clastic soil anomalies,
    274, 276, 282, 284
  geobotany, 483
  geochemical discoveries, 7
  hydromorphic soil anomalies,
    285, 286, 386–391, 437, 441
  provinces, 81, 82
  soil profiles, 153, 155, 204, 205,
    267, 269, 281, 418
  stream-sediment anomalies, 316, 396
    403, 408, 411, 415–418, 438
  termites, 235, 317–318
Zero point of charge, 198–199
Zimbabwe-Rhodesia, *see* Rhodesia
Zinc,
  abundance, 30–32
  adsorption, 200, 202
  in algae, 238
  analytical methods, 55, 68
  anomaly contrast, 251, 359
  associations, 27, 76
  in caliche, 316, 317
  cold-extractable (cxZn), 286, 287
  deposits, *see* Lead–zinc deposits
  epigenetic aureoles,
    104–108, 111–113, 115–121
  general, 580–581
  geobotany, 477–478
  geochemistry, 19–25, 85–86, 206,
    208, 210, 239
  in gossan, 264
  in ground water, 373, 376, 440
  in humus, 281
  in iron–manganese oxides, 282, 412
  in lake sediment,
    365, 367, 420, 423–425
  in lake water, 372, 381
  in magnetite, 45

in marine sediments, 429
ores, 267, 274, 275, 304
in organic matter, 423, 424
in peat, 307, 310, 385, 386
in productive plutons, 93
in productive volcanics, 96
in soil, 268, 272, 273, 277, 280, 282, 314–315
in stream sediment, 204–205, 407, 409, 416, 528, 529
in stream water, 364–367
in till, 296, 300, 302–305
in vegetation, 468, 482
Zinc clays, 196
Zirconium, 75, 91, 581
Zonal soils, 163, 165–167, 170–178
Zoning, 104, 106; *see also* under specific elements